NATO • Challenges of Modern Society

. series of edited volumes comprising multifaceted studies of contem-
orary problems facing our society, assembled in cooperation with NATO
ommittee on the Challenges of Modern Society.

olume 1 AIR POLLUTION MODELING AND ITS APPLICATION I
Edited by C. De Wispelaere

olume 2 AIR POLLUTION: Assessment Methodology and Modeling
Edited by Erich Weber

olume 3 AIR POLLUTION MODELING AND ITS APPLICATION II
Edited by C. De Wispelaere

olume 4 HAZARDOUS WASTE DISPOSAL
Edited by John P. Lehman

olume 5 AIR POLLUTION MODELING AND ITS APPLICATION III
Edited by C. De Wispelaere

olume 6 REMOTE SENSING FOR THE CONTROL OF MARINE POLLUTION
Edited by Jean-Marie Massin

olume 7 AIR POLLUTION MODELING AND ITS APPLICATION IV
Edited by C. De Wispelaere

olume 8 CONTAMINATED LAND: Reclamation and Treatment
Edited by Michael A. Smith

olume 9 INTERREGIONAL AIR POLLUTION MODELING: The State of the Art
Edited by S. Zwerver and J. van Ham

olume 10 AIR POLLUTION MODELING AND ITS APPLICATION V
Edited by C. De Wispelaere, Francis A. Schiermeier, and
Noor V. Gillani

olume 11 AIR POLLUTION MODELING AND ITS APPLICATION VI
Edited by Han van Dop

olume 12 RISK MANAGEMENT OF CHEMICALS IN THE ENVIRONMENT
Edited by Hans M. Seip and Anders B. Heiberg

olume 13 AIR POLLUTION MODELING AND ITS APPLICATION VII
Edited by Han van Dop

olume 14 HEALTH AND MEDICAL ASPECTS OF DISASTER PREPAREDNESS
Edited by John C. Duffy

olume 15 AIR POLLUTION MODELING AND ITS APPLICATION VIII
Edited by Han van Dop and Douw G. Steyn

Air Pollution Modeling and Its Application VIII

Air Pollution Modeling and Its Application VIII

Edited by

Han van Dop

World Meteorological Organization
Geneva, Switzerland

and

Douw G. Steyn

The University of British Columbia
Vancouver, British Columbia, Canada

Published in cooperation with
NATO Committee on the Challenges of Modern Society

PLENUM PRESS • NEW YORK AND LONDON

Library of Congress Cataloging-in-Publication Data

NATO/CCMS International Technical Meeting on Air Pollution Modeling and Its Application (18th : 1990 : Vancouver, British Columbia, Canada)
Air pollution modeling and its application VIII / edited by Han van Dop and Douw G. Steyn.
p. cm. -- (NATO challenges of modern society ; v. 15)
"Published in cooperation with NATO Committee on the Challenges of Modern Society."
"Proceedings of the Eighteenth NATO/CCMS International Technical Meeting on Air Pollution Modeling and Its Application, held May 13-17, 1990, in Vancouver, British Columbia, Canada"--T.p. verso.
Includes bibliographical references and index.
ISBN 0-306-43828-3
1. Air--Pollution--Mathematical models--Congresses.
2. Atmospheric diffusion--Mathematical models--Congresses. 3. Air--Pollution--Meteorological aspects--Mathematical models--Congresses.
I. Dop, H. van (Han van), 1944- . II. Steyn, D. G. III. North Atlantic Treaty Organization. Committee on the Challenges of Modern Society. IV. Series.
TD881.N38 1990
628.5'3'015118--dc20 91-13475
CIP

Proceedings of the Eighteenth NATO/CCMS International Technical Meeting on Air Pollution Modeling and Its Application, held May 13-17, 1990, in Vancouver, British Columbia, Canada

ISBN 0-306-43828-3

A Division of Plenum Publishing Corporation
233 Spring Street, New York, N.Y. 10013

Printed in the United States of America

PREFACE

In 1949, when the North Atlantic Treaty was ratified, one of its articles explicitly noted "that member countries should contribute towards the further development of peaceful and friendly international relations." Specific problems related to the human environment were addressed by the Committee of Challenges of Modern Society (CCMS) of NATO, established in 1969. This provided a framework within which a series of International Technical Meetings (ITMs) on Air Pollution Modelling has been held. This volume documents the proceedings of the 18th meeting in this series.

Science, like the arts and sports, provides an ideal vehicle for "developing peaceful and friendly international relations". National boundaries have never been barriers to the movement of air pollution, and fortunately this has also proved true of scientists studying the transport of air pollution. It is thus satisfying to record that since the mid-seventies it has been commonplace to find Eastern European scientists among attendees at the ITMs which have (in a very modest way) participated in a precursor to the process which has led to historical changes in Europe and which will undoubtedly lead to a tremendous increase in personal and intellectual exchange on a worldwide basis.

The opening address of the 18th ITM was given by Professor M.E. Berlyand who drew the attention to the severe environmental problems faced by the USSR. It is our hope that the ITMs will continue to serve as mechanisms for the transfer of the knowledge required for the abatement of air pollution there and in other East European countries.

The 18th ITM was hosted by the University of British Columbia in Vancouver, elegantly situated between the boreal forest and the North Pacific Ocean. Though fairly remote from the world's major industrial regions, the conference was well attended (some 130 participants).

More than 140 papers of uniformly high quality were submitted for review by the scientific board. This resulted in a very full programme of 103 papers presented in five sessions. The scientific board responsible for paper selection consisted of:

R.M. van Aalst	The Netherlands
R. Berkowicz	Denmark
C. Borrego	Portugal
N. Chaumerliac	France
H. van Dop	Switzerland
N.V. Gillani	USA
J.C.R. Hunt	England
G. Kallos	Greece
W. Klug	FRG
H. Meinl	FRG
M.M. Millán	Spain
J. Saltbones	Norway
S. Sandroni	Italy
G. Schayes	Belgium
F.A. Schiermeier	USA
D.G. Steyn	Canada
J.L. Walmsley	Canada
M.L. Williams	England

The topics were:

Ia Model assessment and policy implication;
Ib Regional acid deposition and oxidant modelling;
II Regional scale modelling;
III Global scale dispersion including sub-grid scale parameterisation of transport and exchange processes;
IV Accidental releases including radioactivity and dense gas dispersion;
V New developments in dispersion modelling and theory.

Papers on the modelling of accidental releases were presented in a poster session.

We are grateful for the excellent facilities and institutional support provided by the UBC conference centre while Elizabeth Ellis and Suzanne Reimer provided outstanding secretarial support before and during the meeting. Marleen Kaltofen deserves explicit acknowledgement for her invaluable assistance in preparation for the meeting, and in compilation of the proceedings.

H. van Dop
D.G. Steyn

September 1990

CONTENTS

OPENING ADDRESS

MODEL ASSESSMENT AND POLICY IMPLICATIONS

REGIONAL ACID DEPOSITION AND OXIDANT MODELLING

REGIONAL SCALE MODELLING

GLOBAL SCALE DISPERSION STUDIES INCLUDING SUB-GRID SCALE PARAMETERIZATION OF TRANSPORT AND EXCHANGE PROCESSES

ACCIDENTAL RELEASES INCLUDING RADIOACTIVITY AND DENSE GAS DISPERSION

POSTER SESSION

VIDEO PRESENTATION

NEW DEVELOPMENTS IN DISPERSION MODELLING AND THEORY

PROBLEMS OF INDUSTRIAL AIR POLLUTION MODELLING CONSIDERING SOVIET EXPERIENCE

M.E. Berlyand

A.I. Voeikov Main Geophysical Observatory
7, Karbyshev Street
Leningrad 194018, USSR

In solving problems of environmental protection against pollution in the USSR, application of atmospheric diffusion modelling methods have for a long time already played a very important, and sometimes the leading role. First of all it is expressed in establishing unified air pollution calculating methods approved by the Government for different types of sources. It is the base for establishing maximum permissible emissions for the planned or operating enterprises and also for implementing air pollution measures. Results of the modelling are also used for designing a rational system and development of air pollution monitoring, in particular in choosing site and time of the observations. At present in the USSR there is a broad network of sites (in about 560 cities and towns) measuring concentrations of several dozens of air pollutants. Results of research in the Main Geophysical Observatory, the Institute of Experimental Meteorology, the Institute of Applied Geophysics and the Computer Centre in Novosibirsk and other Institutes are reviewed and generalized (Berlyand, 1975, 1983, 1985; Byzova, 1974; Israel et al., 1989; Penenko and Aloyan, 1985). These works are carried out in the framework of close international cooperation with CMEA countries and bilateral agreements with the USA, France, Finland and other countries.

Here we have only considered some important problems of industrial pollution modelling taking into account Soviet experience and emphasizing their basic principles and some of the major results of recent studies.

The base for modelling is the solution of the atmospheric diffusion equation:

$$u\frac{\partial q}{\partial x} + w\frac{\partial q}{\partial z} = \frac{\partial}{\partial y}\kappa_y\frac{\partial q}{\partial y} + \frac{\partial}{\partial z}\kappa_z\frac{\partial q}{\partial z} + \alpha q \qquad (1)$$

where q is pollutant concentration; u and w are horizontal and vertical components of the wind velocity respectively; κ_y and κ_z are horizontal and vertical components of the exchange coefficient; α is index of pollutant transformation; $z = 0$ corresponds to the level of the underlying surface.

In the case of uneven terrain instead of x and y used in Eq. (1), flow coordinates $\xi_1 \sim \varphi$ and $\xi_2 \sim \psi$ are used, where φ and ψ are real and imaginary parts of the functions of the conformal transformation of the flow region to a half-plane.

Eq. (1) is solved with appropriate boundary conditions together with a system of other equations determining coefficients and parameters contained in Eq. (1) and in the boundary conditions. Thus, to determine wind velocity u and exchange coefficient κ_z, a system of equations of motion including a turbulent energy balance is solved. For the stationary boundary layer we obtain

$$\frac{d}{dz}\kappa_z\frac{d\vec{U}}{dz} = 2\,\omega_z\,(\vec{U}-\vec{U}_g) \tag{2}$$

where $\vec{U} = u + iv$; u, v are wind velocity components along x and y, ω_z is the vertical component of angular speed of the Earth's rotation and U_g is the geostrophic wind velocity.

In the case of a diabatic boundary layer and a monotone lapse rate, two layers are distinguished: the surface layer ($z \leq h$) where heat flux and friction stress are constant, and the layer above, where an average value $\bar{\kappa} = \kappa_h$ is accepted. According to the turbulent energy balance equation:

$$\kappa_h = \kappa_{h_o}\sqrt{1 - Ri/Ri_{cr}} \tag{3}$$

where κ_{h_o} is the value of κ_h under neutral stability and Ri is the Richardson number.

It follows with sufficient accuracy that:

$$u = u_1\frac{\ln\frac{z}{z_0}}{\ln\frac{z_1}{z_0}}, \quad \kappa_z = \begin{cases}\kappa_1\frac{z}{z_1} \text{ at } z \leq h \\ \kappa_1\frac{h}{z_1} \text{ at } z > h\end{cases}, \quad h = \frac{0.05\,\kappa_1}{\omega_z z_1}. \tag{4}$$

When temperature changes with height are not monotone, additional layers are introduced, in particular for an elevated inversion layer for which the exchange coefficient is also determined using Eq. (3). In the case of a surface inversion, according to similarity theory a logarithmic/linear change with z is taken for u, and a linear-fractional change for κ_z.

For the horizontal component of the exchange coefficient κ_y it follows that $\kappa_y \approx \varphi_o^2\,u$, where φ_o^2 is dispersion of wind direction oscillations for the interval of time for which the concentration q is averaged.

Effects of urban mesoclimate is studied based on non-stationary solutions, on air temperature and humidity daily variation, and also on the daily course of wind velocity and exchange coefficient, taking into account advection of air coming into the city from the surrounding country. The urban mesoclimate model permits to determine the intensity of occurrence of the "heat island" effect (difference between air temperatures inside the city and in surrounding country, $\Delta\theta$, to find out its effects on the variations of the exchange coefficient and wind velocity. E.g., Fig. 1 presents calculated results for a city located at 60° n.l. with the wind velocity U_g = 10 or 20 m/s, under a cloudless sky in June.

Several special features and principles of considered work on modelling can be noted. One of them is the approach to the modelling which is based on the numerical solution of the system of equations including atmospheric diffusion Eq. (1), as well as equations describing variations of determining parameters and coefficients. It has certain advantages when compared with Gaussian models which are to a certain degree empirical-statistical. This approach permits to use the same base to build models for different meteorological and topographical conditions and for different types of sources. That is just how the problem of atmospheric diffusion for light and heavy pollutants has been solved, as well as those for conditions with and without elevated inversions of temperature, for calms and fog, for different types of terrain, building, etc.

As an example, it can be pointed out that from an integrated solution of Eq. (1) and equations describing pollutant rise above industrial sources due to the initial escape and overheating of discharge gases resulted the expression $H_e = H + \Delta H$ for the effective source height H_e, where H is geometrical source height, and

$$\Delta H = \frac{1{,}5\,w_o\,R_o}{u}\left(2{,}5 + \frac{3{,}3\,g\,R_o\,\Delta T}{T_a\,u^2}\right). \tag{5}$$

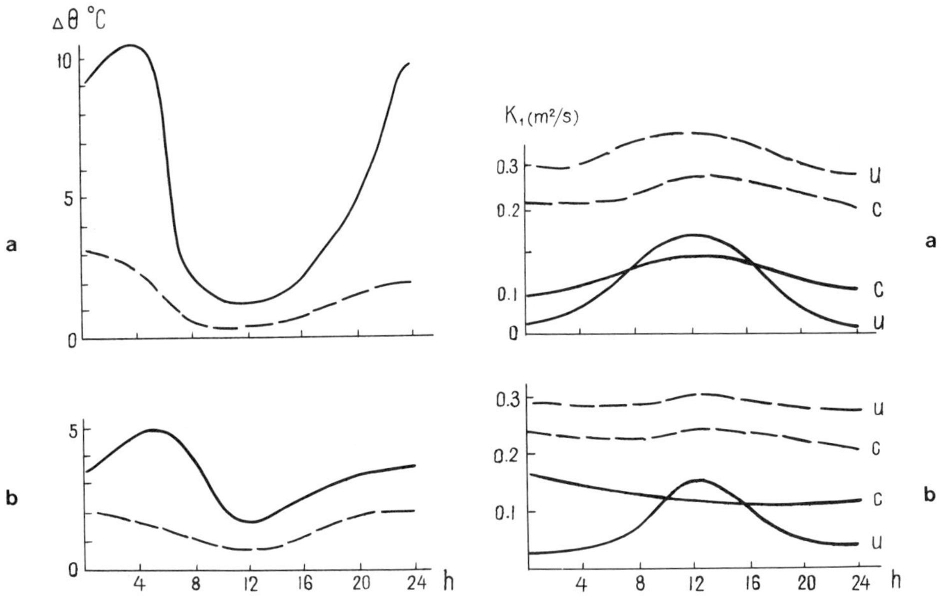

Fig. 1. Urban-rural temperature difference (Δθ) and exchange coefficient (K_1) variation over a day. ——— $U_g = 10$ m/s, - - - - $U_g = 20$ m/s; (a) July, (b) January; (c) city or town, (u) upwind flow.

Here w_o and ΔT are escape velocity and temperature excess of the discharged gases from a stack with orifice radius R_o, g is the acceleration of gravity, u is wind velocity at the height z = 10 m, T_a is temperature in K.

From the integrated solution of Eq. (1), taking into account the initial rise Eq. (5) and the condition that $\frac{\partial^2 q}{\partial x\, \partial u} = 0$ and $\frac{\partial^2 q}{\partial x\, \partial \kappa_z} = 0$, the expression for the maximal value of the short-time average concentration (20 min), C_M, at the distance $x = x_M$ under the critical wind velocity $u = u_M$ and an unfavorable stratification (for elevated sources under unstable conditions) can be calculated. In general we have

$$c = c_M\, r s_1 (x/\rho x_M)\, s_2 (u, |y|/x)$$

$$c_M = \frac{AMFmn\eta}{H^2 \sqrt[3]{V\Delta T}} \qquad (6)$$

where $r = r(u/u_M)$ and $\rho = \rho(u/u_M)$, with $r(1) = \rho(1) = 1$ and $s_1(1) = s_2(u,0) = 1$. M and $V = \pi R_o^2 w_o$ are quantity and volume of gases emitted per unit of time, the coefficient $A = 0.3\ (\kappa_1/\varphi_o u_1)$ is determined from data on spatial-temporal distribution of turbulent parameters (Climatic characteristics, 1983).

If c is expressed in mg/m³, M in g/s, V in m³/s, and H in m, then the value of A over the territory of the USSR will vary from 140-160 for zones with moderate turbulence, located in the Central and Northern halves of the European part of the country to 250 for zones with the most intensive turbulent exchange in subtropics of Central Asia and in Transbaikal region. For other countries it is recommended to establish A according to similarity of climatic conditions.

Coefficient F determines the effects of pollutant deposition velocity. It varies from unity for gases and light pollutants to 2-3 for heavy particles and depends on the efficiency of dust or ash-cleaning.

Table 1

$n_1 = H/h_o$	valley			hill		
	$n_2 = a_o/h_o$					
	4-5	6-9	16-20	4-5	6-9	16-20
< 0.5	4.0	2.0	1.3	3.0	1.5	1.2
1.1 - 2.9	1.8	1.5	1.1	1.4	1.3	1.0
3 - 5	1.4	1.3	1.0	1.2	1.2	1.0

The values m and n depend on auxiliary parameters

$$f = 2.10^3 \frac{w_o R_o}{H^2 \Delta T}, \quad V_M = 0.65 \sqrt[3]{\frac{V \Delta T}{H}}, \quad V'_M = 2.6 \frac{w_o R_o}{H}. \tag{7}$$

The coefficient η incorporates terrain effects. For an even surface $\eta = 1$. For terrain with elongated hills or valleys, maximum values of η depend on source location and the point of concentration measurement. Table 1 shows its maximum values, η_m, as depending on $n_1 = H/h_o$ and $n_2 = a_o/h_o$, where h_o is height difference and a_o is halfwidth of the valley or hill.

Fig. 2 shows values of the analogical index χ, determining correction for the concentration on a smooth surface, which depends not only on the irregularities of terrain (valley or hill), but also on the surface temperature inhomogeneity.

Let us note another principle of the work under discussion, which means definition of the concentration under unfavorable stratification. It permits to minimize the initial meteorological information necessary for the calculations, confining the choice to the indicated values of A for different climatic areas. Often this is sufficient, because to require that the maximum concentration under unfavorable conditions must satisfy air quality standards, also concentration standards under all other meteorological conditions will be satisfied.

Transition from short-time to average annual concentration in the general form is performed by

$$\bar{q} = \int_0^\infty q \, P(q) \, dq$$

where P(q) is probability density of values of q depending in its turn on probability density of wind velocity and direction, as well as of turbulent exchange coefficient. Statistical studies based on observational data show that approximately (in polar coordinates r and φ):

$$\bar{q} = 2 \pi \rho_1(\varphi) \, q_M \, s_1 \, (r/r_M) \, . \tag{8}$$

Here $\rho_1(\varphi)$ is probability of wind direction at an angle φ, determined according to the wind rose; s_1 (r/r_M) is the same function used in Eq. (6), and q_M and r_M are average annual concentration maximum, $\bar{q}$, and the distance, r, where it is reached. q_M and r_M are directly bound to c_M and x_M

$$q_M = c_M \, \Phi_1 \, (A,f) \, r \left(\frac{0{,}8 \, \bar{u}}{u_M} \right), \qquad r_M = x_M \, \Phi_2 \, (A,f) \, \rho \left(\frac{0{,}8 \, \bar{u}}{u_M} \right) \tag{9}$$

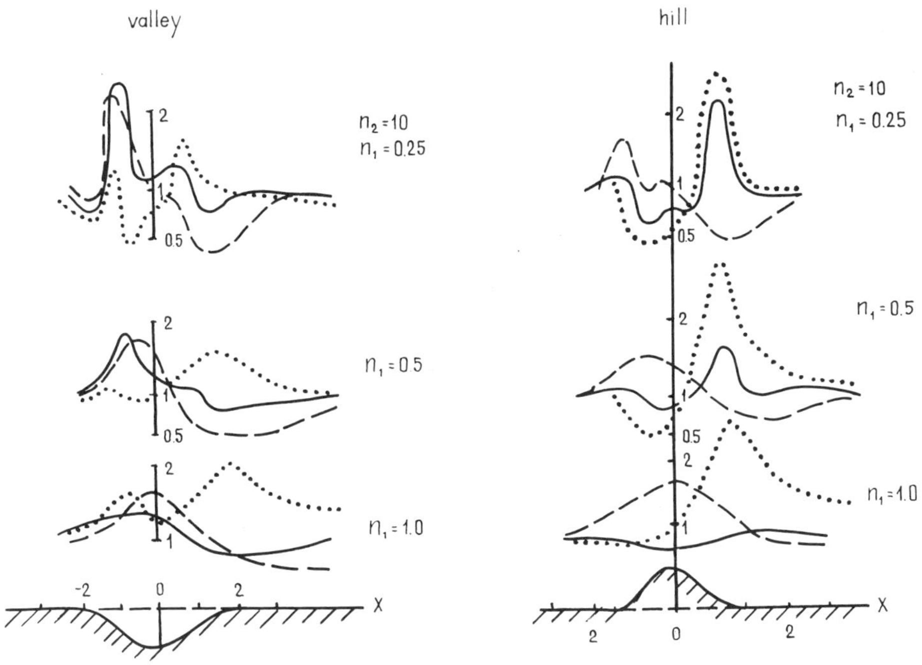

Fig. 2. Correction coefficient χ for dispersion over inhomogeneous terrain include temperature inhomogeneities. ——— homogeneous heating; – – – – heating of windward, cooling of leeward slope; ········ cooling of windward, heating of leeward slope.

where $\bar{u}$ is average wind velocity and r and ρ are the same as in Eq. (6).

One more particularity of considered work carried out in modelling should be noted. When deriving Eqs. (6), (8) and (9), the unfavourable meteorological conditions taken into account are of the "normal" type, so for u and κ_z Eqs. (4) are valid.

Anomalous hazardous conditions are studied separately, among them cases of elevated inversion, calms, fogs, etc. It makes possible a cost-effective implementation of air protection measures. Data calculated according to Eq. (6) for unfavorable cases of normal weather conditions are intended to implement long-term measures. In cases of anomalous hazardous meteorological conditions, usually of short duration, with a probability of occurrence of 1-2%, for which special forecasting methods are developed, it is recommended to take care to reduce emissions for some time.

The concentration of pollutants from multiple sources scattered over a surface is calculated according to the principle of superposition. For this purpose computer programmes have been developed. They permit to take into account hundreds and sometimes thousands of sources characteristic of big cities and industrial areas. Here, the calculating algorithm is constructed in order to greatly save computer time.

The indicated modelling results have been validated by many experiments. Permanent comparison of calculated and experimental data is also one of the principles of considered work. For this purpose observational data have been used which come from about 40 field studies carried out every year in areas with different types of air pollution sources, as well as foreign experimental data, among which those obtained due to international cooperation with CMEA countries, the USA and Finland. For example, Fig. 3 presents the summary graph for normalised maximum concentrations of thermal power stations with a stack height from 30 to 320 m. In particular, it confirms the implication of Eq. (6) that c_M decreases in inverse proportion to H^2. It is

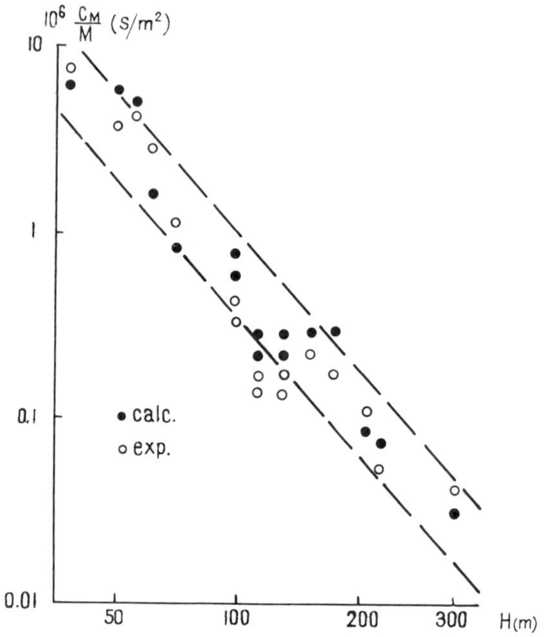

Fig. 3. Maximum concentrations as a function of stack heights.

interesting that there is a possibility to set approximately $c_M \approx \Lambda\ M/H^2$, where $\Lambda = (3\text{-}6)\ 10^{-3}$ with the indicated dimensional representation of c_M, M, H. Good agreement is also obtained for the case of other industrial emissions, for single as well as for multiple sources.

A new all-Union standard document has been written and approved on a governmental level on this base (Methods of calculation, 1987; Berlyand, 1987; Berlyand et al., 1987).

One of the major applications of performed work is the establishment of maximum permissible emissions (MPE). They are determined from the condition:

$$c_M + c_B \leq MPC \qquad (10)$$

and if $c_M = M\Phi$, then $MPE = \dfrac{MPC - c_B}{\Phi}$.

Here, c_B is background concentration and MPC is the maximum permissible concentration.

The work in establishing MPE has already been completed in cities where emissions exceed 80% of total emissions all over the USSR. According to emission inventory data (Annual review, 1989), concentration fields were calculated and MPE established for all major sources in those cities.

One of the new problems in this work is to develop optimization methods for emission reduction (Berlyand et al., 1988; Raputa et al., 1984). The objective is to reduce urban concentration to MPC in the most rational way. E.g., Fig. 4a shows the field of maximum concentrations for unfavourable conditions for one of the cities which is considered with the help of the mentioned programs. Calculation of concentrations have been performed here at the given grid points (i, j), and

$$c_{ij} = \sum_k f_{ijk} M_{ijk} + c_B . \qquad (11)$$

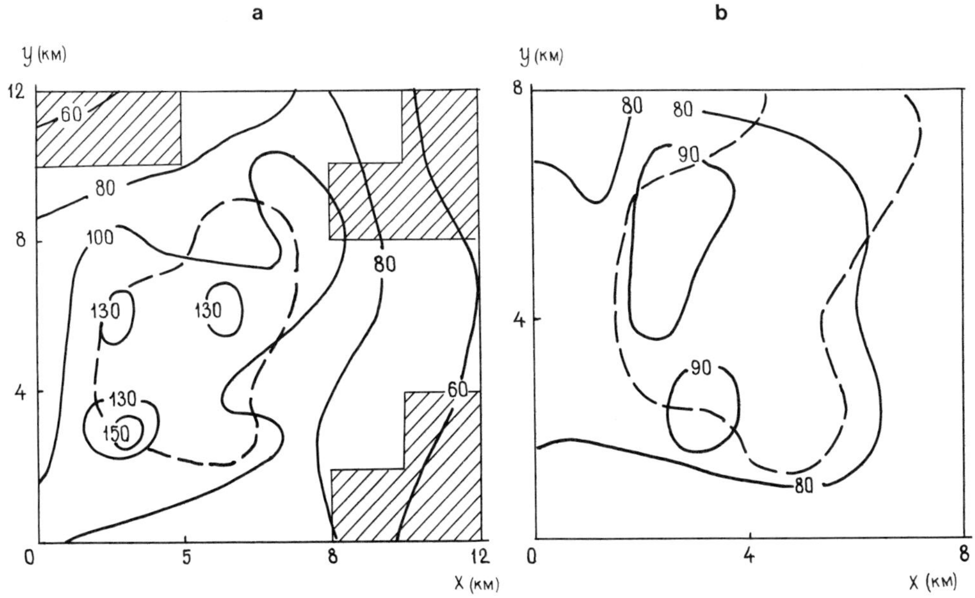

Fig. 4. An example of emission reduction strategy. (a) old situation; (b) new situation, achieved by overall emission reductions of 15%. ——— nondimensional concentrations isolines, 2nd zone; – – – – zone of point sources location, 3rd zone; ▨ stove heating effects.

To reduce emissions optimally, an efficiency function B is introduced, e.g., the total cost of emision reduction (or simply the total emission all over the city) where

$$B = \sum_{\kappa} \frac{a_{\kappa} M_{\kappa} (1 - \lambda_{\kappa})}{\theta_{\kappa}} \qquad \text{with} \quad 0 < \lambda_{\kappa} < 1 . \tag{12}$$

Here a_{κ} is emission reduction cost per mass unit, λ_{κ} is emission reduction coefficient, θ_{κ} is an estimated coefficient of source significance for the city. The following relations must be satisfied simultaneously:

$$B \rightarrow \min,$$

$$\hat{c}_{ij} - \sum_{k} \lambda_{k} \varphi_{ijk} M_{ijk} + c_{B} \leq MPC . \tag{13}$$

The solution of the problem is carried out using linear programming. Fig. 4b shows calculated results for the example in Fig. 4a in the case of a_{κ} = const and $\theta_{\kappa} = 1$. It appeared that for the required reduction of the local concentration with 60% it is sufficient to reduce emissions all over the city only for 15%.

Modelling is also developing in the direction of a more detailed account of different types of sources (Methods of calculating concentrations of air pollutants contained in industrial emissions, 1987). Sources are divided into four classes according to their height, H, and for surface concentration q_o, is taken not the usual value at $z = 0$, but the average concentration, q, over a layer of 2 m adjacent to the surface. So we can introduce two classes of sources: surface sources with $H \leq 2$ m, and low sources with $2 < H \leq 10$, where the difference between q and q_o is significant. Buildings are also taken into account, which can practically be important for the sources with $H \leq 50$ m. Therefore, two more classes of sources are introduced: high sources with $H \geq 50$ m, and medium height sources with $20 < H < 50$ m.

Lower sources, which often have no heat output, frequently cause substantial local pollution. Considering their large number, their total impact can also be felt over great distances. That is why the USSR's initiative in EEC started international cooperation in studying low and non-buoyant sources (LCS).

Among LCS motor vehicles are of special importance, their part in the total emission of numerous cities reaching 80-90%. Much attention was paid to air pollution from cars in the USSR (Annual review, 1990). At the same time for this purpose a number of field studies were carried out, among them the joint Soviet-American experiment "Autoex".

When modelling dispersion of emissions from cars, they were considered as surface sources with values of concentrations and emissions averaged over the lowest layer of the depth $z_h = 2$ m (Berlyand et al., 1990). For a highway of a width d_o with wind velocity u perpendicular to it, pollutant concentration is determined with the formula

$$q = \frac{1{,}24\, M}{d_o \lambda u} \ln \frac{\lambda x \sqrt{z_h^2 + \lambda^2 x^2}}{z_h \left[\lambda(x-d_o) - z_h + \sqrt{z_h^2 + \lambda^2 (x-d_o)^2} \right] \theta(x-d_o)} \qquad (14)$$

where $\lambda = \kappa_1/u_1$ and $\theta(x-d_o)$ is the unit step function.

Formulas have also been obtained and a program for calculating the concentration for the case of arbitrary wind direction has been developed. It follows that the maximal concentration beside a highway is reached when the wind blows along it, and at a sufficient distance x from it when the wind direction is perpendicular to the highway. At intermediate distances, x, hazardous wind directions are those at an angle ϕ. Fig. 5a shows the dependence of this angle ϕ on distance x, and Fig. 5b shows in relative units pollutant concentration distribution from the highway 1,000 m long, when $\phi = 45°$. Fig. 6 presents results of the comparison of calculations with observations near a highway during five field studies. Research has also started on the effects of photochemical transformations of cars exhaust, transformation of NO into NO_2, forming of ozone, etc.

From the requirement for the concentration at the edge of an urban roadway not to exceed MPC, it is possible to establish permissible emissions from the roadway, and from there permissible intensity of traffic on the roadway. For the cases of unfavorable meteorological conditions it is possible to indicate how much to reduce the traffic on the roadway or in other streets, and when necessary all over the city.

Another important type of surface sources are storage ponds which are the subject of recent research in interrelation of air and water pollution (Berlyand et al., 1990). Note that the amount evaporating, M, not only depends on the pond's area, S, but also on the wind velocity, u. Table 2 presents dependence of M on u with a fixed S: $M = a_1 (1 + \beta_1 u) S$, as well as calculated values of c at different distances, x, depending on u. If a sanitary protection zone is established (e.g., extending to 2 km), behind which the concentration c > MPC, it is possible to establish MPE, whose values are also given in Table 2. From here it is possible to determine permissible discharge into the ponds, more precisely to determine permissible area of the pond, S, because S is directly related to M.

Surface sources studies relate to another important problem, modelling of dispersion of accidental emissions. It is explained partly by the fact that many accidental sources are located directly at the surface, e.g. ruptures of gas pipelines, tanks and reservoirs with poisonous substances (chlorine, ammonia, etc.) and the ensuing volatilization of these substances.

One of the major particularities of accidental emissions is the necessity to take into account nonstationarity, that is to introduce the nonstationary term, $\partial q/\partial t$, into Eq. (1). However, if in this case the equation is integrated over time, t, it is possible to retrieve the formula for stationary conditions, where concentration is replaced by dose, $D = \int_0^t q\,dt$. This is the base on which recently a method of forecasting the scale of contamination with poisonous substances following

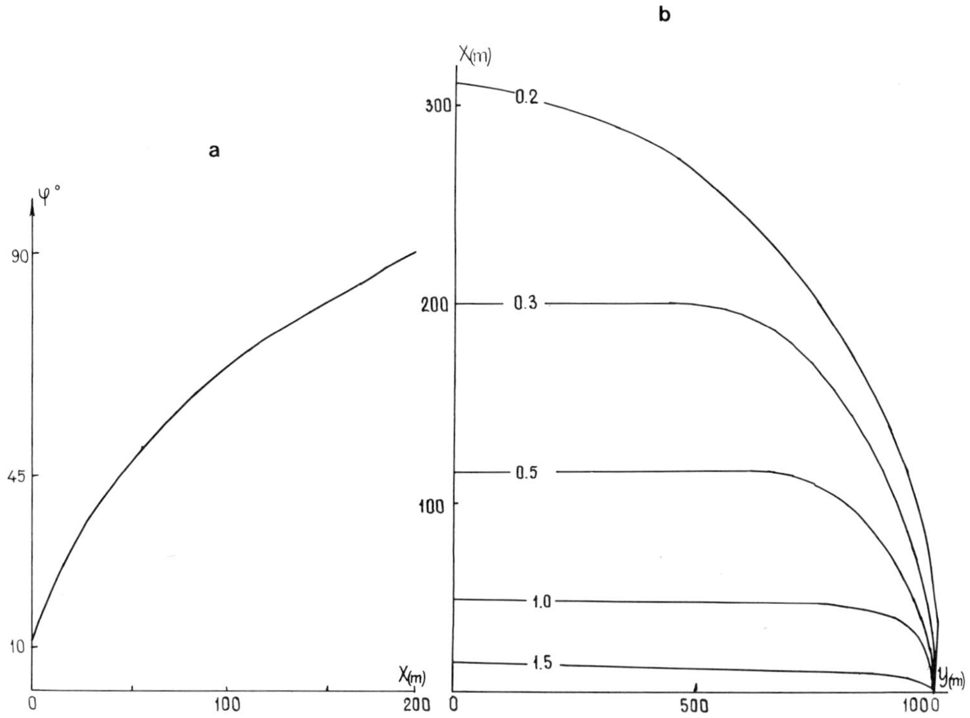

Fig. 5. (a) Location, x, of maximum concentrations as a function of the angle, ϕ, between wind and highway direction; (b) Concentration field due to highway emissions for ϕ = 45°.

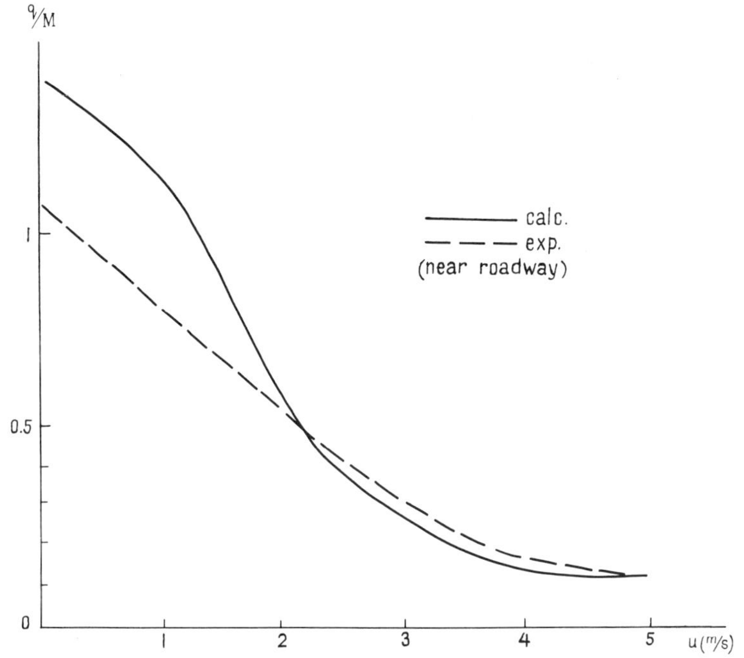

Fig. 6. Normalized concentrations as a function of wind speed.

Table 2

u (m/s)	0.5	1	2	5	10		
M (g/s)	580	605	650	780	1020		
x (km)	Concentration C, (mg/m^3)					c_{Mx}	u_{Mx}
2	26.7	28.0	24.7	26.0	36.0	36.0	10
4	12.8	13.3	11.7	12.0	14.3	14.3	10
8	6.4	6.7	5.6	5.4	6.7	6.7	1.10

accidents on chemically dangerous installations and during transportation has been developed. Public Health Services approved maximum permissible doses for about 25 poisonous substances.

When studying dispersion of LCS emissions located on buildings or near them, sometimes a need arises to take into account recirculation zones (RZ) forming inside built-up areas. Studies permitted to distinguish essential types of such zones. It appeared that their modelling in a wind tunnel does not take into account properly low-frequency wind direction oscillations. In the real atmosphere there is a critical angle between directions of wind and building, when recirculation zones disappear.

Mathematical modelling technique in this case is similar to the one used for uneven terrain. The difference is that the boundary of the recirculation zones (RZ) near buildings is assumed known from experimental data, and in between them a possibility of wind velocity drop to zero with considerable turbulence intensity is admitted. As a result the concentration in the presence of buildings is determined with the following formula (Berlyand et al., 1987; Methods of calculation, 1987):

$$\hat{c} = (1 - \zeta)\, c' + \zeta\, c'' \tag{15}$$

where c' and c" are concentrations averaged over 20 min for cases of absence and presence of the RZ; ζ is the relative RZ life time, given by

$$\zeta = \int_{-\varphi_k}^{\varphi_k} \rho\,(\varphi - \bar{\varphi})\, d\varphi$$

where $\rho\,(\varphi - \bar{\varphi})$ is probability density of wind direction, φ, deviation from its mean value, $\bar{\varphi}$, for the time interval under study (20 min); φ_c is the indicated critical angle (Fig. 7a) which is determined from Fig. 7b. It is also shown that if a variable t_b is introduced,

$$t_b = \begin{cases} \varphi_K \sqrt{u} & \text{when} \quad u \leq 5 \text{ m/s} \\ 2{,}24\, \varphi_K & \text{when} \quad u > 5 \text{ m/s} \end{cases}$$

then ζ is determined from Fig. 7c. In the case of a cluster housing the principle of their recirculation zones superposition is used.

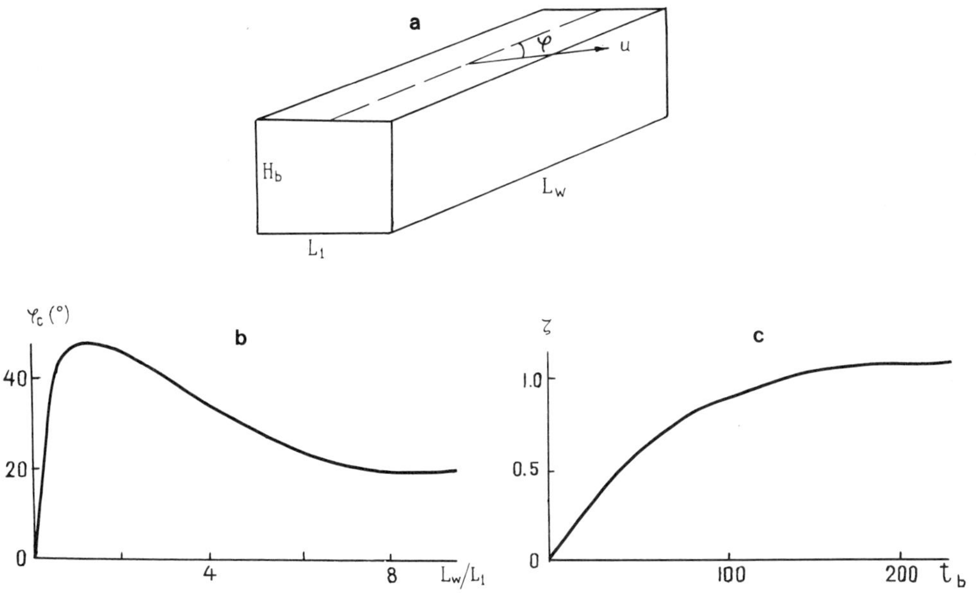

Fig. 7. (a) Geometry of obstacle and mean wind direction; (b) critical angle ϕ_c, as a function of obstacle geometry; (c) relative recirculation zone life time as a function of the parameter t_b.

Much attention is now paid to researches in increasing spatial-temporal modelling scales taking into account deposition and transformation of pollutants. Here the works of the Computer Center in Novosibirsk should be mentioned, where the first in the USSR calculation scheme of photochemical pollutant transformation was proposed (Penenko, Aloyan, 1985). An important role in the international cooperation is played by works of the East European Center at the Institute of Applied Geophysics on long-range pollutant transport modelling (Israel et al., 1989). Their particularity is the possibility to determine transboundary pollutant flows. Application of obtained results permits, in particular, to estimate the part due to sources located at different distances from the boundary in the total transboundary flow. When using data concerning them we were determining the necessary optimal emission reduction from sources on the European USSR territory with the purpose of reducing transboundary flows for 30% according to the accepted international agreement. Above-mentioned method of linear programming can also be useed here with optimization purposes.

We also undertook researches to take into account wash-out and deposition of pollutants from sources like power stations (Berlyand, 1984). It was shown that total amount of deposited pollutant, Π, from a source with capacity M at a distance x from it with a wind velocity u is determined with the formula

$$\Pi = M\left(1 - e^{-\alpha x/u}\right) \tag{16}$$

where α is determined by the structure and intensity of precipitations. Application of Eq. (16) permits to radically change the approach to MPE establishment for powerful sources with high stacks. In fact, MPE are now established with the Eq. (10) and often using the circumstance that according to Eqs. (6) - (10) MPE can be infinitely increased with increasing H. Taking into account hazard criteria for the given locality and accordingly limiting deposition on the soil, Π, it follows from Eq. (16) that MPE are also limited, and from Eq. (6) it is possible to establish maximum height, H, when air quality standard will be satisfied (Eq. (10)). However, it should be noted that the problem of limiting stack height, and consequently capacity of plants, is bound with considerable costs and often concerns interests of several neighboring countries at the same time. So to solve this problem corresponding international discussions are necessary.

At present there are plans to expand work in regional modelling with consideration of interaction of pollutants with the underground and their transformation, with the purpose to estimate effects of urban and industrial emissions into the air over the surrounding territory, in particular on nearby water bodies and soil. It seems that international efforts to organize joint work on determination of necessary parameters and establishment of criteria of hazardous pollutant deposition and air pollution have a special importance in solving these problems. It must promote determination of maximum permissible emissions and wastes into the air, water and on the soil, taking into account their integral efect on the environment.

Such are the problems requiring, as we see it, further development taking into account Soviet as well as foreign experience.

REFERENCES

Annual review on air pollution state and pollutant emissions into the air of cities and industrial centers of the Soviet Union. Volume "Emissions of pollutants" ed. by M.E. Berlyand. Leningrad, GGO, 1989, 370 p.

Berlyand, M.E.: Modern problems of atmospheric diffusion and air pollution. Hydrometeoizdat, Leningrad, 1975, 448 p. (Transl. into English: NEPC US EPA, Raleigh N.C., 1976; into German: Acad. Verlag, Berlin, 1982).

Berlyand, M.E.: Forecasting of air pollution with emphasis on research in the USSR. WMO EMPaRP, No. 19, 1983, Geneva, 208 p.

Berlyand, M.E.: Urgent problems of air pollution due to power plant emissions. Meteorology and Hydrology, 1984, No. 5 (Transl. into English: Soviet Meteorology and Hydrology, 1984, No. 5).

Berlyand, M.E.: Forecast and control of air pollution, Gimiz, Leningrad, 1985, 272 p. (Transl. into English: Kluwer Acad. publ., 1990, in press).

Berlyand, M.E.: New methods of calculating concentrations of air pollutants contained in industrial emissions. Bulletin of construction engineering, 1987, No. 7.

Berlyand, M.E., Genikhovich, E.L., Gracheva, I.G., Onikul, R.I., Filatova, E.N., Khurshudyan, L.G.: On improving air pollution calculation methods. Trudy GGO, v.511, 1987, p. 3-23.

Berlyand, M.E., Genikhovich, E.L., Gracheva, I.G., Zashikhin, M.N., Kulik, A.S., Chicherin, S.S.: Taking into account building when calculating air pollution. Trudy GGO, v.511, 1987, p. 24-38.

Berlyand, M.E., Genikhovich, E.L., Onikul, R.I.: modeling of air pollution due to emissions from low and cold sources. Meteorology and Hydrology, 1990, No. 5, p. 5-19.

Berlyand, M.E., Genikhovich, E.L., Gracheva, I.G., Kanchan, Ya.S., Schtreicher, D.: On improving methodology of standardizing emissions into the atrmosphere. In: Meteorological aspects of air pollution. Gimiz, Leningrad, 1988, p. 10-16.

Byzova, N.L.: Pollutant dispersion in the boundary air layer. Moscow, Gimiz, 1974, 190 p.

Climatic characteristics of air pollution dispersion conditions. Reference book, Bezuglaya, E.Yu, and Berlyand, M.E. (eds.), Leningrad, Gimiz, 1983, 328 p.

Israel, Yu.A., Nazarov, I.M., Pressman, A.Ya., Rovinsky, F.Ya., Ryaboshapko, A.G., Philippova, A.M.: Acid rains. Leningrad, Gimiz, 1989, 270 p.

Methods of calculating concentrations of air pollutants contained in industrial emissions (OND-86), Berlyand, M.E., et al. (eds.), Leningrad, Gimiz, 1987, 94 p.

Penenko, A.V., Aloyan, A.E.: Models and methods concerning environmental protection problems. Novosibirsk, Nauka, 1985, 256 p.

Raputa, V.Ph., Makarov, E.Ya., Skorokhodov, A.A.: Optimization of mode of operation of urban air pollution sources. Computer Center in Novosibirsk, preprint No. 548, Novosibirsk, 1984, 12 p.

DISCUSSION

J.C.R. HUNT

Please explain how you derived your formulae for the effects on dispersion caused by heating and cooling slopes and valleys.

It was a solution of the system of equations of atmospheric diffusion and of motion, of heat transfer and of turbulent energy balance, taking into account the different heating of the terrain surface.

Ia: MODEL ASSESSMENT AND POLICY IMPLICATIONS

chairman: J.D. Shannon

rapporteur: H.J.A. van Jaarsveld

DEVELOPMENTS IN MODELLING AIR POLLUTION FOR REGULATORY USES

J.C.R. Hunt[1,2], R.J. Holroyd[2], D.J. Carruthers[2]
A.G. Robins[3], D.D. Apsley[3], F.B. Smith[4], D.J. Thomson[4]

[1]Dept of Applied Mathematics & Theoretical Physics, University of Cambridge
[2]Cambridge Environmental Research Consultants Ltd
[3]National Power – Technology & Environmental Centre
[4]The U.K. Meteorological Office

1. INTRODUCTION

1.1 Models for 'regulatory' use

Many different types of administrative and operational decisions have to be taken about the prevention of, control of and response to pollutants (e.g. gases, particles, radio-active matter) released into the atmosphere. In part these decisions are taken on the basis of particular kinds of models for calculating the distribution of these pollutants from where they are discharged into the atmosphere to where they are finally deposited on the ground. The transformation of the emissions into other substances by physical and chemical processes also has to be considered. The models have to be of a form that enables them to be widely used for practical purposes which include the laying down of regulations by government agencies - hence the term 'regulatory'.

Most countries of the world now use or have been using models that have a similar basis, and which were pioneered by Pasquill (1961) and Gifford (1961) (see also Pasquill & Smith 1983). However recent research, much faster computers, measurement and communication now enable significant improvements to be made to the models, the data input, the interpretation of their output and its uses for decisions and as inputs to other models. This paper is a review of some of these research developments, and their application with the aid of new systems.

1.2 Requirements

1.2.1 Events

Regulatory models are required for many kinds of situations and emissions, but the most usual fall into three classes:

(i) particular, 'one-off' events, such as Bhopal (Singh & Ghosh 1987) or Chernobyl (e.g. Smith 1988), where all the processes are evaluated in detail for one set of meteorological conditions;

(ii) particular, or 'real time', events where it is necessary to calculate atmospheric processes involving the emissions as *they are happening*, and to make predictions for a few hours or days ahead, using on-line meteorological predictions as they become available, and predictions about the future emission from the source;

(iii) sequences of many hypothetical events, all of which have to be modelled in order to calculate probabilities of risk or harm to people, animals, plants etc. Such calculations are of particular importance for the nuclear industry where many calculations are based on the dispersion models being used in combination with modelling of health effects.

Short-range regulatory models are considered here which are valid up to about 30km from the source. This covers many cases and situations of particular importance. Different kinds of practical models are also available for longer distances.

1.2.2 Form of improved models

Despite the *apparent* similarity in the basis of current models there are substantial differences in how they are used within countries and between countries. Current proposals for improved models provide the opportunity of ensuring a greater uniformity in future, especially as there is a broad scientific consensus about the physical basis of the improvements. The existence of trans-frontier dispersion and the tendency towards international environmental decisions are other reasons why it is essential to work towards greater uniformity.

There are two main criteria that models for regulatory purposes must satisfy: they must be scientifically justifiable and therefore based on the current state of scientific understanding and methods; on the other hand the models must be explicable and reasonably easy to use by non-experts. They must also be *fast* so that they can be used both in real-time situations and also for studying large numbers of scenarios to assess possible environmental consequences.

The technical features that improved modelling systems require are that:

(i) they should be able to use more of the information that is now available from meteorological measurements and models (which current dispersion models cannot use);

(ii) they must deal with complex effects such as buoyant gas emissions (including their interaction with elevated inversion layers) or dense gas emissions and dispersion from arbitrary heights in the atmosphere over irregular terrain, buildings and coastlines. Note that the vertical distributions of concentration must be calculated and not just ground level values;

(iii) they must allow for the complex dispersion processes associated with short time releases, deposition, and concentration fluctuations. (Note that the averaging time for which given levels of pollutants are permissible are steadily decreasing. In the case of SO_2, for example, the mandatory criteria are 70 μg/m^3 for an annual average and 120 μg/m^3 for a daily average, but the World Health Organisation recommended value for 1 hour is 350 μg/m^3 and for 10 minutes is 500 μg/m^3. For toxic gases even shorter periods have to be considered);

(iv) the models and their output must be in forms that can be used in other calculations such as health effects or chemical reactions (for which the fluctuations are significant).

1.3 Recent scientific and technical developments

1.3.1 Basis of current models

Although the present Pasquill-Gifford type of models were designed to be used with simple (slide rule level) calculations and graphs and diagrams they are now generally used with small computer systems. Since such systems are now in general use for air pollution computations and because of the rapid increase in their calculating power, improved 'regulatory' models can be based on recent research which entails lengthier calculations and uses more meteorological information. Small computer systems that are appropriate for these improved models are IBM PS2 or equivalent models which run on MS-DOS and have at least 640 kbytes of random Access Memory.

The essential concept of the Pasquill-Gifford model is that it can determine the line mean dispersion of matter from a source even though the meteorology of the boundary layer is only defined by surface measurements. It was always recognised that this is a simplification, but the approach continues to give estimates that are sufficiently good for many practical purposes. This modelling approach and its range of applicability are discussed in §§2 and 3.

1.3.2 Improved models for boundary layer meteorology

The current proposals for improved models are also based on the assumption that for most of the time the essential features of the atmospheric boundary layer that affect dispersion are determined by a few parameters derived from measurements and a few given functions; however compared to previous models there are more parameters and the detailed structure of the boundary layer is considered. The assumption implies that it is not necessary to compute or measure the state of the boundary layer on every occasion because it is often close to one of several standard forms. Although the concept strictly applies to level terrain in rather steady meteorological conditions it is the first approximation for all conditions and the basis for more complex models that deal with more complex situations.

This idealised state of the boundary layer can be defined by relevant dimensionless groups, namely $(h/L_{MO}), u_*/hf, z_o/h, N_u h/u_*$, and others (Gryning et al 1987, Tennekes 1982, Deardorff 1970). Here the boundary-layer depth is h, L_{MO} is the Monin-Obukhov length based on the surface heat flux F_{θ_0} and friction velocity u_*, f is the Coriolis parameter, z_o is the roughness length of the surface, and N_u is a typical value of the buoyancy frequency of the stably stratified layers above the boundary layer (these parameters are defined in more detail in §2). It would therefore appear that a large number of dimensionless groups would have to be specified to define the state of the boundary layer but fortunately, for the purpose of modelling, it has been found that the first of these groups is by far the most significant.

L_{MO} can be calculated either from widely available measurements (at height z_m) and observations of the surface (principally the mean wind speed U, solar heat flux, cloud cover, roughness length z_o, and vegetation type), or from more advanced instrumentation (e.g. of eddy heat flux). The former, simpler measurements have usually been recorded over longer periods of time, which are needed for obtaining local meteorological statistics. The boundary layer depth h can now be estimated on a routine basis at certain measuring stations, firstly from remote measurements such as radio-sondes or sodar, and secondly by model calculations in which values of h are calculated from surface fluxes or are inferred from those measured at specific times by interpolation and extrapolation. In current practice simple formulæare used (e.g. reviewed by Weil 1985) be produced without introducing another length scale, which may be implicit (as in most schemes) or explicit, as in Smith (1973) and the current U.K. model* (Clarke 1979; Kazanski & Monin 1960), Zilitinkevich (1972) and Weil (1985), where it is derived from the Coriolis parameter f. The differences in this auxiliary length scale is probably one reason for the differences between schemes for defining these dispersion categories.

The other reason for differences is that quite different measurements and calculations are used for determining the surface heat flux and temperature profile, ranging from cloud cover to the two-height temperature difference method of the U.S. Nuclear Regulatory Commission.

In principle there should be very small differences in defining neutral conditions because the surface heat flux is effectively very small. However, an extensive comparison by Kretzschmar & Mertens (1984) showed how different schemes lead to different classifications of the boundary layer in the same meteorological conditions including the neutral case. Using measurements at Mol in Belgium, it was found that the Belgian and U.K. schemes predicted that neutral conditions existed for 20-25% of the time and for 72% of the time respectively! These are substantial differences which should not exist in any internationally agreed scheme.

The current 'regulatory' schemes were designed for sources near the surface, and it was therefore assumed that the turbulence and hence the dispersion near the surface were determined by the variables (U, F_{θ_0}, z_o) at the surface. Research has shown that this is not strictly correct, especially in highly convective conditions.

2.2 Dispersion models

The main features of the calculations of dispersion in the current models (outlined in §1.3.3) are only valid for sources that are within the surface layer, where the relevant velocity field is defined by the surface measurements. Typically the surface layer depth is about $h/10$. (As mentioned at the end of §2.1 there are some errors even in this assumption because σ_y is partly determined by the boundary layer height.) This means that the maximum source height for which the model is valid depends on the depth of the boundary layer. In stable conditions the boundary layer can be as low as 50m, while in highly convective conditions it may be as great as 2000m. Therefore the maximum source height for which these models are valid depends on the relative frequency of very stable or unstable conditions; for north-west Europe, where typically 70% of the time the boundary layer is close to neutral and the boundary layer is about 500m to 1000m thick, the current models are relevant for source heights less than about 50m.

Since emissions from fossil fuel power plants are usually released from heights well above the surface layer, new methods have had to be developed. Some of these are described in the next section.

The advantage of the current surface layer methods is that there is no need to specify the local profiles of turbulence statistics. On the other hand, this is also a disadvantage when the flow is significantly different from the ideal state in which the dispersion coefficients were established experimentally or by computation. This has led to various *ad hoc* adjustments to the models, such as assuming a different stability category for urban areas compared with rural areas, given the other conditions are the same.

---

* The phrase "the current U.K. model" refers to the model developed for the nuclear industry (Clarke 1979). This model has since been used widely outside the nuclear industry, but not so widely as to constitute a true national standard

When dispersion is calculated over terrain that changes from level to hilly, or where the roughness changes, it is necessary to match the dispersion over these two kinds of terrain. Variations in the turbulence, stratification and mean velocity with height are required to calculate dispersion over complex terrain (see §4) but are not used in current surface layer methods over level terrain. Therefore, a satisfactory matching requires a dispersion model for flat terrain that also explicitly accounts for the effects on dispersion of the actual structure of the flow in the boundary layer.

When meteorological conditions are unsteady, or it is necessary to simulate a particular realisation of dispersion, the above models are not appropriate, because they are based on ensemble average statistics for statistically steady conditions during the travel time of the released matter from the source to the receptor. Although with a typical 5m/s wind and conditions steady over an hour the models might be assumed to be valid for only about 20km from the source, in fact they can be used over longer distances because the vertical diffusion is limited by the boundary-layer depths and the variations in the wind direction during the travel time can be allowed for by a suitable form for $\sigma_y(x)$ (e.g. Clarke 1979).

Other approaches have been explicitly developed to account for dispersion that is unsteady and occurs over long times and distances, such as puff models (ApSimon et al. 1985) or random flight models (Lange 1978), but these are not yet simple or fast enough to become 'regulatory ' models.

3. IMPROVED MODELS FOR DISPERSION OVER FLAT TERRAIN

3.1. Meteorological state of the boundary layer

The state of the atmospheric boundary layer flow in steady conditions over level terrain is defined by the dimensionless groups defined in §1.3.2. Modelling and field studies suggest that the broad structure and the essential flow processes are largely determined by h/L_{MO}, although the processes very close to the ground and details of the profiles change as the other dimensionless groups vary. For example the mean velocity profile varies slowly as z_o/h varies.

In the first stage of improved dispersion models various kinds of measurements and computations are used to derive the surface shear stress $\tau(=\rho u_*^2)$, heat flux F_{θ_0}, and boundary layer depth h. The usual method for deriving the surface quantities at meteorological stations (which cannot be assumed to have eddy flux instruments), is to use

(i) measurements of the hourly average wind speed U and the air temperature T_o at a specific height above the ground, the relative humidity and precipitation,

(ii) estimates of the cloud amount C_A, and the state of the ground (i.e. roughness, length z_o, soil type, vegetation),

all recorded at specific times of the day and year.

Since there is a local energy balance the net incoming solar radiation R is equal to the outgoing heat flux from the soil, G, the latent heat flux F_{q_o} and the surface heat flux F_{θ_0} transferred by the turbulence eddies,

$$\text{i.e. } R = F_{\theta_0} + F_{q_o} + G$$

Various approximations have been suggested for estimating R (e.g. van Ulden & Holtslag 1985; Smith & Blackall 1979) in terms of the solar elevation and cloud amount,

C_A. The soil heat flux, G, is generally small (except at certain times of day with rapid variations in R); parameterisations for grass have been suggested based on measurements at Cabauw (the Netherlands) and Cardington (U.K.). Monteith's formula is used to calculate F_{θ_0} and F_{q_o}; this involves assessing various resistances (with units $L^{-1}T$) which determine evaporation. Algorithms for the stomatal resistance (r_s) have been devised which use temperature T_o, moisture, state of the soil and the state of the vegetation. The aerodynamic resistances r_a is determined by u_*^{-1} and the atmospheric stability near the ground (i.e. z_o/L_{MO}), and the climatological resistance r_i is related to the thermodynamic properties of the air. Then

$$F_{\theta_0} = (R - G) \left[\frac{r_a + r_s - r_i}{(1 + \bar{\Delta}/\gamma) r_a + r_s} \right] \tag{3.2}$$

where $\bar{\Delta}$ and γ are properties of the air-water vapour which are determined by the air temperature T_0. $\bar{\Delta} = \partial e_s / \partial T$, where e_s is the saturated vapour pressure, and $\gamma(T)$ is the 'psychometric' parameter (Monteith 1975). Other approaches are required for night time conditions where R is usually negative.

The calculation of u_* from $U(z)$ is connected to that of F_{θ_0}, since in the surface layer

$$U(z) = \frac{u_*}{\kappa} [\ln(z/z_o) - \psi(z/L_{MO}) + \psi(z_o/L_{MO})] \tag{3.3}$$

where $\psi(z/L_{MO})$ are assumed to be universal functions for flat terrain where $z << h$ (or typically $z \lesssim h/10$) and are generally valid over any terrain sufficiently close to the surface. Equations (3.2) and (3.3) have to be solved simultaneously, usually by iteration (but sometimes in stable conditions there are no solutions which correspond to a limit on the amount of heat the turbulence can transport).

Over flat grassland the estimate for F_{θ_0} from (3.2) has been compared with direct measurements of R (with a radiometer), G (by heat flux plates) and F_{q_o} (by lysimeter) and gave a standard error of less than 30 W/m^2, or a typical error in h/L_{MO} of 50% when $h/L_{MO} = 10$. Similarly, u_* calculated from a single $U(z)$ measurement and (3.3) agrees well with u_* derived from a profile of $U(z)$ at several heights (Jiemin 1984).

In convective boundary layers h can be identified from a significant change in the gradient of the mean temperature profile, which is also associated with marked changes in the turbulence and temperature fluctuations. These properties can be detected by different remote sensing apparatus, such as radio-sonde measurements of temperature, FM-CW Doppler measurements of velocity, and by acoustic sodar. There are several simple models appropriate to these conditions for satisfactorily predicting h from surface measurements or from a combination of surface flux measurements and remote sensing measurements at previous times (e.g. Carson 1973). These methods tend to give broadly similar results but computational tests have shown that there can be errors of $\pm$ 30% in h. Such errors can lead to errors of the same order in concentration predictions.

However in neutral and stable boundary layers the objective definition of h is vague. For dispersion models the relevant definition is that there is very limited vertical diffusion above $z = h$. There may be velocity fluctuations but because the atmosphere is stable the diffusion is very slow. h usually corresponds to the height where the shear stress $\overline{-uw}$ tends to zero, even though σ_w is not zero. It also corresponds to where U is close to the geostrophic wind speed, and where the mean potential temperature gradient reaches the value above the boundary layer. Remote measurements of h are therefore not so accurate in these situations, though values can certainly be obtained from sodar which roughly indicate the depth of significant velocity and temperature fluctuations. (These may extend above the region where diffusion is likely to be significant.)

Models to calculate h in neutral and stable boundary layers are used in 'improved' boundary layer models (e.g. Hanna & Paine 1989). However there is some uncertainty about them because the effects of upper layer stratification (with buoyancy frequency N_u) may have just as important effects as the Coriolis parameter f which is commonly used in formulæfor h (e.g. Deardorff 1969; van Ulden & Holtslag 1985).

It is now clear that despite all the factors (each with some uncertainty) involved in practical models for h and L_{MO} and other boundary-layer parameters, values of h/L_{MO} can be computed on a routine basis and statistics of h/L_{MO} can be derived from previous meteorological records, as well as current observations.

In Figs.1 and 2 the current U.K. nomogram (from Smith 1973; Clarke 1979) is shown along with a schematic diagram (adopted from Gryning et al. 1987, Holtslag & Nieuwstadt 1986) to show how the state of the boundary layer is defined by h/L_{MO}. Below the abscissa in Fig. 2 we have plotted the ranges of the Pasquill-Smith categories A-G, based on their U.K. definition and estimates of typical values of h for each category (see Table 1). It can be seen that for *near surface sources* the meteorological state defined by current A-G categories and the h/L_{MO} criteria happen to agree reasonably well (but as mentioned in §1.3.2. different A-G categories could be defined).

The important point about Fig.2 is that it shows how (i) as the boundary layer becomes more convective (i.e. h/L_{MO} decreases), near the surface the effects of convection may affect lateral turbulence before vertical turbulence, so that it is incorrect to use the *same* category for vertical and horizontal diffusion, and (ii) for *elevated* releases the state of the boundary layer is effectively unstable or stably stratified while for near surface releases it is still neutral. This is why during recent years there has been a move away from surface meteorology when calculating the dispersion of exhaust gases from fossil-fuel power plants.

3.2 Boundary layer structure

Given that the state of the idealised flat terrain steady state boundary layer is defined to the first approximation by h, L_{mo}, z_o and the mean wind speed at a reference height, then the profiles of the mean velocity components (U, V), the mean potential temperature $\theta(z)$, the variance of the turbulence components $\sigma_u(z), \sigma_v(z), \sigma_w(z)$ and the integral length scales $\Lambda_u, \Lambda_v, \Lambda_w$ (defined in the direction of the mean wind for example) are determined. For the convective boundary layer it is useful to model the probability distribution of w as a function of z/h, $p(w; z/h)$. It is also necessary to allow for other factors that affect these profiles such as the roughness length, z_o, the structure of an elevated inversion layer (defined by the temperature difference across the layer $\Delta\theta$ and its thickness h_i, which can be parameterised as an inversion Froude number $F_i = \sqrt{TU^2(h)/gh\Delta\theta}$ and the stratification above the boundary layer.

Formulae for these variables have been established by experiment, guided by scaling analysis and some computational and mathematical modelling. There are some differences in the algebraic forms proposed in different models, either in quantities that vary significantly from place to place (e.g. the turning of the wind direction ($\propto (z/h)$ in the stable boundary layer), or in the interpolation between the different principal states of the boundary layer (i.e. convective, neutral and stable), or in regions where several different physical processes compete (e.g. just above the surface layer in the stable boundary layer). For examples of profiles see Hanna & Paine (1989), van Ulden & Holtslag (1985), Nieuwstadt (1984).

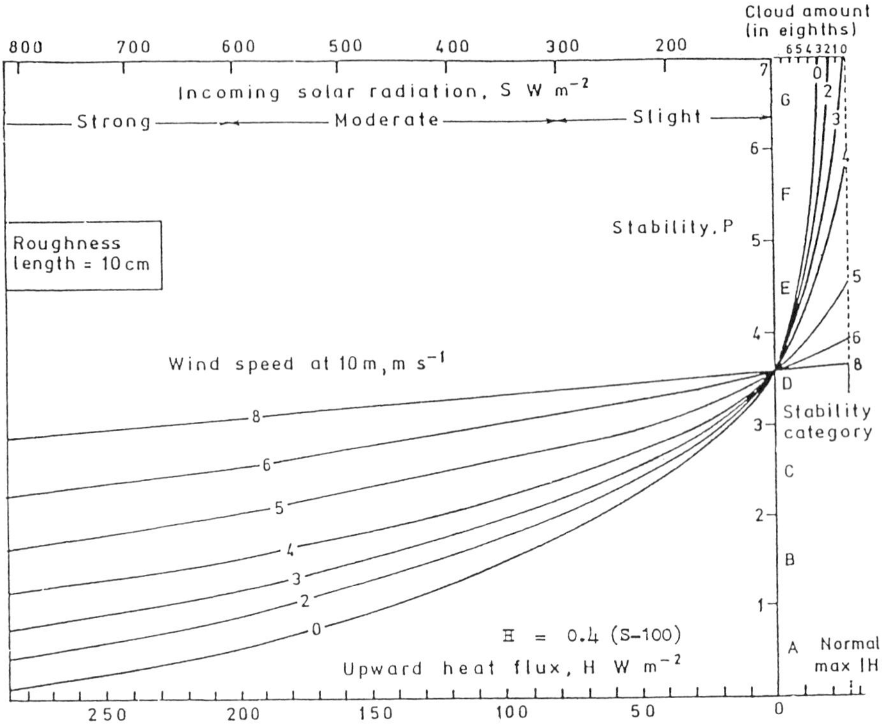

Fig. 1 Estimation of Pasquill stability category from incoming solar radiation, cloud cover and mean wind speed as used in the R91 model (Clarke 1979).

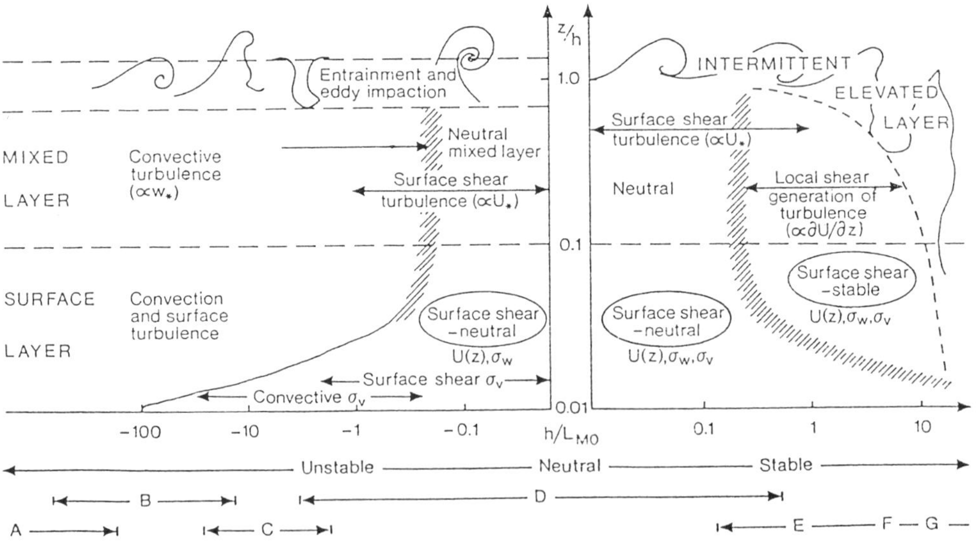

Fig. 2 Origin and nature of turbulence in the atmospheric boundary layer in terms of z/h and h/L_{MO} (developed from diagrams by Holtslag & Nieuwstadt (1986) and Gryning et al (1987)). Approximate ranges of Pasquill categores $A - G$ derived from data in Table 1.

Table 1. Meteorological parameters as a function of stability class (from Clarke (1979, Table A2) with extensions (original parts in italics)).

Pasquill Stability Category P	Smith Numbers	% Frequency	Mean wind speed U_{10} (ms^{-1})	Surface heat flux F_{θ_o} (Wm^{-2})	h(km)	Friction velocity u_* ms^{-1}	Convection velocity w_* ms^{-1}	Monin-Obukhov length L_{MO}(m)	$h/\|L_{MO}\|$	σ_v/U_{10} $z = 10$m
A	*0.5*	*0.125*	*0.625*	*250*	*1.3*	*0.11*	2.2	0.4	2685	1.7
A-B	*1.0*	*1.25*	*1.25*	*180*	*1.08*	*0.17*	1.8	2.5	435	0.4-0.7
B	*1.5*	*3.8*	*2.0*	*150*	*0.92*	*0.24*	1.66	8.4	110	0.5
B-C	*2.0*	*2.6*	*3.37*	*125*	0.5	*0.33*	1.6	26.0	33.7	0.2-0.3
C	*2.5*	*15.0*	*4.12*	*90*	*0.84*	*0.38*	1.36	55.4	15.15	0.26
C-D	3.0	—	5	50	0.6	0.3	0.94	115	7.12	0.2-0.3
D	*3.6*	*62.4*	*4.12*	*0*	*0.5*	*0.36*	0	∞	0	0.17
E	*4.5*	*6.7*	*3.4*	-3.30	*0.4*	0.26	—	476	0.86	0.17
F	5.5	—	2.0	-3.33	0.15	0.16	—	107	1.4	0.19
F-G	*6.0*	*8.4*	*1.2*	-1.39	0.05	0.073	—	25.6	1.95	0.05

Estimates of h are based on he work of Carson (1973) and Smith & Carson (1977).
A surface roughness $z_o = 0.1$m is assumed when calculating u_*.
The standard deviation of h about the mean is expected to be about 250m for each Pasquill category P (A-D).

Sketches of typical profiles plotted in boundary-layer co-ordinates are shown in Fig.3. The different forms of the curves for the different cases of $h/L_{MO} \leq -3$ (convective), $\simeq 0$ (neutral), and ≥ 3 (stable) are examples of how the curves also depend on (h/z_0) and $(N_u h/U(h))$ or F_i (the values of the latter two dimensionless groups are also very important to define the behaviour of plumes at and above elevated inversions). Note that the curves (even over flat terrain in ideal conditions) represent the average of many experiments and therefore the corresponding curves over any given one-hour period differ from them, though the differences are not yet well documented. (Such statistics could be incorporated into these air pollution models.)

3.3 Dispersion models

We now consider the proposed 'improved' methods for calculating the time mean concentration C (i.e. the average over periods much greater than that for a plume to change direction, $\sim 10\Lambda_v/U$, which could be 10 minutes to one hour, depending on the type of boundary-layer flow). The essential new feature of these models is that they incorporate information about detailed structure of the boundary layer (obtained in the previous stage of the calculation). Assuming the mean velocity U is large compared with σ_u, (i.e. ignoring the case of low wind speeds) the concentration downwind from a steady source of strength Q is generally expressed in these models in the form

$$C(x,y,z) \;=\; Q G_y(y, y_k, \sigma_y; x) G_z(z, z_k, \sigma_z; x) \big/ U\left(z_u(x)\right) \tag{3.4}$$

where the functions G_y and G_z represent the horizontal and vertical spread of the plume. They are normalised so that the horizontal flux is preserved (in the absence of deposition)

$$\text{i.e.} \quad \int_{-\infty}^{\infty} \int_{o}^{\infty} C U(z) d(y - y_k) dz \;= Q \tag{3.5}$$

This is sometimes the *only* theoretical constraint on the forms chosen! (It does of course imply that $G_y \to 0$ as $|y - y_k| \to \infty, G_z \to 0$ as $z \to \infty$). The position $(y_k(x), z_k(x))$ is a reference point in the concentration distribution. In current Gaussian plume models over level terrain y_k is the value of y at the source, so the plume moves in a straight line from the source.

The functional form of G_y proposed by Hanna & Paine (1989) and in CTDMPLUS (Perry & Finkelstein 1990) is Gaussian, i.e.

$$G_y \;=\; \frac{\exp\left(-(y - y_k)^2/2\sigma_y^2\right)}{\sqrt{2\pi}\ \sigma_y(x)} \tag{3.6}$$

In the current and improved models G_z is a series of functions

$$G_z \;= \sum_{m=1}^{M} G_{z_m}(z, z_{k_m}, \sigma_{z_m}; x) \tag{3.7}$$

for each of which z_{k_m} is defined. For the current Gaussian models,

$$z_{k_1} = z_s, \quad z_{k_2} = \; -z_s, \quad z_{k_3} = 2h - z_s, \; z_{k_4} = 2h + z_s, \quad \text{etc.} \tag{3.8}$$

where z_s is the source height and $\sigma_{z_1} = \sigma_{z_2} = \sigma_{z_3}$, etc.

The height $z_u(x)$ is the height at which the wind speed should be taken. The specification of $z_u(x), z_{k_m}(x)$, and G_{z_m} are all linked. Different functions could be used to satisfy (3.5) and have similar levels of accuracy. The parameters σ_y, σ_z define the spreading of the profiles relative to the reference points y_k, z_k, and are determined by the structure of the boundary layer.

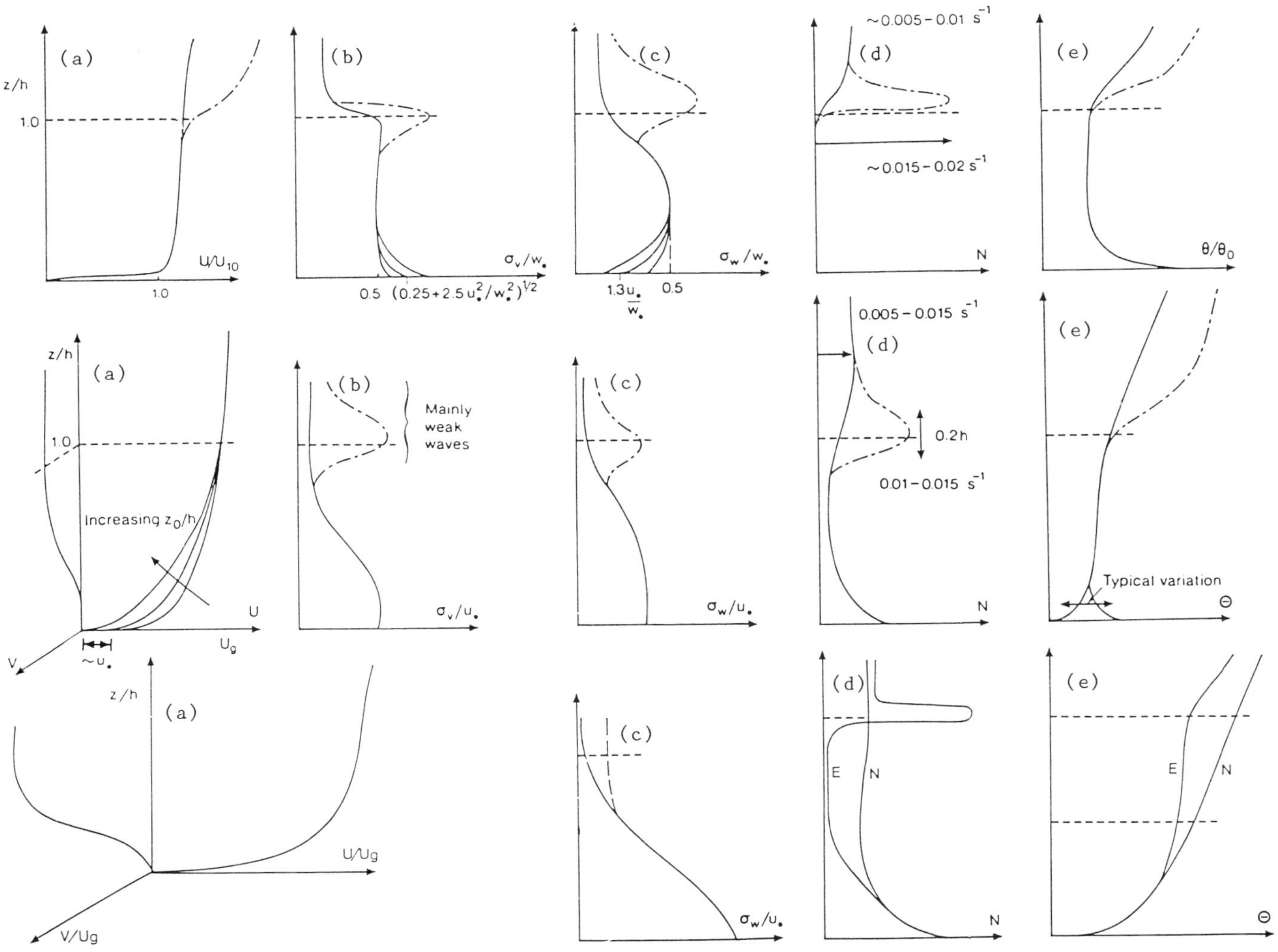

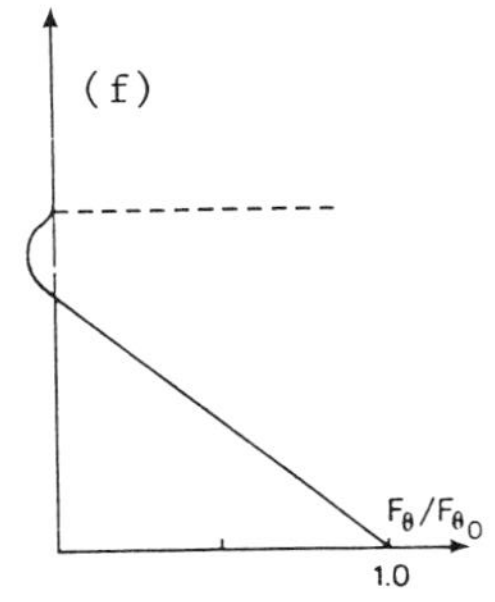

Fig. 3 Boundary layer profiles:-

top – convective
middle – neutral
bottom – stable

(a) U mean velocity
(b) σ_v
(c) σ_w
(d) N buoyancy frequency
(e) θ potential temperature
(f) heat flux F_θ

Ordinate is z/h where h is boundary layer depth.

E early evening
N night time

The reasons why G_z is non-Gaussian in most boundary layers are discussed by Hunt (1985). Different non-Gaussian forms have been proposed for G_z: Moore & Lee (1982) used an asymmetrical Gaussian function that modelled the tendency of elevated plumes to diffuse differently above and below their centre lines because of the vertical variation of turbulence. This model was only developed to predict C on $z = 0$, i.e. at ground level. The improved U.S. models (e.g. CTDMPLUS and HTDM) have also been developed to predict ground-level concentration; however they model G_z using the vertical profile of turbulence. In convective conditions G_z is evaluated from a one-dimensional straight line random flight model of particles released from an elevated source but only the probability distribution $p(w, z)$ is used and the mean profiles are assumed to be uniform.

Explicit profiles for C or G_z in the convective boundary layer should reflect the fact that dz_m/dx is positive but dz_{mx}/dx is *negative* where z_{mx} is the height at which C is maximum. Such a profile cannot be a simple Gaussian profile.

In neutral and stable conditions G_z is a series of Gaussian profiles with the same z_k as in current Gaussian models (3.8), but $\sigma_z(x, z_s)$ is determined from the boundary layer structure at the same height.

The disadvantage of the latter Gaussian plume method is that it does not reflect the pronounced vertical movement of the mean position $z_m(= \int\int zCdydz/\int\int Cdydz)$ of a plume in inhomogeneous turbulence and in shear. This effect can be modelled by letting $z_{k_1}(x)$ in (3.7) be the mean height z_m (which is initially z_s) and estimating dz_m/dx from the approximate eddy diffusion equation, which shows how $\partial z_m/\partial x$ increases with $\partial K_z/\partial z$ and decreases with $\partial U/\partial z$, provided the eddy diffusity has been modified to allow for the time of travel from the source (Hunt 1985).

As yet matching of the convective to neutral diffusion processes has not been satisfactorily modelled.

In some codes the plume spread parameters σ_y, σ_z for given values of h/L_{MO} are given as explicit functions of x/h and of the boundary-layer parameters at the level z_s/h. In other codes σ_z is calculated either implicitly, for example in the random flight method of Hanna & Paine (1989) and CTDMPLUS (Perry & Finkelstein 1990), or at some local value of $z_k(x)$ and a local eddy diffusivity (Smith & Blackall 1979).

The derivation of σ_z from measurements or computed simulations of plume spread depends on the form of G_z. If G_{z_m} are Gaussian profiles and reflections, explicit formulæhave been derived for convective, neutral and stable conditions and for when the source height is above the surface layer, within the surface layer and on the ground supplemented by interpolation functions for intermediate states of the boundary layer and intermediate source heights. A review of such formulæis given by Weil (1985).

A comparison of these improved formulæwith those used in current regulatory models based on Pasquill-Gifford categories is given in Figs. 4a,b,c. (Note that in the calculations for σ_z based on boundary-layer meteorology, it is necessary to state z_s, h, U and L_{MO}.)

The main differences between the two types of model are found in very unstable and stable conditions which are generally associated with low wind speed. Recent field research has shown that there are larger changes in vertical diffusivity which are modelled by the 'improved' schemes, but not by many current models. Also the current models do not adequately represent the change in vertical diffusion from a source in the surface layer when $(h/L_{MO}) > 0.3$; initially it is similar to that in a neutral boundary

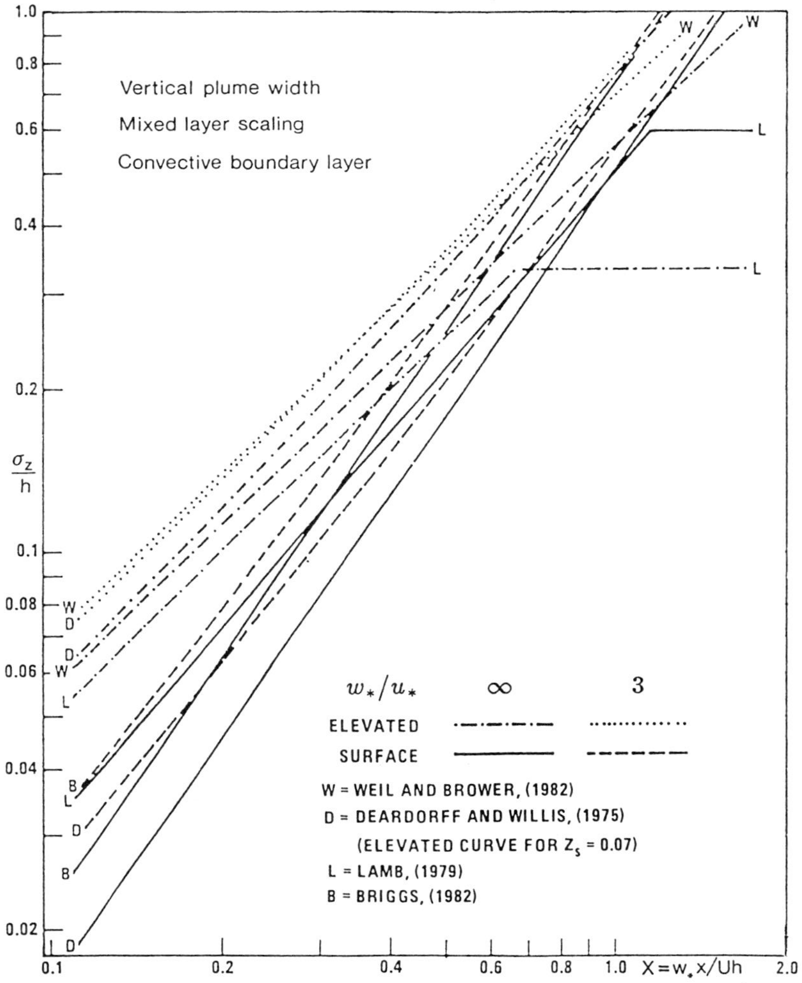

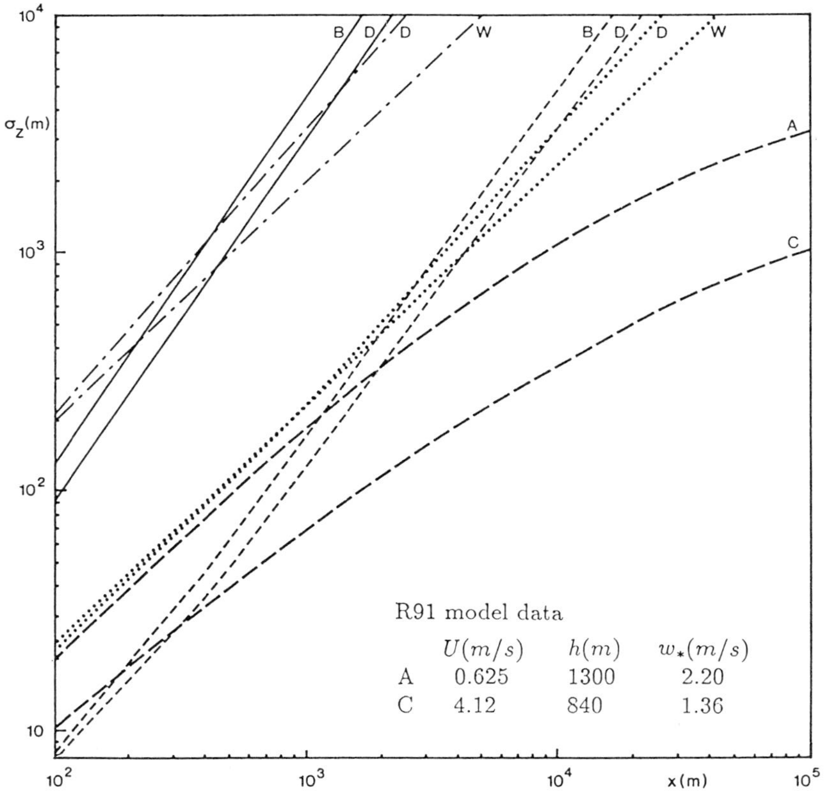

Fig. 4a Variation of σ_z downwind of sources in a convective boundary layer: left – data plotted in boundary layer co-ordinates ($X = w_* x/Uh$, σ_z/h) (from Briggs 1985); right – data plotted in terms of x and σ_z pairing cases $w_*/u_* = \infty$ and 3 with Pasquill categories A and C using data for the R91 model in Clarke (1979; values of U, h, w_* from Table A2; curves labelled A and C from Fig. 8).

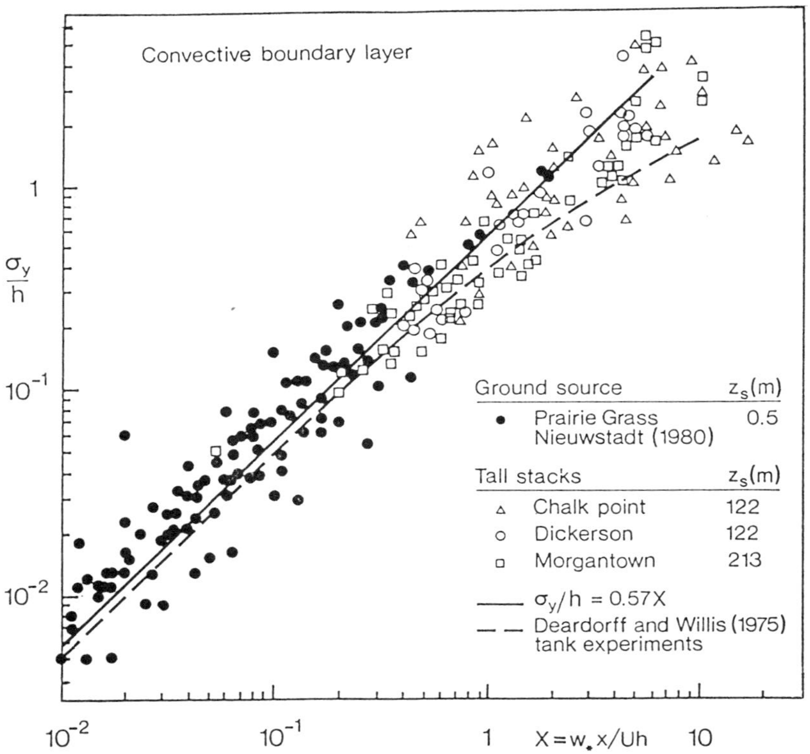

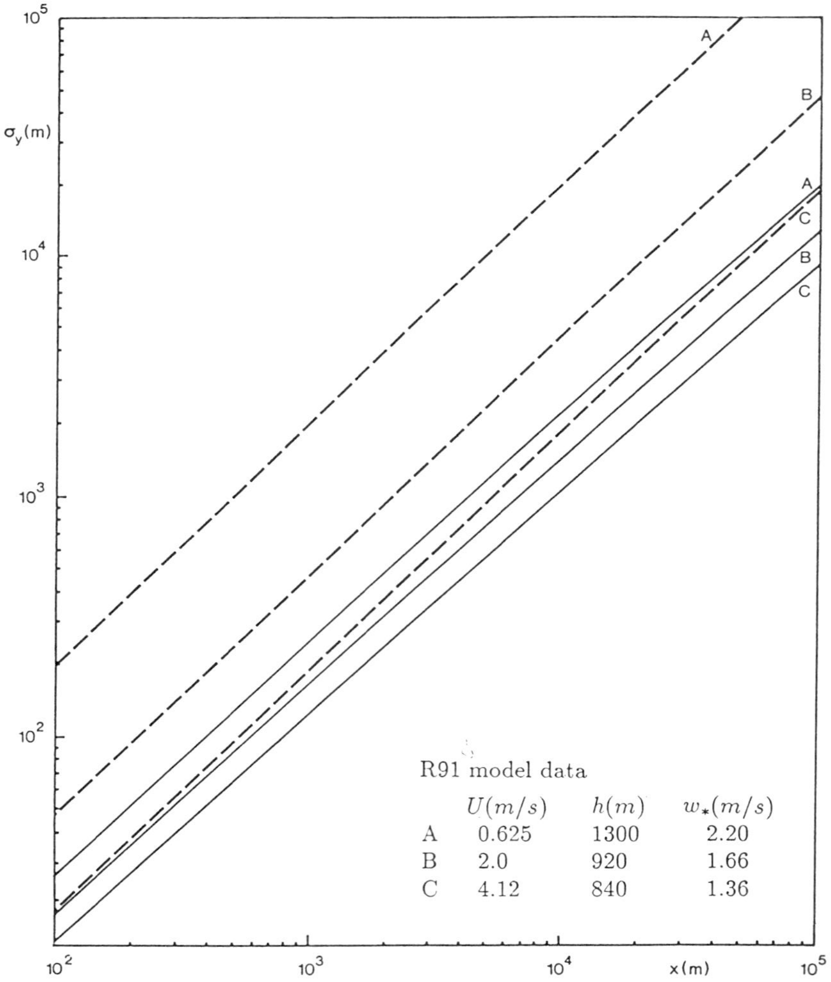

Fig. 4b Variation of σ_y downwind of sources in a convective boundary layer: left – data plotted in boundary layer co-ordinates ($X = w_* x/Uh$, σ_z/h) (from Weil & Brower 1982); right —- $\sigma_y/h = 0.57X$ plotted as σ_y against x (dashed) for Pasquill categories A - C using data for the R91 model in Clarke (1979; values of U, h, w_* from Table A2; curves (solid) from Fig. 10 with allowance for a 30 minute release).

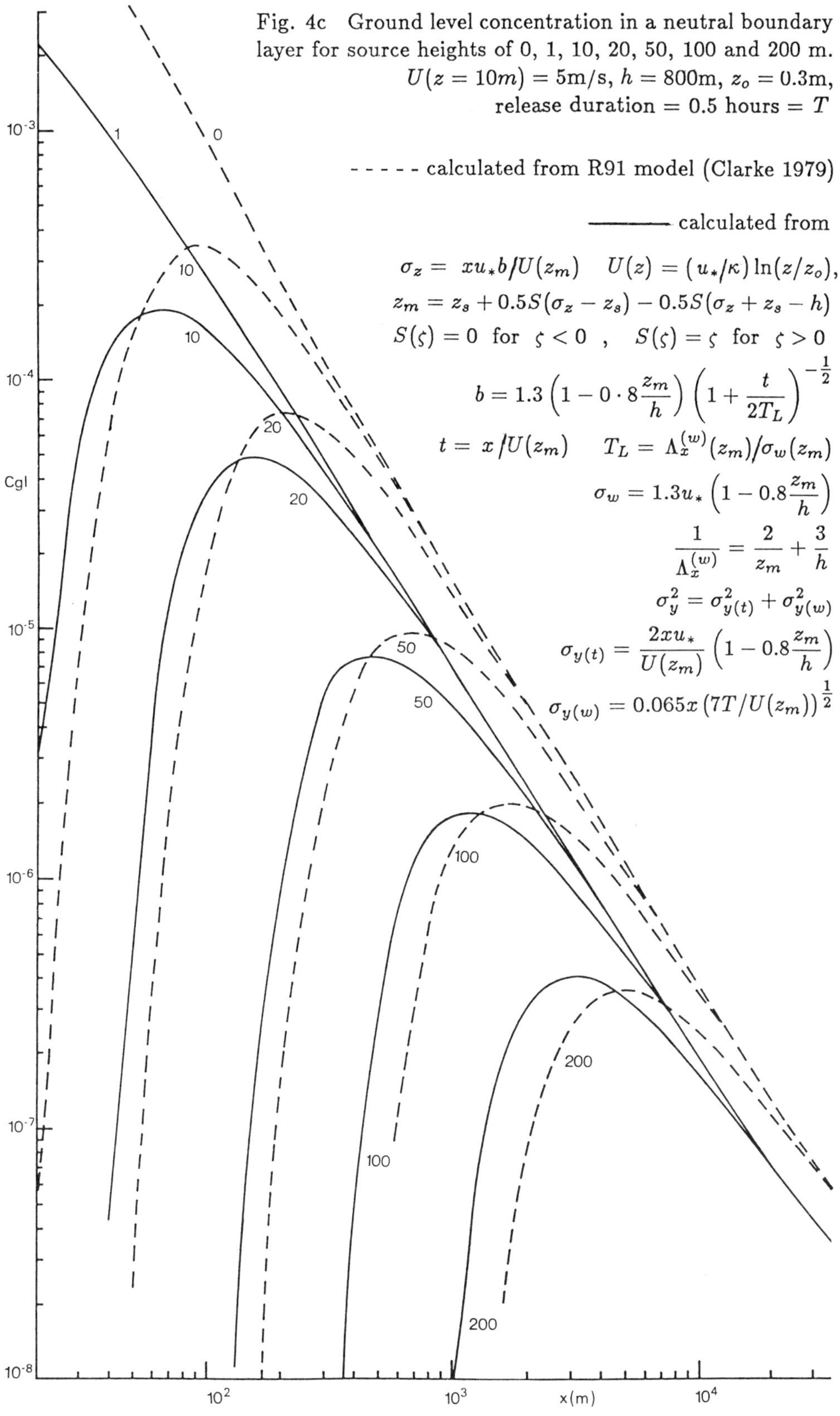

Fig. 4c Ground level concentration in a neutral boundary layer for source heights of 0, 1, 10, 20, 50, 100 and 200 m. $U(z = 10m) = 5\text{m/s}$, $h = 800\text{m}$, $z_o = 0.3\text{m}$, release duration = 0.5 hours = T

- - - - - calculated from R91 model (Clarke 1979)

——— calculated from

$$\sigma_z = xu_*b/U(z_m) \quad U(z) = (u_*/\kappa)\ln(z/z_o),$$

$$z_m = z_s + 0.5S(\sigma_z - z_s) - 0.5S(\sigma_z + z_s - h)$$

$$S(\varsigma) = 0 \text{ for } \varsigma < 0 \;, \quad S(\varsigma) = \varsigma \text{ for } \varsigma > 0$$

$$b = 1.3\left(1 - 0\cdot 8\frac{z_m}{h}\right)\left(1 + \frac{t}{2T_L}\right)^{-\frac{1}{2}}$$

$$t = x/U(z_m) \quad T_L = \Lambda_x^{(w)}(z_m)/\sigma_w(z_m)$$

$$\sigma_w = 1.3u_*\left(1 - 0.8\frac{z_m}{h}\right)$$

$$\frac{1}{\Lambda_x^{(w)}} = \frac{2}{z_m} + \frac{3}{h}$$

$$\sigma_y^2 = \sigma_{y(t)}^2 + \sigma_{y(w)}^2$$

$$\sigma_{y(t)} = \frac{2xu_*}{U(z_m)}\left(1 - 0.8\frac{z_m}{h}\right)$$

$$\sigma_{y(w)} = 0.065x\,(7T/U(z_m))^{\frac{1}{2}}$$

layer, but it changes as the plume enters the upper part of the boundary layer where the turbulence is changed by buoyancy forces.

Although the improved methods lead to some variations in σ_z for different source heights in neutral conditions there is, in fact, little change in the *maximum* ground level concentration compared with current methods. This point is illustrated in Fig.4c.

3.4 Concentration fluctuations

Fluctuations in concentration are important in many dispersion problems. Given our present understanding, estimates of concentration fluctuations are likely to suffer from somewhat larger errors than estimates of mean concentrations. However, by using a combination of theory and experimental data, it is possible to put together a scheme which should give useful guidance on the magnitude of fluctuations.

One approach to the problem of estimating concentration fluctuations is to divide it into two parts; first estimate the mean square concentration, and then parameterize the concentration probability density function (p.d.f.) in terms of the concentration mean and variance. A convenient approach to the first part is to adopt the formalism arising from Sawford's (1983) approximation in two-particle random-walk models of concentration fluctuations in isotropic turbulence. Sawford argued that the displacement of the centre of mass of particle pairs is close to Gaussian and independent of the particle separation. This leads to the result

$$\overline{c^2} = \sigma_c^2 + C^2 = \frac{Q^2}{U^2} G_2(\mathbf{x}\sqrt{2}, \sigma_\Sigma^2 + \sigma_0^2) \int p(\mathbf{r}, \frac{\mathbf{x}}{U}) G_2(\mathbf{r}, \sigma_0^2) d\mathbf{r} \tag{3.9}$$

where Q is the source strength, U is the mean wind speed, $G_2(\mathbf{x}, \sigma^2)$ is an isotropic two-dimensional Gaussian distribution (in the y and z directions) with variance σ^2, σ_0 is the size of the source distribution (assumed Gaussian), σ_Σ^2 is twice the mean square displacement of one component of the centre of mass of two particles, and $p(\mathbf{r}, t)$ is the p.d.f. of $\mathbf{r}$, the separation of two particles divided by $\sqrt{2}$.

The Gaussian assumption for the source distribution is, of course, unlikely to be accurate in practice but an attempt at a more precise description seems unwarranted in view of the other uncertainties associated with source effects. Note that when p is Gaussian, (3.9) takes the same form as in Gifford's meandering plume model (see e.g. Sykes 1988). Theory and random-walk simulations show that p is more peaked than Gaussian, resulting in an additional contribution to $\overline{c^2}$ arising from 'in-plume' fluctuations. Equation (3.9) can be expressed more simply in the form

$$\overline{c^2} = \overline{c^2(0)}(C/C(0))^q \tag{3.10}$$

where $\overline{c^2(0)}$ and $C(0)$ indicate centre-line values (see Fig.5a). In this form the result can take account of the fact that the spread in the y and z directions are different and can be used in conjunction with models which predict non-Gaussian concentration distributions. Of course, there is no reason to think that the departure from the isotropic results will be represented with any great accuracy using (3.10); however, with the right choice for $\overline{c^2(0)}$ and q, (3.10) should give reasonable estimates of $\overline{c^2}$.

For estimating $\overline{c^2(0)}$ and q over short travel times from the source, the turbulence can be regarded as approximately homogeneous. Then random-walk results (Thomson 1990) can be used with some confidence. In particular, they show quite reasonable agreement with the data of Fackrell & Robins (1982) and Mylne & Mason (1990). As a result, q and $\overline{c^2(0)}$ can be parameterized quite easily using scaling arguments, with the

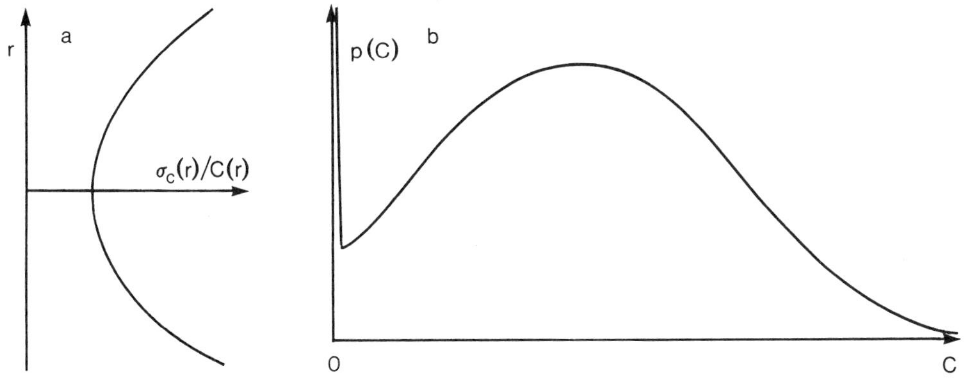

Fig. 5 (a) variation of concentration fluctuations in a plume with distance r at right angles to its axis; (b) a clipped normal probability density function is a good representation of that calulated from concentration measurements at a point in a plume. The delta function at $C = 0$ allows for the intermittent nature of the variable signal i.e. the finite periods of time when only pure air is observed between periods when air with varying traces of pollutant is observed.

constants taken from random-walk simulations. Away from the region where source size is important, $q = 1$, leading to a distribution of σ_c^2 which is rather wider than that of C^2.

At longer times inhomogeneity and shear become important and then the idealised theory is of less help in parameterizing $\overline{c^2}$. However, the pragmatic choice of $\overline{c^2(0)} \simeq 2C(0)^2$ (i.e. $\sigma_c/C \simeq 1$ on the centre line) and $q \simeq 1$ appears to give correct orders of magnitude in comparison with data. For surface sources Fackrell & Robins' (1982) data implies that σ_c decreases as the surface is approached, in contrast to results obtained using (3.10) with $\overline{c^2(0)} = 2C(0)^2$ and $q = 1$. In this situation the error in σ_c is probably not much worse than is expected generally at large times; however, there may be scope here for qualitative improvements.

There are now a number of observations (Lewellen & Sykes 1986; Sawford 1987; Dinar et al. 1988; Mylne & Mason 1990) that indicate that the clipped-normal distribution is an adequate approximation to the p.d.f. of concentration at a point (see Fig. 5b). If we make such an assumption, it is possible to predict the p.d.f. from C and $\overline{c^2}$. The intermittency, defined here as the fraction of time when clean air is observed, can also be derived from such an assumption and is an increasing function of σ_c/C. The "exponential distribution plus intermittency" has also been tested against data and has shown reasonable agreement in some situations. However, the clipped-normal provides much better agreement than the exponential when σ_c/C is small, such as when the averaging time is long enough to remove a substantial part of the concentration variance (Mylne & Mason 1990).

4. Incorporating complex effects into 'regulatory' models

4.1 Meteorology for complex effects:

In almost all practical dispersion calculations, additional effects have to be considered, in particular precipitation, buoyancy of the emitted matter, the disturbed air flow near the source caused by buildings, the non-uniformity of the air flow and stratification in boundary layer caused by variations in the elevation, roughness and temperature of the surface, and unsteadiness of the air flow. With the use of currently available microcomputer systems and more extensive meteorological and topographical information available in computerised data banks it is now possible to incorporate these effects into 'regulatory' models.

For example, many aspects of the spatial and temporal distribution of different types of precipitation are becoming available, especially with the use of radar networks e.g. Browning 1988. There is a large effect of precipitation on the deposition of gases and particles. Given the statistics of precipitation the statistics of concentration and deposition can be estimated (usually with the aid of simplifying statistical assumptions which differ from one model to another) (e.g. Jones 1981).

Calculation of the mean motion and mixing in a buoyant plume requires meteorological input data in the form of the mean velocity profile $U(z)$, the mean potential temperature gradient $d\theta/dz$ (or the buoyancy frequency N if $h/L_{MO} > 0$) and the profiles of integral turbulence quantities (e.g. Briggs 1973, Moore & Lee 1982). If the buoyancy of a plume is strong enough to carry it into any stable regions at the top of the boundary layer, quantities must be defined above as well as in the layer. In most cases of interest this information is only required for $z \lesssim 2h$.

On the other hand, for calculating air flow over hilly terrain it may be necessary to have meteorological data of the mean velocity $U(z)$, its direction $\alpha(z)$, and $d\theta/dz$ up to much greater heights, which are of the order of the horizontal length scale of the hills being considered. For the limited range of models considered here ($\lesssim$ 30km), and where the terrain length scales are of order 10km, it is certainly necessary to specify $U(z)$ and $N(z)$ up to $z \sim$ 10km. For example in the air flow studied by Carruthers & Choularton (1982) over Great Dunn Fell and over the Rockies by Lilly & Klemp (1979), elevated inversion layers at these heights controlled the flow.

In complex terrain the boundary layer structure is not the same as over flat terrain and can only be deduced from surface layer measurements by means of special calculations. The same information is required as for flow over level terrain (e.g. in particular u_* and F_{θ_0}), but it is important to realise that where the mean flow is changing (even by say 20% in 1km) the profiles are markedly changed by these small gradients. Then to estimate the changes in u_* and F_{θ_0} from mean profiles of $U(z)$ or $\theta(z)$, the measurement point would have to be close to the surface (less than 2m) (Hunt, Weng & Carruthers 1988).

Clearly for estimating dispersion in real-time following an accidental release there will be occasions when it is necessary to account for the unsteadiness of the meteorology, even over distances less than 30km. Many current models assume that the plume continually adjusts so as to be a straight line from the source as the direction of the wind changes. This only requires meteorological information at the source site. 'Improved' models (which need not perhaps be as computationally intensive as puff models) would benefit from information about how the flow field changes between the source and the receptor position, in particular the direction and strengths of any fronts.

4.2 Buoyant plume effects

The initial buoyancy and momentum of the matter released into a cross flow have a significant effect on the resulting distribution of concentration, in particular its maximum values in the air and at the surface. The principal basis of practical models and their essential approximation form were defined by the 1950's. These models give the mean vertical displacement at the centre line of the plume, z_p, as a function of down wind distance x, and also the mean radius R_p, mean velocity components u_p, w_p, and mean temperature θ_p. An essential concept of the models is the rate at which fluid from the surrounding flow is 'entrained' into the plume and jet. During the initial stage of the plume this process controls the rate of growth of Rp and how the plume or jet bends over, even in turbulent flow (Hoult & Weil 1972; Coelho & Hunt 1989).

Despite the fact that plumes have a complex non-axisymmetric internal structure (even in some cases splitting into 2 separate plumes) their initial mean motion and growth (i.e. $z_p(x)$, $R_p(x)$) can be modelled by coupled ordinary differential equations, in which the variable coefficients are derived from the mean velocity $U(z)$, potential temperature gradient, and turbulence profiles in the atmosphere. The different models tend to differ in these coefficients. Examples of laboratory experiments by Hunter (1990) of plumes in a neutral boundary layer capped by strong and weak inversion layers are presented in Fig. 6, and are shown to agree approximately with the model of Ooms (1972).

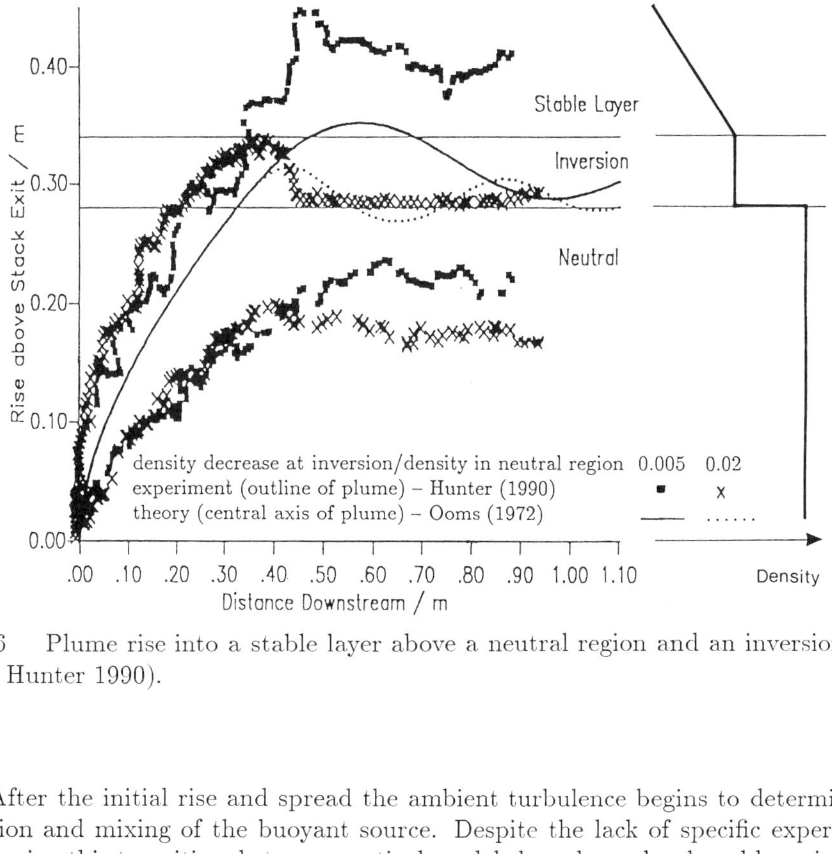

Fig. 6 Plume rise into a stable layer above a neutral region and an inversion layer (from Hunter 1990).

After the initial rise and spread the ambient turbulence begins to determine the diffusion and mixing of the buoyant source. Despite the lack of specific experiments to examine this transitional stage, practical models have been developed by using data from high stacks at power stations (e.g. Moore, Briggs). These models need further laboratory and field testing for low level sources where the turbulence is stronger. Computations using large eddy simulation (e.g. Nieuwstadt & Van Haren 1988) may provide equivalent data, although there may not be sufficient resolution of the flow within the plume. It is particularly important to understand the physics of this process when buoyant plumes are also influenced by nearby buildings or hills.

4.3 Effects of buildings

A number of additional processes affect the dispersion of releases near buildings. The dominant ones are *entrainment* (material caught up and rapidly diffused throughout the recirculating eddy behind a large surface obstacle), *downdraught* (the general downward displacement of streamlines behind the building), *turbulence* (which enhances plume spread in the extensive downstream wake), and *lift-off* (whether a release contains sufficient buoyancy to maintain its tendency to rise after its rapid initial dilution).

The behaviour of emissions for arbitrary source and building configurations is complex. The majority of prescriptions to account for building-affected dispersion have centred either on

- *low-level, full-entrained* releases (e.g. Gifford (1960), or the virtual source model of Barker (1982)) or
- *elevated* releases (either an effective stack height reduction, as in Briggs (1973) or a modification of the σ-curves, as suggested by Huber (1984)).

These correspond to the types of emission generally met at nuclear or fossil-fuelled power stations, respectively.

A general atmospheric dispersion model involving parameterised equations must be flexible enough to describe plumes and dispersion from a variety of sources and configurations. For example, the electricity supply industry in the U.K. is currently applying for permission to build generating plant with lower capital cost. These plants have shorter stacks and a modular form of construction. A *composite* model is needed to describe the effects of such structures.

The aims of a model for building effects should be:

(i) to incorporate building effects within a simple and general dispersion model;

(ii) to account for source height and admit partial entrainment;

(iii) to allow for the extended influence on dispersion due to the downstream building wake;

(iv) to be able to predict concentrations at all downstream positions.

A model that meets these requirements has been developed by Apsley (1988) from earlier work by Robins (1983); its main features are:

(i) a *double plume* structure when the release is only partially entrained (Puttock 1978). This models the thin plume of the non-entrained matter and the wide plume of matter that is entrained and then diffused out into the wake (Fig. 7);

(ii) *uniform near wake concentrations* within a well-mixed recirculating flow region. This only gives an average concentration in this region; more detailed models are needed to predict the distribution (e.g. Turfus 1986, 1988);

(iii) a modified Gaussian plume model in the turbulent wake with the plume centre-line following the streamlines of an analytical small perturbation wake model (Hunt 1982) and modified σ_y, σ_z curves for material diffusing within the turbulent wake (an early version of this model was presented by Robins & Fackrell (1980)).

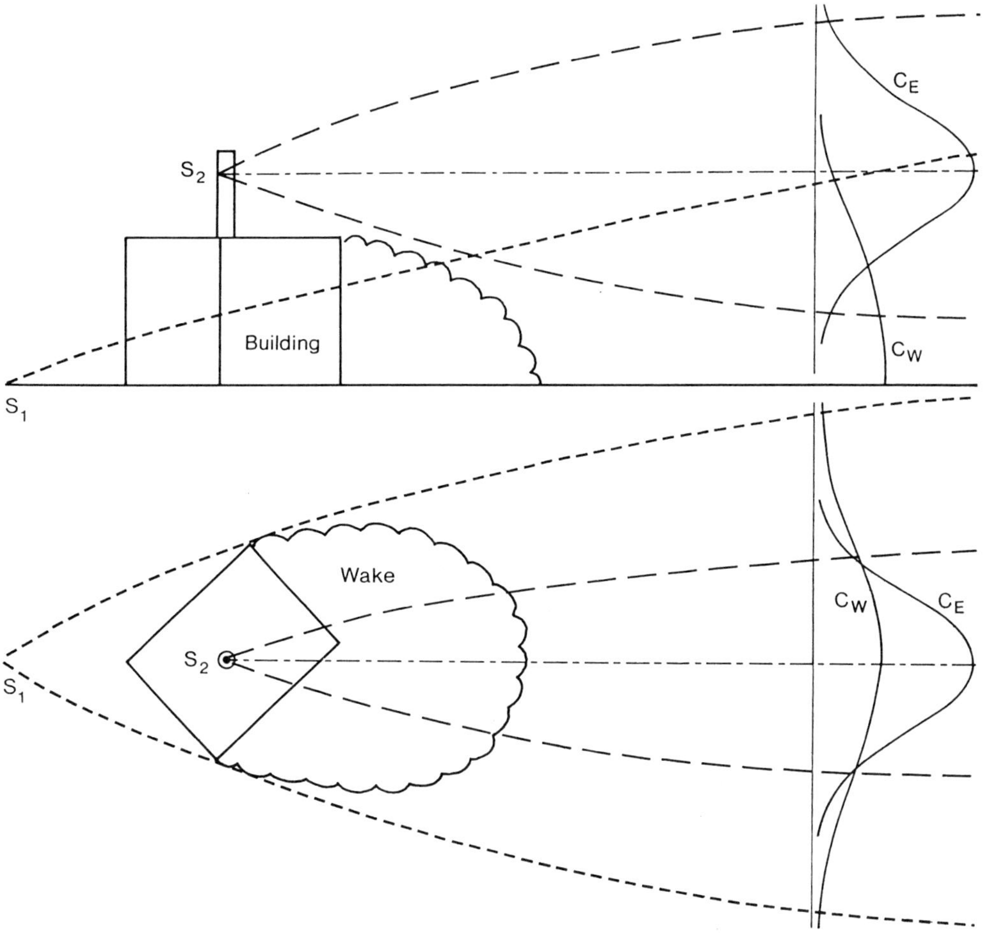

Fig. 7 Modelling the plume from a source of strength Q located on or near a building and partially entrained into its wake by a ground level source S_1 of strength $(1-\alpha)Q_M$ and an elevated source S_2 of strength αQ_M (top-side elevation; bottom-plan). S_1 produces a broad plume. It is assumed that the plume from S_2 is unaffected by the wake. The mean concentration distributions for each plume downstream of the building are as shown (C_W for S_1, C_E for S_2).

4.4 Wind field models for complex terrain

4.4.1 Flat terrain plus warnings

The simplest and most widely used assumption about wind fields over complex terrain is that they are not very different from the flow over flat terrain. Therefore, the height of the mean streamlines above the terrain does not change nor does the turbulence. Some models based on this simple approach employ a 'warning' or 'flagging' procedure, to indicate under what circumstances the model can lead to significant errors (see Jones (1986) from which Table 2 is taken).

4.4.2 Simple deflection/impingement models

A characteristic feature of air flow over hills is that the mean streamlines become closer to the ground. For steep hills and in very stable conditions plumes can impinge onto the hillside which can lead to a significant increase in surface concentration.

Table 2. Criteria for neglecting orographic effects (from Jones (1986, Table 1)).

In neutral and unstable conditions

- the gradient of the surrounding terrain should be less than about 1 in 10

- for a ridge upwind of the source
  either $h > 1.5\ H$
  or $x > 20\ H$ in neutral conditions
  $x > 10\ H$ in very unstable conditions

- for an islated hill upwind of the source
  either $h > 1.5\ H$
  or $x > 7\ H$

- for a hill or ridge downwind of the source
  either $h > H + \sigma_z\ (x)$
  or $\sigma_z\ (x) > H$

In stable conditions

- the gradient of the surrounding terrain should be less than about 1 in 100.

- for an obstacle upwind of the source
  either $h > H$
  or $x > 40\ H$ in slightly stable conditions
  $x > 100\ H$ in very stable conditions

- for an obstacle downwind of the source
  either $h > H + \sigma_z\ (x)$
  or $\sigma_z\ (x) > H$

Notes

The criteria are based on a change of 30% in the wind speed between flat and complex terrain.

The 10 m wind speed must be at least 1 m s^{-1}.

h = the release height.

H = the obstacle height.

x = the distance between obstacle and source.

σ_z (x) = the vertical plume standard deviation.

For some regulatory purposes it may be sufficient just to estimate a typical average affect of the terrain on the dispersion, for example the CRSTER model and its derivatives COMPLEX I, COMPLEX II developed by EPA can be justified on this basis. The dispersion process is assumed to be the same as that over level ground from a source with the same height; in other words, any changes in turbulence are neglected.

To improve the coarse approximation that all streamlines in stable conditions are horizontal, an improved model, RTDM, was developed. This model uses the concept of the 'dividing streamline' of height H_c to discriminate between streamlines that pass over the hill top and impact on the hill side (Snyder et al 1985). For a source of height z_s which lies between H_c and the height of the hill H, the mean streamline passes over the hill. Effectively the flow pattern is similar to that over a hill of height $(H - H_c)$. For z_s below the critical height H_c, the plumes travel in straight lines to the hill. Thus in RTDM, for unstable and neutral conditions, the air flow model is the same as COMPLEX I, II but in stable flow the deflection of the mean streamline through the source depends upon the relative heights of the source z_s, the dividing streamline H_c, and the crest of the hill.

For other regulatory purposes it may be sufficient just to estimate the *largest* effect of the terrain on the surface concentration. In the EPA VALLEY model it is assumed that the worst effect that can occur can be estimated by assuming that a mean streamline is directed *straight* from the source to the hill side in stable conditions. This model has been criticised (e.g. in the review by EPA and the American Meteorological Society; White et al. 1985) for some of the unsatisfactory physics in its assumptions about the air flow and about the dispersion process on to the hill side which can lead to unrealistically high predictions of concentration.

The review by White et al. (1985), comparing models with field experiments, showed that RTDM was a reasonably successful model for predicting the impingement of plumes onto hills in stable flows and for accounting for many neutral and unstable flow conditions. Even so, they recommended that further developments were needed.

The common feature of these deflection/impingement models is that they are based neither on the details of the shape of the hill, nor on any calculations for the flow over the terrain to assess the likelihood of these flows actually occurring.

4.4.3 Interpolation of the wind field using terrain data and stratification

In areas where there are a number of meteorological wind-measuring stations, such as urban areas with nearby mountains and coasts, e.g. Los Angeles or Athens, it is natural to develop methods of modelling wind fields which make optimum use of data available. In such complex wind fields even the most extensive flow field calculations are likely to be in error, and also to require huge computing capacity. This approach has been developed in the U.S., particularly for the Lawrence Livermore National Laboratory by Sherman (1978) and for Rockwell's Rocky Flats Plant by Restrepo (1987); in Europe at Imperial College London by ApSimon et al. (1984), at Karlsruhe by Moussiopoulos & Flassak (1986), and in Greece by Lalas et al. (1983). For reviews and discussions, see Lewellen et al. (1982) and Hunt et al. (1984).

At its basic level, this method requires measurements of the horizontal wind speed and direction at a number of meteorological stations, usually at only one level. The height of the terrain is assumed to be known everywhere in the flow region. Different models require different levels of input about the stratification. In the simplest case, a single parameter roughly equivalent to an inverse Froude number is used but other models require input at more levels and details of the strength of any inversion layers.

The method is to construct a three-dimensional velocity field for the region that is mass consistent (i.e. there are no sources and sinks of fluid within the region) without solving the momentum equations for the flow. It is assumed that values of the constructed velocity field are as close as possible to the values that would be obtained by interpolating between the measured values. So, it is first necessary to interpolate between the measured values and also assume how they vary with z; then the velocity field can be constructed. The essential mathematical point about this method is that if the interpolated velocity field is uniform and the hill symmetrical, then the constructed flow field is also symmetrical and the perturbed flow is just potential flow. No account is taken of wakes behind hills, surface roughness changes, thermal effects, or waves.

The usual computational procedure is to divide up the flow region into cuboids, reduce the differential equations to finite-difference equations, and represent the terrain on this cuboid grid. Typically the depths Δz of the boxes are small near the surface and increase up to the top of the domain ($\sim$ 5km). Moussiopoulos & Flassak (1986) have developed the method to include a terrain-following co-ordinate system. This is also the basis of the NOABL code used widely for wind-energy calculations. For realistic computations over a horizontal scale up to 100km, 30^3 boxes are required, which implies a horizontal length scale of about 3km.

The typical time required for such calculations on a large mainframe computer is a few minutes while storage requirements amount to about seven times the number of grid points. These facts have prevented the use of mass consistent schemes on smaller computers.

The conclusion of many studies (including our own) is that, provided certain conditions are met, mass consistent models can provide a satisfactory and fast approach to mesoscale wind field modelling on scales from about 5 to 100km. Ideally the spacing of the measuring stations should be about the same as the length scale over which the topography and surface conditions change; then comparative tests indicate that these interpolation models are as good as any other system requiring a mainframe computer (see discussion by Hunt et al. 1984). However, for complex flows on scales below 10 km, with sparse data, these methods can be unreliable and it is necessary to compute the flow, without the idealising assumptions used at the larger scales. We consider this scale of motion next.

4.4.4 Predictive computations using terrain data and differential equations

The predictive approach to wind field modelling is similar to that used in weather forecasting and in computational fluid dynamics.

(a) **Equations** It is assumed that the differential equations governing the flow are known everywhere in the flow region R (see Fig. 8). These equations are usually approximations to the full equation of fluid flow because only a few statistics of the flow field are required (for example either the mean flow itself or the mean flow and the second moments of the velocity fluctuations).

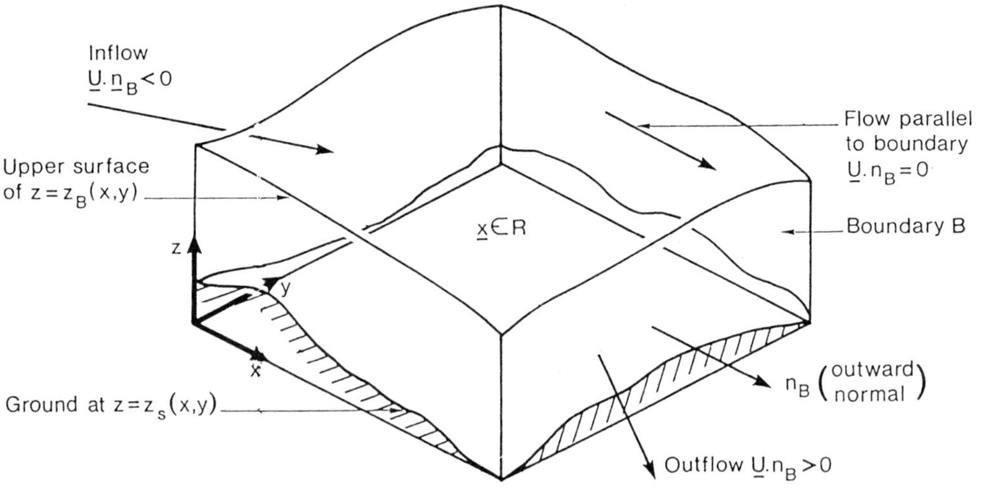

Fig. 8 Notation used when computing flow in a region R above complex terrain.

(b) **Boundary conditions** The equations can be solved to define the flow, i.e. velocity, temperature, pressure field, in the region R after a time t_o provided either the flow is defined everywhere in R at time t_o (for a time dependent problem in a sufficiently large domain), or, more usually, the velocity and fluid properties are defined on the boundaries B of R. In this case the flow has to be defined both at the ground, where $\mathbf{U} = 0$, and an upper surface (e.g. $z = z_B$), and the temperature, surface heat flux, water vapour, radiation need to be defined explicitly or in terms of supplementary equations.

(c) **Simplifications**

(i) **No initial conditions required**: if the elapsed time for the calculation is much greater than either the time for the mean flow to pass over the region or the time over which the upwind surface conditions change, then the *initial* state of the atmosphere does not have to be defined. For example, over a domain of 30km with an average wind speed of 5m/s, neglect of initial conditions would be justifiable after a period of 2 hours.

(ii) **Quasi-steady**: if the length L of the typical variations of surface conditions (e.g. hills, roughness, etc.) are small enough and if the time scales T_B of the approach flow, the upper level air flow and the surface conditions (such as temperature) are long enough for $T_B \geq L/U$ then the unsteady terms, can be neglected. This leads to a considerable simplification. Of course, as the boundary conditions vary with time, so do the flow variables. The assumption means that the flow responds to the instantaneous values of the boundary conditions.

(iii) **Hydrostatic**: For many overall features of mesoscale flows, the hydrostatic approximation is satisfied and it is used in some well-known schemes such as that of Pielke (1984), Pielke et al (1990). The criterion is that $(U/NL)^2(h/L) \ll 1$, so that vertical acceleration can be neglected in the momentum equation. For $U \simeq 10$m/s, $N = 0.01s^{-1}$, $h = 1$km; this implies that $L \gg 1$km.

(d) **Computational methods**

(i) **Constructed non-linear flow model for idealised terrain features**

If the primary goal of an easily computable wind field model is to represent strongly stratified flow (i.e. $F_H < 1$) over and around isolated hills and mountains, it is necessary to have a *non-linear* flow model to account for the large changes in flow speed and direction associated with this kind of flow. The only non-linear stratified flows that can be approximately modelled by analytical methods are flows around idealised shapes such as ellipsoids (Fig. 9). Using this concept, Paine et al. (1987) have developed a computer code CTDM for flow and dispersion impinging into isolated hills and mountains. Its extension, CTDMPLUS, is described by Perry & Finkelstein (1990).

The first step is to approximate the actual terrain by a set of ellipsoids or, in some versions of the code, bell-shaped hills. The second step is to define the Froude number F_H for each hill of height H and if $F_H < 1$, the critical dividing streamline height $H_c = H(1 - F_H)$. Below H_c the flow is modelled as a potential flow in horizontal planes. For streamlines starting upwind above H_c, the flow is modelled approximately as a moderately stratified flow over a hill of height $H - H_c = HF_H$. If $F_H > 1$, the flow is modelled as a potential flow perturbation solution for flow over the idealised shape. There is no matching with the surface boundary-layer flow and wake effects are also ignored.

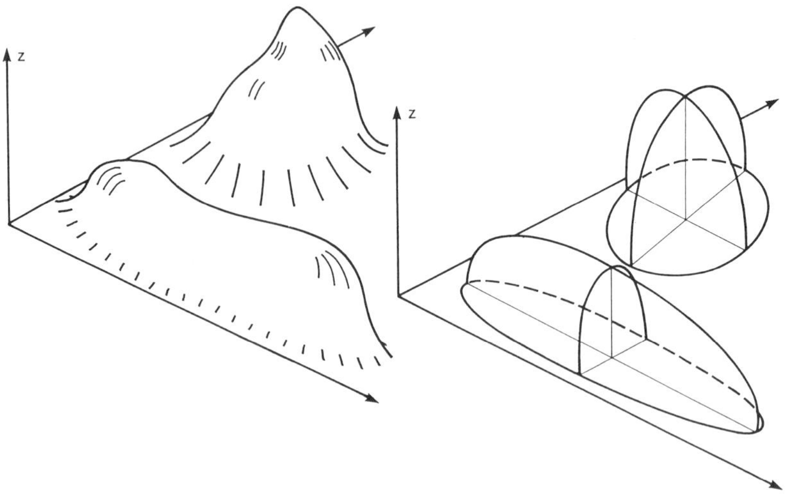

Fig. 9 Representation of arbitrary terrain (left) as a number of isolated semi-ellipsoids in (right) CTDM model.

In general, the CTDM code has been developed to ensure a very good fit with the air flow and dispersion measurements in three field experiments – an isolated 100m hill (Cinder Cone Butte), a long ridge about 85m in height (Hog Back Ridge), and amongst mountains rising up to 1km over distances of 7km near Reno, Nevada (Tracy experiment). The main aim has been to produce a dispersion model for studying the high concentrations associated with plumes impacting on hills; it is less appropriate for the air flow over rolling terrain in near neutral conditions. In addition, it does not allow for interaction between flows over neighbouring hills.

This model is now operational in the U.S.A. and it is likely to replace the streamline impingement model VALLEY.

(ii) **Perturbation/analytical/Fourier methods**

A cursory look at computations or wind-tunnel measurements shows that usually an air flow does not significantly change its direction by more than 20% or its wind speed by more than 100% as it passes over hills or roughness changes. Consequently, one approach to modelling air flow is to assume that the effect of hills and roughness changes is merely to give a small perturbation to the mean flow $U(z)$ of $\Delta\underline{u}(=\Delta u,\ \Delta v,\ \Delta w)$. As in many analyses in engineering and science, perturbation methods gives useful results even when the perturbations are of the same order as the basic flow. Analytical solutions to the linearised equations have been found for certain typical profiles of $U(z), \theta(z)$ and for flow over regular sinusoidal hills, and for flow over sinusoidal variations in surface roughness length.

Earlier authors obtained approximate solutions for neutrally stable flow with logarithmic profiles for $U(z)$ (Jackson & Hunt 1975; Walmsley et al. 1982). Later authors have obtained more accurate solutions for a wider class of profile and have allowed for the effects of stratification over the hills (Hunt, Leibovich & Richards 1988; Hunt, Richards & Brighton 1988). Effects of roughness changes have been introduced by Walmsley et al. (1986) and Belcher, Xu & Hunt (1990).

Because the analysis is *linear*, it means that the perturbation is doubled by doubling the height of a hill; more significantly it also means that perturbations produced by several superposed sinusoidal hills or areas of roughness can simply be added. Therefore, since by the usual method of Fourier analysis, the shape of any hill can be represented by adding many sinusoidal hills, the air flow over any hill can be calculated by adding the solutions for flow over each of its sinusoidal components. Roughness can be handled in the same way. This result is represented schematically in Fig. 10; the

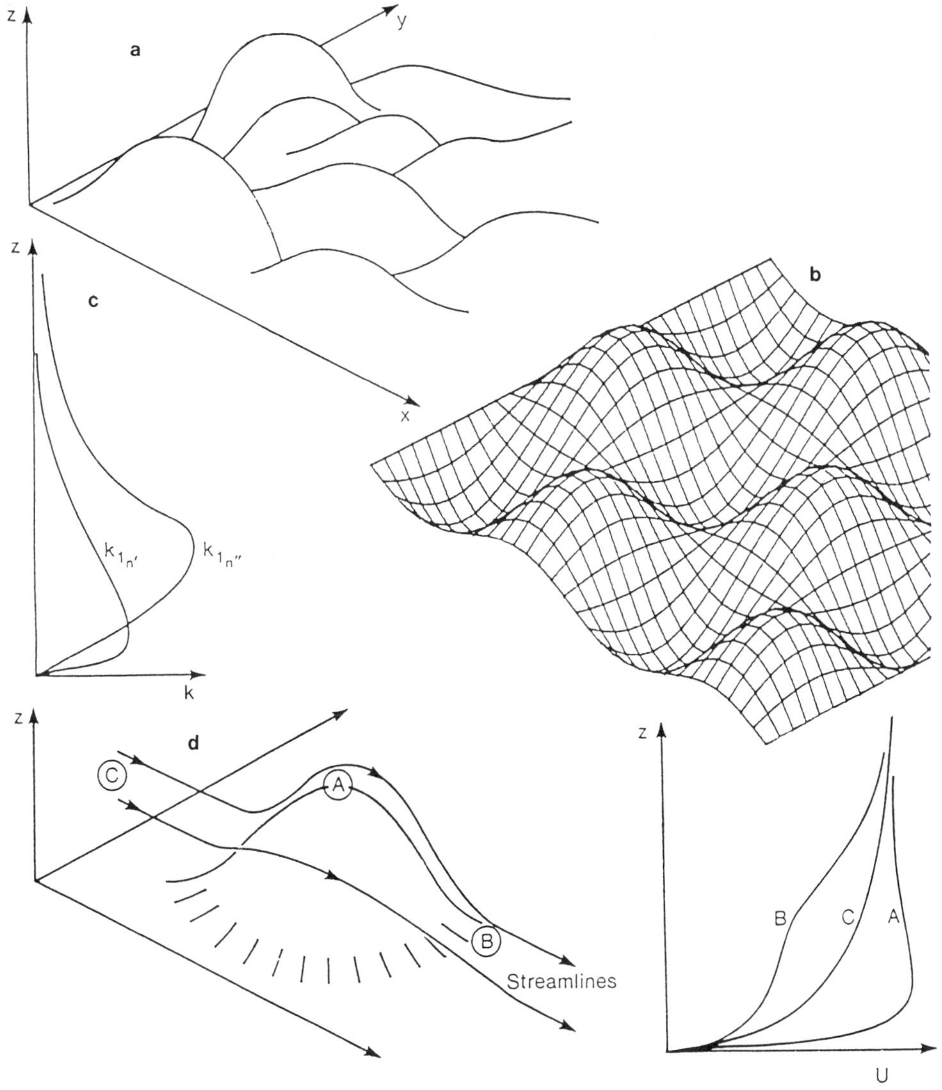

Fig. 10 Illustration of method of solution by linearised Fourier analysis:

(a) actual terrain;
(b) one component of Fourier transform of terrain;
(c) typical vertical profile of amplitude of velocity perturbation for different wave numbers;
(d) typical streamlines (left) and velocity profiles (right) after superposing solutions for each wave number.

terrain is defined on a rectangular $N \times M$ grid in the x and y directions and represented by a finite length double Fourier series with NM components. The flow over each component is calculated and the solutions added to yield the flow over the terrain.

The advantage of this method is that, apart from being very fast on mainframe computers, it can be used on small personal computers. The storage required for three velocity component is $N \times M \times 3$. Thus for a typical region, where the terrain is described by 64×64 points in the horizontal plane only 12500 or so points need to be stored. Since the perturbation quantities are given in terms of analytical functions, the vertical variation of $\Delta \underline{u}$ etc. can be calculated in either great or very little detail. This is useful for streamline calculations for a localised source since only the flow along and near the streamline through the source need be calculated. In the mass consistent interpolation models and full computational methods the whole flow has to be calculated. To compute flow at 3 levels above, along and below a typical streamline over a terrain defined for 64×64 points takes about 15 minutes on an IBM PS2 (or equivalent). The computation of turbulence profiles along these streamlines takes about a further 1 minute. With the new array processors for personal computers now becoming available, it will be possible to employ much larger grids (say 256×256 points) and achieve a much better resolution in the same sort of time.

Computer codes for flow over complex terrain based on the perturbation approach terrain are now available from various organisations in Europe and the U.S.A. but so far only the CERC model FLOWSTAR allows for stratification. This has been compared quantitatively with field experiments of neutral and moderately stratified flow over isolated hills, and with numerical computations and laboratory experiments for many neutral and stratified flows. Some of these are described by Carruthers et al (1988). Recently algorithms to calculate the turbulence field have been incorporated into the model. An example of such predictions is shown in Fig. 11.

The disadvantages of this method are:

- there are errors if the slopes of the hills are greater than about 1/4-1/3. Firstly there are errors in the predictions for the maximum increase in speed over the hills, although this is generally only of the order of 20%. More seriously, the model underpredicts the slow down in the wind speed, the formation of reverse flow, and downwash on the lee slopes and in the wakes of hills. This error, which extends over a larger volume of the flow for isolated hills than for rolling terrain, has been analysed and discussed in detail by Britter et al. (1981) and Mason & King (1985). Formulae or particular computational schemes developed specifically for wake regions have been proposed by Counihan et al. (1974) and Weng et al. (1988). These methods can be incorporated into the perturbation approach as corrections;
- there are errors if the stratification is strong (i.e. if the Froude numbers based on the hill height $F_H < 1$). In this case the air flow below the critical dividing streamline height H_c moves in approximately horizontal planes around the hill. Therefore at each level below H_c there is a point on the hill where the velocity of the approach flow is zero, and a significant zone where the velocity is much less than $U(z)$; the flow, in fact, changes direction by up to 90° and moves parallel to the hill side. Above the critical height the air flow passes over the hill as if it has a height $H^* = (H - H_c)$ with a Froude number $F_{H^*} \simeq 1$. Unlike the mass consistent interpolation methods (such as WAFT) or the equivalent-ideal-shape method (such as CTDM described next), current perturbation codes (such as FLOWSTAR) do not describe this flow regime; but it would be straightforward to incorporate a procedure similar to CTDM into them for individual hills;
- The model does not allow for drainage or other thermally-driven winds.

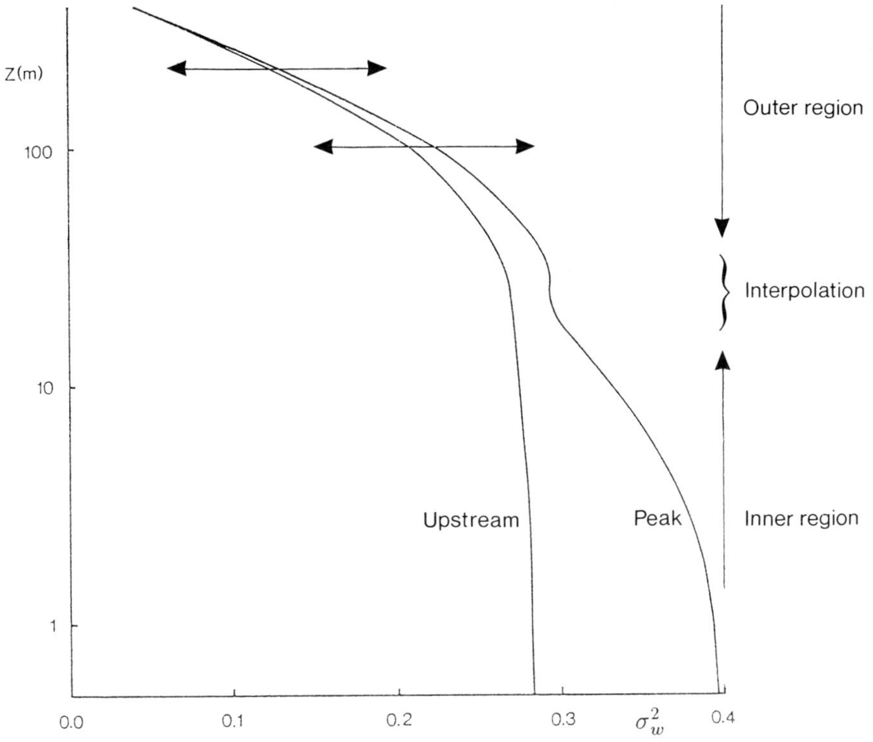

Fig. 11 FLOWSTAR calculation of the variation of σ_w^2 with height in a neutral boundary layer over a sinusoidal ridge $z_s = \frac{1}{2}h(1 + \cos 2\pi x/L)$ with h =100m, L =1000m. The analytical solutions for the inner and outer regions are linked by an interpolation function. The horizontal arrows indicate the range of uncertainty in measurements in the outer layer.

(iii) Finite-difference computations of non-linear equations

Mesoscale models used by meteorological organisations are based on current methods of forecasting synoptic scale weather using the 'numerical' computation of solutions to the governing fluid and thermal equations. Clearly these models incorporate sometimes more and sometimes fewer physical processes than the smaller-scale models described before.

In the widely known and used model of Pielke (1984) the equations are simplified by making the hydrostatic approximation. The topography is 'smoothed' by being represented on an 11km horizontal grid with the whole grid being deformed over the topography, and the flow calculated at height levels with a maximum spacing of 2km at the highest level using equations simplified by making the hydrostatic approximation. The model has been tested for air flow over mountains in Virginia, hills in Israel (Mahrer & Pielke 1985), and for sea breezes off the U.S. Coast. In the latter tests comparison of predicted and radar measurements of convergence zones, cloud growth and rainfall show reasonable agreement.

In the U.K. Meteorological Office model developed by Golding (1987) and others, the horizontal grid comprises 8099 points with an interval of 15km, while vertically there are 16 levels, with Δz ranging from 10m near the ground to about 1.5km at the top of the domain at 12km (even this height may not include all cloud activity in summer.) This mesh is deformed over the filtered topography and the non-hydrostatic form of the equations are used to permit fine resolution. The model covers the whole of the U.K. and is used primarily for forecasting local weather. It is intended to increase the vertical resolution to 32 levels.

Other non-hydrostatic mesoscale models have been developed. For example, the French scheme developed under the auspices of Electricité de France (EDF) (Caneill & Buty 1988) is interesting in that it has a nested system of air flow models. PERIDOT, the French Meteorological Office Model, gives a 24 hour forecast on a 35km horizontal grid spacing. This provides the boundary conditions to the EDF hydrostatic model HERMES which has a grid of 5 to 10km and is used over a regional scale of, say, 300km × 300km. Then for local scale predictions, say over an area of 30km × 30km, they have developed a small-scale numerical model MERCURE, whose grid spacing is 0.5km to 5km. For on-line use (NOWCASTING) they use data from a network of ground-level and remote-sensing measurements in the areas of interest, such as near power stations. These data are used to construct a mass consistent flow field over a region on a scale of about 100-300km using the code MINERVE which employs methods similar to those described in §4.4.2(ii). This flow field may then be combined with the local predictive model MERCURE.

4.5 Dispersion models for complex surface conditions

4.5.1 Methods that avoid flow field calculations

The models of air flow over complex terrain used for regulatory dispersion modelling can be described briefly as

(a) flat terrain plus warning, (e.g. Jones 1986, (see Table 2.)),

(b) corrections to flat terrain concentrations based on graphs, tables from computation and experiments (e.g. Snyder & Thompson 1985, Castro & Snyder 1982),

(c) deflection/impingement models, e.g. CRSTER, COMPLEX I, II.

These flow models are used in conjunction with Gaussian plume models, in the form

$$C = \frac{Q}{\sqrt{2\pi} U_\psi(z_\psi)\sigma_y} \exp\left(-\frac{(y-y_\psi)^2}{2\sigma_y^2}\right) G_z(z, z_\psi) \tag{4.1}$$

where

$$G_z(z, z_\psi) = \frac{1}{\sqrt{2\pi}\sigma_z}\left[\exp\left\{-\frac{(z-z_\psi)^2}{2\sigma_z^2}\right\} + \exp\left\{-\frac{(z-z_{\psi_i})^2}{2\sigma_z^2}\right\}\right] \tag{4.2}$$

and z_ψ is the height of the mean streamline ψ_s through the source. The image source at $z_i = z_\psi - 2n$, where $n = (z_\psi - z_s)$ is the distance from ψ_s to the surface where $z = z_s(x, y)$. Note that the plume is assumed to be straight and parallel to the wind at the source point $y = y_\psi$. Formulae for the height z_ψ are defined without a flow model. The mean speed U_ψ is usually $U(z_s)$.

Models such as COMPLEX I, II can predict the high surface concentration associated with plume impaction approximately but for plumes passing over hills they are unreliable because of the inadequate estimation of the plume height z_ψ, σ_z and σ_y.

4.5.2 Dispersion models for complex surface conditions

Improvements to simple plume deflection models such as COMPLEX I, II, have been developed which make use of the calculated mean flow streamlines, their convergence and divergence and the variation of the turbulence along them and the Gaussian plume models. Usually the plumes are assumed to be 'thin', in the sense that the effects of the variations of mean wind speed and turbulence across the plume are assumed to be small (Hunt 1985). This assumption, which generally is not valid, usually appears to produce only small errors in concentration profiles.

The general form for the mean concentration distribution in a plume that travels over a hill, but does not impact upon it, is given by (4.1) where U_ψ is the mean velocity along ψ_s and $y_\psi(x)$ is the y co-ordinate of the mean streamline. This formula can be used in neutral, weak and moderately stable conditions, but it is only appropriate if the slope is not too great (say less than 1/3). For steeper slopes, local co-ordinates normal and parallel to the surface should be used. When the slope is steep enough for the mean to streamline impact onto the hill, (4.1) is not valid near the point of impaction.

Equation (4.1) was derived from statistical theory by Hunt & Mulhearn (1973) and generalised to three-dimensional flow by Hunt (1985) for the case $n/\sigma_z \gg 1$. In that derivation σ_z, σ_y are calculated in terms of the properties of the turbulence approaching a hill and of the mean flow over the hill. Various simplifying approximations have been developed. If the diffusivity is assumed to be the same as over level ground $\hat{\sigma}_z(t) = \sigma_z^{(o)}(t)$ but a better approximation is

$$\hat{\sigma}_z(t) \simeq \frac{< \sigma_w(t) >}{\sigma_w^{(o)}} \sigma_z^{(o)}(t) , \tag{4.3}$$

where $< \sigma_w(t) >$ is the average value of σ_w along the plume centreline, and $\sigma_z^{(o)}, \sigma_w^{(o)}$ are the plume depth and vertical turbulence in the absence of the hill. Equation (4.4) was found by Britter to give a good estimate of growing plumes impinging onto a body (Hunt, Britter & Puttock 1979).

It is certainly a poor approximation to assume no variation in σ_z over complex terrain, i.e. $\sigma_z(x) = \sigma_z^{(o)}(x)$, and simultaneously to allow for n and U_ψ to vary.

Some results from various models for the general term concentration of a plume passing over a hill are shown in Fig. 12. The recently developed CERC model FLOW-STA-D uses (4.1)-(4.3) together with the air flow calculated by FLOWSTAR to predict the concentration field. (Using an IBM PS2 or equivalent PC it takes about 1 minute per wind direction to calculate the ground level concentrations and about 3 minutes for the whole concentration field.)

The thin plume approximation can be extended by plume splitting, which is necessary when the air flow is strongly stably stratified, i.e. ($F_H < 1$) and has a different structure below and above the dividing streamline where $H_c = H(1-F_H)$. Below it the air flow is approximately homogeneous and therefore a plume approaching a hill can split if some part of the plume is on the stagnation streamline. Therefore downwind of the stagnation region it is necessary to represent the original plume as two separate plumes and these can be modelled as Gaussian plumes with image plumes inside the hill. This approach is used in the EPA code CTDM.

Downwind of a hill in strongly stratified flow the plumes below $z = H_c$ are dispersed across the recirculating, turbulent flow region of the hill. At this stage and further downwind, the plume no longer appears as two plumes, but rather as three – the third being the plume associated with the recirculating region. Sufficiently far downwind these three plumes can be regarded as a single plume again (Snyder & Hunt 1984).

When modelling plumes downwind of hills, the details of the flow over the hill are not usually necessary. Some bulk feature of the hill such as its height, length, breadth, etc. are sufficient to define approximate wake solutions and thence the downwash etc. (cf. the model of Moore & Lee 1982; Hunt, Britter & Puttock 1979). In more complex

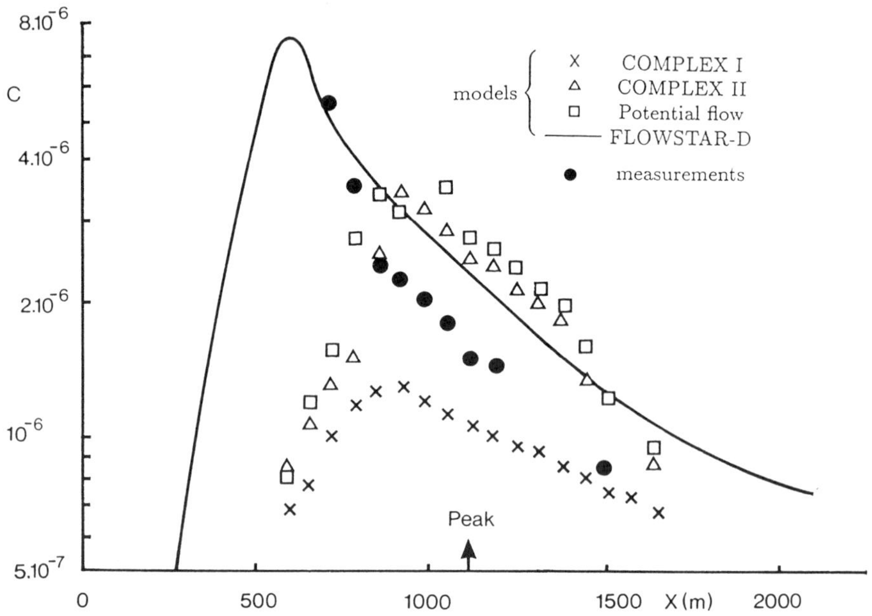

Fig. 12 Measurements of ground level concentration C over Cinder Cone Butte in moderately stable conditions for source height $z_s = 40$m, $u_* = 0.354$ m/s, $U(z_s) = 7.5$ m/s, $N = 0.0423$ s^{-1}, $L_{MO} = 87$m (see Lavery et al 1982). Characteristic height and length of hill in FLOWSTAR-D calculations are 112m and 2000m. Note that all values are derived from contour plots.

terrain, plumes from sources may be dispersed around and over several hills. It is likely that the double-plume approach can also be used in these situations as suggested in §4.3.

Note that in these plume impaction/strongly-stratified flow models, account should be taken of drainage winds. This is an important effect for large hills and mountains and leads to oncoming plumes being carried by the drainage flow down the slope below the source height as well as being carried around the hill by the approach flow. A K-theory model for this process has been developed (Hunt, unpublished).

5. Conclusions about benefits and costs of more complex models

5.1 Improvements in boundary-layer description

Improved boundary-layer 'definition' based on the parameter h/L_{MO} can be implemented for air pollution dispersion models as has already been shown to be possible and practicable in Denmark (Berkowicz et al 1986). The advantages are that:

(i) it would be based on recent and authoritative research and an unusual degree of consensus among researchers. As the review of Kretzschmar & Mertens (1984) makes very clear, the current U.K. classification is not a world-wide or even a European standard. There would be a great advantage in establishing a world standard;

(ii) it would allow better use to be made of local measurements of the boundary-layer depth h, and, perhaps, encourage such measurements to be made (in France they are standard at nuclear power stations);

(iii) it would provide a basis for specifying dispersion parameters as a function of the height of the source;

(iv) it is essential for matching flat terrain models to models for air flow and dispersion over complex terrain which incorporate variations with time of the boundary layer (of great importance for long-range transport) and the effects of precipitation;

(v) the computational cost would not be much more than as for the model based on the Pasquill-Smith approach;

(vi) it would be equivalent to the advanced methods (especially (ii) and (iii)) introduced by D.J. Moore (Moore & Lee 1982) for dispersion from CEGB fossil fuel power stations.

5.2 Improvements in air flow modelling

Practical air-flow models over many kinds of complex surface, such as hills, roughness and temperature changes are currently available and others are being developed. The principal reasons for incorporating these effects are:

(i) complex terrain often gives rise to the largest surface concentrations of pollution, or the most anomalous dispersion. Even quite simple models can account for many *particular* effects as recent field studies in the U.S.A. and U.K. have shown;

(ii) complex terrain effects within 30km of sources also need to be considered in probabilistic risk assessment models. In these models, many situations and cases need to be considered, and simple and fast wind-flow and dispersion models capable of covering most of the likely cases are needed;

(iii) simple and user-friendly schemes for small computer systems are becoming available to model most of the common air-flow situations, to indicate the situations where the models are inadequate, and to model in an *ad-hoc* fashion particular critical situations.

The only disadvantages of using air-flow and dispersion models incorporating real terrain effects is that they are more complex and lengthy to program and operate.

The disadvantage of using simple models is that they may not account for all the effects included in large-scale more computer-intensive models, such as the mesoscale models. However, these models may not be suitable for some of the applications considered here because they do not describe in detail flow over terrain with length scales less than about 15km, which are of interest in site specific studies. It should be noted that most simple models can be 'flagged' to indicate where their use is uncertain.

5.3 Improvements in dispersion modelling

With better models for the air flow in the boundary layer over complex terrain, the present regulatory methods, when used for the prediction of dispersion and surface concentrations, can be improved even when Gaussian plume modelling of the dispersion process is used.

For air flow over flat uniform terrain, the main improvements would be in modelling the extreme conditions of strongly stable and unstable flows, and the changes caused by elevated sources. It would be possible to give some indication of the considerable uncertainty in estimating dispersion in strongly stable boundary layers.

For air flow over complex terrain, better air-flow models would enable systematic changes in the distribution of ground-level concentration, and deposition to be estimated, as well as occasional extremely high concentrations on sloping terrain in highly stable conditions.

Improved air flow and dispersion models for isolated sources would also help improvements in the modelling of related problems, such as plume rise, dense gas dispersion, dispersion from area sources, and more complex processes in the atmosphere, such as chemical change, washout, deposition, etc.

ACKNOWLEDGEMENTS

The writing of this report has benefitted from helpful discussions with many other research workers studying various aspects of pollutant dispersion, but especially Drs. J.A. Jones, P.J. Mason, H.M. ApSimon, M.L. Williams, B.Y. Underwood, and W.H. Snyder, S.R. Hanna and colleagues at the U.S. EPA.

REFERENCES

ApSimon, H.M., Kitson, K., Fawcett, M. & Goddard, A.J.H. 1984. Development of a prototype mesoscale computer model incorporating treatment of topography. Commission of the European Communities, Report EUR 9503 EN.

ApSimon, H.M., Goddard, A.J.H., Wrigley, J. 1985. Long range atmospheric dispersion of radio isotopes. I. The MESOS model. Atmos Env. 19, 99-111.

Apsley, D.D., 1988. A model for dispersion in the wake of large buildings. CEGB Report RD/L/3359/R88.

Barker, C.D. 1982. The application of the virtual source entrainment model in nuclear safety calculations. CEGB Report RD/B/5238/N82.

Belcher, S., Xu, D.P. & Hunt, J.C.R. 1990. The response of the turbulent boundary layer to arbitrary distributed surface roughness. Q.J. Roy. Met. Soc. 116,611-635.

Berkowicz, R., Olesen, H.R. & Torp, U. 1986. The Danish Gaussian air pollution model (OML): description, test and sensitivity analysis in view of regulatory applications. pp. 453-481 Air Pollution Modelling and Its Applications V (eds. C. de Wispelaere, F.A. Schiermeier, N.V. Gillani). Plenum.

Briggs, G.A. 1983. Lift off of buoyant gas initially on the ground. ADTL Contribution File No. 87; Air Resources, Atmospheric Turbulence & Diffusion Lab., NOAA, Oak Ridge, Tennessee.

Briggs, G.A. 1982. Similarity forms for ground-surface surface-layer diffusion. Boundary Layer Met. 23, 489-502.

Briggs, G.A. 1984. Plume rise and buoyancy effects. pp.327-366 (Chap.8), Atmospheric Science and Power Production, ed. D. Randerson. Rep. DOE/TIC-27601, Tech. Info. Center, Office of Sci. & Tech. Info., US Dept. of Energy.

Briggs, G.A. 1985. Analytical parameterisations of diffusion: the convective boundary layer. J. Clim. Appl. Met. 24, 1167-1186.

Britter, R.E., Hunt, J.C.R. & Richards, K.J. 1981. Air flow over a two-dimensional hill: studies of velocity speed up, roughness effects and turbulence. Q.J. Roy. Met. Soc. 107, 91-110.

Browning, K.A. 1989. The mesoscale data base and its use in mesoscale forecasting. Q.J. Roy. Met. Soc. 115, 717-762.

Caneill, J.Y. & Buty, D. 1988. Numerical simulation of the planetary boundary layer over complex topography. Paper 4.3, Proc. 8th Symp. on Turbulence and Diffusion, April 1988, San Diego, California. American Met. Soc.

Carruthers, D.J. & Choularton, T.W. 1982. Air flow over hills of moderate slope. Q.J. Roy. Met. Soc. 108, 603-624.

Carruthers, D.J., Hunt, J.C.R. & Weng, W.-S. 1988. Computational models of air flow over hills. pp.489-492, Proc. Int. Conf. ENVIROSOFT 88, Greece, (ed. P. Zannetti), Springer Verlag.

Carson, D.J. 1973. The development of a dry inversion capped convectively unstable boundary layer. Q.J. Roy. Met. Soc. 99, 450-467.

Castro, I.P. & Snyder, W.H. 1982. A wind-tunnel study of dispersion from sources downwind of three-dimensional hills. Atmos. Env. 16, 1869-1887.

Clarke, R.H. 1979. A model for short and medium range dispersion of radionuclides released to the atmosphere. (First report of a working group on atmospheric dispersion.) National Radiological Protection Board Report NRPB-R91.

Coelho, S.L.V. & Hunt, J.C.R. 1989. On the initial deformation of strong jets in cross flows. J. Fluid Mech. 200, 95-120.

Counihan, J., Hunt, J.C.R. & Jackson, P.S. 1974. Wakes behind two dimensional surface obstacles in turbulent boundary layers. J. Fluid Mech. 64, 529-563.

Deardorff, J.W. 1970. Convective velocity and temperature scales for the unstable planctary boundary layer and for Rayleigh convection. J. Atmos. Sci. 27, 1211-1213.

Deardorff, J.W. & Willis, G.E. 1975. A parameterization of diffusion into the mixed layer. J. Appl. Met. 14, 1451-1458.

Dinar, N., Kaplan, H. & Kleiman, M. 1988. Characterization of concentration fluctuations of a surface plume in a neutral boundary layer. Boundary Layer Met. 45, 157-175.

Fackrell, J.E. & Robins, A.G. 1982. Concentration fluctuations and fluxes in plumes from point sources in a turbulent boundary layer. J. Fluid Mech. 117, 1-26.

Gifford, F.A. 1960. Atmospheric dispersion calculations using the generalized Gaussian plume model. Nuclear Safety 2(2), 56-59.

Gifford, F.A. 1961. Use of routine meteorological observation for estimating atmospheric dispersion. Nuclear Safety 2(4), 47-57.

Golding, B.W. 1987. The UK Meteorological Office mesoscale model. Boundary Layer Met. 41, 97-107.

Gryning, S.E., Holtslag, A.A.M., Irwin, J. & Sivertsen, B. 1987. Applied dispersion modelling based on meteorological scaling parameters. Atmos. Env. 21, 79-89.

Hanna, S.R. & Paine, R.J. 1989. Hybrid plume dispersion model (HPDM) development and evaluation. J. Appl. Met. 28, 206-224.

Holtslag, A.A.M., Nieuwstadt, F.T.M. 1986. Scaling the atmospheric boundary layer. Boundary Layer Met. 36, 201-209.

Hoult, D.P. & Weil, J.C. 1972. Turbulent plume in a laminar cross flow. Atmos. Env. 6, 573-531.

Huber, A.H. 1984. Evaluation of a method for estimating pollution concentration downwind of influencing buildings. Atmos. Env. 18, 2313-2338.

Hunt, J.C.R. 1982. Mechanism for dispersion of pollution around buildings and vehicles. pp.235-266, BMFT/TÜV Coll. on Exhaust Gas Air Pollution caused by Motor Vehicle Emissions. TÜV, Köln.

Hunt, J.C.R. 1985. Turbulent diffusion from sources in complex flows. Ann. Rev. Fluid Mech. 17, 447-485.

Hunt, J.C.R., Britter, R.E. & Puttock, J.S. 1979. Mathematical models of dispersion of air pollution around buildings and hills. pp.145-200, Proc. Conf. Math. Modelling of Turbulent Diffusion in the Environment, Sept. 1978, ed. C.J. Harris. Academic Press.

Hunt, J.C.R., Kaimal, J.C. & Gaynor, J.E. 1985. Some observations of turbulence structure in stable layers. Q.J. Roy. Met. Soc. 111, 793-815.

Hunt, J.C.R., Lalas, D.P. & Asimakopoulos, D.N. 1984. Air flow and dispersion in rough terrain: a report on Euromech 173. J. Fluid Mech. 142, 201-216.

Hunt, J.C.R. & Mulhearn, P.J. 1973. Turbulent dispersion from sources near two-dimensional obstacles. J. Fluid Mech. 61, 245-274.

Hunt, J.C.R., Richards, K.J. & Brighton, P.W.M. 1988 Stably stratified flow over hills. Q.J. Roy. Met. Soc. 114, 1435-1470.

Hunt, J.C.R., Weng, W.-S., & Carruthers, D.J. 1988. Modelling deposition fluxes on hills. pp.623-640, Air Pollution Modelling and its Applications VI, (ed. H. van Dop). Plenum.

Hunter, G.C. 1990. Experimental investigation of a buoyant jet in a stratified cross-wind. Proc. I.M.A. Conf., University of Leeds.

Jackson, P.S. & Hunt, J.C.R. 1975. Turbulent wind flow over a low hill. Q.J. Roy. Met. Soc. 101, 929-955.

Jiemin, W. 1984. Estimation of fluxes from routine meteorological data. Met. Office Turbulence & Diffusion Note No.153.

Jones, J.A. 1981. A procedure to include deposition in the model for short and medium range atmospheric dispersion of radionucleides. (The second report of a working group on atmospheric dispersion.) National Radiological Protection Board Report NRPB-R122.

Jones, J.A. 1986. The uncertainty in dispersion estimates obtained from the working group models. (Seventh report of a working group on atmospheric dispersion.) National Radiological Protection Board Report NRPB-R199.

Kazanski, A.B. & Monin, A.S. 1960. A turbulent regime above the surface atmospheric layer. Izv. Akad. Sci., USSR, Geof. Ser. 1, 110.

Kretzschmar, J.G. & Mertens, I. 1984. Influence of the turbulent typing scheme upon the cumulative frequency distribution of the calculated relative concentration for different averaging times. Atmos. Env. 18, 2377-2393.

Lalas, D.P., Asimakopoulos, D.N., Deligiorgi, D.G. & Helmis, C.G. 1983. Sea breeze circulation and photochemical pollution in Athens. Atmos. Env. 17, 1621-1632.

Lamb, R.G. 1979. The effects of release height on material dispersion in the convective planetary boundary layer. pp.27-33 Proc. 4th Symp. on Turbulence, Diffusion and Air Pollution, Reno. American Met. Soc.

Lange, R. 1978. ADPIC – a three-dimensional transport diffusion model for the dispersal of atmospheric pollutants and its validation against regional studies. J. Appl. Met. 17, 320-329.

Lavery, T.F., Bass, A., Strimaitis, D.G., Venkatram, A., Greene, B.R., Drivas, P.J. & Egan, B.A. 1982. EPA complex terrain modeling program: 1st milestone report (1981). Rep. No. EPA-600/3-82-036.

Lewellen, W.S., Sykes, R.I. & Oliver, D. 1982. The evaluation of MATHEW/ADPIC as a real time dispersion model. Aeronautical Res. Assocs. of Princeton Inc., Rep. No. 442 for Div. Health, Safety and Waste Management, Office of Nuclear Regulatory Res., U.S. Nuclear Regulatory Commission.

Lewellen, W.S. & Sykes, R.I. 1986. Analysis of concentration fluctuations from Lidar observations of atmospheric plumes. J. Clim. Appl. Met. 25, 1145-1154.

Lilly, D.K. & Klemp, J.B. 1979. The effects of terrain slope on non-linear hydrostatic mountain waves. J. Fluid Mech. 95, 241-261.

Mahrer, Y. & Pielke, R.A. 1977. The effects of topography on the sea and land breezes in a two dimensional numerical model. Monthly Weather Rev. 105, 1151-1162.

Mason, P.J. & Derbyshire, S. 1990. Large eddy simulation of the stably stratified atmospheric boundary layer. Boundary Layer Met. (to appear).

Mason, P.J. & King, J.C. 1985. Measurements and predictions of flow and turbulence over an isolated hill of moderate slope. Q.J. Roy. Met. Soc. 111, 617-640.

Monteith, J.L. 1975. Principles of Environmental Physics (second ed.). E. Arnold Ltd.

Moore, D.J. & Lee, B.Y. 1982. An asymmetric Gaussian plume model. CERL Report RD/L/2225 N/81.

Moussiopoulos, N. & Flassak, T. 1986. Two vectorized algorithms for the effective calculation of mass consistent flow field. J. Clim. Appl. Met. 25, 847-857.

Mylne, K.R. & Mason, P.J. 1990. Concentration fluctuation measurements in a dispersing plume at a range of up to 1000m. Q.J.Roy. Met. Soc. (to appear).

Nieuwstadt, F.T.M. 1980. Application of mixed-layer similarity to the observed dispersion from a ground-level source. J. Appl. Met. 19, 157-162.

Nieuwstadt, F.T.M. 1984. The turbulent structure of the stable nocturnal boundary layer. J. Atmos. Sci. 41, 2202-2216.

Nieuwstadt, F.T.M. & van Harem, L. 1988. A large eddy simulation of buoyant and non-buoyant plume dispersion in the convective atmospheric boundary layer. pp. 525-538. Air Pollution Modelling and Its Applications VI (ed. H. van Dop), Plenum.

Ooms, G. 1972. A new method for the calculation of the plume path of gases emitted by a stack. Atmos. Env. 6, 899-909.

Paine, R.J., Strimaitis, D.J., Dennis, M.G., Yamartino, R.J., Mills, M.T. & Insley, E.M. 1987. User's guide to the Complex Terrain Dispersion Model. Vol. 1. Model description and user instructions. Atmos. Sci. Res. Lab., Office of Res. & Development, U.S. EPA.

Pasquill, F. 1961. The estimation of the dispersion of windborne material. Met. Mag. 90, 33-49.

Pasquill, F. & Smith, F.B. 1983. Atmospheric Diffusion (third ed.). Ellis Horwood Ltd.

Perry, S.G. & Finkelstein, P.L. 1990. The U.S. Environmental Protection Agency's model for complex terrain applications. Proc. 10th Int. Conf. of Clean Air Soc. of Australia and New Zealand.

Pielke, R.A. 1984. Mesoscale Meteorological Modeling. Academic Press.

Pielke, R.A., Lyons, W.A., McNider, R.T., Moran, M.D., Moon, D.A., Stocker, R.A., Walco, R.L. & Uliasz, M. 1990. Regional and mesoscale meteorological modelling as applied to air quality studies. Proc. 18th NATO-CCMS Int. Tech. Meeting on Air Pollution Modelling and its Applications.

Puttock, J.S. 1978. Modelling the effects of wakes behind hills and buildings on pollutant dispersion. Proc. 9th NATO-CCMS Int. Tech. Meeting on Air Pollution Modelling and its Applications. Toronto, Canada.

Restrepo, L.F. 1987. Study plan for the sensitivity analysis of the terrain responsive atmospheric code (TRAC). Report RFP-4066, Rockwell Int. N. American Space Op., Rocky Flats Plant, Golden, Colorado.

Robins, A.G. 1983. Recent advances in the modelling of dispersion near buildings. pp.44-56 Proc. I. Mech. Eng. Conf. on Heat and Fluid Flow in Nuclear and Process Plant Safety.

Robins, A.G. & Fackrell, J.E. 1980. Laboratory studies of dispersion near buildings. Proc. CEC Symp. on Radio-active Releases and their Dispersion in the Atmosphere. Risø, Denmark.

Sawford, B.L. 1983. The effect of Gaussian particle-pair distribution functions in the statistical theory of concentration fluctuations in homogeneous turbulence. Q.J.Roy. Met. Soc. 109, 339-354.

Sawford, B.L. 1987. Conditional concentration statistics for surface plumes in the atmospheric boundary layer. Boundary Layer Met. 38, 209-223.

Sherman, C.A. 1978. A mass consistent model for wind fields over complex terrain. J. Appl. Met. 17, 312-319.

Singh, M.P. & Ghosh, S. 1987. Bhopal gas tragedy: model simulation of the dispersion scenario. J. Haz. Mat. 17, 1-22.

Smith, F.B. 1973. A scheme for estimating the vertical dispersion of a plume from a source near ground level. Proc. 3rd meeting of an expert panel on air pollution modelling. Oct. 1972, Paris; Brussels, NATO-CCMS Report 14.

Smith, F.B. & Blackall, R.M. 1979. The application of field experimental data to the parameterisation of the dispersion of plumes from ground level and elevated sources. pp.201-238, Proc. Conf. Math. Modeling of Turbulent Diffusion in the Environment, Sept.1978, ed. C.J. Harris. Academic Press.

Smith, F.B. & Clark, M.J. 1989. The transport and deposition of airborne debris from the Chernobyl nuclear power plant accident with special emphasis on the consequences to the United Kingdom. Met. Office Sci. Paper No.42. HMSO.

Snyder, W.H. 1985. Fluid modelling of pollutant transport and diffusion in stably stratified flows over complex terrain. Ann. Rev. Fluid Mech. 17, 239-266.

Snyder, W.H. & Hunt, J.C.R. 1984. Turbulent diffusion from a point source in stratified and neutral flows around a three-dimensional hill. Part II. Laboratory measurements of surface concentration. Atmos. Env. 18, 1969-2002.

Snyder, W.H., Thompson, R.S., Eskridge, R.E., Lawson, R.E., Castro, I.P., Lee, J.T., Hunt, J.C.R. & Ogawa, Y. 1985. The structure of strongly stratified flow over hills: dividing streamline concept. J. Fluid Mech. 152, 249-288.

Strimaitis, D.G., Lavery, T.F., Venkatram, A., DiChristofaro, D.C., Greene, B.R. & Egan, B.A. (1985) EPA complex terrain model development: 4th milestone report (1984) Rep. No. EPA-600/3-84-110.

Sutton, O.G. 1932. A theory of eddy diffusion in the atmosphere. Proc. Roy. Soc. 135A, 143-165.

Sykes, R.I. 1988. Concentration fluctuations in dispersing plumes. Lectures on air pollution modelling, eds. A. Venkatram, J.C. Wyngaard. American Met. Soc.

Tennekes, H. 1982 Similarity relations, scaling laws and spectral dynamics. pp.37-68 Atmospheric Turbulence and Air Pollution Modelling, eds. F.T.M. Nieuwstadt, H.van Dop. Reidel.

Thompson, R.S. & Snyder, W.H. 1985. Dispersion from a source upwind of a three-dimensional hill of moderate slope. Appendix B.of Strimaitis et al. 1985.

Thomson, D.J. 1986. A random walk model of dispersion in turbulent flows and its application to dispersion in a valley. Q.J. Roy. Met. Soc. 112, 511-530.

Thomson, D.J. 1990. A stochastic model for the motion of particle pairs in isotropic high Reynolds number turbulence, and its application to the problem of concentration variance. J. Fluid Mech. 210, 113-153.

Turfus, C. 1986. Diffusion from a continuous source near a surface in steady reversing flow. J. Fluid Mech. 172, 183-209.

Turfus, C. 1988. Calculating mean concentrations for steady sources in recirculating wakes by a particle trajectory method. Atmos. Env. 22, 1271-1290.

van Ulden, A.P. & Holtslag, A.A.M. 1985. Estimation of atmospheric boundary layer parameters for diffusion applications. J. Clim. Appl. Met. 24, 1196-1207.

Walmsley, J.C., Salmon, J.R. & Taylor, P.A. 1982. On the application of a model of boundary layer flow over low hills to real terrain. Boundary Layer Met. 23, 17-46.

Walmsley, J.C., Taylor, P.A. & Keith, T. 1986. A simple model of neutrally stratified boundary layer flow over complex terrain with surface roughness modulations (MS3 DJH/3R). Boundary Layer Met. 36, 157-186.

Weil, J.C. 1983. Application of advances in planetary boundary layer understanding to diffusion modelling. Proc. 6th Symp. on Turbulence and Diffusion, Boston, MA.

Weil, J.C. 1985 Updating applied diffusion models. J. Clim. Appl. Met. 24, 1111-1130.

Weil, J.C. 1989. Langevin modelling of dispersion in the convective boundary layer. pp.437-449 Air Pollution Modelling and its Applications VII (ed. H. van Dop), Plenum.

Weng, W.-S., Richards, K.J. & Carruthers, D.J. 1989. Some numerical studies of turbulent wakes over hills. p.412-417 Proc. 2nd European Turbulence Conf., Aug. 1988, Berlin, eds. H.H. Fernholz, H.E. Fiedler. Springer Verlag.

White, F.B. (ed.), Ching, J.K.S., Dennis, R.L. & Snyder, W.H. 1985. Summary of complex terrain model evaluation. Atmos. Sci. Res. Lab., Office of Research and Development, US EPA.

Zilitinkevich, S.S. 1972 On the determination of the height of the Ekman boundary layer. Boundary Layer Met. 3, 141-145.

DISCUSSION

M.E. BERLYAND

Is it possible to give some examples of applications for regulatory use?

J.C.R. HUNT

The paper describes in detail a number of models which are used for example when considering the siting of power plant.

D.G. STEYN

In order to provide further insight into turbulent diffusion in complex environments, we may use: a) field observations (with their attendent problems of generalizability), b) wind tunnel experiments (problems of Ri and Re similarity), or c) LES (large computational requirements). Which do you see as the most fruitful approach?

J.C.R. HUNT

A combination of all three. Recent carefully conducted field experiments have yielded more insight than b) or c) - e.g. over hills or using tall towers and measuring all the statistical quantities.

A.A.M. HOLTSLAG

Have you actually compared (or are you planning to compare) concentration predictions with the new improved models based on h/L determination etc., with the "old" models based on Pasquill stability class?

J.C.R. HUNT

We have made a number of comparisons. The new models describe the recent convective diffusion experiments better than the old models. Generally the greatest differences are in "low wind" conditions with strong convection or in stable flows.

A. VENKATRAM

I believe you applied the FLOWSTAR model to Cinder Cone Butte. How do your model estimates compare with corresponding observations of concentrations of SF_6? Can you explain why you underpredict the maximum concentration (important from a regulatory viewpoint) by a factor of two?

J.C.R. HUNT

The results shown in my talk were preliminary and were as Dr Venkatram comments. Subsequent development of the model has rectified this error and the agreement is now satisfactory for this and many other testcases!

MODELED SULFUR DEPOSITION TRENDS SINCE 1900 IN NORTH AMERICA

Jack D. Shannon

Environmental Research Division
Argonne National Laboratory
Argonne, IL 60439

Inventories of historical emissions of acidic deposition precursors in the U.S. have been assembled by Gschwandtner et al. (1985), referred to hereafter as G85, and by Husar (1986). One might expect that such data would be widely applied in studies of possible long-term effects associated with acid deposition, such as lake acidification or forest damage. Use of the inventories in such investigations, however, has been surprisingly limited. This may be because atmospheric deposition rather than precursor emissions is the direct input to ecological systems. Although some researchers have used emission trends as surrogates for deposition trends, the relationships between aggregated emissions and resulting patterns of atmospheric concentrations or deposition have changed over the years as the seasonal, horizontal, and vertical distributions of emissions have changed. In the hope of providing additional information useful to researchers investigating deposition effects, we have simulated wet and dry S deposition and atmospheric concentrations of SO_2 and sulfate over the United States and Canada since 1900 with the ASTRAP model (Shannon, 1985). Source-receptor matrices aggregated to the state or province level have been calculated as well. Simulations are seasonal, at five-year intervals from 1900 through 1960 and at one-year intervals from 1960 through 1985.

Given that one wishes to simulate deposition for such an extensive period, is ASTRAP, a linear Lagrangian model, a proper model to apply? The National Acid Precipitation Assessment Program (NAPAP) of the United States has expended much time and effort in development of an Eulerian model capable of sophisticated treatment of meteorological and chemical processes. However, simulation of deposition with an Eulerian model over such a lengthly period would be prohibitively expensive. In addition, the potential advantages in more detailed treatment of transport, transformation, and removal would be lost because the necessary input data, much more extensive than for a more highly parameterized statistical or Lagrangian model, are inadequate or unavailable. The efficient computational structure of ASTRAP makes multiple simulations of deposition over all of the United States and Canada quite feasible; furthermore, ASTRAP calculations for 1985 deposition compare reasonably well with observations and with the NAPAP Eulerian model, particularly for wet deposition (Dennis, 1990). ASTRAP apparently overestimates dry deposition, but dry deposition can be monitored only indirectly, and the deposition data base is much less extensive than for wet deposition. Spatial analysis of the limited observations (generally sited outside the immediate influence of urban areas and large point sources) and models tuned to such observations may well underestimate dry deposition.

Annual emissions of SO_2 by state from 1900 through 1980, along with nationally aggregated trends in stack heights and seasonal emissions, were taken from G85. Horizontally and vertically detailed seasonal emission data for 1980-1985 for the United States and detailed Canadian emission data for 1980 were obtained from the U.S. Environmental Protection Agency. The Atmospheric Environment Service of Canada provided fairly complete annual emission data for major Canadian point sources for 1970-1985, less complete data for 1955-1970, and extremely sparse data for 1935-1955. No data were obtained for any Canadian emissions earlier than 1935.

In order to maximize the real resolution of emissions input, we first gridded the detailed U.S. emissions seasonally for 1980-1985 with approximately 110-km horizontal resolution and six vertical layers. The G85 data for 1900-1980 were then used to develop scaling factors for annual changes, relative to 1980, in statewide, vertical, and seasonal distributions. The scaling factors were applied to the gridded 1980 detailed emissions to create seasonal emission fields for prior years. Because the detailed 1980 emission inventory differed slightly from the 1980 inventory in G85 when aggregated annually by state, the emission fields input to simulations for years prior to 1980 typically differed from the historical data of G85 by about one percent. The detailed 1980 Canadian emissions were gridded and then adjusted for other years to account for changes in major Canadian point sources and province totals. In order to extend Canadian emissions back to the turn of the century, the earliest Canadian emission estimates (1935 for some point sources, 1955 for other sources) were extrapolated back to 1900 in proportion to the overall emission trend in the United States. Obviously, this makes early Canadian emission estimates more uncertain than those for the United States. The consequences of such increased uncertainty are mainly in calculations for deposition in Canada, because Canadian emissions are calculated to result in significant deposition in only a small portion of the United States. Primary sulfate emissions were available in the detailed 1980 inventory with the same resolution as SO_2 emissions. The scaling factors developed for SO_2 were also applied to sulfate emissions. In Fig. 1 the combined SO_2 and sulfate emissions used in the simulations are shown as equivalent S, aggregated to national totals.

Meteorological data available in a form convenient for use in ASTRAP cover 1960-1983 in one set of meteorological analyses from the University of Michigan (UM) and 1980-1985 in another set of analyses from the Canadian Meteorological Centre (CMC). Both sets of analyses consist of gridded wind and precipitation fields updated several times daily. However, the temporal resolution of precipitation is better in the UM analyses, while the temporal resolution of wind is better in the CMC analyses. The CMC analyses are more reliable in remote areas than are the UM analyses. Comparison of simulations with the alternate data sets during the years of overlap, 1980-1983, indicates that the CMC analysis results average 9.2% less for wet deposition and 5.7% more for dry deposition over the United States and Canada. The main reason for the difference is that rainfall is analyzed in 6-hr time steps in the UM analyses and in 12-hr time steps in the CMC analyses. Because wet removal in ASTRAP is a function of both rainfall amount and rainfall frequency, the different meteorological analyses, even if both were correct, would lead to different results. In addition, some data are

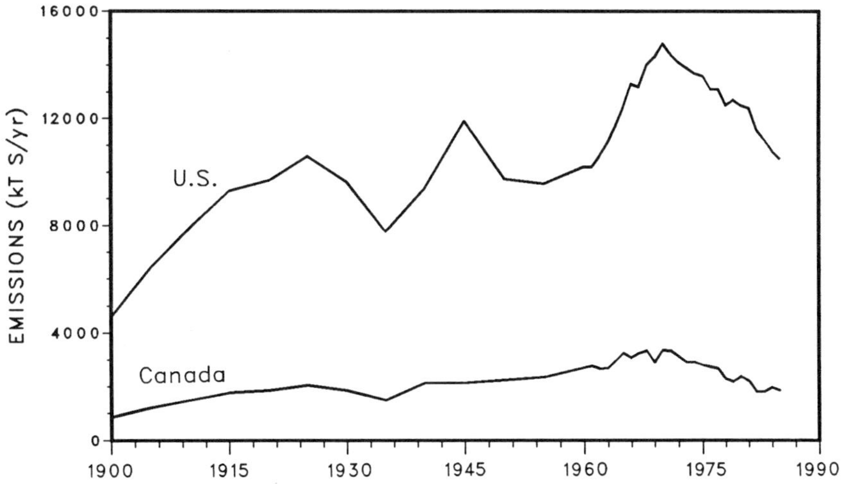

Fig. 1. Nationally aggregated anthropogenic S emissions used in ASTRAP simulations.

missing in the CMC analyses. The differences are well within the uncertainty of the ASTRAP model. For simulations of concentrations and deposition prior to 1960, the average meteorological conditions for the period 1960-1983 were used. The simulations for the pre-1960 period thus omit the effect of interannual meteorological variability and contain additional uncertainty related to the departure of actual meteorological conditions from later climatology.

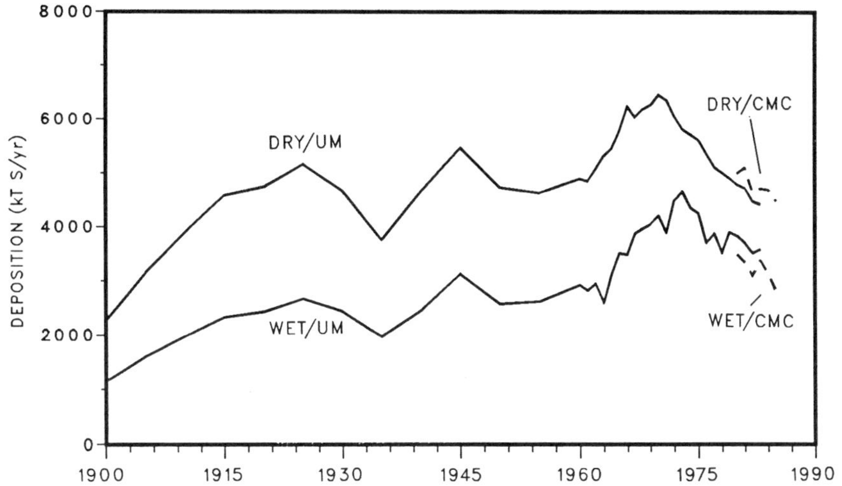

Fig. 2. Trend in modeled wet and dry S deposition integrated across the contiguous United States and Canada south of 60 deg N. UM and CMC indicate the meteorological data set used.

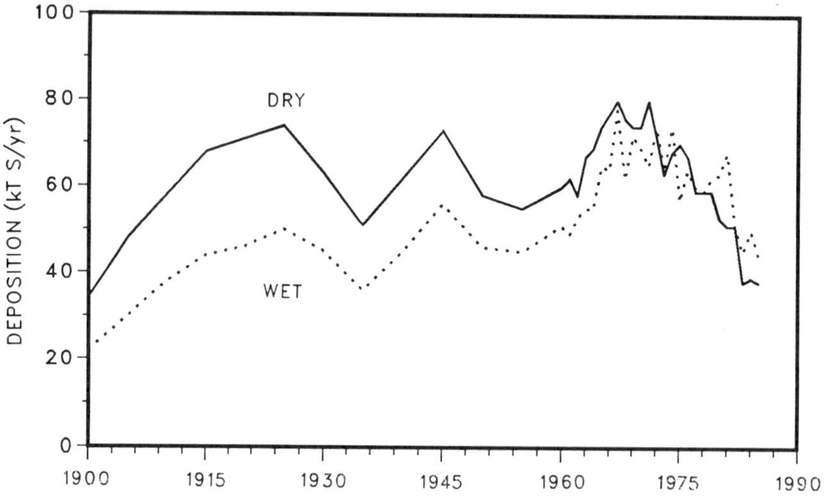

Fig. 3. Trend in modeled wet and dry S deposition integrated across Maine.

The resolution of ASTRAP is determined in part by the theory and structure of the model and in part by the resolution of the emission and meteorological input. The trend in broadly determined variables, such as integrated wet and dry S deposition over North America (Fig. 2), is more reliable than that across a state or substate area, but is potentially less useful to effects researchers who require information on a smaller spatial scale. As one might expect in simulations with a linear model, there is general similarity between the continental-scale emission and deposition trends. However, there are also important differences even over this scale. Initially the ratio of dry:wet was about 2:1, but the ratio decreased to about 4:3 by the 1980s. The increased relative role of wet deposition resulted from several factors, primarily the increase in stack heights (which reduced local dry deposition and thus left more S for later wet removal) and the reduction of the relative emission maximum in winter (when wet removal is much less efficient). On the continental scale, dry and total deposition peaked in 1970, while wet deposition peaked in 1973. Wet deposition is more variable than dry deposition.

The modeled deposition trend across a smaller region, the state of Maine, is shown in Fig. 3. For the years subsequent to 1979, only the results for the CMC meteorological analyses are given, because there was general consistency in results with the alternate analyses during the 1980-1983 overlap. The ratio of dry:wet changed from about 3:2 in 1900 to about 5:6 by 1985. It should be noted that ASTRAP includes neither terrain variations nor droplet deposition from fog or clouds enveloping mountaintops. Much of Maine is elevated terrain, for which ASTRAP typically overpredicts atmospheric concentrations and thus dry deposition. Modeled dry deposition is probably more representative of dry plus droplet deposition. Statewide deposition shows more interannual variability than does continental deposition.

The modeled trend for a single grid cell of about 120 km by 120 km containing the Sudbury, Ontario, smelter sources (Fig. 4) indicates a large decrease after 1971 in total S deposition, mainly in dry

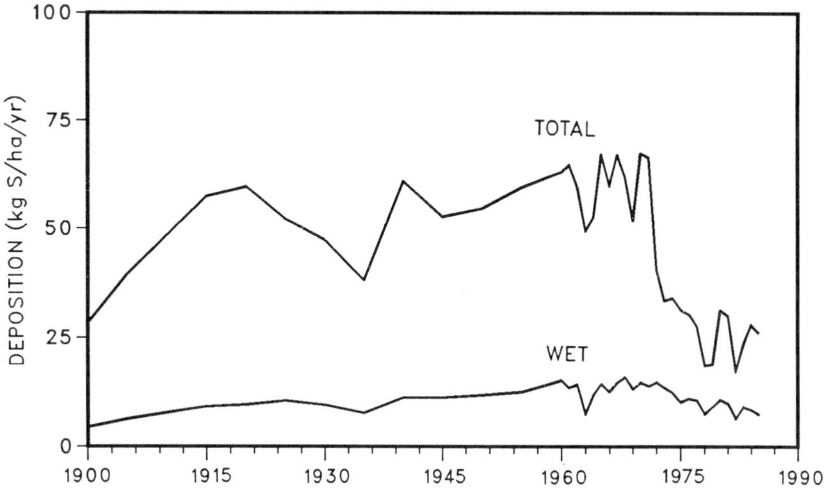

Fig. 4. Trend in modeled wet and total S deposition in the vicinity of Sudbury, Ontario.

deposition, reflecting the introduction of the 381-m stack and a drop in smelter emissions. While it is perhaps an extreme example, the modeled Sudbury trend illustrates the importance of both subregional changes in emission strength and character and greater meteorological variability on the local scale.

Ideally, one would evaluate the accuracy of this modeling study through comparison with equally long-term monitoring data. Unfortunately, no 86-year monitoring record exists. The longest record available, that beginning in the mid-1960s at Hubbard Brook, New Hampshire, is reported as precipitation-weighted concentration (PWC) of sulfate by water year (June-May) and is from a bulk sampler rather than from a wet-only sampler (Likens et al., 1984). Comparisons of bulk and wet-only samples give somewhat mixed results (Stensland et al., 1986). The ratio of the mean of bulk sulfate concentrations to that of wet-only sulfate concentrations ranged between 1.0 and 1.3 for two regions in the northeastern U.S., although a test at Hubbard Brook indicated little difference in sulfate concentrations from the two methods. We have combined the sulfate concentration and precipitation data from Likens et al. to calculate the S deposition trend from the bulk samples. Here we compare the observed deposition trend with a modeled "bulk" trend, which we approximate by wet plus 0.15 times dry S deposition from ASTRAP (Fig. 5). There is a time shift of six months between the two sets of annual deposition, because ASTRAP results represent December through November. Interestingly, a long-term decreasing trend reported by Likens et al. for sulfate concentration is much less evident for S deposition, apparently because annual precipitation was more variable over the period than precursor S emissions. Observed deposition is more variable than is modeled deposition. A much shorter but more direct comparison of trends in modeled and observed wet and total S deposition at Kejimkujik, Nova Scotia (Fig. 6) indicates general agreement between observed and modeled wet deposition, although annual variability of wet deposition is again underestimated. Dry deposition (total minus wet) is apparently overpredicted.

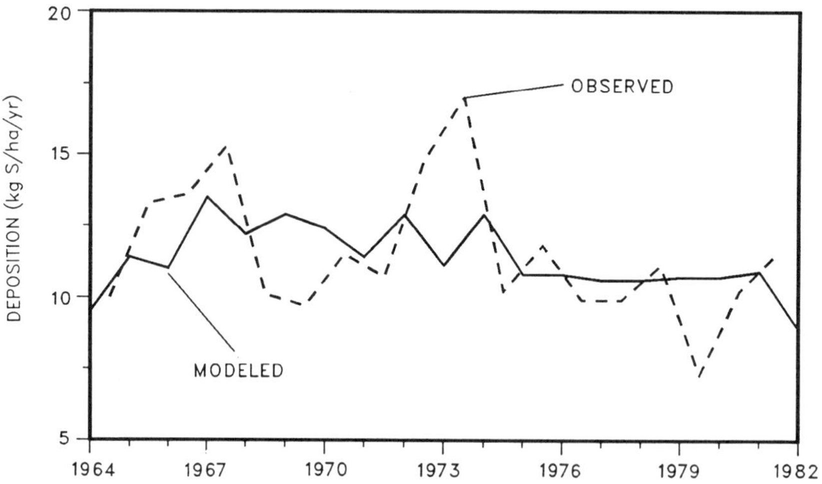

Fig. 5. Comparison of observed total S deposition in bulk samples at Hubbard Brook, New Hampshire, with modeled wet S deposition adjusted by 0.15 times modeled dry S deposition.

Uncertainty in these results arises both from the simplifications in a highly parameterized regional acid deposition model and from factors specific to this application. Although the parameterizations in ASTRAP have been developed without formal model tuning or calibration, informal adjustment and use of parameterizations from other, tuned models have occurred. All such calibration, whether direct or indirect, reflects the atmospheric chemistry conditions of the period since the late 1970s. Extension of the same parameterizations to earlier ambient atmospheric conditions (presumably cleaner regionally but more polluted in some respects in large urban areas) has not been rigorously verified with data, nor with long-term extension of the results of an advanced model.

Uncertainties in the historical emission inventories, in the schemes used to disaggregate the emissions, in the actual meteorological conditions prior to 1960, and in the confounding of meteorological changes with changes in meteorological analysis methods all contribute to model uncertainties in this application. Subsequent to performing all of the model calculations, it was learned that the U.S. SO_2 emission trend had been slightly revised (Gschwandtner et al., 1988, hereafter G88). On a nationally aggregated level, the totals for 1950, 1960, 1970, 1975, 1978, and 1980 were adjusted by 5.9, 0.2, -2.4, -4.3, -1.1, and -6.0%, respectively. Other years were unchanged, although revised estimates were provided only at 5-year intervals (other than 1978). State-level adjustments were generally similar to the adjustment of the nationally aggregated emissions, although not identical. Because linear chemistry is assumed in ASTRAP, rough approximations can be made of how the modeled deposition and concentration fields in the U.S. would differ with scaling factors developed from G88 estimates. The 1950 results would increase by about 12%, 1900 through 1945 results would increase by about 6%, 1955 through 1979 results would increase by about 2-6%, and results for 1980 and subsequent years (which did not use G85 data) would be unchanged. Changes in and near Canada would be smaller. In any case, uncertainty in model results is greater than the relative change between G85 and G88.

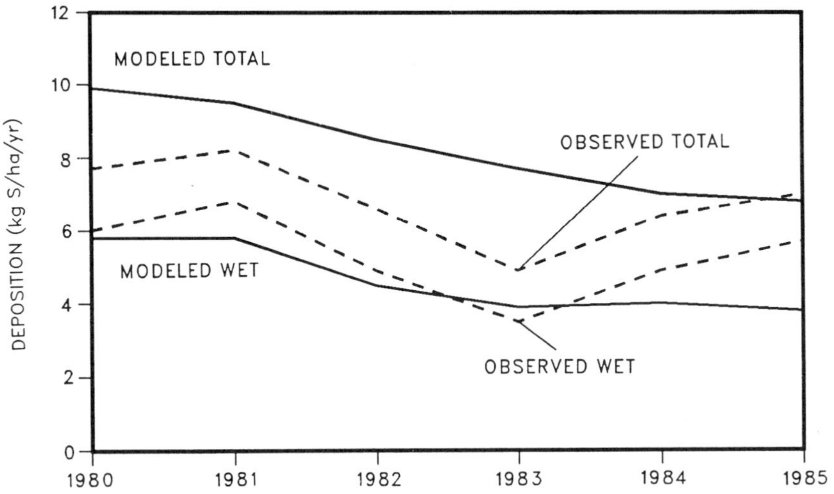

Fig. 6. Comparison of modeled wet and total deposition of S at Kejimkujik, Nova Scotia, with observed data (adapted from Sirois and Summers, 1989).

Because simulations for years prior to 1960 used climatology rather than specific years of meteorology and thus are more uncertain and because revised emission estimates were not made for all years subsequent to 1960, we have not attempted to recalculate deposition trends.

Acknowledgments: This work was supported by the U.S. Department of Energy (DOE), Assistant Secretary for Energy Research, Office of Health and Environmental Research, under contract W-31-109-Eng-38, by the DOE PRECP program, and by an Argonne National Laboratory Computation Science grant. B. Lesht produced graphics and assisted with data preparation. F. Vena of the Atmospheric Environment Service of Canada provided much helpful data.

References

Dennis, R. L., 1990, Evaluation of Regional Acidic Deposition Models, State of Science/Technology Report No. 8, National Acid Precipitation Assessment Program, Washington DC.

Gschwandtner, G., Gschwandtner, K. C., and Eldridge, K., 1985, Historic Emissions of Sulfur and Nitrogen Oxides in the United States from 1900 to 1980, U.S. Environmental Protection Agency Report EPA-600/7-85/009a.

Gschwandtner, G., Wagner, J. K., and Husar, R. B., 1988, Comparison of Historic SO_2 and NO_x Emission Data Sets, U.S. Environmental Protection Agency Report EPA-600/7-88-009a.

Husar, R. B., 1986, Emissions of sulfur dioxide and nitrogen oxides and trends for eastern North America, in "Acid Deposition Long-Term Trends," National Academy Press, Washington DC.

Likens, G. E., Bormann, F. H., Pierce, R. S., Eaton, J. S., and Munn, R. E., 1984, Long-term trends in precipitation chemistry at Hubbard Brook, New Hampshire, Atmos. Environ., 18:2641.

Shannon, J. D., 1985, User's Guide for the Advanced Statistical Trajectory Regional Air Pollution (ASTRAP) Model, U.S. Environmental Protection Agency Report EPA-600/8-85/016.

Sirois, A. and Summers, P. W., 1989, An estimation of acid deposition input of sulfur and nitrogen oxides to the Kejimkujik watershed: 1979-1986, Water, Air, & Soil Pollut., 46:29.

Stensland, G. J., Whelpdale, D. M., and Oehlert, G., 1986, Precipitation chemistry, in "Acid Deposition Long-Term Trends," National Academy Press, Washington DC.

DISCUSSION

A. VENKATRAM

The comparison between model predictions and corresponding observations is poor. Do you have comments on this?

J.D. SHANNON

The model does not reproduce the observed year-to-year variation adequately in the two examples shown here. I feel that this is mainly the result of comparing measurements at a point with the values in a model grid cell of ca. 100 km by 100 km; the within-cell variability may well be much larger than the annual trend. I do not agree that the comparison between model and observations is poor, because the longer-term observed trends are reproduced reasonably well and the magnitudes of simulated and observed deposition are very similar, without model calibration or optimization of coefficients.

R.A. PIELKE

Has ASTRAP been contrasted with RADM results for an episodic event? If ASTRAP cannot be applied in this fashion can a summation of RADM episodes be used to compare with an equivalent sum of periods using ASTRAP?

J.D. SHANNON

ASTRAP cannot be exercised to simulate an episode, because the model includes no short-term horizontal diffusion; instead, long-term horizontal diffusion is parameterized through statistics describing the distribution of multiple trajectories. Trajectories must be calculated for about one month for reliable statistics. Comparisons of annual deposition calculations from aggregated RADM episodic simulations with long-term results produced by ASTRAP and two other linear models for 1985 emissions and 1982-1985 meteorology are underway. Preliminary indications are that ASTRAP and RADM calculations of normalized source-receptor attribution are much alike and their wet deposition fields are reasonably similar, while ASTRAP predicts much more dry deposition than RADM. Estimates of dry deposition from the very limited monitoring data generally lie about midway between the two model estimates. RADM is being extensively evaluated with detailed field data from 1988 and 1989 that include air concentrations as well as wet and dry deposition. It would seem highly desirable to compare RADM with simpler models for this period as well, but at this time I know of no plans to do so.

THE IMPACT OF EMISSION REDUCTIONS ON MESOSCALE ACID DEPOSITION IN THE LOWER OHIO RIVER VALLEY

Rohit Mathur (1), Rick D. Saylor (2), and Leonard K. Peters(1)

(1) Department of Chemical Engineering
University of Kentucky
Lexington, KY 40506

(2) Center for Applied Energy Research
University of Kentucky
Lexington, KY 40511-8433

INTRODUCTION

The growing awareness of atmospheric pollution problems and their consequent effect on human health and the ecosystem have brought about a need for stringent control strategies. Several past studies have been conducted to test model sensitivity to emission reductions and changes in background concentrations. Studies have also focused on the effect of reduction of emissions of the primary species on the eventual deposition patterns over various regional scales. Stockwell et al.(1988) investigated non-linear coupling in the $NO_x - SO_x$ - reactive organic system, and examined the extent of coupling of atmospheric acid formation rates with species that serve as important precursors of photochemical oxidants. Cho et al.(1989) used a one-dimensional time dependent version of the STEM II model to evaluate the effect of reduction in ambient levels of primary pollutants on $SO_4^=$ and NO_3^- wet depositions. Recently, Misra et al. (1989) used the Acid Deposition and Oxidant Model(ADOM) to simulate wet $SO_4^=$ deposition patterns over eastern North America and to study the non-linear response of wet deposition to emission reductions.

This paper presents a series of simulations for various emission reduction scenarios in a mesoscale region. The specific mesoscale region chosen for analysis is the lower Ohio River Valley and consists of Kentucky and portions of the surrounding states (refer to Figure 1). The high density of major pollutant sources in this area makes it an important region for analysis of an emission reduction scenario. In this study, the pollutant transport/transformation/deposition model STEM II/ASM (Carmichael et al., 1990) is used to evaluate the effect of emission reductions on the ambient pollutant concentrations and wet deposition patterns. The time period of the particular deposition episode under consideration is April 29 to May 3, 1985. This period witnessed a slow moving frontal storm that travelled over eastern United States. Details of the meteorology for this specific event are described in Saylor(1989) and Carmichael et al.(1990).

The present work extends previous studies by analyzing the effect of emission reductions on the eventual ambient concentrations of pollutants and their deposition patterns. A major aspect is to investigate the effect of background concentrations of pollutants on the levels of pollutants such as SO_x, NO_x, $SO_4^=$, and NO_3^- in the region under consideration. The results presented in Saylor et al.(1990) are used as a base case for comparing the results of simulations presented in this paper.

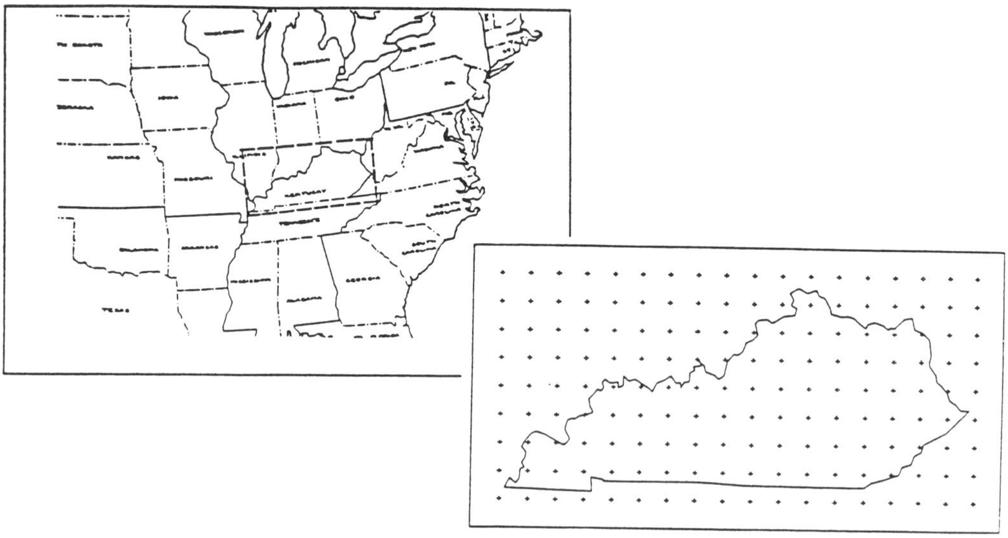

Figure 1. The Mesoscale Region and Grid Domain.

DESCRIPTION OF MODEL AND MODELING REGION

STEM II/ASM is a comprehensive acid deposition model based on an Eulerian formulation; details are presented in Carmichael et al.(1990). Provided with appropriate input meteorological data, STEM II simulates the emission, transport, transformation, and removal of pollutants within the prescribed modeling domain. The Lower Ohio River Valley has a relatively high density of SO_2 and NO_x point sources with over 2000 SO_2 point sources having emissions greater than 10 tons per year. Figure 2 illustrates the distribution of SO_2 point source emissions within the modeling domain.

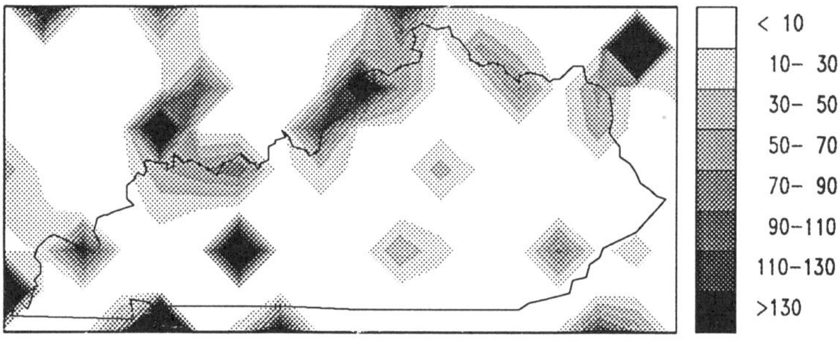

Figure 2. Modeled emissions of SO_2 ($\mu g\ m^{-3}\ hr^{-1}$) for the April 29 to May 3, 1985 base case (KYMET) simulation.

The horizontal region is discretized into 18 x 9 grids of 40 km x 40 km size in the north-south and east-west directions. The vertical extent from ground level to a height of 4 km is discretized into 14 levels. Horizontal and vertical wind fields along with temperature, humidity, eddy diffusivity, surface precipitation, air density and information regarding distribution of clouds constitutes the input meteorological data set used for model computations. Meteorological data were obtained from the National Climatic Data Center. Particulars regarding the gridding of the data to the grid domain are presented by Saylor et al.(1990) and Mathur (1988).

ORGANIZATION OF SIMULATIONS

A series of mesoscale simulations using the STEM II model coupled with the ASM cloud physics module (Hales, 1984) were performed to study the sensitivity of deposition amounts of acidic pollutants to changes in emissions within the region and to changes in background concentrations of pollutants. The base case, which in subsequent discussion will be referred to as KYMET and corresponding to the simulation of the April 29 to May 3, 1985 acid deposition episode, was evaluated against measurements obtained from the KADP/NADP monitoring network. STEM II/ASM predictions of $SO_4^=$ and NO_3^- depositions were found to agree well with network measurements; $SO_4^=$ depositions were predicted more accurately than were NO_3^- depositions (Saylor et al.,1990). These results were analyzed considering uncertainties associated with grid-averaged concentrations as predicted by STEM II/ASM versus observed point measurements, and uncertainties with inadequate representation of subgrid scale chemistry. Simulation results show that in general STEM II/ASM underpredicts NO_3^- wet depositions. One possible explanation of these underpredictions is an inadequate NO_x emissions data base.

Table 1. Simulation Details: Reductions indicated are with respect to Base Case KYMET. Negative percentage reductions indicate an increase in amount deposited.

Simulation Designation	Reductions in Emissions		Reductions in Initial and Background Concentrations	Percentage Reduction in Total Wet Deposition			
	SO_2	NO_x		$SO_4^=$	NO_3^-	$SO_3^=$	HSO_3^-
KYROLL	50 %	40 %	0 %	3 %	16 %	38 %	41 %
KYRSNB	50 %	40 %	SO_2 by 50 %, NO and NO_2 by 40 %	9 %	24 %	46 %	51 %
KYBACK	50 %	40 %	SO_2 and $SO_4^=$ by 50 %, NO, NO_2 and HNO_3 by 40 %	41 %	34 %	36 %	46 %
KYROLB	50 %	40 %	50 % for all advected species	49 %	46 %	8 %	31 %
BDROLL	0 %	0 %	50 % for all advected species	47 %	37 %	-52 %	-22%

Table 1 presents a summary of the conditions for each of the simulations in this study. Simulation KYROLL represents the source reduction scenario where the emissions of SO_2 are reduced by 50% and those of NO_x by 40% of the base case (KYMET) values. The background concentrations for the advected species are the same as the base case. The purpose of the simulation was to investigate the effect of reduction in primary emissions on the total deposition of acidic pollutants in the region. To be consistent with a source reduction scenario the background concentrations of SO_2, NO_2, and NO were reduced for another simulation. This scenario is designated KYRSNB, and in addition to background levels, SO_2 and NO_x emissions were reduced uniformly over the grid domain by 50 % and 40%, respectively, as in KYROLL. The reductions in background concentration were SO_2 by 50 % and NO_2 and NO by 40 % of the base case values.

Simulation KYBACK corresponds to the case where the background concentrations of $SO_4^=$ and HNO_3 were also reduced in addition to those species background levels reduced in KYRSNB. To further investigate the effects, another simulation in which the background concentrations of all advected species were reduced to half of their base case values was performed; this simulation is designated KYROLB. The amount of reduction in emissions for this case is the same as the other simulations; i.e., 50 % for SO_2 and 40 % for NO_x. Hypothetically, this represents a scenario where the regions surrounding the grid domain have ambient concentrations characteristic of a rural environment. Finally, BDROLL corresponds to the scenario where the emissions are the same as the base case, but the background concentrations for all advected species are reduced to half of the base case values.

This simulation was included to determine the extent to which the boundary conditions governed the concentrations of pollutants within the grid domain.

RESULTS AND DISCUSSION

The STEM II/ASM transport/transformation/deposition model was used to investigate deposition of acidic substances for the various emission reduction and background concentration reduction scenarios described in the previous section. Meteorological conditions for the April 29 to May 3, 1985 period have been used. The main purposes of this study are then to investigate and determine the effect of reduction in emissions of primary sources and reduction in background concentrations on the eventual ambient levels and deposition patterns and amounts of major acidic pollutants in the mesoscale region of study.

Figure 3 shows the percentage reductions in ground-level gas-phase SO_2 levels as compared to the base case KYMET. These SO_2 predictions illustrate the predominant effect of emission source strength and location. SO_2 being weakly soluble in aqueous media is not greatly affected by the presence of clouds and rain. As expected, the locations of maximum concentrations of SO_2 are close to the source areas. Direct cutbacks in predicted SO_2 concentrations, especially near the dense source regions are observed in simulations KYROLL, KYROLB, and KYBACK in response to the 50 % reduction in SO_2 emissions.

Figures 4 and 5 present comparisons of simulated $SO_4^=$ and NO_3^- wet deposition amounts over the region for the event for all scenarios; deposition amounts are in mg/m^2. The deposition patterns resulted from the combined effect of SO_2 to $SO_4^=$ downwind transformation and precipitation scavenging of $SO_4^=$. In STEM II/ASM, $SO_4^=$ is formed through both gas- and aqueous-phase reactions, with aqueous-phase reactions more than an order of magnitude faster than gas-phase transformation, provided sufficient amount of oxidants are present in the cloud. On the other hand, HNO_3 (or NO_3^-) is formed from NO_x through gas-phase reactions, and subsequent scavenging results in wet deposition. Since the transformation of gas-phase NO_x to HNO_3 and aqueous-phase formation of $SO_4^=$ are rapid reactions and since cloud uptake of HNO_3 and oxidants are efficient processes, it is expected that wet deposition patterns for $SO_4^=$ and NO_3^- would be similar. Such trends in deposition patterns have also been shown by Misra et al. (1989).

Table 1 also presents the percentage reductions in total region wet deposition amounts for $SO_4^=$, NO_3^-, $SO_3^=$, and HSO_3^-. The reductions are computed in comparison to the base case (KYMET) deposition. Reducing emissions within the region of SO_2 by 50 % and NO_x by 40 % caused relatively small reductions in amounts of $SO_4^=$ or NO_3^- deposited with precipitation. This is reflected in the contour plot for $SO_4^=$ wet deposition for the KYROLL simulation. On this mesoscale, reductions in amount of $SO_4^=$ wet deposited were of the order of 2-3 % . Reductions in NO_3^- deposition were greater being in the range of 15-20 % . Reducing the background concentration of SO_2 by 50 % and NO_x by 40 % (KYRSNB) further reduced the deposition amounts, $SO_4^=$ by 9 % and NO_3^- by 24 %. However, when the background concentrations of $SO_4^=$ and HNO_3 were also reduced (KYBACK), substantial reductions in amounts of wet deposition were observed, $SO_4^=$ by 41 % and NO_3^- by 34 %. These results focus on the dominant effect of background levels of pollutants on the eventual amount of acidic deposition in this mesoscale region. Predictions from simulations KYROLB and BDROLL further illustrated the importance of background pollutant levels, their subsequent advection into the region, and eventual precipitation scavenging. Greater reduction in amounts wet deposited in these two cases resulted from lower oxidant levels which resulted in lesser amounts of $SO_4^=$ and NO_3^- being scavenged.

S(IV) species ($SO_3^=$ and HSO_3^-) serve as precursors to aqueous phase $SO_4^=$. Lesser amounts of S(IV) deposited result from the combined effect of reduced absorption into the clouds and greater amounts being oxidized to $SO_4^=$; these trends are seen for the cases KYROLL and KYRSNB in Table 1. Reducing background levels of $SO_4^=$ and HNO_3 resulted in reduced

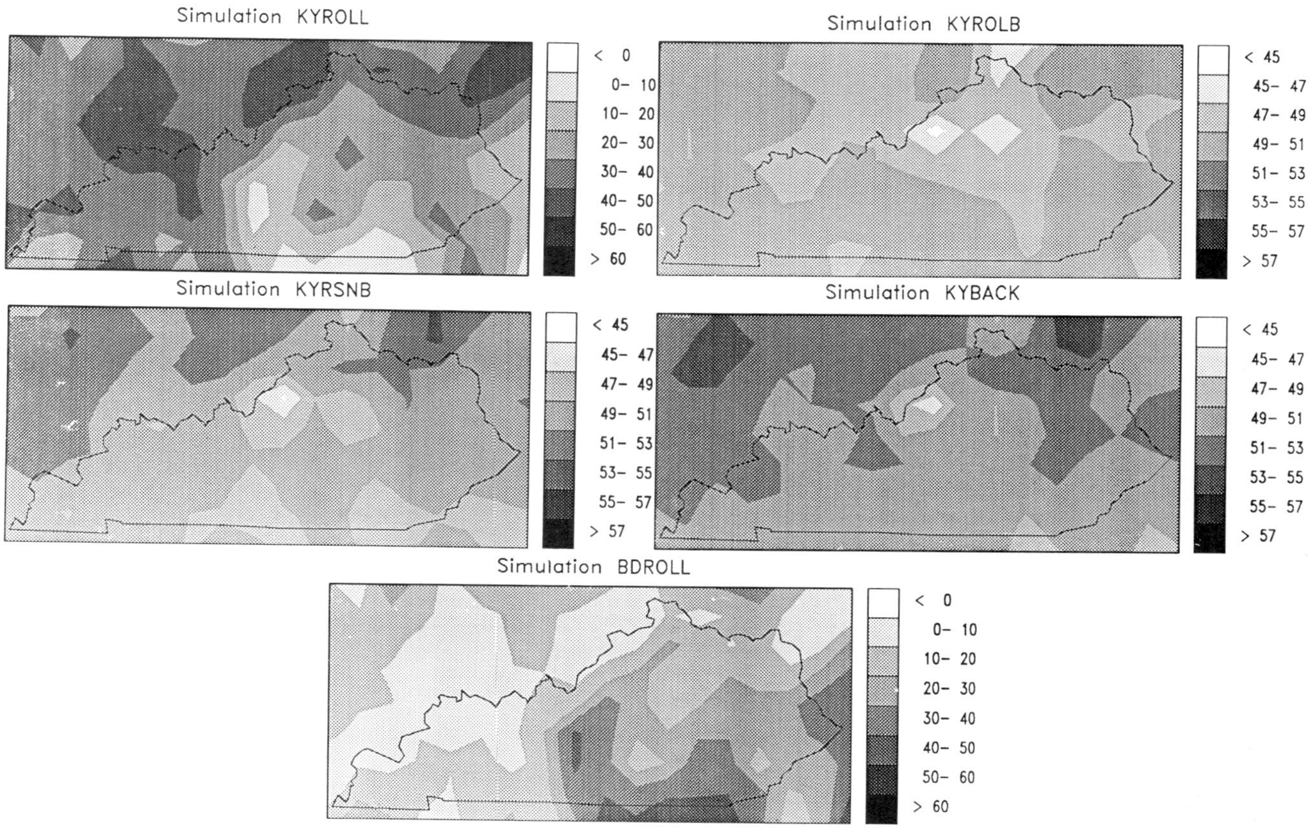

Figure 3. Percentage reduction of event-averaged gas-phase (ground level) SO_2 concentration.

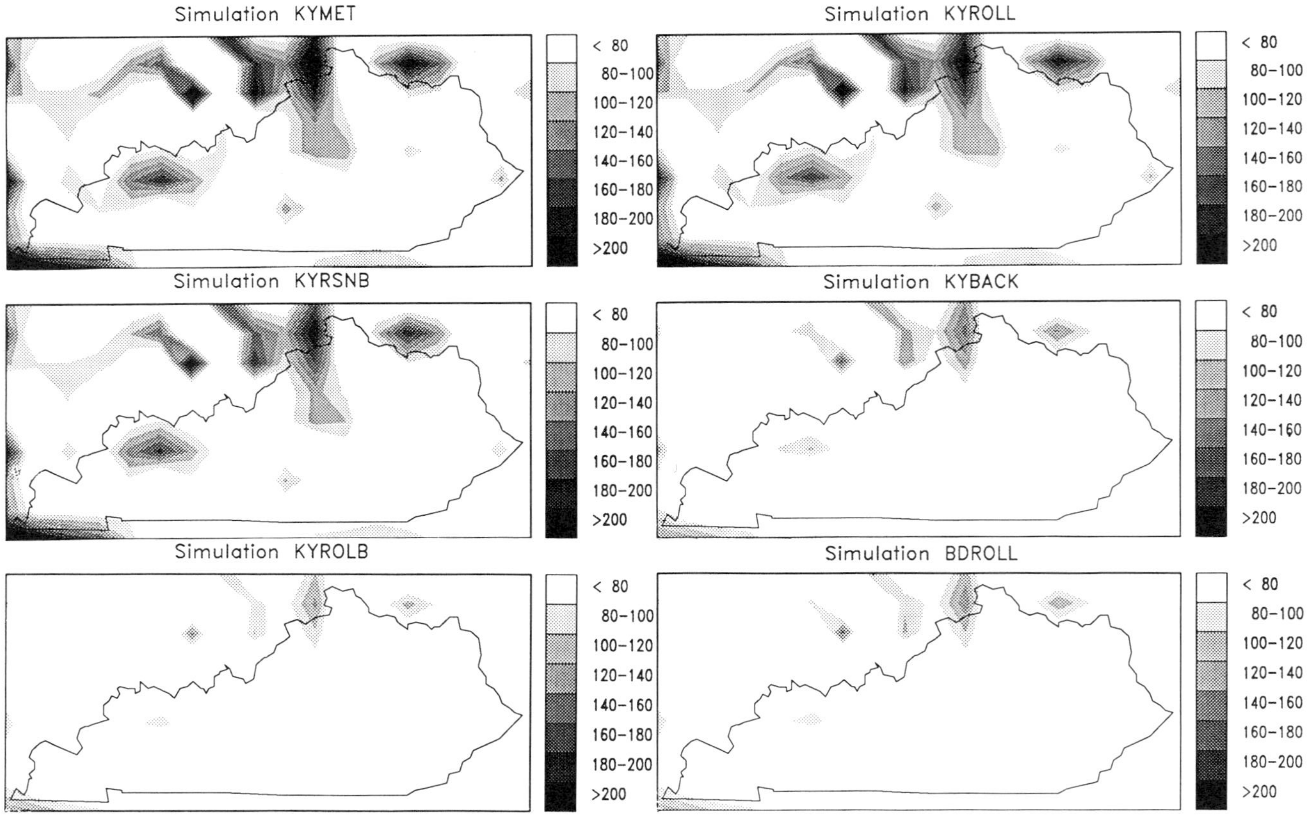

Figure 4. **Predicted total wet $SO_4^{=}$ deposition (mg m^{-2}) for the April 29 - May 3, 1985 deposition episode.**

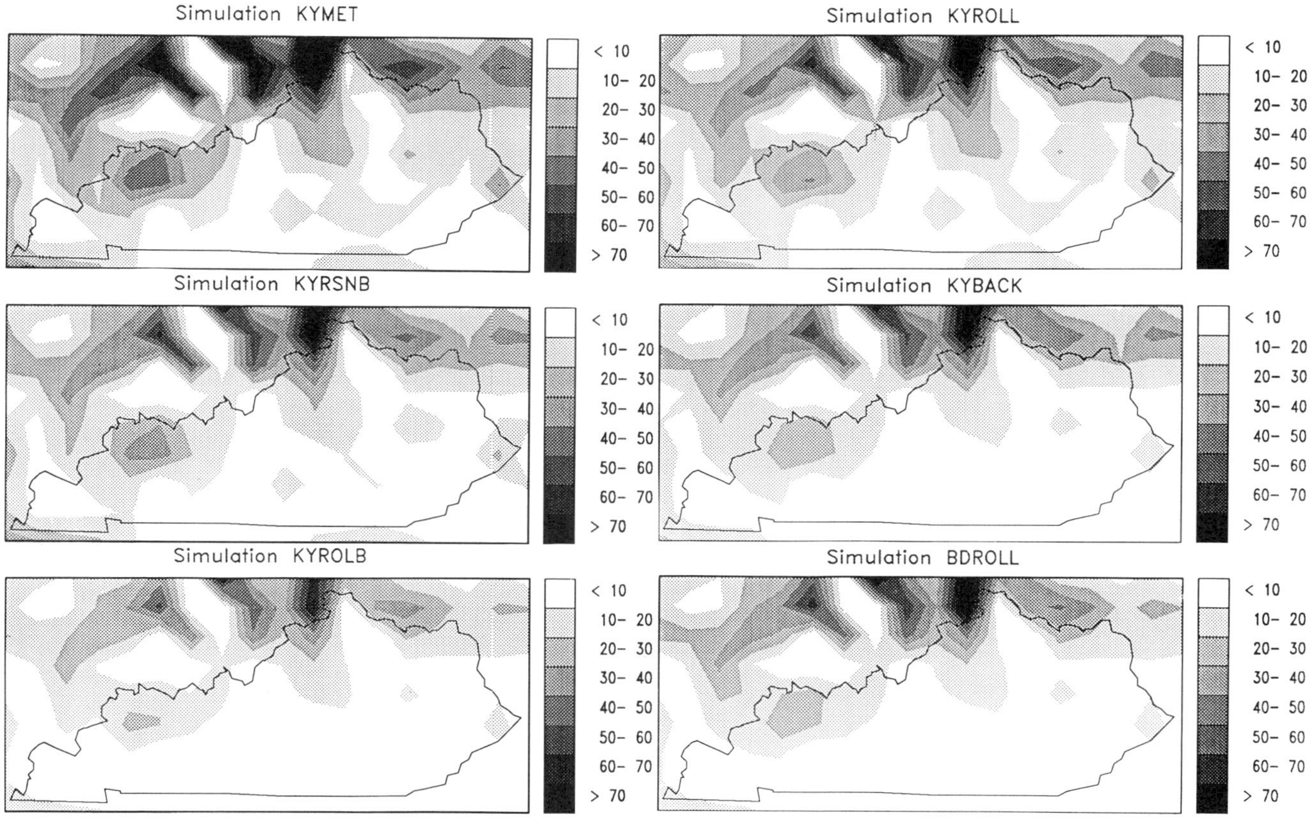

Figure 5. Predicted total wet NO_3^- deposition (mg m^{-2}) for the April 29 - May 3, 1985 deposition episode.

amounts being advected into the region. Consequently, cloud uptake was reduced, with corresponding decreased reductions for S(IV) wet deposition (KYBACK). Greater reductions in amounts of $SO_4^=$ wet deposited in KYROLB and BDROLL illustrate the combined effects of reduced oxidant levels and reduced $SO_4^=$ background levels.

CONCLUSIONS

The results from six simulations using the STEM II/ASM transport/ transformation/ deposition model for a mesoscale region (the lower Ohio River Valley) show that there is significant variation in the gas-phase levels and deposition amounts of major acidic pollutants with varying emissions and background levels of primary species. These simulations, involving various emission reduction and background concentration reduction scenarios, demonstrate the importance of background concentrations of key pollutant species on the final magnitudes and patterns of wet deposition. With relatively high background pollutant levels, reducing SO_x and NO_x emissions does not substantially reduce gas-phase levels of major pollutants or amounts wet deposited. This focuses on the importance of properly representing background concentrations in mesoscale pollutant transport and deposition models. A Nested-Grid approach, wherein the region under study is nested in a larger but coarser grid domain whose boundaries extend far enough so as to preclude any boundary effects, is a technique that has been proposed for computing background concentrations for the mesoscale region. Adaptation of this approach for the given mesoscale region are in progress.

ACKNOWLEDGEMENTS

This research has been supported in part by grants from the U.S. Environmental Protection Agency and U.S. Department of Energy through the NAPAP and PRECP programs. Computer resources and facilities provided by the University of Kentucky Center for Computational Sciences are greatly appreciated.

REFERENCES

Carmichael, G.R., L.K. Peters and R.D. Saylor, "General Model and Overview of STEM II/ASM/RSM", submitted to *Atmos. Environ.* (1990).

Cho, S.Y., Y.S. Chang, and G.R. Carmichael, "An Evaluation of the Effect of Reductions in Ambient Levels of Primary Pollutants on Sulfate and Nitrate Wet Deposition", *Atmos. Environ.*, **23**, 1009 (1989).

Hales, J.M., "An Interim User's Manual for the Advanced Scavenging Module" *Battelle Pacific Northwest Laboratories* (1984).

Mathur, R., "Adjustment of Wind Fields for Application in Air Pollution Modeling", *M.S. Thesis, University of Kentucky*, (1988).

Misra, P.K., R. Bloxam, C. Fung, and S. Wong, "Non-Linear Response of Wet Deposition to Emissions Reduction: A Model Study", *Atmos. Environ.*, **23**, 671 (1989).

Saylor, R.D., "A Study of Mesoscale Acid Deposition in the Lower Ohio River Valley", *Ph.D. Dissertation, University of Kentucky*, (1989).

Saylor, R.D., L.K. Peters, and R. Mathur, "Simulation of a Mesoscale Deposition Episode Using the STEM II Transport/Transformation/Deposition Model", submitted to *Atmos. Environ.* (1990).

Stockwell, W.R., J.B. Milford, G.J. McRae, P. Middleton, J.S. Chang, "Nonlinear Coupling in the NO_x-SO_x Reactive Organic System", *Atmos. Environ.*, **22**, 2481 (1988).

DISCUSSION

J.C.R. HUNT

How does your model allow for the special deposition processes on hills, i.e. the topographic corrections?

R. MATHUR

The deposition velocities used by the model are parameterized to depend on meteorological conditions, land-use type, surface roughness, and time of day. The model does use a terrain-following coordinate system to allow for the topography of the region under study. However, the spatial scales involved in meso- and regional-scale modeling somewhat preclude using detailed topographic features.

M.E. BERLYAND

How did you take the background levels of pollution in the U.S. into account?

R. MATHUR

The Lower Ohio River Valley has a relatively high density of SO_2 and NO_x point sources with over 2000 SO_2 point sources having emissions greater than 10 tons per year. Furthermore, the region accounts for over a quarter of the total point source SO_2 emitted and approximately one-fifth of the total point source NO_x emitted from the contiguous United States. This is indicative of a moderately polluted atmosphere in the region of study. The background pollutant levels used are those characteristic of a moderately polluted atmosphere and are introduced into the model computations through inflow boundary conditions.

J.L. WALMSLEY

How do you account for pollution sources beyond your modelled region? Are you using a nested grid?

R. MATHUR

As indicated earlier, the background levels of pollutants used represent pollutant concentrations in a moderately polluted atmosphere. The results of the various simulations described in this paper are indicative of the sensitivity of final deposition amounts and ambient pollutant levels to the background pollutant levels. Furthermore, the non-linearity introduced by the chemistry often poses a problem when deciding on background levels of pollutants for an emission reduction scenario. One way of addressing this problem is use of nested grids. Our current efforts focus on the implementation of a nested grid scheme to account for pollutant sources beyond the modelled region.

C.J. WALCEK

In one experiment you describe KYROLB, NO_x emissions are reduced by 40%, and nitrate wet deposition is decreased by 46%, an amount greater than the emission reduction. Could you explain more fully how this super linear behavior occurs?

R. MATHUR

Simulation KYROLB corresponds to the case where the background levels of pollutants were reduced by 50% in addition to the primary emissions within the region being reduced. Consequently, the background levels of oxidant species such as O_3 and H_2O_2 were reduced in comparison to the base case. Because of the reduced oxidant levels, the relative conversion rate of NO_x to HNO_3 is reduced, resulting in reductions in nitrate wet deposition to be greater than the amount of reduction in NO_x emissions. These results focus not only on the non-linear relationship between emission reductions and eventual deposition amounts, but also on the importance of the choice of background pollutant levels when analyzing a source-reduction scenario.

DETECTION OF SULPHUR EMISSION REDUCTIONS IN EUROPE

DURING THE PERIOD 1979-1986

Sophia Mylona

The Norwegian Meteorological Institute
P.O.Box 43-Blindern
0313 Oslo 3, Norway

INTRODUCTION

It is well known that in most of the countries of Europe a significant part of the total deposition of sulphur is due to foreign sources. This led, in 1977, to the launch and establishment of a "Co-operative programme for monitoring and evaluation of the long range transmission of air pollutants in Europe" (hereafter referred to as EMEP).

According to the 1985 protocol to the Convention on the Long-Range Transboundary Air Pollution, the countries which are members of EMEP should reduce their national annual sulphur emissions or their transboundary fluxes by at least 30 per cent as soon as possible and at the latest by 1993, using 1980 levels as the basis for the calculation of reductions.

The purpose of this study is to detect and quantify, as far as possible, reductions accomplished in the anthropogenic sulphur emissions in the atmosphere over Europe during the 8-year period 1979-1986, by means of observed and calculated sulphur dioxide and particulate sulphate concentrations, regarding 1980 as the year of reference.

The Norwegian Meteorological Institute (DNMI), acting since 1985 as the Meteorological Synthesizing Centre-West (MSC-W) of EMEP, has at its disposal all of the data which are needed to carry out the above objective.
These data are:

1) Official national emissions of sulphur, annually reported by the countries participating in EMEP.
2) Daily observations from all reporting stations in the EMEP measurement network.
3) Daily computed concentrations at the same sites, derived from the EMEP/MSC-W Sulphur Model.

A first base-case model simulation is used to test the ability of the model to predict the observed concentrations. Subsequently, the investigation is considerably facilitated by the performance of two additional model runs, based on different assumptions:

a) Actual meteorology but the 1980-level emissions employed for all the years.
b) The meteorology of 1980 employed for all the years but emissions annually updated.

By means of the two additional model simulations a) and b), it is possible to isolate the concentration changes due to emission reductions from those due to meteorological factors. This is necessary in order to be able to verify whether the emission reductions have had any quantifiable effect on the measured concentrations. It should be noted however that certain conclusions can be drawn by comparing different model calculations alone. Whilst such conclusions are not influenced by problems related to the station network, they are on the other hand dependent on simplifications in the model design.

THE EMEP SULPHUR MODEL AND INPUT DATA

The calculations presented in the following have been carried out with a receptor-oriented one-layer trajectory model (Eliassen, 1978; Iversen et al., 1989). 850 mb isobaric trajectories are followed 96 hours backward in time from their arrival points, the EMEP measurement sites. Four trajectories arrive at each site every day on a 6-hourly basis. The observation period is therefore covered reasonably well.

Effects such as a variable mixing height inferred from an exchange mechanism between the atmospheric boundary layer and the free troposphere, as well as a variable deposition velocity for sulphur dioxide due to a varying aerodynamic resistance to vertical turbulent transport, are included in the model (Eliassen and Saltbones, 1983).

The operation of the long range transport model requires emission data and meteorological data as input.

Emission data. The current model calculations are based on official national emission data, available up to the time of the sixth session of the Executive Body for the Convention on Long-Range Transboundary Air Pollution, which took place in Sofia, Bulgaria in the autumn of 1988.

For years where no official numbers have been given, data have been interpolated from earlier submissions, taking the officially reported predictions for future years into account. For countries that have not reported their spatial distribution in the 150 km EMEP grid for at least one year, this has been estimated from open information on the population distribution, on the locations of power production facilities and relevant industrial activities.

A rough estimate of the reductions accomplished by each country individually from 1980 to 1986, as estimated from the above data, is presented in Table 1. Recognizing the incompleteness of the emission data and their irregular decrease in some countries, one may still note that the reduction of the total European sulphur emission from 1980 to 1986 seems to be about 4.3 million tonnes or 16.4% down from the 1980 total of about 26 million tonnes. Should the same linear reduction rate continue, the decrease in 1993 would be 9.2 million tonnes or 35.4% of the 1980 value.

Meteorological data. Throughout the year at a 6-hourly time resolution, and at each grid point in the 150 km EMEP grid, the following data are needed:

- Horizontal advection wind for the atmospheric mixing layer,
- surface wind for the calculation of the dry deposition velocity for SO_2,
- vertical velocity at the top of the boundary layer,
- grid-element average rainfall amounts,
- the height of the atmospheric mixing layer at 12 GMT every day. In Europe this will approximate the afternoon maximum mixing height.

Table 1. Emission reductions accomplished by each country individually within the period 1980-1986
Unit: Thousand tonnes of sulphur per annum

Country	1980	1986	Reduction	Red. in %
Albania	25	25	0	0.0
Austria	177	80	97	54.8
Belgium	414	237	177	42.8
Bulgaria	517	535	-18	-3.5
Czechoslovakia	1550	1511	39	2.5
Denmark	219	159	60	27.4
Finland	292	162	130	44.5
France	1756	798	958	54.6
German Dem. Rep	2500	2500	0	0.0
Germ.Fed.Rep.of	1600	1100	500	31.3
Greece	200	180	20	10.0
Hungary	817	710	107	13.1
Iceland	3	3	0	0.0
Ireland	110	77	33	30.0
Italy	1900	1185	715	37.6
Luxemburg	11	7	4	36.4
Netherlands	232	138	94	40.5
Norway	71	45	26	36.6
Poland	2050	2150	-100	-4.9
Portugal	133	125	8	6.0
Romania	100	100	0	0.0
Spain	1625	1592	33	2.0
Sweden	232	122	110	47.4
Switzerland	63	32	31	49.2
Turkey	138	177	-39	-28.3
Soviet Union	6400	5450	950	14.8
United Kingdom	2335	1870	465	19.9
Yugoslavia	588	725	-137	-23.3
SUM	26058	21795	4263	16.4 %

INTERCALIBRATION

In order to ensure comparable results and quantify the existing differences, a high quality of measured data is needed. Since the start of EMEP a series of annual intercomparisons of analytical methods have been performed. For the sampling methods however only one intercomparison test has been organized so far. These tests have shown that the quality of the data from many participants is acceptable and has improved over this period. For some countries however there is clearly room for improvements. (Thrane, 1980, 1981; Hanssen et al., 1983, 1984, 1985, 1987; Nodop, 1985; Nodop and Hanssen, 1986).

In this study the outcome of the intercomparison tests, in both the sampling and analytical methods, has been used as a guideline in determining the relative weight to be given to the data quality from the various stations.

DISCUSSION OF THE STATION NETWORK

Monthly arithmetic mean observed and calculated concentrations for each EMEP measurement site have been produced (Mylona, 1989), but in order to be able to identify any trends, the following discussion will be focused on the results averaged over the whole station network.

In order to obtain comparable results from each site, a set of requirements to be fulfilled by each station has been adopted in this study.

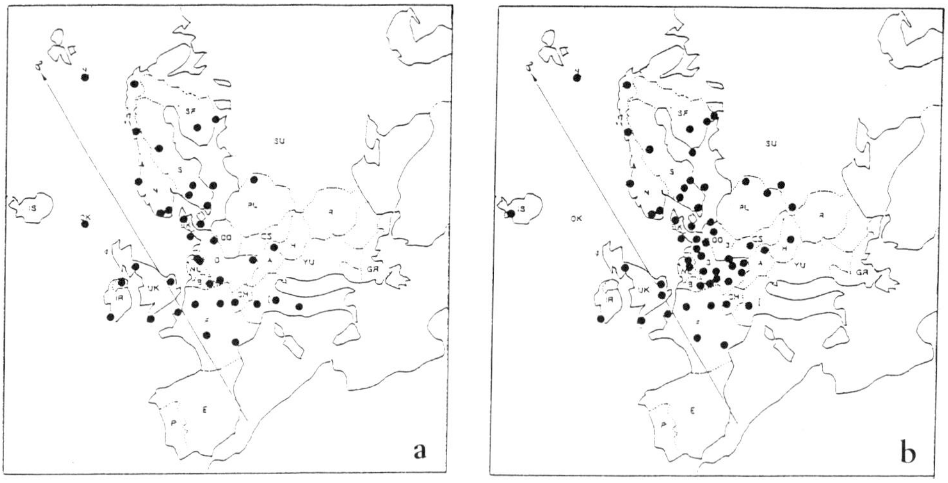

Fig. 1. Distribution of the EMEP station network for sulphur dioxide (a) and particulate sulphate (b) during the period 1979-1986.

According to these a station must:

- not be situated close to the boundaries of the area covered by the model calculations to avoid bias of estimates from certain directions
- not be located very high relative to the surrounding area, since then there is a risk that the sampling occurs above the mixing layer some of the time. In this study a height of 1200 m above sea level has been chosen as the upper limit for the location of a station. A check on the mixing height values for all years has shown that for more than half of the days in a year the mixing height has been below 1200 m. This fact gives some support for the limit chosen.

Even when the sites satisfy the above criteria the expected concentration trends are subject to other problems connected with the station network. These are:

- The spatial distribution of the measurement sites is not uniform with the largest density occuring in central Europe. Consequently the results will over-represent that area.
- The number of reporting stations has increased substantially since the beginning of the period considered.
- Measurement sites from highly polluted areas which started reporting later than 1979, will contribute to the network average concentration more than extra sites from relatively clean areas.
- At several sites monthly measurements are often missing. The gaps so created in the observations sometimes extend to years.

Furthermore, the extent to which a sampling site is representative of its region introduces another difficulty. Subjective inspection of the sites location seems to indicate that some stations do not comply with the requirements a regional station should fulfill (WMO, 1974).

To reduce as much as possible the effect of the above factors on the expected trends, all sites which started reporting after 1980, gave measurements only for a short period, or have shown poor agreement compared with the results from other participants in the sampling and analytical intercomparison tests, have been excluded. The resulting distribution of stations for sulphur dioxide and particulate sulphate is presented in Fig. 1. Unfortunately, by doing so, almost all stations in southern Europe are omitted. It has not been possible to find a fully satisfactory way to deal with this problem. The results will therefore reflect the situation in northern and central Europe.

RESULTS FROM ANALYSIS OF STATION NETWORK AVERAGES

Sulphur dioxide

Monthly observed and calculated concentrations of sulphur dioxide averaged over 42 EMEP stations during the 8-year period 1979-1986 are shown in Fig. 2a. The seasonal variations are generally well captured by the model. The high concentrations in wintertime can be explained by increased consumption of energy used for domestic heating, as well as longer periods with stable stratification resulting from a colder and often snow covered surface. Sinks for SO_2 (i.e. uptake by vegetation) are also less effective in winter than in summer.

Conclusions on emission reductions can more clearly be derived from the annual averages (Fig. 2b). A downward trend appears clearly in both observations and calculations, correlating with the reported reduced emissions.

The dotted curve presents concentrations calculated if the 1980-level emissions are used for all the years. As can be seen there is a marked divergence between the two calculated versions over the years. This can only be ascribed to differences in emissions. The calculations with the reported

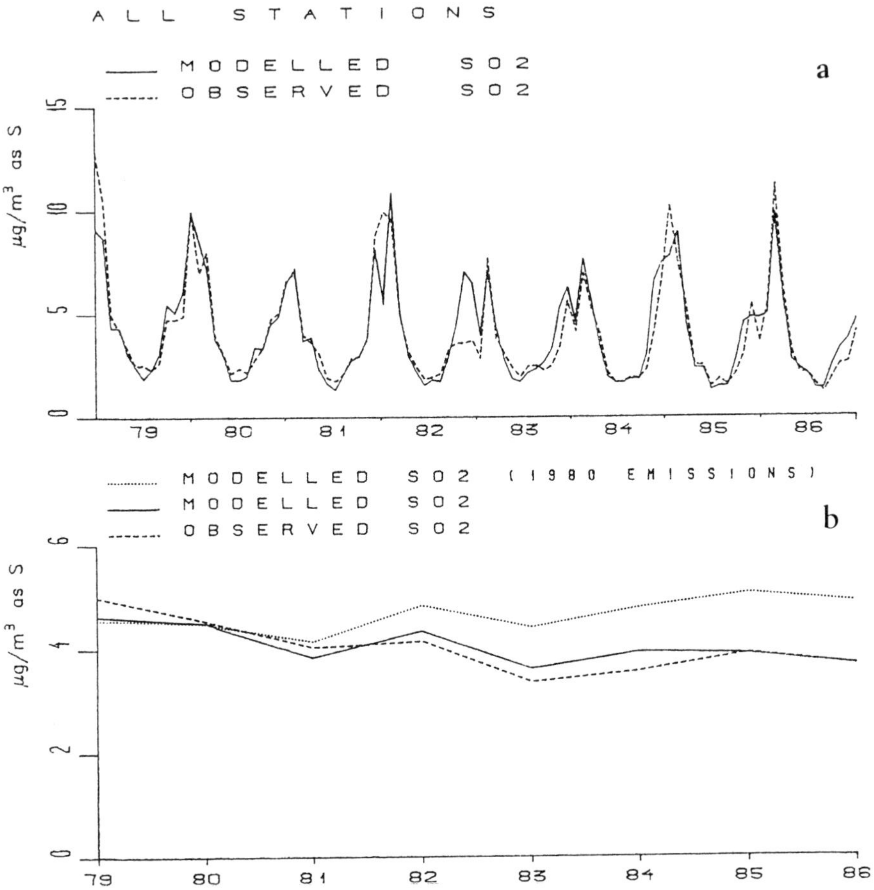

Fig. 2. Monthly (a) and annual (b) mean concentrations of SO_2 over Europe. The dotted curve shows calculated concentrations assuming constant 1980-level emissions over the years.

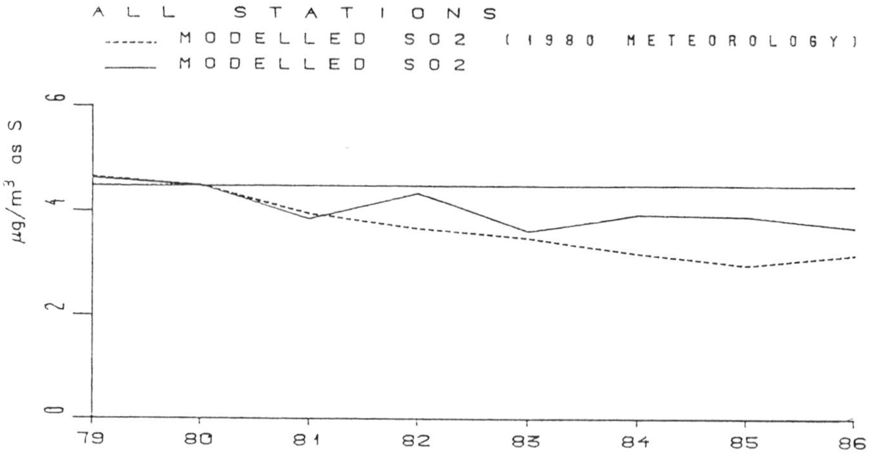

Fig. 3. Annual mean calculated concentrations of SO_2 over Europe. The dashed curve shows concentrations assuming constant 1980-meteorological conditions over the years. The horizontal line shows the calculated 1980 concentrations.

emissions follow the observations quite well and in closer agreement than the calculations based on constant 1980-level emissions.

This approach allows us to state that the expected reduction in SO_2 concentrations in ambient air over Europe is confirmed for the period 1979-1986. A simple estimate shows roughly a decrease of 18.9% in the observed SO_2 concentrations from 4.54% $\mu g/m^3$ in 1980 to 3.68% $\mu g/m^3$ in 1986.

We note that the fluctuations of the dotted curve in Fig. 2b must be attributed to variable meteorological conditions over these years. The effect of the variable number of sites is probably not so important because of the selection process described above.

In order to illustrate the effect of meteorological conditions on the mean concentrations of SO_2 over Europe, the EMEP model has been run once more with annually updated national emissions as reported by the countries, but with the meteorological conditions of 1980 employed for all the years. Thus, any downward trend appearing in these computations will be due to emission reductions only, and any deviation from the calculations will be due to changes in meteorology. The outcome is demonstrated in Fig. 3.

The declining trend of the new curve reflects the reduced emissions fairly well. Average concentrations in 1979, 1981 and 1983 show great similarity with those of the reference year. Positive deviations occur in 1982, 1984, 1985 and 1986. We assume that within these years meteorological conditions were distributed in such a way that high concentrations occured over central and northern Europe more frequently than in 1980.

The most important meteorological factors that influence the concentrations of SO_2 in air are:
- Transport processes caused by the prevailing winds.
- The depth of the mixing layer. The lower the mixing height, the more restricted the dispersion of air pollutants in the boundary layer and the higher the measured concentrations are.
- Deposition processes. Dry deposition is particularly stressed since it is a more efficient scavenging mechanism for SO_2 than wet deposition.

Particulate sulphate

Monthly and annual air concentrations of particulate sulphate averaged over 60 EMEP stations are presented in Figs 4a,b. As can be seen the concentration of sulphate in particles does not show clear seasonal variation. The reason for this is that the production rate of particulate sulphate in air is expressed as the product of the SO_2 concentration and a transformation rate which is a function of the time of the year. During winter the emissions are high, and much of the sulphur is present as sulphur dioxide due to low winter-time transformation rate (95% of the sulphur emission occurs in the form of SO_2). In contrast, during summer emissions are low, but the enhanced summer-time transformation rate accounts for the rapid conversion of sulphur dioxide to particulate sulphate.

The monthly and annual means of observed and calculated concentrations show rather good agreement. A downward trend in the particulate sulphate concentrations can readily be observed (Fig. 4b) but it stops in 1983.

As in the relevant case of SO_2, there is a steady divergence between the calculations with 1980-level emissions and the ones with updated

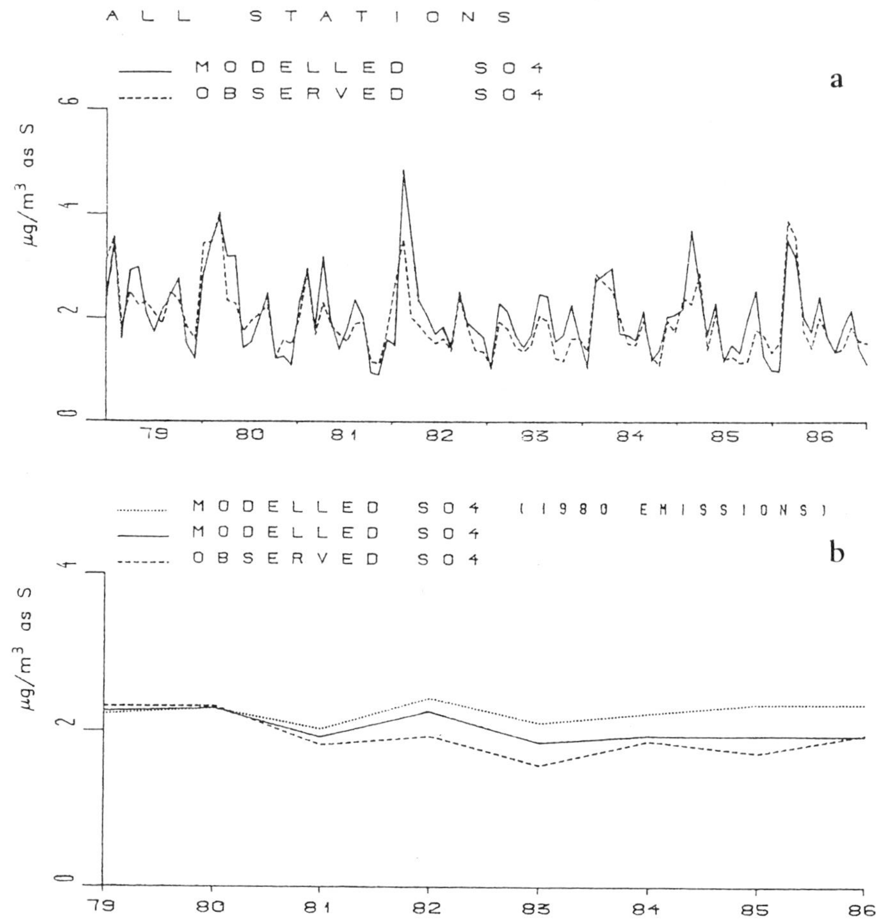

Fig. 4. Monthly (a) and annual (b) mean concentrations of particulate sulphate over Europe. The dotted curve shows calculated concentrations assuming constant 1980-level emissions over the years.

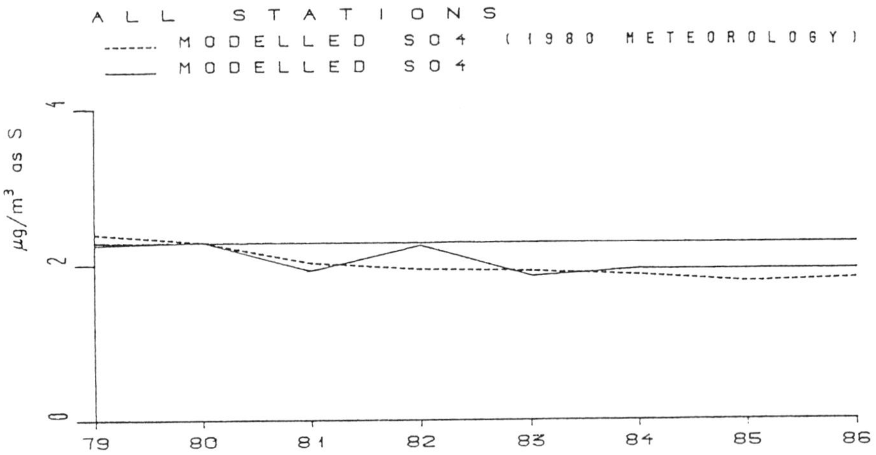

Fig. 5. Annual mean calculated concentrations of particulate sulphate over Europe. The dashed curve shows concentrations assuming constant 1980-meteorological conditions over the years. The horizontal line shows the calculated 1980 concentrations.

emissions over the years. This finding is considered to be a verification of the reduction in emissions.

A rough estimate of the reduction of particulate sulphate observed concentrations accomplished within the period 1980-1986, gives a reduction of 15.5% from the mean value of 2.32 $\mu g/m^3$ in 1980 to 1.96 $\mu g/m^3$ in 1986.

Fig. 5 shows the case where the 1980 meteorological conditions have been employed for all the years, but the emissions have been updated annually. Apart from 1979 and particularly 1982 the calculations seem to be quite consistent with the reduced emissions.

As for SO_2, the same meteorological factors are considered to be the most important in influencing the air concentrations of particulate sulphate, namely: transport processes, deposition processes and the depth of the mixing layer. The first factor is particularly stressed, because particulate sulphate has a longer residence time in the atmosphere than SO_2 and is consequently transported over greater distances. Wet deposition is also very important, because particulate sulphate is much more soluble than SO_2 and is therefore more effectively scavenged by precipitation.

CONCLUSIONS

The EMEP sulphur model has been used to detect reductions in the European sulphur emissions over the 8-year period 1979-1986. The effect of meteorological conditions on the expected trends has also been investigated.

Keeping in mind that the results reflect mainly the situation over central and northern Europe and recognizing that uncertainties in the estimates arise from:
- unsuitability of the monitoring stations location,
- sampling and analytical errors,
- inaccurate emission data,
- limited amount of meteorological information,
- oversimplified assumptions in the model,
- the averaging method used over the station network,

one can still conclude the following:

1) Both monthly and annual average observations are simulated reasonably well by the model.

2) A definite downwards trend is apparent in the annually averaged concentrations of sulphur dioxide. For particulate sulphate the downward trend in the concentrations stops in 1983. These reductions are in line with the sulphur emissions reported by the countries. In particular, it has been found that during the period 1980-1986 there has been a 18.9% reduction in the SO_2 concentrations and a 15.5% reduction in the concentrations of particulate sulphate. These reductions reflect the 16.4% decrease in the sulphur emissions over the same period.

3) Meteorological conditions different from those in 1980 were found to have played an essential role in altering and mainly increasing the air concentrations over the years. It follows that if the meteorological conditions over the years were similar to those of 1980, lower concentrations would have been observed, and consequently the effect of the decreased emissions would be more pronounced.

4) It has to be noted that in deriving the reduction percentages, fluctuations in emissions and concentrations during the period 1980-1986 are disregarded. Furthermore, one should not expect to see the same percentage of reduction in the concentrations as in the emissions. This is mainly due to the fact that areas where the emission reductions have taken place do not always coincide with the location of the measurement stations, and the signal of the decreased emissions is therefore not directly received by them. Moreover, the effect of meteorological conditions on the air concentrations, and consequently on their reductions, must by no means be disregarded since it has been proved to be essential.

ACKNOWLEDGEMENTS

The author wishes to express her appreciation to Jørgen Saltbones and professor Anton Eliassen for their support and helpful discussions.

REFERENCES

Eliassen, A. (1978) The OECD Study of long range transport of air pollutants: long-range transport modelling. Atmospheric Environment 12, 479-487.

Eliassen, A. and Saltbones, J. (1983) Modelling of long-range transport of sulphur over Europe: A two-year model run and some model experiments. Atmospheric Environment Vol. 17, No. 8, pp. 1457-1473.

Hanssen, J.E., Ladegård, N.E. and Thrane, K.E. (1983) The sixth intercomparison of analytical methods within EMEP. EMEP/CCC Report: 2/83 Reference: 0-7729.

Hanssen, J.E. and Ladegård, N.E. (1984) The seventh intercomparison of analytical methods within EMEP. EMEP/CCC Report: 1/84 Reference: 0-7729.

Hanssen, J.E. and Ladegård, N.E. (1985) The eighth intercomparison of analytical methods within EMEP. EMEP/CCC Report: 4/85 Reference: 0-7729.

Hanssen, J.E. and Ladegård, N.E. (1987) The ninth intercomparison of analytical methods within EMEP. EMEP/CCC Report: 4/87 Reference: 0-7729.

Iversen, T., Saltbones, J., Sandnes, H., Eliassen, A. and Hov, Ø. (1989) Airborne transboundary transport of sulphur and nitrogen over Europe-Model descriptions and calculations.
EMEP/MSC-W Report 2/89.

Mylona, S.N. (1989) Detection of sulphur emission reductions in Europe during the period 1979-1986.
EMEP/MSC-W Report 1/89.

Nodop, K. (1985) Sulphur dioxide and particulate sulphate in ambient air. Monthly mean concentrations. January 1978 to december 1982.
EMEP/CCC-Report 1/85.

Nodop, K. and Hanssen, J.E. (1986) Field intercomparison of measuring methods for sulphur dioxide and particulate sulphate in ambient air.
EMEP/CCC-Report 2/86.

Thrane. K.E. (1980) Report on the second and third intercomparison of analytical methods within EMEP.
EMEP/CCC-Report 1/80 Reference: 22977.

Thrane, K.E. (1980) Report on the forth intercomparison of analytical methods within the EMEP.
EMEP/CCC-Report 5/80 Reference: 22977.

Thrane, K.E. (1981) The fifth intercomparison of analytical methods within the EMEP.
EMEP/CCC-Report 2/81 Reference: 22977.

World Meteorological Organization (1974) WMO operations manual for sampling and analysis techniques for chemical constituents in air and precipitation.
WMO-No. 299.

DISCUSSION

H. GEISS

In Germany the SO_2 concentration in 1987 and 1988 dropped by more than a factor of two. This may be due to the mild temperatures causing less SO_2 emission from household and industry.

S. MYLONA

The years 1987 and 1988 are outside the scope of this investigation, but I note the comment with interest. However, in the end of the period considered (1979-1986) the officially reported sulphur emissions by West and East Germany cannot explain a large drop in the air concentrations. Additionally, the occurrence of air pollution episodes in Germany during the winters of 1985 and 1986 seem to have interrupted the downward trend demonstrated for SO_2 for previous years.

D. WILSON

How are the empirical coefficients developed that allow the mixed layer thickness and deposition velocity to be modelled? Are these parameters found by empirical adjustment to fit similar data sets, or are they from other sources of data and theory?

S. MYLONA

First of all, the parameters applied for the modelling of the mixed layer thickness and the deposition velocity have not been adjusted to fit similar data sets, but they are based on scientific reasoning.

The mixing height is variable and is estimated from the information provided by radiosonde data. For each radiosonde report the mixing height is taken as the height up to the lowest stable layer with base above 200 m. A layer was defined as stable in this context if $dT/dz > -0.5 \times 10^{-2}$ km^{-1}, or larger than about half of the dry adiabatic value. No diurnal mixing height variations are captured in this procedure. Instead, it has been attempted to estimate an "enveloping" mixing height, under which both the old and the recently emitted pollution is distributed (Eliassen and Saltbones, 1983).

The dry deposition velocity for SO_2 at a height of 1 m is assumed to vary as a function of time of day, season and latitude (Iversen et al., 1989), with a typical summertime value over land V_{d1} = 0.8 cm/s, which is widely accepted. The introduced variation accounts for the fact that dry deposition of SO_2 over non-melting snow is relatively inefficient (Whelpdale and Shaw, 1975; Dovland and Eliassen, 1976), and that values are considerably lower than 0.8 cm/s outside the main growing season. Parameterization of the lower absorption of SO_2 by plants at night is also included. The deposition velocity used in the model is an adjusted value at 50 m height, calculated from the similarity theory for the constant flux layer. For particulate sulphate the deposition velocity is taken as 0.1 cm/s everywhere all the times. Due to the small value the aerodynamic resistance is ignored, as well as the effect of different surface characteristics.

A. VENKATRAM

The performance of the model is astounding! How sensitive is this performance to model parameters? Has the model been tuned?

S. MYLONA

Experiments performed in 1983 on data from February and June 1979 showed that the best correlation occurred after the introduction of a variable mixing height, a variable deposition velocity, an exchange mechanism between the boundary layer and the free troposphere, the parameterization of the wet deposition by scavenging ratios and the parameterization of the transformation rate fo sulphur dioxide to particulate sulphate. The model results are sensitive to each one of these parameters (Fig. 13, Eliassen and Saltbones, 1983). The sensitivity of the SO_2 concentrations to variations in the transformation rate is small close to the source, while it is large further away, and vice versa for particulate sulphate (Workshop Report on Sources and Evaluation of Uncertainty in Long-Range Transport Models, 1986). The model performance is also very sensitive to the selection of stations for obtaining the network averages, and therefore a very careful screening is required. The principles for screening are mentioned in the article above.

The scavenging ratios are taken from literature (Barrie, 1891; Whelpdale, 1981). The transformation rate is assumed to have a sinusoidal variation over the years. The formulae for the parameterization of the deposition velocity at 1 m height and the transformation rate are based on scientific reasoning and have been developed at DNMI. The mixing height, advection wind, surface wind and precipitation fields are all taken from analyses of measurements and output from the Numerical Weather prediction routine, which is run at DNMI with a 6-hourly data assimilation cycle.

A QUANTITATIVE MODEL ANALYSIS OF YEAR TO YEAR CHANGES IN CONCENTRATION AND DEPOSITION

Hans J.A. van Jaarsveld

National Institute of Public Health and Environmental Protection (RIVM)
P.O. Box 1, 3720 BA Bilthoven, The Netherlands

INTRODUCTION

Concentration and wet deposition of sulphur and nitrogen species show large variations from year to year. On basis of measurements of these species it takes several years to discover significant trends. If it were also possible to measure dry deposition fluxes in an efficient way then it would be easier to determine an overall trend, because one can expect that the sum of dry and wet deposition is much less sensitive to climatological changes than the dry or wet deposition separately. In all cases measurements are required which show no systematic change in calibration over the years. A more sophisticated method do determine trends is to use a long-term long range transport (LRT) model to compensate for the year to year meteorological influences. The model to be used for such a purpose must be able not only to describe the variations in time correctly but also efficiently, because calculations have to be carried out over long periods.

This paper reports on the use of the TREND model for the quantification of the processes and parameters playing an important role in the period to period variations. This model was developed to support the interpretation of data of the National Air Quality Monitoring Network (LML) in The Netherlands. The model will be described briefly, but with some emphasis on the dry deposition velocity parameterization. Model results are compared with measurements and on basis of this comparison an analysis is made of the importance of some parameters. Finally a trend in sulphur and oxidized nitrogen emissions is derived from the combination of measurements and model calculations.

THE MODEL

The model applied is the so called TREND model (Van Jaarsveld and Onderdelinden, 1990). It can be characterized as a long-term Lagrangian model in which the transport equations are solved analytically. Dry deposition, wet deposition and chemical transformation are incorporated as first order processes and independent of concentrations of other species.
To enable calculation of an average concentration at a particular point, the occurring meteorological situations are divided in a number of classes. Next, a representative concentration is calculated for each meteoclass, and after summation of these concentrations weighted by the frequency of occurrence of the classes the period-average concentration is obtained. The analytical approach makes it easy to use the model on different spatial

scales, while the combination of the analytical and statistical approach results in a model which is very efficient with regard to computer resources.

The model described here was originally designed to explain variations in time and space observed in the Netherlands in terms of emission, dispersion and deposition processes. In view of the availability of meteorological data from the LML such as wind direction, wind speed and solar radiation, it was obvious to use these data as far as possible in the characterization of the meteoclasses and in the description of the above-mentioned processes.
The selection of the meteo-classification and the associated parameters to be incorporated into the model is partly based on an analysis of concentration levels observed in the above-mentioned monitoring network.

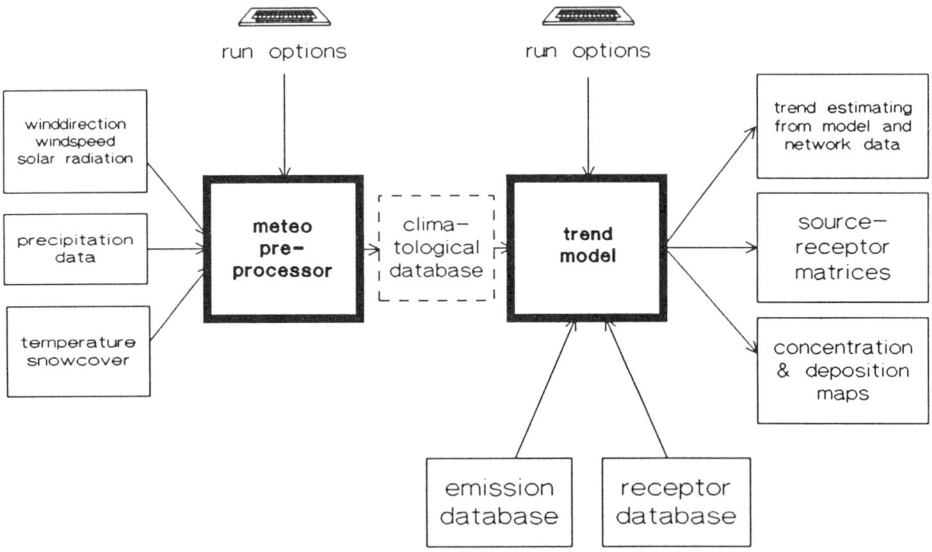

Fig. 1. *Schematic representation of the TREND model system. The meteo-preprocessor is in principle used only once.*

The structure of the model system is presented schematically in figure 1. The collection of meteorological data from various files and the calculation of derived parameters take place in the meteo-preprocessor.
This program produces meteo-statistics for each period (varying from a month to 12 years) and for each area of interest, which are stored in a climatological database. In performing calculations, the model chooses its meteo-input from the climatological database, so that the preprocessor is no longer required for this operation. The remaining model input consists of emission data, data on the locations for which calculations have to be performed, and data on specific properties of the substance to be modelled.

Parameterization of dry deposition velocities

For a model intended to describe temporal changes in concentration and deposition on a certain time scale it is important to emphasize on those processes having a significant role in these changes. Knowledge about these processes can be obtained by studying episodic situations. Periods showing high concentrations are just as important from a scientific point of view as periods characterized by very low concentrations. The LML network, operational since 1977, provides air concentrations of SO_2, NO, NO_2 and others on a hourly basis with a high spatial resolution. Analysis of high SO_2 concentrations for a number of years led to the conclusion that these concentrations could not be explained by phenomena as emission, transport and mixing volume only. Common factor for most of these episodes was the presence of a large scale snowcover in Western Europe. Several authors have published dry deposition velocities for snow-covered soil and most of them come with very low SO_2 deposition velocities for dry snow (Whelpdale and Shaw, 1974; Barrie and Walmsley, 1978; Dovland and Eliassen, 1976; Caddle et al., 1985). Obviously the snowcover (and to some extend also frozen soil) extends the atmospheric residence time of SO_2 , thereby causing accumulation in the lower atmosphere which results in very high concentrations.

Investigating in the same way episodes with very low concentrations led to the conclusion that under certain conditions the dry deposition velocity is very high. From an analysis of the LML-data these conditions appeared to be strongly related with the occurrence of wet soils (Onderdelinden et al.,1984). The latter phenomenon is possibly also related to the presence of high ammonia emissions in the Netherlands. All this was taken as enough evidence to introduce a dry deposition velocity in the model for SO_2 not only depending on atmospheric stability, but also depending on (large scale) soil conditions. This deposition velocity is described with a so called resistance model:

$$v_d = (r_a + r_b + r_c)^{-1}$$

r_a and r_b are aerodynamic resistance and the laminar sublayer resistance, respectively. The surface resistance r_c is parameterized for the different soil conditions with a time resolution of one day. Some values ($s\ m^{-1}$):

snow cover	$r_c = 540$
frozen soil	$r_c = 200$
dry or wet soils	$r_c = 10 + 100\ (1-\exp(-\alpha/6)) + 50\ \exp(-\beta/6)$

with α representing the length of the current dry period in days and β the number of hours with precipation in the present day. This parameterization results in dry deposition velocities which show distinct seasonal and yearly variations (Table 1.)

Table 1. *Effective dry deposition velocities (cm s^{-1}) for SO_2 which follow from the parameterization of the surface resistance r_c*

period	average	range found over 12 years
month	.92	.24 - 1.35
winter	.76	.55 - .87
summer	.98	.95 - 1.04
year	.92	.71 - 1.0

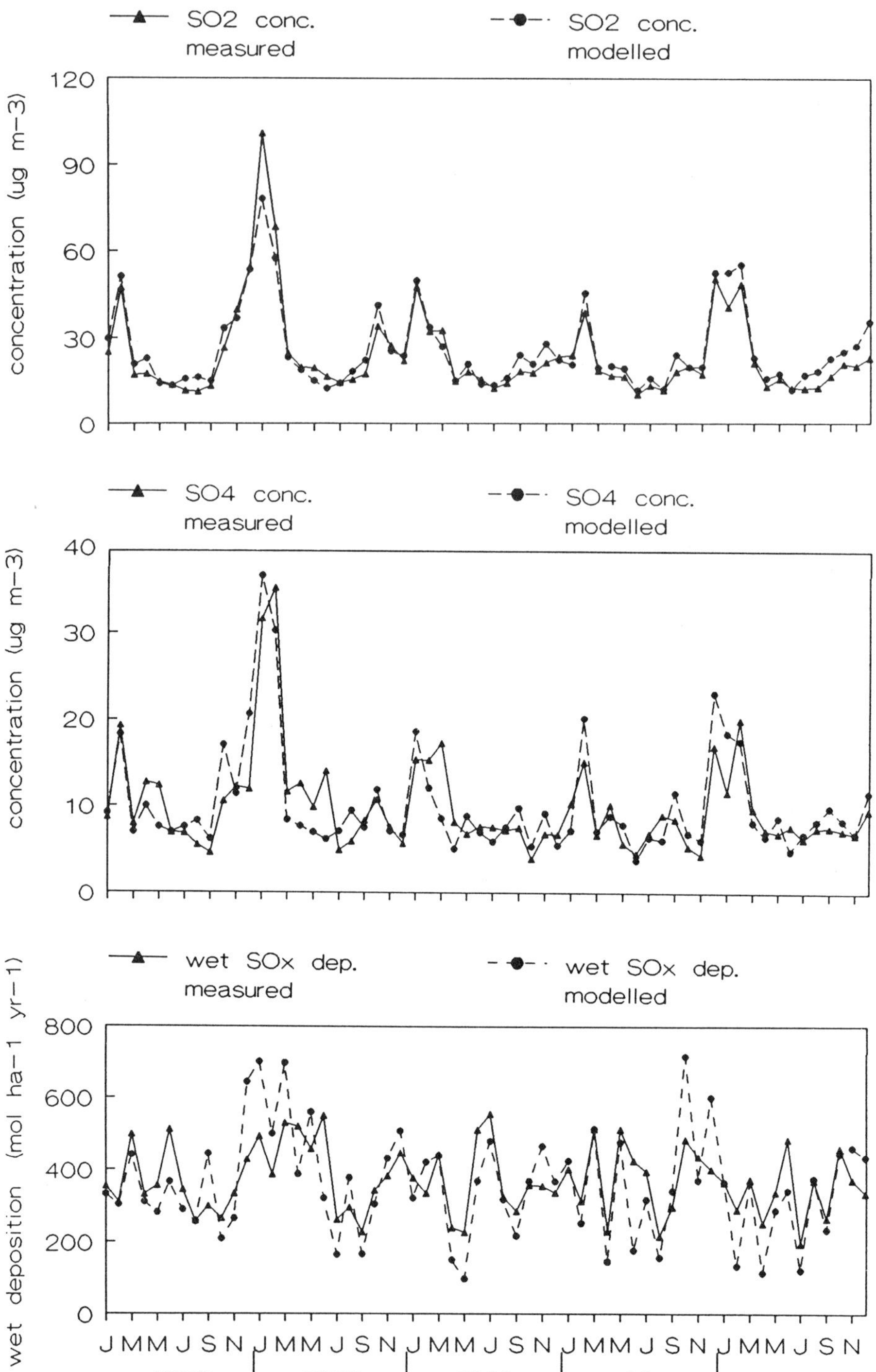

Fig. 2. *Calculated concentrations and depositions versus observations. Values are spatial averages over the Netherlands. The emissions used in the calculations are of the 1980-level for the whole period.*

MODEL PERFORMANCE WITH RESPECT TO SEASONAL VARIATIONS

The capability of the model to describe the monthly variations in concentration and deposition is shown in Fig. 2. In this figure monthly averaged concentrations and depositions, calculated as spatial averages for the Netherlands are compared with observations from the LML network. The number of stations involved in this comparison amount to 99 for SO2, 2-6 for SO4 and 12-18 for the wet deposition. It must be noted that all the calculations have been carried out with emission data representative for the year 1980. Only a temperature dependent correction was applied on the average emissions to account for the effect of variable space-heating sources.

It can be concluded that the modelled and observed concentration values agree very well on this time scale. A similar comparison has been carried out for oxidized nitrogen compounds (Van Jaarsveld, 1988) with comparable results.
The results from this comparison form the basis for a more detailed investigation on the importance of meteorological parameters for the seasonal variations in concentration and deposition.

RESULTS

The impact of climatological differences on time averaged concentrations and depositions

Having introduced a variable dry deposition velocity it was interesting to investigate how this and other parameters as occurring wind-direction, windspeed, temperature, precipitation characteristics and conversion rates affect period-averaged concentrations and depositions. Since winter periods show the largest variations in concentrations, these periods were choosen to demonstrate the effects. The following procedure was followed:

For each of the 12 winters in the period 1977-1989 the concentration and deposition was calculated. Next, concentration and deposition for the winter periods were calculated again but now one of the parameter values was replaced by the average value over the 12 periods. The change in the concentration (or deposition) for a specific winter period due to this parameter value replacement was calculated as:

$$\Delta C = \frac{C_{i,j} - C_i}{C_i} \times 100 \%$$

with: C_i = average concentration period i
$C_{i,j}$ = average concentration in period i when the value of parameter j is replaced by the value of the parameter averaged over 12 periods

Some results of these calculations are presented in Fig. 3. In this figure only the influence of wind-direction/stability/mixing height and dry deposition velocity on SO_2 concentrations is shown. The effect of additional parameters is given Table 2., In this table also results are presented of application of the method to SO_4^{2-} concentration, dry SO_x deposition, wet SO_x deposition and total deposition.

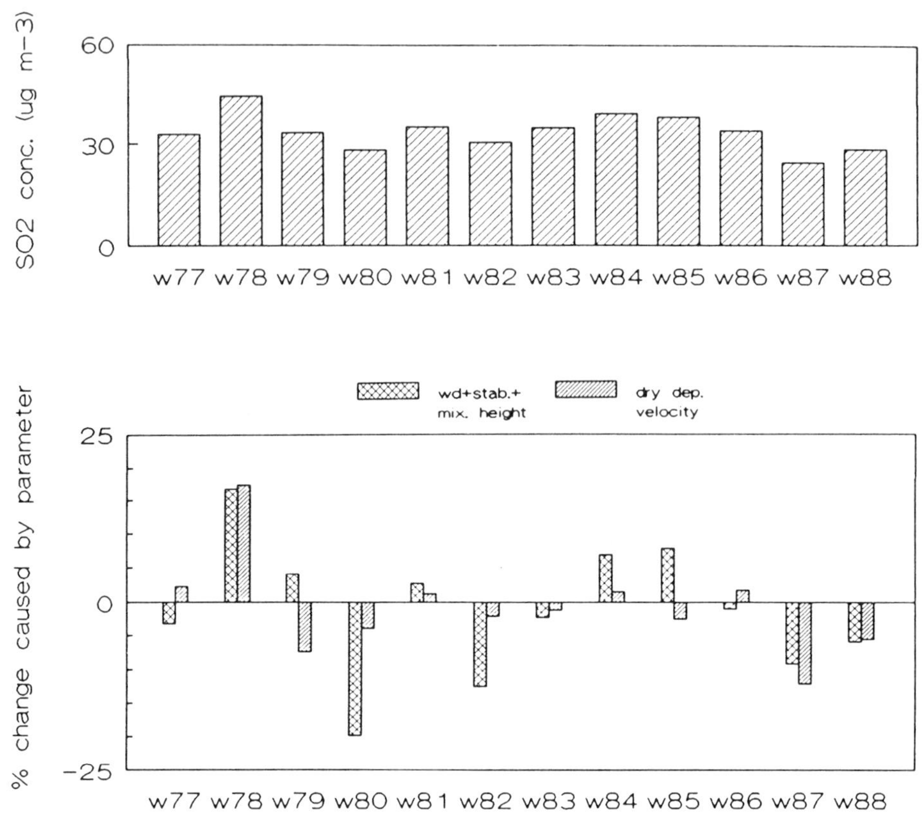

Fig. 3. *Change in* SO_2 *concentrations when period-specific parameter values are replaced by long-term average values. In the upper figure the nominal concentrations as calculated for these winters are shown.*

The parameters wind-direction, atmospheric stability and mixing height cannot be separated in the TREND-model because they are part of the basic meteo-classification as choosen for this model. As expected, this group of parameters dominates the variations in most of the winters. However, for the winter with the highest concentration (w78) the dry deposition velocity appears to be a equally important parameter and this is also the case for the winter with the lowest concentration (w87). In both these winters the two parameters enlarge each others impact on the concentration. There are also winters were the effects have opposite signs, leading to winters with moderate concentrations (w77, w79 and w86). It can be concluded that the variability of the dry deposition velocity plays a major role in explaining SO_2 concentration variations. This conclusion implies that it is not the high SO_2 concentration that causes a high dry deposition flux but that it is the condition of the soil which exhibits or decreases the dry deposition and in that way causing SO_2 accumulation in the atmosphere. From Table 2. it can be seen that the dry deposition flux in the Netherlands is much less sensitive to climatological variations than the concentrations in air or the wet deposition. Total deposition is remarkably constant from winter to winter; the range over 12 winters is only -11 to +8 % from the average value.

Table 2. *Impact of specific meteorological conditions on the winter-average values for 12 consecutive winters. Changes are calculated relative to the nominal values.*

	SO_2 conc.	SO_4^{2-} conc.	SO_x dry dep.	SO_x wet dep.	tot. dep.
average:	33	12	1290	400	1690
range :	24 - 44	8 - 20	1220 - 1400	280 - 540	1500- 1830
	($\mu g\ m^{-3}$	)	(mol $ha^{-1}yr^{-1}$		
		% change caused by parameter:			
winddirection + stability + mixing height	-20 - 17	-20 - 20	-13 - 9	-15 - 13	-13 - 10
wind speed	-1 - 1	-13 - 20	-2 - 2	-9 - 9	-3 - 4
temperature*	-6 - 4	-9 - 4	-6 - 4	-6 - 3	-8 - 3
dry dep. velocity	-12 - 17	-8 - 11	-5 - 6	-6 - 11	-2 - 3
precip. char.	-10 - 4	-10 - 11	-7 - 4	-25 - 19	-2 - 2
conversion rate	-1 - 1	-3 - 2	-1 - 1	-1 - 1	-1 - 1

* effects of temperature variations on emissions only

The same method can easily be applied to other time periods (summer to summer, year to year or month to month) and other components which are incorporated in the model (oxidized nitrogen, reduced nitrogen).

Trends in sulphur and oxidized nitrogen

The extensive set of SO_2 and NO_x concentration data collected in the Netherlands since 1976 offers the possibility to apply trend analysis in various ways. Regression analysis will normally give a good result when applied over such a period. However, as is already shown in Fig.2, the concentrations variations vary strongly from month to month and also from year to year. Average concentrations of SO_2 in the Netherlands appeared to be halved from 1987 to 1988 (Fig.4). It is clear that this sharp decrease is not only due to emission reductions applied in this period. The procedure followed here in order to determine an effective emission-trend from the measurements was:

- average concentrations of SO_2 and NO_x over the Netherlands were calculated for 12 consecutive years, using 1980 - emission levels for all years.
- measured concentrations were averaged over the same area and for the same 12 years.

Both measured and modelled concentrations are given in Fig.4.
As measured concentrations are influenced both by emission variations and meteorological variations, and modelled concentrations only by meteorological variations, the ratio measured and modelled concentration must reflect the effective emission trend with respect to the 1980 situation.

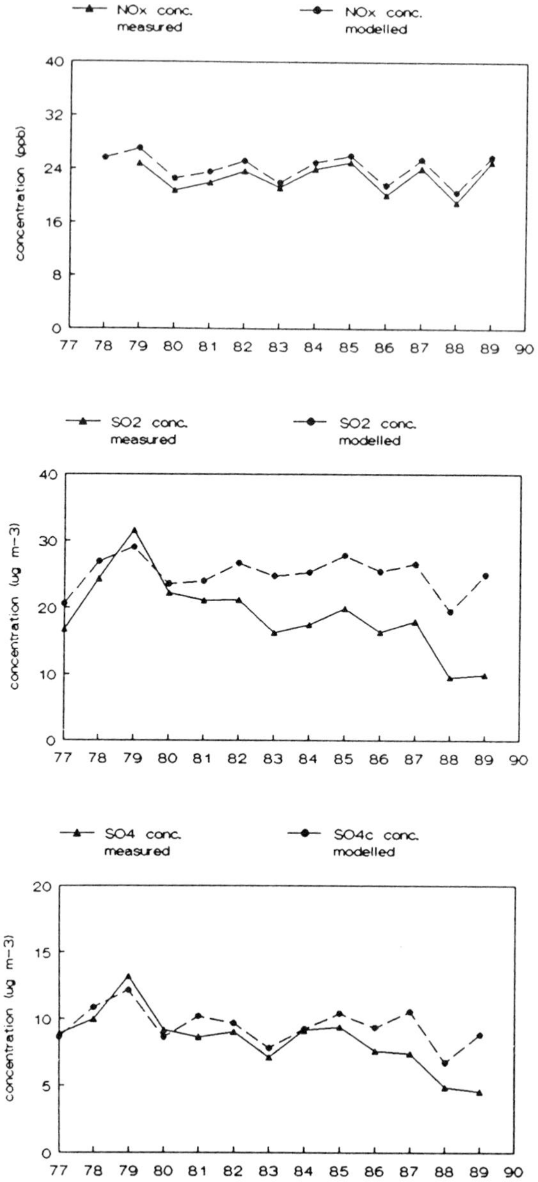

Fig. 4. *Measured and modelled concentrations of* NO_x, SO_2 *and* SO_4^{2-} *on a yearly basis. All model calculations are carried out with emissions of the 1980-level. In the figures on the right hand the ratio measured and modelled is given, representing the relative change in emissions trough the years.*

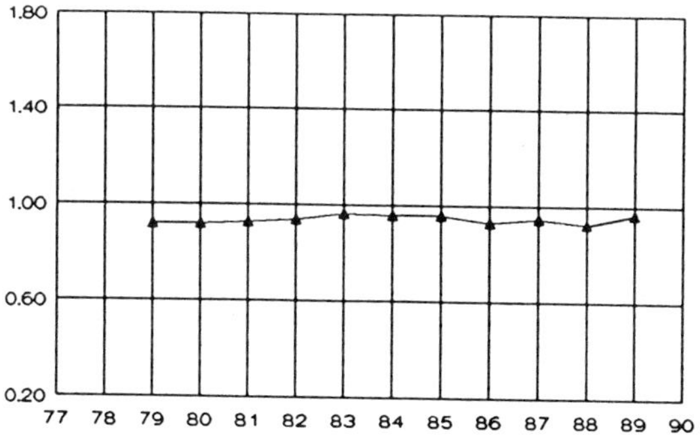

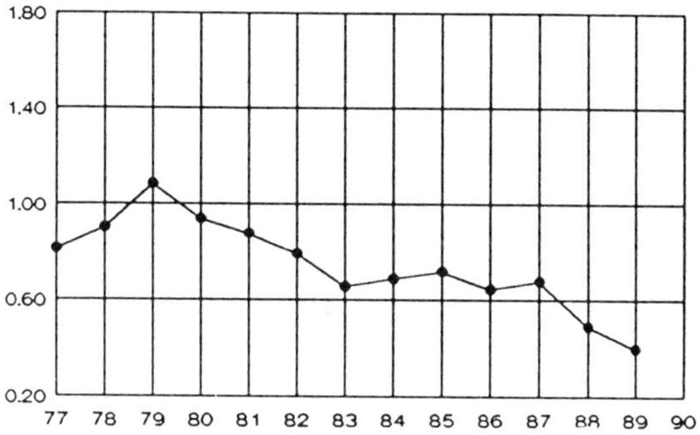

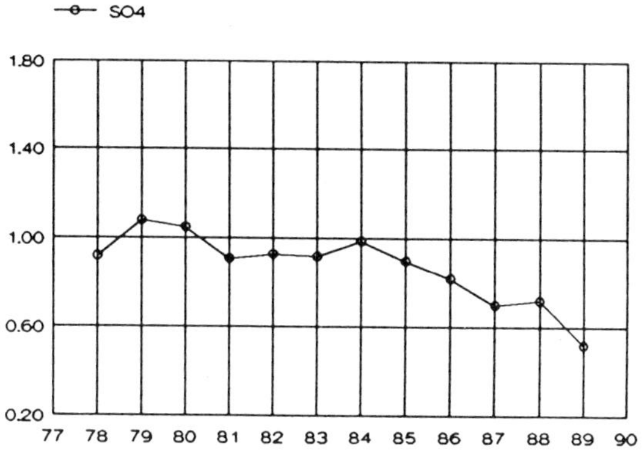

Fig. 4. (Continued)

From Fig. 4 it can be seen that this ratio shows a significant downward trend over the period 1980-1989 of almost 50% for SO_2. In the NO_x case there is no trend at all or less than ± 4%. As surrounding countries are responsible for more than 70% of the SO_2 concentrations and more than 50% of the NO_x concentrations in The Netherlands (Derwent et al., 1989), it can be postulated that these trends represent the general emission situation for an area including The Netherlands, Belgium, Western Germany, France and the United Kingdom. A rather strong decrease of SO_2 emissions is suggested for the period after 1987. Fact is, that in Western Germany the so called 'Grossfeuerungsanlage-Verordnung' came into effect in July 1988. This law demands de-sulphurization equipment on all larger furnaces, resulting in a 64% emission reduction in 1988 compared to 1983.

If we had not used the assistance of a model but based the trend estimate on the bare measurements, then we came out possibly at the same overall trend for SO_2. However, it would not have been possible to interprete for example the situation of 1989 as still a down going trend. The conclusion is that with a method based on measurements and model calculations it is possible to detect trends in an early stage.

The same method applied on sulphate concentrations or wet deposition does not necessarily lead to the same SO_2 emission trends. Possible reasons for differences are:

- SO_4^{2-} concentrations and also wet SO_4^{2-} depositions have their origin in sources located in a much larger area than SO_2 concentrations, because they are secondary formed products.

- Measurements are less available and often less reliable.

- Model results for these components are less reliable because a linear conversion rate is maintained between SO_2 and SO_4^{2-}. This is especially a problem in case of large emission changes.

The emission trend derived from sulphate aerosol concentrations has the same value as the trend derived from SO_2 concentrations when considered over the total period 1980-1989. Most of the decrease in emissions is suggested for the period after 1984. Considering the fact that an important part of the sulphate concentration in the Netherlands originates from sources in Eastern Europe and that reported emission data for these countries show no decrease at all (Mylona, 1989), a somewhat less pronounced trend was expected. Probably the limited number of monitoring locations (only 2 in 1989) and a change in position of the stations did had some impact on the derived trend.

CONCLUSIONS

Meteorological variations have a large impact on year to year concentrations and wet depositions. Variations in the dry deposition velocity due to large scale meteorological phenomena as snow-covered or wet soils play an very important role in the occurrence of extreme SO_2 concentrations in Western Europe. In such cases the dry deposition parameter is as important as parameters like wind-direction and mixing depth. Therefore, if high SO_2 concentrations are partly induced by low deposition rates, it is not correct to calculate dry deposition fluxes for episodes from measured concentrations and average deposition velocities. As a consequence of the above, the total deposition is much less sensitive to climatological changes, when it is considered over a larger area.

A model which is able to describe period to period changes in concentrations and depositions, can be successfully used to detect emission trends in an early stage. A trend for NO_x emissions derived from NO_x concentrations shows that these emissions have not changed since 1980 more than ± 4 %. This applies for the area in Europe with the highest emission density. For sulphur a significant downward trend can be detected of 50% since 1980.
Effects of recent emission reduction measures in Western Germany can be evidenced with the applied method.

REFERENCES

Barrie, L.A., and J.L. Walmsley (1978), A study of sulfur dioxide depositon velocities to snow in Northern Canada, Atmospheric Environment, 12, 2321-2332

Cadle, S. H., Dash, J.M. and Mulawa,P.A.(1985), Atmospheric concentrations and the deposition velocity to snow of nitric acid, sulfur dioxide and various particulate species, Atmospheric Environment, 19, 1819-1827.

Derwent, R.G., Hov, O., Asman, W.A.H., Van Jaarsveld, J.A. and De Leeuw, F.A.A.M. (1989) An intercomparison of long-term atmospheric transport models; The budgets of acidifying species for the Netherlands. Atmospheric Environment, 23, 1893-1909.

Dovland, H. and Eliassen, A. (1976) Dry deposition on a snow surface, Atmospheric Environment, 10, 783-785.

Granat, L. and Johansson C.(1983) Dry deposition of SO2 and NO2 in winter, Atmospheric Environment, 17, 191-192.

Van Jaarsveld, J.A. and Onderdelinden, D. (1990) TREND; An analytical long term deposition model for multi-scale applications. Report 228603009. RIVM, To be published.

Van Jaarsveld, J.A. (1988) Modelling of NOy budgets over The Netherlands. In: Extended abstracts of papers presented at the (ECE/EMEP) Workshop on modelling transformation processes and transport of air pollution with special reference to nitrogen oxides, Potsdam, GDR, March 21-25,1988, Meteorologischer Dienst der DDR, Potsdam.

Mylona N., (1989) Detection of sulphur emission reductions in Europe during the period 1979-1986. EMEP MSC-W Report 1/89 The Norwegian Meteorological Institute P.O.Box 43-Blindern, N-0313 Oslo 3, Norway

Onderdelinden D., Van Jaarsveld, J.A. and Van Egmond N.D. (1984) Bepaling van de depositie van zwavelverbindingen in Nederland (Determination of the deposition of sulphur compounds in the Netherlands.) Report 842017001, RIVM, P.O. Box 1, 3720 BA Bilthoven, The Netherlands.

Whelpdale, D.M. and Shaw, R.W. (1974) Sulfur dioxide removal by turbulent transfer over grass, snow and water surfaces, Tellus, 26, 196-204.

DISCUSSION

H. GEISS — Did you take into account all stations or only the rural stations?

H.J.A. VAN JAARSVELD — The basic configuration of the National Air Quality Network is a regular grid over the Netherlands. In addition there are stations in industrial and (sub)urban areas. For the comparison with model results the data of the gridded stations were used. Most of these stations are situated in rural areas, but there are also some stations in (or close to) industrial and urban areas.

C.J. NAPPO — Is your statement that SO_2 deposition is small over snow based on actual measurements or on model results? Is it not a question of wet or dry snow cover?

H.J.A. VAN JAARSVELD — The statement of low SO_2 deposition velocities to snow is based on the results of measurements as reported in open literature. It is indeed a question of dry or wet snow cover. All statements about snow cover related properties refer to the situation of a dry snow cover.

AN AIR QUALITY FORECAST SYSTEM FOR PHOTOCHEMICAL SMOG EPISODES

H.Jetske van Rheineck Leyssius and Frank.A.A.M. De Leeuw

National Institute of Public Health and Environmental Protection
P.O.Box 1,3720 BA Bilthoven, The Netherlands

INTRODUCTION

One of the tasks of the RIVM (National Institute of Public Health and Environmental Protection) is to inform the County Governor on the actual and expected air pollution concentrations during smog episodes. The County Governor has the authority to order measures for emission reduction according to the Air Pollution Act. The measured and expected values of oxidant (Ox, the sum of the O_3 and NO_2 concentrations) form the basis on which the County Council decides to order these measures. For this purpose three threshold values are defined, based on the expected daily average oxidant concentrations (Table 1).

Table 1. Threshold values for photochemical smog episodes. Based on daily averaged oxidant concentrations. Actions taken by the County Council.

threshold value	phase	actions
63 ppb	pre-warning	no measures advice and information to the public
75 ppb	warning	press communiqué measures: voluntary
100 ppb	alert	active information to the public measures: obligatory

To perform this task during photochemical episodes the RIVM:
- gives on-line (hourly) information on the concentrations measured by the National Air Quality Monitoring Network (LML) to the County Council and the public (ViewData system);
- has developed a statistical method to estimate the Ox concentrations (prognosis of 0-24 hours) based on observations of the LML;
- has developed a method to predict the air quality (prognosis of 0-3 days) based on the weather forecast of the ECMWF (European Centre for Medium Range Weather Forecast, Reading, U.K.) and a Lagrangian long-range transport model with non-linear chemistry (MPA model).

A description of both prediction methods will be presented. The applicability of the methods in forecasting the photochemical episodes of the summer of 1989 will be discussed. The possibility of using both methods to form a smog prognosis on a routine basis in coming summers will be discussed as well.

Air Pollution Modeling and Its Application VIII, Edited by H. van Dop and D. G. Steyn, Plenum Press, New York, 1991

PHOTOCHEMICAL SMOG EPISODES IN THE NETHERLANDS IN 1989

Several photochemical smog episodes occured in The Netherlands in the summer of 1989 (Schokkin et al., 1989a, 1989b). The pre-warning level of 63 ppb for oxidant was exceeded on 20% of the days, the 75 ppb level on 17 days and the 100 ppb level on 1 day in the period from May until October 1989.

The public interest in the environment is large in The Netherlands. The traffic emissions in The Netherlands contribute considerably (50-60%) to the total NOx and VOC emissions. In the discussion on the possible abatement measures during episodes, a general ban on private transport is one of the options. For most Dutchmen their car is extremely important. So, the discussion arised whether and if so when and where to stop the traffic. In this discussion the need for an accurate smog prognosis becomes evident. As a consequence two prediction methods were developed and used in the summer of 1989. During smog episodes the smog prognosis was formulated by a group of experts of the RIVM using both tools.

High oxidant concentrations occur due to oxidant production in a mixture of organic substances and nitrogen oxides under the influence of solar radiation. In the Netherlands this event occurs in specific meteorological situations consisting of a stagnant anticyclone over Europe. The weather is sunny and the temperatures are above 25°C. The chemical reactions which play a role during photochemical episodes are numerous, but may be represented schematically (reaction R1, R2). The oxidant concentrations are independent of the position of the photostationary equilibrium:

$$NO_2 \rightleftarrows NO + O_3 \qquad R1$$

Photochemical production of oxidant takes place by:

$$RH + 2NO + 2O_2 \rightarrow aldehyde + 2NO_2 + H_2O \qquad R2$$

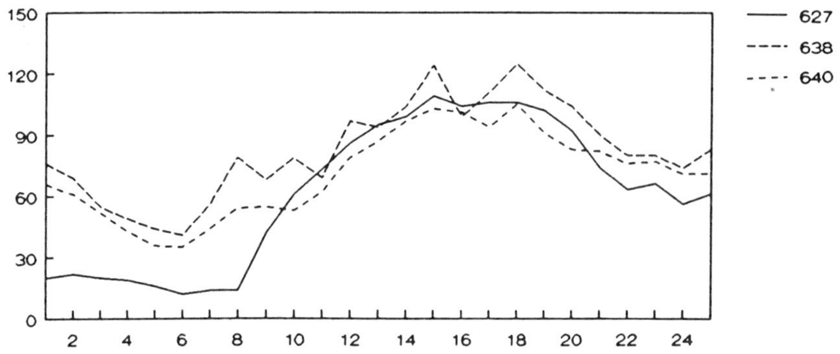

Figure 1 Oxidant concentrations on September 7, 1989 at three different types of stations of the LML: 627 (urban situation) and (street level).

The oxidant concentrations show a specific daily pattern (Figure 1). The concentrations are relatively low in the early morning. The concentrations rise due to both photochemical production and fumigation. The maximum concentration is reached around 18 hours MET (Middle European Time). The concentration drop at the end of the day is caused by dry deposition in combination with the stop of photochemical production processes and fumigation. The influence of local (traffic) emissions on the oxidant concentrations is shown in the three different station types in Figure 1. The urban concentrations are higher than the rural concentrations due to local oxidant formation from traffic emissions. The concentrations at street-level are slightly higher than the urban concentrations and show more scatter, again due to local traffic emissions.

A STATISTICAL PREDICTION METHOD (< 24H)

The assessment of the daily average oxidant concentration at a time when only a limited number of hourly measurements is available is not straight foreward. The daily concentration profile depends on the meteorological situation as the processes of photochemical production, fumigation and dry deposition are influenced by meteorological parameters. A statistical method (OXPRO) has been developed to solve this problem. The method is based on the oxidant data from the past and the actual meteorological situation. The relative daily concentration profile is determined for several meteorological classes (daily average) for each station of the LML:

$$F(IU) = \frac{C(IU)}{C(8)} \qquad \text{for } IU = 9 - 24$$

Where: IU = hour of the day (MET)
C = oxidant concentration
F = relative oxidant concentration (relative to the concentration of 8 MET)

When measurements are available up to time KU the daily average concentration can be estimated by estimating the meteorological class and finishing the daily concentration profile:

$$C(IU) = C(KU) * \frac{F(IU)}{F(KU)} \qquad \text{for } IU = KU+1 - 24 \qquad (4)$$

The meteorological classification is based on routinely available information:
- Month of the year;
- Cloud cover (< 0.75; > 0.75);
- Wind direction; two sections: - flow from the continent
  - flow from sea
- Wind velocity (< 4 m/s; > 4 m/s).

These parameters (including global radiation for the estimation of the cloud cover) are measured at the LML stations. With this information the meteorological class can be estimated.

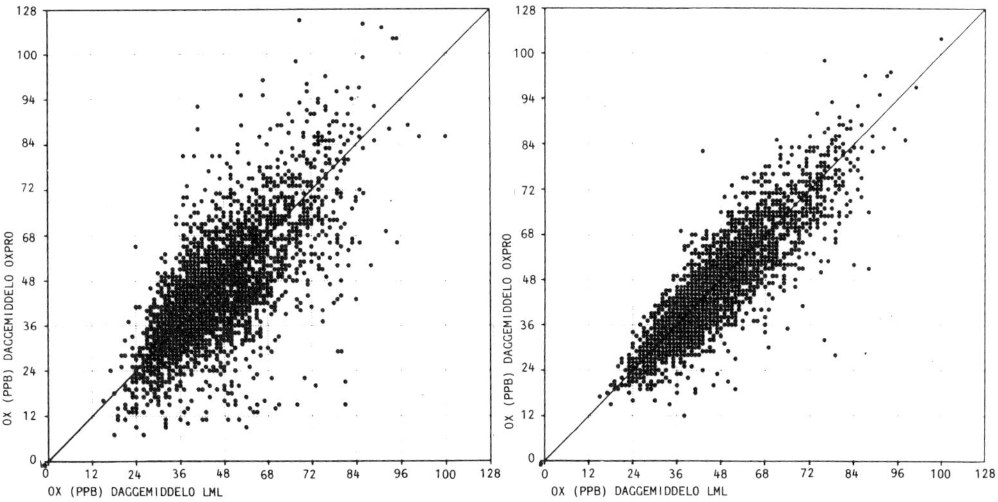

Figure 2 Scatterplots LML (measured) vs OXPRO (forecasted) (daily averaged oxidant concentration). May - September 1989, left observations available up to 8 MET, right observations available up to 11 MET.

Figure 2 shows the measured daily average oxidant concentrations vs. the OXPRO results (measurements available up to 8 MET (OXPRO 8 MET) and 11 MET (OXPRO 11 MET) respectively; all LML stations).

The prediction of OXPRO 11 MET is better than of OXPRO 8 MET. This may be explained by a larger part of the measurements within the daily average concentrations and a better estimation of the meteo class. The prediction of the exceedance of a threshold value may be indicated by the percentage false alarm (exceedance predicted but not measured) and the percentage alarmed (exceedance measured was also predicted). The overall performance may be indicated by the correlation coefficient r and the root mean square error RMSE:

	% false alarm		% alarmed		r	RMSE
	63 ppb	75 ppb	63 ppb	75 ppb		(ppb)
OXPRO 8 MET	38	52	57	39	.72	10
OXPRO 11 MET	24	41	71	60	.88	6

Generally, the exceedance of the 75 ppb level was less well predicted than the exceedance of the 63 ppb level. There was no significant difference in the prediction quality for different station types or different regions within the Netherlands. Ninety percent of the measurements was within 30 % of an OXPRO 8 MET prediction and 15 % of an OXPRO 11 MET prediction.

A PROGNOSTIC LONG RANGE TRANSPORT MODEL (< 72H)

The prognosis for up to three days is based on a long range transport model, the MPA model (De Leeuw et al., 1990). The MPA model is a trajectory model including anthropogenic and natural emissions, non-linear atmospheric gas-phase chemistry, dry deposition, exchange between boundary layer and free troposphere, and fumigation between mixed layer and aged smog layer. The trajectories are calculated from windfields of the ECMWF which are obtained via the Royal Netherlands Meteorological Institute (KNMI). The windfields of the 1000- and 850 mbar level prognostic up to 96 hours are available. Other meteorological parameters along the trajectory are estimated from the measurements of the LML. For a prognostic calculation this information is extrapolated from the last available measurements.

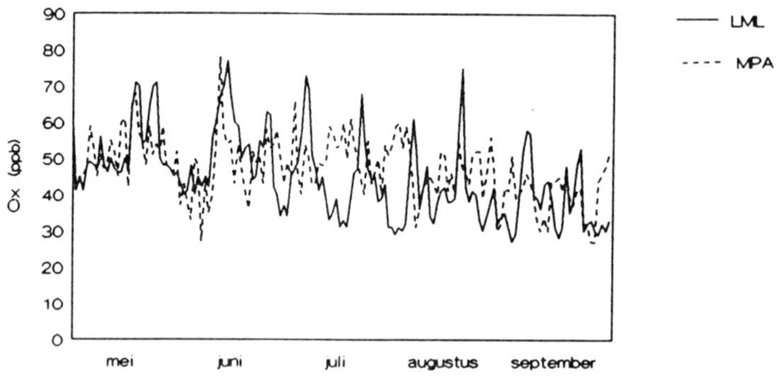

Figure 3 Measured (LML) and calculated (MPA) oxidant concentrations for the summer of 1989. Receptor-area Middle Netherlands. Average transport height trajectories 800 m.

The modelled concentrations are sensitive to the selected transport height. A vertical interpolation to a transport height of 800 m for the assessment of the trajectory showed the best results for the episode of

May 1989 (Schokkin et al., 1989a). The results of this method for the summer of 1989 were sometimes good (May/June) and sometimes they were not (July, see Figure 3). The correlation coefficient r was for the whole period .43 and in the months May and June .75.
Figure 4 shows the 1000- and 850 mbar trajectories for a number of days in the summer of 1989 (arrival 16 MET). The measured and predicted oxidant concentrations for these days are given in table 2. On the basis of this information the model performance in specific meteo situations may be evaluated.

Table 2. Observed (LML) and modelled (MPA) oxidant concentrations (ppb, daily averaged) in the middle-Netherlands for a number of days in the summer of 1989.

date	LML	MPA
89 5 26	50	51
89 5 27	48	59
89 8 20	75	48
89 9 7	58	46

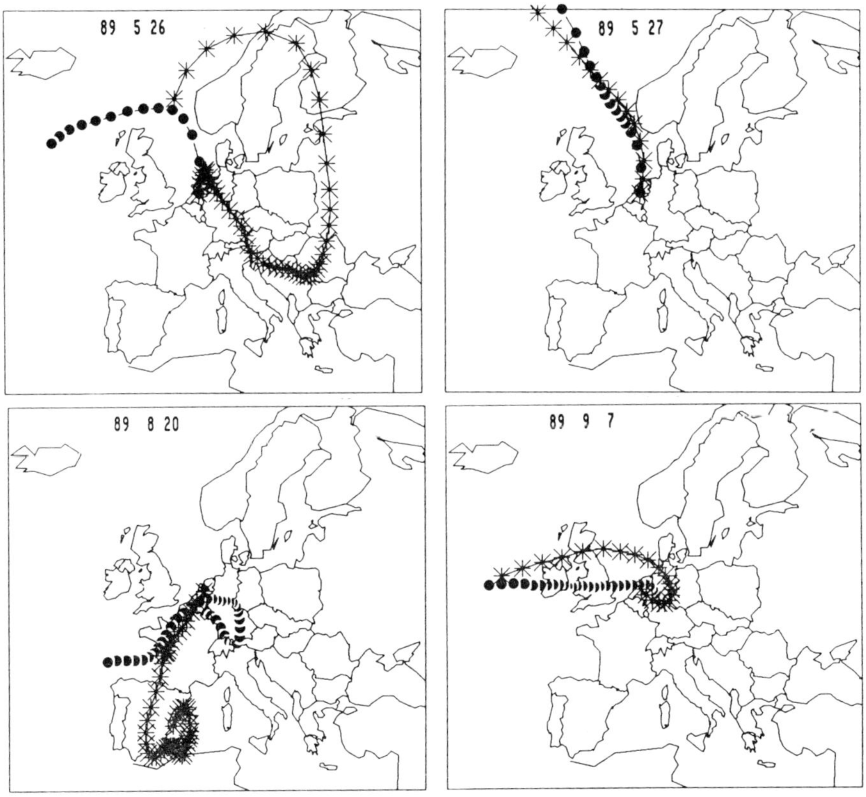

Figure 4 1000 mbar (dot) and 850 mbar (star) trajectories for a number of days in the summer of 1989 (arrival 16 MET). Receptor-area Middle Netherlands. Every two hours are marked.

A North-Western flow (i.e. transport from sea) was maintained above the Netherlands in July. Above sea the dry deposition of O_3, NO and NO_2 is negligible. In this case the boundary conditions of the model become important. The initial concentrations at the start of the trajectory and the concentrations in the free troposphere are obtained from a 2-D tropospheric model (Isaksen and Rodhe, 1978, Isaksen and Hov, 1987). In the version of the Isaksen model applied in the prediction method the calculated tropospheric O_3 concentrations are probably too high. As a result, the oxidant concentrations will be overestimated when there is transport from sea (see also May 27, Table 2, Figure 4).

Recently new results of this 2-D model have come available (Roemer and Van den Hout, 1989). The tropospheric O_3 concentrations are lower than in the original results. The MPA model results when these new boundary conditions were used improved but in cases of transport over sea the concentrations were still slightly overestimated (Figure 5). This overestimation of the concentrations is no problem for the prediction of smog episodes. The oxidant concentrations will not be high in The Netherlands in case of a flow from sea.

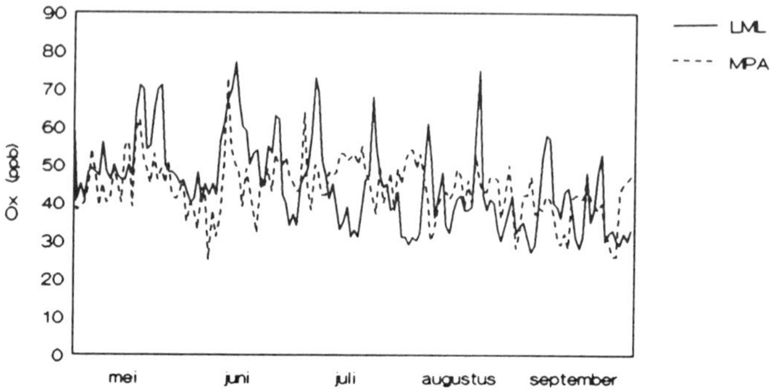

Figure 5 Measured (LML) calculated (MPA) oxidant concentrations for the summer 1989. Receptor-area Middle Netherlands. Improved boundary conditions (tropospheric O_3).

In times of extreme low wind velocities (< 2.5 m/s; for example August 20, September 7) the concentrations may be underestimated (Figures 3 and 4, Table 2). On such days there is a large difference between the wind velocities at different heights (850 mbar, 1000 mbar, 20 m. (meteo stations LML). When using trajectories at the 1000 mbar level the simulation of these peak concentrations may be improved. The high concentrations at August 20 could be simulated, the peak concentrations at September 7 were still underestimated (Figure 6). The wind velocity at 1000 mbar may still be an overestimation of the real transport velocity. This will be a problem when predicting smog episodes: an episode day like this may not be forecasted by the system. The results of May 18 are good (Table 2, Figure 4).

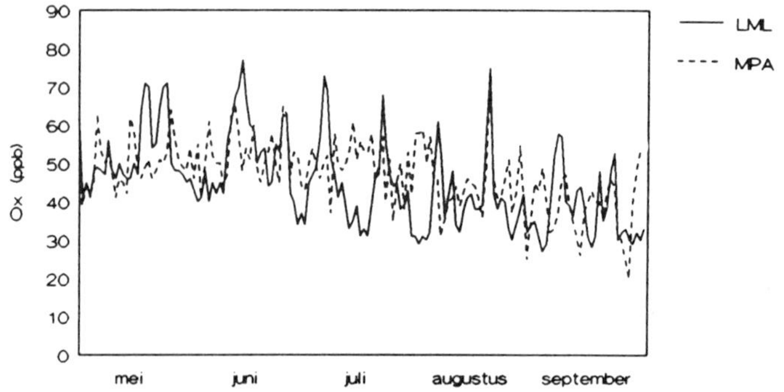

Figure 6 Measured (LML) and calculated (MPA) oxidant for the summer 1989. Rector-area Middle Netherlands Trajectories at 1000 mbar level.

DISCUSSIONS

OXPRO

A deviation of 30% of the results of OXPRO 8 MET and a deviation of 15% of the results of OXPRO 11 MET should be taken into account. Then 90% of the exceedances will be predicted with an acceptable percentage false alarm.
The oxidant statistical data available in 1989 consisted of three rather moderate summers (1986,1987,1988). In the summer of 1990 the oxidant data may be supplied by the sunny (photochemically active) summer of 1989. This may improve the method, in particular as far as the prediction of high concentrations (exceedances of threshold values) is concerned.
The estimation of the meteo class may be improved by using the weather forecast to estimate the cloud cover, wind direction and wind velocity. In this way, weather changes which take place later in the day may be included.

THE MPA MODEL

The model results of the MPA model should be used very carefully. The meteorological situations in which the model performs less well are well known. The presentation of the 1000- and 850 mbar trajectories may give more information about the reliability of the model results (Figure 4). During transport over sea the concentrations may be overestimated. If there is a large difference in wind velocity or wind direction at different heights the model results may be less valid. There is the possibility of an overestimation of the transport velocity, even when 1000 mbar trajectories are used. In this case the concentrations may be underestimated.
In this case qualitative predictions only may be recommended.

The computing time of the MPA model is 15 minutes per trajectory calculation (hourly concentrations for a receptor-area of 2000 km^2). To assess the gradient over the Netherlands and the daily average oxidant concentrations for the next 3 days several trajectory calculations are necessary. Since the ECMWF wind fields are available at 6 MET and the prognosis should be made at 10 MET a choice must be made for a limited number of trajectory calculations.

The method can be improved at several points:
- using the weather forecast for estimating parameters like cloud cover, temperature, relative humidity etc.;
- improving the estimation of the boundary conditions by using more recent results of the Isaksen model;
- improving the trajectory calculation by implementing an iterative method where the transport height is fixed at the middle of the mixing height.

CONCLUSIONS

Both methods proved to be useful tools when making a smog prognosis in the summer of 1989. The method OXPRO may be reliable enough to be used on a routine basis to form the smog prognosis for "today". The interpretation of the results of the MPA model is more difficult and the method may - in this version - not directly be used to form a smog prognosis for the next 3 days.

ACKNOWLEDGEMENTS

The authors wish to thank H.Daan, S.Kruisinga and A.P.van Ulden of the KNMI for their support and stimulating discussions on this subject.

REFERENCES

De Leeuw, F.A.A.M., Van Rheineck Leyssius, H.J. and Builtjes, P.J.H., 1990.
Calculation of long term averaged ground level ozone concentrations. *Atmospheric Environment* **24A** 185-193.

Isaksen, I.S.A. and Rodhe, H., 1978.
A two-dimensional model for the global distribution of gases and aerosol particles in the troposphere. Report AC-47, Department of Meteorology, University of Stockholm.

Isaksen, I.S.A. and Hov, O.,1987.
Calculation of trends in the tropospheric concentrations of O_3, CO, CH_4 and NOx. *Tellus* **39B**, 271-285.

Roemer, M.G.M.and Van den Hout, K.D., 1989.
De invloed van Europese en Mondiale emissiereducties op mondiale ozonconcentraties. Report R 89/073, TNO, Delft, The Netherlands.

Schokkin, G.J.H., Van Rheineck Leyssius, H.J. and Onderdelinden, D.(1989a).
Luchtverontreinigingsepisode van 17-27 mei 1989, oxidantconcentraties (metingen en modelresultaten) in Nederland. Report nr. 228702018, Rijksinstituut voor Volksgezondheid en Milieuhygiene, Bilthoven, The Netherlands.

Schokkin, G.J.H., Van Rheineck Leyssius, H.J. and Onderdelinden, D.(1989b).
Luchtverontreinigingsepisoden in de maanden juni t/m september 1989. Oxidantconcentraties (metingen en modelresultaten) in Nederland. Report nr. 228702019, Rijksinstituut voor Volksgezondheid en Milieuhygiene, Bilthoven, the Netherlands.

ANALYSIS OF OZONE AIR QUALITY OVER THE NEW YORK METROPOLITAN AREA

S.T. Rao and G. Sistla

Division of Air Resources
New York State Department of Environmental Conservation
Albany, New York 12233-3259

K. Schere* and J. Godowitch*

Atmospheric Research and Exposure Assessment Laboratory
U.S. Environmental Protection Agency
Research Triangle Park, North Carolina 27711

INTRODUCTION

The New York Metropolitan Area (NYMA), comprising portions of the States of New Jersey, New York, and Connecticut, lies within the emissions-rich Northeastern urban corridor with significant inter-urban transport of ozone and its precursors. Peak ozone concentrations during the summer months in NYMA are significantly higher than the National Ambient Air Quality Standard (NAAQS) for ozone. The multi-day character of these elevated ozone concentrations over NYMA (see Rao, 1988) is primarily associated with summertime meteorological conditions and south-southwesterly flows over the Northeastern United States which favor transport of ozone and its precursors along the Northeast urban corridor.

This paper attempts to assess (a) the temporal and spatial behavior of measured ozone concentrations over the tri-state area, and (b) the variability of the modeled ozone concentrations due to the variations in meteorology and emissions. To this end, measured concentrations of ozone and its precursors from different monitoring stations in the NYMA during the years 1980 through 1989 are assembled and analyzed. Peak ozone concentrations predicted by an urban-scale photochemical model are also examined to delineate the effects of varying emissions and meteorology on ozone air quality.

The results of the analysis of the air quality data in NYMA over the ten-year period indicate that despite a distinct downward trend in the ozone levels over Connecticut, no trend is evident for the ozone concentrations over the New Jersey/New York City portions of the tri-state region. Further, there is little difference in the peak ozone concentrations

*On assignment from the National Oceanic and Atmospheric Administration, U.S. Department of Commerce.

between weekday and weekend despite the weekday to weekend differences in the pollutant emissions pattern over NYMA. The modeling results suggest that the variability in meteorology can overwhelm the changes in ozone air quality resulting from weekday to weekend differences in the emissions pattern.

DATA AND ANALYSIS PROCEDURES

Hourly concentrations of ozone during summer months (June, July, and August) for the period 1980 to 1989 from the monitoring stations (see Figure 1a) in the tri-state area, and hourly concentrations of NO_2 and CO from selected monitoring sites were obtained from the EPA's Aerometric Information Retrieval System (AIRS) database. In this study, the NO_2 and CO concentration data were from stations located within the central business district such as Mabel Dean Bacon in New York City and those located downwind of the emission-rich region, namely, Eisenhower Park, NY, or Bridgeport, CT. In addition, the daily maximum 1-hr ozone concentrations over the ozone monitoring network in each state for the summer months were assembled along with data for Wednesdays and Sundays. The air quality data used here were subjected to thorough quality assurance/quality control checks to insure the integrity of the data.

The number of days exceeding the level of the ozone NAAQS (0.12 ppm) in each year for each region is determined to examine the severity of the ozone non-attainment problem over the region. Average diurnal profiles of ozone and a precursor specie (NO_2) for Wednesday and Sunday were obtained by averaging concentrations for a given hour over the summertime to study the weekday/weekend effects. The Urban Airshed Model (UAM), an Eulerian photochemical grid model (Ames, et al., 1985) with the Carbon Bond II chemical mechanism (Whitten, et al., 1980), is employed to simulate ozone air quality over NYMA under different meteorological conditions (i.e., 5 high ozone days) and emission scenarios. The weekday/weekend behavior of measured ozone concentrations over NYMA is examined in light of model predicted variability in the ozone concentration field.

RESULTS AND DISCUSSION

Figure 1b shows the number of days exceeding the ozone NAAQS of 0.12 ppm over NYMA during the period 1980 to 1989. It is evident from Figure 1b that whereas the highest number of ozone violations occurred over Connecticut during 1980 through 1984, from 1985 onwards the highest number of ozone exceedances is being recorded over the New Jersey portion of NYMA. This reversal of the trend in the ozone exceedances may be due to either year-to-year changes in the prevailing meteorology or changes in the ozone precursor emissions over NYMA. Since this trend in ozone violations has persisted since 1985, year-to-year variability in meteorology can most likely be ruled out as the primary cause for the observed reversal of the trend. Hence, change in the emissions over the NYMA might be a more plausible explanation for this trend reversal.

In order to further examine the trend reversal for the ozone exceedances, the median of the daily maximum 1-hr concentration from all sites in each region along with its 95% confidence interval, determined from Type II extreme value distributional fit (see Tabony, 1983 and Rao, et al., 1988) to the summertime daily maxima, for each year is presented in Figure 1c. Examination of the temporal behavior of the three-year running median of the measured maximum ozone concentration reveals the

Figure 1. (a) Location of the monitoring stations over the tri-state area used in this study.

(b) The number of days exceeding the ozone NAAQS of 0.12 ppm in each year for each region in the tri-state area.

(c) The median of the hourly maximum ozone concentrations for the summer months and its 95% confidence bounds based on Type II extreme value distribution fit to the daily maxima of 1-hour ozone concentrations. The dashed line represents the ozone trend based on the three-year running median of the measured median hourly maximum ozone concentration in each region.

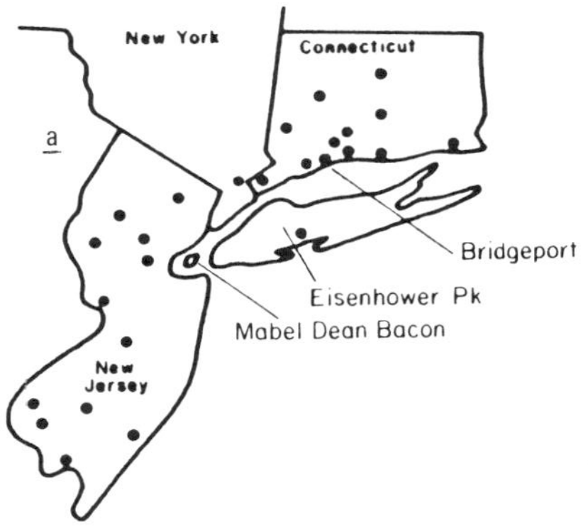

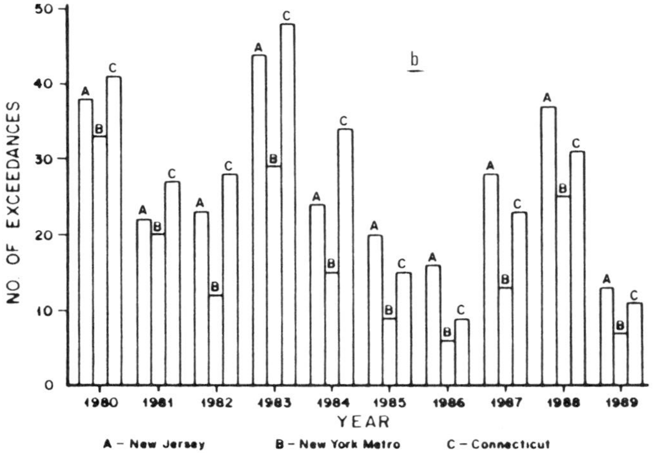

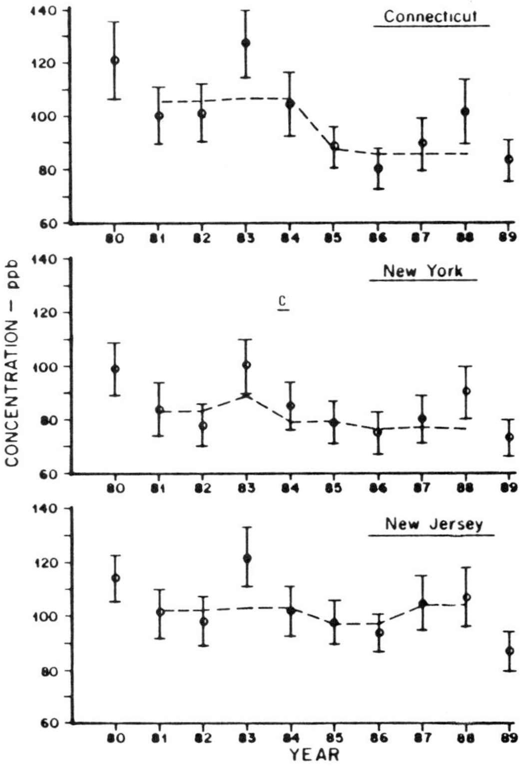

lack of a downward trend in the ozone levels over New Jersey and New York City despite a distinct downward trend in the ozone levels over Connecticut during this ten-year period. Whereas the median daily 1-hr maximum ozone concentration over Connecticut was about 105 ppb through 1984, the median concentration had dropped to about 85 ppb from 1985 onwards. This downward trend in the ozone levels over Connecticut complements the observed reversal of the trend in the ozone exceedances discussed above. Since the year-to-year variability in meteorology is smoothed or moderated somewhat when the temporal behavior of the running three-year median is considered, this 20% improvement in the ozone levels over Connecticut during the period 1980-1989 may be attributable primarily to the change in the emissions from the sources located in the tri-state region.

Having examined the long-term changes in the ozone air quality over NYMA, we now will attempt to assess the influence of the weekday/weekend differences in the precursor emissions on the ozone concentrations. The diurnal profiles of the ratio of the average NO_2 concentration on Wednesday to the average NO_2 concentration on Sunday (a typical example shown in Figure 2a) for the years 1984 to 1987 at Mabel Dean Bacon, New York, (a downtown site) indicate that the mid-morning and evening peak NO_2 concentrations on Sunday are considerably lower than those on Wednesday for all years. A similar trend is also evident in the case of CO concentrations (see Figure 2b for a typical example) at the Eisenhower Park, New York, monitoring location. The diurnal profile of the ozone concentration ratio at the Mabel Dean Bacon station (not shown here) indicates that the morning ozone concentrations on Sunday are generally higher than those on Wednesday. Since this monitoring site is located within New York City's central business district, higher ozone concentrations on Sunday than on Wednesday can be expected during the morning hours since NO_2 concentrations due to area and mobile source emissions are reduced on Sundays. However, examination of the cumulative frequency distributions of daily maximum 1-hr ozone concentrations over all years from all locations within New York City, shown in Figure 3a, indicates little change in the peak ozone levels between Wednesday and Sunday.

Although ozone in the Northeast is primarily a regional-scale problem, a previous urban-scale modeling study (see Rao and Sistla, 1988) indicated that the change in the emissions from sources located within the NYMA affects the magnitude of the peak ozone concentration over Connecticut. Therefore, the diurnal profile of the ratio of Wednesday to Sunday ozone concentrations for the Connecticut domain are also examined. Although the average NO_2 (6-10 AM) concentrations between Wednesday and Sunday are significantly different at the Bridgeport, Connecticut location (see Figure 3b), there is little difference in the ozone concentrations between these days at this location (see Figure 3c). The ozone concentrations ratios over Connecticut in Figure 4 indicate lower ozone on Sundays in some years and higher ozone on Sundays in other years. Thus, a consistent pattern between Wednesday and Sunday ozone levels is not evident despite the notable difference in the levels of ozone precursors over NYMA for these days. One would expect to observe lower ozone concentration levels farther downwind of the New Jersey/New York City metropolitan area (that is, over Connecticut) on Sundays if emissions reduction was substantial. The lack of a notable difference between Wednesday and Sunday peak ozone levels over Connecticut in Figure 3d implies that either the emissions reduction during the weekend from this urban area is not substantial enough to cause a reduction in the ozone levels or the meteorological influence is so dominant that it is obscuring the improvement in ozone

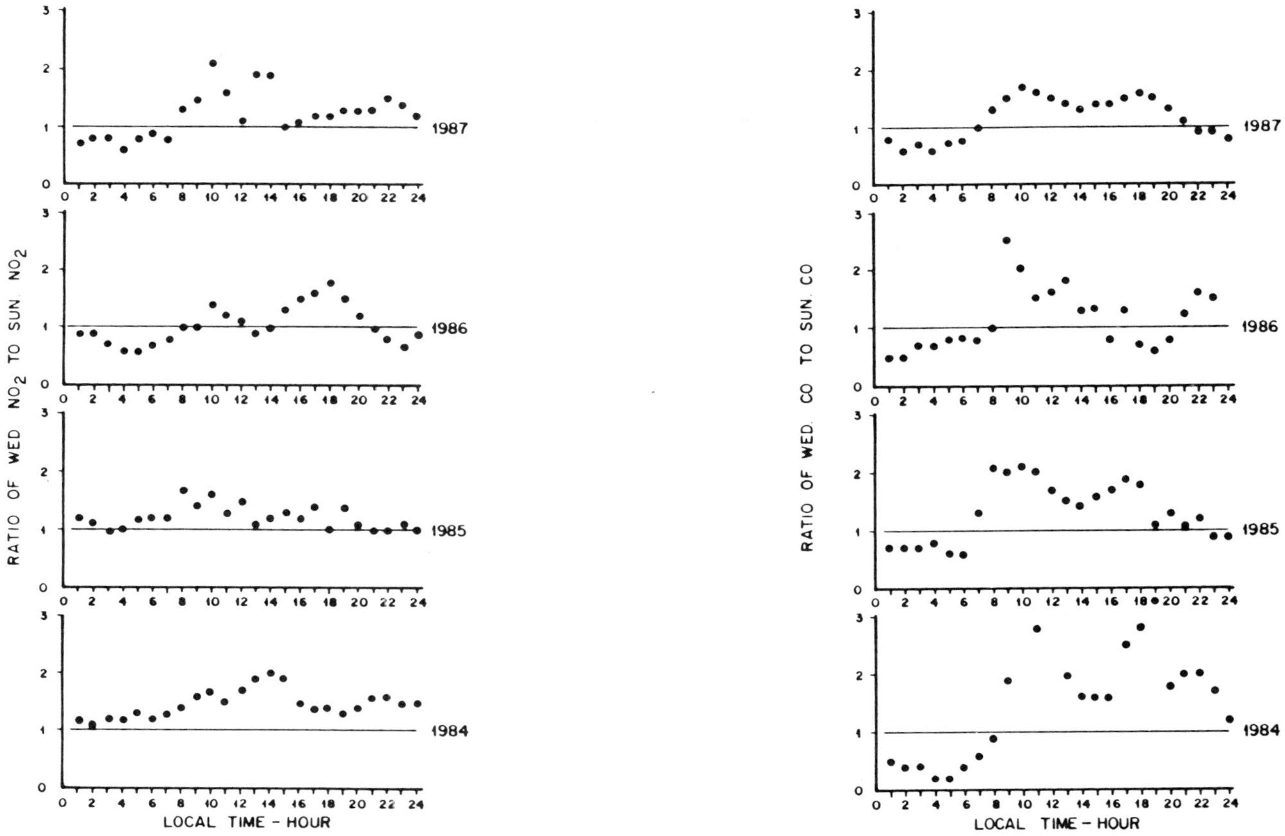

Figure 2. (a) The diurnal profiles of the ratio of the average wednesday NO_2 concentration to the average Sunday NO_2 concentration for different years at the Mabel Dean Bacon, NY station.

(b) Same as (a) except that the ratio is for CO concentrations at Eisenhower Park, NY station.

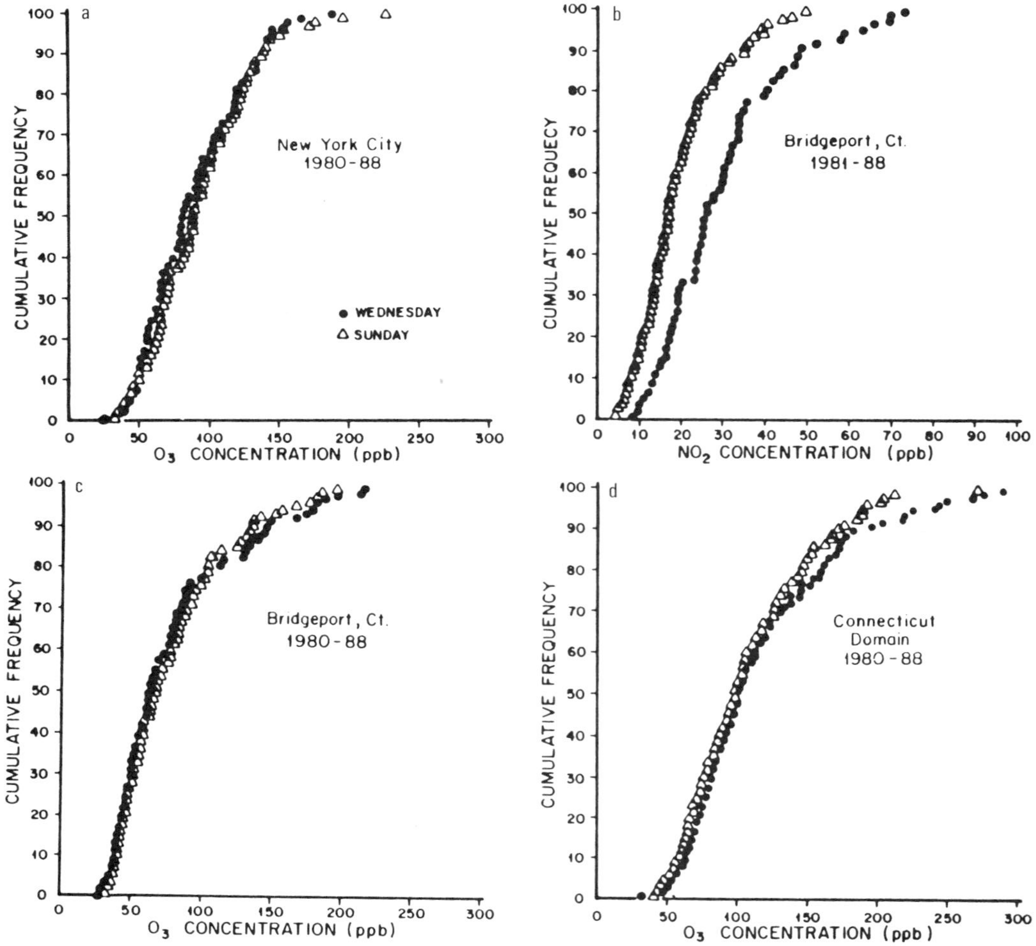

Figure 3. (a) Cumulative frequency distributions of domain maxima of hourly ozone concentrations for Wednesdays and for Sundays over the New York City area for the years 1980 through 1988.

(b) Cumulative frequency distributions of the morningtime average NO_2 concentrations for 1981 through 1988 at the Bridgeport, CT site.

(c) Same as (b) except for the daily maximum hourly ozone concentrations.

(d) Same as (a) except for the Connecticut domain.

air quality that can be expected from a reduced level of precursor emissions on Sundays. This feature is further examined through a model assessment of the sensitivity of the peak ozone concentrations to the variations in the emissions and meteorological conditions as outlined below.

The peak ozone concentrations over the Connecticut domain predicted by the Urban Airshed Model (UAM) with the Carbon Bond II (CB-II) chemical mechanism for different meteorological conditions (see Rao, 1987) are presented in Figure 5. Given the same emission inventory, the UAM predicts a variability of about 80 ppb (see Figure 5a) in the peak ozone concentration over Connecticut in response to the differences in the meteorological conditions occurring on 5 different days during the summer of 1980. When the UAM was exercised for a single day (August 8, 1980) with the same emissions, air quality, and other meteorological variables but with different methodologies for generating the wind fields as described in Rao, et al. (1989a), the model predicted a variability of about 60 ppb in the peak ozone concentration (see Figure 5b). Thus, the model predicts that variations in meteorological conditions alone could produce about 30% change in the predicted peak ozone concentration over Connecticut. Keeping the same meteorological conditions, and reducing VOC and NO_x emissions in the modeling domain by about 50% and 23%, respectively, the UAM predicted a change of about 65 ppb or about 30% change (see Figure 5c) in the peak ozone concentration over Connecticut (see Rao, et al., 1989b). These simulations suggest that the change in the peak ozone concentration over Connecticut due to the variability in meteorology is comparable to the change in the ozone peak due to a 50% and 23% reduction in the VOC and NO_x emissions, respectively, from sources located within the modeling domain. Since the models are deterministic while the atmospheric transport and diffusion processes which contribute to peak ozone concentrations are stochastic, the models tend to underestimate the variability in the predicted maximum ozone concentration due to the variability in meteorology. Thus, these results suggest that unless the emissions reduction is substantial, variations in prevailing meteorology may mask the improvement in ozone air quality that can be expected from modest reductions in the emissions.

CONCLUSIONS

Analysis of measured ozone concentrations during the 1980-1989 period indicates little change in the ozone air quality over New Jersey and New York City despite a distinct downward trend in the ozone levels over Connecticut. The lack of a clear downward trend in the ozone air quality over the New Jersey/New York portions of NYMA suggests that the emissions reduction from sources located upwind of NYMA may not be significant enough to influence the ozone concentrations in this region. Photochemical modeling simulations reveal that the peak ozone concentrations over Connecticut are strongly influenced by the meteorological conditions. The modeling results indicate that unless emissions reduction is substantial, the variability in meteorology can obscure the improvement in the ozone air quality. Hence, the change in ozone air quality due to the weekday/weekend differences in the emissions may not always be discernible because of the dominance of meteorology in dictating the ozone levels over NYMA.

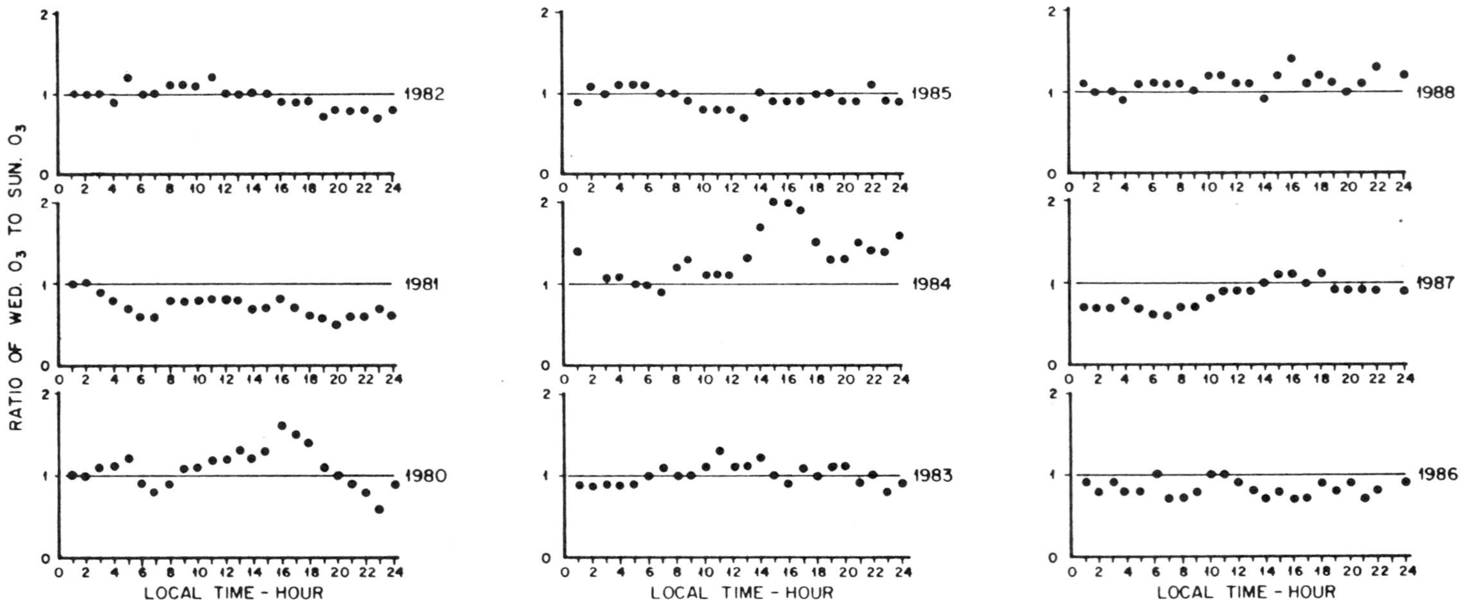

Figure 4. The diurnal profile of the ratio of the average ozone concentration on Wednesday to the average ozone concentration on Sunday over the Connecticut domain for the years 1980 through 1988.

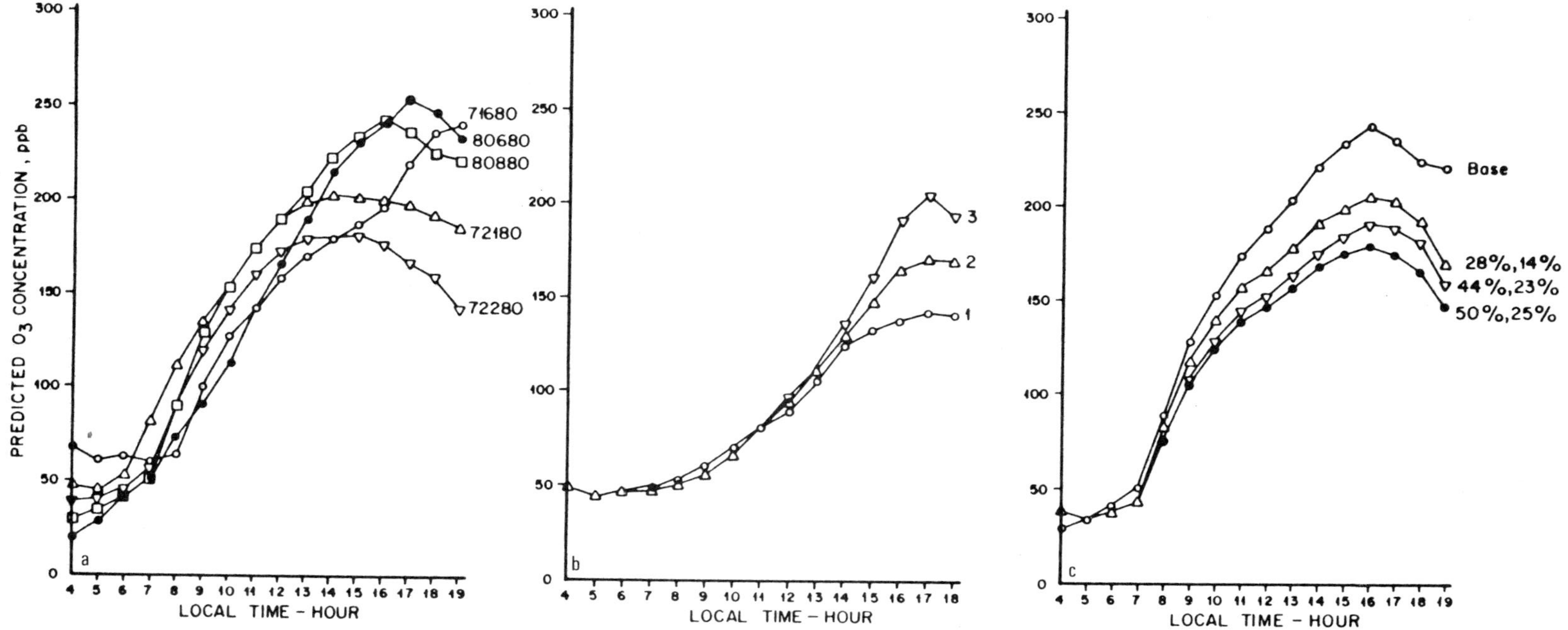

Figure 5. (a) The diurnal profile of maximum ozone concentrations over Connecticut predicted by the Urban Airshed Model for same emissions scenario on different days with differing meteorology.

(b) Same as in (a) except that different wind fields were used in the modeling simulations.

(c) Same as in (a) except that the meteorological conditions were now fixed while the VOC and NO_x emissions reduction were varied. The percentages shown represent the reduction in the VOC and NO_x emissions, respectively, from the base year.

ACKNOWLEDGEMENTS

The authors are grateful to Stephanie Liddle for typing the manuscript and to Gary Lanphear for drafting of the diagrams. The authors appreciate the encouragement of Thomas M. Allen, Director of the NYS Department of Environmental Conservation's Division of Air Resources, for this study.

REFERENCES

Ames, J., T. Meyers, L. Reid, D. Whitney, S. Golding, S. Hayes and S. Reynolds, "SAI Airshed Model Operation Manual Vol. I - Users Manual," EPA-600/8-85/1007A, PB85-191567, NTIS, Springfield, VA 22161 (1985).

Rao, S.T., "Application of the Urban Airshed Model to the New York Metropolitan Area," EPA-450/4-87-011 (1987).

Rao, S.T., "prepared discussion: Ozone air quality models," JAPCA 38:1129 (1988).

Rao, S.T. and G. Sistla, "Sensitivity Analysis of the Urban Airshed Model," The Scientific and Technical Issues Facing Post-1987 Ozone Control Strategies, Air & Waste Management Association, Pittsburgh, PA (1988).

Rao, S.T., G. Sistla, and J.Y. Ku, "Temporal and Spatial Variability of Ozone Concentrations in the New York Metropolitan Region," The Scientific and Technical Issues Facing Post-1987 Ozone Control Strategies, Air & Waste Management Association, Pittsburgh, PA (1988).

Rao, S.T., G. Sistla, J.Y. Ku, K. Schere, R. Scheffe and J. Godowitch, "Nested Grid Modeling Approach for Assessing Urban Ozone Air Quality," 82nd Annual Meeting of AWMA, Anaheim, CA, 89-42A-2 (1989a).

Rao, S.T., G. Sistla and R. Twaddell, "Photochemical Modeling Analysis of Emissions Control Strategies in the New York Metropolitan Area," EPA 230/2-89/026 (1989b).

Tabony, R.C., "Extreme Value Analysis in Meteorology," Met. Mag., 112:78 (1989).

Whitten, G.Z., J. P. Killus, and H. Hogo, "Modeling of Simulated Photochemical Smog with Kinetic Mechanism," EPA-600/3-80-028, PB 80-175102, NTIS, Springfield, VA 22161 (1980).

DISCLAIMER

DISCUSSION

R.A. BROST

Do you believe that the step change in observed O_3 over Connecticut between 1984 and 1985 was produced by a step change in meteorology?

S.T. RAO

NO! The diagram shows the temporal behavior of the median of the three year medians of measured daily maximum 1-hour ozone concentrations over Connecticut. This allows us to look at the trends in the ozone concentrations without being unduly biased by the year-to-year changes in meteorology. Note, the meteorological variability is moderated or filtered out by this data analysis procedure. Hence, the observed change in the median ozone concentration over Connecticut between the periods1980-1984 and 1985-1989 is attributable primarily to the change in the precursor emissions over the New York metropolitan area rather than the change in meteorology.

Ib: REGIONAL ACID DEPOSITION AND OXIDANT MODELLING

chairman: R.L. Dennis

rapporteur: A.A.M. Holtslag

DELINEATION OF THE INFLUENCE OF MAJOR EMISSIONS REGIONS ON THE ANNUAL ACIDIC DEPOSITION AT SENSITIVE AREAS

Robin L. Dennis,[1] John N. McHenry,[2] Dae Won Byun,[2] Terry L. Clark,[1] and Steven D. Reynolds,[3]

[1]U. S. Environmental Protection Agency, Atmospheric Sciences Modeling Division, Research Triangle Park, North Carolina

[2]Computer Sciences Corporation, Research Triangle Park, North Carolina

[3]Systems Applications Incorporated, San Rafael, California

INTRODUCTION

One central question for atmospheric processes research of the National Acid Precipitation Assessment Program (NAPAP) of the United States has been identification of the fraction of species deposition at a sensitive receptor region that is attributable to a particular emissions source region. The concern centers on distinguishing between effects of distant and local sources of emissions on sensitive ecological regions. The prevalent means for distinguishing the effects of sources is to develop source-receptor relationships. However, the debate about the nonlinearity in the atmospheric processes raises doubts about the ability of linear models to realistically portray source-receptor relationships. Also, questions have been raised regarding the loss in precision of describing three-dimensional transport (as compared to use of dynamic models) due to the lack of resolution inherent in interpolation techniques and due to the exclusion of cloud-influenced vertical transport.

To provide answers for the debate and to reduce uncertainty regarding source-receptor relationships, NAPAP developed a comprehensive model, the Regional Acid Deposition Model (RADM). This model was intended to represent the state-of-the-science of incorporating atmospheric processes of pollutant transport and transformation related to acidic deposition.

ANALYSIS BACKGROUND

The Comprehensive Model, RADM

The RADM is a mathematical model that incorporates known major physical and chemical atmospheric processes related to acidic deposition. Specifically, the RADM is designed to represent mathematically (1) the nonlinear dynamics of oxidant formation from precursor emissions of nitrogen oxides and volatile organics (reactive hydrocarbons); (2) the nonlinear dynamics of wet scavenging of sulfur oxides; and (3) the three-dimensional dynamics of transport, transformation, and deposition--including the effects of cloud processes. The RADM uses an Eulerian framework to carry out these representations (see Chang et al., 1987 and Chang et al., 1990).

The Mesoscale Meteorological Model, MM4, is used to provide the meteorological and dry-deposition velocity fields required by RADM. MM4 is a primitive-equation model that uses the hydrostatic assumption. To control error growth in the numerical simulation, MM4 is adapted to assimilate observations throughout the simulation period (Four-Dimensional Data Assimilation, FDDA). See Seaman and Stauffer (1989) for a description. The advantages of the MM4/FDDA system are that the model is able to use the plentiful observations available after the initial time of each simulation, while retaining MM4's numerical integration framework to develop mesoscale details, vertical motions, and an intervariable consistency that are not contained in the analyzed observations. In this way it is possible to generate input data sets whose accuracy, particularly regarding transport and dynamic consistency, is superior to both direct interpolation of synoptic-scale analyses of observations and to a purely predictive-mode model result. The MM4/FDDA model is capable of simulating nonlinear interactions that cannot be represented by direct interpolation between analyses based on synoptic observations separated by 12 hours (h).

The RADM model is used to represent changes in acidic deposition over a geographic scale of 1000 to 2000 kilometers (km). The domain under study is eastern North America. The model's horizontal grid is 80 km. The vertical grid has six layers, nonuniformly spaced, between the ground and 100 millibars (mb) (approximately 15 km in altitude). The three lowest RADM layers are typically in the planetary boundary layer during the day, and the remaining three layers are in the free troposphere. Emissions are input hourly. The simulation steps are of the order of seconds for the chemistry, minutes for the transport process, and an hour for cloud processes. The model requires 1.8 h of computing time on a CRAY XMP to simulate 5 days.

<u>The Tagged Sulfur Engineering Model</u>

A set of simpler models, based on the RADM, have been developed for applications to sulfur-only emissions changes (see Chang et al., 1990). These models solve the three-dimensional conservation equations for gas- and aqueous-phase airborne sulfur dioxide and sulfate in a manner identical to the RADM's. The only difference between these models and the RADM occurs in the gas-phase chemistry; the algorithms for the other processes are the same as in the RADM. One of these models, the Tagged RADM/EM, was specifically designed for use in source-receptor applications. This model allows sulfur dioxide emissions and products from specified emissions sources to be tagged and tracked through the entire simulation. This procedure allows source attribution to be examined in situ in a nonlinear system without having to invoke the assumption of linear superposition of the effects of individual sources.

The underlying principle of the Tagged RADM/EM is the assumption of a chemically well-mixed grid cell. This assumption ignores subgrid-scale differences where, for example, distinct plumes of tagged material might not be well mixed with surrounding nontagged material.

Development of Annual Averages

The development of annual averages is described in a companion paper by R. L. Dennis et al., "Characterization of the Nonlinear Change in Annual Sulfur Deposition for a Change in Emissions." The annual averages produced by RADM are designed to represent the 1982-85 period, a reasonably typical period.

Analysis Approach

With the 80-km horizontal grid of the RADM, the influence of single sources cannot be resolved very well. It is best to consider the influence of all of the sources in a cell. If the sources are not dispersed over the cell, estimates of their influence across space will still contain uncertainty, especially near the sources. The use of several cells should result in a canceling of directional errors and a reduction of the uncertainty. The authors are primarily interested in studying the influence of utility emissions, which tend not to be uniformly placed throughout a RADM grid. These emissions tend to be large in a few cells that are clustered in space. Therefore, blocks of several grid cells that cover regions of major emissions density were defined for tagging. Only the utility sources were tagged because they are the sources of greatest interest.

Three subregions of major emissions, source regions, were defined for this study: the cluster of sources around the upper Ohio River (Upper Ohio), the cluster of sources around the lower Ohio River (Lower Ohio), and the cluster of sources in the southeastern United States below Tennessee (Southeast). These clusters of sources represent three dominant subregions of sulfur dioxide emissions density east of the Mississippi River. The RADM grid cells for which utility emissions were tagged and the respective emissions densities of these cells are shown in Figure 1. The tagged emissions of the Upper Ohio, Lower Ohio, and Southeast represent 17%, 12%, and 7.2%, respectively, of the major point source emissions in the RADM domain. Note that the emissions densities within each source region vary substantially and that a few RADM cells dominate each source region with their emissions.

RESULTS

Percentage Contributions to Total Deposition

The percentage contribution of tagged emissions sources to total annual sulfur deposition predicted by RADM is shown in Figures 2, 3, and 4 for the Upper Ohio, Lower Ohio, and Southeastern tagged source regions, respectively. Comparisons with Figure 1 indicate that the "pattern" of the percentage contribution is controlled by the largest sources (the grid cells with highest emissions) in the tagged regions. The influence of the meteorology is also apparent in the patterns. The pattern is stretched along a northeastern axis for the Upper Ohio sources and is stretched in an arc from north to east for the Lower Ohio sources. The pattern is fairly symmetric for the Southeastern sources.

In all three tagged source regions, the tagged sources contribute more than 50% of the total deposition to a noticeable portion of the area including and immediately "downwind" of the largest tagged sources. The radius of this major influence is approximately 100 km for the Southeastern and Lower Ohio sources and 100-140 km for the Upper Ohio sources.

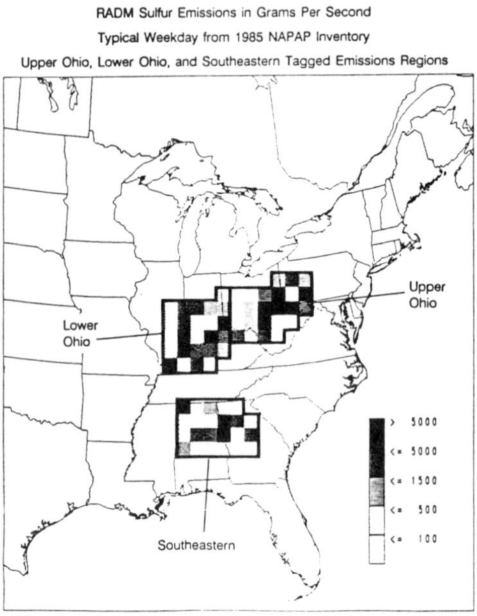

Fig. 1. Map of the three tagged source regions in the RADM domain. The cells for which utility emissions are tagged are shown, and the emissions densities in these cells are subdivided into five groups.

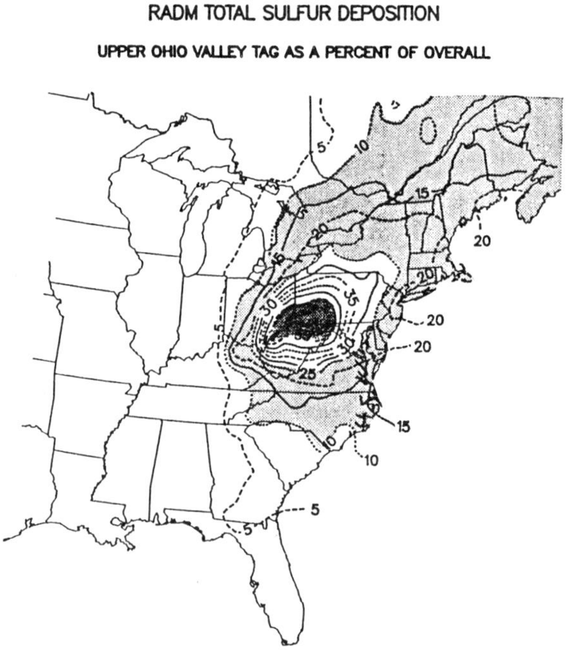

Fig. 2. Percentage contribution of tagged Upper Ohio Valley utility emissions to total 1982-85 annual sulfur deposition (wet plus dry) predicted by RADM.

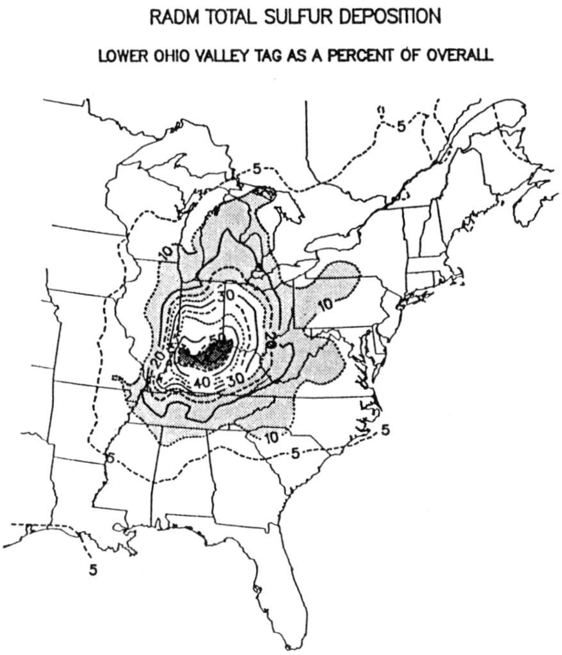

Fig. 3. Percentage contribution of tagged Lower Ohio Valley utility emissions to total 1982-85 annual sulfur deposition (wet plus dry) predicted by RADM.

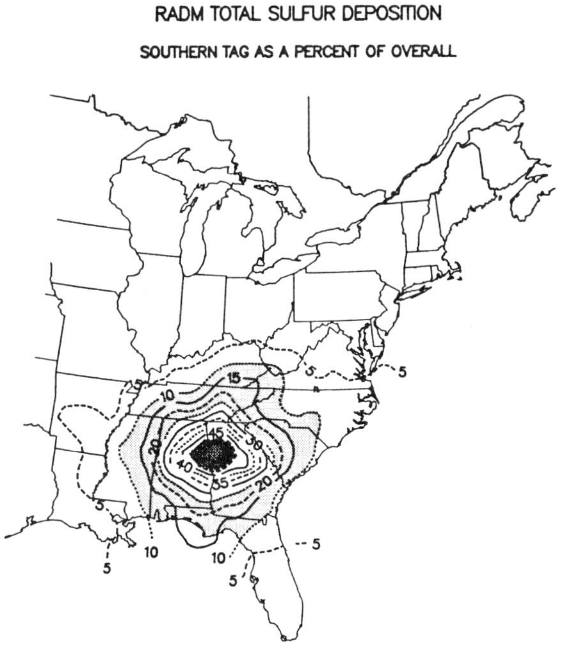

Fig. 4. Percentage contribution of tagged Southeastern utility emissions to total 1982-85 annual sulfur deposition (wet plus dry) predicted by RADM.

One identifiable measure of the range of important influence for these tagged source regions is the distance from the dominant emissions grid cells at which the percentage contribution to total deposition drops below 25%. Considering primarily the distances that appear to constitute "downwind," as noted above, the range of importance for the Upper Ohio, Lower Ohio, and Southeastern sources is approximately 400 km, 350 km, and 250 km, respectively. The range of the percentage contribution is influenced by several factors, including the magnitude of the emissions included in each source region, the meteorology, and the magnitude of the emissions upwind that are mixing with the source region's emissions.

Absolute Contribution to Deposition

A second way to analyze the range of influence of the source regions is to examine the area impacted by different levels of total sulfur deposition from each of the source subregions. The peak deposition produced by the regions ranges from 14 to 18 kg sulfur(S)/hectare(ha) (15, 14, and 18 kg S/ha for Southeastern, Lower Ohio, and Upper Ohio, respectively). The differences in peak deposition do not correspond to the differences in emission magnitudes (7.2%, 12%, and 17% of the domain-wide emissions for Southeastern, Lower Ohio, and Upper Ohio, respectively). Source density is important, and the source densities of the dominant cells in each source region are similar.

A measure that is less sensitive to the immediate source density is the area enclosed by a rate of deposition defined so that it is normalized for the differences in emissions tonnage between the subregions. The emissions in each source region were divided by those of the Southeastern region's and the quotient multiplied by 1 kg/ha to obtain a set of normalized contour levels. A level of 1 kg/ha was chosen so that the deposition would be approximately a factor of 10 lower than the peak. The areas encompassed by the 1-kg-S/ha contour (Southeastern), the 1.7-kg-S/ha contour (Lower Ohio), and 2.4-kg-S/ha contour (Upper Ohio) are roughly similar, implying that the "unit of emissions" range of influence is approximately the same for the source regions.

Even though the "unit of emissions" area of deposition coverage is similar, the magnitude of the deposition decreases more rapidly from the Southeastern sources than it does for the Lower Ohio or Upper Ohio sources. The distances over which the deposition drops to one-third of its peak value are 150 to 170 km, 240 to 270 km, and 250 to 320 km for the Southeastern, Lower Ohio, and Upper Ohio source regions, respectively.

DISCUSSION

The contribution of the source regions to local deposition is substantial, approximately 50% of the deposition in all of the source areas examined. The long-range influence of source regions decreases fairly rapidly, over a range of several hundred kilometers, but it is still important. For example, the contribution of the Upper Ohio sources to deposition at sensitive regions is estimated to be 21%, 30%, and 40% for the Adirondacks, the Poconos, and the Mid-Appalachian sensitive regions, respectively.

The long-range influence depends on the meteorology (or transport), the surrounding emissions, and the source strength of the tagged sources. For example, at a distance of approximately 300 km, the percentage of the deposition contributed by the Upper Ohio sources to the Poconos region is 30%, whereas the percentage contributed by the Southeastern source region to the Southern Blue Ridge sensitive area is 18%. If one were to simplistically account for the difference in emissions density between the two regions by increasing in the deposition from the southern sources, based on the difference in emissions total, then the deposition contributed by the Southeastern source region to the Southern Blue Ridge would increase to 34%. This relative contribution is much more comparable to the influence of the Upper Ohio on the Poconos region.

The RADM is presently undergoing a rigorous performance evaluation in two phases, with the first phase currently in progress. Preliminary results indicate that the model is capturing the day-to-day variation in surface concentrations as synoptic systems pass through the model domain. Also, the model is replicating the differences in magnitude of air concentrations between subregions. These results are considered weak evaluation tests, but they provide insight into performance related to source attribution. The model appears to be functioning well.

The MM4/FDDA has also undergone model verification/evaluation on the assimilated and nonassimilated primitive variables. The assimilated variables are important to transport. Examples of assimilated variables are surface level pressure, 500-mb winds, and 850-mb temperatures. None of the mean or root-mean-square errors for the sampled assimilated variables are large, compared to typical values expected for objective analyses. While the model does not return fields that are identical to the analyses (by design), it appears to have only very small systematic and random error characteristics.

The aggregated annual estimates of source attribution produced by the RADM are qualitatively similar to those produced by linear models using interpolated fields when the emissions inputs are the same; however, the quantitative details are different. The 1985 source contributions of the Upper Ohio and Southeastern source regions are shown for the Regional Lagrangian Model of Air Pollution in Figures 5 and 6. Work is continuing to examine what improvement in skill is produced by the more advanced models such as RADM.

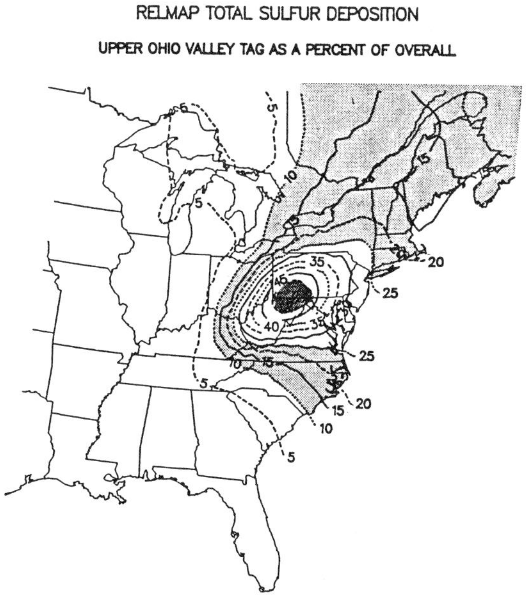

Fig. 5. Percentage contribution of tagged Upper Ohio Valley utility emissions to total 1985 annual sulfur deposition (wet plus dry) predicted by RELMAP.

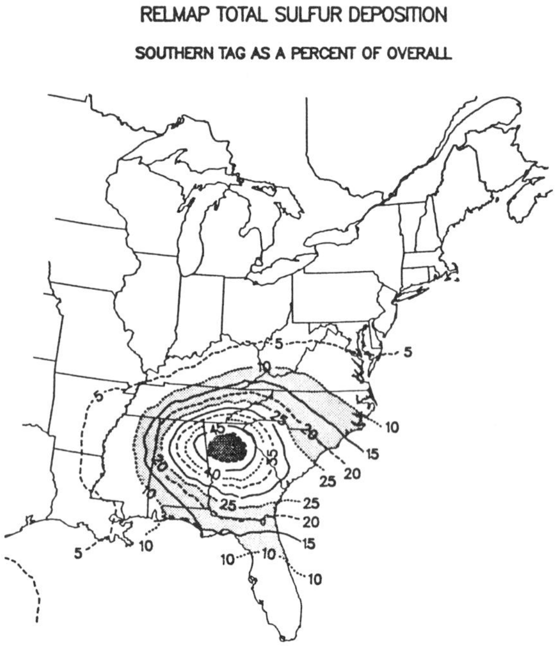

Fig. 6. Percentage contribution of tagged Southeastern utility emissions to total 1985 annual sulfur deposition (wet plus dry) predicted by RELMAP.

REFERENCES

Chang, J. S., Brost, R. A., Isaksen, I. S. A., Madronich, S., Middleton, P., Stockwell, W. R., and Walcek, C. J., 1987, A three-dimensional Eulerian acid deposition model: Physical concepts and formulation, J. of Geophysical Res., 92:14,681.

Chang, J. S., Binkowski, F. S., Seaman, N. L., McHenry, J. N., Byun, D. W., Samson, P. J., Stockwell, W. R., Walcek, C. J., Madronich, S., Middleton, P., Pleim, J. E., and Lansford, H. H., 1990 (in press), "The Regional Acid Deposition Model and Engineering Model," National Acid Precipitation Assessment Program, Washington, D.C.

Seaman, N. L., and Stauffer, D. R., 1989, "Development of Four-Dimensional Data Assimilation for Regional Dynamic Modeling Studies," Final Report to U.S. EPA, Contract CR-814068-01-0, The Pennsylvania State University, University Park, PA.

DISCUSSION

J.C.R. HUNT

How does your model account for the fact that nearby sources do not mix as well as simple mixing models suggest? The reason is that plumes "flap" in phase with each other, as shown by laboratory studies, stochastic models and field studies in La Trobe Valley, Victoria, Australia.

R.L. DENNIS

Eulerian models do not account for this observation at all. The emissions in a grid cell are instantaneously mixed. For RADM, this means all sources within an 80 km grid cell mix together. You raise what we term a sub-grid-scale issue. Particularly with models incorporating non-linear chemistry there is a question of what error is introduced by mixing first and then "doing" chemistry rather than doing chemistry and then mixing. We expect to be exploring this latter sub-grid-scale issue in the future.

A. VENKATRAM

You talked about contributions of various source areas to sulfur deposition. I do not know whether you made the point that these contributions are a function of the emission level (you are dealing with a non-linear system).

R.L. DENNIS

This is a good point that was not made clearly in the paper. As emissions levels increase in a given area to where oxidant limitation (the cause of the nonlinearity) becomes important, competition for limited oxidants will cause emissions to transport farther before depositing, thus changing the relative contributions of the sources to deposition across space. As a result, linear superposition of source influence is no longer a reasonable approximation. This is why source attribution needs to be calculated "in situ" for a non-linear system, as done here.

CHARACTERIZATION OF THE NONLINEAR CHANGE IN ANNUAL SULFUR DEPOSITION FOR A CHANGE IN EMISSIONS

Robin L. Dennis,[1] John N. McHenry,[2] Steven K. Seilkop,[3] and Julius S. Chang[4]

[1]U. S. Environmental Protection Agency, Atmospheric Research and Exposure Assessment Laboratory, Research Triangle Park, North Carolina

[2]Computer Sciences Corporation, Research Triangle Park, North Carolina

[3]Analytical Sciences, Inc., Durham, North Carolina

[4]State University of New York-Albany, Atmospheric Sciences Research Center, Albany, New York

INTRODUCTION

A central question for atmospheric processes research posed by the National Acid Precipitation Assessment Program (NAPAP) of the United States has been: What degree of nonproportionality in the response in sulfur deposition is expected to be associated with a specified change in precursor emissions? Nonlinearity (or nonproportionality) exists where the fractional changes in emissions of a primary pollutant are not matched by a commensurate (proportional) change in the deposition of its primary or secondary product. Nonlinearity is only expected to occur in wet deposition processes. When sulfur dioxide in the aqueous phase is converted to sulfate, the most reactive oxidant, hydrogen peroxide, can become depleted in the cloud (oxidant limited). During the in-cloud-water conversion if there is more sulfur dioxide than is needed to react fully with available hydrogen peroxide, the hydrogen peroxide can be depleted and unreacted sulfur dioxide left over (an excess). Subsequently, if emissions of sulfur dioxide are reduced, the "excess" sulfur dioxide is reduced, leading to less than proportional (nonlinear) reductions in sulfate deposition. One concern is that if emission controls are implemented, deposition may not reduce in proportion to emissions reductions.

Both prior to and since the 1983 National Research Council report (NRC, 1983), an intense debate has been carried out regarding nonlinearity. To provide answers for the debate and to ascertain the degree of nonlinearity, NAPAP developed a comprehensive model, the Regional Acid Deposition Model (RADM). This model was intended to represent the state-of-the-science in incorporating atmospheric processes of pollutant transport and transformation related to acidic deposition.

ANALYSIS BACKGROUND

The Comprehensive Model, RADM

The RADM is a mathematical model based on the time-dependent, three-dimensional equations describing the conservation of mass of chemically reactive pollutants in the atmosphere. The model incorporates known major physical and chemical atmospheric processes related to acidic deposition. Specifically, the RADM represents mathematically (1) the nonlinear dynamics of oxidant formation from precursor emissions of nitrogen oxides and volatile organics (reactive hydrocarbons); (2) the nonlinear dynamics of wet scavenging of sulfur oxides; and (3) the three-dimensional dynamics of transport, transformation, and deposition--including the effects of cloud processes. The RADM, of necessity, uses an Eulerian framework to carry out these representations. For a description of the model, see Chang et al. (1987) and Chang et al. (1990).

A preprocessor, run independently of the RADM, is used to provide the meteorology and dry-deposition-velocity fields required by the model. This preprocessor, the MM4, is a mesoscale meteorological model based on the primitive equations that use the hydrostatic assumption. The MM4 uses a four-dimensional data assimilation technique to incorporate surface relative humidity and temperature measurements taken every 3 hours (h) as well as upper-air wind and temperature measurements taken every 12 h.

The RADM is used to represent changes in acidic deposition over a geographic scale of 1000 to 2000 km. The domain under study is eastern North America. The model's horizontal grid is 80 km. The vertical grid has six layers, nonuniformly spaced, between the ground and 100 millibars (mb) (approximately 15 km in altitude). The three lowest RADM layers are typically located in the planetary boundary layer during the day, and the remaining three layers are in the free troposphere. The RADM is an episodic model; that is, meteorology and emissions are input hourly. The simulation steps are of the order of seconds for the chemistry, minutes for the transport processes, and an hour for cloud processes. The model requires 1.8 h of computing time on a CRAY XMP to simulate 5 days on a 35 x 38 x 6-grid array.

Development of Annual Averages

Because the RADM is an episodic model requiring detailed setup of meteorological and emissions inputs to simulate several days, it would be cost- and time-prohibitive to simulate an entire year to estimate annual deposition values. Instead, a sample of meteorological events was taken, and model predictions for these events were aggregated to estimate annual totals of deposition.

Conceptually, the aggregation approach is based on grouping meteorological cases with similar 850-mb windflow patterns (strata). A stratified, random sample is then developed. The classification of these windflow patterns into 19 sampling strata was achieved through cluster analysis. The period 1982-85 was so classified. A random selection of meteorological cases from within a stratum was performed, selecting approximately the same proportion of cases relative to the number of cases which occurs in each of the strata of windflow patterns (Samson, 1990). The deposition values from the model predictions for each meteorological case were then weighted and aggregated according to the strata sampling frequencies.

Thirty meteorological cases were used to aggregate RADM predictions to annual totals. The totals are representative of average annual deposition for the period 1982-85. Of the 30 cases, 19 were chosen from nine strata that contribute roughly 75% of the total, annual accumulated wet deposition (sulfate and nitrate). The 30-case sample has a sampling uncertainty of approximately +/- 25%, a significant reduction over simple, random sampling from the entire 1982-85 period. The performance of the aggregation sample plus the RADM model is shown in the discussion section.

Analysis Approach

Nonproportionality exists if the percentage of reduction in emissions results in an other than equal percentage of reduction in deposition. In an oxidant-limited region (where sulfur dioxide concentrations are much greater than hydrogen peroxide concentrations, a less-than-proportional change in wet deposition is expected, based on relationships derived from the stoichiometry of the chemical transformations. That is, the amount of sulfur dioxide converted to wet sulfate would be limited by the availability of hydrogen peroxide. The change in sulfur dioxide concentrations associated with an emissions change would not be reflected in the amount of sulfur dioxide converted by the same availability of hydrogen peroxide.

However, nonproportionality can arise, even for a linear system, if the emission reductions are not uniform, and especially if they vary greatly across space. Consider a simple example. Source region A, far upwind, contributes 20% to the deposition in region B and its emissions are unchanged. Sources in region B contribute 80% to the deposition and its emissions are reduced 50%. In subregion B the emissions are reduced by 50%, but the deposition is only reduced by 40% (half of 80%).

For this study only emissions reductions that occurred equally across the modeling domain were investigated--in particular, a 50% "across-the-board" reduction. By keeping the emissions reduction spatially simple, it was possible to avoid having to distinguish and identify the contributions of the two sources of nonproportionality. A 50% reduction is also the approximate emissions reduction currently being considered by the U.S. Congress.

RESULTS

The 30 cases were simulated for a 50% reduction in sulfur emissions everywhere in the RADM domain. The sulfur boundary and initial conditions were also reduced by 50%. The RADM simulations were then aggregated to annual totals, and the change in percentage relative to the 1982-85 base deposition was calculated.

Annual Wet Sulfur Deposition

The percentage of reduction in annual wet sulfur deposition across the RADM domain resulting from the 50% emissions reduction is shown in Figure 1, using isopleth intervals of 2%. The percentage of change in this deposition is nonlinear on an annual basis. The regions of greatest nonlinearity, i.e., those having reduction percentages of 38 to 40%, are associated with the large source of sulfur emissions of the upper and lower Ohio River areas, near Atlanta, GA, and the Noranda and Sudbury smelters in Canada. The 40% reduction contour surrounds the regions of major sulfur dioxide emissions.

The effect of oxidant limitation is strongest in winter, as expected, and is weakest, although still evident, in summer. In winter, the deposition is only reduced by 35% over Ohio, western Pennsylvania, and western New York in the United States and around the two largest Canadian smelters. In summer, the depression of percentage reduction due to nonlinearity surrounds the Ohio River, surrounds a corridor between Detroit and Toronto, and circles the largest Canadian smelters.

The effect of nonlinearity is to reduce the effectiveness of the emissions reduction by 20% in the region of highest emissions. Farther out from the source regions, near sensitive regions, the effectiveness of the emissions decrease is reduced by 10%.

Annual Dry Sulfur Deposition

The percentage of reduction in annual dry sulfur deposition across the RADM domain is shown in Figure 2, using isopleth intervals of 2%. Over the RADM domain, the change in the percentage of dry deposition is essentially linear everywhere over the continent.

Some researchers have speculated that because of the nonlinearity in the wet deposition, a greater-than-linear response in the dry deposition might be found. Such behavior is simulated by the RADM for spring cases

RADM WET SULFUR DEPOSITION
REDUCTION PERCENTAGE FROM 1985 BASE CASE

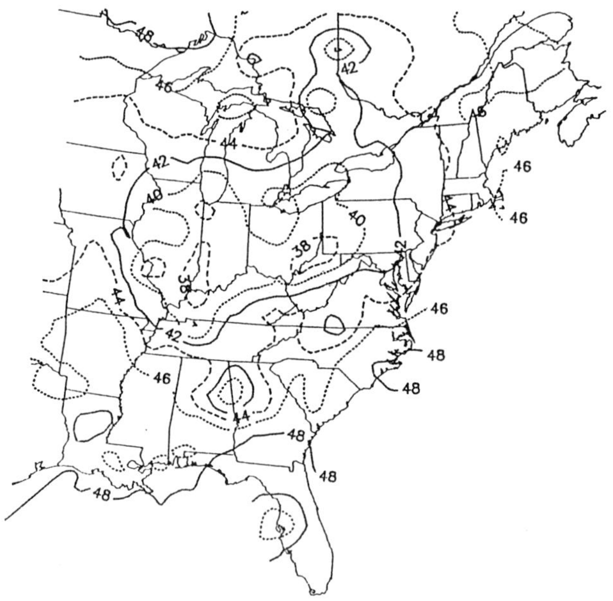

Fig. 1. Percentage change in annual wet sulfur deposition predicted by RADM for a 50% uniform reduction in sulfur dioxide emissions from all sources (boundary and initial concentrations also reduced by 50%).

for the Sudbury plume, an isolated plume, but nowhere else in the domain. As seen in Figure 2, a greater than proportional response in dry deposition does not occur in the annual average over the continent. However, it does seem appear to appear far out over the ocean, once emissions are no longer being injected into the system.

<u>Annual Total Sulfur Deposition</u>

The percentage of reduction in annual total sulfur deposition is shown in Figure 3, using isopleth intervals of 2%. The percentage of reduction in deposition for a 50% reduction in emissions is nonlinear on an annual basis. As one would expect, the pattern of the percentage of reduction is similar to that of wet deposition, with smaller departures from proportionality. In this case the percentage reduction contour of 44% surrounds the major source regions.

The degree of nonproportionality over the source regions is roughly half of that for wet deposition. The effectiveness of a 50% emissions decrease is reduced by 10% to 45%. Away from the source regions, the degree of nonlinearity of total deposition is two-thirds that of wet deposition because dry deposition contributes only approximately one-third of the total deposition.

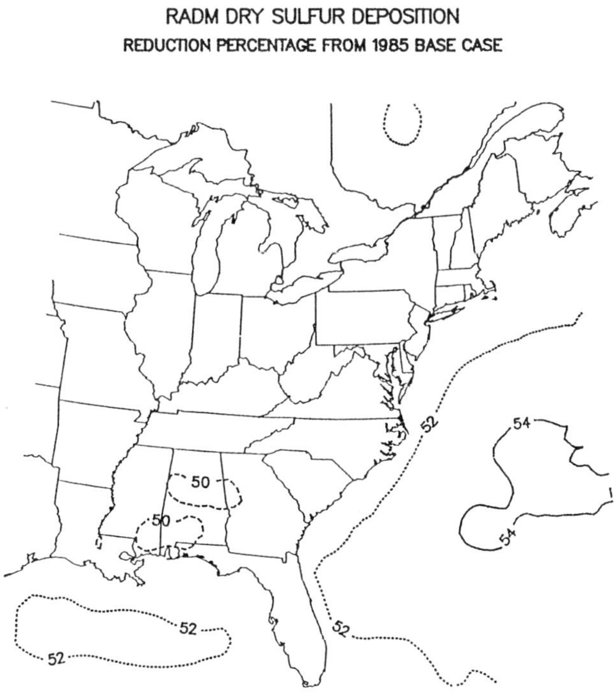

Fig. 2. Percentage change in annual dry sulfur deposition predicted by RADM for a 50% uniform reduction in sulfur dioxide emissions from all sources (boundary and initial concentrations also reduced by 50%).

DISCUSSION

The percentage of reduction of total, annual sulfur deposition integrated over the eastern North American continent is 45.4%. For dry deposition the reduction average is 50.6%, and for wet it is 43.0% Thus, the system is still nonproportional over the spatial scale of eastern North America. Because the initial and boundary conditions were reduced 50%, the results are slightly more linear than if a constant "global" background of sulfur had been assumed.

A diagnosis of the oxidation processes occurring in RADM for several spring meteorological cases indicates that in the source regions, where oxidant limitation is expected to be pronounced due to high levels of sulfur dioxide, hydrogen peroxide is responsible for about two-thirds of the oxidization of sulfur dioxide in the aqueous chemistry, and oxidation by ozone accounts for basically the rest of the oxidized sulfur dioxide. The pH levels for eastern North America are low enough that other oxidation pathways, such as iron catalysis, do not come into play. Downwind, where oxidant limitation is much less pronounced or has ceased, hydrogen peroxide is responsible for more than 90% of the oxidation of sulfur dioxide to sulfate.

RADM TOTAL SULFUR DEPOSITION
REDUCTION PERCENTAGE FROM 1985 BASE CASE

Fig. 3. Percentage change in annual total sulfur deposition (wet plus dry) predicted by RADM for a 50% uniform reduction in sulfur dioxide emissions from all sources (boundary and initial concentrations also reduced by 50%).

The average wet sulfur deposition, aggregated annually, compares very well with measured deposition (see Figure 4). The average dry sulfur deposition, aggregated annually, compares reasonably well with empirically derived deposition, although the comparison for dry deposition has a bias and the comparison for wet deposition does not (see Figure 5).

The RADM is presently undergoing a rigorous two-part performance evaluation, with the first part currently in progress. Initial model evaluation comparisons against September 1988 field data have indicated two model components needing modest refinement. An iteration of model refinement has been completed and the refined RADM is expected to be evaluated against the 1988 field data during the NAPAP phase of the RADM evaluation.

After completion in 1990 of this first, rigorous evaluation, the results presented here can be more accurately appraised. The results described here should be considered preliminary. Based on sensitivity studies with the RADM, however, little change in the results is expected.

With this work, the authors have attempted to demonstrate that RADM, the new, comprehensive model, is providing answers to primary questions regarding the degree of nonlinearity for annual averaging times. Significantly, the comprehensive, regional acidic deposition models have improved greatly in a short time and continue to mature rapidly. As model developers and users obtain an increased understanding of RADM and its ability to make predictions, their confidence in the model will continue to grow.

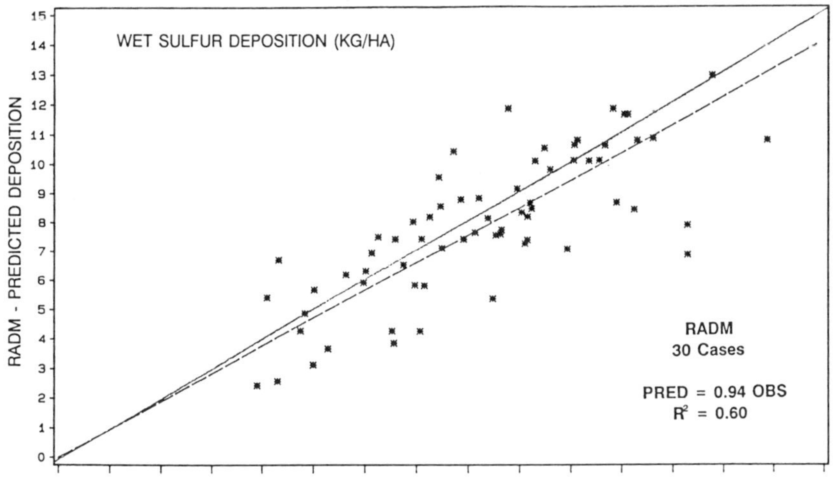

Fig. 4. Comparison of 30-case aggregated RADM predictions and observations of annual average sulfur wet deposition for the 1982-1985 period.

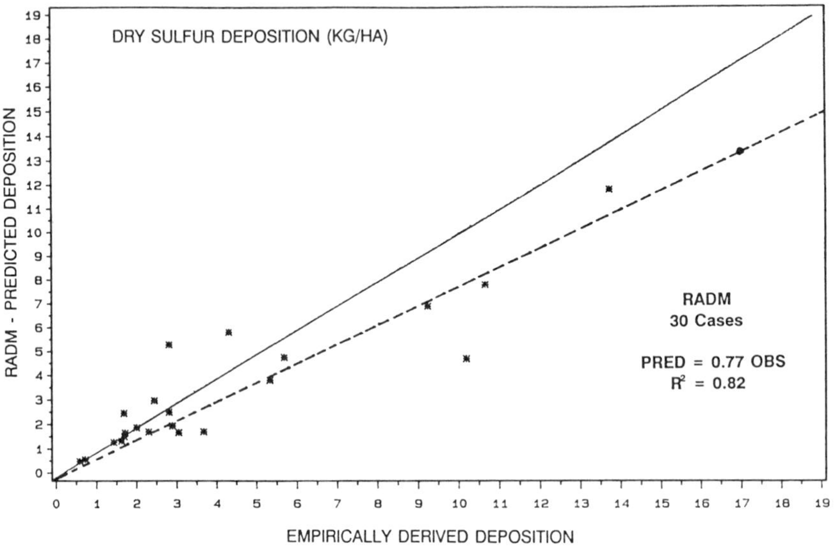

Fig. 5. Comparison of 30-case aggregated RADM predictions and empirically derived annual average dry deposition for the 1982-1986 period.

REFERENCES

Chang, J. S., Brost, R. A., Isaksen, I. S. A., Madronich, S., Middleton, P., Stockwell, W. R, and Walcek, C. J., 1987, A three-dimensional Eulerian acid deposition model: Physical concepts and formulation, J. of Geophysical Res., 92:14,681.

Chang, J. S., Binkowski, F. S., Seaman, N. L., McHenry, J. N., Byun, D. W., Samson, P. J., Stockwell, W. R., Walcek, C. J., Madronich, S., Middleton, P., Pleim, J. E., and Lansford, H. H., 1990 (in press), "The Regional Acid Deposition Model and Engineering Model," National Acid Precipitation Assessment Program, Washington, D.C.

National Research Council, 1983, "Acid Deposition: Atmospheric Processes in Eastern North America: A Review of Current Scientific Understanding," National Academy Press, Washington, D.C.

Samson, P. J., Brook, J. R., and Sillman, S., 1990, "Estimation of Seasonal and Annual Acidic Deposition Through Aggregation of Three-day Episodic Periods," EPA Report No. EPA/600/3-90/000, Atmospheric Research and Exposure Assessment Laboratory, Research Triangle Park, NC, 27711.

DISCUSSION

J.D. SHANNON

Relative to the integration over the eastern US of observations of dry S deposition by Hicks, RADM is 50% lower. Have you resolved the discrepancy?

R.L. DENNIS

The evaluation of RADM just completed for the National Acid Precipitation Program, NAPAP, (including use of Hicks's data) indicates that RADM calculations of dry S deposition are a factor of

20% low; hence, an integration over the eastern US using RADM predictions would be expected to be 20% low. I do not know where you obtained the number 50%. My best judgment at this time is that the underprediction of dry S deposition is due primarily to RADM's underprediction of ambient sulfate at the surface. We still have not resolved the underprediction of sulfate at the surface.

MODELLING PHOTOCHEMICAL POLLUTION BY AN EULERIAN LONG-RANGE TRANSPORT MODEL

Jesper H. Christensen#, Øystein Hov* and Ruwim Berkowicz#

#National Environmental Research Institute
Frederiksborgvej 399, 4000 Roskilde, Denmark

*Geophysics Institute, University of Bergen
N-5007 Bergen, Norway

1. INTRODUCTION

It is a well-known fact now that photochemical oxidants are produced, transported and deposited on continental scales. The large-scale distribution of photochemical pollution implies that adequate abatement strategies must have a more global character. Suitable transport models can give substantial support to the development of adequate control strategies.

Modelling formation and transport of photochemical oxidants requires a complex description of both chemical and meteorological processes.

For models to be able to serve as scientific tools for decision-making, it must be ensured that at least the most significant processes are properly taken into account. A critical evaluation of models is thus necessary in order to guarantee that right answers are given for right reasons. Testing a model on available data is a powerful tool to achieve this security, but it is not sufficient. It is also important to verify how the model responds to changes in different driving parameters and whether this response is in accordance with the known physical and chemical processes.

Some results from an Eulerian long-range transport model for photochemical pollution are presented here. The main emphasis is placed on modelling the formation and transport of ozone. Of special interest in this report is the sensitivity of the predicted ozone and NO_2 concentrations on the parametrization of exchange processes between the boundary layer and the free troposphere.

2. DESCRIPTION OF THE MODEL

The model is based on a set of coupled continuity equations described symbolically by

$$\frac{\partial c_i}{\partial t} = \text{Advection+Diffusion+Deposition+Emission+Chemistry} \qquad (1)$$

where c_i is the concentration in the air of the i'th chemical species.

The model equations are solved on an Eulerian grid system consisting of 32x32 equally spaced grid points. The distance between the grid points is 150 km.

Air Pollution Modeling and Its Application VIII, Edited by H. van Dop and D. G. Steyn, Plenum Press, New York, 1991

The grid system is compatible with that used by the Lagrangian Long-Range Transport Model developed at the Meteorological Synthesizing Centre - West (MSC-W) of the European Monitoring and Evaluation Program (EMEP) (Eliassen and Saltbones, 1983).

The basic version of the model is two-dimensional, i.e. only horizontal advection and diffusion is considered and all pollutants are assumed to be homogeneously distributed through the whole mixing layer. A three-dimensional version of the transport model has also been developed, but is not yet operational (Zlatev et al., 1983). In the present paper we also present a simple extension of the model to a two-layer version, but it is done with the sole purpose of studying the significance of the exchange processes for the calculated concentrations of ozone.

The mathematical treatment of the model equations is described in several publications and will not be discussed here (Zlatev, 1988; Zlatev et al., 1984). We will only mention that a very efficient pseudo-spectral method is used for space discretization, which leads to a significant reduction of numerical errors arising from space discretization. The pseudo-spectral method for long-range transport modelling was first introduced by Prahm and Christensen (1977).

The model has previously been applied for calculations of transport and deposition in Europe of sulphur and nitrogen compounds (Zlatev et al., 1985a,b; Zlatev and Christensen, 1989). In that model version only a simple, parameterised linear chemistry was used for both sulphur and nitrogen. Although the model results compare very well with measurements, the greatly simplified treatment of chemistry prohibits use of the model for evaluation of control strategies, at least with regard to nitrogen pollution. This restriction, as well as particular interest in photochemical pollution, especially ozone, motivated the extension of the model to include much more complex chemistry.

2.1. The chemistry in the model

The basic structure of the model is the same as in the aforementioned version with the simplified linear chemistry, but instead of only two compounds, (one primary and one secondary pollutant), now 35 chemical species are treated simultaneously.

The chemical scheme used in the model is a slightly modified version of the Carbon Bond Mechanism IV (CBM-IV) (Whitten et al., 1980). It is the same scheme as that used in the Regional Transport Model (RTM III) within the framework of the Dutch-German PHOXA project (Photochemical Oxidant and Acid Deposition Model Application in the Framework of Control Strategy Development) (Pankrath et al., 1988). In the model presented here, production of particulate nitrate is also described, including reaction involving ammonia, as described in Iversen et al. (1989).

The matching of the advection part and the chemical part of the continuity equation creates difficult numerical problem when the chemistry involved is non-linear. Some of these problems and solutions that have also been applied in the present work are discussed in Hov et al. (1989).

The chemistry is the most time consuming element in the model. However, due to the highly efficient and vectorized computer code it is possible on a vector processor to perform long-term model calculations (e.g. one year) with a modest consumption of computer time.

2.2 Emissions

The model requires emissions for sulphur dioxide, nitrogen oxides, carbon monoxide, non-methane hydrocarbons and ammonia.

The data used for sulphur and nitrogen are those reported by Iversen et al. (1989). Emissions of ammonia are from Asman and Janssen (1987) with some recent modifications (W.A.H. Asman - privat communication). The hydrocarbon emissions are from Hov et al. (1986). The emissions of carbon monoxide are assumed to be proportional to emissions of hydrocarbons.

In the model a seasonal variation is imposed for sulphur dioxed, with winter emission twice the summer emission. For nitrogen oxides, hydrocarbons and carbon monoxide a diurnal cycle is imposed with daytime emission being three times larger than the nighttime emission.

The very coarse spatial resolution of the emissions (150km x 150km) is one of the major sources af uncertainty in the model. Also the considerable uncertainty in emissions of hydrocarbons and their split-up must be emphasized. Only anthropogenic hydrocarbon emissions are accounted for in the present version of the model.

2.3. Meteorological data

The meteorological data used in the model runs presented here have been supplied by the MSC-W of EMEP at the Norwegian Meteorological Institute.

The meteorological fields consist of gridded values of horizontal winds (1000 mb and 850 mb), average rainfall and height of mixing layer. The latter is given twice daily (00 GMT and 12 GMT), while the time resolution for all other data is 6 hours.

Presently only the 850 mb wind field is used in the model.

Recently a more extensive set of meteorological data became available, including vertical velocity at mixing layer top, cloud cover, temperature and humidity. Those data are, however, not yet adapted for model calculations.

2.4. Deposition

Both dry and wet deposition are modelled as first order linear removal processes.

The dry deposition velocity for SO_2 and O_3 has a seasonal and diurnal variation but no explicit dependence on atmospheric stability has yet been introduced. A procedure as that outlined in Iversen et al. (1989) would be a significant improvement.

The daytime deposition velocity for ozone vary from 0.15 cm/s in winter to 0.6 cm/s in summer. The nightime deposition velocity is set to one fourth of the daytime value, in order to simulate the decrease in uptake of ozone by plants due to the closing of their stomata at night. Furthermore, the dry deposition of ozone over water is assumed to be negligible. In Eliassen et al. (1982) it is shown that the dry deposition has a significant impact on the calculated ozone concentrations. A realistic parametrization of the dry deposition of ozone is therefore important.

The removal by wet deposition is modelled using species specific scavenging coefficients. Modelling wet removal as a linear process is clearly a strong simplification. Introduction of liquid phase chemistry would be a significant improvement, but presently the most significant limitation is the very coarse timeresolution of precipitation (6 h intervals). Fortunately, the wet removal processes have only a minor significance for the photochemical pollutants, such as NO_2 and O_3.

3. MODELLING EXCHANGE PROCESSES

The approaches presented in the following should be considered only as attempts to explore the different methods used to simulate the exchange of ozone between the mixing layer and the free troposphere and the sensitivity of model results with respect to this process.

Three model versions are presented here:

Basic version. In this case the model is run as a one-layer model and the exchange processes are not simulated at all.

One layer version with linear exchange. In this case the model is still run as a one layer model, but the following linear exchange term is added to the continuity equation (1).

$$v_e \frac{c_f - c}{h_{mix}} \tag{2}$$

Here, v_e is the prescribed exchange velocity, while c_f and c are the concentrations of ozone in the free troposphere and the mixing layer, respectively. The additional exchange term (2) is used only for ozone. The concentration of ozone in the free troposphere is a model input parameter.

Two layer version. In this case the model consists of two layers. The first layer is the mixing layer while the second layer is a reservoir layer. The two layers are considered as parallel layers. Chemistry and advection is calculated in both layers, but deposition is calculated only in the mixing layer. All emissions are assumed to take place only in the mixing layer. This is certainly a very crude simplification, as pollutants from high sources can be emitted above the mixing layer, especially during nighttime. Exchange between the two layers takes place only when the mixing height is rising (from sunrise until 12 GMT) or when the daytime mixing layer colapses after sunset. The diurnal cycle of the mixing height is shown in Fig. 1.

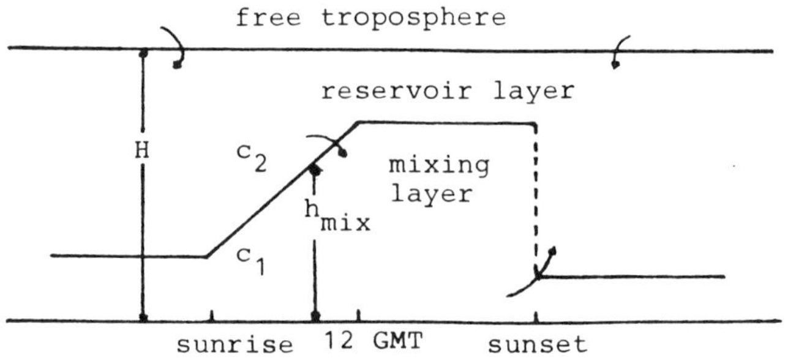

Fig. 1. Illustration of the vertical structure of the two-layer model version.

The amount of all chemical species exchanged between layers is calculated in the following way:

When the mixing height is rising, the concentrations in the mixing layer are changed by

$$\Delta c_1 = (c_2 - c_1) \frac{\Delta h_{mix}}{h_{mix}} \tag{3}$$

where c_1 and c_2 are concentrations in the mixing layer and the reservoir layer, respectively. Δh_{mix} is the increase of the mixing height.

After sunset the reverse process takes place, i.e. the concentrations in the reservoir layer are changed according to the amount of pollution released from the mixing layer. The height H of the top of the reservoir layer is set to 3000 m, which is a rather arbitrary choice.

A linear exchange term is added to the continuity equation for the reservoir layer in order to simulate the exchange of ozone between the free troposphere and the reservoir layer. The exchange velocity and concentration in the free troposphere are input parameters.

A similar multi-layer structure is now widely used in many regional and mesoscale models, among these the Regional Transport Model (RTM III) (Liu et al.,1989) and the Dutch mesoscale model by Egmond and Kesseboom (1985).

The two-layer model version presented here should be considered as a test version only, because such parameters as e.g. the height of the reservoir layer, the exchange velocity and the free troposphere ozone concentrations suffer from the lack of reliable physical justification. A significant drawback is also that all pollutants are assumed to be emitted only in the mixing layer.

4. RESULTS

We show here results of some model experiments performed with data from 1985. The data used for comparison with model results are from the OXIDATE database (Grennfelt et al., 1987). Model calculations are available for the entire year 1985, but only results for July and December 1985 are presented here. For presentation of the detailed results on hourly basis we have selected three measuring stations: Deuselbach and Schauinsland in the Federal Republic of Germany and Wray in the United Kingdom.

4.1. Winter conditions

The model experiments show that concentrations of ozone in winther are almost entirely controlled by the exchange processes. This is illustrated in Fig. 2, where a comparison of measured and calculated ozone concentrations at Wray is shown. In the case of the one-layer version with exchange we use: v_e = 0.8 cm/s and c_f = 30 ppb while for the two-layer version: v_e = 0.8 cm/s and c_f = 48 ppb. Concentrations of ozone calculated without exchange (Fig. 2a) are significantly lower than the observed concentrations. Much better results are obtained with the two-layer version (Fig. 2c). Good correlation in time variation is observed, but the calculated concentrations are here slightly higher than the observed. The calculated concentrations of ozone are very sensitive to the background concentrations.

Concentrations of ozone in winter are strongly correlated with NO_2. Measurements of NO_2 and model results for Wray are shown in Fig. 3.

Comparing concentrations of NO_2 and O_3 in Wray, one can see that they are almost inversely correlated. Maxima in NO_2 correspond to minima in O_3. Comparing the one-layer version without exchange (Fig. 3a) with the two-layer version (Fig. 3c), it is seen that increase of ozone concentrations also leads to increased NO_2 concentrations. This happens when the production of NO_2 is limited by the amount of available ozone, due to the reaction between O_3 and NO. Increase of ozone concentrations can, however, also result in decrease of NO_2 concentrations. This is the case when ozone is no longer the limiting factor for formation of NO_2, but reacts with NO_2 leading to formation of nitrate through the following reaction sequence:

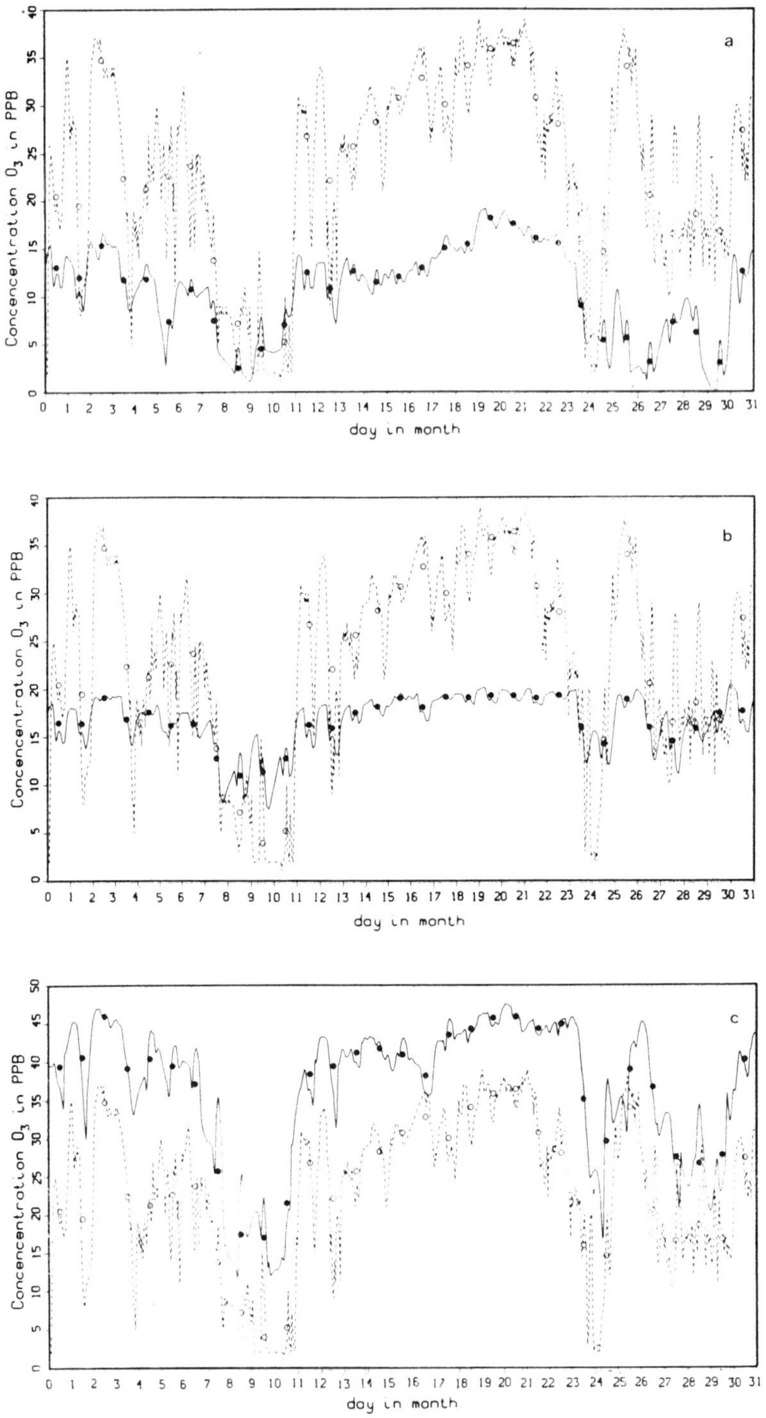

Fig. 2. Calculated and measured hourly ozone concentrations at Wray. December 1985. ——— model results, ------ measurements, • calculated daily averages, o measured daily averages. a) one-layer model without exchange; b) one-layer model with exchange; c) two-layer model.

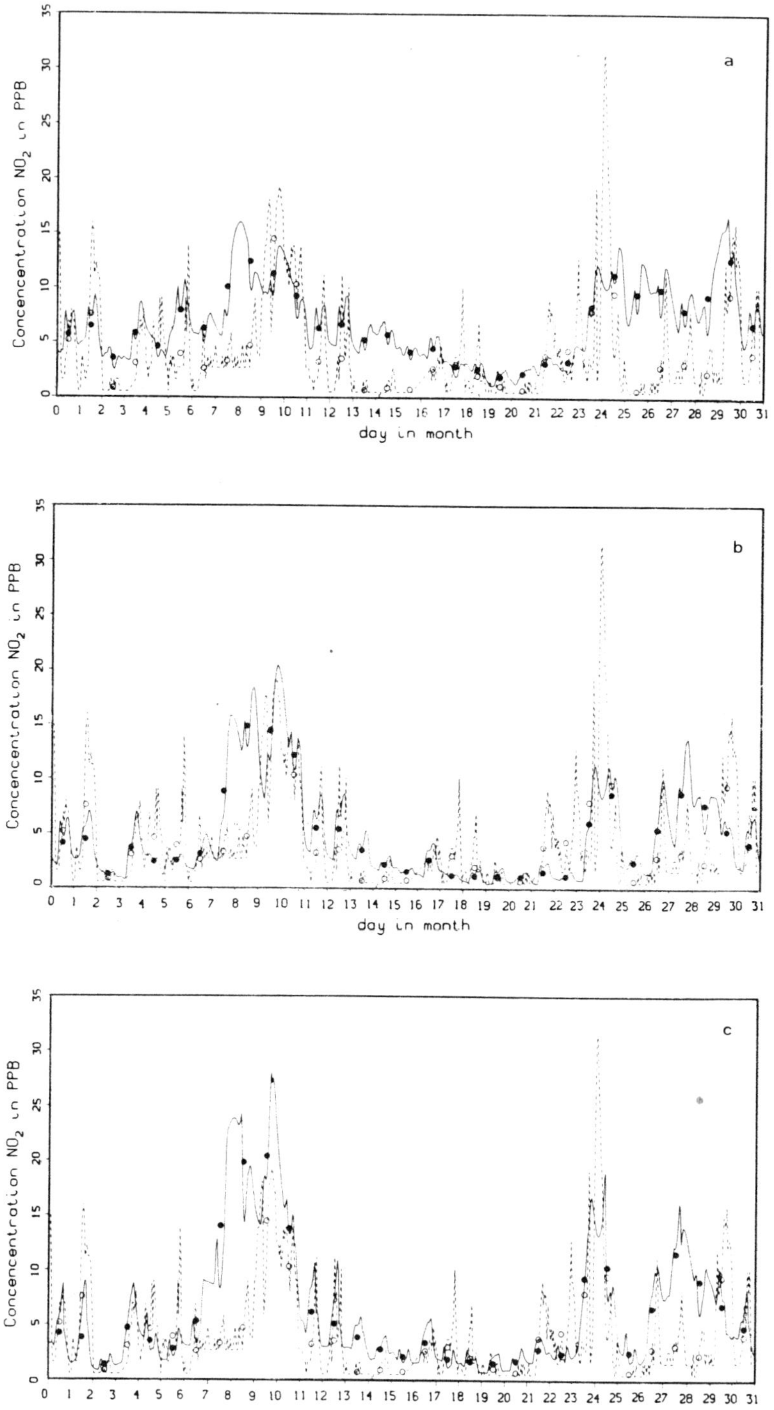

Fig. 3. Calculated and measured hourly NO_2 concentrations at Wray. December 1985. ——— model results, ------ measurements, ● calculated daily averages, o measured daily averages. a) one-layer model without exchange; b) one-layer model with exchange; c) two-layer model.

$$NO_2+O_3 = NO_3+O_2$$
$$NO_3+NO_2 = N_2O_5$$
$$N_2O_5+H_2O = 2NO_3^-+2H^+$$

This reaction sequence is only effective in winter and during the night. In the presence of sunlight, N_2O_5 and NO_3 quickly dissociate and NO_2 is restored. Also in NO-rich areas the reaction between NO_3 and NO will lead to reproduction of NO_2. We have only shown here results from Wray. Other stations reporting hourly values of NO_2 show in general much higher concentrations than predicted by our model. This is particularly valid for such stations as Gent in Belgium, Eibergen in the Netherlands and Bottersford in England. NO_2 concentrations measured at these stations show very great hourly variation. This phenomena and the observed concentrations levels indicate significant influence from local sources.

4.2. Summer conditions

A comparison of measured and calculated ozone with the basic version (without exchange) for July 1985 is shown in Figs. 4a, 5a and 6a. Corresponding results using the one-layer version with exchange and the two-layer version are shown in Figs. 4b, 5b, 6b and Figs. 4c, 5c and 6c, respectively. The exchange velocity is 0.4 cm/s in the case of the one-layer version, and 0.2 cm/s in the case of the two-layer version. In both cases c_f=32 ppb. The general impression is that the exchange processes are not so significant for ozone concentrations in summer. This result might be due to the rather low value of the free tropospheric ozone chosen in this study. A model study by Derwent and Hov (1987) indicates that also in summer the ozone concentrations might be very sensitive towards aloft concentrations. In the case of the examples shown here the photochemical production of ozone in the boundary layer leads to higher concentrations than the prescribed background value and model runs with exchange predict in general slightly lower concentrations of ozone than without exchange.

Results for Wray indicate quite good agreement with the measurements. The model tends, however, to overestimate the concentrations of ozone at night. This is probably because the model predicts average concentrations in the mixing layer, while the measurements represent concentrations in the surface layer. Significant vertical gradients can occur in stable atmospheric dispersion conditions. Stable conditions are expected during clear sky nights, typical for high pressure situations, which usually are connected with summer ozone episodes (Grennfelt et al., 1987).

Overestimation of ozone concentrations in nighttime has almost no consequences for the calculated daytime ozone concentrations. This fact is important when evaluating the model performance. Considering e.g. the adverse effects of ozone on vegetation, only the daytime concentrations are of particular interest.

Model results and measurements of NO_2 in Wray are shown in Fig. 7. Considering the three model versions, one can conclude that the two layer version results in slightly higher concentrations of NO_2 than the one layer version. Diurnal variation of the mixing height, introduced in the two layer version, has the most pronounced effect on NO_2 concentrations. Ozone is seldom the limiting factor for formation of NO_2 in summer, and therefore the concentrations of NO_2 are not sensitive to small changes in ozone concentrations.

In general, the modelled NO_2 concentrations are smaller than the measured concentrations. Contributions from local sources might have a significant effect on the model performance because of the limited spatial resolution of the model. Strong variations of NO_2 concentrations from hour to hour, also observed in Wray, are an indicator of influence from some local sources. Underestimation of NO_2 concentrations in summer is, however, also evident at some remote locations, where the influence from local sources must be

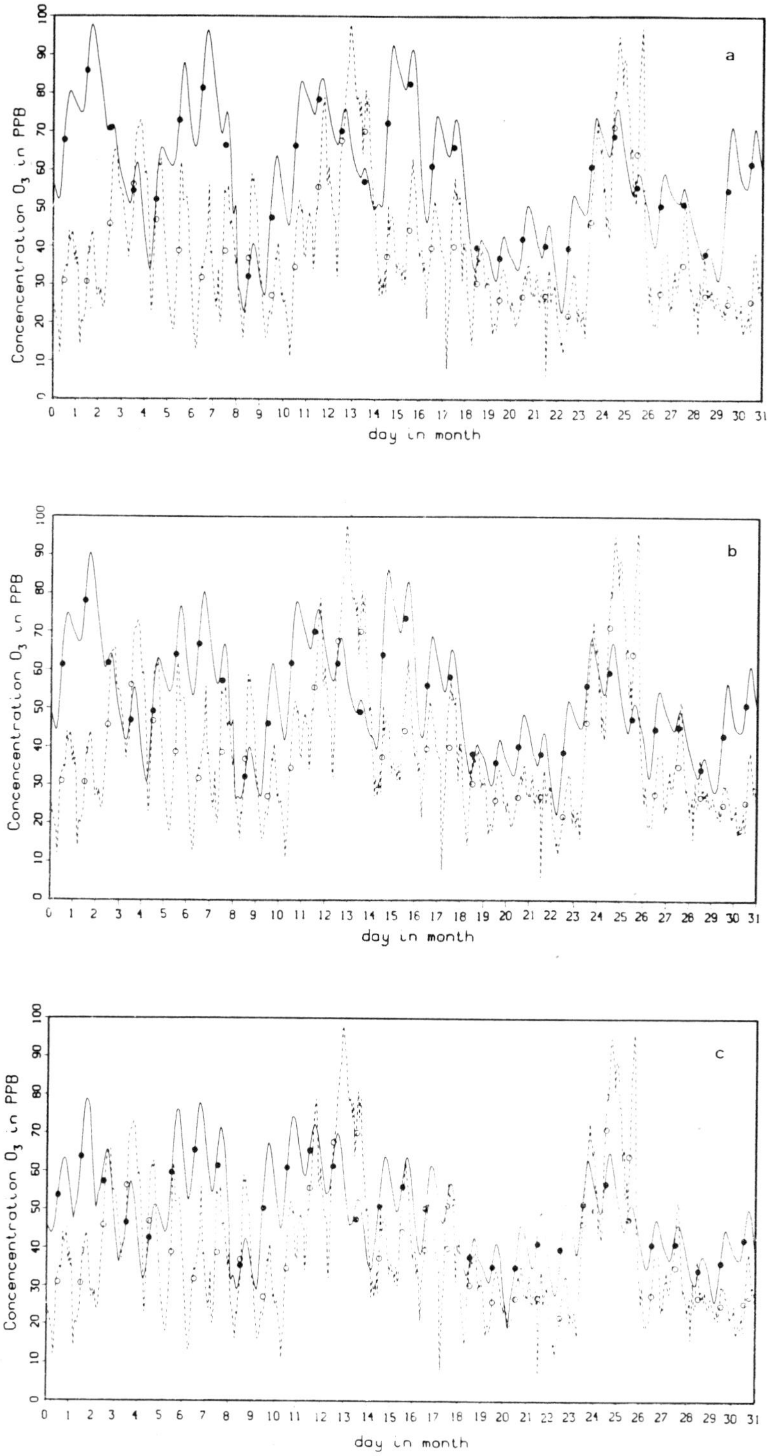

Fig. 4. Calculated and measured hourly ozone concentrations at Deuselbach. July 1985. ——— model results, ------ measurements, • calculated daily averages, o measured daily averages. a) one-layer model without exchange; b) one-layer model with exchange; c) two-layer model.

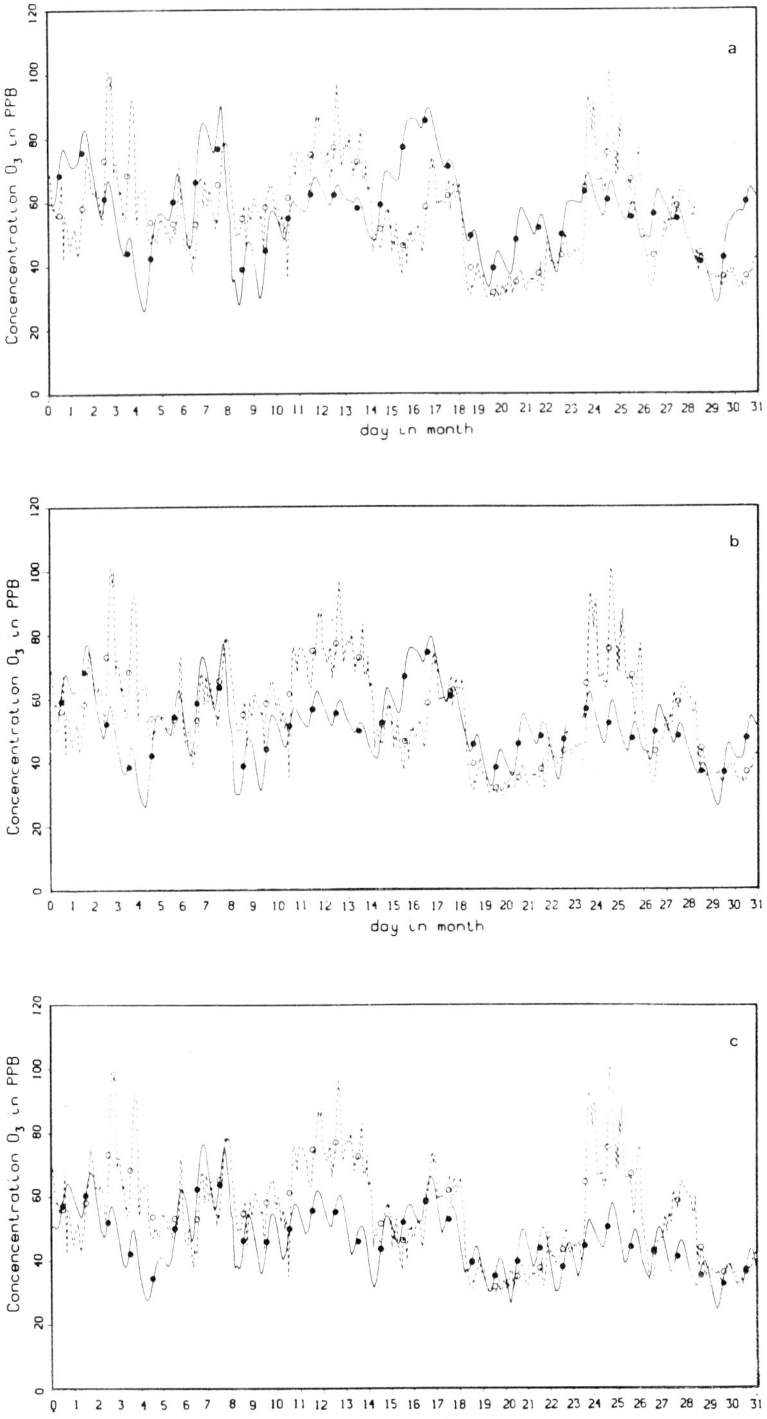

Fig. 5. Calculated and measured hourly ozone concentrations at Schauinsland. July 1985. ——— model results, ------ measurements, • calculated daily averages, o measured daily averages. a) one-layer model without exchange; b) one-layer model with exchange; c) two-layer model.

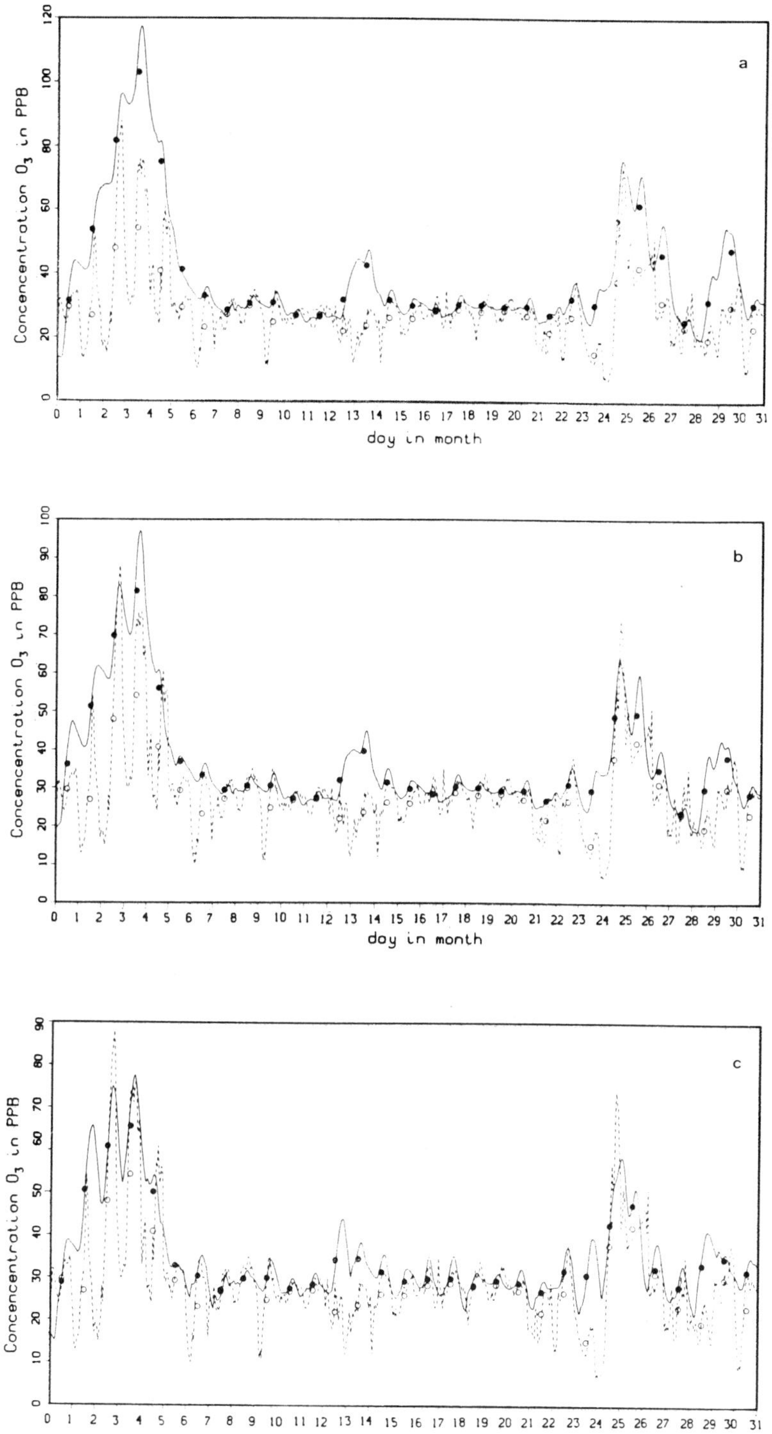

Fig. 6. Calculated and measured hourly ozone concentrations at Wray. July 1985. ——— model results, ------ measurements, • calculated daily averages, o measured daily averages. a) one-layer model without exchange; b) one-layer model with exchange; c) two-layer model.

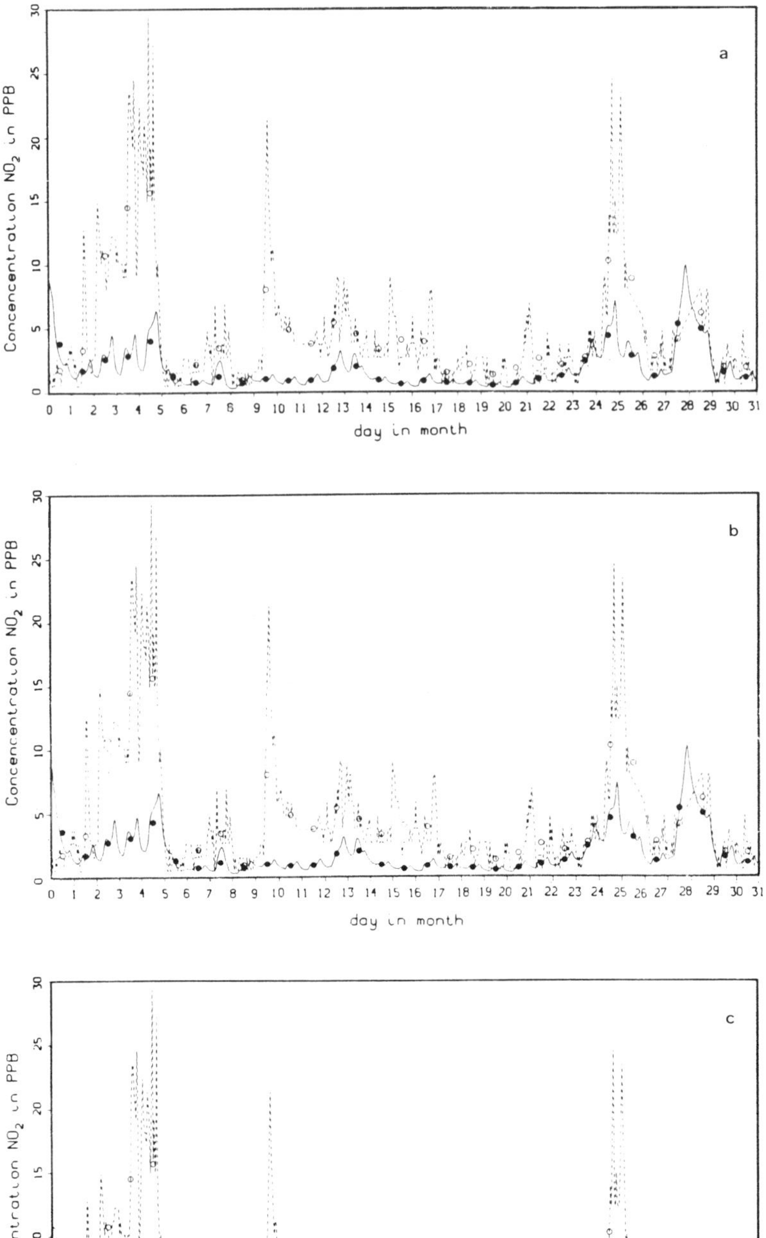

Fig. 7. Calculated and measured hourly NO_2 concentrations at Wray. July 1985. ——— model results, ------ measurements, • calculated daily averages, o measured daily averages. a) one-layer model without exchange; b) one-layer model with exchange; c) two-layer model.

negligible. The process which is most effective in removing NO_2 from the air is the reaction with OH radicals:

$$NO_2+OH = HNO_3$$

A comparison of model results with some available measurements of total nitrate shows that the modelled concentrations are significantly higher than the measured. This might indicate that the concentrations of OH, which controls the formation of HNO_3 and removal of NO_2, are overestimated. OH effectively reacts with CO and hydrocarbons producing HO_2 and organic radicals, which in turn oxidate NO to NO_2. Underestimation of the concentrations of CO and hydrocarbons will thus result in overestimation of the concentration of OH and consequently in a too high removal rate of NO_2. It is well known that the present estimations of hydrocarbons and CO emissions are very uncertain. New model experiments are required in order to quantify the sensitivity of the model to changes in emissions. We believe that this exercise will help to explore the model behaviour and contribute to future improvements.

5. CONCLUSIONS

An Eulerian Long-Range Transport model for photochemical pollutants was presented here. The model is still under development and the results shown are only preliminary. The chemical scheme used in the model is a slightly modified version of the Carbon Bond Mechanism IV. Due to a very effective numerical code, long term model runs are possible and the model behaviour could be examined through a whole year. Results for two months only, December and July 1985, are shown here.

It is shown that the concentrations of ozone in winter are almost entirely controlled by the exchange processes. In summer, this effect is not so pronounced, but this might be due to the prescribed value of the background ozone.

The model reproduces reasonably well the observed ozone concentrations in summer. However, the diurnal variation is not well reproduced. This is probably because the model predicts average concentrations in the mixing layer, while the measurements represent conditions in the surface layer, where significant vertical gradients can occur.

The comparison of model results with measurements of NO_2 is less satisfactory. Many of the measuring stations used in this study are affected by contributions from local sources. Due to the coarse spatial resolution these contributions cannot be resolved by the model.

Concentrations of NO_2 in winter are sensitive to ozone concentrations. In the case when ozone is the limiting factor for formation of NO_2, the increase in ozone concentrations leads to increased NO_2 concentrations. When ozone is not the limiting factor, the increase in ozone concentrations results in more intense oxidation of NO_2 to HNO_3 and thus consequently to decrease of NO_2 concentrations.

The model in general underpredicts concentrations of NO_2 in summer. This is most likely due to overprediction of the transformation of NO_2 to HNO_3 by OH radicals. It is necessary to examine the sensitivity of OH concentrations with respect to emissions of CO and hydrocarbons. This study is in progress.

ACKNOWLEDGEMENTS

This work has been supported by the Nordic Council of Ministers and by the Meteorological Synthesizing Centre - West of EMEP.

REFERENCES

Asman, W.A.H. and Jannsen, A.J., 1987. A long-range transport model for ammonia and ammonium for Europe. Atmos. Env., 21, 2099-2119.

Derwent, R. and Hov, Ø., 1987. The Application of Sensitivity and Uncertainty analysis Techniques to a Photochemical Ozone Model. AERE R 12615, Harwell Laboratory, Oxfordshire, OX11 ORA, England.

Egmond, N.D. and Kesseboom, H., 1985. A numerical mesoscale model for long-term average NO_x and NO_2-concentration. Atmos. Env., 19, 587-595.

Eliassen, A., Hov, Ø., Isaksen, I.S.A., Saltbones, J. and Stordal, F., 1982. A lagrangian long-range model with atmospheric boundary layer chemistry. J. Appl. Meteor., 21, 1645-1661.

Eliassen, A. and Saltbones, J., 1983. Modelling of long-range transport of sulphur over Europe: A two year model run and some model experiments. Atmos. Env., 17, 1457-1473.

Grennfelt, P., Saltbones, J. and Schjoldager, J., 1987. Oxidant Data Collection in OECD-Europe 1985-87 (OXIDATE). NILU OR: 22/87. Norwegian Institute for Air Research, Norway.

Hov, Ø., Becker, K.H., Builtjes, P., Cox, R.A. and Kley, D., 1986. Evaluation of the photooxidants-precursor relationship in Europe. COST 611 - Task Force. Commission of the European Communities. Brussels, Air Pollution Research Report 1.

Hov, Ø., Zlatev, Z., Berkowicz, R., Eliassen, A. and Prahm, L.P., 1989. Comparison of numerical techniques for use in air pollution models with non-linear chemical reactions. Atmos. Env., 23, 967-983.

Iversen, T., Saltbones, J., Sandnes, H., Eliassen, A. and Hov, Ø., 1989. Airborne transboundary transport of sulphur and nitrogen over Europe - Model description and calculations. EMEP MSC-W Report 2/89. MSC-W, Norwegian Meteorological Institute, Norway.

Liu, M.K., Morris, R.E. and Killus, J.P., 1984. Development of a regional oxidant model and application to the Northeastern United States. Atmos. Env., 18, 1145-1161.

Pankrath, J., Stern, R. and Builtjes, P., 1988. Application of long-range transport models in the framework of control strategies: Example of photochemical air pollution. In: "Environmental Meteorology". (Grefen, K. and Löbel, J. eds.), Kluwer Academic Publishers, 585-611.

Prahm, L.P. and Christensen, O., 1977. Long-range transmission of pollutants simulated by a two dimensional pseudospectral dispersion model. J. Appl. Met., 16, 896-910.

Whitten, G.Z. Hogo, H. and Killus, J.P., 1980. The carbon bond mechanism for photochemical smog. Env. Sci. Technol., 14, 690-700.

Zlatev, Z., 1988. Treatment of some methematical models describing long-range transport of air pollutants on vector processors. Parallel Computing, 6, 87-98.

Zlatev, Z., Berkowicz, R. and Prahm, L.P., 1983a. Three dimensional advection-diffusion modelling for regional scale. Atmos. Env., 17, 491-499.

Zlatev, Z., Berkowicz,R. and Prahm, L.P., 1984. Implementation of a variable stepsize variable formula method in the time-integration of a code for treatment of long-range transport of air pollutants. J. Comp. Phys., 55, 278-301.

Zlatev, Z., Berkowicz, R. and Prahm, L.P., 1985a. Studying the sulphur pollution over Europe. MST LUFT-A98. NERI, 4000 Roskilde, Denmark.

Zlatev, Z., Berkowicz, R. and Prahm, L.P., 1985b. Studying the nitrogen pollution over Europe I. Description of the model and some simple tests. MST LUFT-A99. NERI, 4000 Roskilde, Denmark.

Zlatev, Z. and Christensen, J.H., 1989. Studying the sulphur and nitrogen pollution over Europe. In: "Air Pollution Modelling and its Applications VII". (H. van Dop, ed.), Plenum Press, New York, 351-360.

DISCUSSION

A.A.M. HOLTSLAG

How do you derive the boundary layer height in the two-layer version?

J. CHRISTENSEN

The boundary layer height used in the model is derived from radiosoundings. We are aware that the night time stable boundary layer height might be misinterpreted using radiosoundings data. We are planning in the future to relate the night time boundary layer height to wind speed or friction velocity.

R.L. DENNIS

The model underpredicted NO_x in the summer, but not in the winter. Also, only anthropogenic emissions are input into the model. The underprediction in the summer was noted as a puzzle. Emissions inventory work of NAPAP is currently finding that soil NO_x might be a large source in summer, after fertilizer application. It might be worth investigating biogenic sources of NO_x emissions as an explanation of summer NO_x underprediction.

J. CHRISTENSEN

The biogenic emissions of NO_x may contribute to NO_x concentrations in summer, but it is not likely that this contribution is large compared with European anthropogenic emissions.

THEORETICAL PREDICTION OF LOCAL OZONE MAXIMA ASSOCIATED WITH THE PASSAGE OF COLD FRONTS

R. J. Barry and J. S. Chang

State University of New York at Albany
100 Fuller Rd., Albany, NY 12205 USA

Introduction

To better understand the processes controlling the distribution and time evolution of trace chemical species such as ozone within the troposphere, comprehensive and coupled transport, transformation, and deposition models have been developed. An example of this kind of modeling system is the Regional Acid Deposition Model Version 1.0 (RADM 1.0) developed by Chang *et al.* (1987). Though specifically produced to study the relationship between sources and receptors of acid deposition, RADM 1.0 is equally well suited for studying the processes potentially responsible for the evolution of ozone in the troposphere. Many of the physical and chemical processes that affect ozone are explicitly included in RADM 1.0. These include long range transport, turbulent diffusion, both gas and aqueous phase chemistry, cloud scavenging and mixing, and dry deposition.

During a 72 hour simulation run of RADM 1.0 over the eastern United States an ozone maximum was observed to form along a surface cold front and then propagate eastward. To understand why this is occurring we have analyzed the first 48 hours of this simulation for the mechanisms that were acting within the model to produce and maintain this feature. The benefit of this analysis is it points out potential means by which localized ozone maxima are created and sustained within the troposphere. With this information measurement programs and observational data analysis can be tailored towards verifying that this feature does occur in the real atmosphere, and if observed, that the mechanisms acting within the model are consistent with those occurring in nature.

Model Overview

RADM 1.0 is an Eulerian transport, transformation, and deposition model developed to study the relationship between sources and receptors of acids by simulating episodic chemical concentrations and dry and wet deposition (Chang *et al.*, 1987). RADM 1.0 has a 30 by 30 horizontal lattice with each grid cell approximately 80 x 80 km^2 in size. The horizontal domain of RADM 1.0 covers the eastern half of the United States, southeastern Canada, and part of the western Atlantic Ocean. In the vertical RADM 1.0 has six layers. The approximate heights of these layers are: 0 - 0.15 km, 0.15 - 0.55 km, 0.55 - 1.3 km, 1.3 - 3.8 km, 3.8 - 8.5 km, and 8.5 - 15 km. Meteorology for RADM 1.0 is produced separately by a hydrostatic, primitive equation meteorological model (Anthes and Warner, 1978).

RADM 1.0 operates by solving a set of chemical conservation equations of the form

$$\frac{\partial C}{\partial t} = - \nabla \cdot VC + \nabla \cdot (K_e \nabla C) + \left(\frac{\partial C}{\partial t}\right)_{chem} + \left(\frac{\partial C}{\partial t}\right)_{cld} + \left(\frac{\partial C}{\partial t}\right)_{dry\ dep} + E\,, \qquad (1)$$

Air Pollution Modeling and Its Application VIII, Edited by H. van Dop and D. G. Steyn, Plenum Press, New York, 1991

where C is the chemical species volume mixing ratio, V is the three-dimensional velocity vector, K_e is the eddy diffusivity used to represent the subgrid-scale fluxes of trace species due to turbulent motions other than those due to clouds, $(\partial C/\partial t)_{chem}$ represents production and destruction due to gas phase chemical interactions, $(\partial C/\partial t)_{cld}$ is the change in concentration due to cloud effects (i.e. subgrid-scale vertical mixing, aqueous phase chemistry, and scavenging), $(\partial C/\partial t)_{dry\ dep}$ is the change in concentration due to dry deposition, and E is the emissions rate. RADM 1.0 solves (1) for 24 transported species using an operator splitting technique. An 11 additional species are computed by assuming a local chemical steady state exists.

Description of the Phenomenon

The data analyzed in this paper is from a 72 hour model simulation of the time period starting 00 UTC on 22 April 1981. This was a simulation of the fourth period of the Oxidant and Scavenging Characteristics of April Rains (OSCAR IV) field measurement experiment (Benkovitz *et al.*, 1984). The OSCAR IV period was characterized by a well developed wave cyclone moving through the RADM 1.0 domain. At the initial hour of the simulation a ridge of high pressure was running from central Ontario, through the eastern Great Lakes and into the southeastern United States. The wave cyclone of interest had yet to enter the RADM 1.0 domain. Within the lowest layer of the model ozone was maximized in the Southern States and along the Appalachian Mountains.

At approximently 11 UTC on 22 April the cold front and cyclone center were just beginning to enter the model domain. Precipitation associated with the frontal system was occurring from Minnesota southward to Mississippi and Alabama. This area also extended into central Illinois and Indiana. Model generated cloudiness was rather extensive in the southerly flow out ahead of the surface cold front. Ozone was increasing along the western edge of the domain east of the cold front from western Iowa southward to central Mississippi and southern Alabama . Maximum values of ozone in this area reached 35 to 40 ppb. This was an increase of approximately 20 ppb over the past twelve hours.

During the day on 22 April the wave cyclone continued to moved eastward through the domain. Ozone maxima with magnitudes of up to 50 ppb were forming in areas of precipitation and cloudiness located within the warm sector. These maxima were advancing northward in the southerly flow out ahead of the cold front. The ozone maximum along the

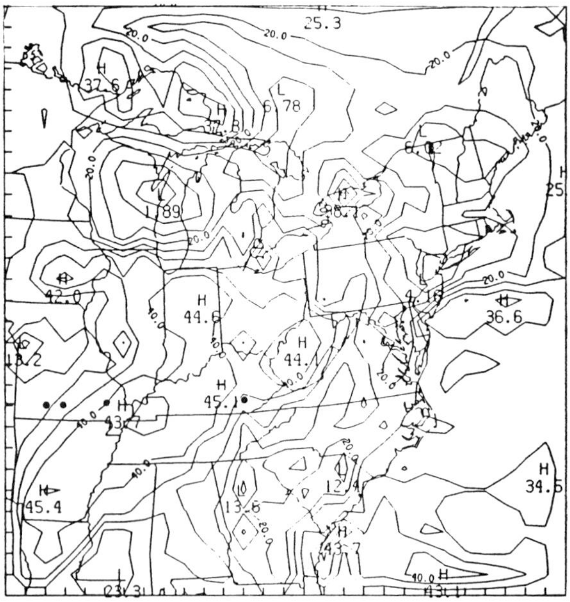

Figure 1. Ozone concentration (ppb) for the lowest model layer at 11 UTC 23 April 1981. Contours are every five ppb. The black dots indicate grid locations used for the budget analysis presented later in this paper.

western edge of the domain remained relatively stationary. By the end of the day at (00 UTC 23 April) the wave cyclone was located such that the cold front extended from a low center in northeastern Minnesota and ran southward through Wisconsin, western Illinois, southeast Missouri, and northwestern Arkansas.

On the night 23 April high ozone air was observed to accumulate along the eastward moving surface cold front. By 11 UTC, with the storm center in northern Michigan and the frontal zone running southward through Indiana, the Ohio River Valley, then into Arkansas, a distinct ozone maximum associated with the surface cold front was evident (Fig. 1). The magnitude of the maximum was about 45 ppb. This was a decrease in the high ozone values observed the previous day at the surface. On the previous day it was also observed that certain isolated surface maxima of ozone were forming in areas of precipitation. During the night this did not seem to be the case as the ridge of maximum ozone was beginning to fall slightly behind the precipitation.

During the day on 23 April the low center had moved only slightly eastward. On the other hand, by the end of the day the cold front had pushed well eastward into the area of the Appalachian Mountains . The ozone maximum maintained its relative position with the eastward moving cold front. High ozone air was also being advected northward along the cold front wrapping into the cyclone center. During this time the precipitation pattern and maximum ozone were once again almost exactly coincident. The magnitude of the ozone maximum along the frontal zone had increased, ranging from approximately 54 to 58 ppb at its highest value.

Analysis

Analysis of the physical and chemical processes during this event points to two mechanisms that are significant in the formation and maintenance of the ozone maximum along the surface cold front. The first of these mechanisms was the advection of high ozone air into the frontal zone. The other mechanism was the limitation of the destruction of ozone by dry deposition in the vicinity of the surface front, or more specifically in regions of precipitation.

Advection by model resolved motions

During the simulation gas phase ozone maximized along the surface cold front between

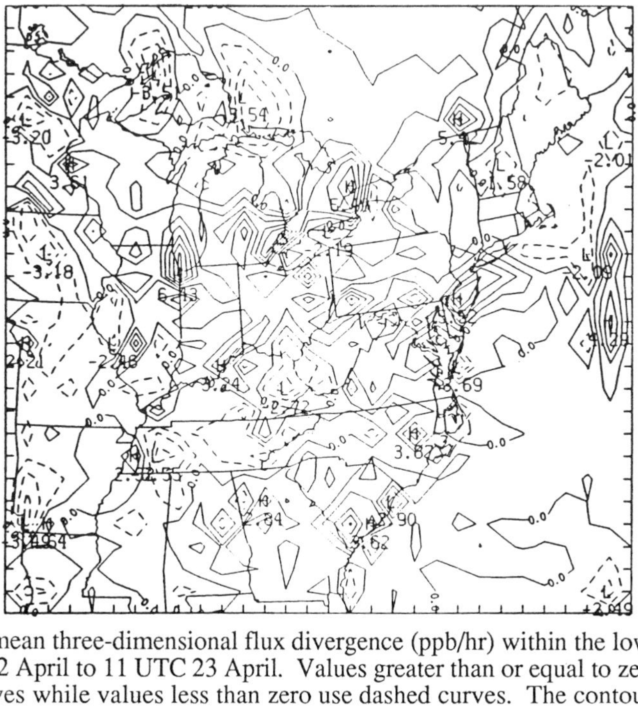

Figure 2. The mean three-dimensional flux divergence (ppb/hr) within the lowest model layer from 23 UTC 22 April to 11 UTC 23 April. Values greater than or equal to zero are contoured using solid curves while values less than zero use dashed curves. The contour interval is one ppb/hr.

approximately 23 UTC on 22 April and 11 UTC on 23 April. These being nighttime hours only chemical destruction of ozone is possible. Without a chemical source, some transport process (large scale transport, turbulent mixing, cloud mixing) and/or the localized limitation of the loss of ozone relative to surrounding areas (chemical destruction, dry deposition) must have been responsible for the development of this relative maximum in ozone along the front.

The analysis of large scale transport is based on the three-dimensional flux divergence of ozone. During the twelve hour period during which we saw ozone accumulating along the front the flux divergence was calculated hourly for every grid location and then an average for each grid location was obtained for the time period.

Figure 2 is a plot of the twelve hour mean three-dimensional flux divergence from 23 UTC 22 April to 11 UTC 23 April within the lowest model layer. From this figure we see advection bringing high ozone air into the region of the surface cold front (Indiana through the Ohio River Valley and into eastern Arkansas). These time averaged rates of increase in ozone concentration along the front are as high as 6 ppb/hr but generally range from 1-2 ppb/hr. Advection of high ozone air into the region of the frontal zone is therefore one mechanism contributing to the organization of this feature.

Having shown there is a positive flux divergence into the region of the frontal zone, we need to investigate whether this local increase was due to lateral entrainment or the downward advection (i.e. stratospheric injection) of high ozone air by motions on this scale. Figure 3 is a vertical cross-section of ozone concentration and vertical velocity at 5 UTC on 23 April. This cross-section runs through southern Missouri, Kentucky, Virginia, and northern North Carolina. On the large scale rising motion is occurring above the surface ozone maximum therefore eliminating the downward vertical flux of high ozone air into the area of the surface maximum. As the frontal zone moves into the model domain post-frontal subsidence can be observed but this downward motion cannot be made responsible for the low level ozone already present near the surface. The magnitude of this subsidence ranges up to approximately 5 cm s^{-1}. Even with this maximum value of downward motion a parcel of air of stratospheric origin would take approximately two days to traverse the troposphere. Based on the structure

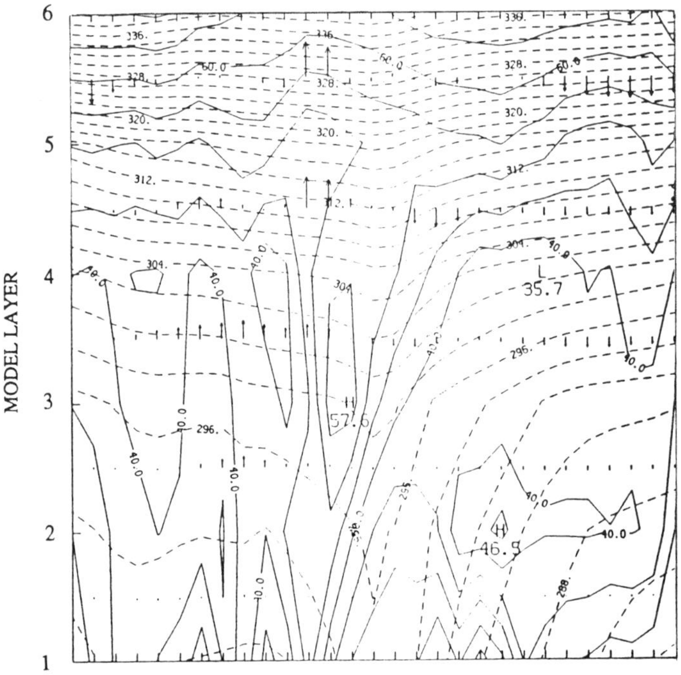

Figure 3. A vertical cross-section of ozone concentration (ppb) and vertical motion (m s^{-1}) through southern Missouri, Kentucky, Virginia, and northern North Carolina at 5 UTC 23 April. Ozone is contoured every five ppb with solid curves and vertical motion is represented by vertical vectors.

and magnitude of the model simulated vertical motion stratospheric injection can be ruled out as a mechanism for the formation of this maximum. Instead we see convergence and deformation associated with frontogenesis acting to entrain and stretch preexisting maxima into and along the surface front.

Dry deposition

Dry deposition is calculated within the model by multiplying the ozone concentration of the lowest model layer by a spatially and temporally varying deposition velocity, v_d. This product is then divided by the thickness of the lowest model layer (Sheih *et al.*, 1986). Using this concept of a deposition velocity, processes affecting dry deposition are represented. Specifically, the degree of turbulent mixing within the surface layer, molecular diffusion within the interfacial layer, and the ability of the underlying surface to allow a chemical species to be dry deposited are parameterized.

Within RADM 1.0 precipitation limits the sink of ozone to the surface during the day. Figures 4 is a plot of the rate of loss of ozone due to dry deposition at 17 UTC on 22 April. In precipitating regions, such as over Mississippi and Alabama, loss is limited to values less than 1 ppb/hr. In non-precipitating regions ozone is being depleted at rates as high as 8 ppb/hr. This lack of a sink to the surface coupled with some source of ozone (e.g. photochemistry, downward transport through cloud or turbulent mixing) is creating low level ozone maxima. It is this high ozone air that is later entrained into the cold frontal zone forming the frontal maximum later that night.

During the day the lack of dry deposition in precipitating areas within the model aids in the formation of local ozone maxima. At night this is not the case. Instead it is observed that the greatest amount of dry deposition occurs in the regions of highest ozone concentration. Precipitation still limits dry deposition to the surface. But because such effects as turbulent mixing and photosynthetic activity decrease domain wide during the night, deposition velocities in non-precipitating areas are reduced to values similar to those in precipitating regions. Dry depostion therefore becomes greatest in high ozone areas limiting the formation of local ozone maxima at night.

On the second day of the simulation the effect of limited dry depostion in precipitating

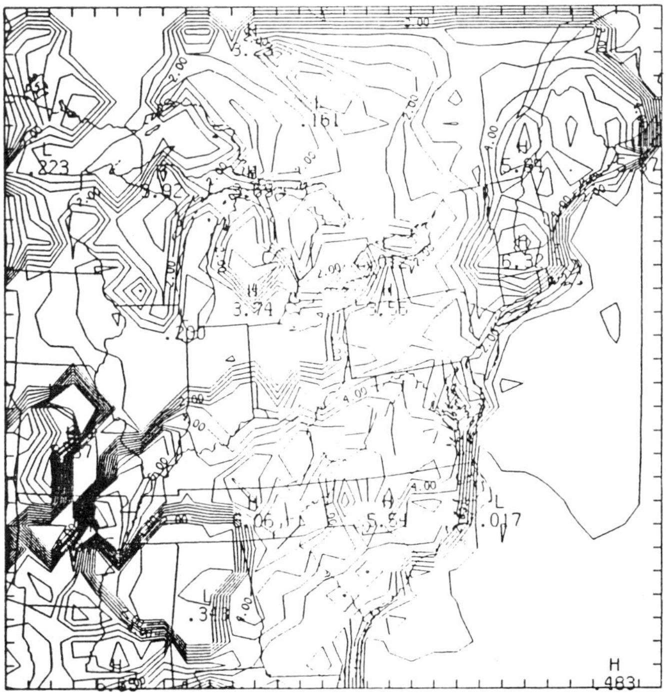

Figure 4. The rate of loss of ozone due to dry deposition (ppb/hr) across the domain at 17 UTC 22 April. Contours are every .5 ppb/hr.

regions once again played a positive role in this phenomenon. The ozone maximum was almost exactly coincident with the precipitation associated with the cold front. By not allowing ozone to be lost to the surface, the ozone maximum is maintained in this precipitating region and therefore along the front.

Ozone Budget Analysis

During the RADM 1.0 simulation the hourly change in ozone due to each relevent physical or chemical process as well as the overall change were analyzed for four different grid locations at the surface at four different times. The species conservation equation for ozone and these data are contained in Table 1. The grid points analyzed are indicated on figure 1.

During the first night of the model run cloud and turbulent mixing transported high ozone air from aloft downward to the lower model layers. This was generally along the western edge of the domain just ahead of the surface front where increased cloudiness and enhanced shear were found. High ozone air to the south was also being advected northward in the southerly flow ahead of the front. This is evident in the data presented in Table 1. The grid volume in southwest Missouri at 5 UTC on 22 April is representative of what is occurring along the western edge of the domain during the first night of the simulation. Overall ozone is increasing at a rate of 1.6 ppb/hr. Contributing to this increase as mentioned previously are advection of high ozone air from the south (+1.5 ppb/hr), cloud mixing (+0.9 ppb/hr), and vertical diffusion (+0.3 ppb/hr). Advection and cloud mixing are the dominant mechanisms acting to increase ozone at this time. Chemical destruction and dry deposition are limiting the increase of ozone. It is not clear that the ozone that built up along the western edge of the domain actually contributed to the maximum that we saw approximately twenty four hours later, but these transport processes associated with the wave cyclone would contribute to the correlation of the low level ozone maximum and the frontal zone later in the simulation.

With daylight photochemical production began to take place. In areas of precipitation, where the sink to the surface is limited, ozone began to accumulate forming local ozone maxima, generally in the west and southwest portion of the domain. The source of this ozone was in situ photochemical production (though this was limited by the presence of clouds) and the continued downward mixing of high ozone air by both clouds and smaller scale turbulence. The grid volume located in south central Missouri is ahead of the surface front at 17 UTC on 22 April. Ozone is increasing at a rate of 1.5 ppb/hr with chemical production (+1.1 ppb/hr), cloud mixing (+0.7 ppb/hr), and vertical diffusion (+0.5 ppb/hr) contributing to this increase. What is not reflected in this table is the limitation of the sink of ozone to the surface by precipitation wetting the ground. If this grid cell was not within an area of precipitation, dry deposition would be larger than the observed loss rate of 1 ppb/hr and ozone would not be accumulating as rapidly, if at all.

As the wave cyclone moved towards the center of the domain on the second night of the model run the local ozone maxima formed previously were horizontally entrained into and stretched along the surface cold front. The large scale motions associated with frontogenesis (i.e. convergence, deformation) were responsible for ozone being maximized along the frontal zone. Enhanced cloud and turbulent mixing along the frontal zone at this time also continued to be a source of ozone. Data from Table 1 illustrates these points. The grid cell located in southeast Missouri is generally in the area where the maximum organized into the feature we observed. At 5 UTC on 23 April ozone is increasing at a rate of 2.0 ppb/hr at this location. Contributing to this increase of ozone is advection (+1.4 ppb/hr), cloud mixing (+1.1 ppb/hr), and turbulent mixing (+0.3 ppb/hr) Both chemical destruction and dry deposition are depleting ozone at this point.

On the second day of the simulation the ozone maximum maintained itself along the surface frontal zone almost exactly coincident with the precipitation pattern. In this region ozone was introduced to low levels by cloud mixing, turbulent mixing, and photochemical production, but it was the limitation of the sink to the surface by the precipitation along the front that allowed the ozone maximum to maintain itself in this area. At 17 UTC on 23 April the grid cell in southeast Kentucky is located in the middle of the ozone maximum along the surface cold front. Overall ozone is increasing at a rate of 2.1 ppb/hr. The two major sources at this time are chemical production and cloud transport (+2.2 ppb/hr and +1.9 ppb/hr respectively). Once again the limitation of the sink of ozone to the surface is not made totally evident until the rate of loss at this location is compared to locations where precipitation is not occurring.

Summary

This analysis involved the examination of a 72 hour simulation using RADM 1.0. During this simulation ozone was observed to maximize along a surface cold front during the second night of the simulation and then propagate eastward. This study has given some insight into potential mechanisims operating in the natural environment that could be working to produce and/or maintain high concentrations of ozone. During the first day of the model run ozone maxima were formed and maintained in areas where dry deposition was limited by precipitation wetting the ground. Photochemical production was the major contributor of ozone but cloud mixing of high ozone air downwards and vertical diffusion also supplied ozone to these areas. On the second night of the simulation with the wave cyclone approximately in the center of the domain the ozone maxima formed previously were entrained into the area of the frontal zone. Enhanced cloudiness along the cold front and increased vertical diffusion relative to areas away from the front also contributed to the increase of ozone along the front. During the second day of the simulation the ozone peak maintained itself along the front where precipitation was occurring. With dry deposition being limited, photochemical production, cloud mixing, and vertical diffusion served as the sources of ozone allowing the maximum to maintain itself and in fact increase in amplitude. These processes identified as being responsible for the accumulation of ozone at low levels within the model could also be operating in a similar manner in the lower troposphere. The extent to which these processes are operating and the observational manifestation of this phenomenon would of course be expected to vary greatly form episode to episode depending on the initial conditions of ozone in the vicinity of the frontal zone.

As with any model of the natural environment there is of course a degree of uncertainty in the results acheived. Therefore it is necessary to confirm that these mechanisms are operating in the real atmosphere in a similar manner to that which was observed during the simulation. This can be done through well planned measurement programs involving both chemical and meteorological measurements designed to potentially isolate the phenomenon in question. Observations made during the OSCAR measurement program do not allow for the isolation of ozone features on the scale that we observed during this model simulation. For this reason using observations to confirm that this particular phenomenon did occur cannot be accomplished . It has been shown that RADM 1.0 is able to produce realistic surface ozone concentrations. Middleton and Chang (1990) have compared surface ozone concentrations from the RADM 1.0 OSCAR IV simulation with data from selected ozone observation sites. They found the model data and observations were in agreement to better than a factor of two, and 50% of the locations were in agreement to a factor of 1.1 or better.

Acknowledgements - Although the research described in this article has been funded by the U.S. Environmental Protection Agency under contract 68D80016 to the Research Foundation of the State University of New York, the results have not been subject to the Agency's peer and policy review and therefore do not necessarily reflect the views of the Agency and no official endorsement should be inferred.

References

Anthes, R. A., and T.T. Warner, 1978: Development of Hydrodynamical Models Suitable for Air Pollution and Other Mesometeorological Studies. *Mon Wea. Rev.*, 106, 1045-1078.

Benkovitz, C. M., V. A. Evans, J. L. Tichler, J. V. Hayes, and K. J. Norden, 1984: Oxidizing and Scavenging Characteristics of April Rains - OSCAR Data Report. BNL 51881. Brookhaven National Laboratory, Brookhaven, NY. 240 pp.

Chang, J. S., R. A. Brost, I. S. A. Isaksen, S. Madronich, P. Middleton, W. R. Stockwell, and C. J. Walcek, 1987: A Three Dimensional Eulerian Acid Deposition Model: Physical Concepts and Formulation. *J. Geophys. Res.*, 92, 14681-14700.

Middleton, P., and J. S. Chang, 1990: Analysis of RADM Gas Concentration Predictions Using OSCAR and NEROS Monitoring Data. *Atmos. Environ.*, in press.

Pleim, J., A. Venkatram, and R. Yamartino, 1984: ADOM/TADAP Development Program, Volume 4, The Dry Deposition Module. ERT Document No. P-B980-520, 98 pp.

Seinfeld, J. H., 1986: Atmospheric Chemistry and Physics of Air Pollution. John Wiley and Sons, 738 pp.

Sheih, C. M., M. L. Wesley, and C. J. Walcek, 1986: A Dry Deposition Module For Regional Acid Deposition. EPA/600/3-86/037 U.S. Environmental Protection Agency, 63pp.

Warneck, P., 1988: Chemistry of the Natural Atmosphere. Academic Press, 757 pp.

DEVELOPMENT OF AN URBAN GRID MODEL FOR THE SANTIAGO, CHILE REGION

P. K. Karamchandani, A. Venkatram and K. Kashanian

ENSR Consulting and Engineering
1220 Avenida Acaso
Camarillo, CA 93012

M. Tubino

Intec Chile
Casilla 19002-Correo 19
Santiago de Chile

and P. Ulriksen

Facultad de Ciencias
Fisicas y Matemáticas
Universidad de Chile
Santiago de Chile

INTRODUCTION

The city of Santiago frequently experiences episodes of high air pollutant concentrations, particularly during the winter months. These episodes are a consequence of the unique geographic setting of the Santiago region and emissions of large quantities of pollutants emitted by various surface sources such as diesel buses, automobiles, and wood burning, and point sources such as power plants and smelters. Concern about present and future air quality in the Santiago metropolitan region has motivated the Intendencia of the city of Santiago to sponsor the development of a state-of-the-science Air Quality Modeling System.

The modeling system consists of a set of air quality models to be used for air quality evaluation and planning in the region. The principal model in the system is a 3-dimensional terrain-following Eulerian model which incorporates a state-of-the-art gas-phase chemical mechanism to simulate the photochemical production of oxidants. The model predicts the transport, deposition and chemistry of sulfur and nitrogen oxides, reactive organic compounds, carbon monoxide, oxidants such as ozone and PAN, and particulates. A key component of the modeling system is a diagnostic wind field generation scheme to drive the grid model. The diagnostic wind field scheme also operates in terrain-following coordinates. This paper describes the special features of the modeling region, the diagnostic wind field preprocessor and the Eulerian model. Results from preliminary evaluation studies conducted with the model will be described in a subsequent paper.

SPECIAL FEATURES OF THE SANTIAGO AREA

The Santiago region is in a long north-south valley, with an eastern boundary formed by the Andes mountain range which exceeds 6000 meters in elevation. This range effectively blocks the flow of pollutant-laden air to the east. The Maipo River Valley opens up the city to marine influences of the Pacific Ocean at the western boundary. Figure 1 shows a map of the Santiago region. Most of the pollutants (SO_2, NO_x, CO, organic compounds, and particulates) are emitted in downtown Santiago and the surrounding suburbs. A large number of small area sources (such as diesel buses and automobiles) contribute to the pollution in these areas. Figure 1 shows the approximate area covered by downtown Santiago. In addition, a molybdenum smelter, located approximately southwest of the Santiago region (see Figure 1), is the largest SO_2 point source in the region.

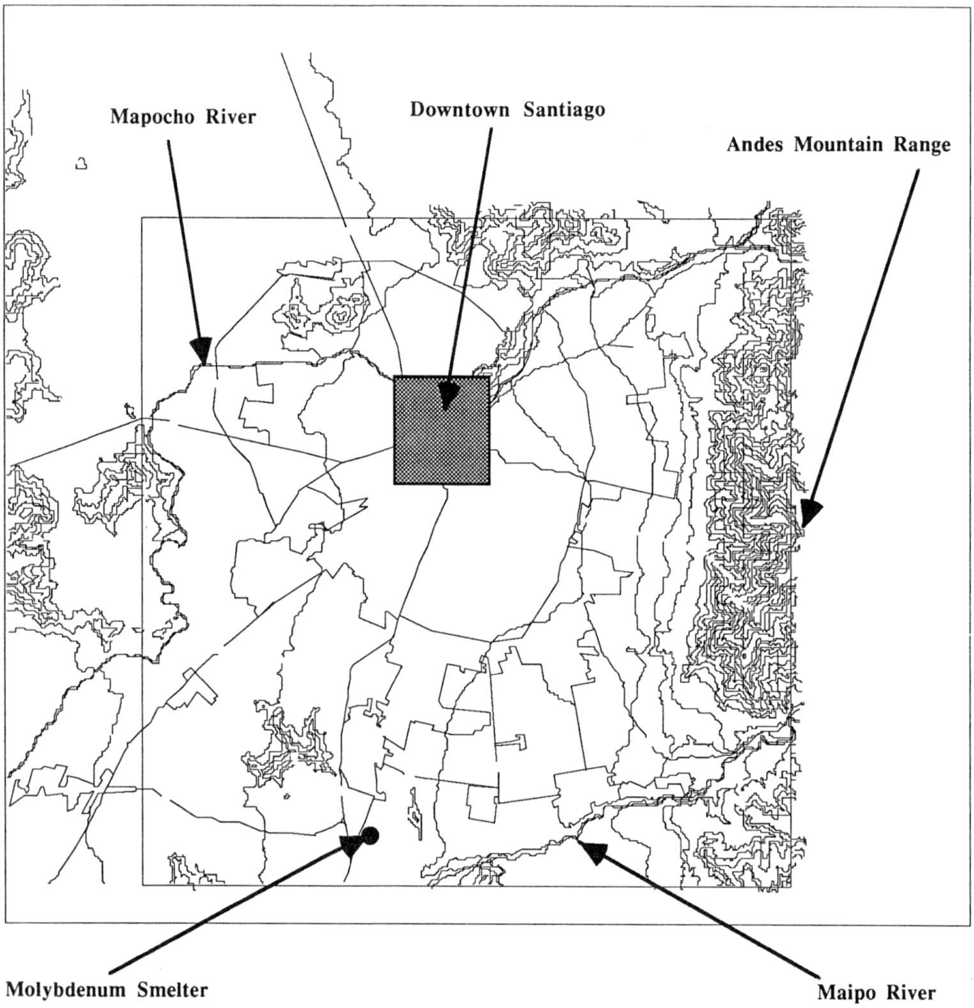

Figure 1. Map of the Santiago region. The inner box shows the boundaries of the modeling domain.

Meteorological conditions in Santiago are strongly dependent on space and time. Data reported by Ulriksen (1980) show that winds are generally calm at night and early morning hours and have their greatest values in the late afternoon and early evening. Afternoon winds are much higher in summer as compared to winter. The daily cycle of wind directions is also very pronounced. In winter, prevailing winds at night are from the east, representing drainage flow from the Andes. Conversely prevailing winds at mid-afternoon are from the west, indicating up-slope flow and a possible synoptic component. Overall, a system of mountain-valley winds is quite evident. Within the city, different characteristics are exhibited due to local features and the urban heat island effect. Winter months are characterized by relatively low mixing heights. These conditions restrict dispersion and lead to a build-up of pollutants in the region.

GRID SYSTEM AND GOVERNING EQUATIONS

The modeling domain used in the analysis encompasses a 34 x 34 km^2 area surrounding metropolitan Santiago, with a horizontal resolution of 2 x 2 km^2. The modeling domain is defined by the inner box shown in Figure 1. The vertical extent of the domain is 4 km, which accommodates both the high terrain in the region as well as the mixed layer heights during summer. The actual vertical coordinate system is terrain following. Because the largest vertical gradients of the important variables such as concentrations, turbulence, and winds occur near the ground, the vertical resolution of the grid system used is finer near the ground than aloft. The terrain following coordinate system requires the formulation of a vertical coordinate ρ, defined by,

$$\rho = [z - h(x,y)]/\pi(x,y) \tag{1a}$$

$$\pi(x,y) = H - h(x,y) \tag{1b}$$

where x, y, z is the Cartesian coordinate system, h(x,y) is the height of the terrain, and H is the specified top of the model domain.

In the x, y, ρ system, the mass conservation equation that forms the basis for the transport model can be written as

$$\frac{\partial \hat{C}}{\partial t} + \frac{\partial}{\partial x}(u\hat{C}) + \frac{\partial}{\partial y}(v\hat{C}) + \frac{\partial}{\partial \rho}\left(\frac{W}{\pi}\hat{C}\right) = \frac{\partial}{\partial \rho}\left(\frac{K}{\pi^2}\frac{\partial \hat{C}}{\partial \rho}\right) + S\pi + R\pi \tag{2a}$$

$$\hat{C} = \pi C \tag{2b}$$

$$W = w - u\frac{\partial h}{\partial x}(1-\rho) - v\frac{\partial h}{\partial y}(1-\rho) \tag{2c}$$

where C is the pollutant concentration, K is the vertical diffusivity, S is a source term, R is the chemical reaction term, and u, v, and w are the velocity components in the Cartesian coordinate system x, y, and z. W is the vertical velocity component in the terrain following coordinate system, x, y, and ρ. Note that horizontal diffusion is ignored in the derivation of Equation (2a). The justification for this assumption is that the numerical or artificial diffusion associated with solving the horizontal advection terms of the mass-conservation equation is usually much larger than the horizontal diffusion associated with the unresolved wind field.

DIAGNOSTIC WIND FIELD MODULE

The diagnostic wind field module generates mass-consistent three-dimensional gridded wind fields in a terrain-following coordinate system using available surface and upper air data. The wind fields are generated in two steps. In the first step, observed wind data are horizontally and vertically interpolated to the grid system. In the second step, variational calculus is used to adjust the interpolated wind field to satisfy the mass continuity equation.

The horizontal interpolation scheme used by the module is an improved inverse-distance-squared (IDS) method suggested by Ross and Smith (1987). This scheme is specially suitable for complex terrain. It improves upon the standard IDS method by incorporating the mean of the observed wind field into the interpolation routine via a prescribed radius of influence. Incorporation of the radius of influence into the interpolation routine is a means by which the randomness or uniformity of the observed wind field is taken into account in the interpolation process. If the observed wind field is highly uncorrelated, more weight is given to the mean of the observations in the interpolation process. This prevents the extrapolation of a single observation — which, due to the randomness of the observations, may not be representative of the prevailing wind field — into remote regions. On the other hand, if the observations are uniform, the interpolation scheme reduces to the standard IDS method.

The vertical interpolation scheme employed by the module depends on the number of locations where vertical soundings are available. If vertical soundings are available at only one location, the scheme assumes that the slopes of the vertical profiles are the same throughout the grid at each hour. The slopes of the vertical profiles are first estimated at each grid level using a linear interpolation scheme. The user also has the option of choosing a cubic spline interpolation scheme. The slopes are then linearly interpolated in time to provide 24 hours of data. Then, the hourly slopes, assumed to be uniform throughout the grid, are used to integrate upwards from the surface using the previously interpolated hourly surface velocities to obtain the upper-air velocities at the grid cell centers.

When vertical soundings are available at more than one location, the module linearly interpolates the observed velocities vertically to the grid heights. If desired, a cubic spline interpolation technique can be used instead of linear interpolation. The velocities at each grid height are then linearly interpolated in time to obtain 24 hours of data. The improved IDS interpolation routine is then used to interpolate the hourly upper-air velocities at each grid level horizontally to the grid cell centers.

Once the interpolation process is complete it is necessary to adjust the interpolated velocity field for mass consistency. This is accomplished by defining an integral function whose optimum solution minimizes the variance of the difference between the given and the adjusted velocity components. The integral function is solved under the constraint that the solution must satisfy the continuity equation. The integral function is defined in the weighted least-squares sense and a Lagrangian multiplier is used to incorporate the continuity constraint within it as follows:

$$F(U,V,W,\lambda) = \iiint_V \{\alpha_1^2(U-U_o)^2 + \alpha_1^2(V-V_o)^2 + \alpha_2^2(W-W_o)^2 + \lambda(\nabla.\vec{V})\}\, dxdyd\rho \quad (3a)$$

$$\nabla.\vec{V} = \frac{\partial U}{\partial x} + \frac{\partial V}{\partial y} + \frac{\partial W}{\partial \rho} \quad (3b)$$

where U_o, V_o, and W_o are the observed wind velocity components; U, V, and W are the adjusted wind velocity components; $\lambda(x,y,\rho)$ is the Lagrangian multiplier; and α_1 and α_2 are weights associated with the horizontal and vertical velocity components, respectively. All velocity components are in terrain-following coordinates.

The solution of the integral equation leads to the following differential equation for the Lagrangian multiplier, λ:

$$\frac{\partial^2\lambda}{\partial x^2}+\frac{\partial^2\lambda}{\partial y^2}+\left(\frac{\alpha_1}{\alpha_2}\right)^2\frac{\partial^2\lambda}{\partial\rho^2}=-2\alpha_1^2\left(\frac{\partial U_o}{\partial x}+\frac{\partial V_o}{\partial y}+\frac{\partial W_o}{\partial\rho}\right) \quad (4)$$

with the following boundary conditions:

$$\frac{\partial\lambda}{\partial\rho}=0 \text{ at the terrain surface} \quad (5a)$$

$$\lambda=0 \text{ at the open boundaries} \quad (5b)$$

The model uses an alternating direction implicit (ADI) method to solve Equation (4). The method involves solving three one-dimensional systems of finite difference equations in the x, y, and ρ directions at half time steps over the grid until a steady state solution is obtained. Each system of equations is expressed in the form of a tridiagonal matrix and an efficient tridiagonal solver is used to solve for the Lagrangian multipliers.

URBAN-SCALE EULERIAN AIR QUALITY MODEL

The Eulerian model, referred to as SANMOD (*SAN*tiago *MOD*el), consists of the following major components: a horizontal transport (advection) module; a vertical transport module; a dry deposition module; and a gas-phase chemistry module. The model solves the mass-conservation equation [Equation (2a)] for pollutants using the operator splitting method. In this method, the horizontal advection is first simulated for half a model time step:

$$\frac{\partial\hat{C}}{\partial t}+\frac{\partial}{\partial x}(u\hat{C})+\frac{\partial}{\partial y}(v\hat{C})=0 \quad (6)$$

The numerical solution of the above equation is based on the Smolarkiewicz (1983) scheme. The scheme has the advantage of being computationally efficient and of maintaining the positivity of concentrations. At the same time, the numerical diffusion associated with the scheme is reasonably small. As currently implemented in the model, the one-dimensional version of the scheme is used — the x and y-direction components of Equation (6) are solved separately.

Next, the vertical component of Equation (2a) is solved for half a time step:

$$\frac{\partial\hat{C}}{\partial t}+\frac{\partial}{\partial\rho}\left(\frac{W}{\pi}\hat{C}\right)=\frac{\partial}{\partial\rho}\left(\frac{K}{\pi^2}\frac{\partial\hat{C}}{\partial\rho}\right)+S\pi \quad (7)$$

The source term in the above equation, S, represents the emissions of the pollutant. Equation (7) is solved using a hybrid time differencing scheme. When the vertical eddy diffusivity, K, is small or moderate, the Crank-Nicholson semi-implicit method is used. When K is large, a fully implicit method is used.

Dry deposition is handled as a boundary condition at $\rho = 0$ in Equation (7). The concept of dry deposition velocity is used to calculate the deposition flux to the ground,

$$F_i = V_{di}\, C_i \quad (8)$$

where V_{di} is the dry deposition velocity, and F_i and C_i are the flux and concentration at a reference height for species i. For the parameterization of dry deposition velocity, the electrical resistance analog is used. The deposition velocity for gases is represented by three resistances in series:

$$V_d = \frac{1}{r_a + r_d + r_s} \tag{9}$$

where

r_a = aerodynamic resistance
r_d = deposition layer resistance
r_s = surface (canopy) resistance

The aerodynamic resistance, r_a, is given by:

$$r_a = \frac{1}{ku_*}\left[\ln\left(\frac{z}{z_o}\right) - \psi_h\left(\frac{z}{L}\right)\right] \tag{10}$$

where ψ_h is the nondimensional heat flux (Businger, 1973), z_o is the surface roughness of the given locality, u_* is the friction velocity, and L is the Monin-Obukhov length.

The deposition layer resistance for gases is estimated from the following relation (Wesely and Hicks, 1977):

$$r_d = \frac{2}{ku_*}\left(\frac{K_D}{D_g}\right)^{2/3} \tag{11}$$

where K_D is the thermal diffusivity of air, and D_g is the molecular diffusivity of the gas in air. The surface resistance, r_s, is estimated from a lookup table compiled by (Walcek et al., 1986).

For particulates (e.g., sulfate), the surface deposition velocity is written as:

$$V_{ds} = \frac{1}{r_a + r_s} \tag{12}$$

The parameterization for V_{ds} is given by Wesely et al. (1985) as:

$$\begin{aligned} V_{ds} &= u_*/500 && z/L \leq 0 \\ V_{ds} &= (u_*/500)[1 + (-300/L)^{2/3}] && z/L > 0 \\ V_{ds} &= 0.0009u_* (-z_i/L)^{2/3} && z_i/L < -30 \end{aligned} \tag{13}$$

The estimated value of V_{ds} is limited by the observed maximum value.

After the solutions of Equations (6) and (7) for half time steps, the chemistry equations are integrated for a full time step:

$$\frac{\partial \hat{C}}{\partial t} = R\pi \tag{14}$$

The gas-phase chemical mechanism used in the urban model is based on a condensed mechanism developed by Lurmann et al. (1987). The mechanism is an updated and refined version of an older mechanism (Lurmann et al., 1986) and is suitable for use in urban-scale atmospheric modeling. The gas-phase chemistry module includes the photochemical reactions that yield radical pools for species conversions. In addition, it also treats the chemistry during dark conditions. The mechanism uses the "lumped-molecule" and surrogate species approaches to partition the hundreds of different types of hydrocarbons present in the atmosphere into representative categories. The mechanism consists of 110 reactions among 45 species, 28 of which are transported (i.e., integrated). The remaining species are treated as steady-state or constant.

The stiff system of differential equations describing the chemical kinetics is solved by using a hybrid asymptotic predictor-corrector integration scheme (Young and Boris, 1977). This scheme uses a standard formula for the non-stiff equations and a simple asymptotic formula for the stiff equations.

APPLICATION OF THE MODEL

The model described in the previous sections is currently being applied to the Santiago region for two episodes during which the region experienced high pollutant concentrations: a winter episode from August 1 to August 5, 1988, and a summer episode from January 9 to January 15, 1989. Air quality measurements were made at 6 locations during these episodes. 4 measurement sites were located in the downtown area while the remaining 2 sites were located in the southwest and northeast corners of the modeling domain.

The winter episode was characterized by light winds and low mixing heights during the night, leading to a build-up of pollutants. Observed wind speeds did not exceed 2 to 3 m/s at most locations. Nighttime mixing heights were less than 100 m, while maximum daytime mixing heights were typically of the order of 500 m. Maximum SO_2 concentrations measured in downtown Santiago ranged from 80 to 150 ppb. Maximum CO concentrations from 17 to 26 ppm were also measured. Ozone was not a major problem during this episode. For all but one day of the episode, maximum ozone concentrations were below 90 ppb. However, a maximum ozone concentration of 130 ppb was measured on August 5 near downtown Santiago.

In contrast, ozone was the major pollutant for the summer episode. Maximum ozone concentrations ranged from 110 to 160 ppb for the first 6 days of the episode. Maximum SO_2 concentrations observed during this episode ranged from 30 to 65 ppb, while maximum CO concentrations ranged from 2 to 7 ppm.

Results from the application of the model to these episodes will be presented in a subsequent paper.

ACKNOWLEDGEMENTS

The work described in this study is being sponsored by the Intendencia of the city of Santiago, Chile.

REFERENCES

Businger, J. A., 1973, Turbulent transfer in the atmospheric surface layer, in: " Workshop on Micrometeorology", D. A. Haugen, ed., American Meteorological Society, Boston, MA.

Lurmann, F. W., Lloyd, A. C., and Atkinson, R., 1986, A chemical mechanism for use in long-range transport/acid deposition computer modeling, J. Geophys. Res., 91:10905.

Lurmann, F. W., Carter, W. P. L., and Coyner, L. A., 1987, "A Surrogate Species Chemical Reaction Mechanism for Urban-Scale Air Quality Simulation Models, Vols. I and II", EPA Contract No. 68-02-4104, U.S. EPA, Research Triangle Park, NC.

Ross, D. G., and Smith, I. N., 1987, "Diagnostic Wind Field Modeling: Development and Validation", 5/87, Center for Applied Mathematical Modelling, Chisholm Institute of Technology, Victoria, Australia.

Smolarkiewicz, P. K., 1983, A simple positive definite advection scheme with small implicit diffusion, Mon. Wea. Rev., 111:479.
Ulriksen, P., 1980, Factores meteorologicos que influyen en la contaminacion atmosferica y conditiones existentes en Santiago, Trakla 1, 2:189.
Walcek, C. J., Brost, R. A., Chang, J. S., and Wesely, M. L., 1986, Sulfate and HNO3 deposition velocities computed using regional landuse and meteorological data, Atmos. Environ., 20:949.
Wesely, M. L., Cook, D. R., Hart, R. L., and Speer, R. E., 1985, Measurements and parameterizations of particulate sulfur dry deposition over grass, J. Geophys. Res., 90:2131.
Wesely, M. L., and Hicks, B. B., 1977, Some factors that affect the deposition rates of sulfur dioxide and similar gases on vegetation, J. Air Pollut. Control Assoc., 27:1110.
Young, T. R., and Boris, J. P., 1977, A numerical technique for solving stiff ordinary differential equations associated with the chemical kinetics of reactive flow problems, J. Phys. Chem., 81:2424.

DISCUSSION

M.E. BERLYAND

What are the main features of your urban atmospheric model?

P.K. KARAMCHANDANI

The model consists of the following modules: horizontal and vertical transport, dry deposition, gas-phase chemistry. The model operates in terrain-following coordinates and receives its meteorological and micrometeorological input from a state-of-the-art processor which includes a diagnostic wind field module for complex terrain. Further details on the model can be found in the paper.

THE GERMAN AIR POLLUTION OBSERVATION NETWORK AS A BASE FOR THE VALIDATION OF EURAD

H. Geiss, A. Volz, D. Kley

Institut für die Chemie der belasteten Atmosphäre
Forschungszentrum Juelich
P.O. Box 1913, D-5170 Juelich, FRG

Introduction

Complex three-dimensional regional models like the **E**uropean **A**cid **D**eposition Model EURAD yet may contribute to understand the behaviour of atmospheric compounds like ozone. The regional distribution and the temporal evolution of the ozone concentration due to changes in meteorological conditions as well as the influence of emissions of urban areas on the ozone concentration in the vicinity are of interest. Another important point is the valuation of the antropogenic contribution to the total ozone balance in the troposphere. Before such a model can be used operativly one has to compare the model results with observations, besides a test on internal consistence (like mass balance). This model evaluation can be carried out in different ways for models like EURAD which are composed of modules describing the various meteorological and chemical processes. On the one hand the results of each single modul can be tested using the observations of especially designed experiments. The main assumption for this validation step is, that there does not exist any nonlinear coupling between the individual processes.
Accordingly, recently two series of experiments with chemicaly non reactive tracers have been conducted in the US to proof the performance of transport and diffusion modules in various long range transport models.
During the Cross-Appalachian Tracer Experiment (CAPTEX) (Ferber et al.,1986) 3-hour and 6-hour averages of the tracer concentrations have been measured at the surface within 1100 km of the release point up to seven days after the release. Vertical profiles of the tracer have been measured too by aircraft.
Although this data set enabled modelers for the first time to proof the performance of their models, the low number of the releases prevented from comprehensive statements to qualify the accuracy of the models. In addition the experiments have been restricted to synoptic situations with homogenoues meteorological conditions in time and space to keep the tracer in narrow sampling sector to get a measurable plume. Anyway the observations have been used to validate various transport models and the evaluation procedure provided important information about model performance. (Draxler, 1987; Haagenson et al. 1987;Kao und Yamada, 1988; Brost et al., 1988).
The experimental deficiencies of the CAPTEX-experiment in ANATEX mostly have been avoided. During ANATEX more single releases (66) over a period of three month under various meteorological conditions have been conducted, the network has been extended to 3000 km against 1100 km during CAPTEX and the avarage distance of the measuring sites was about 250 km. This data base is to be used to assess the performance of ten long-range transport and diffusion models (Clark et al., 1989).
In Europe up to this date no such long-range tracer experiments have been carried out. But comprehensive measurements of radioactivity in air and on the ground after the accident in Chernobyl provide the possibility to proof the transport and diffusion modul of EURAD as well as the modules for dry and wet deposition (Hass et al., 1990). In this special case chemical reactions during the

Air Pollution Modeling and Its Application VIII, Edited by H. van Dop and
D. G. Steyn, Plenum Press, New York, 1991

transport can be neclected. Although there are uncertainties in the amount of radioactive material released there is a good agreement of important model parameters with measurements like the arrival time of the radioactive cloud and the spatial distribution of wet deposited radioactivity.
For the modules describing the gas phase chemistry no comparable atmospheric data sets exist. Therefore the results of the gas phase chemistry modul in RADM/EURAD for the present has been verified using the results of smog chamber experiments (Stockwell, 1986), despite the fact that atmospheric conditions cannot be simulated adequatly in a smog chamber. In the real atmosphere one finds conditions nearly like in a smog chamber only under stable meteorological conditions with low cloudiness and low wind speed. During such conditions the ozone concentration in north-east of the US has been simulated by the ADOM/TADAP model and the results have been compared with observations (Karamandchani, 1989).
Validating single modules is no longer wise if nonlinear processes are coupled in a model as it is the case of acid rain formation. Here one has complex interactions including horizontal and vertikal transport and the formation of clouds as well as gas phase and heterogenoues chemistry. Wet deposition modeling has been performed with RADM using the observations during the OSCAR-experiments in 1981. Interpreting the difference between observations and calculations a problem associated with integral validation has been indicated. The deviation of the results of the whole model cannot be assigned definitly to errors in a single modul (Middleton et al., 1988).
During OSCAR no other meteorological parameter than precipitation has been measured. Even the results of the OSCAR-measurements show clearly that the following criteria have to be taken into account for the choice of the date base for model validation.
On the one hand the network has to be designed problem oriented. That means, that besides the allready mentioned completness of the parameters to be measured the distance between the observational stations and the temporal resolution have to be chosen in a way, that all processes and phenomena of interest can be recorded. On the other hand the structur of the network and the structur of the model have to be adjusted. In particular this has to be done with grid size, the temporal resolution and the kind and number of species to be studied. A higher resolution of the network data compared to the model results may be convenient to achieve a higher statistical certainty.
Taking the spatial and temporal variation of ozone in the Federal Republic of Germany (FRG) before, during and after the passage of a cold front we will demonstrate, that routine measurements of an air pollution network can provide informations which may serve as a base for model validation.

Air pollution monitoring in Germany

In the FRG, at present, there exist about 350 monitoring stations. These are operated by the German Environmental Protection Agency (UBA) the federal states and the German Weather Service (DWD). Air pollution observation in the FRG started in the 60's with a few scientific stations of the Deutsche Forschungsgemeinschaft (DFG) but was operating only periodicly. In 1971 the German Weather Service began with continoues measurement of Ozone at the Hohenpeissenberg Observatory. In 1975 the UBA took over five sientific stations from the DFG, located at rural sites near the German border from Westerland in the northern FRG to Brotjacklriegel in the south-eastern part of the FRG. They are designed to record the long term background pollution in the FRG and the trans boundary transport. Since then these so called pilot stations are measuring daily mean concentration of SO_2, NO_2 and dust. In 1978 ten more of those stations - most of them at rural sites - have been added. These UBA-stations together with the Hohenpeissenberg Observatory of the DWD form the backbone of the air pollution network. In 1980 the UBA started with the continoues monitoring of ozone at the five pilot stations with a temporal resolution of one hour.
In addition since the middle of the 70's the federal states each have built there own network situated mainly in polluted areas and urban regions. At these stations 30-min. means of O_3, SO_2, NO_2, CO and dust are measured as well as meteorological parameters. Because these stations in the first place serve as air pollution warning systems the spatial density in urban areas is quite high. The distance of them sometimes is smaller than two kilometers.
From the beginning of the eighties the networks have been completed by several stations in rural regions, especially where the so called "Neuartige Waldschaeden" appeared.
The spatial distance of those stations is 30 to 80 km with larger gaps in the north-west and south-east of Germany (fig. 1). A detailled analysis using the ozone measurements of more then five years shows that about 40 sites may be addressed as rural. Another 40 can be characterized as rural with little urban influence (Kley et al., 1989). All others are definitly located in high polluted areas.

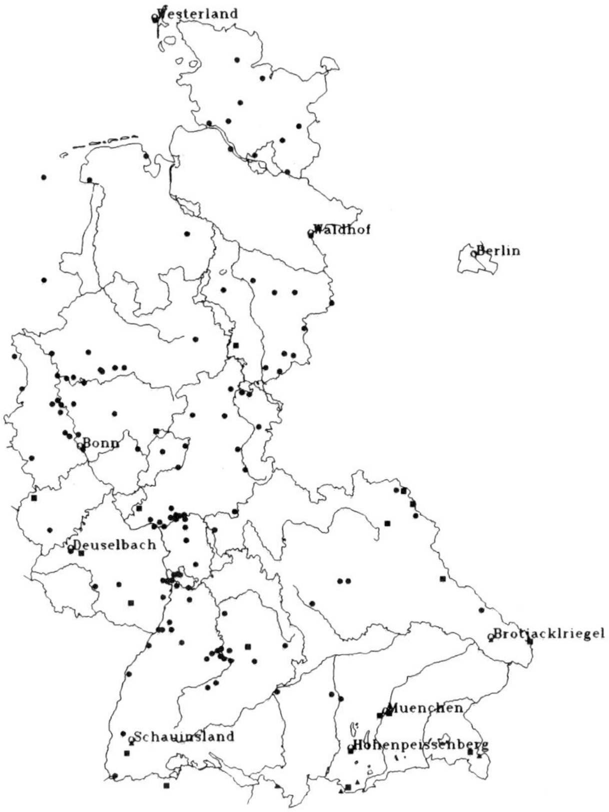

Fig. 1. The German Air Pollution Monitoring Stations (for the clearness of the presentation some of the urban stations have been omitted; dot points : altitude below 500 m, square points : 500 - 1000 m, triangle : above 1000 m)

All measurements inclusive those recorded by the federal states are central registered and stored by the UBA.

Variation of the ozone concentraion during a cold front passage; two examples

Analyzing the temporal evolution of ozone at many rural stations in the FRG one frequently identifies a drop of the concentration to less than 20 ppb with a time lag between the stations of a few hours to two days. As a typical example for such an event we look at the synoptic situation between July 13, and July 18, 1986 in some more detail.
Between July, 13 and July,17 a ridge of high pressure expands from the Atlantic Ocean to south-east Europe. In Germany partly cloudy skies at the very beginning of the episode turned to sunny weather on 14th. Winds were light changing there directions. With increasing sunshine photochemical production of ozone is indicated by both the general increases as well as the diurnal cycle.
During the 15th a low pressure system was developing west of Ireland and was moving north-east over Scotland towards Norway. The associated cold front crossed Great Britain and Middle Europe during the 16th, 17th and 18th heading the Alpes in the evening of the 18th (fig. 2). A part of the air mass behind the cold front originating north-west of Scotland in the North Atlantic Ocean crossed the industrialized regions with large NO_x-emissions in Great Britain, Belgium, the Netherlands and the FRG. During the 16th and 17th the ozone concentration rises at all stations exceeding 80 ppb at most sites (fig. 3).
Maximum values have been reached in the western part (Deuselbach) and in the middle (Eggebirge) of the FRG at one moment in the afternoon of the 17th, but in the south-west (Schauinsland) and the south-east (Brotjackriegel) not before evening of July 17 (fig. 4) During and after the cold front

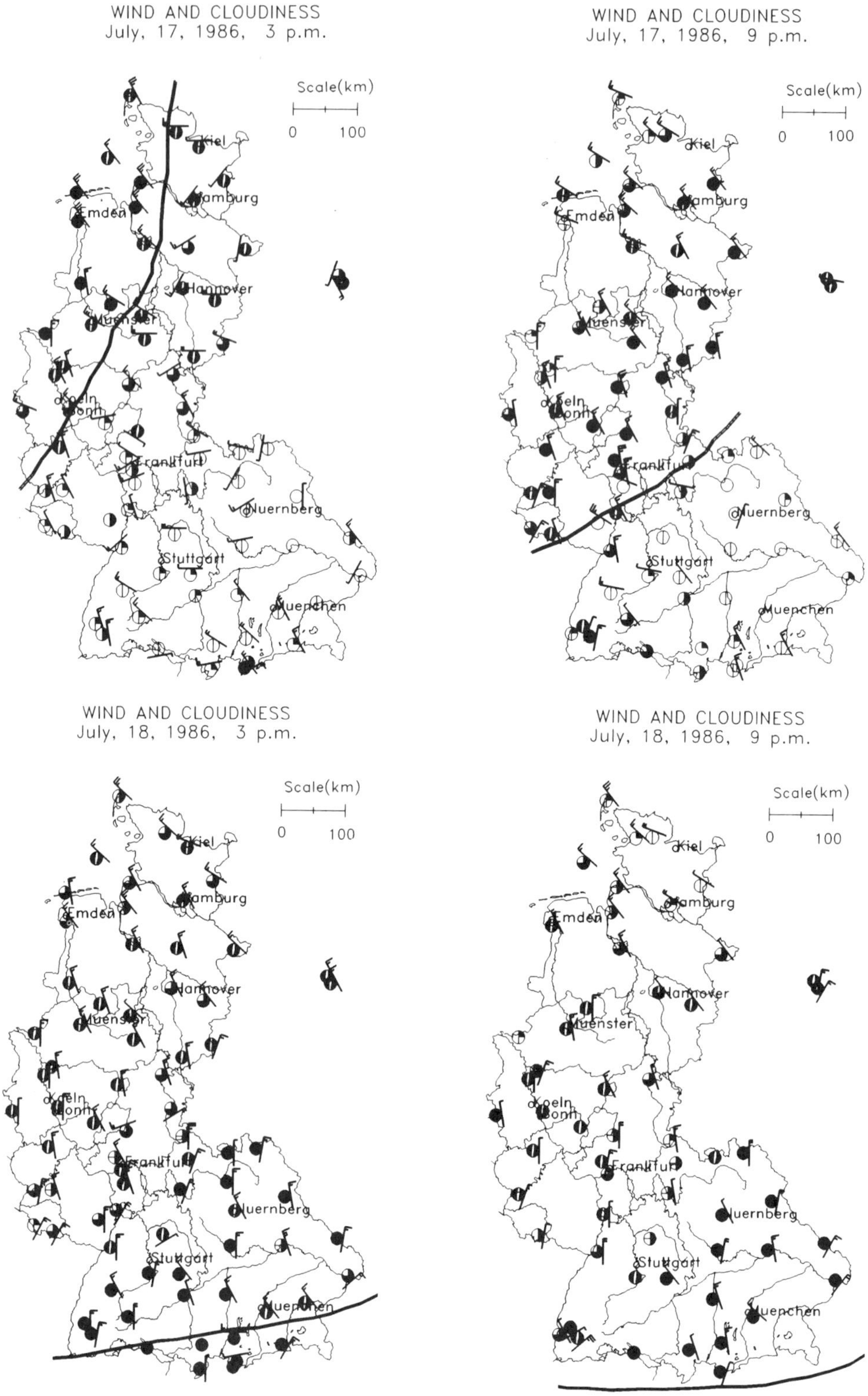

Fig. 2. Meteorological situation July 17, 3 and 9 p.m. and July 18, 3 and 9 p.m. Wind direction and wind speed are indicated by the arrows, filled circels show high cloudiness, open circels clear skies. The position of the cold front is marked by the solid lines.

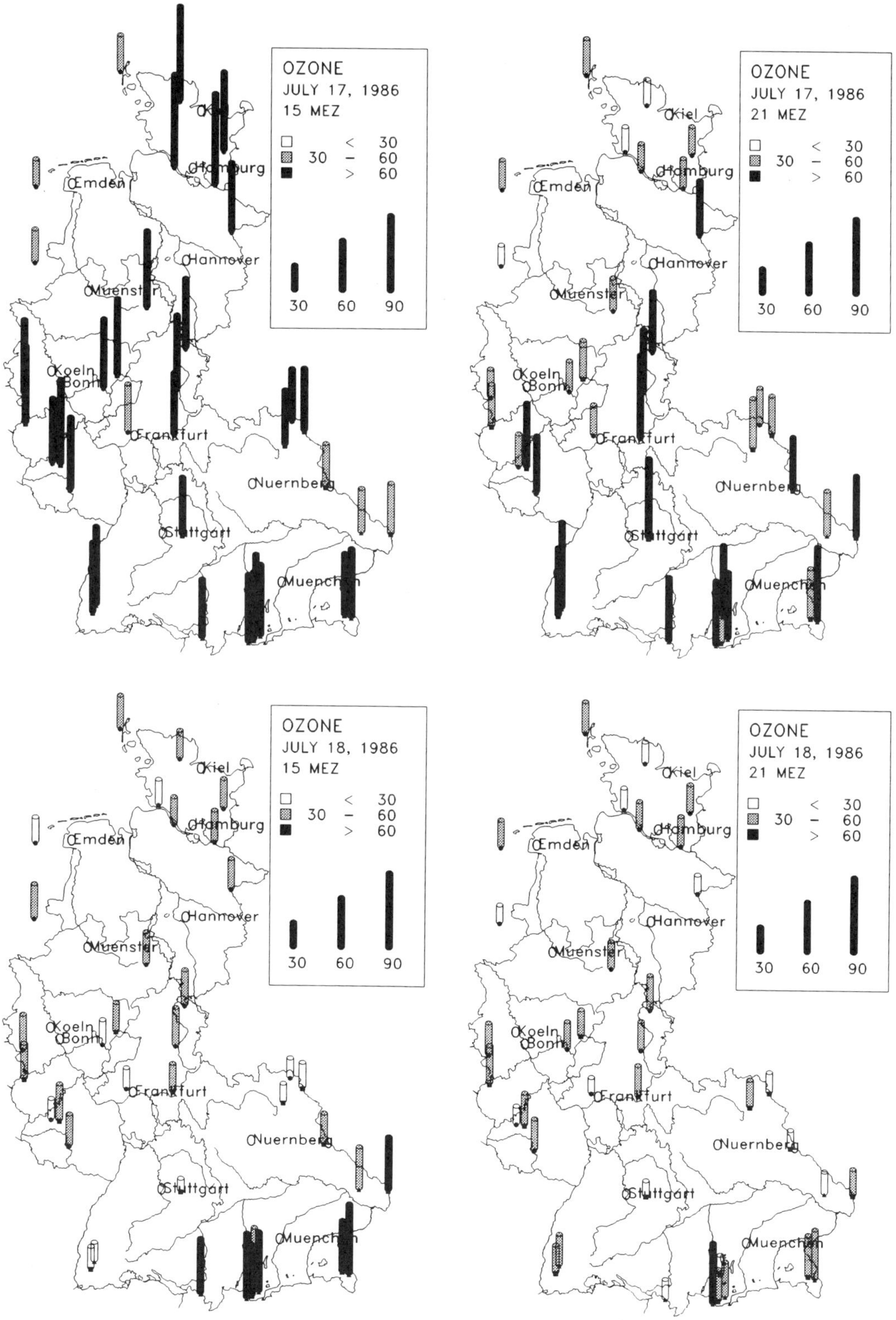

Fig. 3. O_3 in Germany on July 17, 3 and 9 p.m., and July 18, 3 and 9 p.m.

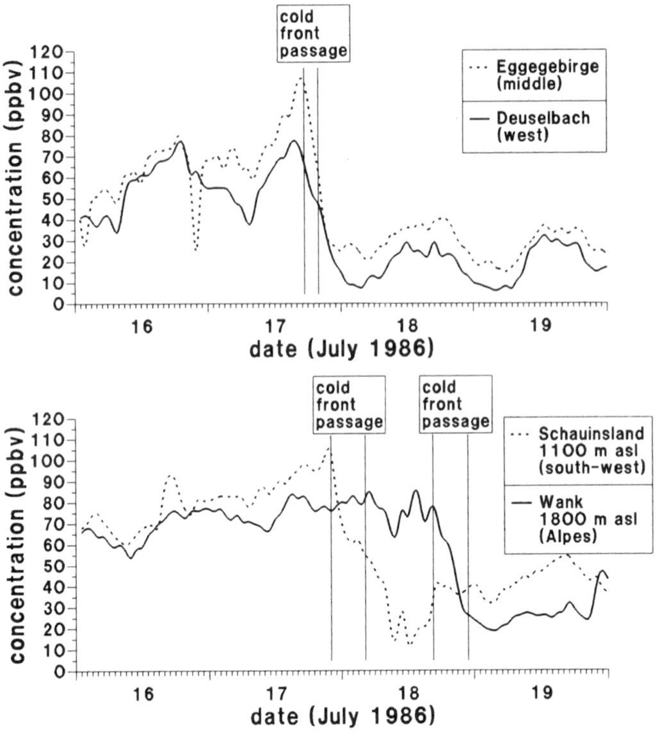

Fig. 4. Time evolution of O_3 at four different sites in Germany before, during and after a cold front passage (Eggegebirge (middle), Deuselbach (western border), Schauinsland (south-west), Wank (Alpes))

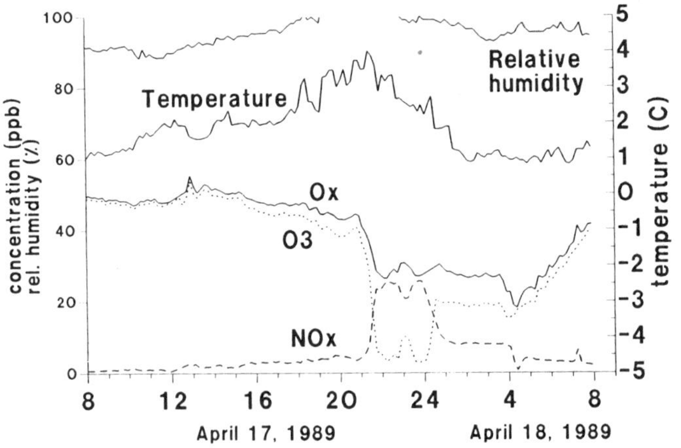

Fig. 5. Time variation of different species at Schauinsland before, during and after a cold front passage.

passage at all measuring stations the ozone concentration dropped drasticly with a rate of 20 ppb to less than 20 ppb except the coastel sites in the north, where the concentration fell to 30 ppb. During the 17th the region with low ozone extended over whole Germany to the Alpes (fig. 3), stopped during the 18th and headed the Alpes in the evening of the 18th. Even the elevated sites in the Alpes like Wank (1800 m asl) show the same drop of the ozone concentration as the other rural stations.
The temporal variation of ozone at four of the rural stations is shown in fig. 4. Those stations are located in different regions of the FRG more than 200 km seperated from each other. The passage of the cold front is indicated by vertical lines in fig. 3. From the available synoptic data the cold front passage may be fixed not more accurate than within a time intervall of 3 to 6 hours. The time difference of the maximum ozone concentration reached first in the middle and later on in the south corresponds exactly to the speed of the moving front. Measurement over the Atlantic Ocean show that the ozone concentration of fresh atlantci air reaches 30 to 35 ppb (Smit et al., 1990). Crossing Europe a net destruction of ozone within the air mass behind the cold front must have been happened.
Besides dry deposition of ozone the reaction with NO from antropogenic emissions forming NO_2 may be one reason for the low values. But the reaction $O_3 + NO \rightarrow NO_2$ does not lead to a net loss of ozone. This can be explained only by the reaction chain

$$O_3 + NO \rightarrow NO_2 + O_2$$
$$O_3 + NO_2 \rightarrow NO_3 + O_2$$
$$NO_2 + NO_3 \rightarrow N_2O_5$$

followed by heterogenoues removal of NO_3 and N_2O_5 (Volz et al., 1988).
An other example displaying a cold front passage at the Schauinsland station of the "Institut für die Chemie der bealsteten Atmosphäre" (ICH) of the Research Center Jülich is shown in fig. 5. Since the beginning of 1989 at this station continouesly ozone and its precursors has been measured within the framework of the Tropospheric Ozone Research (TOR).
The development of humidity and temperature fixes the cold front passage to 9 p.m.. At this time the ozone concentration suddenly drops. At the same time NO_x increases with the same rate as ozone decreases indicating the advection of a polluted air mass. Despite the fact that NO_x is increasing the decrease of O_x is an other hint to the heterogenoues loss of ozone.
The above specified situations are not singular. Averaged over several years in Middle Europe one can expect 8 to 10 times a year a cold front with air originating over the North Atlantic and crossing Great Britain and Middle Europe towards south-east. Within the 6-years series of measurements of ozone at 120 stations in the FRG since 1984 one may find a large number of cases to validate EURAD.
Events like cold front passages above all are suitable for the validation as within a period of a few days several processes, which are important for ozone chemistry, occur one after the other. Photochemistry, transport, destruction of ozone by the emission of NO_x as well as heterogenoues losses can be recorded successively. So the speed of the cold front and the arrival time can be fixed accurate to a time interval of one hour or less using the ozone records. With that, the transport modul of EURAD can be validated.
But the data of the same episode can be used to verify other modules like that for heterogenoues chemistry too. Heterogenoues chemistry in the first place is determined by meteorological processes at the cold front and in the air behind. In this case above all the formation of clouds and precipitation are the most important processes. The value of the ozone concentration as well as the temporal evolution after the front passage point on the way and the intensity of the ongoing heterogenoues processes. If EURAD or other models are able to simulate such phenomena that may strengthen the credibility of such models.

Summary

The german air pollution monitoring stations fulfil the most important needs for a network to validate models :

- they provide continouces time series of ozone for six or more years at a great number of stations, which guerrantees the results of the validation to be statistically significant
- the stations have an average distance of about 40 km sufficiently small to explore regional scale processes
- the time resolution of the measurements is small enough to resolve the variations within the spatial scale.

The number of species which are recorded at this time is not sufficient. To be able to clarify some open questions upon heterogenouces destruction of ozone there are additional measurements with high accuracy necessary as they are carried out allready at the Schauinsland station of the ICH.

Acknowledgments

The supply of data from the following agencies and research institutions is gratefully acknowledged :

Bayerisches Landesamt für Umweltschutz, München; Landesanstalt für Umweltschutz Baden-Wüttemberg, Karlsruhe; Landesamt für Umweltschutz und Gewerbeaufsicht, Oppenheim; Hessische Landesanstalt für Umwelt, Wiesbaden; Landesamt für Immissionschutz Nordrhein-Westfalen, Essen; Niedersächsisches Landesamt für Immissionsschutz, Hannover; Gewerbeaufsichtsamt, Itzehoe; Umwetbundesamt, Berlin; UBA Pilotstation, Offenbach; Deutscher Wetterdienst, Meteorologisches Observatorium, Hohenpeissenberg; Fraunhofer Institut für Umweltforschung, Garmisch-Partenkirchen

Literature

Brost R.A. et al., 1988
The effect of Diffusion on Tracer Puffs Simulated by a regional scale model in : Journal of Geophysical Research, Vol. 93, D3, pp. 2389-2404, 1988
Clark, T.L. et al., 1989
Comparison of modelled and measured tracer gas concentrations during Across North America Tracer Experiment (ANATEX) in : Air Pollution Modelling and its Application VII, Volume 13, pp. 307-321, Plenum Press, 1989
Derwent, R.G., Kay P.J.A.
Factors Influencing the Ground Level Distribution of Ozone in Europe Environ.Poll.,0269-7491/88/$03-50, 1988
Draxler, R.R., 1987
Sensitivity of a trajectory model to the spatial and temporal resolution of the meteorological data during CAPTEX, J,Clim.and App. Meteor., 26:1577
Draxler, R.R. et al., 1988
Modeling the CAPTEX vertical tracer concentration profiles, J.Appl.Meteor., 27:617.
Ferber, G.J. et al., 1986
Cross-Appalachian Tracer Experiment (CAPTEX '83) Final Report,NOAA Technical Memorandum ERL-ARL-142, National Oceanic and Atmospheric Administration, Silver Spring, Maryland
Hass, H. et al., 1989
Simulation of the Chernobyl Radioactive Cloud over Europe using the EURAD-Model, Atm.Env., Vol. 24A, No.3,pp. 673-692
Kley,D. et al., 1989
Concentration, trend and distribution of reactive trace gases in rural Germany in the context of forest damage

Smit, H. et al., 1990
The Meridional distribution of ozone and water vapor over the Atlantic Ocean between 30° S and 52° N in September/October 1988, in : Physico-Chemical behaviour of atmospheric pollutants, No. 23, Kluwer Academic Publishers, Dordrecht, Netherlands

Stockwell, W..R. ,1986
A homogenoues gas phase mechanism for use in a regional acid deposition model, Atm.Env. 20, pp. 1615-1632
Volz, A. et al., 1988
Correlation of Ozone and solar radiation at Montsouris and Hohenpeissenberg : Indications for photochemical influence Proc.Quadr.Ozone Symp., Göttingen, FRG, 1988

ROUTINE MODEL FOR THE LONG-RANGE TRANSPORT CALCULATIONS OF SULPHUR AND NITROGEN COMPOUNDS FROM CONTINUOUS SOURCES ALLOWING FOR NON-LINEAR EFFECTS

Michael Galperin

Meteorological Synthesizing Centre - East EMEP
Moscow

INTRODUCTION

The air pollution long-range transport can be considered as an aggregate of four processes:

1. Horizontal transport and diffusion.
2. Vertical diffusion and dry deposition.
3. Chemical and physical transformation of pollution.
4. Washout.

In order to make allowance for the non-linearity of the two last processes and wind variation along the vertical it is necessary to use calculation schemes of horizontal transport and diffusion dealing with integral pollution amounts in grid cells (Eulerian models). The problem of the long-range transport requires schemes with small pseudodiffusion as well as with adequate description of impulsive effects on pollution concentrations (precipitations, point sources). Such schemes are sophisticated and require big computer resources, therefore they fit little to routine calculations covering long periods of time.

In this context the MSC-E has developed a hybrid model described below. The model can use any scheme for the calculation of vertical diffusion and dry deposition. To save computer resources it is desirable to have a model of vertical diffusion and dry deposition and wind input data at several levels without detail calculations of pollution vertical distribution. The report presents a model of vertical diffusion which meets the requirements and agrees with the hybrid model developed for horizontal transport and diffusion.

HORIZONTAL TRANSPORT AND DIFFUSION

There is a rectangular grid with gridsize $N_x \times N_y$ square cells and $k = 1,\ldots N_x$, $l = 1,\ldots N_y$ - whole-number co-ordinates. Explicit calculational schemes operating with pollution amounts in grid cells have zero coefficient of artificial viscosity if the conditions

$$\Delta x/t_b = u; \quad \Delta y/t_b = v \tag{1}$$

Air Pollution Modeling and Its Application VIII, Edited by H. van Dop and D. G. Steyn, Plenum Press, New York, 1991

are met, where Δx and Δy - mass centre shift along axes x and y - while the pollution is being exchanged between cells, t_b - time which is taken for the exchange, u and v - wind components.

To fulfil equality (1) a virtual trajectory from the cell mass centre is considered in each cell. The mass is transferred to a neighbouring cell when the virtual point, the trajectory end, is transported to cell-receiver. Horizontal dispersion is simulated only at the moment of transfer.

Assume that the mass centre of cell A at moment t is in point 1 (fig.1). Under the influence of wind the virtual point is displaced from point 1 to points 2 and 3 at moments $t_2 = t_1 + \Delta t$ and $t_3 = t_2 + \Delta t$. During this period pollution is accumulated in cell A due to emission in A and input from other cells and it is removed due to dry deposition, washout and chemical transformation. At moment t_4 when virtual point comes to cell B (position 4), the part of mass accumulated in cell-donor A is transferred to cell-receiver B and its centre is prescribed to point 4. At this moment the dispersion from point 4 is calculated. If in cell B at moment t_4 there is no pollution but coming from A then the virtual point belonging to B starts from point 4 and occupies position 5 at moment t_5. At the same time a new virtual point in cell A starts from a new mass centre of this cell (it is not shown in the figure). Assume that the virtual point of cell B is found in cell D at moment t_6 (cell B is a donor for cell D). Accordingly the part of mass from B is transferred to D (point 6) and the dispersion from cell D is calculated.

Now it is obvious that pollution transport velocity from cell to cell in the scheme is exactly equal to wind speed, just as it is in Lagrangian models. Dispersion can be assumed arbitrarily. For example it should be proportional to the distance passed corresponding to the known empirical dependences.

Masses coming to a cell are summed up and the co-ordinates of their common centre are calculated taking into account the mass centre conservatism. Further this mass centre prescribes only the starting position of a virtual point, but from now the centre mass bias does not influence on the virtual point.

According to said above, if a virtual point remained in a cell at the i-th step, then at the (i+1)-th step the virtual point co-ordinates are

$$X_{Li+1} = X_{Li} + u\Delta t \qquad (2)$$

$$Y_{Li+1} = Y_{Li} + v\Delta t \qquad (3)$$

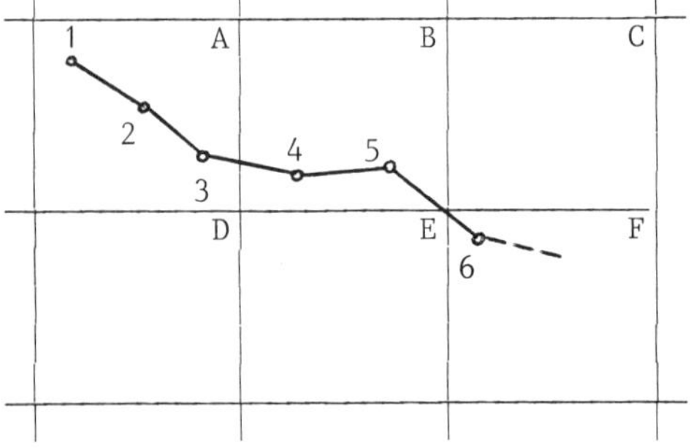

Fig.1. Consecutive positions of virtual points within grid cells.

where u and v - wind components, Δt - time step.

If a virtual point left a cell at the i-th step, then $X_{Li} = X_c$ and $Y_{Li} = Y_c$, where X_c and Y_c - mass centre co-ordinates for the moment of the virtual point exit from a cell.

Further calculations depend on the fulfilment of inqualities

$$X_{min} \leqq X_{Li+1} < X_{max}, \tag{4}$$

$$Y_{min} \leqq Y_{Li+1} < Y_{max} \tag{5}$$

where X_{min}, Y_{min}, X_{max}, Y_{max} - cell limits. If (4) and (5) are satisfied then at the end of the (i+1)-th step the cell pollution mass is:

$$m_{i+1} = m_i + q_{i+1} + \Sigma m_{i+1,j} + \Delta m_{i+1}, \tag{6}$$

where q_{i+1} - emission, $m_{i+1,j}$ - masses coming from neighbouring cells; Δm_{i+1} - mass transformation caused by deposition and chemical transformation. While calculating mass centre co-ordinates it is assumed that the emission occurs in the geometric cell centre x_c, y_c.

If (4) and/or (5) are not satisfied, then after the (i+1)-th step

$$m_{i+1} = \Sigma m_{i+1,j} + (1-d_t)\ (m_i+q_i+\Delta m_{i+1}) \tag{7}$$

where d_t - transport coefficient.

Coefficient d_t is chosen in such a way that pollution amount in a cell-donor should correspond on the average to continuous advection.

This condition is of the form: $d_t = 2(|w|\Delta t+1)/(3+|w|\Delta t) \cong 2/3+|w|\Delta t/3$, where $|w| = (v^2 + u^2)^{1/2}$ and $|u|\Delta t$, $|v|\Delta t$ are less than a grid step. Mass

$$m_L = (m_i + q_i + \Delta m_{i+1})d_t \tag{8}$$

is distributed into two parts. The first is

$$m_0 = m_L(1 - d_h), \tag{9}$$

where d_h - horizontal dispersion coefficient, it is transferred to a cell-receiver where the virtual point appeared. Mass centre m_0 is located in point X_{Li+1} and Y_{Li+1}. A fraction of mass m_L, equal to $d_h m_L$, is displaced to cells adjoining both the cell-receiver and cell-donor, as shown by arrows in fig.2. Here E - cell-donor and any other cell can turn out to be a receiver (depending on wind direction). If cell A is a receiver then diffusion goes to B and D; if B is a receiver then diffusion goes to A and C; if C is a receiver then diffusion goes to B and F; if F is a receiver then diffusion goes to C and I, etc. Diffusion masses are:

$$m_{dj} = d_h m_L (1/L_j^2)/(1/L_1^2 + 1/L_2^2),$$

where L_j - distance from point X_{Li+1}, Y_{Li+1} to the border of appropriate cell (j = 1,2).

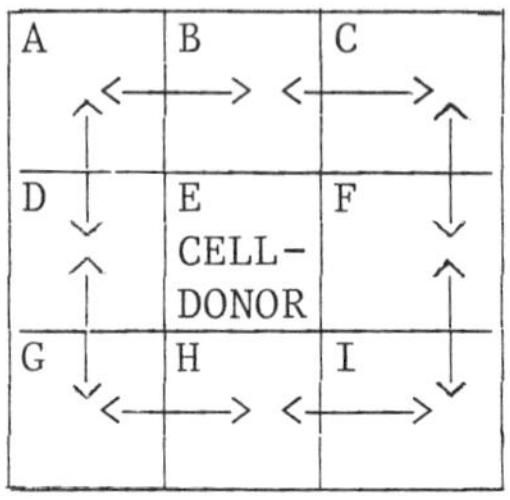

Fig.2. Mass distribution at horizontal diffusion

Masses m_0 and m_{dj} are just masses $m_{i+1,j}$ in equations (6) and (7). Correspondingly they and their centre co-ordinates are included to the equation determining X_c and Y_c of each cell at each time step.

The choice of d_h coefficient depends on the assignment of wind along the vertical. If the mean wind value is taken (one-layer model) then $d_h \cong 0.25$ corresponds to empirical dependence $\delta \cong 0.1\ r$ (δ^2 - horizontal dispersion, r - pollution track, see Pressman, Galperin, Mikhailova, 1986).

At long-range transport horizontal diffusion is caused mainly by the wind turn along the vertical. Therefore at multilayer wind assignment the task for each layer is solved by the above method at $d_h \leqq 0.1$ ($d_h = 0$ can be used).

In case of rough wind assignment (for example two layer wind assignment at 850 and 1000 mbar) it is expedient to use simplified schemes for vertical distribution calculation, one of which is considered below.

Mass centres m_{dj} have the following co-ordinates (see fig.2): for corner cells-receivers A, C, G and I - x_c, y_c of adjoining cells (B and D, B and F, etc.); for cells-receivers B and H - x_c of adjoining cells, Y_{Li+1}; for cells-receivers F and D - X_{Li+1} and y_c of adjoining cells.

The crossing of grid border by masses m_0 and m_{dj} is checked up. If any of the masses is found outside one of the four grid borders then it is assumed that it has left the grid forever. Integral mass crossing each border is calculated. In case some mass crossed the grid corner it is added to the mass which crossed the nearest counter-clockwise border part. Thus pollution amounts leaving the grid are fixed by the model. If mass $m_i + q_i$ in a cell turned out to be less of limit value m_{min}, then the calculation at the (i + 1)-th step is not made. Mass $m_i + q_i$ is added to PVP, which is used for the evaluation of undecided deposition over the grid. Value m_{min} is chosen to be correspondent to concentration known beforehand to be lower than the background (for sulphur this value is 0.01 $\mu g/m^{-3}$). Thereby the mass budget is observed for the calculated period.

This scheme for calculations of horizontal transport and diffusion being apparently complex possesses a number of beneficial properties:

a) it can make practically any description of vertical diffusion;

b) it operates with masses in cells similar to Eulerian models and allows to simulate non-linear effects (chemical reactions and washout);

c) it does not produce, similar to Lagrangian models, pseudodiffusion in the horizontal plane and allows to select a model value for horizontal dispersion coefficient which fits the experiment;

d) calculations run with high speed and it is simple with programming;

e) it is absolutely stable to both coefficient variations and to impulse effects (sources, precipitations).

Simplification of calculations and above all reduction of the necessary on-line storage is due to the emergence of a random dispersion component in mass distribution inside the deposition track and assignment of (6) and (7) mainly near a source. Model runs confirmed that this error is negligible (< 10%) when daily averaging is made and source is continuous. This error is decreasing proportional to the time of averaging.

VERTICAL DIFFUSION AND DRY DEPOSITION

Pollution distribution with the vertical from unit point source is (Monin, 1956):

$$\psi(z,\tau) = \frac{\exp(-z^2/4K_z\tau)}{\sqrt{\pi K_z\tau}} - \frac{V_d}{K_z}\exp V_d \frac{z + V_d\tau}{K_z}\,\mathrm{erfc}\,\frac{z + 2V_d\tau}{2\sqrt{K_z\tau}}, \qquad (10)$$

where z - vertical co-ordinate, τ - pollution age, K_z - vertical diffusion coefficient, V_d- dry deposition velocity. It is assumed in (10), that source height h << $V_d\tau$ and h << $\sqrt{K_z\tau}$, i.e. a situation observed far from a source is considered. Air pollution amount of τ age is (see Pressman, Galperin, Mikhailova, 1986):

$$\zeta(\tau) = \int_0^\infty \psi(z,\tau)\,dz = \exp V_d^2 \frac{\tau}{K_z}\,\mathrm{erfc}\,\frac{V_d\sqrt{\tau}}{\sqrt{K_z}}. \qquad (11)$$

Formulae (10) and (11) do not include processes of volumetric removal - washout and chemical transformation. The impact of wash-out will be considered below and pollution transformation does not affect the vertical distribution. Comparing (10) at z = 0 with (11) we obtain that pollution amount at the surface layer is

$$\psi_o(\tau) = 1/\sqrt{\pi K_z\tau} - V_d\zeta(\tau)/K_z \qquad (12)$$

At the long-range transport K_z and V_d are varying versus conditions in cells, i.e. they are functions of co-ordinates x, y and astronomical time and consequently of τ age. At the same time in (10)-(12) productions $K_z\tau$ and $V_d\tau$ should be substituted by $\int_0^\tau K_z(\tau)d\tau$ and $\int_0^\tau V_d(\tau)d\tau$. Represent $\psi_o(\tau)$ as

$$\psi_o(\tau) = \psi_{oo}(\tau) - V_d\zeta(\tau)/K_z, \qquad (13)$$

where $\psi_{oo}(\tau) = 1/\sqrt{\pi\int_0^\tau K_z(\tau)d\tau}$ - pollution amount at the surface layer, when V_d = 0 along the whole trajectory. The second term in (13) represents decrease $\psi(\tau)$ due to dry deposition. Adding equation

$$d\zeta(\tau)/d\tau = - V_d(\tau)\psi_o(\tau), \qquad (14)$$

we obtain a description of dry deposition. Relationships (10)-(14) can be used for a description of dry deposition allowing for vertical diffusion, if the pollution age is known (the process is not autonomic). In calculation schemes operating with integral masses in grid cells this is practically unacceptable. In order to devise a scheme, which can provide local (autonomic) description of a process with variable K_z and V_d, we introduce an equivalent mixing layer:

$$H(\tau) = \zeta(\tau)/\psi_o(\tau) \tag{15}$$

and express components of a full differential:

$$dH = (\partial H/\partial \tau)d\tau + (\partial H/\partial V_d)dV_d + (\partial H/\partial K_z)\, dK_z, \tag{16}$$

through local values H, V_d and K_z. By definition we have:

$$d\zeta(\tau)/\zeta(\tau)d\zeta = -\, V_d(\tau)/H(\tau). \tag{17}$$

To find $\partial H/\partial \tau$ we use an approximate equality

$$\exp x^2 \mathrm{erfc}\, x \cong (2/\sqrt{\pi})/(x + \sqrt{x^2 + 4/\pi}) \tag{18}$$

After calculations from (11), (17) and (18) we obtain an estimation:

$$\partial H/\partial \tau \cong 2V_d + 2K_z/\pi H. \tag{19}$$

From (13) we find values $\partial H/\partial V_d$ and $\partial H/\partial K_z$. Values $\psi_{oo}(\tau)$ and $\zeta(\tau)$ do not depend on local variations K_z and V_d, therefore

$$\partial H/\partial V_d \cong H^2/K_z, \tag{20}$$

$$\partial H/\partial K_z \cong -\, V_d H^2/K_z^2 \tag{21}$$

Equalities (19), (20) and (21) give a possibility to find dH at each step.

While using scheme of item 2 surface concentrations C_o, pollution mass m and $H = m/C_o$ are calculated in each cell. When pollution mass is transferred from cell to cell, then appropriate C_o are summed and the mixing layer is calculated from relationship:

$$H = \Sigma m_j/\Sigma C_{oj} \tag{22}$$

As an initial value for H we assume $H \cong 2h$, where h - source height. Equivalent and physical mixing layers are equal to each other as far as the order of magnitude is concerned. This approach can be used for rough two- and three- layer models.

MODEL PARAMETERS AND DESCRIPTION OF WASHOUT

To evaluate dry deposition velocity V_d information of two types was used. First, measurement data presented by Nowicki, 1987. The data exhibited sufficiently complete picture of underlying surface effect on dry deposition velocity. In particular these data show that the soil humidity extent makes a decisive effect on V_d for SO_2, but these data are concerned only with positive temperatures. Second, V_d estimations obtained from budget assessments over the whole area under EMEP. It was shown that V_d mean value is varied spasmodically at the temperature of about 0^oC, i.e. when water is freezing. Synthesizing data of direct measurements and budget estimations we obtained the following formula for V_d: $V_d = V_{do}K_w(\Theta, P, S)$ where V_{do} - minimum dry deposition velocity relevant to dry surface; it is assumed that $V_{do} = 0.32$ cm s^{-1} for SO_2 and 0.14 cm s^{-1} for sulphates-aerosols;

$K_w(\Theta, P, S)$ - coefficient dependent on temperature Θ, precipitation amount P and surface type S (S = 2 for water basins, S = 1 for grid squares with mixed surface types, S = 0 for land): $K_w(\Theta,P,S) = 1+[2P_s+S(1-P_s)+1]\Theta_s$; $P_s = 0$, if $P<0.1$ mm h^{-1}, and $P_s = 1$, if $P \geq 0.1$ mm h^{-1}; $\Theta_s = 0$ if $\Theta < 0^oC$, and $\Theta_s = 1$, if $\Theta \geq 0^oC$.

For vertical turbulent diffusion coefficient model uses Θ dependence, obtained by a regression of balance estimations of K_z for different temperatures: $K_z \cong 2.2[m^2\ s^{-1}] \exp 0.14\Theta\ [^oC]$.

Chemical transformation rate $SO_2 \rightarrow SO_4^{2-}$ λ_T and its dependence on temperature was evaluated by statistical filtration of time series of SO_2 and SO_4^{2-} concentrations (Galperin, 1988 (1)). Using these data dependence $\lambda_T(\Theta)$ was reproduced: $\lambda_T \cong 0.009[h^{-1}] \exp 0.06\Theta[^oC]$.

The study of dependences of pollution concentrations in precipitation p on their air concentrations a (Wallen, 1986; Galperin, 1988 (2) and a number of other papers) shows that washout is a non-linear process, p is function of a and dependence p(a) is (Galperin, 1988 (2)):

$$p = p_o[1 - \exp(-a/a_o)] \tag{23}$$

where p_o - saturation concentration and a_o - specific parameter. For sulphur compounds $p_o \cong 3.2$ mg S l^{-1} and $a_o \cong 3$ µg S m^{-3}, for nitrogen compounds (NO_2, NO_3^-) $p_o = 1.5$ mg N l^{-1} and $a_o \cong 1.2$ µg N m^{-3} if precipitation intensity is 1 mm h^{-1}. Value p_o depends on precipitation intensity I as $I^{-0.25}$. It follows from (23) that washout coefficient should be considered as a function of pollution amount in an air column. Mean precipitation value within a cell cannot be used for calculations because of the strong variability of precipitation intensity over a cell. To solve this problem two-staged calculation of washout coefficient is carried out:

1. Mean linear wash out coefficient in a cell is calculated (at $a \rightarrow 0$): $W = 1 - [\Sigma \exp(- I_i \Lambda t)]/N$, where I_i - precipitation intensity at the i-th site , N - number of sites in a cell.
2. Real washout coefficient in a cell is estimated:

$$W_R = p_o SIW/(Wm + p_o SI), \tag{24}$$

where m - total mass in a cell, S - cell square, I - precipitation amount during Δt in a cell. Λ depends on precipitation type (i.e. temperature). $\Lambda \cong 1$ mm^{-1} for rain and $\Lambda \cong 0.8$ mm^{-1} for snow if $I \cong 1$ mm h^{-1}. Function $\Lambda(I)$ is similar to $p_o(I)$. As shown by Galperin, 1988 non-linearity of washout processes creates pollution redistribution along the vertical which influences greatly on deposition and transport processes. Washout mainly takes place in the upper layers, but not in the lower one. Thereby there is usually no washout impact on the surface concentration. The equivalent mixing layer scheme allows for this effect very simply: as far as C_o at washout does not vary and m is decreasing then H is decreasing correspondingly.

MODEL VERIFICATION

The Table presents comparison results of model with measurements (mean monthly concentrations) for two months, when model had large errors caused by unusual meteorological conditions. SO_2, SO_4^{2-} and NO_3^- data from 46 and NO_2 data from 16 EMEP stations were used.

Since modelling processes may be non-linear ones and errors have big multiplicative component, regrission equation should be used in form:

$$\ln(y + b) = \ln a + \gamma \ln x, \tag{25}$$

where x and y - measurement and modelling data respectively. It is obvious that b = 0, γ = 1 are ideal values. Monthly averages for all stations x and y in $\mu g\ S\ m^{-3}$, $\mu g\ N\ m^{-3}$, $mg\ S\ l^{-1}$ and $mg\ N\ l^{-1}$ are given in table as well as a, γ and b from (25). Additional parameter for model quality evaluation is frequency of its being in "factor 2" relative to measurement (c/m in table). It can be compared with the probability of more than twice deviation of singular measurement from mean value for a cell square (m/m in table). The last was evaluated by comparison of data, obtained from sites, disposed in the distance less then 150 km from each other.

Table. Comparison of model results with EMEP network measurement data for January and July 1985

Substance	Month	x	y	Parameters of eq.(25) a	γ	b	"Factor 2" c/m	m/m
SO_2 air	1	18.0	14.4	0.96	0.81	1.12	0.85	0.91
SO_4^{2-} air	1	2.5	2.3	1.11	0.80	0.00	0.79	0.90
SO_4^{2-} pr.	1	1.17	1.31	1.10	1.02	0.03	0.71	0.68
SO_2 air	7	2.3	2.8	1.00	0.98	0.21	0.67	0.88
SO_4^{2-} air	7	1.36	1.11	0.79	0.81	0.09	0.85	0.88
SO_4^{2-} pr.	7	1.22	1.24	1.11	1.00	-0.10	0.90	0.75
NO_2 air	1	4.4	3.9	0.85	0.99	-0.22	0.59	*
NO_3^- pr.	1	0.70	0.54	0.71	0.88	0.02	0.65	0.67
NO_2 air	7	1.60	1.60	1.10	1.00	-0.16	0.60	*
NO_3^- pr.	7	0.48	0.49	0.93	0.99	0.04	0.75	0.67

* Representative data for estimation is absent.

REFERENCES

Galperin, M.V., 1988 (1), An assessment of the atmospheric chemical transformation coefficients of sulphur and nitrogen compounds by time series filtration of measured concentrations, in: "Papers on Workshop of modelling transformation processes and transport of air pollution," Potsdam, p.89-91.

Galperin, M.V., 1988 (2), Adsorption-kinetic non-linear wash-out model of sulphur and nitrogen compounds from the atmosphere, in: "Air Pollution Modelling and Its Application VII," H. van Dop, ed., Plenum Press, N.Y. and London, p.475-484.

Monin, A.S., 1956, Half-empiric theory of turbulent atmospheric diffusion, Proc. Geophysical Inst. of the USSR Academy of Science, 33:3-47.

Nowicki, M., 1987, Measurements of sulphur dioxide dry deposition velocity on the Earth's surface, EMEP/PL/01/87, Inst.of Environ.Protection, Warsaw.

Pressman, A.Ya., Galperin, M.V., Mikhailova, J.E., 1986, Trajectory models of long-range air pollution transmission, in: "Acidification and its policy implication," Elsevier, Amsterdam.

Wallen, C.-C., 1986, Sulphur and nitrogen in precipitation an attempt to use BAPMoN and other data to show regional and global distribution, WMO/TD-N 103, GEMS, Geneva.

THE EFFECT OF SIZE-DEPENDENT DRY DEPOSITION VELOCITIES IN AN EULERIAN REGIONAL-SCALE PARTICULATE MODEL

O. Russell Bullock, Jr.*

Atmospheric Research and Exposure Assessment Laboratory
U.S. Environmental Protection Agency
Research Triangle Park, North Carolina

INTRODUCTION

Atmospheric aerosol concentrations are frequently observed at levels that may adversely affect health and welfare. As a first step toward developing an Eulerian model of the physical and chemical processes which affect atmospheric aerosol concentrations, EPA's Atmospheric Research and Exposure Assessment Laboratory (AREAL) has developed the Engineering Aerosol Model (EAM). The EAM's framework was derived from the Regional Acid Deposition Model (RADM) which is described by Chang et al. (1987). Specifically, the EAM was developed from the RADM's Engineering Model, Version 2 (EM2) by the addition of various aerosol chemistry and kinetics mechanisms. These mechanisms were originally developed by Systems Applications, Inc. (SAI) (Hudischewskyj et al., 1987), and were subsequently corrected and improved for use in the RADM model framework.

Internally, the EM2 uses all of the same computational routines as the RADM, except that the EM2's gas-phase chemistry mechanism treats directly only those reactions related to sulfur. The EM2 uses the RADM's chemical output files, along with the same meteorological files the RADM uses to produce those output files, to recompute the sulfur deposition quantities that result from using differing sulfur emission inventories. Kleinman (1988) has shown mathematically that the governing equations can be split in this manner. It has also been demonstrated that the results obtained from the EM2 match the RADM results when the same sulfur emissions inventory is used, and differ only by a few percent when the sulfur emissions are increased by a factor of four in both models. When the sulfur emissions are decreased by a factor of four, the differences are only about one percent. The EM2, rather than the RADM, was chosen as the framework on which to develop AREAL's first Eulerian particulate model because of its much faster execution time. AREAL is currently using the knowledge gained from the development and testing of the EAM to put aerosol chemistry and kinetics mechanisms into the full-scale RADM, thereby producing a full-scale Regional Particulate Model (RPM).

The aerosol mechanisms in the EAM simulate the gas- and aqueous-phase processes of emission, nucleation, coagulation, condensational growth, evaporational shrinkage, oxidation, transport, and deposition, in seven

*On assignment from the U.S. Department of Commerce, NOAA

Table 1. EAM Size Sections

Section Number	Diameter Range (μm)
1	$0.010 \leq d < 0.022$
2	$0.022 \leq d < 0.048$
3	$0.048 \leq d < 0.106$
4	$0.106 \leq d < 0.230$
5	$0.230 \leq d < 0.520$
6	$0.520 \leq d < 1.130$
7	$1.130 \leq d < 2.500$
8	$2.500 \leq d < 5.000$
9	$5.000 \leq d \leq 10.000$

size sections spanning the range of diameters between 0.01 μm and 2.50 μm, and simulate emission, transport, and deposition in two size sections between 2.50 μm and 10.00 μm. Table 1 shows the range of diameters used for each of the EAM's 9 size sections. When the aerosol mechanisms were originally developed by SAI (Hudischewskyj et al., 1987), the use of additional size sections for a regional-scale Eulerian model was considered impractical due to the number of calculations that would be required to accurately model the coagulation process. Like the EM2 and RADM, the EAM can simulate a variable number of vertical layers, but is normally operated with 6. The EAM, EM2, and RADM each have a horizontal resolution of approximately 80 km in each dimension.

THE NUMERICAL DIFFUSION PROBLEM

One of the most common problems encountered in using Eulerian modeling frameworks is the artificial diffusion caused by finite element advection schemes. The EAM uses the iterated upstream finite difference scheme of Smolarkiewicz (1984) to minimize the numerical diffusion from the advection calculations in all three orthogonal directions. However, the sectional representation of the continuous size distribution of aerosol particles causes numerical diffusion of a different nature. The processes of coagulation, condensation, and evaporation act to change the size of the simulated aerosol particles. This change in size is represented in the size-sectional framework as a transfer of mass from one size-section to another, similar to the advective transfer of mass from one grid volume to another. Most sulfate particle size distributions from EAM simulations resemble the well documented "accumulation mode" (Whitby, 1978), having a geometric mean diameter by volume in the 0.15 μm to 0.5 μm range. However, the particle size range of the typical EAM distribution is somewhat larger that is generally found in atmospheric aerosol samples.

Warren and Seinfeld (1985) examined "size-diffusion" in the Expanded Sectional Multicomponent Aerosol Package (ESMAP) by simulating the growth of a monodisperse aerosol from an initial diameter of 0.01 μm to a diameter of 1.01 μm over a time period of 100 s. At the end of the simulation the conserved fraction of the analytic value for the sectional mass using 9, 18, and 36 section per decade was 0.656, 0.821, and 0.909, respectively. To evaluate the EAM in a similar test environment, a one-cell version of the EAM has been used to simulate the rapid growth of a monodisperse aerosol from a diameter of 0.016 μm (the average for size section 1) to a diameter of 0.825 μm (the average for size section 6). A rate of sulfate production was specified such that condensation on existing particles would

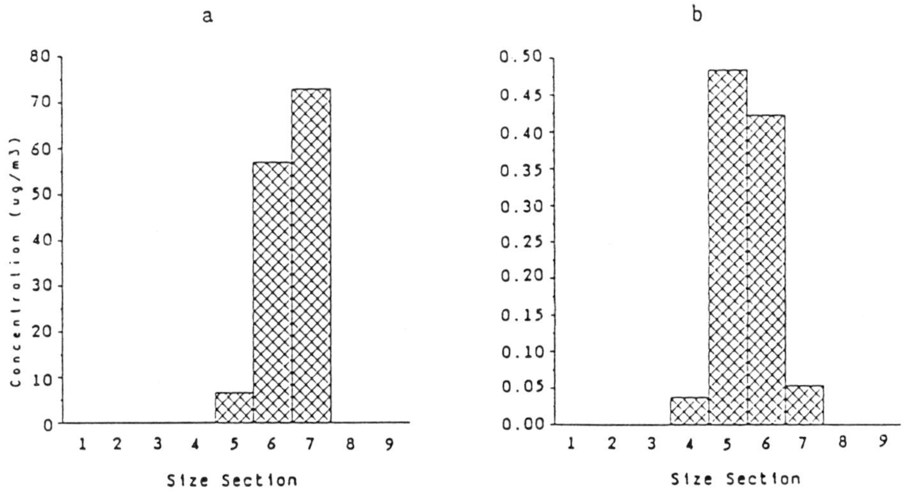

Fig. 1. Result from a one-cell version of the EAM simulating the rapid growth of a monodisperse aerosol from size section 1 to size section 6 showing (a) the mass distribution and (b) the normalized number distribution at the end of the simulation.

result in the growth of the initial particle population of from 0.016 µm to 0.825 µm in 100 s. The initial sulfate concentration of 0.001 µg m^{-3} was contained in size section 1 only. Wet and dry deposition were set to zero, but the coagulation algorithms in the EAM could not be deactivated for this test. Due to the rapid growth rate and monodisperse nature of the test population of particles, the coagulation effects were not significant. Figure 1 shows the final size distribution of sulfate mass and normalized particle number from the simulation. The transfer of mass into larger size sections appears to be somewhat over-predicted, while the transfer of particle number is somewhat under predicted. Size section 6 contains 0.417 of the analytic value for the sectional mass. Compared to the results of Warren and Seinfeld, this measure of size diffusion is reasonable considering that the EAM uses only 3 sections per size decade.

To test the EAM under more normal conditions of sulfate aerosol development with nucleation and coagulation, a second test with the one-cell version was performed. Here, the initial sulfate concentrations for each of the size sections was set to zero, and a moderate sulfate production rate was specified such that with no removal the total sulfate concentration would reach 150 µg m^{-3} after 72 h. Wet deposition was set to zero, however dry deposition of sulfate was modeled using a deposition velocity (v_d) of 0.3 cm s^{-1}. The resulting size distribution of sulfate concentration after 72 h is shown in Figure 2(a). These results still do not show the amount of size distribution spreading that is typically found in regional-scale EAM simulation results. The total sulfate concentration after 72 h was 32.9 µg m^{-3}, indicating that about 78 percent of the sulfate produced during the simulation had been removed through dry deposition by that time.

The transport of a sulfate particle through regimes of differing concentrations of water vapor, or any other volatile aerosol component,

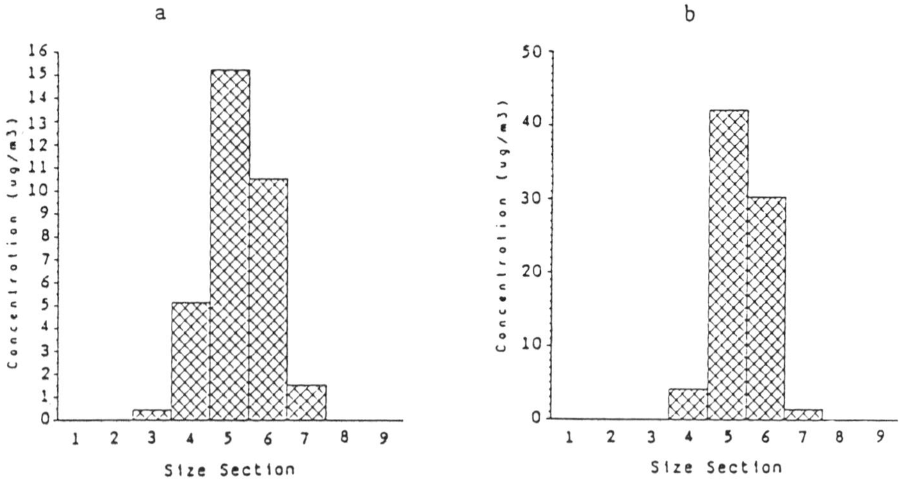

Fig. 2. Size distributions from the one-cell EAM after 72 h of sulfate production in an initially clean atmosphere using, (a) a v_d of 0.3 cm s^{-1}, and (b) Sehmel-Hodgson size dependent v_d estimates.

would result in many growth and shrinkage episodes during the lifetime of that particle. The daily thermal cycle creates alternating growth and shrinkage conditions for aerosol particles due to changes in the relative humidity. Such changing growth/shrinkage conditions appear to be the cause for the numerical size diffusion observed from regional application of the EAM. Increasing the number of sections per size decade in the EAM would reduce the numerical diffusion, but the time and funding resources for a significant modification of the EAM are not currently available, and the resulting increase in computational burden makes the possibility less attractive in a regional model.

SIZE DEPENDENT DRY DEPOSITION VELOCITIES

In order to mitigate the effects of numerical size diffusion in the EAM, the use of size dependent v_d estimates for particulate model species has been investigated. Investigators of dry deposition of particulate matter have determined from wind tunnel studies that the v_d for aerosol particles is size dependent (Sehmel and Hodgson, 1974; Sehmel and Hodgson, 1978; Slinn, 1982). For particle diameters greater than 1 μm, v_d becomes larger with increased eddy diffusion and gravitational settling. For particle diameters less than about 0.1 μm, v_d tends to increase with decreasing diameter due to increased Brownian diffusion across the laminar sub-layer between the deposition surface and the turbulently mixed atmosphere. By using these relatively larger v_d values for the largest and smallest particle sizes in the EAM, it was hoped that the speading of the simulated size distribution would be reduced.

The EAM was first developed with the v_d model used in both the RADM and the EM2 (Walcek et al., 1986). This technique considers meteorological variables and land use type in the estimation of a single v_d value for particulate sulfate, and does not address the particulate nature of any other chemical species. In addition to sulfate, the aerosol mechanisms in

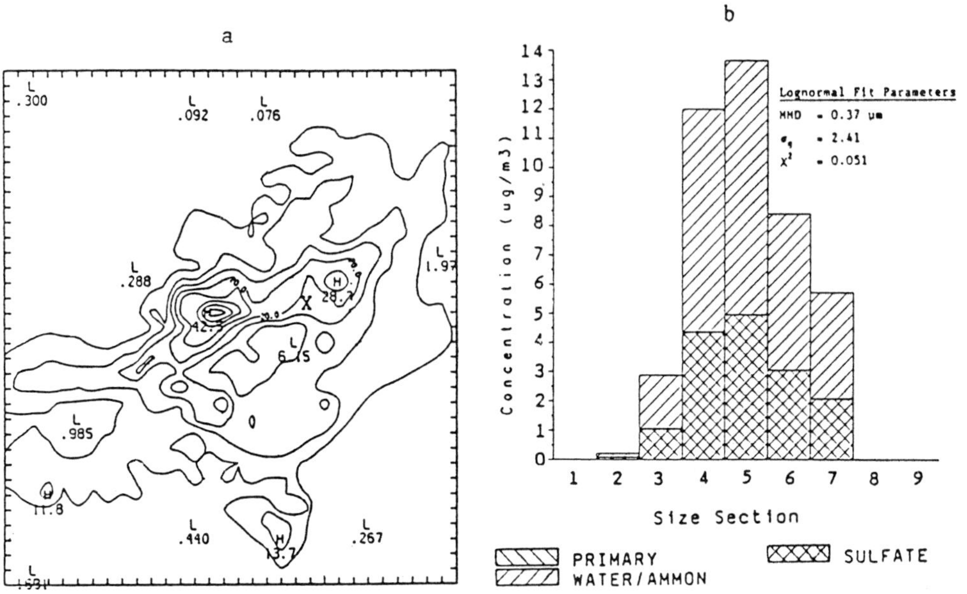

Fig. 3. Standard EAM output for layer 1 at hour 72 of the 17-20 July 1985 simulation, showing (a) the total sulfate aerosol concentration ($\mu g\ m^{-3}$) field, and (b) the aerosol size distribution and chemical composition at the location marked "X" in (a).

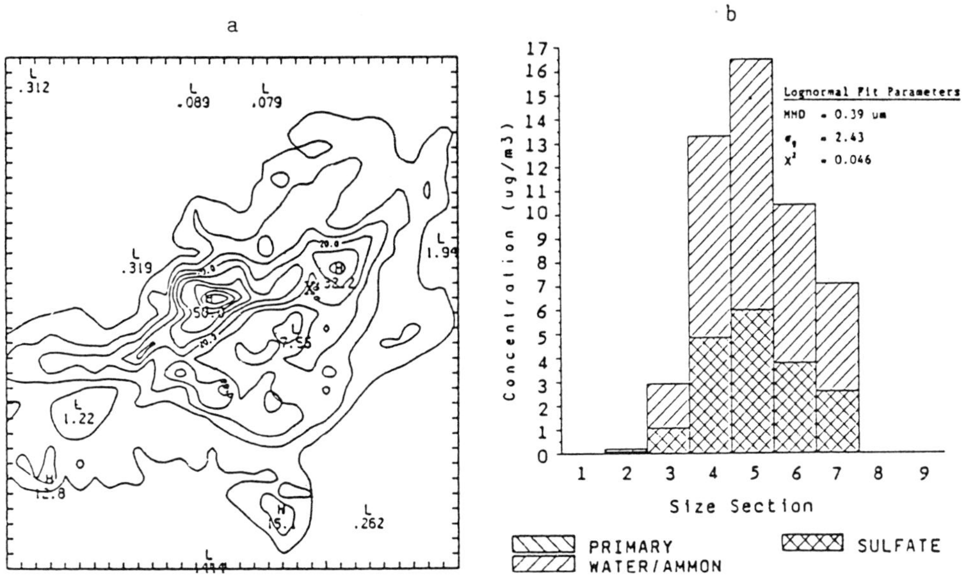

Fig. 4. Sehmel-Hodgson EAM output for layer 1 at hour 72 of the 17-20 July 1985 simulation, showing (a) the total sulfate aerosol concentration ($\mu g\ m^{-3}$) field, and (b) the aerosol size distribution and chemical composition at the location marked "X" in (a).

the EAM also consider nitrate, ammonium, and organic compounds as possible chemical components of the modeled particles, along with water and primary aerosol material such as soil derived and metallic dusts. It was therefore technically necessary to add to the EAM a v_d model which considers the conglomerate nature of the modeled aerosol particles.

The Air Quality Modeling Section of the California Air Resources Board (CARB) has produced a computer program which provides a refined analysis of v_d for particles based on the method of Sehmel and Hodgson (1978). This method considers the processes of Brownian diffusion, turbulent diffusion, and gravitational settling when estimating v_d for particles of specified diameter and density. The FORTRAN 77 program from CARB has been incorporated into the EAM as a subroutine providing v_d estimates for a large number of particle diameters evenly spaced across the range of all 9 size section used by the EAM. A trapezoidal approximation is used to obtain the area under the curve for each size section, and an integral average v_d is obtained.

The CARB subroutine requires as input the following variables; air temperature (T_a), surface roughness length (z_0), reciprocal of the Monin-Obukhov length (1/L), friction velocity (u_*), and the density of the aerosol particle (ρ_p). The value for ρ_p is supplied from the EAM aerosol mechanisms, while T_a, z_0, 1/L, and u_* are all obtained from the meteorological input files used by the EAM, EM2, and RADM.

To test the effect of the Sehmel-Hodgson v_d estimates on the size distribution spreading, a second 72 h test of the one-cell version of the EAM was performed using the Sehmel-Hodgson v_d values corresponding to the conditions used for the first 72 h test, specifically a temperature of 298 K, z_0 of 0.1 m, 1/L of -0.01 m^{-1}, u_* of 1 m s^{-1}, and a particle density of 1.5 g cm^{-3}. The results shown in Figure 2(b) show that the spreading of the size distribution is reduced, and that the maximum size-sectional concentrations are approximately doubled.

The effect of the Sehmel-Hodgson v_d estimates on regional EAM results was evaluated by comparison with results obtained from the unmodified EAM. Standard and Sehmel-Hodgson versions of the EAM were used to simulate the period from 0000Z 17 July 1985 to 0000Z 20 July 1985, over a model domain covering all of the eastern United States and parts of southern Canada. Early in this test period, a relatively pristine continental air mass began to cover the modeled region and eventually stalled over the area. This air mass was gradually modified such that widespread haze was reported over the test region by the end of the simulated period.

Figure 3(a) shows the total sulfate aerosol concentration field for layer 1 at the end (hour 72) of the standard EAM simulation. Figure 3(b) shows the corresponding layer 1 size-distribution statistics and chemical composition of the modeled aerosol at the location marked by the "X" in Figure 3(a). The site marked by the "X" was found to be representative of the rural areas in the modeled region that are regularly impacted by urban plumes. Figure 4 shows similar results from the Sehmel-Hodgson EAM simulation. The Sehmel-Hodgson results show larger concentrations, but no significant change in the geometric standard deviation (σ_g) of the best log-normal fit of the size distribution. The proportion of aerosol mass in the form of volatile components (water and ammonium) is nearly the same for each simulation, indicating that the thermodynamic balance has not been significantly modified by the Sehmel-Hodgson v_d estimates.

The larger sulfate concentrations from the Sehmel-Hodgson version of the EAM are due to its generally smaller v_d estimates. This has been demonstrated by repeating the 17-20 July 1985 simulation with a version of

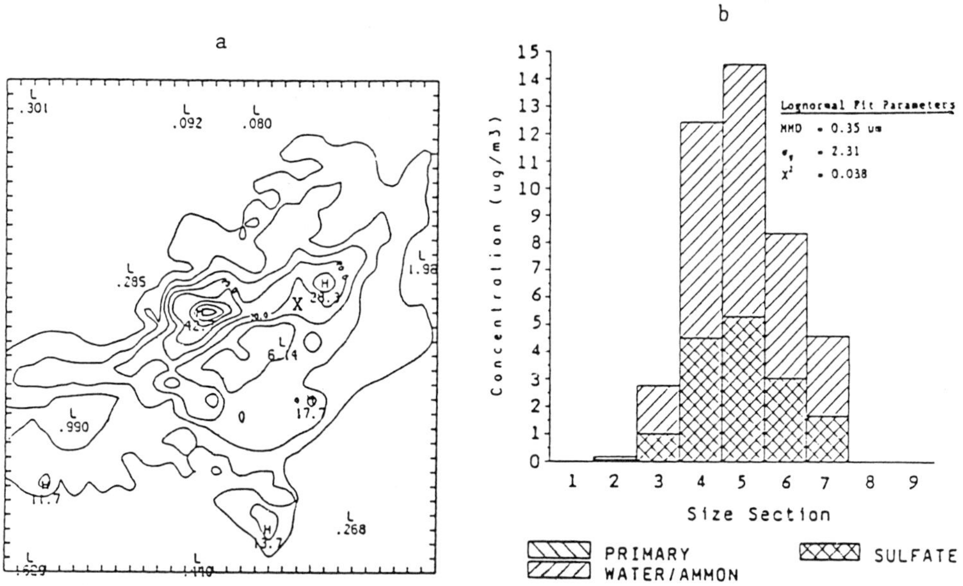

Fig. 5. "Scaled" Sehmel-Hodgson EAM output for layer 1 at hour 72 of the 17-20 July 1985 simulation, showing (a) the total sulfate aerosol concentration (μg m^{-3}) field, and (b) the aerosol size distribution and chemical composition at the location marked "X" in (a).

the EAM in which, for each grid cell and time step, the Sehmel-Hodgson v_d estimates are scaled so that the resulting total dry deposition rate for sulfate is equal to that obtained from applying the single v_d value used by the standard EAM. Figure 5(a) shows the simulated field of total sulfate aerosol concentration at hour 72 of the "scaled" Sehmel-Hodgson simulation. Figure 5(b) shows the layer 1 size-distribution statistics and chemical composition of the modeled aerosol at the location of the "X" in Figure 5(a). Figure 5(b) shows that the stronger influence of the scaled Sehmel-Hodgson v_d estimates resulted in a smaller σ_g and a better log-normal fit to the size distribution of mass.

CONCLUSIONS

The typical EAM simulated particle size distribution for locations outside of areas with large sulfur emissions has a σ_g value of around 2.4 for the best log-normal fit. This figure for σ_g is somewhat larger that the average continental value of 2.03 found by Whitby (1978). In a rapid particle growth simulation with a one-cell EAM, the magnitude of the size distribution spreading is reasonable when compared to the test results of Warren and Seinfeld (1985), considering that the EAM uses 3 sections per size decade. One-cell tests under more realistic conditions of sulfate production, dry deposition, coagulation, and balance between nucleation and condensation show that the size-dependent v_d estimates of Sehmel and Hodgson (1978) reduce the spreading of the particle size distribution and lead to significantly higher total sulfate aerosol mass concentrations. The use of Sehmel-Hodgson v_d values in a regional-scale application of the EAM has very little effect on the value of σ_g for the best log-normal fit of the resulting size distributions. When the Sehmel-Hodgson v_d estimates are scaled to match the total sulfate deposition rates obtained from the original EAM v_d values, σ_g values are reduced to approximately 2.3.

It is hypothesized that the strength of the numerical size-diffusion seen in regional EAM applications is attributable to repeated cycles of particle growth and shrinkage forced by a daily relative humidity cycle in the lowest model layers and by vertical advection and diffusion between higher relative humidity regimes in the upper mixed layer and lower relative humidity regimes in the lower mixed layer. An aerosol mechanism is now in development at AREAL which uses multiple, variable σ, log-normal modes to define the distribution of aerosol mass as a function of particle diameter. With this methodology, particle growth and shrinkage can be modeled as changes in the mean and standard deviation of the log-normal modes, thus eliminating the numerical diffusion associated with size-sectional aerosol characterization.

REFERENCES

Chang, J. S., R. A. Brost, I. S. Isaksen, S. Mandonich, P. Middleton, W. R. Stockwell, and C. J. Walcek. 1987. A three-dimensional Eularian acid deposition model: physical concepts and formulation. J. of Geophys. Res., 92:14,681-14,700.

Hudischewskyj, A. B., P. Saxena, and C. Seigneur. 1987. Development of computer modules of particulate processes for regional particulate model. EPA/600/3-87/021 (PB87-227 278), Atmospheric Sciences Research Laboratory, Research Triangle Park, NC, 270 pp.

Kleinman, L. I. 1988. Evaluation of SO_2 emission scenarios with a nonlinear atmospheric model. Atmos. Environ., 22:1209-1219.

Sehmel, G. A., and Hodgson, W. H. 1974. Predicted dry deposition velocities. Proc., Atmospheric Surface Exchange of Particulate and Gaseous Pollutants, USAEC Report CONS-74-0921.

Sehmel, G. A., and Hodgson, W. H. 1978. Model for predicting dry deposition of particles and gases to environmental surfaces. PNL-SA-6721, Battelle Pacific Northwest Laboratories, Richland, WA.

Slinn, W. G. N. 1982. Predictions for particle deposition to vegetative canopies. Atmos. Environ., 7:1785-1794.

Smolarkiewicz, P. K. 1984. A fully multidimensional positive-definite advection scheme with small implicit diffusion. J. Comput. Phys., 54:325-362.

Walcek, C. J., R. A. Brost, and J. S. Chang. 1986. SO_2, sulfate, and HNO_3 deposition velocities computed using regional land use and meteorological data. Atmos. Environ., 5:949-964.

Warren, D. R., and J. H. Seinfeld. 1985. Simulation of aerosol size distribution evolution in systems with simultaneous nucleation, condensation, and coagulation. Aerosol Sci. and Tech., 4:31-43.

Whitby, K. T. 1978. The physical characteristics of sulfur aerosols. Atmos. Environ., 12:135-159.

DISCUSSION

J. ALCAMO

Could you explain why you think the Sehmel-Hodgson model underestimates dry deposition velocity?

O.R. BULLOCK, JR.

The underestimation of dry deposition velocity by the Sehmel-Hodgson model in the EAM is due primarily to the fact that its parameterization was developed from wind tunnel studies with roughness elements of less than 1 cm in height. The model's basic concept of a laminar sub-layer of fixed thickness over the deposition surface works well as long as the surface roughness elements do not penetrate the sub-layer. The FORTRAN code that we used from the California Air Resources Board was formulated to use a maximum sub-layer thickness of 10 cm. The actual height of the surface roughness elements on the Earth's surface is often much larger than 10 cm. Therefore, I think it is safe to assume that in nature the Brownian diffusion across this laminar sub-layer is often short-circuited, leading to higher observed deposition rates than the model would estimate.

A. VENKATRAM

A great deal of the gas-phase sulfate is produced in the aqueous phase. How do you specify the initial size distribution for the aqueous-phase produced sulfate?

O.R. BULLOCK, JR.

Since the cloud model obtained from the RADM does not realistically simulate cloud droplet size, the size of the aerosol particles produced from cloud evaporation cannot be defined in a physically based manner. Currently, the quantity of sulfate produced in cloud water is passed to the EQL and MARS routines in the EAM and treated as an additional quantity of sulfate produced by the aqueous reactions on wet aerosol particles. When a cloud model is developed for the EAM which actually models the droplet size distribution, then a more realistic treatment of evaporative aerosol particles will be possible.

SIMULATION OF A WET DEPOSITION CASE IN EUROPE USING THE EUROPEAN ACID DEPOSITION MODEL (EURAD)

H. Hass, H.J. Jakobs, M. Memmesheimer, A. Ebel, J.S. Chang*

Institut für Geophysik und Meteorologie, Universität zu Köln
Projekt EURAD, Salierring 48, 5000 Köln 1, FRG

*Atmospheric Sciences Research Center, State University of New York, 100 Fuller Road, Albany 12205, NY, USA

INTRODUCTION

EURAD is a comprehensive multi-level Eulerian regional transport and deposition model. The modelling system contains two major modules, a mesoscale meteorological model (the PSU/NCAR model MM4; Anthes et al., 1987) and a chemistry-transport model (CTM, based on RADM; Chang et al., 1987; Hass et al. 1990). EURAD is applied for pollutant transport and deposition during an episode in February/March 1982. This episode was also simulated by the TADAP model within the PHOXA program (Scherer and Stern, 1989).

The major difference between the TADAP and the EURAD application is the generation of the meteorological fields. For the TADAP model short time forecasts of 12 or 18 hours were combined to a quasi diagnostic data set using a nudging technique. EURAD used prognostically derived data in a way which will be described later. Further differences between the two applications exist in geophysical input data which are constant for the simulated time, i.e. roughness lengths, landuse categories etc. and the preparation of the emission data set.

THE EURAD MODELING SYSTEM

The episode which had to be covered according to the EMEP air quality network was 20 February 1982, 6 GMT to 11 March 1982, 6 GMT. Hence, starting from 17 February 12 GMT this time period was covered by 8 consecutive forecasts with 3 or 2 days duration. The period 17 - 20 February is used for initialization. The horizontal resolution on a polar stereographic projection is 63.5 km for both x- and y-direction. The integration domain shown in Fig. 1 is covered by 59 x 44 grid points. In the vertical 15 layers of unequal thickness span the height between ground and 100 hPa using the σ-system. Further details how MM4 is used to predict the meteorological fields including precipitation is described by Hass et al. (1990).

The versions of CTM used in EURAD are based on RADM (Chang et al., 1987) but have several modifications. For this application version 2 was used and the differences to Chang et al. will be explained shortly.

Air Pollution Modeling and Its Application VIII, Edited by H. van Dop and D. G. Steyn, Plenum Press, New York, 1991

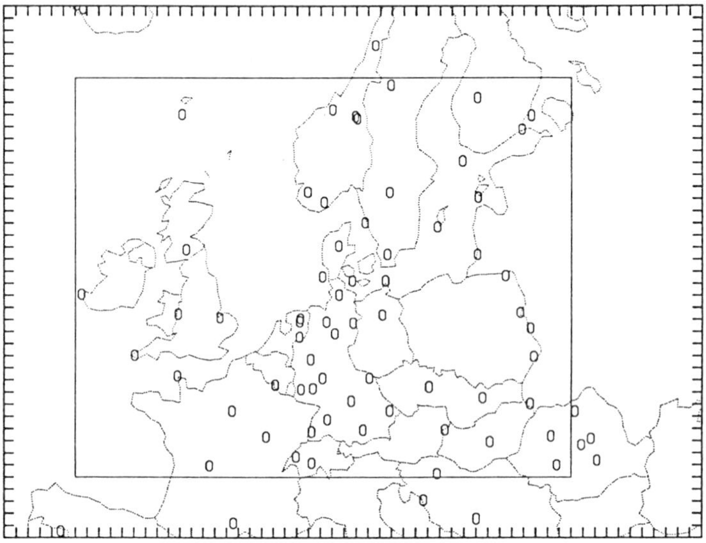

Fig. 1. EURAD modeling area for the February/March 1982 episode. EMEP measuring stations are indicated. The inner rectangle defines the TADAP modeling area (Scherer and Stern, 1990). Emissions are given only on this domain so that the rest of the area is more or less an extended boundary.

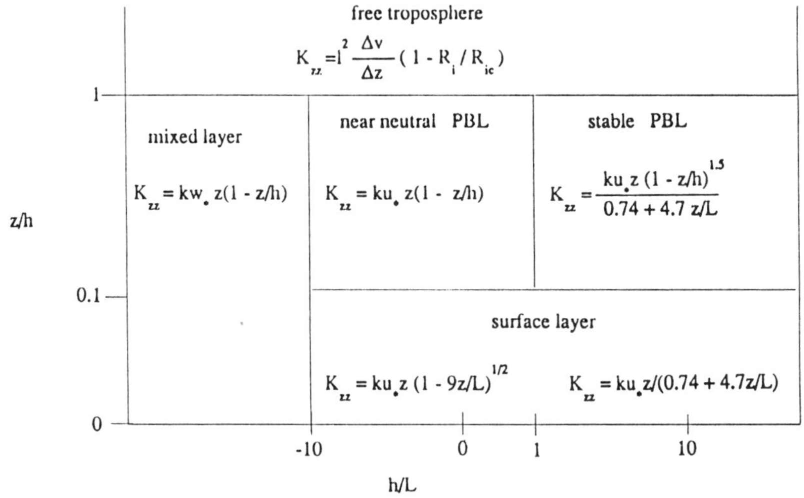

Fig. 2. Scaling regions of the atmospheric boundary layer as defined within CTM (simplyfied from Holtslag and Nieuwstadt, 1986). Applied K_{zz} formulations are also given. z is height, h is the PBL height, L is the Monin-Obukhov lenght, k is the von Karman constant, u_* is the friction velocity and w_* is the convective velocity scale.

Chang et al. (1987) used the method of fractional steps in a way which formally is of first order accuracy in time. To accomplish second order accuracy in time the operator splitting realized in CTM2 is the following for each transported constituent C (Mc Rae et al., 1982)

$$C^{n+1} = T_{cld}\ T_{a2}\ T_{d}\ T_{c}\ T_{c}\ T_{d}\ T_{a1}\ C^{n-1}. \qquad (1)$$

The operator that approximated advection numerically is denoted by T_a, diffusion by T_d and chemistry by T_c. Cloud processes are treated by T_{cld} and are incorporated in the integration schemes once per hour. All other operators use Δt = 150 s. Further, T_a is split into $T_{a1} = T_x\ T_y\ T_z$ and $T_{a2} = T_z\ T_y\ T_x$ in order that the operators commute on successive time steps. In CTM2 the advection scheme of Smolarkiewicz (1983) in the 1-D formulation is implemented and usually two corrective steps to remove most of the numerical diffusion are used.

Trace gases concentrations and transformations are predicted with the homogeneous chemical reaction mechansism of RADM2 (Stockwell et al., 1990). Different to Chang et al. (1987) we used a vertical subgrid-scale turbulent diffusion parameterization based on scaling regions of the atmospheric boundary layer (Holtslag and Nieuwstadt, 1986). Fig. 2 gives a simplified picture of the scaling regions together with the appropriate K_{zz} equations used by the model. The equations for K_{zz} are based on the specific turbulent structure of the regimes and are proposed by Panofsky and Dutton (1984), Brost et al. (1988) and Anthes et al. (1987). The solution of the diffusion finite differences scheme can be varied from fully implicit to a Crank-Nicholson method, the latter being used for this application.

INITIALIZATION, EMISSIONS

The simulation was started on 17 February 1982 with initial conditions being horizontally homogeneous but exhibiting vertical dependencies for some of the species. The used profiles represent clean conditions (Chang et al., 1987). These background concentrations are also used at the boundaries and are kept constant during the simulation. Following the initialization CTM is run for two days prior to the actual simulation period with realistic meteorology. For the considered episode this period is sufficient to change the conditions from clean background values to a strongly polluted situation over central Europe.

The emission data base created within the PHOXA program (Builtjes, 1989; Scherer and Stern, 1989) was slightly modified as to match the representative species in the gase-phase chemical mechanism. The applied splitting of the total VOC emissions into the hydrocarbon surrogate species is described by EURAD (1989).

RESULTS OF A "BASE CASE" SIMULATION

Within the episode the meteorological situation changes from a blocking situation to flow patterns with frontal systems crossing Europe rapidly. During the blocking situation pollutants can accumulate in the weak anticyclonic flow over Central Europe. Then westerly flow disperses pollutants northeastwards and due to washout by precipitation in the frontal systems the concentrations decrease to near background values.

This picture is shown in Fig. 3 for the observed SO_2 and sulfated concentrations in air at Deuselbach (FRG). Accumulation is preceeding until February 27 followed by the major frontal passage during March 1 and 2. The 24-hourly SO_x concentrations predicted by the model are in a comparable

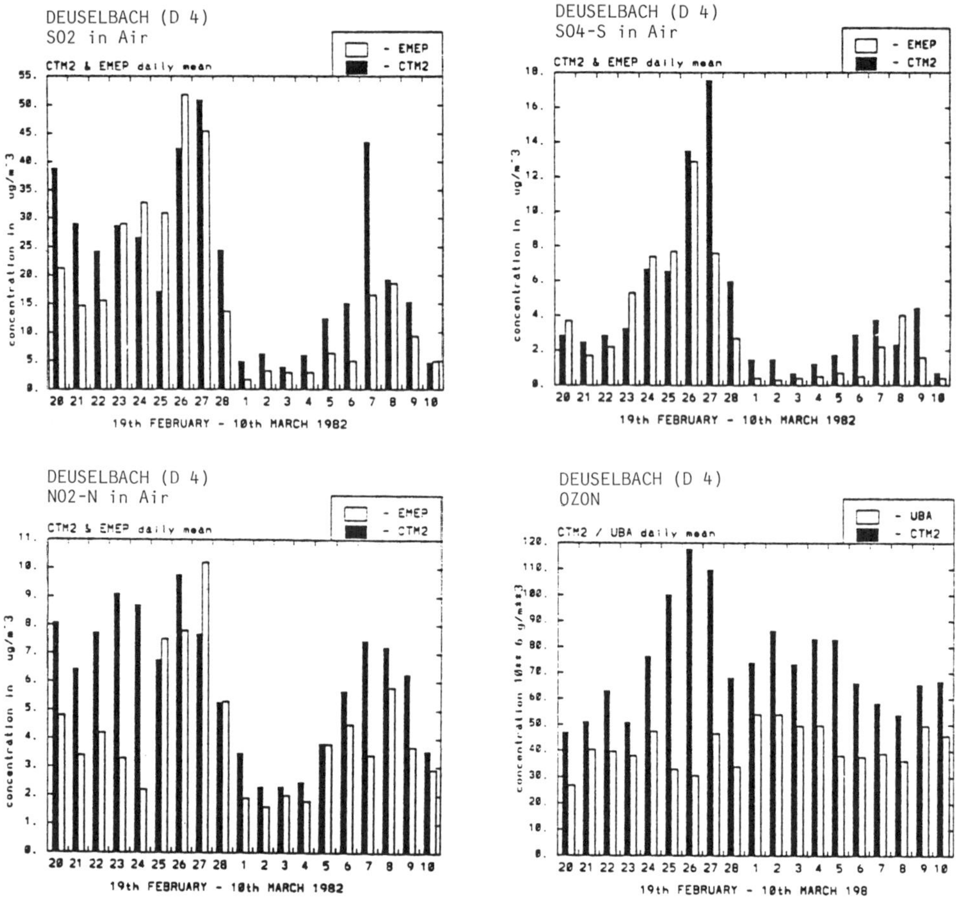

Fig.3. Predicted and measured 24-hourly (6 GMT to 6 GMT) concentrations of SO_2, SO_4, NO_2 and O_3 at Deuselbach (FRG).

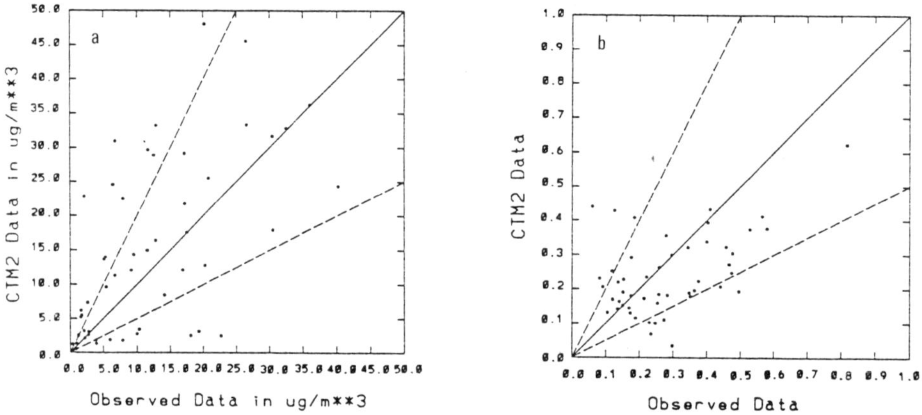

Fig. 4. Scatter diagramms of the observed versus predicted SO_2 episode mean air concentrations (a) and the SO_4/SO_x -ratio (b). Dashed live indicates the factor of 2 deviation.

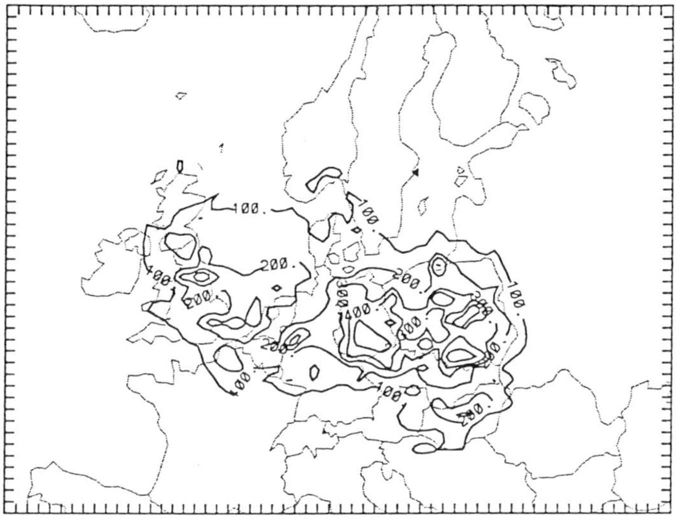

Fig. 5. Accumulated SO_2 dry deposition in mgS/m^2. Accumulation period: 19 Feb. 12 GMT - 11 March 12 GMT.

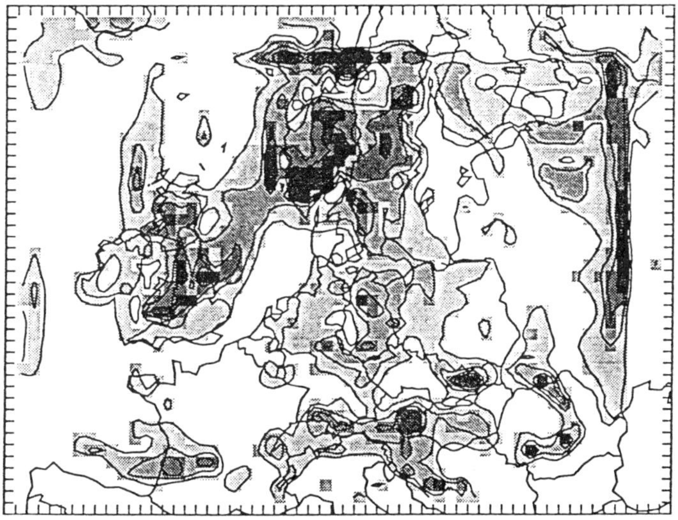

Fig. 6. As Fig. 5, but for sulfate wet deposition. Shading intervals: 20, 30, 40 mgS/m^2.

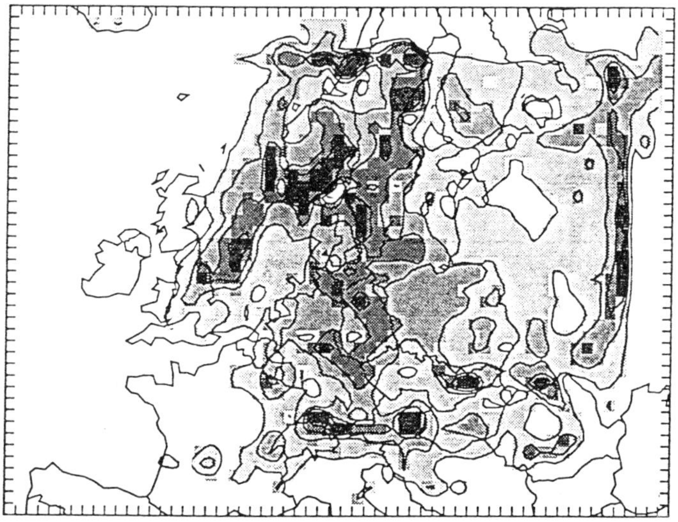

Fig. 7. As Fig.5, but for nitrate wet deposition. Shading intervals: 4,8,12,16 mgN/m^2.

agreement with observations. The same general trend is also visible for the NO_2 concentrations at Deuselbach. However, NO_2 is overpredicted by the model for the first five days of the episode. Ozone at Deuselbach does not exhibit the pronounced trend as observed and modeled SO_x and NO_2. There is even an indication for an opposite behaviour during February 25 to March 4. The model is overpredicting ozone for this rural area for all days of the episode. The inspection of similar figures for the rest of the EMEP measuring stations gives an overprediction of SO_2 concentrations especially for the first 3 days in the westerly part of the modeling region. Similar results were also gained by Scherer and Stern (1989).

The scatter diagramms of Fig. 4 gives the general performance of the model concerning the SO_x air concentrations at the measuring stations. Here the episode mean SO_2 and SO_4/SO_x-ratio of observation versus prediction is displayed. Most of the predicted ratios agree within a factor of two with the observed ratios. Displaying the scatter of the individual daily ratios would just fill the picture with more points but not change the overall view.

Accumulated SO_2 dry deposition is depicted in Fig. 5 and is reviewing the results obtained by Scherer and Stern (1990). Two maxima are clearly seen: near source deposition cause the high values in eastern Europe while the maxima over the southern North Sea are related to transport. The specific meteorological condition explains this deposition maxima (see Scherer and Stern (1990)).

The accumulated sulfate wet deposition pattern is given in Fig. 6. Distinct areas with more than 30 mg^N/m^2 are found over Great Britain, the North Sea, Scandinavia, the eastern alpine region and the GDR. Transport is the reason for the enhanced deposition over the North Sea, Scandinavia and the alpine region whereas near source deposition is the process causing most of the remaining spots.

The main sulfur deposition process in the central modeling region during this episode is SO_2 dry deposition (compare Fig. 5 and 6). Sulfate wet deposition is comparable to SO_2 dry deposition over southern Scandinavia and the alpine region. There are only minor SO_x sources in these areas so that they receive deposition following long range transport.

In general, the accumulated nitrate wet deposition distribution is exhibiting comparable structures as can be found in the sulfate deposition (Fig. 7). Sulfate wet deposition is a result of in-cloud oxidation of SO_2 with accompanying wash-out. Aqueous nitric acid is formed by scavenging of HNO_3 which first must be produced in the gasphase by NO_2 oxidation. The distribution of sulfate and nitrate deposition indicates a simular transportation path of SO_2 and HNO_3.

One reason for the development of state-of-the-art long range transport models was the question of how emission control affect the changes in deposition. Models like EURAD are most suited for analysing the effects of emission scenarios. Hence, a model simulation was performed with 50 % SO_x emission reduction for the whole emission area (Fig. 1). In another simulation NO_x emissions were reduced by 50 % across the emission domain.

Fig. 8 shows the percentage reduction in wet deposition due to the 50 % emission reduction. Reduction of sulfate wet deposition due to SO_x emission reduction is depicted in Fig. 8a and the respons in nitrate wet deposition due to NO_x emission reduction is shown in Fig. 8b. For both deposition distributions the change in the central modeling region is non-linear. For the SO_x-reduction this response is not only found over the areas with high emissions but also over the receptor areas like Scandinavia. Receptor areas have a more linear response with respect to NO_x reduction. Oxidant limitation

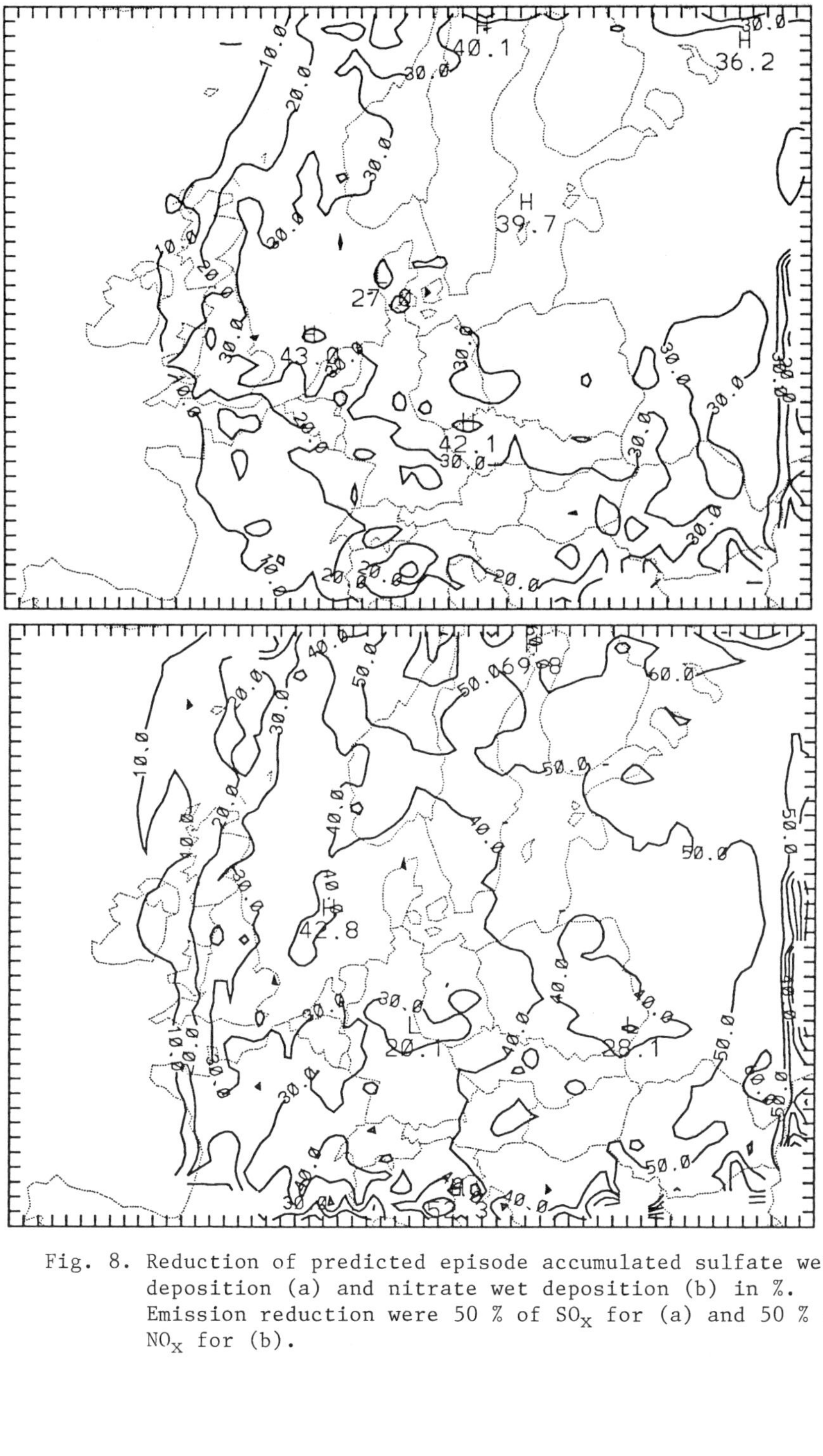

Fig. 8. Reduction of predicted episode accumulated sulfate wet deposition (a) and nitrate wet deposition (b) in %. Emission reduction were 50 % of SO_x for (a) and 50 % of NO_x for (b).

of the in-cloud SO_2 oxidation is likely the process controlling the receptor response to SO_x emission reduction for most areas.

SUMMARY

EURAD has been applied to a wintertime acid deposition episode in central Europe. Comparison with SO_x measurements showed that the episodic trends were modeled reasonalbly well. The distribution of the deposition patterns in general agree with observations too. A more detailed validation with respect to other pollutants like ozone is difficult as only a few observations other than SO_x are available for Europe for this episode.

ACKNOWLEDGEMENT

This work was funded by the Federal Ministry of Research and Technology (BMFT) of the FRG. Support concerning the emission data base came form the Umweltbundesamt. EURAD is a contribution to the European environmental research project EUROTRAC.

REFERENCES

Anthes, R.A., Hsie, E.Y. and Kuo, Y.H., 1987, Description of the Pennstate/ NCAR Mesoscale Model Version 4 (MM4), NCAR Technical Nore, NCAR/TN--282+STR., 66 pp.

Brost, R.A., Haagenson P.L., and Kuo, Y.-H., 1988, The effect of diffusion on tracer puffs simulated by a regional scale Eulerian model, J. Geophys. Res., 93, 2389-2404.

Builtjes, P.J.H., 1989, The use of emission data bases in air pollution dispersion modelling. Air pollution modeling and its application VII, H. van Dop (Editor) Plenum Press, 620 pp, New York and London.

Chang, J.S., Brost, R.A., Isaksen, I.S.A., Madronich, S., Middleton, P., Stockwell, W.R., and Walcek, C.J., 1987, A threedimensional Eulerian acid deposition model: Physical concepts and formulation, J. Geophys. Res., 92, 14681-14700.

EURAD, 1989, Das EURAD-Modell: Aufbau und erste Ergebnisse, Report 61, Institute of Geophysics and Meteorology, University of Cologne.

Hass, H., Memmesheimer, M., Geiß, H., Jakobs, H.J., Laube, M., and Ebel, A., 1990, Simulation of the Chernobyl radioactive cloud over Europe using the EURAD model, Atmos. Environm., in press.

Holtslag, A.A.M., and Nieuwstadt, F.T.M., 1986, Scaling the atmospheric boundary layer, Boundary-Layer Met., 36, 201-209.

McRae, G.J., Goodin, W.R., and Seinfeld, J.H., 1982, Numerical solution of the atmospheric diffusion equation for chemically reacting flows, J. Comp. Phys., 45, 1-45.

Panofsky, H.A., and Dutton, J.A., 1984, Atmospheric turbulence, John Wiley & Sons, New York, 397 pp.

Scherer, B., and Stern, R., 1989, Simulation of an acid deposition episode over Europe with the TADAP/ADOM Eulerian regional model, Air pollution modeling and its application VII, H. van Dop (Editor), Plenum Press., New York.

Smolarkiewicz, P.K., 1983, A simple positive definit advection scheme with small implicit diffusion, Mon. Wea. Rev., 111, 479-486.

Stockwell, W.R., Middleton, P., Chang, J.S., and Tang, X., 1990, The RADM2 chemical mechanism for regional air quality modeling, J. Geophys. Res., in press.

DISCUSSION

A. VENKATRAM

How does your error in the photolytic rates affect sulfate production rates? I noticed that your sulfate concentrations were estimated very well. Could you comment.

H. HASS

Only the first 3 days of the episode were simulated with an error in the photolysis data set (using sec instead of min). Thus, sulfate production rates are too large during these days. The model will adjust from this wrong "initialization" after about 2 more days of simulation.

NUMERICAL MODELLING OF THE ATMOSPHERIC TRANSPORT, CHEMICAL TRANSFORMATIONS AND DEPOSITION OF MERCURY

G. Petersen[1], B. Schneider[1], D. Eppel[1], H. Grassl[2]
A. Iverfeldt[3], P.K. Misra[4], R. Bloxam[4], S. Wong[4]
W.H. Schroeder[5], E. Voldner[5], J. Pacyna[6]

[1] GKSS Research Centre, Institute of Physics
Max-Planck-Str. 1
D-2054 Geesthacht, Federal Republic of Germany
[2] University of Hamburg, Meteorological Institute and
Max-Planck-Institute of Meteorology, Bundesstr. 55
D-2000 Hamburg 13, Federal Republic of Germany
[3] Swedish Environmental Research Institute
P.O. Box 47086, S-40258 Goeteborg, Sweden
[4] Ontario Ministry of the Environment
125, Resources Road, Rexdale, Ontario, Canada, M9W 5LW
[5] Atmospheric Environment Service
4905 Dufferin Street, Downsview, Ontario, Canada M3H 5T4
[6] NILU, Norwegian Institute for Air Research
P.O. Box 64, N-2001 Lillestrom, Norway

INTRODUCTION

During the last two decades a large number of surface waters in Scandinavia and North America have been impacted by mercury pollution, which has resulted in an increase of the mercury content of their fish. (Lindqvist et al., 1984; National Academy of Sciences, 1978; National Research Council Canada, 1979). This wide-spread pollution problem cannot be connected to local discharges, but is correlated with atmospheric transport and deposition of other air pollutants.

A very cost-effective means of delineating the atmospheric transport and deposition pathways of environmental contaminants is through numerical modelling. This paper presents the framework for applying a simplified atmospheric mercury chemistry scheme in a comprehensive Eulerian model (Misra et al., 1989) to be applied over eastern North America. A Lagrangian trajectory model (Eliassen and Saltbones, 1983; Petersen et al., 1989) with linear chemistry will be initially applied over Europe. The potential importance of natural emissions of mercury, as well as the development of anthropogenic mercury emission data bases, is also described in this paper.

ATMOSPHERIC MERCURY CHEMISTRY

The simplified scheme used in the models to describe the atmospheric chemistry of mercury is based on the present knowledge about the occurrence of various physical/chemical forms of mercury in ambient air and possible transformation processes involving mercury species in the gas (air) and/or the liquid phase (rain, cloudwater).

Air Pollution Modeling and Its Application VIII, Edited by H. van Dop and D. G. Steyn, Plenum Press, New York, 1991

Table 1. Concentration ranges (c) and Henry's law constants ($H = c_{gas}/c_{aq.}$, 20-25 °C) for mercury species in continental air masses.

Substance	c [ng m^{-3}]	H [v/v]	
Hg^0	2 - 4	0.29	(Sanemasa, 1975)
$(CH_3)_2Hg$	0 - 0.01	0.31	(Talmi and Mesmer, 1975)
CH_3HgCl	0 - 0.1	$1.9 \cdot 10^{-5}$	(Iverfeldt and Lindqvist, 1982)
$HgCl_2$	0 - 0.1	$2.9 \cdot 10^{-8}$	(Iverfeldt and Lindqvist, 1984)
particulate Hg	0 - 0.1		

Different species of atmospheric mercury have been identified by measurements. The reported concentration ranges (Bloom and Fitzgerald, 1988; Brosset, 1987; Iverfeldt and Rohde, 1988; Schroeder and Jackson, 1987) and their respective Henry's law constants are given in Table 1. Dimethylmercury is not considered in the model as the concentrations are very low and the high Henry's law constant indicates that the deposition of this compound is negligible. Monomethylmercury is also not taken into account because its concentrations in rainwater (Lee and Iverfeldt, 1990; Bloom and Watras, 1989) show that its contribution to the mercury deposition is of minor importance. Inorganic gaseous mercury ($HgCl_2$) makes up only a small fraction of the total mercury in air, but may contribute significantly to the deposition of Hg as its Henry's law constant is extremely low. However, $HgCl_2$ could not be included into the model at the present time since no data on the emission of this species is available. Hence, the model calculations presented here are confined to elemental Hg and to the fraction that is associated with particles.

Simplifications were also necessary with respect to chemical and physical transformations of atmospheric mercury. It is assumed that particulate Hg is chemically inert and that no exchange of Hg between the gas phase and particles occurs. For elemental Hg, an oxidation reaction has to be considered as this process controls the removal of Hg^0 from the atmosphere. Ozone is assumed to be the major oxidizing agent and may react with Hg^0 either in the gas phase (Pyankov, 1949) or in the liquid phase (Iverfeldt and Lindqvist, 1986). However, field measurements on gaseous oxidized Hg species show that the gas phase oxidation is probably less important than the oxidation in rain or cloudwater. For the aqueous phase reaction

$$Hg^0 + O_3 + H_2O \rightleftarrows Hg^{2+} + O_2 + 2\ OH^-$$

evidence was found experimentally (Iverfeldt and Lindqvist, 1986). By this process dissolved Hg^0 is removed from the solubility equilibrium resulting in a permanent flux of Hg^0 from the air into rain/cloudwater. The kinetics of the oxidation is described on the basis of a second order reaction:

$$\frac{d}{dt}[Hg^{2+}] = k_1 \cdot [Hg^0]_{aq.} \cdot [O_3]_{aq.} \quad (1)$$

It is assumed that the transfer of Hg^0 and O_3 from the air into water droplets is fast compared to the oxidation reaction. Consequently, $[Hg^0]_{aq.}$ and $[O_3]_{aq.}$ are close to equilibrium concentrations and may be expressed by the corresponding concentrations in air and the Henry's law constants:

$$\frac{d}{dt}[Hg^{2+}] = k_1 \cdot H_{Hg^0} \cdot H_{O_3} \cdot [Hg^0]_{gas} \cdot [O_3]_{gas} \quad (2)$$

A first guess for k_1 by Iverfeldt and Lindqvist (1986) covers several orders of magnitudes (10^5-10^8 dm^3 mol^{-1} s^{-1}) and is based on laboratory studies of the flux of gaseous elemental Hg into an ozone containing water phase. Hence, mass transfer processes as well as aqueous phase oxidation are involved. Experiences from homogeneous aqueous phase kinetics of mercury oxidation are also included in this estimate.

MERCURY EMISSION INVENTORIES

Anthropogenic Hg emissions have been independently estimated for both Europe and eastern North America. Emission factors are calculated separately for various source categories.

For Europe the most important source types are:

1. Combustion of coal in power plants and other industrial, commercial and residential boilers (68 %)
2. Chlor-alkali production (18 %)
3. Waste incineration (7 %)
4. Roasting and smelting operations in non-ferreous metal smelters (6%).

The emission estimates were then distributed spatially on the basis of geographical coordinates for point sources and information on the population density for area sources. The 1982 Hg emissions were distributed on the EMEP grid of 150 x 150 km^2 (Figure 1). Total anthropogenic emissions for Europe were estimated at ~ 400 t (Dornier, 1990). This

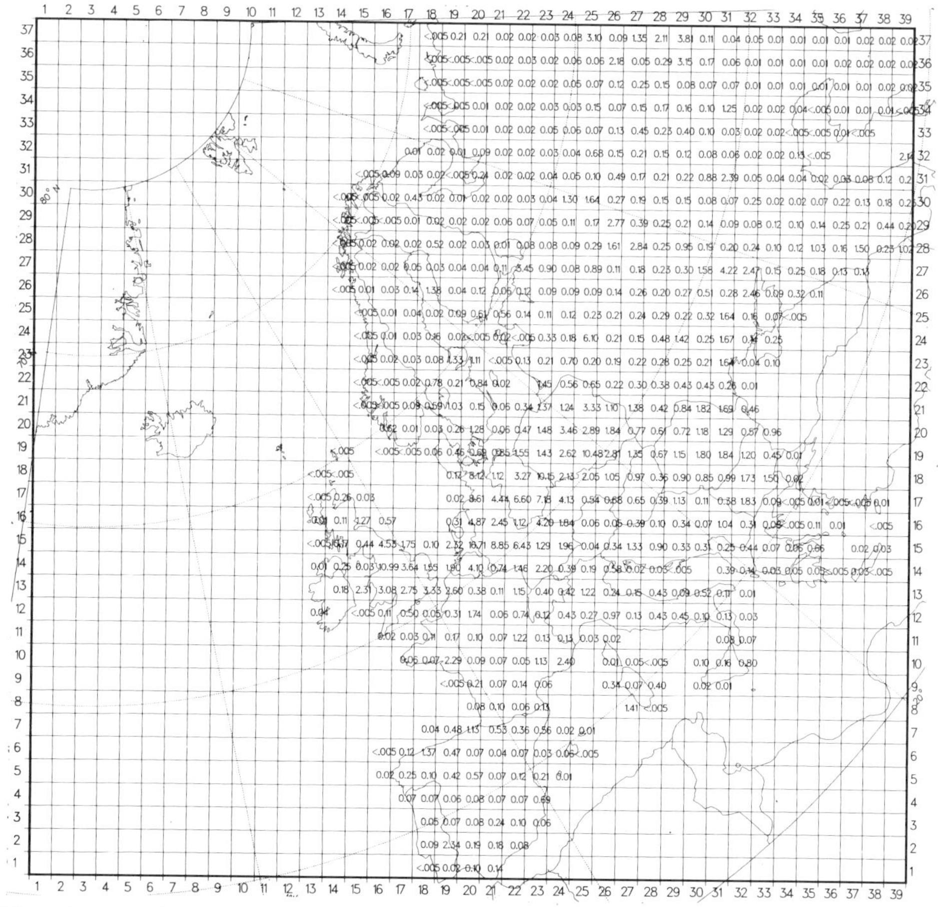

Fig. 1. Total anthropogenic mercury emissions for Europe - 1982 [tonnes/year] (Dornier, 1990)

inventory does not include Hg emissions from oil combustion nor its release from paint production/application and battery manufacturing.

Annual anthropogenic emissions in Canada (Jacques, 1984) and in the U.S.A. (Voldner and Smith, 1989) are estimated at about 31 and 650 tonnes, respectively. The Canadian emission rates refer to 1982. The base year for the U.S. source estimates varies from 1980 to 1985. The most important source sectors in the U.S.A. include fossil fuel combustion at over 20 %, use of consumer goods such as paints and electrical equipment at 55 %, non-ferreous smelters at about 10 %, and sewage sludge and municipal waste incineration at approximately 15 %. In Canada, base metal recovery is the largest contributor to the atmospheric emissions at over 45 %, followed by fuel combustion from stationary sources at 26 %. The third highest source of emissions is paint volatilization, at about 15 %. Incineration accounts for roughly 6 %, while chlor-alkali plants and agricultural chemicals production contribute about 3 % each.

A survey of the natural Hg emissions is presently underway in Europe but a very preliminary inventory is available for North America. For Canada, natural Hg emissions have been estimated at 3500 tonnes. Approximately 2000 tonnes are from vegegation and about 1400 tonnes from outgassing of soils and rocks, with small amounts of Hg emissions estimated from fresh water surfaces and fires (Environmental Protection Service, 1981). In the U.S. an estimate of natural emissions from soils and rocks of 1000 tonnes has been produced (Van Horn, 1975). More recent measurements indicate that these rates are too high (Schroeder, 1989; Canada/FRG Workshop, 1989). For comparison global natural emissions of

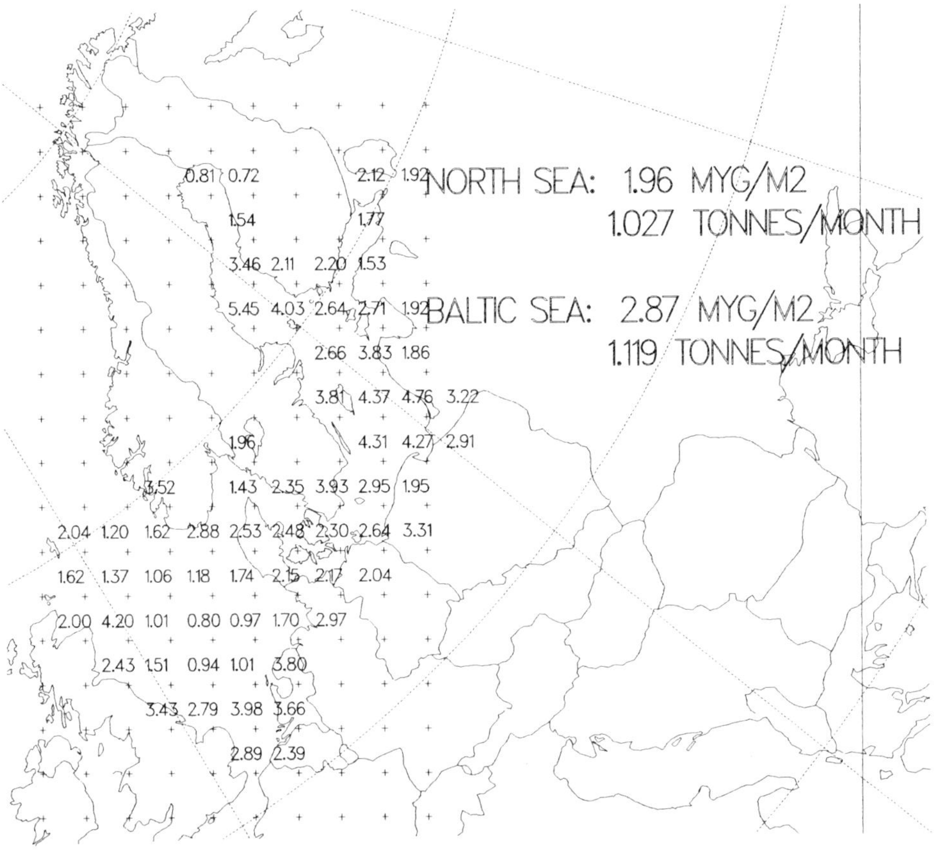

Fig. 2. Monthly total deposition of mercury [µg m^{-2} $month^{-1}$], January 1985

mercury have been estimated to be in the range 100 to 4900 tonnes with a median value of 2500 tonnes annually (Nriagu, 1989). In this study, we have assumed the natural emissions to be roughly 20 % of the official North American values.

RESULTS

Europe

A simplified 2 species mercury system (elemental mercury in the gas phase Hg^0 and particulate mercury, Hg(part.)) has been applied in the EMEP model. It is assumed that 95 % of the emitted mercury (Fig. 1) is Hg^0 and 5 % is in particulate form. Based on flux measurements taken in the 1970's a constant natural Hg^0 flux of 25 ng m^{-2} h^{-1} from land areas has been used in the model runs together with a global Hg^0 background concentration of 1.5 ng m^{-3}. Since Hg^0 has a very low water solubility the dry deposition of this species is zero. For Hg(part.) a constant dry deposition velocity of 0.2 cm s^{-1} is applied. Wet removal is parameterized by scavenging ratios, i.e. the ratio of mercury concentration in air to mercury concentration in precipitation. Values of 10^4 for Hg^0 (Lindqvist and Rodhe, 1985) and $5 \cdot 10^5$ for Hg(part.) have been used. The latter value is based on measurements of the Hg mass size distribution.

For a detailed investigation of deposition processes over the North Sea and the Baltic Sea a special receptor area consisting of grid elements which totally or partly cover those areas, was implemented into the model. An example for the calculated total monthly deposition of mercury into the North Sea and the Baltic Sea is shown in Fig. 2. The calculated fluxes as shown in the grid elements strongly depend on the precipitation rates, since about 90 % of the mercury is wet deposited. The average mercury flux into the Baltic Sea is about 50 % higher than for the North Sea due to higher precipitation amounts over the Baltic in this month. The calculations have been extended to the entire year 1985. The results for the North Sea and the Baltic Sea are summarized in Table 2 and Table 3, respectively. The calculated monthly means of concentrations in air fit well into the range of measured concentrations in Scandinavia (Iverfeldt and Rodhe, 1988). The calculated annual input of ~ 10 tonnes to the North Sea and ~ 12 tonnes to the Baltic correspond to 18.8 μg m^{-2} yr^{-1} and 30.2 μg $m^{-2}yr^{-1}$, respectively. These fluxes are in reasonable agreement with a measured annual flux of 27 μg m^{-2} a^{-1} at Roervik in southern Sweden in 1987 (ATMOS, 1988).

North America

As a first step for modelling the transport and deposition of mercury over North America the Eulerian model ADOM (Acid Deposition and Oxidant Model) has been adopted to examine a simple Hg^0 mercury system. The ADOM model includes 12 levels in the vertical. Modules for transport and diffusion, cloud mixing, aqueous phase chemistry, and wet and dry deposition are included in ADOM (Misra et al., 1989).

The emission inputs used in the model include anthropogenic and natural emissions over eastern North America as described earlier. At present the model is restricted to aqueous phase conversion of Hg^0 to soluble oxidized mercury (Equation 2). Ozone aqueous phase concentrations determined by the ADOM model are used in the model runs.

Since observations indicate a global Hg^0 background concentration, a boundary concentration field of Hg^0 in the range of 0.5 to 2.0 ng m^{-3} is employed for the lowest 2 km of the atmosphere. This provides a flux of mercury into the domain throughout the simulation period.

Fig. 3 shows the ground level air concentrations of Hg^0 for April 14, 1981 when the mercury transport and mixing algorithms were run for the period from April 10 to 14, 1981. The boundary concentrations of Hg^0

Table 2. Results from the trajectory model for the North Sea - 1985

	Concentration of elemental mercury ng m^{-3} gaseous	part.	deposition ta^{-1} dry	wet	total	precip mm
JAN	2.89	0.038	0.105	0.922	1.027	40.4
FEB	2.48	0.035	0.087	0.291	0.378	17.5
MAR	3.05	0.039	0.107	0.762	0.869	36.2
APR	1.85	0.014	0.038	0.668	0.706	54.7
MAY	2.18	0.029	0.082	0.430	0.512	23.2
JUN	2.08	0.023	0.062	0.711	0.773	41.3
JUL	2.37	0.032	0.090	1.086	1.176	53.1
AUG	1.81	0.014	0.040	0.979	1.019	78.5
SEP	1.89	0.024	0.054	0.797	0.851	57.2
OCT	2.50	0.044	0.121	0.444	0.565	27.6
NOV	2.17	0.025	0.067	0.732	0.799	56.9
DEC	2.23	0.020	0.055	1.168	1.223	88.2
MEAN	2.29	0.028	-	-	-	-
SUM	-	-	0.908	8.99	9.89	574.8

Table 3. Results from the trajectory model for the Baltic Sea - 1985

	Concentration of elemental mercury ng m^{-3} gaseous	part.	deposition ta^{-1} dry	wet	total	precip mm
JAN	3.49	0.054	0.111	1.008	1.119	53.2
FEB	3.97	0.070	0.130	0.551	0.681	23.1
MAR	4.64	0.068	0.140	1.092	1.232	40.7
APR	2.50	0.036	0.071	0.802	0.873	55.0
MAY	2.43	0.041	0.085	0.479	0.563	34.8
JUN	2.43	0.035	0.070	0.827	0.897	56.7
JUL	2.80	0.043	0.090	1.193	1.283	69.1
AUG	2.07	0.024	0.051	0.967	1.017	78.2
SEP	2.23	0.026	0.052	0.929	0.981	73.5
OCT	2.64	0.042	0.087	0.551	0.637	38.1
NOV	2.55	0.035	0.069	0.894	0.963	68.0
DEC	3.39	0.047	0.096	1.440	1.536	86.6
MEAN	2.92	0.043	-	-	-	-
SUM	-	-	1.052	10.733	11.785	677.0

Table 4. Comparison of model results and measurements for the Roervik site (annual means).

	concentrations in air [ng m^{-3}]	concentrations in precipitation [ng dm^{-3}]
calculated	2.6	36
measured	3.3	37

used for this model was 1.5 ng m^{-3} from ground level to 400 m, 0.5 ng m^{-3} to 1.5 km and 0.06 ng m^{-3} to 10 km. The concentrations shown in fig. 3 for Hg^0 are in the range of observed values.

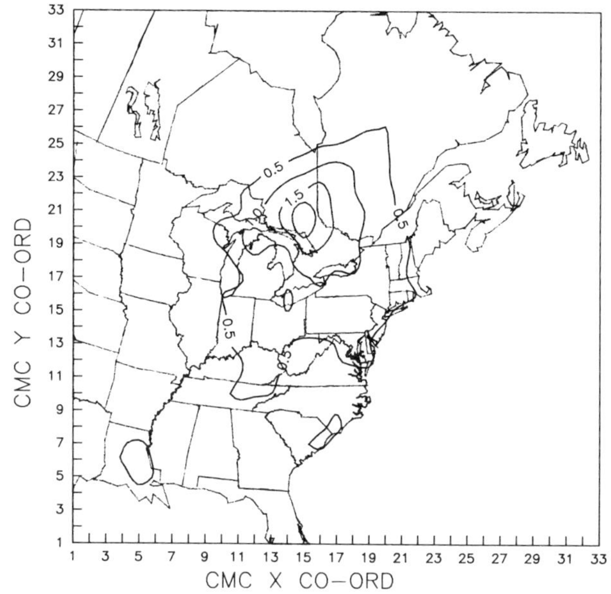

Fig. 3. Hg^0 ground level air concentrations April 14, 1981

DISCUSSION AND FUTURE WORK

The mercury concentrations in air calculated with the EMEP model for the Roervik site agree reasonably with the results from corresponding measurements (Table 4). However, this is partly due to the fact that the background concentration (1.5 ng m^{-3}) superimposed to the transported Hg makes up a major fraction of total Hg in air at this location. Also the agreement between calculated and measured concentrations in precipitation (Table 4) is to a certain degree not surprising as the Roervik precipitation data was used besides other data from Sweden to estimate the scavenging ratio applied in the model. Consequently, the model has to be validated also at other sites especially in regions close to anthropogenic source areas. But this requires that Hg species which are rapidly removed from the atmosphere (e.g. $HgCl_2$) have to be included into the model. Therefore, an attempt has been made to establish an emission inventory for gaseous $HgCl_2$ and to describe the emissions of particulate Hg in a more realistic way for future model calculations.

Moreover, a new chemical module for the liquid-phase chemistry of Hg has recently been developed which takes into account the complexity of oxidation, reduction, and adsorption processes in rain and cloudwater. To validate this chemical model and to determine the kinetic and thermodynamical constants involved, a field measurement campaign has been inititiated.

ACKNOWLEDGEMENTS

The Eulerian model studies on the mercury transport and deposition over eastern North America are carried out within the framework of the German/Canadian Cooperation Agreement in Science and Technology. The application of the long-range transport models over Europe is funded by the German Umweltbundesamt under contract 104 02 726.

REFERENCES

ATMOS (1988), Interim Report on Atmospheric Input to the North Sea. Convention for Prevention of Marine Pollution from Land-based Sources. Sixth Meeting of the Working Group on the Atmospheric Input of Pollutants to Convention Waters, Paris, November 15-17, 1988 (ATMOS 6/Info 19 E).

Bloom, N.S., Fitzgerald, W.F., 1988, Determination of Volatile Mercury Species at the Picogram Level by Low Temperature Gas Chromatography with Cold-Vapor Atomic Fluorescence Detection, Anal. Chim. Acta, 208: 151-161.

Bloom, N.S., Watras, C.J., 1989, Observations of methylmercury in precipitation, Sci Total Env., 87/88, 199.

Brosset, C., 1987, The Behavior of Mercury in the Physical Environment, Water, Air and Soil Pollut., 34: 145-166.

Canada/FRG Workshop, 1989, Modelling the Atmospheric Transport and Deposition of Mercury, Geesthacht, FRG, September 18-19, 1989.

Dornier, 1990, Europäische Modell-Emissionsdatenbasis Quecksilberkomponenten - 1982, F+E Vorhaben im Rahmen des UBA-Forschungsvorhabens 104 02 726 'Belastung von Nord- und Ostsee durch ökologisch gefährliche Stoffe am Beispiel atmosphärischer Quecksilberverbindungen'.

Eliassen, A., Saltbones, J., 1983, Modelling of Long-range Transport of Sulphur over Europe: a Two Year Model Run and Some Model Experiments, Atmos. Environ., 17: 1457-1483.

Environmental Protection Service, 1981, National Inventory of Natural Sources and Emissions of Mercury Compounds, Air Pollution Directorate, Environment Canada.

Iverfeldt, Å., Lindqvist, O., 1982, Distribution Equilibrium of Methyl Mercury Chloride between Water and Air, Atmos. Environ., 16: 2917-2925.

Iverfeldt, A., 1984, "Structural, Thermodynamic and Kinetic Studies of Mercury Compounds; Application within the Atmospheric Mercury Cycle", Ph.D. Thesis, Chalmers University of Technology, Goeteborg, Sweden.

Iverfeldt, Å, Lindqvist, O., 1984, The transfer of mercury at the air/ water interface, in W. Brutsaert and G.H. Jirka (eds.), Gas transfer at water surfaces, pp. 533-538, D. Reidel.

Iverfeldt, Å., Lindqvist, O., 1986, Atmospheric Oxidation of Elemental Mercury by Ozone in the Aqueous Phase, Atmos. Environ., 20: 1567-1573.

Iverfeldt, Å., Rodhe, H., 1988, Atmospheric Transport and Deposition of Mercury in the Nordic Countries, progress report prepared for the Nordic Council of Ministers, Swedish Environmental Research Institute, Goeteborg, No. L 87-285.

Jacques, A.P., 1987, Summary of Emissions of Antimony, Arsenic, Cadmium, Copper, Lead, Manganese, Mercury and Nickel in Canada, Internal Report, Environmental Analysis Branch, Conservation and Protection, Environment Canada, Ottawa, Ontario.

Lee, Y.H., Iverfeldt, Å., 1990, Measurement of methylmercury and mercury in stream-, lake-, and rainwater, in manuscript.

Lindqvist, O., Jernelöv, A., Johannson, K., Rodhe, H., 1984, Mercury in the Swedish Environment. Global and Local Sources, National Swedish Environmental Protection Board, Solna, Sweden, Report PM 1816.

Lindqvist, O., Rodhe, H., 1985, Atmospheric Mercury - a Review, Tellus, 37B: 136-159.

Misra, P.K., Bloxam, R., Fung, C., Wong, S., 1989, Non-linear Response of Wet Deposition to Emissions Reduction: A Model Study, Atmos. Environ., 23: 671-683.

National Academy of Sciences, 1978, An Assessment of Mercury in the Environment, U.S. National Academy of Sciences, Washington, DC.

National Research Council of Canada, 1979, Effects of Mercury in the Canadian Environment, National Research Council Canada, Ottawa, Ontario, Report No. 16739.

Nriagu, J. O., 1989, Natural versus Anthropogenic Emissions of Trace Metals to the Atmosphere, in: "Control and Fate of Atmospheric Trace Metals", J.M. Pacyna and B. Ottar, eds., Kluwer Academic Publishers, Dordrecht/Boston/London.

Petersen, G., Weber, H., Grassl, H., 1989, Modelling the Atmospheric Transport of Trace Metals from Europe to the North Sea and the Baltic Sea, in: "Control and Fate of Atmospheric Trace Metals", J.M. Pacyna and B. Ottar, eds., Kluwer Academic Publishers, Dordrecht/Boston/London.

Pyankov, V.A., 1949, Kinetics of a reaction between mercury vapor and ozone, Zhur. Obschev Khim. (J. gen. Chem.), 19: 224-229

Sanemasa, I., 1975, The Solubility of Elemental Mercury Vapor in Water, Bull. Chem. Soc. Jpn., 48: 1795-1798.

Schroeder, W.H., Jackson, R.A., 1987, Environmental Measurements with an Atmospheric Mercury Monitor Having Speciation Capabilities, Chemosphere, 16: 183-199.

Schroeder, W.H., Lindqvist, O., Munthe, J., 1989, Volatilization of Mercury from Natural Surfaces, Proc. 7th International Conference of Heavy Metals in the Environment, Geneva, Switzerland, September 12-15, 1989, Edt. Vernat, J.P., Publ. C.E.P. Consultants Ltd., Edinburgh, U.K., pp. 480-484.

Stevens, R.D.S., Reid, N.W., Schroeder, W.H., McLean, R.A.N., 1982, "The Chemical Forms and Lifetimes of Mercury in the Atmosphere", 65th Canadian Chemical Conference (CIC), Toronto, Ontario, Canada.

Talmi, Y., Mesmer, R.E., 1975, Studies on Vaporization and Halogen Decomposition of Methyl Mercury Compounds Using GC with a Microwave Detector, Water Res., 9: 547-552.

Van Horn, 1975, Material Balance and Technology Assessment of Mercury and the Compounds on a National and Regional Basis, EPA-560/3-75-007.

Voldner, E.C., Smith, L., 1990, Appendix 2 from the Workshop on Great Lakes Atmosperic Deposition, International Joint Commission, Windsor, Ontario, in press.

DISCUSSION

J.C.R. HUNT

1. Have you considered Switzerland as a source of Hg? I believe they incinerate a high proportion of their solid wast.

2. Please clarify the figure for the number of tonnes/year of Hg on the North Sea.

G. PETERSEN

1. Switzerland has been considered as a source of Hg, but the emissions in this country are of minor importance for the deposition of mercury to the North Sea.

2. The numbers in the grid squares are total mercury deposition fluxes in micrograms per square meter and month. The fluxes have been converted to the total input into the North Sea in tonnes per month by calculating an average monthly flux and multiplying this value by the North Sea areas of 525,000 km^2.

H.J.A. VAN JAARSVELD

What is known about the emission process of mercury? In which form is mercury emitted; is it mainly emitted as elemental mercury?

G. PETERSEN

In Europe about 70% of the anthropogenic mercury emissions is elemental mercury in the vapor phase, 25% are divalent inorganic mercury compounds and 5% is particulate mercury.

EFFECTS OF SPATIAL RESOLUTION ON TRACER FIELD SIMULATIONS WITH AN EULERIAN LONG-RANGE TRANSPORT MODEL (EURAD)

H.J. Jakobs, H. Hass, M. Memmesheiser, and Y.H. Kuo*

Institut für Geophysik und Meteorologie, Universität zu Köln
Projekt EURAD, Salierring 48, 5000 Köln 1, FRG

*National Center for Atmospheric Research
P.O. Box 3000, Boulder, CO 80307, USA

INTRODUCTION

The reliability of transport calculations of atmospheric constituents is highly dependent on the transport mechanism and obviously on the choice of the horizontal and vertical resolution of the model domain. To examine this effect several numerical simulation experiments have been conducted with the European Acid Deposition (EURAD) model. The investigated tracer field used for this study was the release of nuclear particles during the Chernobyl accident (April 25 - May 10 1986).

MODEL DESIGN

The EURAD model consists of two main components: The Mesoscale Meteorological Model MM4 (NCAR/PSU mesoscale model, version 4; Anthes et al., 1987) and the Chemical Transport Model (CTM), which is based on RADM (Regional Acid Deposition Model; Chang et al., 1987). In the forecast mode of EURAD, the preprocessors of MM4 provide the initial and boundary conditions for the meteorological fields, using the ECMWF global analysis. In addition the terrain and landuse data are incorporated for the forecast. Then MM4 provides the prognostic variables for the Chemical Transport Model. For this study CTM is designed only for transport and both wet and dry deposition calculations using a point source of emitted radioactive particles at Chernobyl. No losses through chemical transformation, but only through radioactive decay of the elements have been considered.

SIMULATION EXPERIMENTS

Several simulation experiments have been conducted with the EURAD model for the above mentioned episode after the accidental release from the nuclear power plant in Chernobyl (April 25 - May 10 1986). The radioactive constituents Iodine (I-131) and Cesium (Cs-137) are treated as released particles. The emission scenario was taken from the International Atomic Energy Association (IAEA) (see Hass et al. (1990)).

The EURAD model can run in two different modes. In the forecast mode

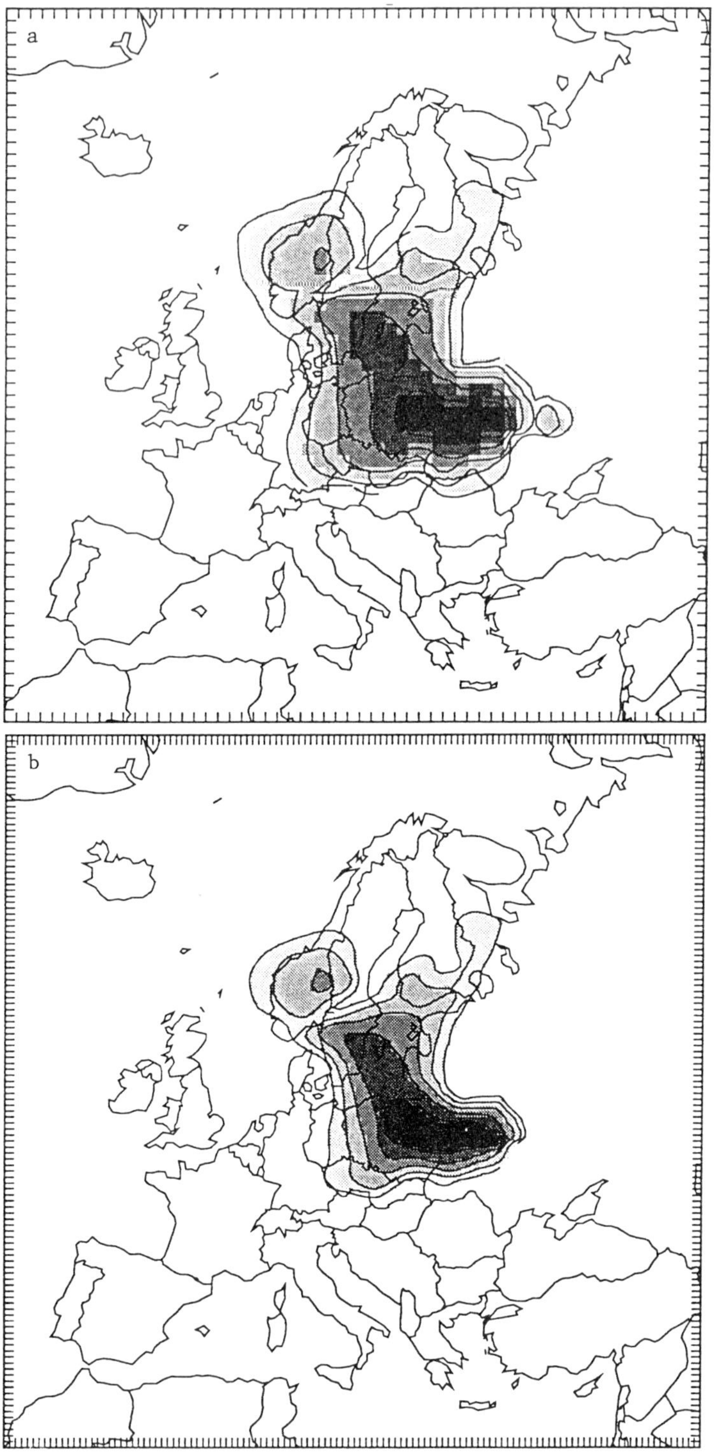

Figure 1. Horizontal distribution of the Cs-137 concentration (Bq/m^3) in the lowest model layer at April 28 1986 1200 UTC a) for the experiment with 80 km horizontal grid size and b) for the experiment with 40 km grid size. Results for the analysis mode of EURAD. Contour lines are 0.001, 0.01, 0.1, 1 and 10 Bq/m^3.

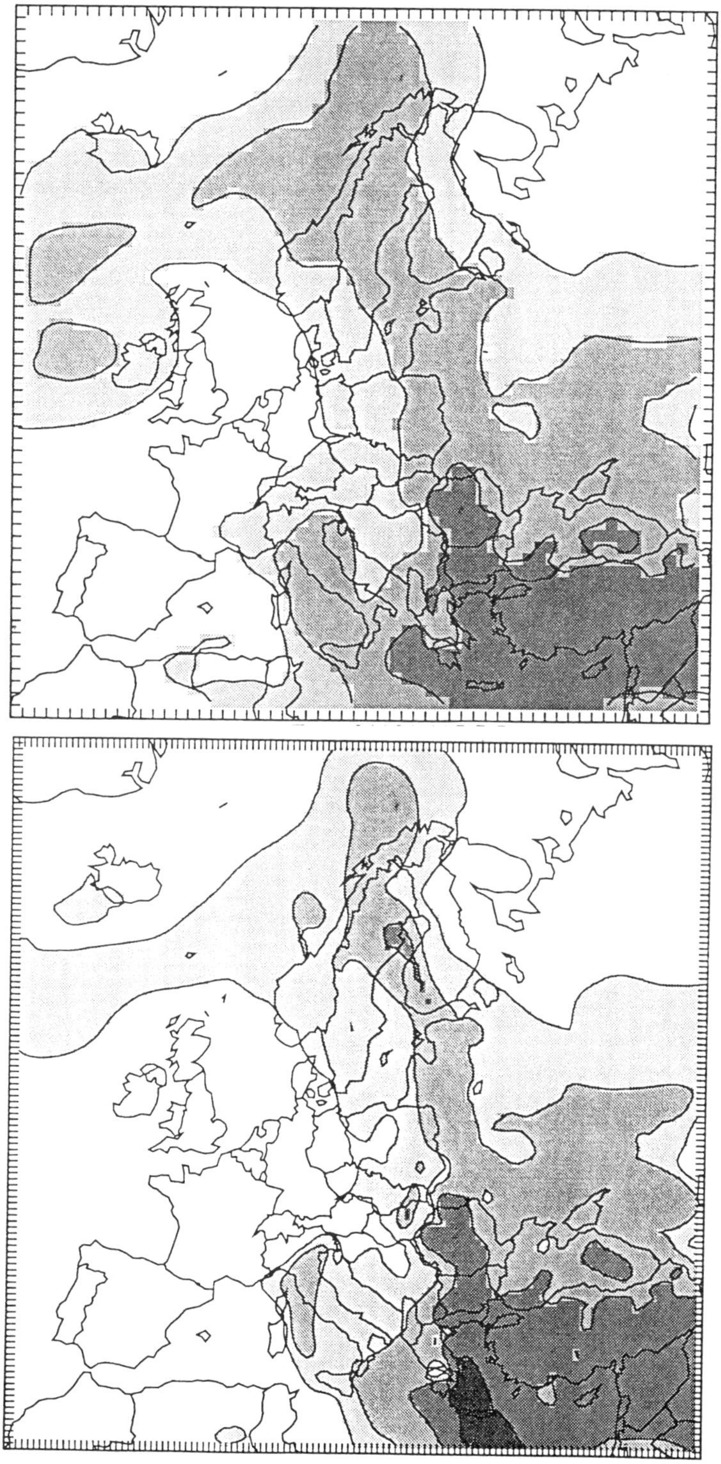

Figure 2. Same as Fig. 1, but at May 10, 1986 1200 UTC

MM4 provides the meteorological fields (including precipitation) as prognostic variables (usually a 3 day forecast). In the analysis mode the meteorological fields from the 6 hour ECMWF global analysis are interpolated on the selected model domain for the selected vertical layers. For this mode the precipitation data are taken from the observed data base from the German Weather Service. The simulation experiments have been conducted in both modes to demonstrate the skill of the forecast.

In addition the effect of horizontal grid resolution was investigated. Therefore two horizontal grid sizes are incorporated for both modes of EURAD in the simulation, i.e. 80 km and 40 km (61 x 61 and 121 x 121 gridpoints in x- and y-direction).

For both modes the following additional parameters and physical processes are considered for the experiments: A projection on a Lambert conformal plane, 15 vertical layers on σ-coordinates. The numerical techniques and physical processes of CTM and MM4 (in the forecast mode) are described in more detail by Hass et al. (1990).

RESULTS

Four different simulation experiments are conducted for the period April 25 1200 UTC - May 10 1200 UTC (in the forecast mode only the first 9 days with 3 consecutive 3 day forecasts). First the results from the experiments using the analysis mode are presented.

Figure 1 shows the horizontal distribution of Cs-137 in the lowest vertical layer at April 28 1986 1200 UTC for the experiments with the analysis mode of EURAD. The coarse structure of both transport patterns is nearly the same. But in the simulation experiment with higth resolution a more detailed structure is indicated. Also the experiment with 80 km gridsize demonstrates a higher dispersion pattern due to higher degree of numerical diffusion. The tracer field reaches southern parts of Germany. It is notable that the analysis for the meteorological fields is the same for both experiments. This different transport behaviour is more significant at the end of the investigated episode (May 10 1986 1200 UTC; Figure 2). The experiment with 40 km gridsize shows again a much more detailed structure, whereas in the low resolution experiment the spreading of the tracer field is much more evident after 15 days of time integration.

Figure 3 displays the accumulated wet deposition of Cs-137 on May 1 1200 UTC. The deposition also exhibits a slightly different pattern for both horizontal resolution experiments. Due to the spreading effect demonstrated earlier, the areas of deposited radioactive material is larger for the experiment with 80 km resolution.

The experiments using the EURAD model in the forecast mode reveal quite different results for the transport of the radioactive tracers. Figure 4 shows the horizontal distribution of Cs-137 at April 28 1200 UTC. In comparison to the experiments in the analysis mode (e.g. Fig. 1), these experiments exhibit a highly different transport behaviour for the tracer field. It seems that in the forecast mode the wind velocities are much stronger than in the analysis mode, i.e. MM4 predicts stronger pressure gradients than the analysis. The reason for that finding may be that MM4 and other mesoscale meteorological models tend to have a cold bias. The different dispersion behaviour and horizontal structure for the two different horizontal resolutions are more pronounced for this mode. In addition in the forecast mode the flow over the Baltic Sea is more southerly whereas in the analysis mode a more south-easterly flow is observed. These differences must lead to different arrival times (earlier for the forecast mode) of

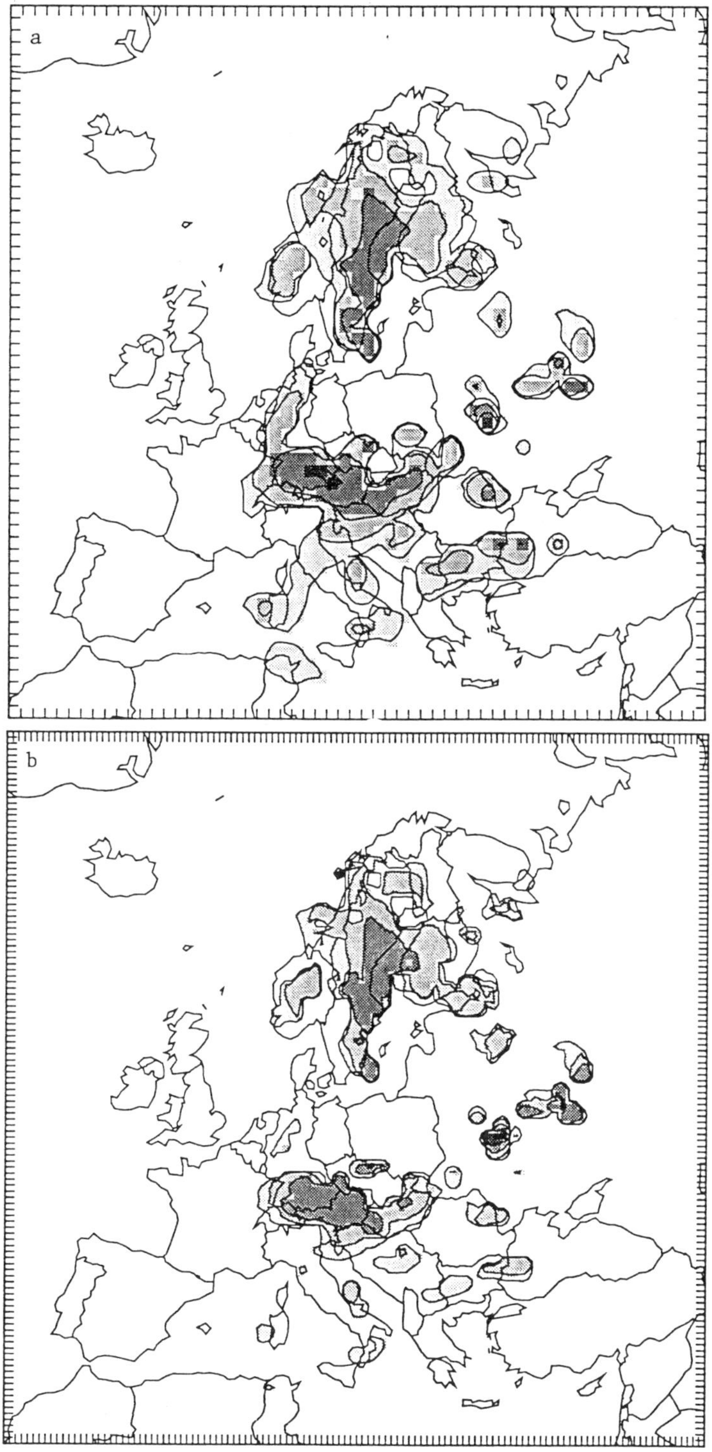

Figure 3. Horizontal distribution of the accumulation wet deposition of Cs-137 (kBq/m^2) at May 01, 1986 1200 UTC a) for the experiment with 80 horizontal grid size b) for the experiment with 40 km horizontal grid size. Results for the analysis mode of EURAD. Contour line are 1, 10, 100 and 1000 kBq/m^2.

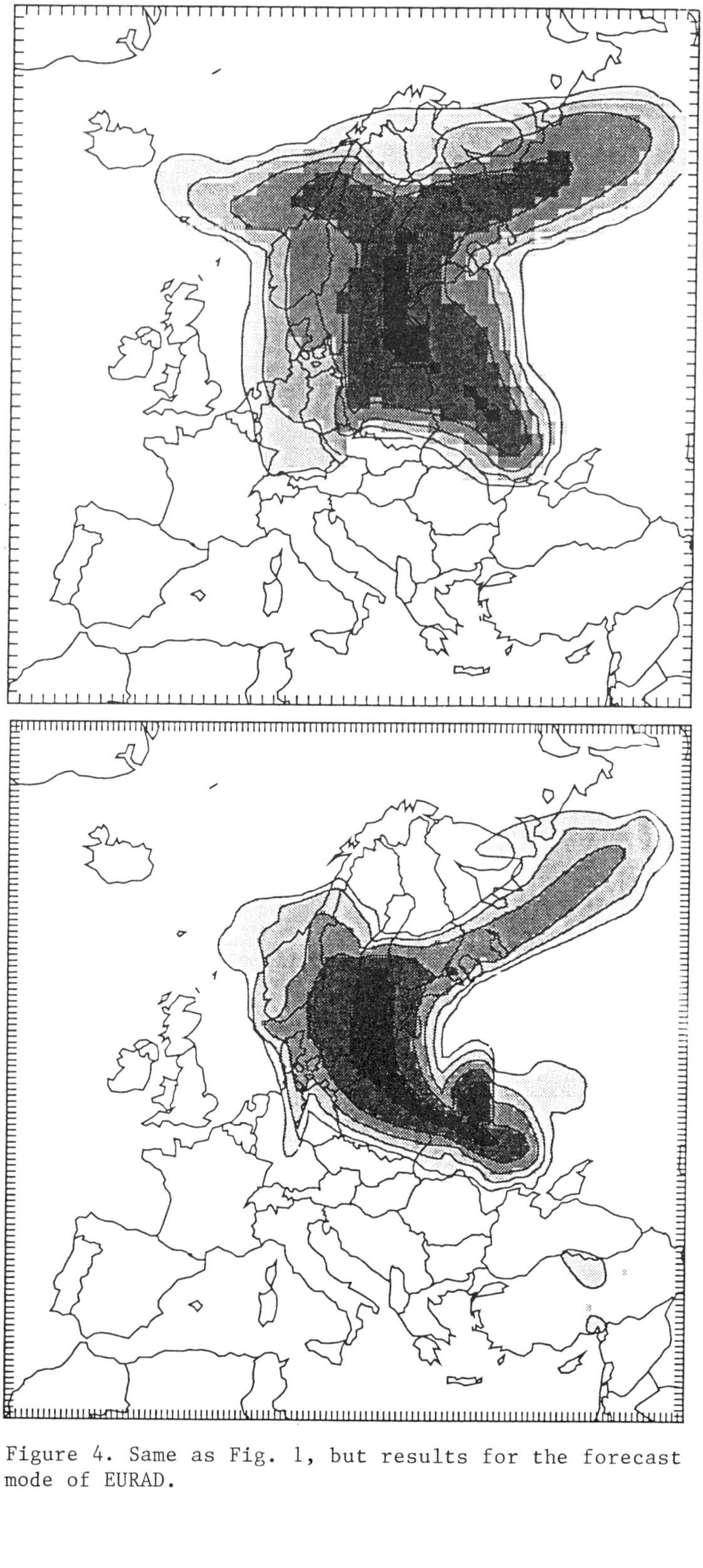

Figure 4. Same as Fig. 1, but results for the forecast mode of EURAD.

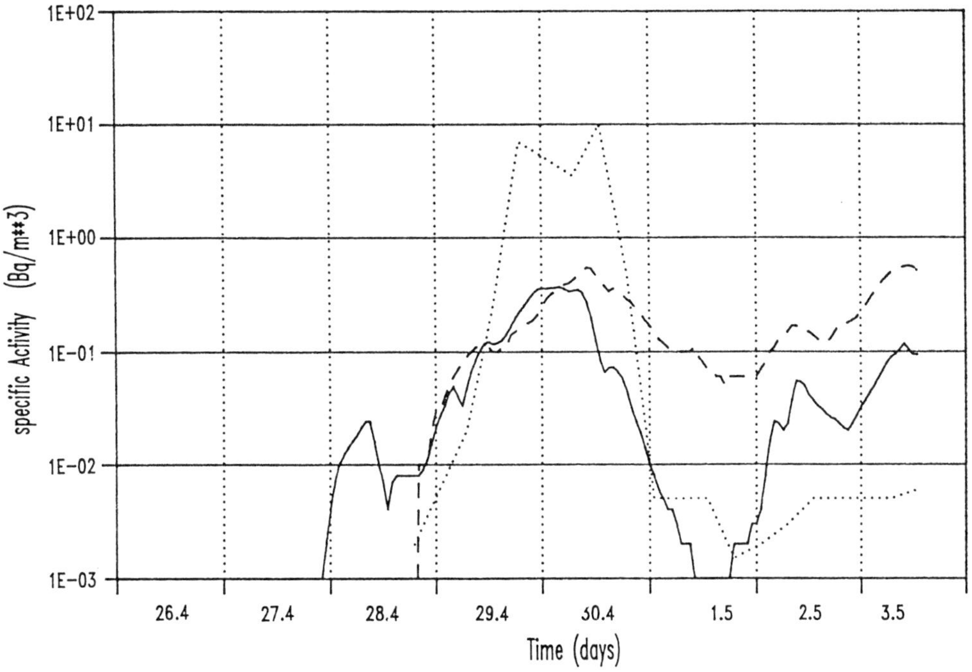

Figure 5. Comparison of the simulated Cs-137 concentration with observation at the station Berlin. Dotted curve: measurement, dashed curve: simulation with 80 km horizontal grid size, solid curve: simulation with 40 km horizontal grid size. Results for the analysis mode of EURAD.

radioactive material in Scandinavia and later to Central Europe.

But preliminary investigations concerning the evaluation of the simulation experiments indicate that the results for the forecast mode show a more realistic arrival time and magnitude of the measured radioactive cloud at least for the first three days of the forecast. At this point it should be noted that in principle the EURAD model could run in a forecast mode for operational applications like emergency cases.

Figure 5 shows the comparison of our simulations in the forecast mode with different horizontal resolution with measurements at a selected station. This plot clearly demonstrates that the experiment with high resolution (40 km) reproduces better the observed trend of Cs-137 at this station in Berlin. Strong differences appear after the maximum on April 30.

CONCLUSIONS

The different simulation experiments clearly demonstrate the different transport behaviour of the tracer field for different horizontal resolution. It seems that the experiments with high resolution display a more realistic dispersion of the radioactive cloud in comparison with observation. Additional comparisons with ground-based observation will be done in connection with the ATMES study for evaluation the Chernobyl case.

REFERENCES

Anthes, R. A., Hsie, E. Y., and Kuo, Y. H., 1987, Description of the Penn State/NCAR mesoscale model version 4 (MM4), NCAR Technical Note, NCAR/TN-282+STR.

Chang, J. S., Brost, R. A., Isaksen, I. S. A., Madronich, S., Middleton, P., Stockwell, W. R., and Walcek, C. J., 1987, A three-dimensional Eulerian acid deposition model: Physical concepts and formulation, J. Geophys. Res., 92, 14681-14700.

Hass, H. Memmesheimer, M., Geiß, H., Jakobs, H. J., Laube, M., and Ebel, A., 1990, Simulation of the Chernobyl radioactive cloud over Europe using the EURAD model, Atmos. Environm., in press.

ESTIMATION OF THE ANNUAL CONTRIBUTION OF US SULFUR EMISSIONS TO CANADIAN ACIDIC DEPOSITION AND VICE VERSA

Francis S. Binkowski[1], Julius S. Chang[2], John N. McHenry[3]
Steven D Reynolds[4], Richard D. Cohn[5]

[1]U.S. Environmental Protection Agency, Atmospheric Sciences Modeling Division, Research Triangle Park, North Carolina (On Assignment from NOAA)

[2]Atmospheric Sciences Research Center
State University of New York at Albany, Albany, New York

[3]Computer Sciences Corporation
Research Triangle Park, North Carolina

[4]System Applications, Inc.
San Rafael, California

[5]Analytical Sciences, Inc.
Research Triangle Park, North Carolina

INTRODUCTION

The determination of a source-receptor relationship for acidic deposition has been one of the major thrusts of the National Acid Precipitation Assessment Program (NAPAP). The specific source-receptor relationship considered here is that showing the relative contribution of U.S. and Canadian sulfur emissions to deposition amounts. The results presented here are from a comprehensive Eulerian mathematical modeling system consisting of a core model, the Regional Acid Deposition Model (RADM) (Chang *et al.*, 1987; Chang *et al.*, 1990), a meteorological processor, MM4, (Seaman, 1989) and several interpretative tools. RADM consists of components which represent various atmospheric processes including: gas-phase photochemical production of ozone, hydrogen peroxide, and other oxidants from emissions of reactive organic compounds and oxides of nitrogen; oxidation of sulfur dioxide in both clear air and within cloud water; three-dimensional transport, mixing by clouds, and wet and dry deposition. The meteorological processor, MM4, which provides the information required as input for RADM is a mesoscale numerical weather prediction model using four-dimensional data assimilation (Seaman et al., 1990). The RADM is used episodically, that is, emissions and meteorological information for a specific weather event of three days duration are used for simulating the air concentration of important photochemical oxidants and other chemical species as well as the accumulated material which is deposited to the surface. The output files are then archived for future analysis using the interpretative tools. For the present discussion, the tool of choice is the Tagged Sulfur Engineering Model (TSEM).

TAGGED SULFUR ENGINEERING MODEL (TSEM)

The TSEM, developed to examine source-receptor relationships and to study sulfur reduction strategies, uses the archived RADM output files along with the same meteorological information used for the original simulation. These are combined with a sulfur emissions inventory which may be reduced, tagged, or both. The chemistry component of TSEM uses archived values of the oxidants (ozone, and the hydroxyl and hydroperoxyl radicals). Thus, sulfur oxidation by hydroxyl in clear air and by hydrogen peroxide and ozone in cloud water, as well as the production and loss of hydrogen peroxide, is explicitly represented in TSEM. Kleinman (1988) has shown that the governing mathematical equations of TSEM can be split in the manner described. He also demonstrated that the results are exact within machine accuracy for the same sulfur input inventory and differ slightly, on the order of a few percent , when the sulfur emissions are *increased* by a factor of four. With a *decrease* of sulfur emissions by a factor of four, the difference becomes much smaller, on the order of one percent or less.

The reason that this splitting of the chemistry is possible is that the the most important sulfur oxidant in the gas-phase is the hydroxyl radical. Due to the nature of the sulfur chemistry itself, the hydroxyl radical concentrations are not affected by their involvement in the oxidation of gas-phase sulfur dioxide. This means that for the gas-phase, variations in sulfur concentration do not imply variations in the hydroxyl radical concentration. This situation presents a fortuitous opportunity to decouple the sulfur chemistry from the more complex nitrogen and volatile organic compound (VOC) chemistry modeled in the RADM. Hence, hydroxyl radical concentrations may be precomputed by the RADM for a given NOx/VOC scenario, and then read in by the TSEM. The concentrations may then be used to compute gas-phase sulfur chemistry for a wide variety of sulfur concentration scenarios and different distributions of tagged sources, under the presumption that the NOx/VOC chemistry is unchanged in all cases. Other NOx/VOC scenarios require further RADM runs, but each offers the opportunity to generate a variety of sulfur scenarios using only the TSEM.

Aqueous-phase conversion of S(IV) compounds to S(VI) compounds involve a variety of oxidants, the most important of which are ozone and hydrogen peroxide. It is generally recognized that hydrogen peroxide is likely to be the dominant oxidant and that it also may be limited in its availability to oxidize aqueous-phase S(IV) to S(VI). Ozone is thought to be relatively plentiful and thus unlimited in its availability to act as an in-cloud oxidant, although its oxidation rates are considerably slower than peroxide's under cloudy, aqueous-phase atmospheric conditions . It is thus possible that limitation of hydrogen peroxide in cloud-based reactions may create a non-linear, emissions-reduction-to-deposition-reduction relationship.

TSEM accounts for this important possibility by including a rate equation for hydrogen peroxide because the consumption of hydrogen peroxide in the aqueous-phase will differ from that in the original RADM simulation when the sulfur dioxide emissions are reduced in the cases of control strategy simulations. Thus, having a dynamic response of hydrogen peroxide to these reductions allows a more realistic simulation of the aqueous chemistry than if the hydrogen peroxide levels from RADM were used directly. The levels of hydroxyl and hydroperoxyl are set by photochemical production from the NOx and VOC levels which are held constant in the sulfur control scenarios. The changes in hydrogen peroxide production do not greatly influence these concentrations as has been verified in separate calculations (Chang *et al*, 1990). The aqueous-phase cloud and chemical submodels are incorporated in the TSEM exactly as they are in the RADM. Wet and dry deposition of sulfur dioxide, sulfate, and hydrogen peroxide are modeled in the TSEM according to the same deposition schemes incorporated in the RADM.

The TSEM uses this split chemistry approach along with specifically tagged sulfur sources. The tagging is done by treating the sulfur dioxide from tagged and nontagged sources as well as the sulfate which is derived from the tagged and untagged sulfur dioxide as four separate species. The total deposition for a given event is then composed of tagged and untagged sulfur, and a ratio of tagged to total can be obtained. This has been done in studies of source apportionment and is presented by Dennis *et al.*, (1990). For examining the relative contribution of U.S. and Canadian sulfur emissions, all emissions in Canada were tagged and all U.S. emissions were treated as untagged.

Fourteen individual events were chosen randomly from a five year collection of events of sulfur deposition over eastern North America. The events in the collection were grouped into clusters corresponding to synoptic weather system type. Using a weighted summation technique based upon the frequency of occurrence of a particular type of event, the deposition values from each of the fourteen events are combined into an estimate of the annual average deposition (Samson, *et al.*, 1989).

RESULTS

Figure 1 shows the contribution of Canadian sources to total (wet plus dry) annual sulfur deposition. Note that the U.S. contribution is 100 minus the isopleth value shown. Thus, a value of 15% means that 85% of the sulfur was of U.S. origin. Over southern Ontario, most of the sulfur is of U.S. origin, while north of the Sudbury smelter, more than 50% of the sulfur is of Canadian origin.

Canadian sources contribute less than 5% to total annual deposition in the U.S. except over northern New England, upstate New York, northern Michigan and northern Wisconsin where the contribution is between 5% and 15%. Figure 2 shows the contribution of Canadian sources to dry sulfur annual deposition. The 5% isopleth is much farther south. This is consistent with the interpretation that dry deposition would be the dominant deposition mechanism with a prevailing northerly component to the wind under anticyclonic conditions. Figure 3 shows the annual wet sulfur deposition. This pattern more closely resembles the total annual deposition and appears consistent with a prevailing southerly component to the wind during the cyclonic conditions associated with precipitation.

Table 1 shows the average value and range (minimum and maximum) for selected receptor areas. The tabular entries are arranged from north to south and show the expected systematic decrease of values with increasing distance from Canadian sources. In the Adirondacks only 13% of the total sulfur deposition originates in Canada, this includes 24% of the dry and 9% of the wet deposition. For the Mid-Atlantic region, only 1% of the total sulfur deposition originates in Canada. This is made up of 1% of the wet and 2% of the dry deposited sulfur. Only 0.4% of the total sulfur deposited in the Southern Blue Ridge comes from Canadian sources with 0.8% of the dry and 0.3% of the wet being of Canadian origin. For Toronto, only 22% of the total sulfur deposition is of Canadian origin, but 43% of the dry, and 15% of the wet is of Canadian origin. For Montreal, the values are 37% total, 65% dry, and 28% wet deposited sulfur is of Canadian origin, while for Quebec City , the values are 31% total, 53% dry, and 27% wet. Olson (1989) has summarized results from previous studies which showed values for total sulfur deposition in Eastern Canada from Eastern U.S. sources in the range 37% to 72% of the total sulfur deposition in Eastern Canada.

Both the values displayed in the Figures and those in Table 1, are consistent with the following physical interpretation. Wet deposition in southern Ontario and southern Quebec is primarily from long range transport of U.S. emissions, since the 50% line is well north of the U.S.-Canadian border. Dry deposition, on the other hand, is of more local origin since the dry deposition maxima would be downwind of the major source areas. Thus, the 50% line is nearly along the border nearer the Canadian sources.

The range of deposition values in Table 1. is the smallest for the Poconos This is true for total, wet, and dry deposition. The largest range of values is for the Mid-Atlantic deposition, and is a factor of 10 for wet deposition, a factor of 4 for dry deposition, and a factor of 5 for total deposition. . It is not easy to assign a physical reason for this behavior without further careful analysis of the actual trajectories followed by the air. A tentative interpretation is that the Poconos are dominated primarily by the Ohio Valley sources and secondarily by the sources in the industrialized southern part of the York Peninsula of Ontario. Wet deposition in the Mid-Atlantic area can occur with a large variety of flow patterns. Thus, a ten-fold range in the small Canadian contribution is easily possible. Another way of saying this is that the Mid-Atlantic region is at the edge of any significant Canadian influence..

Toronto shows the largest variability of the Canadian locations with about a factor of 2 in wet deposition and a factor of 3 in dry deposition. The wet deposition is dominated by U.S. sources and as in the case of the Mid-Atlantic deposition, differences in trajectories account for the range. For dry deposition, the values shift from being dominated by U.S. sources to being dominated by Canadian sources. The simplest explanation for this is a shift in the low-level winds. Southerly flow will bring air contaminated by U.S. sources over the Toronto area. Westerly through northerly to easterly flow will bring air contaminated primarily by Canadian sources.

Finally, these results represent only the averaged impact of Canadian emissions (or conversely that of the U.S. emissions) over the region. They do not clarify the sub-regional source-receptor relationships. Because of the complexity and variability of synoptic-scale flow, the direction of pollutant transport and the amount transported across any boundary region between the U.S. and Canada fluctuate on a daily basis. This can be studied by examining the transboundary flux of sulfur for each individual event. For example, because of the predominant southwesterly flow much of northeastern Canada receives sulfur from the U.S. while New York and Pennsylvania receive a mixture of relatively fresh Canadian and leftover U.S. emissions (flowing over Canada) along their western boundaries, The Midwestern U.S. emissions move into Canada mostly across the Michigan-Ontario boundary. Farther west, the flows become more random depending on the passage of storms and the origin of synoptic scale disturbances. More detailed analysis of each of the aggregated cases will be needed to further quantify the interpretation given here.

Acknowledgement

The authors wish to thank Terry L. Clark of the Atmospheric Sciences Modeling Division for his perceptive review of the manuscript of this paper.

Table 1. Percentage Contribution of Canadian Sources to Sulfur Deposition in Selected Areas

Area	Wet Deposition	Dry Deposition	Total Deposition
Adirondacks	**9** (7-14)	**24** (16-37)	**13** (9-19)
Poconos	**4** (3-5)	**7** (5-7)	**5** (4-6)
Mid-Atlantic	**1** (0.2-2)	**2** (1-4)	**1** (0.6-3)
Southern Blue Ridge	**0.3** (0.1-0.4)	**0.8** (0.7-1.1)	**0.4** (0.3-0.6)
Toronto	**15** (9-23)	**43** (21-68)	**22** (12-34)
Montreal	**28** (22-34)	**65** (55-76)	**37** (27-49)
Quebec	**27** (18-43)	**53** (42-70)	**31** (22-48)

Top entry is the average in the area, values in parentheses are the maximum and minimum values for grid cells in the area. The U.S. contribution is 100 minus the table entry.

RADM TOTAL SULFUR DEPOSITION

CANADA TAG AS A PERCENT OF OVERALL

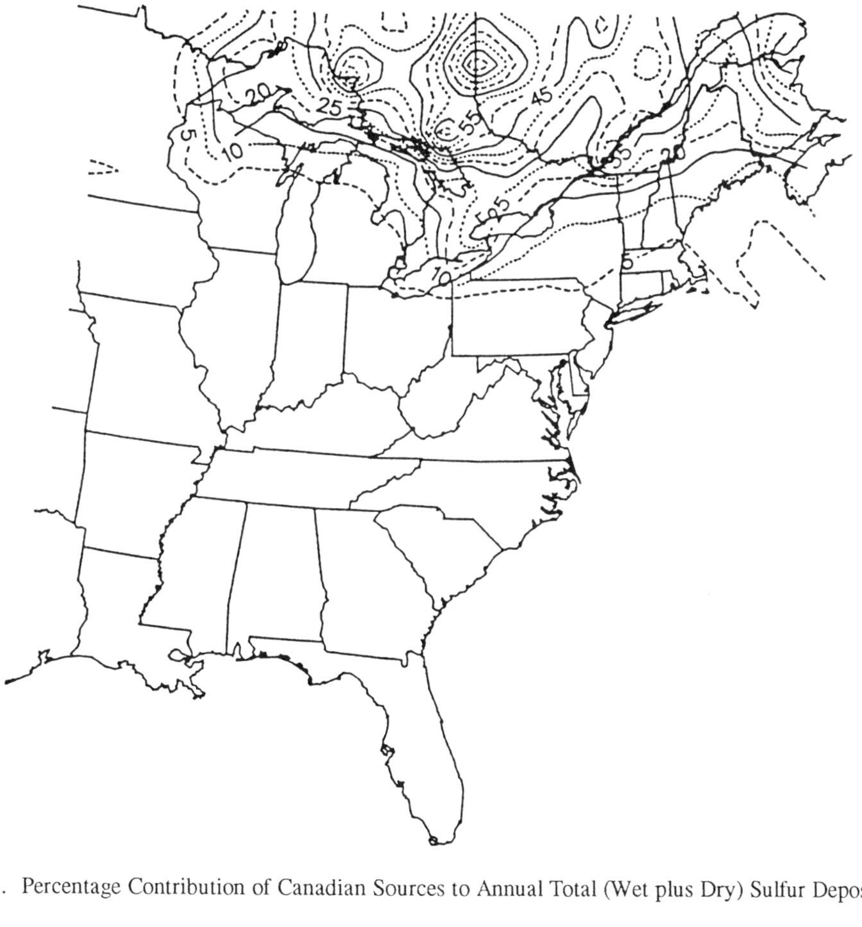

Fig 1. Percentage Contribution of Canadian Sources to Annual Total (Wet plus Dry) Sulfur Deposition

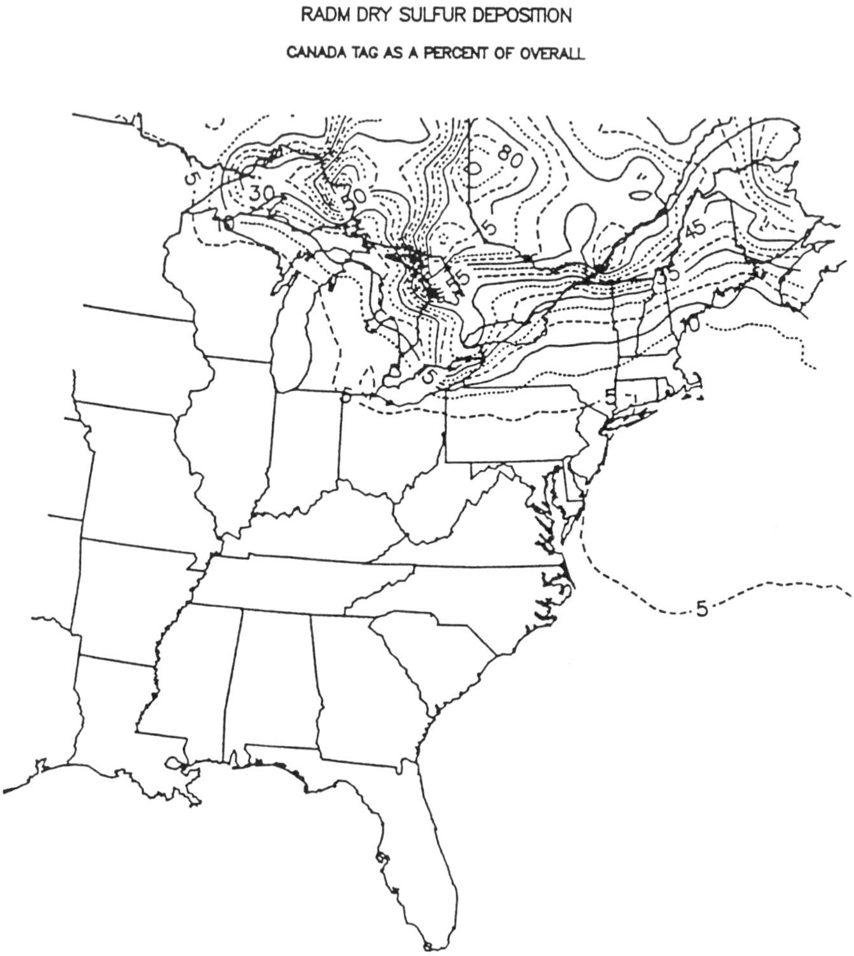

Fig 2. Percentage Contribution of Canadian Sources to Annual Dry Sulfur Deposition

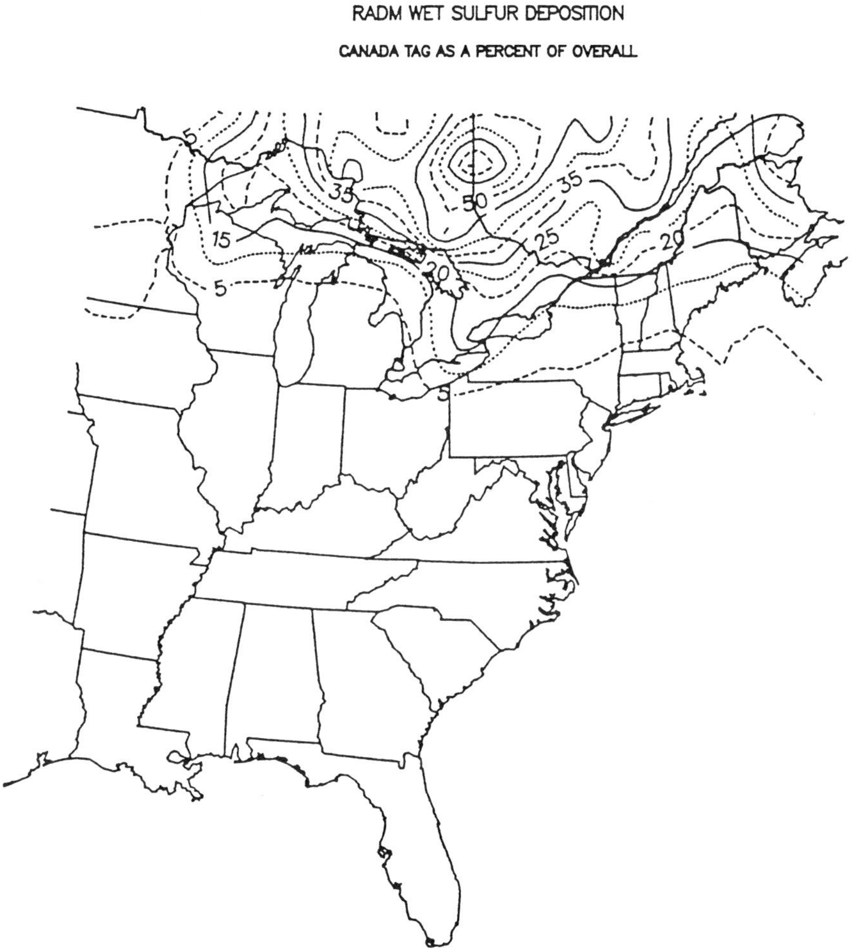

Fig 3. Percentage Contribution of Canadian Sources to Annual Wet Sulfur Deposition

REFERENCES

Chang, J.S., R.A. Brost, I.S.A. Isaksen, S. Madronich, P.Middleton, W.R. Stockwell, and C.J. Walcek, 1987, A Three-dimensional Eulerian acid deposition model: physical concepts and formulation, Journal of Geophysical Research,, 92: 14681.

Chang, J.S. , F.S. Binkowski, N.L. Seaman, J.N. McHenry, D.W. Byun, P.J. Samson, W.R. Stockwell, C.J. Walcek, S. Madronich, P. Middleton, J.E. Pleim, and H.H. Landsford, 1990,(in press) "The Regional Acid Deposition Model and Engineering Model" National Acid Precipitation Assessment Program, Washington, D.C.

Dennis, R.L, J.N. McHenry, D.W.Byun, T.L. Clark, and S.D. Reynolds, 1990 , Delineation of the influence of major emissions regions on the annual acidic deposition at sensitive areas, in press, these proceedings

Kleinman, L,I., 1988, Evaluation of SO2 emissions scenarios with a nonlinear atmospheric model, Atmospheric Environment 22: 1209-1219.

Olson, M.P. 1989, Author's Reply, Atmospheric Environment 23: 2864-2867.

Samson, P.J., 1989, "Aggregation of Pollutant Deposition Episodes into Seasonal and Annual Estimates", EPA Report, Atmospheric Research and Exposure Assessment Laboratory, Research Triangle Park, N.C.

Seaman, N.L., 1990, "Description and Evaluation of the Meteorological Cases Used in RADM NAPAP Evaluation and Assessment", EPA Report , Atmospheric Research and Exposure Assessment Laboratory, Research Triangle Park, N.C.

THE ROLE OF GRID-BASED, REACTIVE AIR QUALITY MODELING IN POLICY ANALYSIS: PERSPECTIVES AND IMPLICATIONS, AS DRAWN FROM A CASE STUDY

Philip M. Roth[1], Charles L. Blanchard[2], Steven D. Reynolds[3], and Robin L. Dennis[4]

[1]21 Catskill Court
San Anselmo, California

[2]956 Kains Avenue
Albany, California

[3]Systems Applications, Inc.
San Rafael, California

[4]U.S. Environmental Protection Agency
Research Triangle Park, North Carolina

INTRODUCTION

During the 1980s the National Acid Precipitation Assessment Program (NAPAP) supported the development of the Regional Acid Deposition Model (RADM). While sound performance evaluation was to be a part of the development process, concern existed that policy makers may hold overly optimistic expectations of RADM's performance, and of the time to RADM's acceptability for unrestricted use. A primary objective of this study is to gain an improved understanding of the role of quality of performance in determining a model's acceptability and usefulness to the policy maker, and thus to aid in developing soundly-based expectations of the modeling process. The vehicle for pursuing this objective is examining the historical evolution of the Urban Airshed Model (UAM), a grid-based photochemical model that is similar in basic formulation to RADM, and its application to policy analysis in the South Coast Air Basin (SOCAB) of California.

In this study we surveyed both the performance evaluation history of the UAM and the degree of acceptance achieved by the model in the policy making and scientific communities. We first examined and interpreted performance statistics for the model over the history of its development. We then interviewed three categories of interested and informed parties--members of the contemporaneous scientific community familiar with the UAM and its development history, industry representatives, and regulatory agency staff and policy makers who have had to assess the merits of using the results of the UAM. Complete documentation of this work is contained in the report by Roth, Blanchard, and Reynolds (1989).

Air Pollution Modeling and Its Application VIII, Edited by H. van Dop and D. G. Steyn, Plenum Press, New York, 1991

THE URBAN AIRSHED MODEL

The formulation of the UAM is based on the time-dependent, three dimensional equations describing the conservation of mass of a chemically reactive pollutant in a turbulent atmosphere. To solve the resulting system of coupled nonlinear partial differential equations, it is necessary to utilize numerical integration techniques, which require that the modeling domain be partitioned into an array of three dimensional grid cells. These cells have fixed horizontal dimensions of 2 km to 10 km and time-varying vertical dimensions of 50 meters to several hundred meters. The overall modeling region may extend from 50 km to 200 km in the horizontal and from less than 1 km to 3 km in the vertical. Ideally, the geographical extent of this region should contain all areas with sources of reactive hydrocarbon and NO_x precursor emissions that contribute to ozone formation and the downwind areas where significant ozone levels are likely to occur.

Development of the UAM was initiated by Systems Applications, Inc. in 1970 under contract to the National Air Pollution Control Administration (incorporated into the EPA in 1971). Initial model development activities were accompanied by simulation of six single day ozone events in 1969. Evaluation of the model's performance and its components showed that improved treatments of some physical and chemical processes were necessary. Sponsored by the EPA, another series of research efforts was carried out between 1973 and 1977 to improve model components. Improvements to the software and documentation as well as application of the model to St. Louis using Regional Air Pollution Study (RAPS) data were supported by the EPA from 1977 through 1979.

The first serious interest in applying the UAM was expressed by the California Department of Transportation (CALTRANS) and the California Air Resources Board (CARB), who jointly sponsored a study in 1974 and established a set of model inputs for 26 and 27 June 1974. These inputs were a point of departure for several UAM applications to the SOCAB over the next ten years.

In the mid-1970s, the Southern California Edison Company (SCE) sponsored a study to examine issues associated with the implementation of NO_x emission controls on power plants in the SOCAB. Work also focused on simulating a multiple day episode, to avoid problems connected with single day simulations. Four other UAM applications to the SOCAB were conducted in the late 1970s--studies sponsored by the Southern California Association of Governments, the EPA Office of Air Quality Planning and Standards, the National Commission on Air Quality, and the CARB. In the early 1980s, a major application of the UAM to the SOCAB was sponsored by the Western Oil and Gas Association (WOGA). This study examined the relative effectiveness of reactive hydrocarbon and NO_x controls in reducing ozone and NO_2 concentrations.

A significant body of experience has also been acquired in applications to other areas of the U.S., Europe, and Far East. In Europe, simulations have been conducted for The Netherlands, the Rhine-Ruhr area of West Germany, and Turin, Italy. Recently, the model has been applied in Japan and to the cities of Taipei and Kaohsiung in the Republic of China (Taiwan). A more complete description of the UAM development and application history is provided by Roth, Blanchard, and Reynolds (1989).

THE PERFORMANCE OF THE URBAN AIRSHED MODEL

Of particular interest to this study is the evolution of model performance over time, especially in Los Angeles applications, since the model has been used most extensively in this area. In addition, the model's ability to simulate atmospheric processes in other areas provides further insight into its range of applicability.

A review of UAM performance indicates that the model has tended to underpredict ozone concentrations in all but the most recent applications. Furthermore, the bias in ozone results has decreased over time, presumably because the model now contains better representations of atmospheric processes and model users have become more skilled in preparing inputs. The overall error in ozone results has also decreased over the past several years, as shown in Figure 1; currently, discrepancies between calculated and observed hourly-averaged values are in the range of 20% to 40%. Uncertainties in ozone measurements are of the order of 15% to 20%. These figures suggest that overall discrepancies between calculated and measured values of less than 20% may be difficult to achieve.

Over the past 15 years, only limited efforts have been made to demand good performance for NO_x and to test NO_x performance with the same rigor as ozone. In general, the overall error in hourly-averaged NO_2 results is in the range of 30% to 50%. The poorer level of performance achieved for NO_2 has contributed to concerns about the accuracy of calculated future year air quality values.

Note that inadequacy of model formulation can be masked during evaluation by compensating biases among the outputs of the model components (modules), such as emissions estimates, predictions of wind speeds or mixing depths, estimated deposition rates, and chemical rates of formation and consumption of ozone. Historically, the limitations of available data have not permitted adequate examination of the possible presence and extent of compensating biases. Also, the range of meteorological conditions associated with high ozone episodes--the conditions under study--is rather circumscribed (i.e., high temperatures, low winds, high solar insolation, and a strong inversion layer aloft). Thus, while a significant number of episodic events have been studied for several cities, the range of conditions examined is somewhat limited. This concern is central in identifying the potential for inadequate performance evaluation, for misgivings on the part of the policy community, and for uncertainties among the model developers.

THE UAM AND POLICY-RELATED STUDIES

To assess the extent to which UAM results were utilized as technical input to selected air quality control decisions, we interviewed several people who participated in such decision making activities. The broad questions posed in these interviews were:

Was the model acceptable?

Did its degree of acceptability change over time?

How was model acceptability related to model performance?

Four applications of the UAM were selected for discussion with interviewees; each application involved a major policy decision pertaining to air quality. Since the policy events were of a varied nature, we generally refer to them simply as "activities". The four activities were all associated with decisions pertaining to the SOCAB. By focusing on the SOCAB, we were able to assess model acceptability over a six-year period during which UAM estimates were submitted as technical input to air quality decisions, and during which most of the technical and policy issues were relatively unchanged.

The first activity was the adoption of Rule 1135.1 by the CARB in 1979, which established emissions requirements for electric utilities. The next activities were the 1979 Air Quality Management Plan (AQMP) and its 1982 revision; both plans were drafted by the South Coast Air Quality Management District (SCAQMD) as part of California's SIP. The final activity was a series of hearings in 1985 by a subcommittee of the SCAQMD board on the need for additional NO_x controls in the SOCAB. In addition, some individuals chose to discuss two recent activities involving air quality policy decisions in two other areas of California, namely Santa Barbara and Kern counties.

Individuals participating in the survey were selected from five distinct backgrounds: regulatory affiliation (board members, management staff, and technical staff), industry affiliation, and scientific peer group (others involved in air quality modeling). Individuals with a regulatory affiliation represented the EPA, CARB, and the SCAQMD. Those with an industry affiliation represented SCE and WOGA. Peer group representatives were university faculty. Each individual surveyed had substantive involvement with modeling application in one or more of the activities cited.

A questionnaire was developed and mailed to 38 individuals. Each individual was then contacted by telephone, at which point we requested a telephone interview (30 individuals agreed to be interviewed). The interviews were conducted after each person had an opportunity to review the questionnaire. The number and affiliation of respondents are summarized in Table 1.

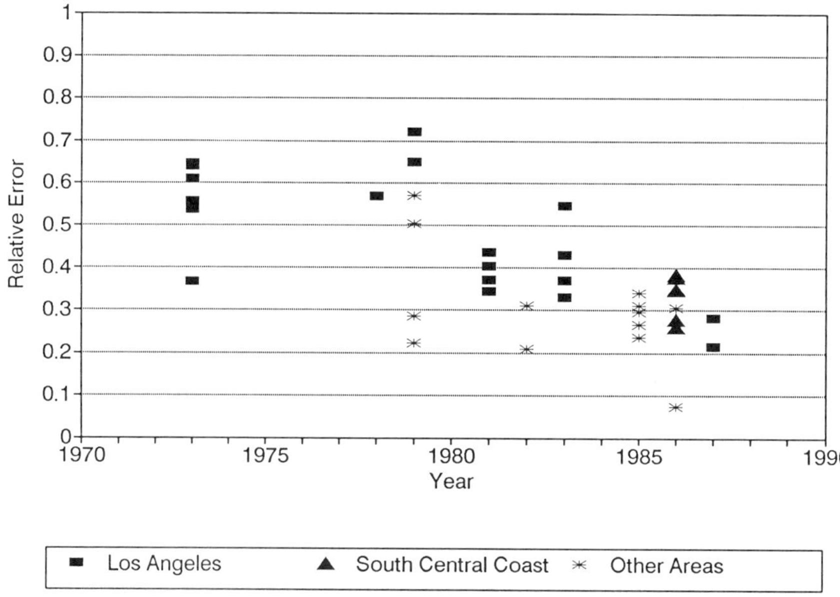

Fig. 1. Time history of the average absolute relative error in UAM ozone results. The absolute relative error is expressed as: |calculated concentration - observed concentration|/observed concentration.

The principal conclusions that emerged from the discussions with interviewees are as follows. The extent to which UAM results were utilized as technical input to air quality decisions increased over the period from 1979 to 1986 (see Table 2), as indicated by the increase in the proportion

Table 1. Number of individuals participating in each activity by affiliation

Affiliation	Activity: Rule 1135	1979 AQMP	1982 AQMP	1985 NO_x	Santa Barb.	Kern Cty.
Regulatory: board	4	4	5	3	0	0
Regulatory: management	6	7	7	4	4	4
Regulatory: technical	2	2	3	3	1	1
Industry	3	3	5	5	1	1
Scientific peer group	2	1	2	2	0	0
Total	17	17	22	17	6	6

Table 2. Influence of UAM on respondents' assessments of control measures

Influence	Number of respondents by activity: Rule 1135	1979 AQMP	1982 AQMP	1985 NO_x	Santa Barb.	Kern Cty.
Little or none	5	2	3	1	0	0
Marginal or moderate	3	2	5	2	0	0
Substantial	5	5	6	6	3	3
Most substantial	3	4	5	5	0	4
Question not applicable	14	14	8	14	24	23
Answer missing	0	3	3	2	3	0

of respondents who indicated that the influence of the UAM was "substantial" or "most substantial". We note that lengthy summary comments were obtained from individuals. The extent of change in acceptance was more clearly reflected in these subjective comments than in the summary statistics presented here (see Appendix D of Roth, Blanchard, and Reynolds,

1989). Throughout this period, interviewees qualitatively ranked the performance of the UAM as "good enough for estimates to be considered" but not "good enough to provide a high degree of confidence in the model" (see Table 3). Most respondents stated that they had originally believed that quantitative assessment would be possible, but with improvements in their understanding of the model's capabilities they now concluded that only qualitative comparisons of alternatives was possible.

The increase in acceptance of UAM results was not related to improved performance (or to perceived performance), since the interviewees' qualitative rankings of performance were unchanged during the 1979 to 1986 period. The interviewees reported that their understanding of the model's capabilities and limitations increased. This improvement in understanding focused on key aspects of the model to which policy-relevant technical findings are sensitive, e.g., the dependence of the NO_x-ozone relationship on the ratio of hydrocarbons to NO_x. Thus, the interviewees' understanding of the dependence of model results on key technical aspects of model formulation was the critical element in facilitating their increased degree of acceptance of the model with time.

Most interviewees believed that better input data, preferably from a data base specifically designed for use as model inputs, would be necessary to improve the model's reliability and thus its acceptability. They favored the collection of comprehensive data bases designed specifically for supporting model evaluation, with attention given to stressing the model to improve the model's acceptability. In addition to performance evaluations, sensitivity analyses were also considered to be important in assessing the accuracy and reliability of the model.

THE INFLUENCE OF MODEL RESULTS ON POLICY

While model acceptance increased and performance improved over the period 1979 to 1986 and thus appear correlated, a causal relationship cannot be inferred. In fact, information on time trends in performance were generally unavailable, and individuals interviewed probably were not aware of the nature of the trends. Also, the interviewees' qualitative rankings of performance were unchanged over the period 1979 to 1986.

Table 3. Respondents' assessments of UAM performance levels

	Number of respondents by activity				
	Rule 1135	1979 AQMP	1982 AQMP	1985 AQMP	Kern/Santa Barb. Cty.
UAM estimates were:					
Too poor to consider	1	1	0	0	0
Questionable	2	0	1	1	0
Good enough to consider	9	9	11	10	4
Good enough to use for policy decisions	0	0	1	2	0
Question not applicable	16	14	10	13	23
Answer missing	2	6	7	4	3

Increased acceptance with time appears to derive in part from a combination of (1) limited influence on policy of model results at the beginning of the period under study and (2) greater comfort with modeling as understanding of the models--their strengths and limitations--increased. Interviewees indicated that there was an early reluctance to accept modeling results because of the polarization that existed over issues that modeling was to address, the limitations imposed on the value of results due to inadequacies in the supporting data base, and concerns about the quality of performance of the model. With time came considerably improved understanding of the specifics of models and modeling, including the interrelationships among key variables, and of the capabilities of the UAM in general. Thus, greater confidence accompanied judgments, and a greater willingness to take advantage of the strengths of models emerged.

Two primary factors have favored acceptance of the UAM: (1) its performance, as gauged by statistical measures, compares favorably with that of other models, and (2) its testing, in a formal sense, has been more extensive than that of comparable models.

Interviewees nevertheless indicated that in their view the model results are of value primarily in a relative or qualitative sense. However, they have little confidence that a model can be used to reliably estimate (1) the actual peak concentrations (or other, more robust measures of high concentrations) likely to result as a consequence of adopting a control strategy, or (2) the extent to which an air quality standard is likely to be approached or met. The current 20% to 40% discrepancies between calculated and measured values, the inconsistencies between estimated emission rates and ambient air quality levels for hydrocarbons, the quality of ambient NO_2 concentration estimates, the uncertainties in emissions models and inventories and other factors provide the basis for this reservation. In effect, the interviewees believe that the average errors associated with model estimates are unacceptably large relative to the magnitude of concentration changes that they are attempting to effect. Based on these findings, we conclude that "competence" in calculated values is clearly a factor in establishing acceptance.

In summary, the UAM is judged to be valuable, and performance evaluation has been an essential element in establishing confidence in its value. The time required by the regulatory and industrial communities to gain familiarity with and an understanding of the model--its strengths and shortcomings--is a key factor in the process of acceptance. The UAM achieves its highest value when it is used in a way that takes advantage of its strengths and recognizes its limitations.

A PERSPECTIVE ON EVALUATING RADM PERFORMANCE: LESSONS FROM THE UAM EXPERIENCE

A primary purpose of conducting this study was to learn some lessons from the UAM experience and present them for consideration in planning and executing RADM development, evaluation, and application activities.

A Model of the Modeling Process

Modeling is most appropriately conceived as a more or less continuing process of model development, collection of data, testing, diagnostic analyses, modification and improvement of the model, and repetition of the cycle. This "process model" contrasts with the "product model"--that of model development to produce a "one shot" or final product.

Modeling is a codification of current knowledge. When a series of gains in knowledge is made, the model is modified to reflect this new information or understanding. Experience suggests that, for complex atmospheric systems, the time needed to accommodate successive cycles of advancement to the point of diminishing returns is of the order of ten to twenty years. Expectations of policy makers should be based on an understanding of the development process and its associated time frame.

This "model" of the modeling process suggests that a goal of "completing" RADM (i.e., producing a final "product") and applying it in a policy context--with the model producing reliable results and attaining credibility in the eyes of the scientific and policy communities in the short term--would be optimistic and possibly naive. About seven years have elapsed since the inception of RADM's development. By 1990 to 1991, the second full round of evaluation will have been completed. Past experience suggests that additional cycles of development and testing will be needed, and that credibility and acceptance may be attained only after perhaps three to four cycles.

The Role of Performance Evaluation in Model Development

Performance evaluation is an essential step in model development. Interviewees were unanimous in the view that, if confidence in a model is to be achieved, field monitoring programs designed to support evaluation of model performance over a range of conditions must be carried out, and models must be tested using these data. The converse, of course, does not hold; that is, data collection and testing do not assure technical acceptance of a model.

What, then, are the implications for RADM? It seems clear that the model is unlikely to supply unambiguous, accurate answers to the questions of policy makers in the short term. Rather, the modeling process, as described, will provide gains in knowledge and improved versions of the model which will be of value to the policy maker. Moreover, the modelers become expert and are valued as technical advisers in interpreting findings, despite the limitations of the models themselves. Thus, emphasis should be placed on the process itself, as discussed earlier, rather than on the product, if expectation is to reflect reality.

Technical Acceptance of RADM

The main factor likely to inhibit RADM's future technical acceptance is the limited understanding of certain key processes that govern acidic deposition. The main factor that will promote RADM's technical acceptance is the collection and thoughtful use of a well designed data base in performance evaluation, diagnosis, and further development. If the commitment to obtaining such a data base is compromised, so will be technical acceptance.

The Acceptance of RADM in Policy Analysis

Acceptance of a model by the policy community is influenced by technical arguments and other factors as well. These factors include (1) degree of demand for the model by decision makers and by managers who fund model development and application and (2) availability and attractiveness of alternative models.

In the case of UAM, demand for the model during the first ten years of its development and application were relatively low. This was attributable to (1) the relative novelty of grid-based models for the policy community; (2) the availability and relative simplicity (and thus the attractiveness) of the Empirical Kinetics Modeling Approach (EKMA); and (3) the geographical imbalance in demand for the model in California as compared with the rest of the country.

In contrast, acidic deposition received intensive attention more than a decade after urban ozone was the focus of regulatory action. Familiarity with models and model use was much higher in the early 1980s, and the pros and cons of employing simpler versus more sophisticated models were clearer in the minds of policy makers and managers. Moreover, a strong national interest existed in addressing the acidic deposition issue. RADM is a centerpiece of NAPAP. Thus, the policy community badly wants and needs a sound model for use in analysis. This "mindset" could be conducive to promoting acceptance of an "adequate" or "best available" model more rapidly than might occur (or be justified) based on scientific assessment alone.

A complicating factor in policy acceptance is what might be termed "the involvement syndrome". History suggests that model acceptance is clearly enhanced if interested parties are participants in or financial supporters of model development and testing, and/or actively participate in a review capacity. Nonparticipants and policy or technical representatives of groups, agencies, or interests that might hold different views or have different missions than the supporting agency or group tend to be more skeptical and less accepting of the model. Thus, the model can become the focus of differences, which in turn heightens questioning and delays acceptance.

A number of issues have been raised in this paper. A number of speculations have been made. Analysis of historical events and policy issues is clearly subjective. Yet, from our perspective further analysis and investigation is warranted. The art and science of modeling is likely to benefit from a fuller understanding of past practice and its impacts of the technical and policy communities.

REFERENCES

Roth, P.M., C.L. Blanchard, and S.D. Reynolds. 1989. The role of grid-based reactive air quality modeling in policy analysis: perspectives and implications, as drawn from a case study. EPA/600/3-89/082, U.S. Environmental Protection Agency, Research Triangle Park, North Carolina.

ACKNOWLEDGEMENTS

The work reported in this paper was funded by the United States Environmental Protection Agency under Contract No. 68-01-6849.

DISCUSSION

F.L. LUDWIG

What about the importance of user documentation?

R.L. DENNIS

I think the central point we learned is that users are uncomfortable in putting trust in a model that appears as a "black box". User documentation is a very important means to dispel the black box problem. Documentation of the science (the "why" and the "how") will, therefore, be far more important than computer code documentation. Documentation of sensitivity analyses is another form of documentation that our study suggests would be very useful.

J. ALCAMO Do you think model user workshops are important for getting a model used?

R.L. DENNIS I think the issue is communication -- what the model can do, what its strengths and limitations are and what the analyst can do with the model. I see here communication being a process that occurs over time with the user, especially with the complex, Eulerian models the paper focuses on. A user workshop would be one of many communication tools, and maybe not the best for these models. For more "transparent", self-contained models, workshops may be a good tool and workshops probably represent a good way to initiate the communication process.

THE SCIENTIFIC CREDIBILITY OF REGIONAL ATMOSPHERIC DEPOSITION MODELS USED FOR CONTROL STRATEGY DEVELOPMENT

Thomas F. Lavery

Environmental Science & Engineering, Inc.
P.O. Box 1703
Gainesville, Florida 32602-1703

INTRODUCTION

The achievement of the scientific credibility of models like RADM (Chang, 1987) and ADOM/TADAP (Venkatram et al, 1988) comes from the gradual acceptance by the scientific and user community that these models are appropriate tools to simulate atmospheric processes and emission scenarios. The gradual acceptance results from a process of consensus among the modelers, the scientific community, sponsors, decision-makers who use the model results, and other interested groups (i.e., regulated industry or environmental groups). These disparate groups must be involved throughout the model development, evaluation, and application processes. Only through experiencing the actual process will confidence be built in the predictions or use of the model. The concerns and needs of these groups can conflict. Every effort must be made, however, to satisfy any concerns before the models can be used for policy making and the development of emission control strategies.

Confidence in the model and model acceptance is obtained by:

1. Knowledge and understanding of the models, model applications, and support data bases;
2. Rigorous tests, evaluations, and model refinements/improvements;
3. Documentation in the scientific literature and to the sponsors; and
4. Various presentations to the interested groups.

Recommendations for the scientific evaluation and other activities in support of RADM, ADOM/TADAP, and similar models are presented in this paper.

MODEL EVALUATION PROCESS

The model evaluation process involves several related steps:

1. Description of the intended model applications,
2. Presentation of the field measurement data bases used to evaluate the models,

3. Description of the atmospheric phenomena important to the intended applications,
4. Model evaluation activities,
5. Quality assurance,
6. Data exchange and chain-of-custody, and
7. Communication and coordination.

For models like RADM and ADOM/TADAP, intended applications include:

1. Investigating source attribution,
2. Developing and evaluating the effectiveness of emissions control strategies, and
3. Calculating transboundary fluxes for acidifying pollutants.

Future applications could include development of ozone abatement strategies or investigation of the effects of air toxics emissions. Model acceptance involves how well a model performs given the intended applications of the model (i.e., do model performance limitations inhibit the application of the model for certain key issues?).

Quality assurance (QA) is an important aspect of model evaluation. QA activities could include verification of the model software, audits of the models by an independent modeling team, estimates of the uncertainty of the model input (e.g., emissions) data, establishing protocols for data exchange, and chains-of-custody.

A large part of model credibility could be achieved by the proper display and documentation of the model evaluation process and the model results. The basic guiding principle for documenting the evaluation process should be to enable independent duplication of the work. The tests performed to test the model and its modules should be assembled in a consistent framework to provide review and help in future studies.

Improvement in the display and manipulation of the input data sets and output from the acid deposition models is necessary. In particular:

1. All data should be manipulated by standard data base management systems; there should be minimal special purpose software. The use of these systems would simplify the task of transferring information between modeling systems.

2. Use of data base management systems simplifies the task of documenting the data sets because pointers could be added to narrative discussions (e.g., day-specific points source data).

3. More attention should be given to the presentation of model output. Animated graphic presentations of the fields should be used. These displays could be used for both diagnostic purposes and for presentations to policy makers.

PRESENTATION AND ANALYSIS OF EMEFS DATA BASE

RADM and ADOM are currently being evaluated using data collected during the Eulerian Model Evaluation Field Study (EMEFS) (see Section 5 of the report by Dennis et al, 1989). Presentation and interpretation of the EMEFS data are important components of model credibility. Descriptions of the EMEFS data base should include:

1. Simple displays
   - Isopleth maps,
   - Cross-sections, and
   - Time series;
2. Statistics of measurements;
3. Uncertainties;
4. Seasonal variability;
5. Diurnal behavior;
6. Relationships among variables;
7. Statistical analysis (e.g., principal component analysis); and
8. Description of unusual or unexpected behavior; and
9. Description of concentration or deposition events that are related to the intended applications.

MODEL EVALUATION TESTS

Model evaluation tests include:

1. Operational,
2. Mechanistic-diagnostic,
3. Integrated-diagnostic,
4. Comparative,
5. Sensitivity, and
6. Diagnostic simulations.

Operational evaluation tests involve comparing model output with observations (e.g., annual average wet sulfur deposition or annual average sulfur dioxide concentrations). Operational evaluation has been the most widely used (Clark et al, 1989) approach to model testing and has most often involved using seasonal and annual average depositions and concentrations.

Diagnostic testing can involve the full model (integrated-diagnostic) or individual modules (mechanistic-diagnostic) (e.g., the MM4 [Anthes and Warner, 1978] meteorological model used to develop the meteorological input to RADM). Diagnostic testing involves the specification of several technical questions that are answered during the evaluation process. The technical questions are designed to stress the model and identify any flaws quickly.

Integrated-diagnostic evaluation tests should focus on special measurements which will stress the models. For example, a main component of the RADM evaluation involves using the EMEFS intensive aircraft measurements. Selected ground-level measurements (e.g., hourly ozone concentrations or hydrogen peroxide measurements) could also be useful (Lavery et al, 1980).

The diagnostic evaluation of modules should include:

1. Scientific Review
   - Are the scientific details of the module sufficient given the intended applications and any practical constraints?
   - Is the geographic resolution adequate?
   - Are the numerical methods representative?
   - Has the computer software been verified?

2. Comparison of modules with parent models or other, more complete models

3. Evaluation with specialized data bases (ANATEX, PRECP, PEPE/NEROS)

4. Direct comparison of module output with observations
   - Aircraft meteorological output versus model data
   - Smog chamber data

5. Sensitivity Studies (McRae et al, 1982)

6. Diagnostic Simulations
   - Mass conservation
   - "Normal" behavior
   - Advection tests
   - Vertical mixing with and without clouds

7. Documentation of the results

Comparative evaluation tests involve comparing various modules used in RADM and ADOM. For example, the gas-phase chemistry modules from RADM and ADOM have recently been compared. The transport and mixing models will be compared using the ANATEX (Draxler and Heffter, 1989) data base.

Sensitivity evaluation tests are designed to determine the response of a model or module to changes in an input variable or a model parameterization. Diagnostic simulations are performed to isolate certain physical features of the integrated model. For example, running a model with an inert tracer through a cloud field could be used to examine the reasonableness of vertical mixing produced by clouds.

Model credibility will be enhanced when the focus of the evaluations is diagnostic. Today, model evaluation is progressing from operational tests to the more diagnostic evaluations being undertaken for RADM and ADOM. The achievement of credibility is a process of consensus which is enhanced by documentation and communication of all aspects of the model evaluation process.

REFERENCES

Anthes, R.A. and Warner, T.T., 1978, Development of hydrodynamic models suitable for air pollution and other mesometeorological studies, Monthly Weather Review, 106:1045-1078.

Chang, J.S., Brost, R.A., Isakson, I.S.A., Madronich, S., Middleton, P., Stockwell, W.R. and Walcek, C.J., 1987, A three-dimensional Eulerian acid deposition model: Physical concepts and formulation, Journal of Geophysical Research, 92(D12):14,681-14,700.

Clark, T.L., Voldner, E.C., Dennis, R.L., Seilkop, S.K., Alvo, M, and Olson, M.P., 1989, The evaluation of long-term sulfur deposition models. Atmospheric Environment, 23:2267-2288.

Dennis, R.L., Barchet, W.R., Clark, T.L., Seilkop, S.K., and Roth, P.M., 1989, NAPAP SOS/T Report No. 5, Evaluation of Regional Acidic Deposition Models.

Draxler, R.R. and Heffter, J.L., 1989, Across North American Tracer Experiment (ANATEX), Vol 1: Description, ground-level sampling at primary sites, and meteorology, NOAA Technical Memorandum ERL ARL-167, National Oceanic and Atmospheric Administration, Air Resources Laboratory, Silver Spring, MD, 83 pp.

Lavery, T.F., Baskett, R.L., Thrasher, J.W., Lordi, N.J., Lloyd, A.C., and Hidy, G.M., 1980, Development and validation of a regional model to simulate atmospheric concentrations of sulfur dioxide and sulfate, Second Joint Conference on Applications of Air Pollution Meteorology, American Meteorological Society, Boston, MA, 869 pp.

McRae, G.J., Tilden, J.W., and Seinfeld, J.H., 1982, Global sensitivity analysis - A computational implementation of the Fourier Amplitude Sensitivity Test (FAST), Computers and Chemical Engineering, 6:15-25.

Venkatram, A., Karamchandani, P.K., and Misra, P.K., 1988, Testing a comprehensive acid deposition model, Atmospheric Environment, 22(4):737-747.

II: REGIONAL SCALE MODELLING

chairmen: G. Kallos
F.A. Schiermeyer

rapporteurs: J.L. Walmsley
N. Chaumerliac

REGIONAL AND MESOSCALE METEOROLOGICAL MODELING AS APPLIED TO AIR QUALITY STUDIES

R.A. Pielke[1], W.A. Lyons[2], R.T. McNider[3], M.D. Moran[1], D.A. Moon[2]
R.A. Stocker[1], R.L. Walko[1], and M. Uliasz[1]

[1]Dept. of Atmospheric Science, Colorado State University, Fort Collins Colorado 80523
[2]R*SCAN Corporation, Business and Technology Center, 511 Eleventh Avenue South, Minneapolis, MN 55415
[3]Dept. of Mathematics and Statistics, Univ. of Alabama at Huntsville Huntsville, Alabama 35899

SHORTCOMINGS OF CURRENT REGULATORY MODELS

Current air quality regulatory models are summarized in Turner et al. (1989). Unfortunately, these models have been applied more to conserve legal consistency over time than to accurately represent atmospheric flow, particularly in complex terrain and coastal zones [see discussions in Moran et al. (1986) and Lyons et al. (1983)]. These regulatory models are also not generally validated on a case-by-case basis but rather by using rank-order correlations (e.g., Fox, 1981).

The need for accurate air quality modeling is essential if we are going to address atmosphere-related environmental issues adequately. The National Park Service, for example, has concluded that air quality is the most serious threat to many U.S. national parks (Everhart, 1983). The Department of Energy needs realistic short-range predictions of air flow in order to more accurately evacuate population centers during emergency response events.

Major shortcomings of current regulatory models include:

- Absence of an adequate spatial and temporal characterization of the wind and turbulence field;
- Neglect of wind recirculation on the accumulation of pollution;
- Lack of model validation on a case-by-case basis.

This paper presents examples of a modeling approach which can remedy these shortcomings for episodic air quality studies. Another paper in this volume (Moran et al., 1990) addresses related issues of model resolution.

METEOROLOGICAL MODELING CAPABILITIES

Current Computer Power

For many years, computer power has advanced at a steady and rapid pace. Today's most advanced supercomputers such as the CRAY Y-MP out-perform the fastest machines of 15 or 20 years ago by over two orders of magnitude in both processing speed and central memory. In addition, supercomputers are becoming more widely available in universities, government labs, and other research facilities. This has opened the door to a much broader class of computational problems, including complex 3-D atmospheric simulations (Cotton et al., 1990).

Perhaps more important has been the introduction of the "super-mini" computers or graphics superworkstations. Machines of this class perform to within a factor of 10 of the speed of the most advanced supercomputers but cost only one-hundredth as much. They are also very easy and inexpensive to maintain, and require no special cooling systems or other restrictive environmental conditions. The affordability of these machines makes them accessible to a much wider range of users, and makes feasible their dedication to specific tasks which, if performed on supercomputers, would normally compete with other jobs for throughput. Thus machines of this class are highly suited to all but the largest computer jobs, and acquiring them is well within the means of many organizations.

Regional-Scale and Mesoscale Meteorological Models

In a recent report, Pielke (1989) reviewed mesoscale meteorological modeling capabilities in the United States as related to air quality. Some other regional-scale National Weather Service operational models and European models, which have the potential to be used in air quality assessments, are discussed in Anthes (1983) and NCAR (1988). Seigneur (1989) prepared a similar review to that of Pielke (1989) for atmospheric chemistry models. In the Pielke report, in which each active modeling group was solicited for their input and later reviewed and commented on the drafts, the NCAR/Penn State (PSU) regional-scale meteorological model and the Colorado State University (CSU) Regional Atmospheric Modeling System (RAMS) were judged to be the most general meteorological modeling tools currently available in the United States. An earlier version of the NCAR/Penn State model was used as the Regional Atmospheric Deposition Model (RADM) meteorological preprocessor (Chang et al., 1987). The RADM version, however, does not contain several significant improvements found in the current NCAR/Penn State meteorological model. Most of the other models reviewed in the report could be generalized to be as comprehensive as the NCAR/PSU and CSU RAMS models, but do not at present have several of their features.

Tables 1 and 2 summarize the capabilities and needed improvements in each of the seven North American models evaluated by Pielke (1989). Models outside of North America should also be compared in a similar way. Pielke (1984, Appendix B) presents a somewhat dated overview and comparison of international mesoscale model capabilities. In addition, NCAR (1988) describes the first side-by-side comparison of a group of regional-scale meteorological models for a common case study.

The CSU Regional Atmospheric Modeling System is a highly versatile tool that can be used to simulate a wide variety of atmospheric phenomena. It is the successor by merger of a cloud model developed by W.R. Cotton and a mesoscale model developed by R.A. Pielke. Continued development of RAMS has followed the philosophy of including a broad range of capabilities within one modeling system. The code is constructed with a front-end preprocessor that allows the user to construct a custom-designed model whose features can be selected from a menu of model options. This structure is facilitated by the use of modularization of the code into groups of subroutines which perform specialized tasks. Some RAMS features include:

- Two-way interactive nesting with any number of either telescoping or parallel fine nest grids.
- Cartesian or polar stereographic horizontal coordinates with terrain-following coordinate surfaces.
- Non-hydrostatic or hydrostatic time-split time differencing.
- Various levels of microphysical complexity from no microphysics to a mixed phase system including cloud water, rainwater, pristine ice crystals, graupel or hailstones, and aggregates of ice crystals.

- Parameterizations of shortwave and longwave radiative transfer through clear and cloudy atmospheres.
- Cumulus and turbulence parameterization schemes.
- Surface-layer parameterization with a prognostic soil model and vegetative canopy.
- Several options for upper and lateral boundary conditions (e.g., radiative boundaries) and for finite difference operators.

Table 1. Comparison of Mesoscale Meteorological Model Capabilities (from Pielke, 1989).

	San Jose Model	MASS Model	NCAR/PSU/SUNY Model*	Drexel Model	CSU RAMS Model	ERT CMC/PBL Model Combination	Los Alamos Model
Governing Equations	i) vorticity framework ii) hydrostatic iii) incompressible	i) primitive equation framework ii) hydrostatic iii) anelastic	i) primitive equation framework ii) hydrostatic iii) anelastic	i) primitive equation framework ii) hydrostatic	i) primitive equation framework ii) hydrostatic or nonhydrostatic iii) anelastic	i) primitive equation ii) hydrostatic iii) anelastic	i) primitive equation framework ii) hydrostatic iii) incompressible
Dimensionality	3-D	3-D	3-D	3-D	3-D	3-D/1D	3-D
Grid	staggered	Arakawa–A staggered	staggered	non-staggered	Arakawa–C staggered grid	staggered	staggered
Horizontal Grid Spacing	i) minimum 1.6 km	i) minimum 7 km	i) 80 km *ii) as small as 1 km*	i) minimum ~25 km	i) minimum ~100 m	i) 127 km	i) minimum 380 m
Vertical Grid Spacing	stretched grid	stretched grid	stretched grid	stretched grid	stretched grid	stretched grid	stretched grid
Model Domain	i) several hundred kilometers on a side	i) arbitrary ii) one-way interactive grid	i) arbitrary *ii) two-way interactive nested version*	i) arbitrary	i) arbitrary ii) two-way interactive iii) telescoping grid	i) northern hemisphere / eastern North America	i) arbitrary ii) one-way interactive grid
Initialization	i) dynamic initialization	i) objective analysis scheme using synoptic[1] analyses including satellite data; removal of integrated mass divergence ii) multivariate variational adjustment scheme planned for December 1989	i) dynamic initialization from synoptic[1] data	i) objective analyses of synoptic[1] data	i) objective analysis scheme of synoptic[1] data	i) normal mode initial analysis of synoptic[1] data/ dynamic initialization with nudging towards CMC fields in PBL	i) objective analysis of synoptic[1] data/ requirement for wind field to satisfy mass conservation
Solution Technique	finite difference	finite difference	finite difference	finite difference	finite difference	spectral (CMC)/ finite element (PBL)	finite difference
Coordinate System	$x-y-z$	$x-y-\sigma_p$	$x-y-\sigma_p$	$x-y-\sigma_z$	$x-y-\sigma_z$	$x-y-\sigma_p$	$x-y-\sigma_z$
Lateral Boundary Conditions	open	sponge from synoptic data	from synoptic data	from synoptic data	from synoptic data	from synoptic data/ from CMC analyses	from synoptic data and solution of 1-D primitive equations
Top Boundary Conditions	from synoptic specifications; up to present, model top confined to below mid-troposphere	absorbing layer	$\frac{d\sigma_p}{dt}=0$ at $\sigma=0$ *absorbing layer*	$\frac{dp}{dt}=0$	absorbing layer	specified from CMC analyses; PBL model top at 2 km	zero vertical gradient
Surface Boundary Conditions	i) heat energy and moisture surface budgets	i) heat energy and moisture surface budgets ii) specified from satellite iii) representation of vegetation effects	i) heat energy and *moisture surface budgets* *ii) representation of vegetation effects*	i) heat energy budget	i) heat and moisture surface budgets ii) representation of vegetation effects iii) specified from satellite	i) heat and moisture surface budgets ii) representation of vegetation effects	i) heat energy budget ii) representation of vegetation effects
Parameterization of Subgrid Mixing	1st and 2nd order closure	1st order closure	1st order closure	1st order closure	1st or 2nd order closure	1st order closure; a simplified 2nd order closure in the PBL model	2nd order closure
Cumulus Parameterization	none	1-D cloud models	1-D cloud models	1-D cloud models	1-D cloud models	1-D cloud models	1-D cloud models
Radiation Parameterization	i) solar and long-wave radiation at the surface ii) solar and long-wave radiative flux divergence in the atmosphere	i) solar heating of surface; long-wave radiation for longer than 24 hour runs (December 1989)	i) solar and long-wave radiation at the surface *ii) long-wave radiative flux divergence in the atmosphere*	i) solar and long-wave radiation at the surface ii) long wave radiation flux divergence in the atmosphere	i) solar and long-wave radiation at the surface ii) solar and long-wave radiative flux divergence in the atmosphere	i) solar and long-wave radiation at the surface ii) long wave radiation flux divergence in the atmosphere	i) solar and long-wave radiation at the surface ii) long-wave radiative flux divergence in the atmosphere

* Italicized information corresponds to NCAR/PSU version of the code.
[1] Only upper air synoptic data is used.

Table 1. Continued.

	San Jose Model	MASS Model	NCAR/PSU/SUNY Model*	Drexel Model	CSU RAMS Model	ERT CMC/PBL Model Combination	Los Alamos Model
Stable Precipitation Algorithm	none	i) rainout for relative humidity greater than 100% ii) ice microphysics representation in cirrus anvil clouds	i) rainout for relative humidity greater than 100%	i) includes equations for cloud liquid water and rain water – 1st order parameterization	i) includes equations for cloud ice, cloud liquid water, snow, and rain – 1st and 2nd order parameterizations	i) gradual rainout based on a dewpoint depression threshold dependent upon large-scale environment	none
Algorithms to Link to Other Models	i) plume models ii) Eulerian chemical models iii) Eulerian dispersion models iv) cloud models	i) trajectory model ii) 3-D nonhydrostatic cloud scale model	i) Eulerian chemical model (RADM) ii) trajectory model	none	i) advection diffusion (Eulerian dispersion) model ii) SAI chemistry model iii) Lagrangian dispersion model	i) Eulerian chemistry model (ADOM)	i) Lagrangian particle dispersion model
Phenomena Studied of Relevance to Complex Terrain	i) polluted urban coastal boundary layer	i) extratropical cyclones ii) frontal system iii) mesoscale convective circulations iv) sea breezes v) plume transport and acid rain deposition vi) squall line	i) extratropical cyclones ii) *sea-land breezes* iii) *forced airflow over rough terrain* iv) *frontal circulations* v) *mesoscale convective systems*	i) frontal zones ii) mesoscale convective circulations iii) cyclogenesis iv) heavy precipitation events	i) orographic precipitation ii) topographically-induced gravity waves iii) turbulence structure in the boundary layer iv) extratropical squall lines v) transport and diffusion of pollutants vi) sea-land breezes vii) mountain- valley flows	i) acid deposition ii) oxidant formation	i) diurnal variation of planetary boundary layer ii) sea breezes iii) nocturnal drainage flow iv) formation of plumes over a cooling pond v) turbulence in cloud in marine boundary layer vi) long-range transport

* Italicized information corresponds to NCAR/PSU version of the code.

Table 2. Summary of Model Capabilities (adapted from Pielke, 1989).

	San Jose Model	MASS Model	NCAR/PSU/SUNY Model*	Drexel Model	CSU RAMS Model	ERT CMC/PBL Model Combination	Los Alamos Model
1. Able to explicitly represent clouds and fog (i.e., to prognose or diagnose cloud water or cloud ice)	no	no	no	yes (water clouds)	yes	no	yes, using a statistical procedure for nonprecipitating cumulus water clouds
2. Appropriate tool which could be used to estimate transport over complex terrain	not unless terrain-following coordinate system is added	yes	*yes*	yes	yes	no	yes
3. Able to represent boundary-layer processes in complex terrain and coastal environments	no (no terrain in model)	yes	*yes*	yes	yes	no	yes
4. Nested grid versions	no	yes	*yes*	no	yes	no	yes
5. Satisfactory numerical solution algorithms as based on prior performance	yes	yes	yes	yes	yes	yes	yes
6. Needed data for initialization	i) synoptic	i) synoptic upper air ii) terrain iii) vegetation/soil	i) synoptic upper air ii) terrain iii) *vegetation/soil*	i) synoptic upper air ii) terrain	i) synoptic upper air ii) terrain iii) vegetation/soil	i) synoptic upper air ii) terrain iii) vegetation/soil	i) synoptic upper air ii) terrain iii) vegetation/soil
7. Needed data for mesoscale validation	mesoscale resolution wind, turbulence, and temperature data; at minimum of 20 equispaced surface stations and sounding data	mesoscale resolution wind, turbulence, and temperature data; at minimum of 20 equispaced surface stations and sounding data	mesoscale resolution wind, turbulence, and temperature data; at minimum of 20 equispaced surface stations and sounding data	mesoscale resolution wind, turbulence, and temperature data; at minimum of 20 equispaced surface stations and sounding data	mesoscale resolution wind, turbulence, and temperature data; at minimum of 20 equispaced surface stations and sounding data	mesoscale resolution wind, turbulence, and temperature data; at minimum of 20 equispaced surface stations and sounding data	mesoscale resolution wind, turbulence, and temperature data; at minimum of 20 equispaced surface stations and sounding data
8. Summary of ease of application to complex terrain (i.e., need for major model development)	major development work needed	relatively minor development work needed	major development work needed in SUNY version; relatively minor development work needed for NCAR/PSU research version	major development work needed	relatively minor development work needed	very major development work needed	relatively minor amount of development work needed (for nesting algorithm)

* Italicized information corresponds to NCAR/PSU version of the code.

- Isentropic data analysis package for model initialization.
- A suite of model-output analysis routines including vorticity, divergence, streamfunction, energy, momentum flux, variances and covariances, power spectrum analysis, layer-averaged quantities, and source/sink terms in physical routines.

RAMS is a highly portable modeling system which is currently being run on the following computers: CRAY X-MP under COS and UNICOS, CRAY Y-MP and CRAY-2 under UNICOS, CYBER 205, DEC MicroVax II under VMS, Ardent Titan under UNIX, COMPAQ-386, and others.

DISPERSION MODELING CAPABILITIES

Current Computer Power

The increased availability of inexpensive, high performance computers has made possible the replacement of simpler Gaussian plume and Lagrangian puff models of atmospheric dispersion with more computationally intensive (and more accurate) 3-D Lagrangian particle dispersion models (LPDMs) and Eulerian advection-diffusion models. LPDMs simulate the transport and diffusion of tracer releases by following the mean-wind- and turbulence-driven motion of a large number of individual particles, each of which represents a given amount of the tracer. These models frequently employ tens of thousands and sometimes more than one hundred thousand particles. Current computers are fast enough to simulate the motion of this many particles at a rate much faster than real time, and the results are thus useful for emergency response applications. Advection-diffusion models, on the other hand, simulate the emission and spread of pollutant from area or volume sources in an Eulerian framework. Computational cost for this type of model depends on grid resolution and domain size rather than number of particles released.

An essential component of modern LPDM and Eulerian advection-diffusion model simulations is the ability to display the output in a manner which rapidly communicates the results to the modeler. Machines of the graphics superworkstation class are well suited to this purpose as they have sophisticated color graphics packages for plotting 2-D and 3-D fields and for animating the time variation of the fields. For example, output from the RAMS LPDM is routinely animated on the computer screen in a 3-D perspective. The user can rotate the domain and view it from any direction as well as zooming in on (i.e., magnifying) smaller-scale features, thus acquiring a good understanding of the location and structure of the simulated particle plume. Such a capability is invaluable for emergency response applications where emergency personnel must quickly learn where toxic releases are predicted to go.

Lagrangian Particle Dispersion and Eulerian Advection-Diffusion Models

Modeling mesoscale dispersion, the result of both differential horizontal and vertical advection and turbulent diffusion, is an even more difficult problem than modeling shorter-range atmospheric dispersion. Both mean and turbulent flow components are nonstationary and inhomogeneous on the mesoscale, precluding the use of small-scale diffusion models based on Gaussian distributions or similarity theory. The two main classes of 3-D episodic dispersion models which have been employed on this scale are (1) Eulerian concentration moment models, which require the numerical solution of some form of the Reynolds-averaged advection-diffusion equation on a fixed grid, and (2) Lagrangian particle dispersion models (also known as Lagrangian statistical or stochastic models, Lagrangian Monte Carlo models, stochastic dispersion models, Markov chain models, random walk models, and random flight models). The latter will be emphasized in this paper because of their flexibility and because RAMS employs an LPDM for air quality studies (e.g., McNider et al., 1988).

K-theory dispersion models are the simplest of the Eulerian concentration moment models and have been widely used (e.g., van Egmond and Kesseboom, 1983; Brost et al., 1988). However, they suffer from several limitations. First, use of the gradient-transfer hypothesis limits these models to time scales much larger than the turbulence integral time scale and to pollutant spatial scales much larger than the turbulence integral spatial scales (e.g., Corrsin,

1974; Pasquill and Smith, 1983). Second, the pollutant mass being modeled must have a spatial extent at the very least equal to four or more horizontal and vertical grid increments in order that gradients be adequately defined and advection phase errors minimized (Pielke, 1984; Avissar et al., 1990). Third, the advection schemes used in these models may suffer from computational diffusion and from Gibbs oscillations at pollutant boundaries (which gives negative concentrations). These Eulerian model characteristics make point and line sources particularly difficult to treat. On the other hand, Eulerian models are well suited to handle complicated physical processes such as reactive nonlinear chemistry and wet deposition. A few modelers have also used higher-order Eulerian models (see Lamb, 1982; Pasquill and Smith, 1983). Eulerian advection-diffusion models have been applied to urban-scale, meso-β-scale, and regional-scale air quality problems. More information may be found in Klug et al. (1985), Chang et al. (1987), Thomson et al. (1987), Dronamraju et al. (1988), and Venkatram et al. (1988) among many others.

In LPDMs, turbulent dispersion is modeled by simulating the release of a large number of pollutant point-particles either simultaneously or sequentially and then advecting this set of particles in response to a resolved mean flow and to parameterized turbulent fluctuations. Concentration field moments can then be estimated based on the spatial distribution of the set of predicted particle-positions through estimation of the univariate or multivariate particle-position probability density function $P(z, t)$ (e.g., Sawford, 1985; de Baas, 1988). Single-particle models, in which particles travel independently of one another, can only be used to estimate the (ensemble) mean concentration field. Estimation of higher-order moments requires the use of multiple-particle models in which groups of particles are released whose turbulent velocities are correlated in space. Note that many particles can be released simultaneously in a single-particle model but such a release is still equivalent to an ensemble of one-particle realizations for identical flow conditions.

LPDMs have a number of advantages over Eulerian advection-diffusion models. First, a Lagrangian framework is more natural for modeling turbulent diffusion, itself a Lagrangian phenomenon. Second, LPDMs automatically conserve mass. They also do not suffer from computational phase dispersion or computational diffusion. Third, diffusion over small space and time scales, including emissions from point and line sources, can easily be treated. Fourth, LPDMs can be shown to have an equivalent Eulerian form, the Fokker-Planck equation, which reduces to the Eulerian K-theory advection-diffusion equation for large travel times (e.g., Monin and Yaglom, 1971; Thomson, 1987). Finally, LPDMs are flexible, conceptually simple, and, when used with a reasonable number of particles, computationally inexpensive. In principle, complex flows should be as amenable to this approach as simple ones provided that the mean flow, Lagrangian time scales, and turbulence statistics are known. With respect to disadvantages, LPDMs are not well suited for modeling the dispersion of nonlinear reacting pollutants (Lamb, 1980). Questions of statistical sampling error must also be addressed when estimating pollutant concentrations in areas of low particle density, sometimes requiring the release of very large numbers of particles. Other theoretical and practical issues concerning the use of LPDMs for modeling mesoscale dispersion are discussed in the material that follows.

The origins of LPDMs lie in the early random-walk models developed by Einstein (1905) and Langevin (1908) to describe Brownian motion and molecular diffusion. Most LPDMs treat particle velocity as a stochastic process and use a discrete form of the Langevin equation to describe particle velocities as subject to a damping force and a random acceleration. LPDMs were first applied to turbulent diffusion in the surface layer and vegetation canopies, and then more recently to the full planetary boundary layer as insight was gained in handling inhomogeneous turbulence and stratification effects. Studies of small-scale dispersion with LPDMs include those of Hall (1975), Hanna (1978), Lamb (1978), Reid (1979), Wilson et al. (1981a,b), Ley (1982), Legg and Raupach (1982), Thomson (1984), van Dop et al. (1985), de Baas et al. (1986), Sawford and Guest (1987), and Luhar and Britter (1989). Early applications of the LPDM approach to mesoscale dispersion studies include McNider et al. (1980, 1982), McNider (1981), and Garrett and Smith (1982). More recent applications include those of Arritt (1985), Etling et al. (1986), Schorling et al. (1986), Yu and Pielke (1986), Gross et al. (1987), Kao and Yamada (1988), McNider et al. (1988), Segal et al. (1988), Yamada and Bunker (1988), and Uliasz (1990b). All of the above studies employed one-particle models. Durbin (1980), Lamb (1981), Sawford (1982, 1984), and Thomson (1986b) have used two-particle Lagrangian models to study concentration fluctuations and relative dispersion.

Mesoscale LPDMs have generally used the coupled, single-particle Markov-chain equations

$$X_i(t + \Delta t) = X_i(t) + [u_i(t) + u_i'(t)]\Delta t \ , \ i = 1,2,3 \tag{1}$$

and

$$u_i'(t) = u_i'(t - \Delta t)R_i(\Delta t) + u_i''(t) + w_c(t)\delta_{i3} \ , \quad i = 1,2,3, \tag{2}$$

to determine the position of each of the independent virtual tracer particles, where $X_i(t)$ is the i-th component of the particle location at time t, Δt is the time step, and $u_i(t)$ is the i-th component of the resolvable or grid-scale wind vector at time (t) and location $X_i(t)$. $u_i'(t)$ is a turbulent velocity fluctuation component determined by means of the discrete Langevin equation (2), where $R_i(\Delta t)$ is the i-th velocity component Lagrangian autocorrelation coefficient for lag Δt and $u_i''(t)$ is a random, normally-distributed turbulent component which is independent of $u_i'(t)$. Term $w_c(t)$ in the vertical velocity component equation is the so-called drift correction.

If $w_c(t)$ is uniformly zero, then Equations (1) and (2) are appropriate for homogeneous, stationary turbulence. However, in the real boundary layer, a particle moving away from the ground experiences motions with longer Lagrangian time scales and varying turbulent energy. Legg and Raupach (1982) introduced the drift correction term as a way of dealing with the model-predicted but unrealistic accumulation of particles in low-turbulent-energy flow zones. Since then, a number of papers have explored the appropriate formulation of Equation (2) for inhomogeneous and nonstationary turbulence (Wilson et al., 1983; Ley and Thomson, 1983; Thomson, 1984, 1986a, 1987; van Dop et al., 1985; de Baas et al., 1986; Sawford and Guest, 1987, 1988; and Luhar and Britter, 1989). The best form for $w_c(t)$ is still under debate but some combination of terms based on higher-order turbulence moments and their temporal and spatial derivatives are required for an accurate treatment of inhomogeneous, nonstationary turbulence.

However, several studies have pointed out practical difficulties with some proposed forms of $w_c(t)$. First, some of the required statistical quantities (e.g., third-order velocity moments) may not be available from either observations or model predictions (Brusasca et al., 1989). Second, accurate treatment of some of these quantities may require very small model time steps to be used (Hurley and Physick, 1990). The primary constraint on the time step Δt used in an LPDM is that it be considerably smaller than the Lagrangian integral time scale T_L (e.g., Hall, 1975; Legg and Raupach, 1982). However, in mesoscale applications where particles may have to be tracked for days, computational cost will be sensitive to the choice of Δt and tradeoffs may have to be made between inaccuracies in the trajectory calculations and the number of particles released and followed. Note that if Δt is set to a value much larger than T_L, then the Langevin stochastic model reduces to a true random-walk model where velocity is treated as white noise and has no memory (e.g., Hall, 1975; Runchal, 1980). This is equivalent to the Eulerian K-theory model (Durbin, 1983).

Equation (1) is written very generally in that turbulent diffusion is considered in all three flow directions. In practice, cross-stream and streamwise diffusion are often neglected and only vertical diffusion is considered. However, a number of investigators have treated the more general case, including velocity covariance terms (Ley, 1982; Davis, 1983; Legg, 1983; Ley and Thomson, 1983; Zannetti, 1984, 1986; Thomson, 1986a). Longitudinal velocity variations tend to have a small but not insignificant effect (5–10%) in the near field (Davis, 1983; Legg, 1983) but their importance at longer distances has not been evaluated. Figure 1 shows particle plumes from a fumigation simulation in a coastal environment using the Mesoscale Dispersion Modeling System (MDMS) of Uliasz (1990a). Three variants of turbulent diffusion parameterizations were used: I – only vertical diffusion of particles; II – both vertical and horizontal diffusion (Eq. 2); III – vertical and horizontal diffusion including velocity component covariances (Zannetti, 1984). In the near field, use of Scheme I results in a narrower plume than the other two schemes but this difference decreases with travel time.

The turbulent velocity covariances appear to contribute little for this case.

One very important but frequently overlooked aspect of LPDMs is the method by which meteorological information is supplied. Many of the assumptions and approximations underlying LPDMs arise in this step. As Lagrangian models, LPDMs require Lagrangian meteorological inputs which are seldom available. However, such quantities may be estimated from Eulerian measurements if certain simplifying assumptions are made. The most fundamental of these is that particles are fully caught up in the mean flow so that a particle's deterministic velocity can be interpolated based on its position within the mean flow field. Second, particle turbulent velocity variances and covariances are assumed to be equal to the local Eulerian values, e.g., $< W^2 >= \sigma_w^2 = \sigma_{wE}^2$, although this relation does not always hold (Legg, 1983; Sawford, 1985). Third, various expressions are assumed for the Lagrangian integral time scales T_{Lu}, T_{Lv}, and T_{Lw}, often in terms of the Eulerian integral time scales T_{Eu}, T_{Ev}, and T_{Ew} (Hanna, 1982; Davis, 1983; Legg, 1983; Sawford, 1985; McNider et al., 1988). Fourth, a functional form must be assumed for the Lagrangian autocorrelation functions $R_{Lu}(\Delta t)$, $R_{Lv}(\Delta t)$, and $R_{Lw}(\Delta t)$ (Durbin, 1983; Sawford, 1985). Fifth, when mesoscale models are used to supply meteorological input to the LPDM, expressions for such Eulerian turbulence quantities as σ_{wE}^2 and T_{Ew} must be constructed in a fashion consistent with the turbulence parameterization employed by the mesoscale model, including level of parameterization complexity (first-order, second-order) and averaging type (ensemble, volume) (e.g., McNider et al., 1988; Uliasz, 1990a).

Most LPDMs to date have assumed that the Lagrangian particles are passive and non-buoyant but several investigators have considered buoyant particles (Cogan, 1985; Schorling et al., 1986; Gaffen et al., 1987; Walklate, 1987). Applications include LPDM simulations of crop spraying, cooling tower plume dispersal, heavy gas dispersion, dry deposition, and battlefield visibility. As has been reported for other regional-scale and long-range transport models (e.g., Kuo et al., 1985; Martin et al., 1987), LPDMs are also sensitive to the value of large-scale vertical velocity imposed. Uliasz (1990b) has presented the results from one such sensitivity test using three different large-scale vertical velocities.

Several approaches have been employed to estimate ensemble mean concentration fields based on particle positions. The simplest is to count the number of particles in a sampling volume to arrive at a particle number density or to weight the particles by their mass and obtain a mass density (i.e., concentration). This cell-average approach has several weaknesses. First, there is the question of what size the sampling volume should be. If the size of the sampling volume is too small, the resulting concentration estimates become too noisy as a result of random sampling fluctuations. If the size of the sampling volume is too large, then the concentration distribution is oversmoothed and significant smaller-scale features may be lost. Lamb (1982) and Davis (1983) have suggested several approaches to this problem. Second, the cell-average method is discrete in the sense that the concentration estimate is implicitly assumed to be constant everywhere within the sampling volume rather than varying smoothly with position.

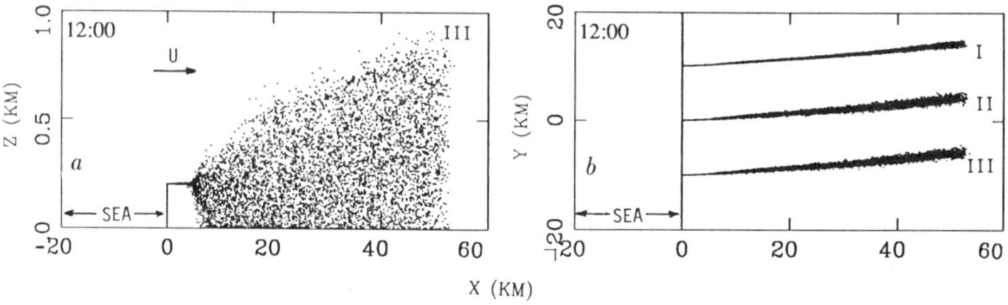

Figure 1. 3-D particle simulation of fumigation case over heated land with weak onshore flow using 2-D meteorological simulation: (a) xz-projection of particle plume at 1200 LST for scheme II; (b) plan view of plumes at 1200 LST for all three schemes. Emission source (z_e = 200 m) was translated along the coastline in (b) to demonstrate differences between LPDM turbulence parameterizations I, II, and III. The particles are released at a rate of 1200 particles per hour beginning at 0600 LST.

A new approach which makes use of a statistical technique called kernel density estimation (KDE) has been applied to this concentration estimation problem (Lorimer, 1986; Lorimer and Ross, 1986; Yamada and Bunker, 1988). The kernel density estimation technique overcomes the two weaknesses of the cell-average approach just mentioned and may require fewer tracer particles to make equally accurate concentration estimates (Grossman, 1989). It also provides a firmer statistical foundation on which to calculate estimation error bounds. However, influence bandwidths must be specified for this technique, and these present much the same problem as the specification of sampling volume mentioned above, especially for continuous plume releases where the particle distribution is elongated (Uliasz, 1990b). Figure 2 shows a comparison of four different KDE implementations for the fumigation case pictured in Figure 1. Choice of bandwidth can lead to oversmoothed or undersmoothed concentration estimates.

Source-Oriented Modeling

The determination of the impact of a particular source upon its surroundings is the basic concept behind source-oriented modeling. The two basic frameworks within which source impacts can be considered are the Lagrangian and Eulerian frameworks that were discussed in the last section. As previously stated, the Eulerian approach is most applicable to area or volume releases while the Lagrangian approach is commonly used when point or line sources are being considered.

Table 3 presents LPDM mesoscale dispersion estimates from two instantaneous volume sources. This table can also be used to estimate source attribution assuming an inert tracer release with no deposition velocity (i.e., the maximum potential impact). This dispersion estimate was calculated based on a number of different initial particle source counts. The two source areas are denoted in Table 3 as "NORTH" and "NGS" (Navajo Generating Station; see Figure 6); the receptor of interest is Hopi Point in the Grand Canyon (see Figure 6). The "NGS" source was approximately 125 km away from the receptor and the "NORTH" source was approximately 350 km away from the receptor, about three times more distant. The dispersion calculations were made for both daytime (mid-morning) and nighttime (mid-evening)

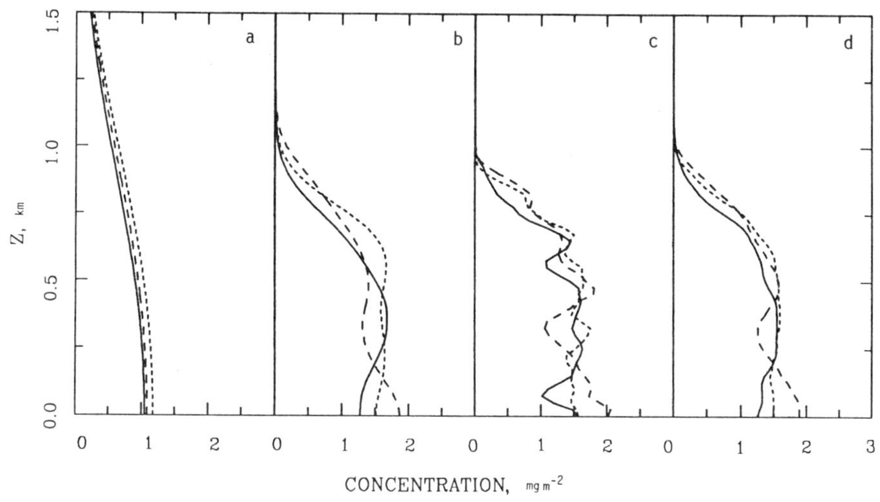

Figure 2. Vertical profiles of cross-wind-integrated concentration calculated by different kernel density estimators for the plumes shown in Figure 1b at $x = 50$ km using: (a) Gaussian kernel (Yamada and Bunker, 1988); (b) same as (a) but with an upper limit for the vertical bandwidth $h_z \leq 100$ m; (c) uniform kernel with bandwidths $h_x = h_y = 1000$ m, $h_z = 50$ m; and (d) same as (c) but $h_z = 100$ m. The three profiles in each panel correspond to the three LPDM parameterizations: I – dotted line, II – dashed line, and III – solid line.

Table 3. Maximum impact at Hopi Point receptor site from two instantaneous volume sources ("NORTH" and "NGS"). See Figure 6 for site locations ("NORTH" refers to the location on the northern border above the Canyonlands site).

	NUMBER OF PARTICLES RELEASED	MAXIMUM NUMBER OF PARTICLES AT HOPI POINT	ELAPSED TIME (hrs) AT MAXIMUM IMPACT	DISPERSION ESTIMATE
Day				
NGS	240	5	12.1	0.0208
NORTH	840	2	22.7	0.0024
Night				
NGS	240	8	8.5	0.0333
NORTH	840	2	20.6	0.0024
Day				
NGS	1500	51	11.5	0.0340
NORTH	1500	6	22.8	0.0040
Night				
NGS	1500	70	7.3	0.0467
NORTH	1500	8	17.0	0.0053
Day				
NGS	10000	363	11.6	0.0363
NORTH	10000	65	22.9	0.0065
Day				
NGS	20000	688	11.6	0.0344
NORTH	60000	320	23.0	0.0053
Night				
NGS	20000	1135	7.2	0.0578
NORTH	60000	464	17.0	0.0077

Day – large amount of dispersion; night – small amount of dispersion

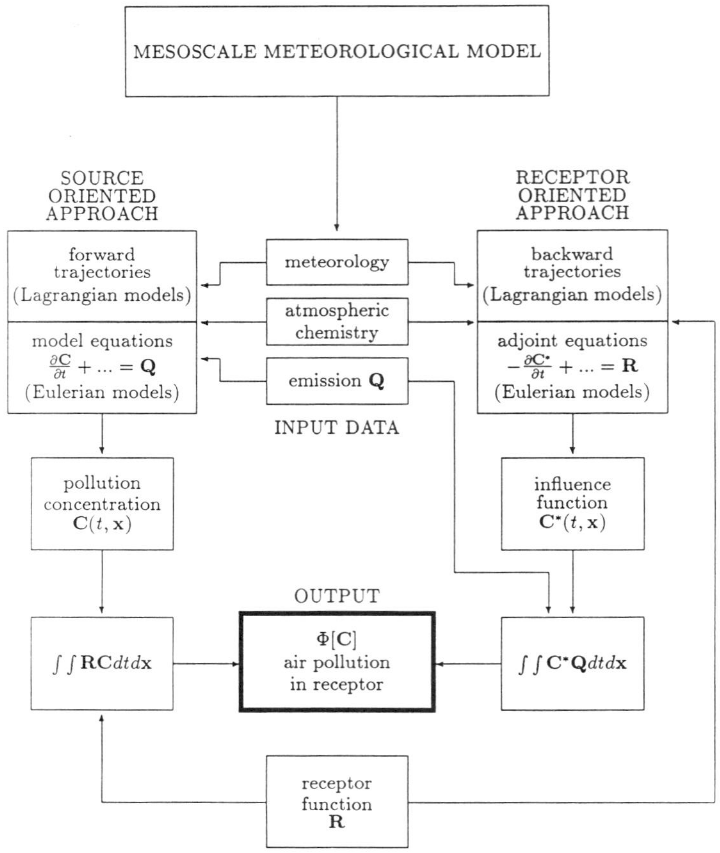

Figure 3. Source-oriented and receptor-oriented approaches in air pollution modeling.

release times. Comparing the daytime and nighttime simulations, dispersion is highest during the day for the "NGS" release and about equal for the "NORTH" release. This apparent discrepancy can be explained by the travel times to the receptor. The "NGS" release has a travel time of between 8-12 hours so that particles released at night are influenced almost entirely by nighttime conditions (low dispersion), whereas for the "NORTH" source, since travel times are on the order of a day, almost a full diurnal cycle has been completed for both daytime and nighttime releases before the receptor is reached.

The variability in the dispersion estimate based on the number of particles released is also illustrated by this table. By using only a small amount of particles for each source, the probability that any one of the particles will end up at the receptor is quite variable. This demonstrates that releasing a larger number of particles from a given source will produce a better dispersion estimate and reduce sampling fluctuations. Also, the further the source is away from the receptor, the greater the number of particles that are needed to account for the increased dispersion with distance. For the "NGS" source, the dispersion calculation started to level off for a much lower total particle count than for the "NORTH" source. The table provides an indication as to the number of particles that need to be released to obtain accurate dispersion figures when using a Lagrangian framework.

Receptor-Oriented Modeling

A traditional source-oriented approach in dispersion modeling consists in solving dispersion model equations forward in time for given sources of pollutant. As a result a time- and space-dependent concentration field C is obtained. In order to investigate another emission scenario Q, the solution of the model equations must be repeated. In many practical applications, air pollution at a given receptor is of primary interest and the alternative of receptor-oriented modeling should be considered as a more effective approach. Figure 3 presents a comparison of the source-oriented and receptor-oriented modeling approaches which can be used as complementary tools in air quality studies. Air quality at the receptor is characterized by an integral of pollution concentration over the modeling domain and time of simulation, i.e., $\Phi = \int\int RCd\mathbf{x}dt$. The receptor function R determines the location and shape of the receptor (point, area, volume), and the required averaging time of the concentration. The integral Φ is equivalent to averaging the concentration field with R as a weighting function. In the receptor-oriented modeling method, an influence function C^* is determined instead of concentration. The influence function can be calculated from backward trajectories of particles or puffs if Lagrangian dispersion models are used. Eulerian models governed by partial differential equations are subject to a variational formulation (e.g., Marchuk, 1982) and then the influence function is obtained as a solution of adjoint equations backward in time with the receptor function R as a source term. The influence function C^* calculated for a given receptor depends on meteorology and transformations of pollutant in the atmosphere but is independent of emission sources. The air quality integral can now be calculated in the alternative way by averaging the emission field Q with the influence function as a weight function, $\Phi = \int\int C^*Qd\mathbf{x}dt$. In a more general form, the integral Φ is expressed as a sum of contributions from (1) local sources within the modeling domain, (2) distant sources outside the modeling domain (in terms of pollutant fluxes across lateral boundaries, and (3) the initial concentration field. It is possible to simply calculate Φ for any distribution of pollutant sources (e.g., to investigate different heights of a planned stack) without repeating the solution of model equations. Receptor-oriented modeling is especially relevant to such applications as emissions control, planning locations of new emission sources, and assessing contributions from different sources to air pollution in a given region.

The same numerical algorithm and the same meteorological input created by a mesoscale meteorological model is used in the source-oriented and receptor-oriented modeling approaches. In the second case, meteorological fields must be read backward in time. Receptor-oriented modeling is computationally more effective than the source-oriented method if the number of receptors considered is less than the number of emission scenarios to be investigated. Applicability of receptor modeling is limited to linear dispersion models. However, this approach based on the variational formulation of model equations may still be useful to perform sensitivity analysis of the Eulerian dispersion models with nonlinear chemistry (Uliasz, 1983, 1985). Receptor-oriented modeling is discussed in more detail in a companion paper (Uliasz and Pielke, 1990), where some examples of calculations for mesoscale dispersion in a coastal zone of the Baltic Sea are also presented.

EXAMPLES OF APPLICATIONS

Mt. Isa, Australia

Realistic treatment of dispersion on the 20 km to 2000 km range requires the correct representation of the atmospheric temporal and spatial energy spectra. Because of the interest in microscale turbulence in air pollution, the high-frequency, planetary-boundary-layer (PBL) turbulent part of the spectrum is generally well treated in models. However, present rawinsonde systems spaced 300 km apart with observations taken at 12 h intervals cannot provide the necessary spatial and temporal resolution to incorporate energy at wavelengths below 600 km or periods below 24 h. Figure 4 shows the behavior of a simulated plume using an LPDM driven by a 1-D PBL dynamical model in which various parts of the atmospheric energy spectrum are included and excluded, respectively. Horizontal variations of the meteorological field are not considered. The major differences in the behavior of the plume are due to either PBL shear or synoptic (thermal wind) shear. This demonstrates the connectivity between scales in that microscale turbulence must be operating to allow the plume to interact with the vertical shear which is produced by much larger and longer scale processes.

Observations also appear to support the importance of temporal energy. Figure 5 shows observations of plume width taken at mesoscale distances downwind from Mt. Isa, a smelter in Queensland, Australia. Note the sustained diffusion rate out to nearly 1000 km. Superimposed on the observations are LPDM results from two simulations of the Mt. Isa plume. The model results, which include PBL temporal variations, show reasonable agreement with the observations. The study is reported in more detail in McNider et al. (1988).

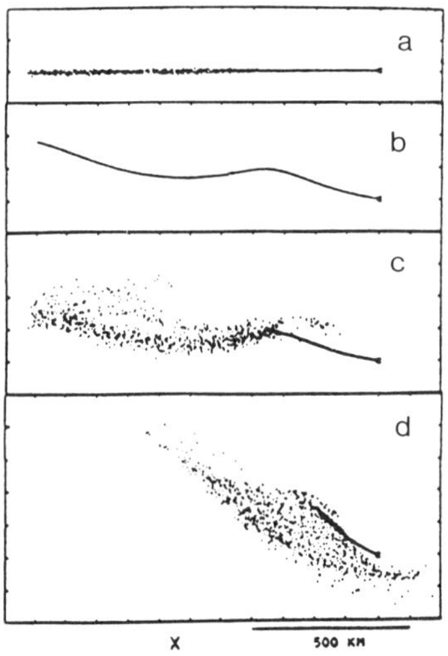

Figure 4. Depiction of an LPDM-simulated plume showing the response of the plume to various scales of atmospheric processes. Includes (a) only microscale turbulence; (b) Coriolis force but not boundary-layer shear or turbulence; (c) microscale turbulence, boundary-layer shear, and Coriolis force; (d) same as (c), but also including synoptic-scale geostrophic shear (from McNider et al., 1988).

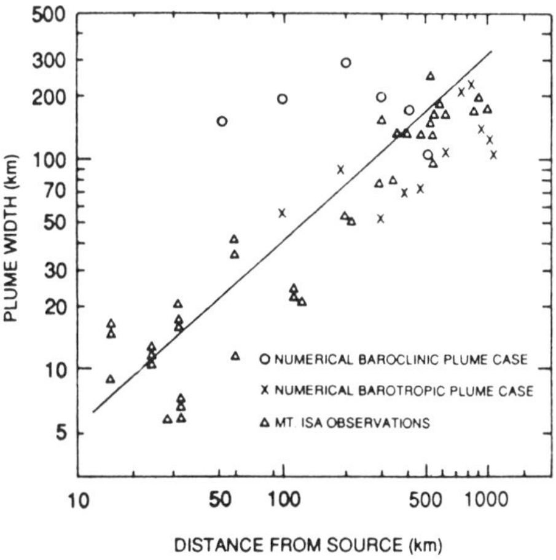

Figure 5. Horizontal spread of plume vs. downwind distance for the Mt. Isa observations and the barotropic and baroclinic LPDM simulations (from McNid-er et al., 1988).

Regional or mesoscale dynamical models with high resolution grids can provide some atmospheric flow information missed by the synoptic-scale observing network. In general, the high temporal resolution of these models can preserve most of the diurnal and inertial energy. However, even fairly high resolution regional models (e.g., $\Delta x = 80$ km) cannot provide much information on flow energy below approximately $4\Delta x$ because of filtering and/or numerical smoothing (Pielke, 1984; Avissar et al., 1990). Thus, even these high resolution dynamical models are not correctly handling the spatial energy spectrum below about 320 km. Even if an advection-diffusion grid model has finer horizontal resolution than the meteorological model providing it with wind and turbulence fields, it does not receive the proper spatial representation of the wind and turbulence fields from the meteorological model. This question of model resolution is discussed in greater detail by Moran et al. (1990).

WHITEX

Another example of the application of a mesoscale meteorological model (Mahrer and Pielke, 1977) coupled with an LPDM (McNider, 1981) is a simulation carried out for a period within the Winter Haze Intensive Tracer Experiment (WHITEX). This six-week study (January – mid-February, 1987) was conducted to determine the impact of regional and local sources on the haze production within certain Class I areas in northern Arizona and southern Utah (Malm, et al., 1989). The time period being modeled (February 11-12, 1987) represented a time period in which there was a particularly strong haze episode around the Grand Canyon in northern Arizona. The synoptic conditions for February 11 and 12 and for the five previous days indicated stagnant conditions due to the influence of a surface high-pressure system centered over the WHITEX area.

Figure 6 shows the model domain used for this simulation. The area of the domain was approximately 380 km^2 with a horizontal grid spacing of 5.5 km and a model top at 11 km. The terrain is very complex within this region as can be seen by examining Figure 6. The mesoscale thermally-forced circulations which are generated during strong stagnant conditions in this area are correspondingly complex.

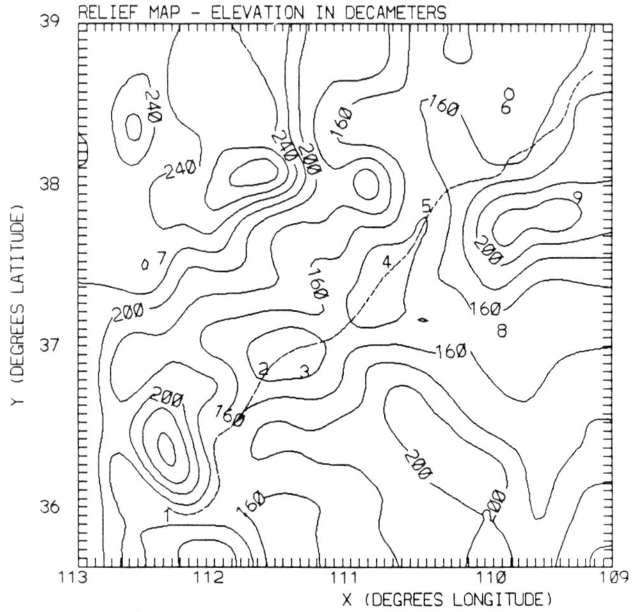

Figure 6. Topography used in the mesoscale meteorological model to represent the WHITEX region: 1 – Hopi Point; 2 – Page; 3– Navajo Generating Station (NGS); 4 – Bullfrog; 5 – Hite; 6 – Canyonlands; 7 – Bryce Canyon; 8 – Mexican Hat; and 9 – Monticello. The Colorado River is indicated by a southwest/northeast-oriented dashed diagonal line through the domain.

Figures 7a-c present several of the meteorological fields for 1500 LST, 11 h after the start of the simulation. Figure 7a shows a plan view of the vector surface winds, which are strongly influenced by the mesoscale circulations. Notice that during the afternoon, the winds along Marble Canyon (just to the south of Page, Arizona; see Figure 6) and towards the southeast are primarily out of the northeast which indicates down-valley flow. This is in agreement with surface winds obtained in this area for WHITEX during this time period (Malm et al., 1989). This northeast flow is mostly due to the heating of the Kaibab Plateau (located just north of Hopi Point on an elevated plateau; see Figure 6) which acts as a local heat source during the day and drives daytime upslope winds. In contrast, the 500 m (~ the effective stack height of the NGS plume) vector wind field (Figure 7b) also suggests northeast transport, but these winds are primarily being driven by the synoptic pressure gradient force imposed at the start of the simulation. Another interesting feature in the 500 m flow is the cyclonic eddy circulation which develops in the mountain/valley region around Monticello. This circulation would tend to block any flow coming east along the San Juan River basin (Mexican Hat is located within this basin; see Figure 6). This feature is evident at both the surface and 500 m levels. Figure 7c shows the thermal structure along a y-z cross section centered at 111.7°W. In this figure, the region around Page and NGS (~36.9°N) shows a PBL approximately 700 m in depth. The deepest PBL occurs in the mountain range around Bryce Canyon, where the PBL exceeds 1.5 km due to strong mesoscale circulations in the area. Observations obtained from

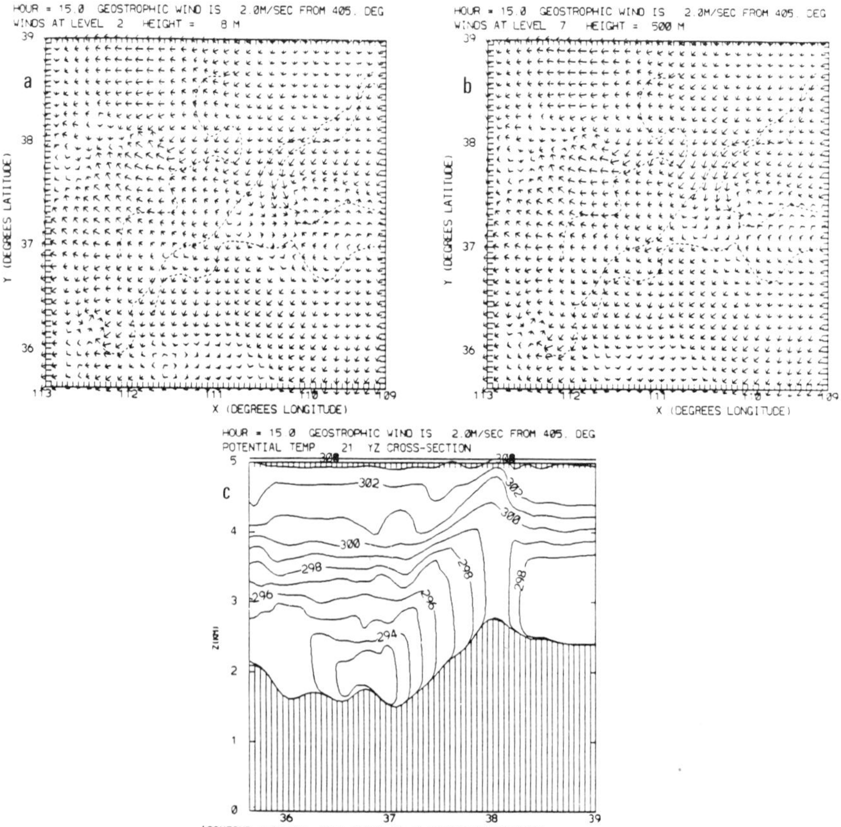

Figure 7. (a) Mesoscale meteorological model vector wind field at 8 m above ground at 1500 LST; (b) same as (a) except at 500 m; and (c) potential temperature at 1500 LST on a y-z cross section centered at 111.7° W. The Colorado River is indicated in the first two panels by a dashed line as in Figure 6. The 1600 m contour interval is also marked and shows the confluence of the Colorado River and San Juan River valleys.

the Canyonlands and Page soundings for this time period suggest that the maximum PBL depth around these two areas was approximately 800-900 m. The model produces maximum PBL depths in these two areas on the order of 1 km.

Figures 8a-d show plan views of particles representing the NGS tracer with the particles being released at a rate of 18 particles per hour. This continuous point source release was initialized with a release height of 600 m (the estimated effective stack height). These figures show that with the inclusion of thermally-forced mesoscale circulations, the NGS plume impacts Hopi Point in about 12 h. The plume tends to travel on the east side of the canyon rim, which agrees with the time-lapse sequences of this event (Malm et al., 1989). At 0200 LST, notice that the last half of the plume is very tightly clustered due to the lack of substantial turbulence at night. Therefore, this part of the plume stays above the surface during the night due to the decoupling of the surface flow from the winds around 500 m but starts to be mixed down to the surface by 0900 LST the next morning. This work is reported on further in Stocker and Pielke (1990).

As can be seen from this case study, mesoscale thermally-forced circulations in complex terrain can create very complex flow patterns. The visualizations of these flow patterns by particle releases serves not only to determine the impact of different sources on a particular receptor but also emphasizes how poorly, in general, any one particular meteorological measurement site can characterize the flow field in the entire region.

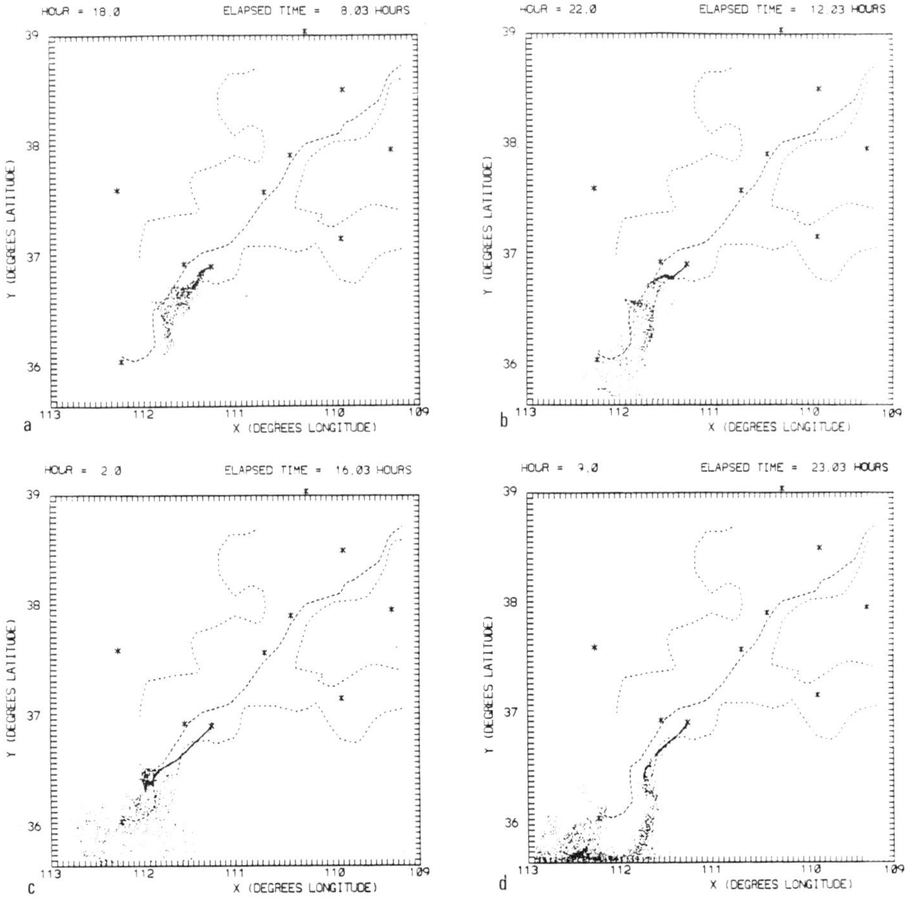

Figure 8. (a) Location of tracer particles at 1800 LST from a continuous release starting at 1000 LST from NGS; (b) same as (a) except at 2200 LST; (c) same as (a) except at 0200 LST; and (d) same as (a) except at 0900 LST.

South Florida

The CSU mesoscale dispersion modeling system has also been used to make estimates of worst-case maximum ground-level concentrations of sulfur dioxide (SO_2) for two National Park Service management areas in southern Florida due to SO_2 emissions from four major point sources located 60 to 80 km away from the park boundaries. The two park areas are Everglades National Park and Big Cypress Preserve. They are situated at the southern end of the Florida peninsula relatively close to a number of rapidly growing urban areas (Miami, Fort Lauderdale, West Palm Beach, Fort Meyers). Three of the large, coal-fired power plants of interest are located on the southeast coast of the peninsula (Fort Lauderdale, Port Everglade, Turkey Point) and one on the southwest coast (Fort Meyers).

Over the last 15-20 years, it has become clear from a number of observational and modeling studies that the lower atmosphere over the southern Florida peninsula is strongly influenced by local sea- and land-breeze circulations, especially during the summer (e.g. Frank et al., 1967; Pielke and Mahrer, 1978; Lopez et al., 1984; McQueen and Pielke, 1985). Such mesoscale circulations will significantly affect the mesoscale transport and diffusion of pollutants in this area and thus must be considered in any dispersion calculations.

An air pollution potential climatology was first constructed for southern Florida in order to identify synoptic patterns which are likely to be associated with mesoscale transport from one or more of the four coastal sources to the two parks and which occur at least once a year. Two seasonal worst-case scenarios, a summer case and a winter case, were chosen and were used to initialize the mesoscale model. The latter case is discussed in Segal et al. (1986). The summer case will be summarized here and is described in detail in Segal et al. (1988).

The worst-case summertime synoptic scenario consists of moderate easterly flow over southern Florida associated with the Bermuda High. Under these conditions, the low-level

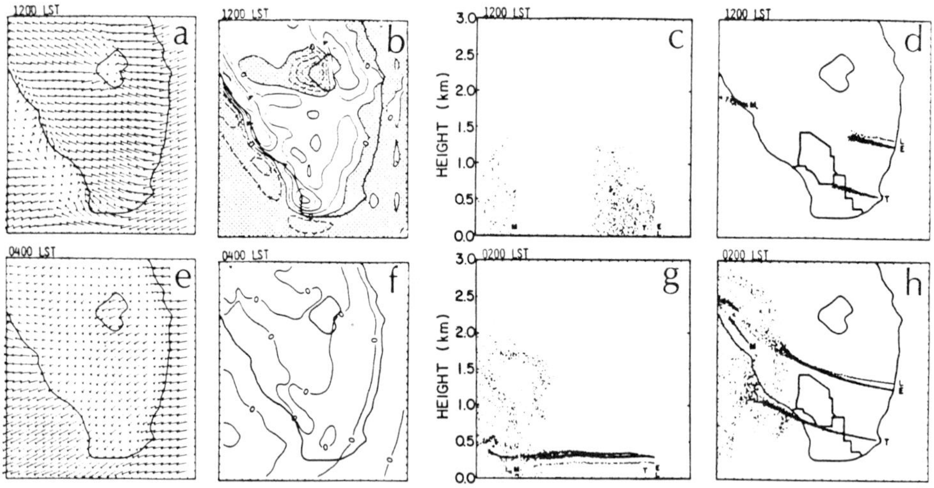

Figure 9. Model-predicted fields from meteorological model and LPDM over southern Florida domain: (a) horizontal wind field at 25 m at 1200 LST; (b) vertical velocity field at 500 m at 1200 LST (contour interval is 2 cm s^{-1}; stippled areas indicate sinking motion); (c) xz-projection of particle positions at 1200 LST looking northward (M – Fort Meyers source; T – Turkey Point source; E – Port Everglades source; L – Fort Lauderdale source); (d) xy-projection of particle positions at 1200 LST; (e) same as (a) except for 0400 LST; (f) same as (b) except for 0400 LST; (g) same as (c) except for 0200 LST; (h) same as (d) except for 0200 LST. The east-west extent of all panels is 264 km (24 grid squares times 11 km/grid square). The two park areas are indicated at the bottom of the Florida peninsula in panels (d) and (h); Lake Okeechobee is near the top center of these panels.

flow will carry pollutants from the three east coast sources towards the park areas of concern. Such a weather pattern occurs about 15% of the time during the summer months. The three-dimensional version of the mesoscale meteorological model was run for this scenario for 30 h beginning at 0800 LST. Horizontal grid spacing was 11 km and 14 vertical levels were employed to a height of 5 km. Low-level wind speed was 5.8 m s^{-1}. The surface parameterization included both a multilayer soil model and vegetation canopy.

Figures 9a and 9e show the simulated 25 m horizontal flow field at 1200 LST and 0400 LST, 4 and 20 h after the start of the simulation. During the day the east-coast sea breeze augments the easterly synoptic flow while the west-coast sea breeze generates an opposing low-level westerly flow. A well-defined west-coast convergence zone forms and moves inland during the afternoon (see Figure 4c of Segal et al., 1988). The decay of the sea breeze towards the evening and the establishment of a stable surface layer over land leads to the weakening of the inland surface flow (in contrast to the over-water flow). During the night, the land breeze along the west coast is noticeable through an acceleration of the flow as it moves offshore (Figure 9e). Figures 9b and 9f show the 500 m level vertical velocity field at the same times as Figures 9a and 9e. During the day, rising motion generally occurs over land and sinking motion occurs over water. However, note the subsidence zone in the lee of Lake Okeechobee and the enhanced upward motion in the west-coast convergence zone (Figure 9b). Vertical velocities diminish at night across the model domain.

In the LPDM simulation, particles were released at regular intervals from the four power plant sites beginning at 0800 LST. The number of particles emitted per source per unit time was scaled by the relative magnitude of each source in order to provide a realistic visual perception of the source contributions in the plots of particle dispersion. Diurnal variations in load factor were accommodated by adjusting the mass of each virtual particle as it was emitted. Since the power-plant stack gases are heated and buoyant but LPDM particles are assumed to be nonbuoyant, a bent-over plume model described by Hanna et al. (1982) was used to calculate the effective stack height. The particles were then released at this height, which varied diurnally with changes in atmospheric conditions, especially thermal stability (see Figure 10 of Segal et al., 1988). This approximation was felt to be justified because far-field rather than near-field surface concentrations were the focus of this study.

Figures 9c and 9g present features of the particle dispersion simulation at 1200 LST and 0200 LST as viewed when the particles are projected onto an xz-plane. During the daytime, the particles are well mixed within the growing boundary layer (Figure 9c). Following sunset and the onset of stable thermal stratification, however, the east coast plumes (T, E, L) are layered at the effective stack height level (Figure 9g). The deep layer of daytime-mixed pollutant is advected offshore along the west coast during the night. Fumigation of the elevated stable plumes over land then occurs the next morning. Figures 9d and 9h provide the following information: (i) a narrow plume indicates a plume emitted into and dispersing within a thermally stable layer (note that this situation does not include light wind conditions where plume meander can be important); (ii) a wide plume indicates a plume which has been emitted into (or at some point has resided in) a thermally less stable layer. A wider plume may also be due to variations of wind direction with height, i.e., to dispersion due to differential advection. Only narrow plumes are evident in Figure 9d while both narrow and wide plume segments are present in Figure 9h. In the evening and during the night, as the lower atmosphere stabilizes, plumes emitted from the east coast sources exhibit less lateral dispersion except over the western section of the peninsula, where significant changes in the flow direction resulted after the cessation of the sea breeze.

Concentrations were computed as averages for the lowest 25 m within each horizontal grid cell. This depth was chosen in order to enclose most of the vegetative canopy. The averaging area (11 km × 11 km) was felt to be sufficiently small to permit estimation of concentrations at grid cells well removed from the source of an advecting, diffusing plume. The mass of particles resident in each cell was summed and then divided by the grid cell volume to obtain concentrations in units of $\mu g\ m^{-3}$ of air. The average concentration for the previous 3 h was then calculated hourly, beginning 3 h after the start of the simulation. SO_2 depletion due to wet or dry deposition and chemical transformation have not been considered. The computed SO_2 concentrations are thus overestimated to a certain degree. However, this simplification is not as limiting as it would be for longer travel times.

Figure 10 presents the predicted 3-h-average SO_2 concentration fields for the lowest 25 m. Maximum 3-h-average computed values in the areas of the parks (well removed from the sources) are less than 25 $\mu g\ m^{-3}$, which is the maximum incremental 24-h average SO_2 value permitted for new sources to impact legislatively-designated clean air Class I areas. On Day 1 of the simulation, as the plumes are advected westward with a slight northward component, only the Turkey Point plume (southernmost east-coast source) affects Everglades National Park in its northern section, with a concentration $< 10\ \mu g\ m^{-3}$ (Figures 10a, 10b). During the night as (i) emissions occur into a stable layer, and (ii) directional wind shear due to frictional decoupling reduces the daytime low-level concentrations of pollutants which were advected westward, there is a gradual reduction of computed surface concentrations to a very low level (Figure 10d). Fumigation resulting from the development of the boundary layer during the morning hours, however, leads to significant increases in the computed near-surface concentrations (Figures 10e, 10f). Again, only the Turkey Point plume affects the Everglades with peak concentrations about 20 km from the source and a second peak in the western coastal area (in both cases the surface concentration peaks are about 20 $\mu g\ m^{-3}$). While the first peak results directly from trapping conditions (i.e. a low-lying capping inversion), the west coast peak is attributable to trapping following a fumigation event. The reduction of wind speed and the change of its direction along the west coast area during the morning hours contributed further to an increase of computed pollutant concentration there.

Kennedy Space Center

The Cape Canaveral Air Force Station (CCAFS), adjacent to the Kennedy Space Center in Florida, handles, transports, and stores large quantities of hazardous materials which are potential atmospheric contaminants. Other concerns include exhaust plumes from spacecraft and debris clouds from launch mishaps, which could include aerosol radionuclides from on-board nuclear-powered electrical generators. One of the more sophisticated Emergency Response Dose Assessment Codes (ERDACs) available today, MARSS, the Meteorological and Range Safety Support System (Taylor and Schumann, 1986) can employ either the Ocean Breeze/Dry Gulch algorithms or a somewhat more versatile puff-advection scheme (LOMPUFF) driven by the surface-layer wind field from the regional mesonetwork. While systems

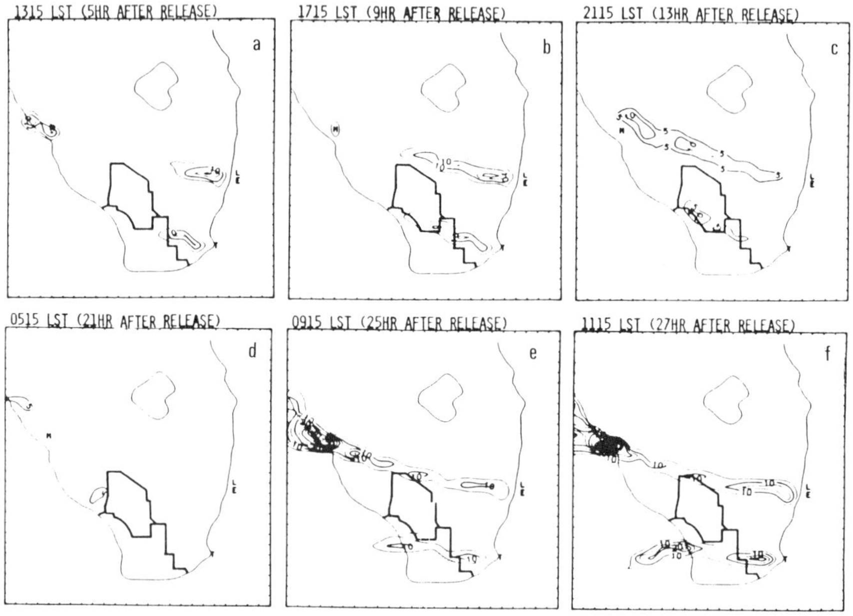

Figure 10. Three-hour-averaged SO_2 concentrations in the lowest 25 m in $\mu g\ m^{-3}$ as computed from the Lagrangian particle distribution for periods ending at (a) 1315 LST; (b) 1715 LST; (c) 2115 LST; (d) 0515 LST; (e) 0915 LST; and (f) 1115 LST. The averaged concentrations were computed in the period prior to the stated time. Contour intervals are 5 $\mu g\ m^{-3}$ (from Segal et al., 1988).

such as MARSS represent significant improvements over earlier systems, they share at least two common failings with those used frequently in emergency response: (1) the use of near-surface winds only, which are incapable of resolving recirculating mesoscale flows associated with sea/land breezes, and (2) the use of time-independent winds. Even if a perfect 3-D description of the transport winds were available over the entire domain of interest, the rapidly evolving flow field associated with many mesoscale weather regimes mandates using *predicted* winds valid during the next several hours to determine the location of the severely impacted receptors.

An increasing number of military, nuclear power, and chemical facilities worldwide are installing ERDACs on a variety of computational platforms. The ultimate futility of attempting to apply Gaussian steady-state straight-line plume models and their numerous segmented plume and puff-advection progeny to all meteorological regimes has been widely reported, yet emergency response planners have had little choice. While considerable sophistication may be desired for such ERDACs, real-world requirements demand a system that can (1) be initialized with data sources either on hand or easily available; (2) be run using affordable computational resources; (3) generate output very quickly, typically within 15 minutes; and (4) produce results requiring minimal interpretation to provide useable response guidance. The prediction of mesoscale dispersion for emergency response situations is critically dependent upon many factors, including accurately describing and predicting the 3-D wind and turbulence fields within the plume domain. The difficulties inherent in describing, much less predicting, mesoscale dispersion within coastal zones have been extensively detailed in the past. A 1981 DOE Model Validation Workshop held at Savannah River Lab concluded that (1) Gaussian trajectory models make assessments at travel distances where a single wind at the source can no longer be representative of the free transport of pollutants; (2) further improvement in the numerical accuracy of these models is not nearly as important as the accurate calculation of the winds; and (3) more effort should be directed at improving the analyses of the wind fields driving the models. The DOE/AMS Air Pollution Model Evaluation Workshop at Kiawah, SC in 1985 noted that (1) model performance can be improved by using more accurate three-dimensional analyses of wind fields; (2) particle-in-cell methods should yield results superior to Gaussian models when coupled with 3-D wind data and ample computing resources; and (3) within several years, as new graphic supercomputers or "desktop" supercomputers become available and affordable, computer limitations will no longer be an impediment to the use of mesoscale prognostic models to drive predictive ERDACs. The issue will shift to the availability of properly tested and evaluated software. R*SCAN Corporation and ASTeR Inc. have teamed to provide a commercial version of the CSU RAMS model called ARAMS – the Advanced Regional Atmospheric Modeling System. ARAMS, when combined with an LPDM, appears to provide the basis for a new generation of ERDACs better suited for emergency response applications in complex coastal wind fields. An evaluation conducted at CCAFS was aimed at assessing the potentials of this new technology.

The first phase of the Kennedy Atmospheric Boundary Layer Experiment (KABLE) was conducted during November, 1988 (Taylor et al., 1989). In addition to over four dozen monitoring sites in the operational surface mesonetwork (wind, σ_θ, precipitation), special rawinsondes ascents were taken at the easternmost tip of the Cape and a continuously operating Doppler sodar was installed near Titusville (TIX) on the west bank of the Indian River. On 7 November 1988, there were fair skies in the region as a weak high-pressure system drifted eastward through the eastern Gulf of Mexico. At daybreak, there were moderate (5 m s^{-1}) northwest surface winds. Temperatures over land were approximately 10-13°C as compared to the 22°C to 26°C water surface temperatures offshore. Solar insolation was near the maximum possible for the date, yet inland high temperatures only reached about 3°C warmer than those over the ocean. This was sufficient to allow development of an Atlantic Sea Breeze (ASB). It was first detected, as is common, along the west shore of the Indian River at 1500 GMT. It very slowly pushed inland, reaching 9 km west of the river shore by 2100 GMT. Though insolation was relatively weak due to the proximity to the winter solstice, surface heating over the adjacent Merritt Island and Cape Canaveral land masses resulted in distinct convergence onto these "heated islands". This was accompanied by strongly divergent flow off the Indian and Banana Rivers, sometimes referred to locally as the "river breezes". Figure 11a shows the complex model-predicted surface-layer wind flows at 1700 GMT. The ASB frontal position and the island convergence zones are shown to persist for most of the day. These perturbations in the wind field were located near major Space Shuttle and missile launch complexes, as well as storage and handling facilities for highly toxic oxidizers and other chemicals.

ARAMS was employed to perform a hindcast of the conditions on this day. By using a nested inner grid over the CCAFS/KSC region as fine as 1000 m, it was possible to simulate with considerable fidelity, the wind, temperature, and turbulence patterns. Figure 11a shows the surface-layer wind streamlines predicted by ARAMS at 1700 GMT. In addition to an accurate presentation of the surface-layer winds, a full three-dimensional flow field was generated. Figure 11b shows the contours of Pasquill-Gifford (PG) stability class in the lowest 30 m (determined by temperature lapse rate). Over the cooler water PG class E prevailed, whereas over the heated land, PG classes A and B predominated. Inland, the mixing depth reached 1300 m to 1500 m. The vertical section east-west through the center of the domain (Figure 11c) shows the three distinct circulations associated with the ASB, Merritt Island, and Cape Canaveral. Turbulence and stability patterns were also generated. In the sea-breeze inflow, a well formed TIBL was present, with afternoon mixing depths over most of the CCAFS/KSC complex being less than 400 m. The predicted onset time of the ASB at the sodar site was within several minutes of the observed. The inflow depth ranged between 200 and 400 m. The predicted inflow, averaging 234 m was somewhat shallower than the observed 322 m, but the basic patterns were similar. Through mid-afternoon, the predicted and observed inland penetrations were nearly identical. Of considerable interest were the peak vertical motions resulting from the convergence patterns which formed. Since these features are very narrow (on the order of 1 km) the predicted vertical motions are sensitive to the horizontal grid scale

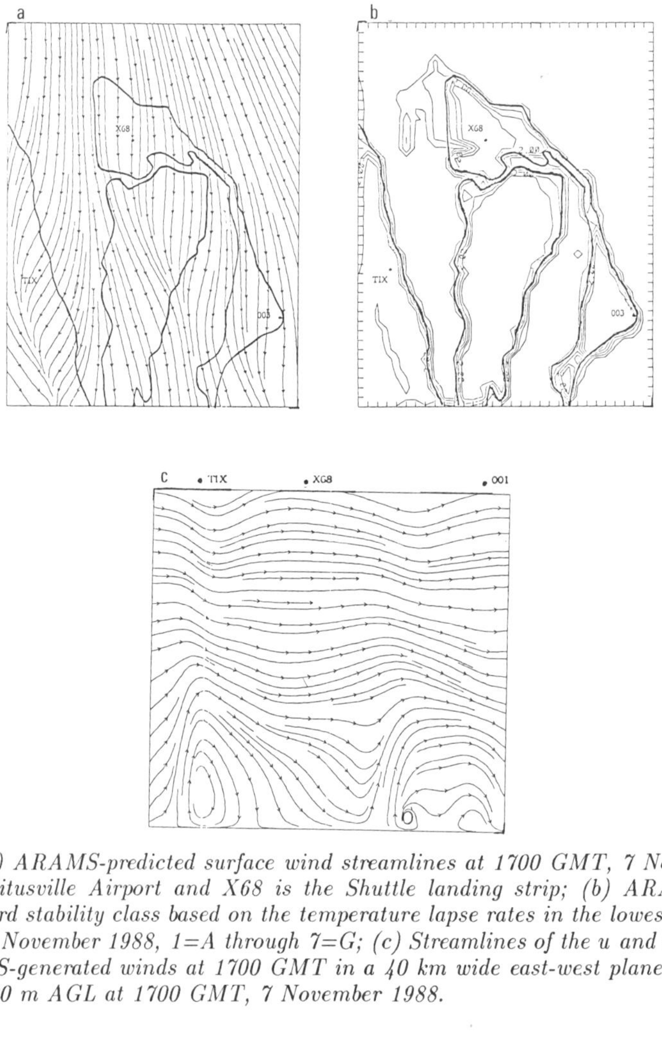

Figure 11. (a) ARAMS-predicted surface wind streamlines at 1700 GMT, 7 November 1988. TIX is the Titusville Airport and X68 is the Shuttle landing strip; (b) ARAMS-predicted Pasquill-Gifford stability class based on the temperature lapse rates in the lowest 30 meters at 1700 GMT, 7 November 1988, 1=A through 7=G; (c) Streamlines of the u and w components of the ARAMS-generated winds at 1700 GMT in a 40 km wide east-west plane through TIX, surface to 3000 m AGL at 1700 GMT, 7 November 1988.

used in the model. As shown in Figure 12, maximum predicted updrafts in the ASB frontal zone were resolved to be only 30 cm s^{-1} using only the 9 km coarse grid. This is comparable to most other mesoscale model results. Yet recent field observations suggest much stronger vertical motions are likely. Indeed, the 1 km ARAMS nested configuration produced a peak updraft of 170 cm s^{-1} during mid-day. Figure 13 shows the surface-layer convergence patterns. The main convergence zone was associated with the ASB, yet strong convergence was also found over the two heated islands. Vertical motions in excess of 30-40 cm s^{-1} were noted over southern Merritt Island much of the mid-day period. Thus contaminants released within this domain and advecting into the organized updraft zones could be expected to undergo substantial vertical transport. Any resultant dispersion is likely to be at considerable variance with that indicated by a two-dimensional diagnostic wind field model driven by surface-layer anemometers.

Figure 14 shows a very simple application of the LPDM. In this case, an accidental spill of 1000 gallons of nitrogen tetroxide (N_2O_4) was simulated to occur near launch pad 41 at 1700 GMT. The resultant pool was approximately 30 meters across, and the evaporation rate was assumed to be constant, completed within 15 minutes of the spill. Thus an elongated puff was released (from 2 m effective height) into the north-northeast flow down the Banana River. Gaussian models assume that the σ_x and σ_y dispersion coefficients are equivalent. In this case, using the Lagrangian particle dispersion model, as the puff advected down the river, the strong difluent flow onto the adjacent heated island caused the puff to be elongated in its transverse direction. By the time it reached the center of Cape Canaveral (15 km downwind), it had reoriented into an almost east-west front. Maximum mixing over the water was confined to under 400 m, although portions of the puff were beginning to become entrained into a updraft over the Cape after about an hour.

A data management and display subsystem is being developed which allows retrieval of the peak surface concentration as a function of downwind distance (which represents a complex non-linear shape). In addition, at each receptor on the plume "centerline" the temporal history of the passage of the puff or plume can be displayed (Figure 15). Figure 16 further illustrates the complexity of the dispersion patterns over the CCAFS/KSC region on 7 November. In a second hypothetical release, a tank car derailment on a railroad bridge over the Indian River was simulated to occur at 1700 GMT. The release (assumed at 2 m) continued at a constant rate for 8 hours. A plan view of the particle cloud revealed that the plume drifted inland a short distance, and intersected the ASB frontal position. The cloud then drifted south. After several hours, the cloud appears to bifurcate into two major "arms", one heading south along the convergence zone, the other moving almost due east over southern Merritt Island. Since the Stardent 2000, used to display the figures in this section, allows full image rotation and perspective viewing, an elevated view from the southwest clarifies the processes involved. The plume was actually convected upwards in the strong updraft zone. Part of the plume matter moved towards the sea near the top of the return flow layer. The lower particles subsided in the downdrafts over the Indian River and were re-entrained into the ASB inflow. There they underwent fumigation for a second time, reaching the surface in significant concentrations some 20 km south of the source. This confirms the possibility of phenomena such as alongshore quasi-helical trajectories and first and second trip fumigation.

The same basic meteorological modeling system can be used for a variety of purposes, thus amortizing costs over several activities. For instance, a utility can use the improved mesoscale temperature predictions as an input to their daily load forecasting algorithms. Improvements in local wind, cloud, and thunderstorm forecasts can be of considerable benefit to military operations and missile launching ranges, as demonstrated by Lyons et al. (1987) in their sea-breeze thunderstorm forecasting experiment at the Kennedy Space Center.

MODEL VALIDATION

There have been few validation studies of mesoscale and longer-range dispersion. Only limited mesoscale dispersion data sets such as CAPTEX (Ferber et al., 1986) and WHITEX (Malm et al., 1989) have been collected. Therefore, confidence in model simulations of dispersion have relied on accurate meteorological simulations in which comparisons are made between observed wind and temperature, and equivalent model predictions (to the extent possible since the model results are grid-volume values while observations are point values; e.g., see Pielke, 1984, Chapter 4). Since observational data are spatially limited, the assumption is made that the model simulation accuracy, which is determined from comparisons where observations and model data are coincident, will be the same throughout the modeled domain. This assumption is justified because the meteorological model uses physically consistent conservation equations to simulate the temperature, velocity, and moisture fields. Unfortunately, less information is normally available to estimate the turbulent motion field or other scales of air movement smaller than the model grid resolution. Examples of meteorological model simulations of this type which have been quantitatively compared with observations include Pielke and Mahrer (1978), Segal and Pielke (1981), Snow (1981), Segal et al. (1982), and Steyn and McKendry (1988).

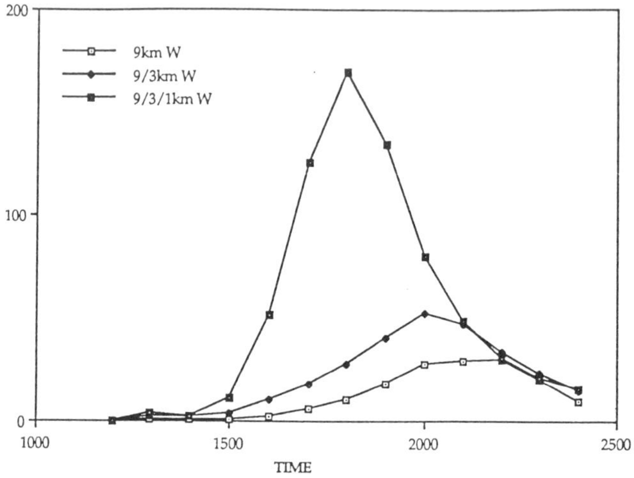

Figure 12. ARAMS-predicted maximum vertical motions (cm s^{-1}) in the sea-breeze convergence zones around CCAFS/KSC, 7 November 1988, as a function of the size of the finest nest horizontal grid cell.

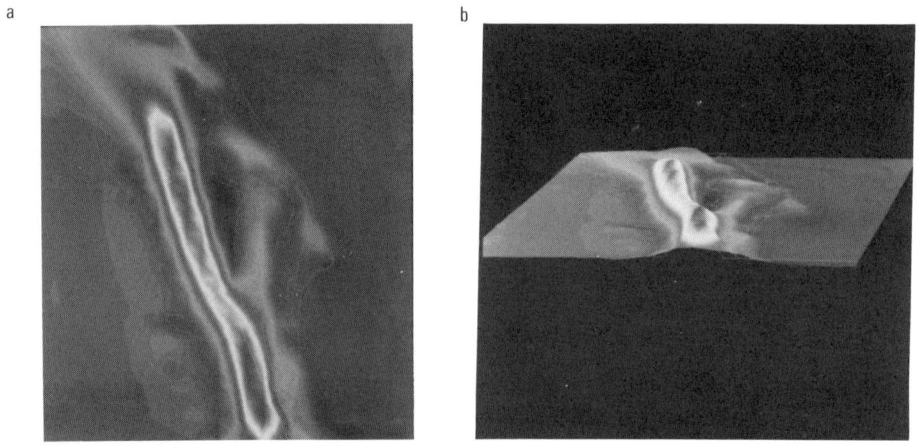

Figure 13. ARAMS-predicted convergence of surface winds in CCAFS/KSC region, 2000 GMT, 7 November 1988, visualized using the Stardent 2000 graphics package, in plan (a) and perspective (b) views (originals in color).

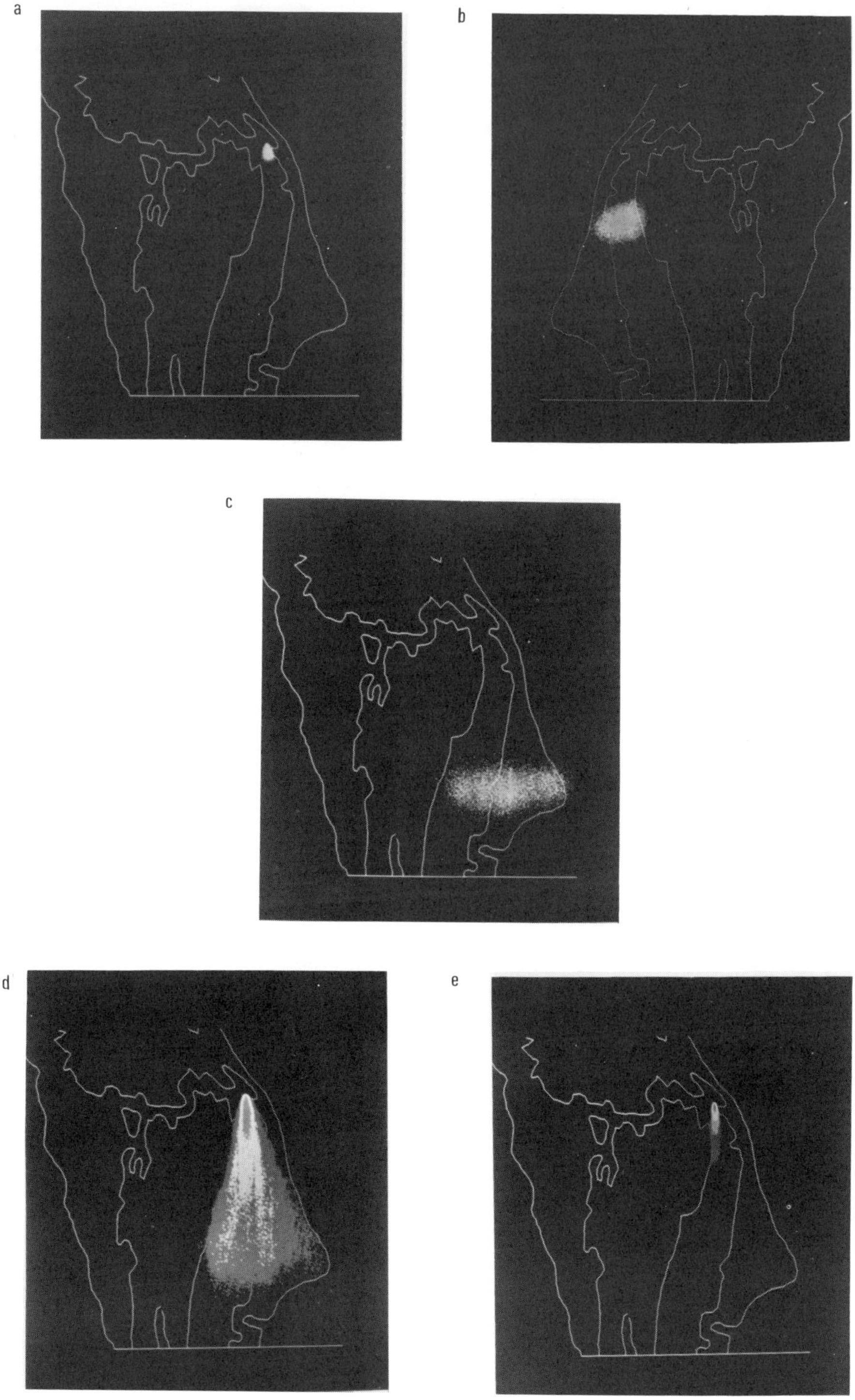

Figure 14. Particle cloud at (a) 10 minutes, (b) 60 minutes, and (c) 120 minutes after start of 15-minute duration spill. Surface-layer maximum concentrations experienced during (d) passage of puff, and (e) time-integrated concentrations. Colors in original correspond to exposure limits for nitrogen tetroxide.

CONCLUSIONS

This paper gives an overview of current regional-scale and mesoscale meteorological modeling capabilities as related to air quality studies. Shortcomings of current regulatory models are discussed, and several examples are presented to demonstrate the capability of the new assessment tools. Computer speed and memory, particularly with respect to graphics superworkstations, have advanced sufficiently that meteorological models on this scale, coupled to Eulerian advection-diffusion and/or Lagrangian particle dispersion models can be used for routine air quality assessments and emergency response. Both source- and receptor-oriented approaches can be applied.

Examples of the use of these modeling approaches are presented for Mt. Isa, Australia; northern Arizona and southern Utah; south Florida; and the Kennedy Space Center for specific applications using versions of the modeling capabilities developed at Colorado State University, R*SCAN Corporation, and ASTeR, Inc. Other models in the United States, as listed in Table 1, could similarly be applied, as well as a range of excellent models in this scale in Japan, Great Britain, the Federal Republic of Germany, France, and elsewhere.

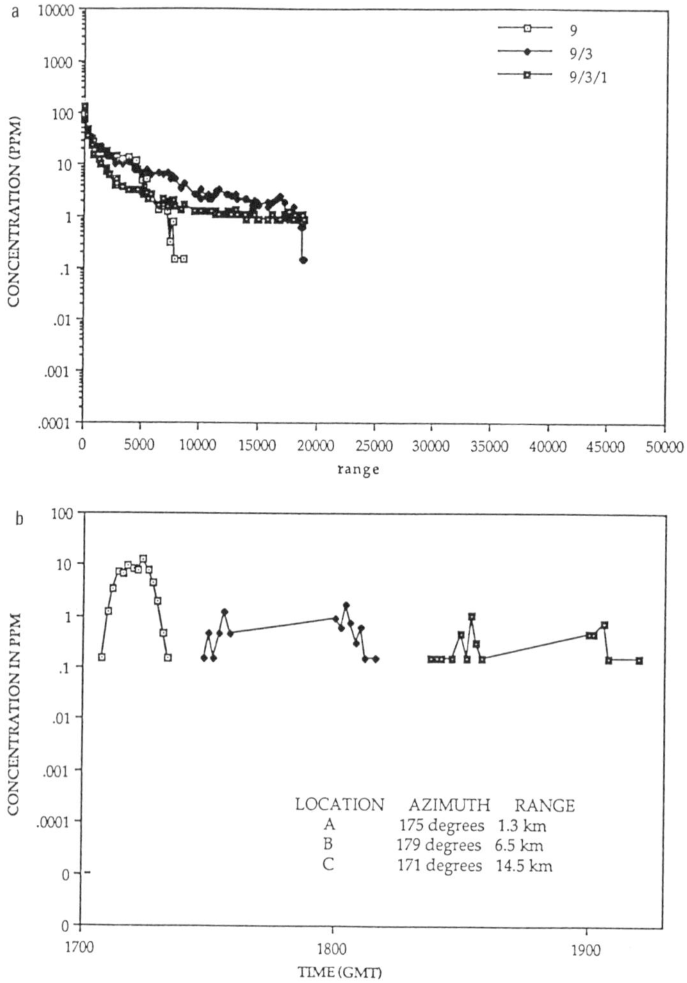

Figure 15. (a) Concentrations (ppm) as a function of range downwind from spill site using ARAMS runs at three different resolutions as input into the LPDM; (b) temporal variation of N_2O_4 concentrations as the cloud passed three receptors at 1.3, 6.5, and 14.5 km downwind from source.

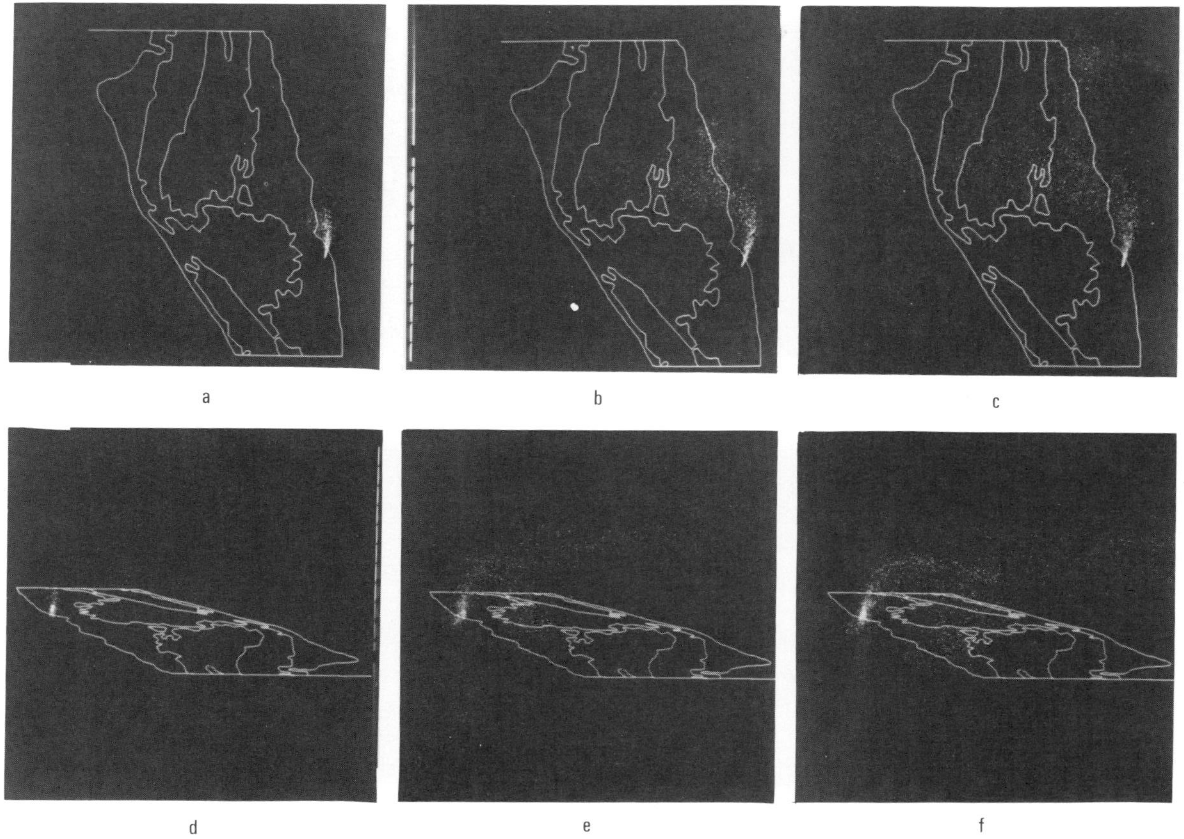

Figure 16. *LPDM visualization of a continuous release at 2 m above the Indian River starting at 1700 GMT, 7 November 1988, showing plan views of the particle cloud at (a) one, (b) three, and (c) five hours after the onset. Perspective views of the plume seen from the southwest at (d) one, (e) three, and (f) five hours after the onset reveal the bifurcation of the plume and the impact of the surface on the recirculated plume matter as it undergoes "second trip" fumigation (originals in color).*

ACKNOWLEDGEMENTS

The authors gratefully appreciate the editorial assistance of Dallas McDonald in coordinating and completing this paper. The research reported in this paper was supported by the National Park Service under Contract NA81RAH00001, Amendment 17, Item 15, the U.S. Air Force, Space Systems Division, Los Angeles, California under Contract F04701-89-C-0052, the Electric Power Research Institute, Inc. under Contract Nos. RP-1630-53 and RP-2434-6, and the Illinois Department of Nuclear Safety under Contract No. 1245. A portion of the calculations were completed at the National Center for Atmospheric Research (NCAR). NCAR is supported by the NSF.

REFERENCES

Anthes, R.A., 1983: Regional models of the atmosphere in middle latitudes. *Mon. Wea. Rev.*, **111**, 1306-1335.

Arritt, R.W., 1985: Numerical studies of thermally and mechanically forced circulations over complex terrain. Final Report, Contract No. NA81RAH00001, Amendment 17, Item 15, National Park Service, U.S. Department of the Interior, Denver, Colorado, 201 pp.

Avissar, R., M.D. Moran, G. Wu, R.N. Meroney, and R.A. Pielke, 1990: Operating ranges of mesoscale numerical models and meteorological wind tunnels for the simulation of sea breezes and land breezes. *Bound.-Layer Meteor.* (In press).

Brost, R.A., P.L. Haagenson, and Y.-H. Kuo, 1988: Eulerian simulation of tracer distribution during CAPTEX. *J. Appl. Meteor.*, **27**, 579-593.

Brusasca, G., G. Tinarelli, and D. Anfossi, 1989: Comparison between the results of a Monte Carlo atmospheric diffusion model and tracer experiments. *Atmos. Environ.*, **23**, 1263-1280.

Chang, J.S., R.A. Brost, I.S.A. Isaksen, S. Madronich, P. Middleton, W.R. Stockwell, and C.J. Walcek, 1987: A three-dimensional Eulerian acid deposition model: physical concepts and formulation. *J. Geophys. Res.*, **92**, 14681-14700.

Cogan, J.L., 1985: Monte Carlo simulation of buoyant dispersion. *Atmos. Environ.*, **19**, 867-878.

Corrsin, S., 1974: Limitations of gradient transport models in random walks and in turbulence. *Adv. Geophys.*, **18A**, 25-60.

Cotton, W.R., C.J. Tremback, M.P. Meyers, M.E. Nicholls, and T.S. Pedersen, 1990: Prospects for short-term and mesoscale numerical forecasting – A basic researches perspective. *Proc. Third AES/CMOS Workshop on Operational Meteorology*, Atmospheric Environment Service, Toronto, Canada.

Davis, P.A., 1983: Markov chain simulations of vertical dispersion from elevated sources into the neutral planetary boundary layer. *Bound.-Layer Meteor.*, **26**, 355-376.

de Baas, A.F., 1988: Some properties of the Langevin model for dispersion. Report No. Risø-M-2627, January, Risø National Laboratory, DK-4000, Roskilde, Denmark, 250 pp.

de Baas, A.F., H. van Dop, and F.T.M. Nieuwstadt, 1986: An application of the Langevin equation for inhomogeneous conditions to dispersion in a convective boundary layer. *Quart. J. Roy. Meteor. Soc.*, **112**, 165-180.

Dronamraju, M., L.K. Peters, G.R. Carmichael, P. Kasibhatla, and S.-Y. Cho, 1988: An Eulerian transport/transformation/removal model for SO_2 and sulfate – III: Comparison with the July 1974 SURE database. *Atmos. Environ.*, **22**, 2003-2011.

Durbin, P.A., 1980: A stochastic model of two-particle dispersion and concentration fluctuations in homogeneous turbulence. *J. Fluid Mech.*, **100**, 279-302.

Durbin, P.A., 1983: Stochastic differential equations and turbulent dispersion. NASA Ref. Pub. 1103, National Aeronautics and Space Administration, Washington, D.C., 73 pp.

Einstein, A., 1905: Uber die von der molekularkinetischen Theorie der Warme geforderte Bewegung von in ruhenden Flussigkeiten suspendierten Teilchen. *Annalen der Physik*, **17**, 549.

Etling, D., J. Preuss, and M. Wamser, 1986: Application of a random walk model to turbulent diffusion in complex terrain. *Atmos. Environ.*, **20**, 741-747.

Everhart, W.C., 1983: *The National Park Service.* Boulder, CO, Westview Press, 197 pp.

Ferber, G.J., J.L. Heffter, R.R. Draxler, R.J. Lagomarsino, F.L. Thomas, R.N. Dietz, and C.M. Benkovitz, 1986: Cross-Appalachian Tracer Experiment (CAPTEX '83) final report. NOAA Tech. Memo. ERL ARL-142, January, Air Resources Laboratory, National

Oceanic and Atmospheric Administration, Silver Spring, Maryland, 60 pp.
Fox, D.G., 1981: Judging air quality model performance. *Bull. Amer. Meteor. Soc.*, **62**, 599-609.
Frank, J.L., P.L. Moore, and G.E. Fisher, 1967: Summer shower distribution over the Florida Peninsula as deduced from digitized radar data. *J. Appl. Meteor.*, **6**, 309-316.
Gaffen, D.J., C. Benocci, and D. Olivari, 1987: Numerical modeling of buoyancy dominated dispersal using a Lagrangian approach. *Atmos. Environ.*, **21**, 1285-1293.
Garrett, A.J. and F.G. Smith, III, 1982: A two-dimensional dynamical drainage flow model with Monte Carlo transport and diffusion calculations. ASCOT Report 82-6, Savannah River Laboratory, Aiken, South Carolina.
Gross, G., H. Vogel, and G. Wippermann, 1987: Dispersion over and around a steep obstacle for varying thermal stratification – numerical simulations. *Atmos. Environ.*, **21**, 483-490.
Grossman, P.A., 1989: Kernel density estimation applied to a Lagrangian particle dispersion model. Final report, Chisholm Institute of Technology, Melbourne, Victoria, Australia, 36 pp. + App.
Hall, C.D., 1975: The simulation of particle motion in the atmosphere by a numerical random-walk model. *Quart. J. Roy. Meteor. Soc.*, **101**, 235-244.
Hanna, S.R., 1978: A statistical diffusion model for use with variable wind fields. *Proc. Fourth AMS Symp. on Turbulent Diffusion and Air Pollution*, American Meteorological Society, Boston, 15-18.
Hanna, S.R., 1982: Lagrangian and Eulerian time-scale relations in the daytime boundary layer. *J. Appl. Meteor.*, **20**, 242-249.
Hanna, S.R., G.A. Briggs, and R.P. Hosker, Jr., 1982: *Handbook on Atmospheric Diffusion*, DOE/TIC-11223 (DE2002045). Technical Information Center, U.S. Department of Energy, 102 pp.
Hurley, P. and W. Physick, 1990: A fumigation model incorporating a Lagrangian particle approach. *Atmos. Environ.* (Submitted).
Kao, C.-Y.J. and T. Yamada, 1988: Use of the CAPTEX data for evaluations of a long-range transport numerical model with a four-dimensional data assimilation technique. *Mon. Wea. Rev.*, **116**, 293-306.
Klug, W., P.J.H. Builtjes, H. van Dop, N.D. van Egmond, H. Glaab, D. Gömer, B.J. de Haan, K.D. v.d. Hout, N. Kesseboom, R. Röckle, and C. Veldt, 1985: Comparison between four different interregional air pollution models. In *Interregional Air Pollution Modelling: The State of the Art*, S. Zwerver and J. van Ham, Eds., Plenum Press, New York, 149-273.
Kuo, Y.-H., M. Skumanich, P.L. Haagenson, and J.S. Chang, 1985: The accuracy of trajectory models as revealed by the observing system simulation experiments. *Mon. Wea. Rev.*, **113**, 1852- 1867.
Lamb, R.G., 1978: A numerical simulation of dispersion from an elevated point source in the convective planetary boundary layer. *Atmos. Environ.*, **12**, 1297-1304.
Lamb, R.G., 1980: Mathematical principles of turbulent diffusion modeling. In *Atmospheric Planetary Boundary Layer Physics*, A. Longhetto, Ed., Elsevier Scientific Publishing Co., New York, 173-210.
Lamb, R.G., 1981: A scheme for simulating particle pair motions in turbulent fluid. *J. Comp. Physics*, **39**, 329-346.
Lamb, R.G., 1982: Diffusion in the convective boundary layer. In *Atmospheric Turbulence and Air Pollution Modelling*, F.T.M. Nieuwstadt and H. van Dop, Eds., D. Reidel Publishing Co., Boston, 159-229.
Langevin, P., 1908: Sur la théorie du mouvement brownien. *Comptes rendus, Académie des Sciences*, **146**, 530-533.
Legg, B.J., 1983: Turbulent dispersion from an elevated line source: Markov chain simulations of concentration and flux profiles. *Quart. J. Roy. Meteor. Soc.*, **109**, 645-660.
Legg, B.J. and M.R. Raupach, 1982: Markov-chain simulation of a particle dispersion in inhomogeneous flows: the mean drift velocity induced by a gradient in Eulerian velocity variance. *Bound.-Layer Meteor.*, **24**, 3-13.
Ley, A.J., 1982: A random walk simulation of two-dimensional turbulent diffusion in the neutral surface layer. *Atmos. Environ.* **16**, 2799-2808.
Ley, A.J. and D.J. Thomson, 1983: A random walk model of dispersion in the diabatic surface layer. *Quart. J. Roy. Meteor. Soc.*, **109**, 847-880.
Lopez, P.E., P.T. Gannon, Sr., D.O. Blanchard, and C.C. Balch, 1984: Synoptic and regional circulation parameters associated with the degree of convective shower activity in south Florida. *Mon. Wea. Rev.*, **112**, 686-703.

Lorimer, G.S., 1986: The kernel method for air quality modelling – I. Mathematical foundation. *Atmos. Environ.*, **20**, 1447-1452.

Lorimer, G.S. and D.G. Ross, 1986: The kernel method of air quality modelling – II. Comparison with analytic solutions. *Atmos. Environ.*, **20**, 1773-1780.

Luhar, A.K. and R.E. Britter, 1989: A random walk model for dispersion in inhomogeneous turbulence in a convective boundary layer. *Atmos. Environ.*, **23**, 1911-1924.

Lyons, W.A., C.S. Keen, and J.A. Schuh, 1983: Modeling mesoscale diffusion and transport processes for releases within coastal zones during land/sea breezes. U.S. Nuclear Regulatory Commission, NUREG/CR-3542, 202 pp.

Lyons, W.A., J.A. Schuh, D. Moon, R.A. Pielke, W.R. Cotton, and R.W. Arritt, 1987: Short range forecasting of sea breeze generated thunderstorms at the Kennedy Space Center: A real-time experiment using a primitive equation mesoscale numerical model". *Symposium on Mesoscale Analysis and Forecasting, Incorporating Nowcasting*, August 17-19, 1987, Vancouver, British Columbia, Canada, 503-508.

Mahrer, Y. and R.A. Pielke, 1977: A numerical study of the airflow over irregular terrain. *Beitrage zur Physik der Atmosphare*, **50**, 98-113.

Malm, W., K. Gebhart, D. Latimer, T. Cahill, R. Eldred, R. Pielke, R. Stocker, and J. Watson, 1989: Winter Haze Intensive Tracer Experiment, NPS Final Report, December, Air Quality Division, National Park Service, Fort Collins, Colorado.

Marchuk, G.I., 1982: Mathematical issues of industrial effluent optimization. *J. Meteor. Soc. Japan*, **60**, 481-485.

Martin, D., C. Mithieux, and B. Strauss, 1987: On the use of the synoptic vertical wind component in a transport trajectory model. *Atmos. Environ.*, **21**, 45-52.

McNider, R.T., 1981: Investigation of the impact of topographic circulations on the transport and dispersion of air pollutants. Ph.D. Dissertation, University of Virginia, Charlottesville, 210 pp.

McNider, R.T., S.R. Hanna, and R.A. Pielke, 1980: Sub-grid scale plume dispersion in coarse resolution mesoscale models. *Proc. Second AMS/APCA Conf. on Applications of Air Pollution Meteorology*, March 24-27, New Orleans, Louisiana, American Meteorological Society, Boston, 424-429.

McNider, R.T., K.J. Anderson, and R.A. Pielke, 1982: Numerical simulation of plume impaction. *Proc. Third AMS Conf. on Applications of Air Pollution Meteorology*, January, San Antonio, Texas, American Meteorological Society, Boston, 126-129.

McNider, R.T., M.D. Moran, and R.A. Pielke, 1988: Influence of diurnal and inertial boundary layer oscillations on long-range dispersion. *Atmos. Environ.*, **22**, 2445-2462.

McQueen, J.T. and R.A. Pielke, 1985: A numerical and climatological investigation of deep convective cloud patterns in south Florida. Atmospheric Science Paper No. 389, Department of Atmospheric Science, Colorado State University, Fort Collins, Colorado.

Monin, A.S. and A.M. Yaglom, 1971: *Statistical Fluid Mechanics, Vol. 1*, MIT Press, Cambridge, Massachusetts, 769 pp.

Moran, M.D., R.W. Arritt, M. Segal, and R.A. Pielke, 1986: Modification of regional-scale pollutant dispersion by terrain- forced mesoscale circulations. *Transactions of the APCA Second International Specialty Conference on the Meteorology of Acidic Deposition*, Albany, New York, March 17-20, 1986, Air Pollution Control Association, Pittsburgh, Pennsylvania, 136-157.

Moran, M.D., R.A. Pielke, and R.T. McNider, 1990: Temporal and spatial resolution requirements for regional-scale dispersion models. *Proc. 18th NATO/CCMS Int'l Tech. Mtg. on Air Pollution Modeling and Its Application*, May 13-17, Vancouver, Canada [This volume].

NCAR, 1988: Proceedings of Workshop on Limited-Area Modeling Intercomparison. Nov. 15-18, Mesoscale and Microscale Meteorology Division, National Center for Atmospheric Research, Boulder, Colorado, 244 pp.

Pasquill, F. and F.B. Smith, 1983: *Atmospheric Diffusion*, Third Edition, John Wiley and Sons, New York, 437 pp.

Pielke, R.A., 1974: A three-dimensional numerical model of the sea breezes over south Florida. *Mon. Wea. Rev.*, **102**, 115-139.

Pielke, R.A., 1984: *Mesoscale Meteorological Modeling.* Academic Press, New York, 612 pp.

Pielke, R.A., 1989: Status of mesoscale and subregional models: Volume 2. Mesoscale meteorological models in the United States. Report No. EN-6649, October, Electric Power Research Institute, Palo Alto, California.

Pielke, R.A. and Y. Mahrer, 1978: Verification analysis of the University of Virginia three-dimensional mesoscale model prediction over south Florida for July 1, 1973. *Mon. Wea. Rev.*, **106**, 1568-1589.

Reid, J.D., 1979: Markov chain simulations of vertical dispersion in the neutral surface layer for surface and elevated releases. *Bound.-Layer Meteor.*, **16**, 3-22.

Runchal, A.K., 1980: A random walk atmospheric dispersion model for complex terrain and meteorological conditions. *Proc. Second AMS/APCA Conf. Applications of Air Pollution Meteorology,* March 24-27, New Orleans, American Meteorological Society, Boston, 430-437.

Sawford, B.L, 1982: Lagrangian Monte Carlo simulation of the turbulent motion of a pair of particles. *Quart. J. Roy. Meteor. Soc.*, **108**, 207-213.

Sawford, B.L., 1984: The basis for, and some limitations of, the Langevin equation in atmospheric relative dispersion modelling. *Atmos. Environ.*, **18**, 2405-2411.

Sawford, B.L., 1985: Lagrangian statistical simulation of concentration mean and fluctuation fields. *J. Climate Appl. Meteor.*, **24**, 1152-1166.

Sawford, B.L and F.M. Guest, 1987: Lagrangian stochastic analysis of flux-gradient relationships in the convective boundary layer. *J. Atmos. Sci.*, **44**, 1152-1165.

Sawford, B.L. and F.M. Guest, 1988: Uniqueness and universality of Lagrangian stochastic models of turbulent dispersion. *Proc. Eighth AMS Symp. on Turbulence and Diffusion*, April 25-29, San Diego, American Meteorological Society, Boston, 96-99.

Schorling, M., H.-U. Mast, and J. Schmidt, 1986: Lagrangian simulation of neutral and heavy gases in complex terrain and streets. Unpublished manuscript, Hauptabteilung Systemstudien, Industrieanlagen-Betriebsgesellschaft, Ottobrunn, West Germany.

Seigneur, C., 1989: Status of subregional and mesoscale models. Volume 1: Air quality models. Report No. EN-6649, March, Electric Power Research Institute, Palo Alto, California.

Segal, M. and R.A. Pielke, 1981: Numerical model simulation of biometeorological heat load conditions – summer day case study for the Chesapeake Bay area. *J. Appl. Meteor.*, **20**, 735-749.

Segal, M., R.A. Pielke, and Y. Mahrer, 1980: Quantitative assessment of air quality in the greater Chesapeake Bay area using a three-dimensional mesoscale atmospheric model. *Proc. Symposium on Intermediate Range Atmospheric Transport Processes and Technology Assessment*, October 1-3, Gatlinburg, Tennessee, 105-115.

Segal, M., R.T. McNider, R.A. Pielke and D.S. McDougal, 1982: A numerical model simulation of the regional air pollution meteorology of the Greater Chesapeake Bay area - summer day case study. *Atmos. Environ.*, **16**, 1381-1397.

Segal, M., R.A. Pielke, R.W. Arritt, M.D. Moran, C.-H. Yu, and D. Henderson, 1986: Southern Florida air pollution climatology study and selected episodic impacts. Report prepared for Air Quality Division, National Park Service, Department of the Interior under Contract NA81RAH00001, Amendment 17, Item 15, Denver, Colorado, 237 pp.

Segal, M., R.A. Pielke, R.W. Arritt, M.D. Moran, C.-H. Yu, and D. Henderson,, 1988: Application of a mesoscale atmospheric dispersion modeling system to the estimation of SO_2 concentrations from major elevated sources in southern Florida. *Atmos. Environ.*, **22**, 1319-1334.

Snow, J.W., 1981: Wind power assessment along the Atlantic and Gulf coasts of the U.S. Ph.D. Dissertation, Dept. of Environmental Sciences, University of Virginia, Charlottesville.

Steyn, D.G. and I.G. McKendry, 1988: Quantitative and qualitative evaluation of a three-dimensional mesoscale numerical model simulation of a sea breeze in complex terrain. *Mon. Wea. Rev.*, **116**, 1914-1926.

Stocker, R.A. and R.A. Pielke, 1990: Source attribution during WHITEX – A modeling study. *Proc. AWMA/EPA International Speciality Conference on Visibility and Fine Particles*, Estes Park, Colorado, October 15-19, 1989, Air Waste Management Association, Pittsburgh, Pennsylvania (In press).

Taylor, G.E. and R.A. Schumann, 1986: A description of the Meteorological and Range Safety Support (MARSS) System. *Preprints, Fifth AMS/APCA Conference on Applications oj Air Pollution Meteorology*, American Meteorological Society, Boston, 133-136.

Taylor, G.E., C.R. Parks, M.K. Atchison, G.W. Drape, A. Dianic, D. Nuggent, B.Jubach, and H. Firstenberg, 1989: KABLE data analysis and EMERGE sea breeze algorithm evaluation. Final Report ARS-MPR-89-23, ENSCO, Inc., 66 pp.

Thomson, D.J., 1984: Random walk modelling of diffusion in inhomogeneous turbulence. *Quart. J. Roy. Meteor. Soc.,* **110** 1107-1120.

Thomson, D.J., 1986a: A random walk model of dispersion in turbulent flows and its application to dispersion in a valley. *Quart. J. Roy. Meteor. Soc.,* **112**, 511-530.

Thomson, D.J., 1986b: On the relative dispersion of two particles in homogeneous stationary turbulence and the implications for the size of concentration fluctuations at large times. *Quart. J. Roy. Meteor. Soc.*, **112**, 890-894.
Thomson, D.J., 1987: Criteria for the selection of stochastic models of particle trajectories in turbulent flows. *J. Fluid Mech.*, **180**, 529-556.
Thomson, R.B., R.P. Angle, and S. Sakiyama, 1987: Selecting air quality–acid deposition models for mesoscale application. *J. Air Pollut. Control Assoc.*, **37**, 260-265.
Turner, D.B., L.W. Bender, T.E. Pierce, and W.B. Petersen, 1989: Air quality simulation models from EPA. *Environ. Software*, **4**, 52-61.
Uliasz, M., 1983: Application of the perturbation theory to sensitivity analysis of an air pollution model. *Zeitschrift für Meteorology*, **33**, 355-362.
Uliasz, M., 1985: Comparison of the sensitivity analysis methods for meteorological models. *Zeitschrift für Meteorologie*, **35**, 340-348.
Uliasz, M., 1990a: Development of the mesoscale dispersion modeling system using personal computers. Part I: Models and computer implementation. *Zeitschrift für Meteorologie*, **40**, No 2, 104-114.
Uliasz, M., 1990b: Development of a mesoscale dispersion modeling system using personal computers. Part II. Numerical simulations. *Zeitschrift für Meteorologie* (In press).
Uliasz, M. and R.A. Pielke, 1990: Application of the receptor oriented approach in mesoscale dispersion modeling. *Proc., 18th ITM on Air Pollution Modeling and Its Application*, May 14-17, Vancouver, Canada [This volume].
van Dop, H., F.T.M. Nieuwstadt, and J.C.R. Hunt, 1985: Random walk models for particle displacements in inhomogeneous unsteady turbulent flows. *Phys. Fluids*, **28**, 1639-1653.
van Egmond, N.D. and H. Kesseboom, 1983: Mesoscale air pollution dispersion models. – I. Eulerian grid models. *Atmos. Environ.*, **17**, 257-265.
Venkatram, A., P.K. Karamchandani, and P.K. Misra, 1988: Testing a comprehensive acid deposition model. *Atmos. Environ.*, **22**, 737-747.
Walklate, P.J., 1987: A random-walk model for dispersion of heavy particles in turbulent air flow. *Bound.-Layer Meteor.*, **39**, 175-190.
Wilson, J.D., G.W. Thurtell, and G.E. Kidd, 1981a: Numerical simulation of particle trajectories in inhomogeneous turbulence, I: Systems with constant turbulent velocity scale. *Bound.-Layer Meteor.*, **21**, 295-313.
Wilson, J.D., G.W. Thurtell, and G.E. Kidd, 1981b: Numerical simulation of particle trajectories in inhomogeneous turbulence, II: Systems with variable turbulent velocity scale. *Bound.-Layer Meteor.*, **21**, 423-441.
Wilson, J.D., B.J. Legg, and D.J. Thomson, 1983: Calculation of particle trajectories in the presence of a gradient in turbulent-velocity variance. *Bound.-Layer Meteor.*, **27**, 163-169.
Yamada, T. and S. Bunker, 1988: Development of a nested grid, second moment turbulence closure model and application to the 1982 ASCOT Brush Creek data simulation. *J. Appl. Meteor.*, **27**, 562-578.
Yu, C.-H. and R.A. Pielke, 1986: Mesoscale air quality under stagnant synoptic cold season conditions in the Lake Powell area. *Atmos. Environ.*, **20**, 1751-1762.
Zannetti, P., 1984: New Monte Carlo scheme for simulating Lagrangian particle diffusion with wind shear effects. *Appl. Math. Modelling*, **8**, 188-192.
Zannetti, P., 1986: Monte-Carlo simulation of auto- and cross-correlated turbulent velocity fluctuations (MC-LAGPAR II MODEL). *Environ. Software*, **1**, 26-30.

DISCUSSION

K. AYOTTE

What is your feeling for the effect of vertical resolution on regional air quality modeling?

R.A. PIELKE

Vertical resolution must be sufficient to resolve the atmospheric structure at the initial time and as the flow evolves. This means, for example, that greater vertical resolution is required when the atmosphere is stably stratified and long-wave radiative flux divergence is important. Enhanced resolution is generally more critical near the surface and near inversions. Observations can be used as guidance for the needed vertical grid resolution.

C.J. WALCEK

You suggested that 3 or 9 km resolution could not adequately represent motions occurring on a 1 km scale. Using even higher resolution, perhaps 1 km resolution can be shown to be inadequate. How fine a resolution is "adequate" and how do you determine this?

R.A. PIELKE

I suggest that a spectral analysis of the spatial distribution of surface heating and roughness distribution be evaluated. This is a necessary condition for realistic simulations since it will be necessary to resolve spatial scales of sufficiently large contributions to the forcing or to parameterize their effects. Below a certain spatial scale, our work suggests that the variability can be averaged, but a more sophisticated representation is needed for larger scales.

A.A.M. HOLTSLAG

How are you dealing with inhomogeneity effects on the subgrid scale?

R.A. PIELKE

Our recent papers (Avissar, R. and R.A. Pielke, 1990: The impact of plant stomatal control on mesoscale atmospheric circulations, *Agric. and Forest Meteor.*, (in press) and Pielke et al., 1990: Influence of landuse on weather and climate and its representation for use in larger scale models) summarize our current work on this topic. For small enough scales we can average the surface fluxes for the separate landuse areas; for larger spatial scales we need to include the influence of subgrid scale coherent circulations.

M. SCHATZMANN

When you presented your examples, you did not indicate whether you used the hydrostatic or non-hydrostatic version of your model. I assume you used most times the hydrostatically approximated model. Are you not afraid of non-hydrostatic effects in cases of high resolution and highly structured terrain?

R.A. PIELKE

We generally have applied a non-hydrostatic model in our recent work.

A. ANDRÉN

On the smaller mesoscale, say 10-200 km, it might be of interest to be able to provide calculations of dispersion for different sampling times. How do you relate your particle calculations to different sampling times? What part of the spectrum of eddies is resolved in your model and what part is parameterized?

R.A. PIELKE

This is an important question. In our model, motions with a spatial scale greater than 4Δ are reasonably well resolved - thus, dispersion due to differential advection on that scale is explicitly represented. For smaller scales, the motions are parameterized using a statistical representation in which observationally-based formulations are used to obtain probabilistic estimates of subgrid-scale velocities associated with diffusion. At present, our "sampling time" is limited to the time-period of averaging from which the observationally-based formulas are obtained. Moreover, since our procedure involves spatially-uncorrelated particles in the stochastic part of the Lagrangian particle model, our dispersion results are a "most-likely pattern" rather than an individual realization. Since the observationally-based formula presents horizontally-homogeneous, steady state turbulence effects only, any dispersion which results from small-scale unresolved coherent circulations which are not associated with such turbulence will be missed.

NOCTURNAL FLOW ON A WESTERN COLORADO SLOPE*

John M. Leone, Jr. and Paul H. Gudiksen

Lawrence Livermore National Laboratory

Box 808, Livermore, California 94550

INTRODUCTION

The Department of Energy sponsored Atmospheric Studies in Complex Terrain (ASCOT) program has conducted a research program designed to increase our knowledge and understanding of terrain-dominated flows with specific emphasis on nocturnal flows within mountain valleys. ASCOT has sponsored both field studies and numerical modeling efforts to improve our understanding of the wind, temperature and turbulence structure of nocturnal drainage flows. One of the most recent ASCOT sponsored field studies involves a study within the Mesa Creek Basin in western Colorado to investigate the seasonal frequency of occurrence of drainage flows along the sloped surfaces and within the basin, and to evaluate the effect of the ambient meteorology on their development.

The Mesa Creek Basin, situated on the north slope of the Grand Mesa, encompasses a roughly 10 × 20 km area that is approximately 30 km east of Grand Junction. The observational segment of the study was undertaken jointly by the Lawrence Livermore National Laboratory and the NOAA Wave Propagation Laboratory, and involved the operation of a network of eight meteorological towers and a monostatic sodar within the Mesa Creek study area over a period of one year that extended from December 1988 through November 1989. These measurements were augmented by tethersonde observations to define the vertical wind and temperature structure during a few nights. The preliminary results of the field study will be reported in King, Gudiksen, and Russell (1990). The modeling portion of the study is being undertaken by Lawrence Livermore Laboratory using a three-dimensional prognostic boundary layer model to gain further insight into the dynamics of the seasonal variations and the effect of cloud cover on the development of the drainage flows. It is the purpose of this paper to present preliminary results from a numerical simulation done as part of this study.

Work performed under the auspices of the U.S. Department of Energy by the Lawrence Livermore National Laboratory under contract No. W7405-Eng-48.

THE MODEL

The model used for the simulations reported here solves the 3-dimensional, nonhydrostatic, Boussinesq equations of motion shown below.

$$\frac{\partial u}{\partial t} + u\frac{\partial u}{\partial x} + v\frac{\partial u}{\partial y} + w\frac{\partial u}{\partial z} = -\Theta_0\frac{\partial \pi}{\partial x} + f(v - v_g) + \frac{\partial}{\partial x}(K_H\frac{\partial u}{\partial x}) + \frac{\partial}{\partial y}(K_H\frac{\partial u}{\partial y}) + \frac{\partial}{\partial z}(K_V\frac{\partial u}{\partial z})$$

$$\frac{\partial v}{\partial t} + u\frac{\partial v}{\partial x} + v\frac{\partial v}{\partial y} + w\frac{\partial v}{\partial z} = -\Theta_0\frac{\partial \pi}{\partial y} - f(u - u_g) + \frac{\partial}{\partial x}(K_H\frac{\partial v}{\partial x}) + \frac{\partial}{\partial y}(K_H\frac{\partial v}{\partial y}) + \frac{\partial}{\partial z}(K_V\frac{\partial u}{\partial z})$$

$$\frac{\partial w}{\partial t} + u\frac{\partial w}{\partial x} + v\frac{\partial w}{\partial y} + w\frac{\partial w}{\partial z} = -\Theta_0\frac{\partial \pi}{\partial z} + (\frac{\theta - \theta'}{\Theta_0}g) + \frac{\partial}{\partial x}(K_H\frac{\partial w}{\partial x}) + \frac{\partial}{\partial y}(K_H\frac{\partial w}{\partial y}) + \frac{\partial}{\partial z}(K_V\frac{\partial w}{\partial z})$$

$$\frac{\partial u}{\partial x} + \frac{\partial v}{\partial y} + \frac{\partial w}{\partial z} = 0$$

$$\frac{\partial \theta}{\partial t} + u\frac{\partial \theta}{\partial x} + v\frac{\partial \theta}{\partial y} + w\frac{\partial \theta}{\partial z} = \frac{\partial}{\partial x}(K_H\frac{\partial \theta}{\partial x}) + \frac{\partial}{\partial y}(K_H\frac{\partial \theta}{\partial y}) + \frac{\partial}{\partial z}(K_V\frac{\partial \theta}{\partial z})$$

where the usual meteorological notation has been used with u,v,w are the x,y,z components of the velocity respectively; is the deviation potential temperature, and is the deviation Exner function. The constants f, g, u_g, and v_g are the Coriolis parameter, the gravitational acceleration, and the x and y components of the large scale geostrophic wind, respectively. The equations have been linearized about a motionless hydrostatic basic state defined by:

$$\Theta(x, y, z, t) = \Theta_0 + \theta(x, y, z, t)$$

$$\Pi(x, y, z, t) = \Pi_0(z) + \pi'(z) + \pi(x, y, z, t)$$

where $\frac{\partial \Pi_o}{\partial z} = \frac{g}{\Theta_0}$ and $\frac{\partial \pi'}{\partial z} = g\frac{\theta'}{\Theta_0^2}$

with $\theta'(z)$a horizontally averaged potential temperature.

To close the system of equations, we must define the turbulent diffusivities, K_H and K_V. For the simulations reported herein, the horizontal eddy diffusivity has been taken to be constant with a value of 100 m^2/s, while the vertical diffusivity, K_V, was modeled using a parameterization developed by McNider and Pielke (1981).

The spatial discretization of the equations is accomplished via the finite element method using tri-linear basis functions for the velocity components and the temperature and piecewise constant basis functions for the pressure. For efficiency, mass lumping and 1 point quadrature were employed. The time integration scheme is a variant of the forward Euler scheme. The forward backward scheme of Sun (1980) is used on the buoyancy term, and the advection terms are stabilized via the implementation of the balancing tensor diffusivity. (The model is described in detail in Leone and Lee, 1989.)

THE SITE

The topographic features of the Mesa Creek drainage basin resolved within the modeled terrain and the locations of six of the meteorological towers are shown in Fig. 1. The study area is bounded on the south and west sides by the Grand Mesa, a ridge on the east side and the Plateau Valley toward the north. The terrain varies in height from 3200 m on the Grand Mesa to about 1500 m within the basin. The slope generated flows drain into the basin where they merge with those generated within the Plateau valley whose drainage area is largely situated east of the study area. The flows within the Plateau Valley subsequently drain westward into the Colorado River drainage area.

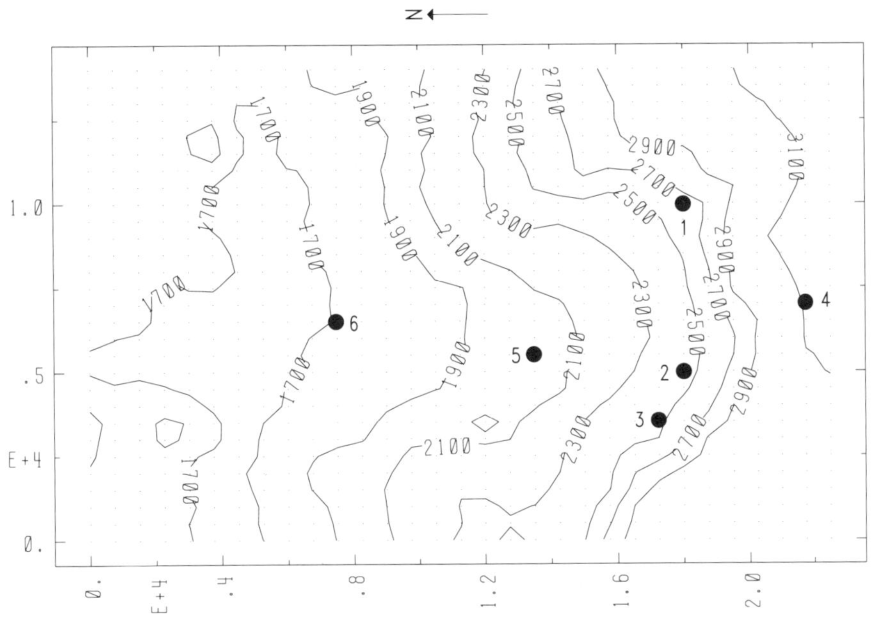

Fig. 1. The topography within the computational domain, contours every 200 m. Also shown are the locations of 6 of the observational towers.

THE SIMULATION

The horizontal domain used for the simulations reported here is the area shown in Fig. 1 and encompassed a distance of 14 km in the cross basin direction and 22.5 km in the down basin direction. The top of the domain was taken to be at 4500 meters above mean sea level giving a vertical domain size of 1469 meters above the highest point and 2976 meters above the lowest point. The domain was divided into twenty-eight (28) 500 m long elements in the cross basin direction, thirty (30) 750 m long elements in the down basin direction, and sixteen (16) elements in the vertical direction which were 7.5 m high at the ground and increased in size toward the top of the domain.

The simulation reported here was for a simple undisturbed nocturnal flow development. The large scale geostrophic wind was assumed to be zero as was the initial velocity field. The initial potential temperature field was slightly stable, with

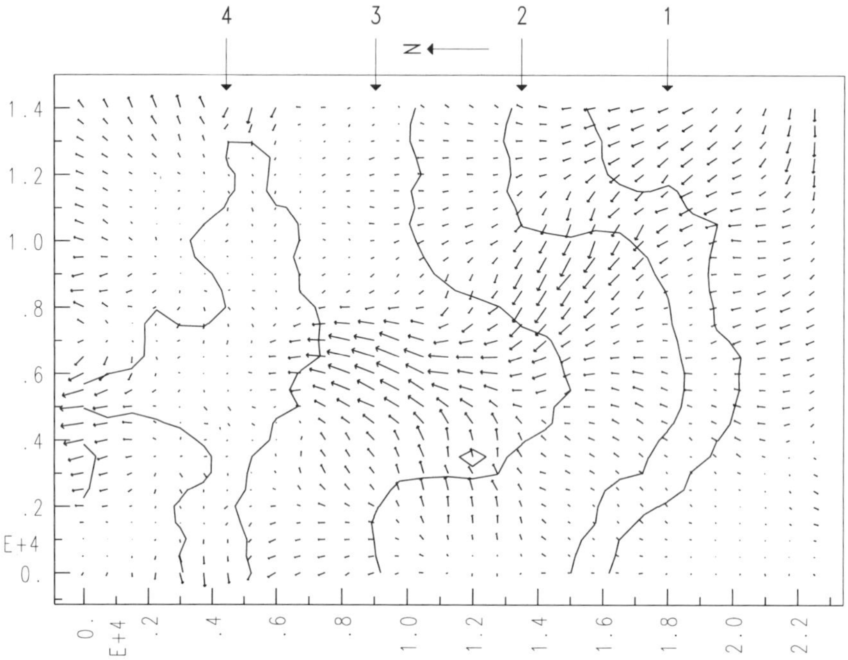

Fig. 2. The horizontal wind field after 2 hours of cooling at 7.5 m AGL. Terrain contours every 400 m are provided for reference. The numbered arrows at the top of the figure indicate the positions of the vertical cross sections presented in Fig. 3.

a vertical gradient of 1 K/km, and horizontally uniform. The driving force was a surface heat flux of $-40\ \mathrm{Wm}^{-2}$ applied uniformly on the domain for a period of two hours. The velocity boundary conditions were the no-slip condition on the surface and at the top of the domain the velocity was set equal to the large scale geostrophic wind (zero in this case). The lateral boundary conditions were:

$$f_n = -\pi + K_H \frac{\partial u_n}{\partial x_n} = 0.0 \ ;$$

$$f_{\tau_1} = K_{\tau_1} \frac{\partial u_{\tau_1}}{\partial x_n} = 0.0; \ f_{\tau_2} = K_{\tau_2} \frac{\partial u_{\tau_2}}{\partial x_n} = 0.0 \ ;$$

$\frac{\partial \theta}{\partial x_n} = 0.0$where n $\rightarrow$ normal and $\tau_i \rightarrow$ the ith tangent to the boundary.

THE RESULTS

The results presented here will be for a simulated time of 2 hours. While the details of the flow were not at steady state, the main features were well established and steady. Also, it was felt that continuing the simulation with a constant surface heatflux as a forcing would be unrealistic. Figure 2 is a snapshot of the flow at 7.5 meters above ground level (AGL) after two hours of cooling. Three distinct drainage regions can be seen; one originating on the southeast rim of the basin, a second coming from the center of the mesa, and the third flowing off the west wall of the basin. These merge into a single down basin flow through the center of the basin.

Figure 3 presents a series of cross-basin vertical cross sections of the northerly velocity component. Figure 3a, located on the upper slope (arrow 1 in Fig. 2), shows a significant down-basin flow across 3/4's of the basin with two distinct centers representing the northeast corner flow and the central flow. The maximum down-basin speed is 2.9 m/s within the central flow. Fig. 3b, half way down the basin (arrow 2 in Fig. 2), shows a continuation of this pattern. However, the central flow is now clearly dominant with the maximum speed increasing to 3.4 m/s. Figure 3c, just before the entrance to the Plateau River Valley (arrow 3 in Fig. 2), shows a single strong core flow with a maximum speed of 5.1 m/s. The flow is seen to be vertically confined to a shallow layer. Figure 3d, within the Plateau River Valley (arrow 4 in Fig. 2), shows a significantly altered pattern. Here the maximum speed in the northerly direction is only 2.5 m/s and the height of the jet core has elevated significantly. We believe that this anomaly is a numerical artifact of two weaknesses in the simulation. The first is the lack of significant terrain within the model domain north of the Plateau River Valley, and the second is the inability of the model to reproduce the flow down the Plateau River Valley due to upstream influences. Each of these contribute to the lack of turning of the local flows into and down the Plateau River Basin. We believe however, that these effects are very local and do not significantly perturb the solution on the basin slopes.

As a further aid in understanding the structure of the flow and the representativeness of the observation sites, we examined the flow in more detail at the six tower locations marked on Fig. 1. Sites 1, 2, and 3 represent an across basin section, while sites 4, 5, and 6 depict a down basin section. The vertical profiles of down slope winds across the basin are plotted in Fig. 4. Site 1 which is in the southeast rim flow and site 2 which is in the central flow are seen to be very similar in both speed and vertical structure. Site 3 which is between the central flow and the west wall flow is significantly weaker and shallower. These results seem to indicate that the difference is caused by the character of the upwind terrain. Site 3 is down wind of a relatively flat region that does not appear to drain as readily as the more sloped regions upwind of the other two sites. In contrast, the vertical temperature change profiles shown in Fig. 5 for the same site show a more uniform change with the eastern site having cooled the most, the western site cooling the least, and the central site midway between.

The down basin variation of the vertical profiles of downslope winds is presented in Fig. 6 using sites 4 on the mesa top, 5 about half way down the basin, and 6 near the entrance to the Plateau River Valley. The flow on the mesa top is weak and relatively shallow. The flow at site 5 is seen to be deeper than on the mesa top, but only very slightly stronger. The slow speed is a function of the site's position just to the west of the core of the flow which has a maximum speed of 3.4 m/s. This suggests that the tower data may underestimate the speed of the drainage flow at mid-basin. The flow at site 6 is significantly faster than at the other two sites but the flow depth is approximately the same as at site 5 indicating that the drainage flow depth is not strictly a function of mass flux but also a strong function of thermal structure which is shown in Fig. 7. As expected, the maximum cooling has taken place at site 6 at the foot of the basin and the least has taken place at site 4 on the mesa top. The depth of the cooling reaches a maximum at approximately site 5 and is constant down to the foot of the basin.

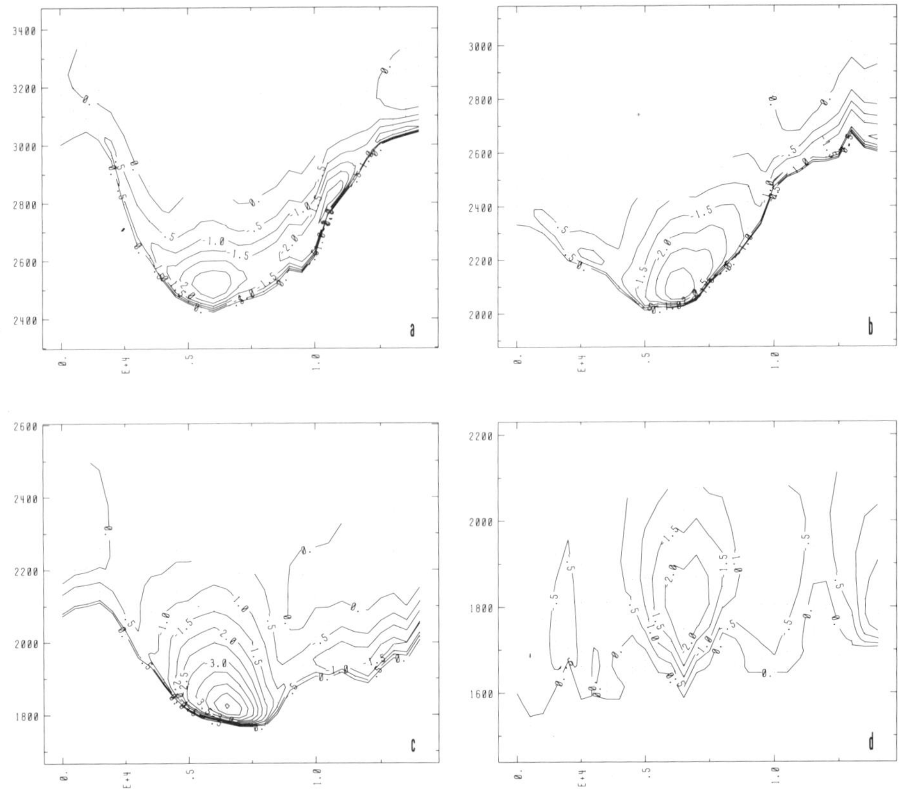

Fig. 3. Vertical cross sections of the northerly component of the wind at the 4 locations marked in Fig. 2. The contour interval 0.5 m/s. a) is along the slope (arrow 1 in Fig. 2) b) is in the middle of the basin (arrow 2 in Fig. 2) c) in near the entrance to the Plateau River Valley (arrow 3 in Fig. 2) d) is in the Plateau River Valley.

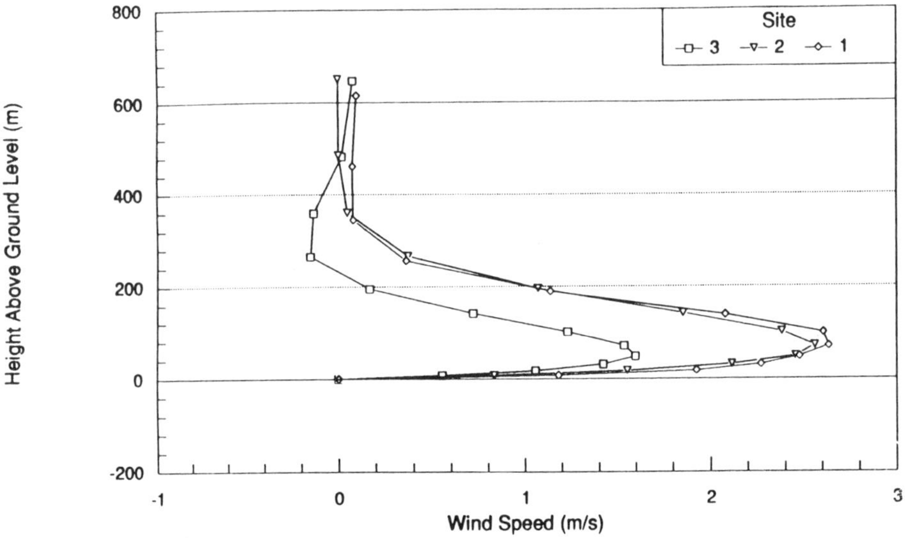

Figure 4. Vertical profiles of downslope velocity after two hours of cooling at the sites marked as 1, 2, and 3 in Fig. 1.

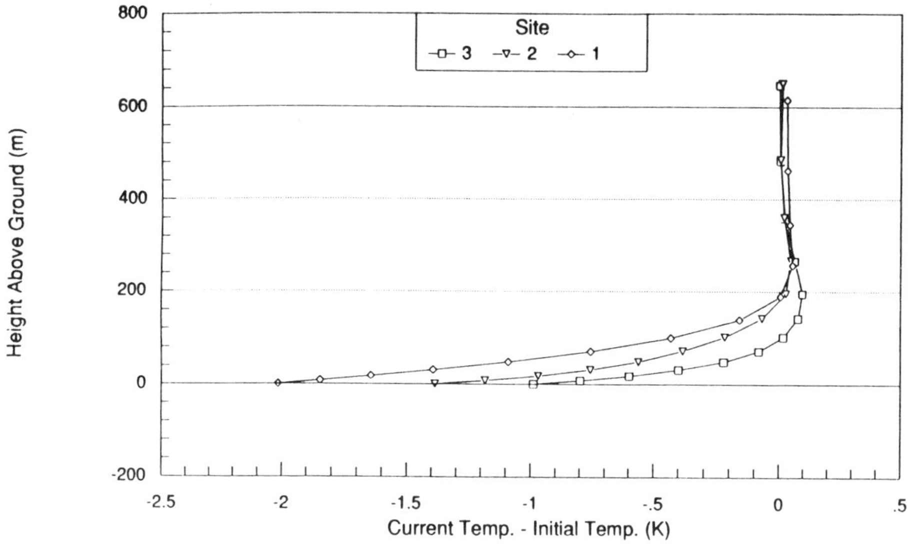

Figure 5. Vertical profiles of the total temperature change over two hours at the sites marked as 1, 2, and 3 in Fig. 1.

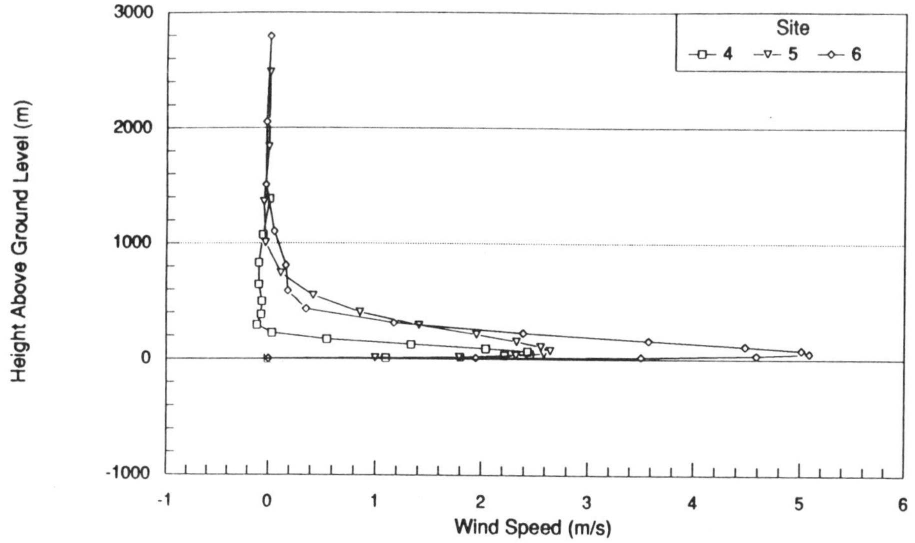

Figure 6. Vertical profiles of downslope velocity after two hours of cooling at the sites marked as 4, 5, and 6 in Fig. 1.

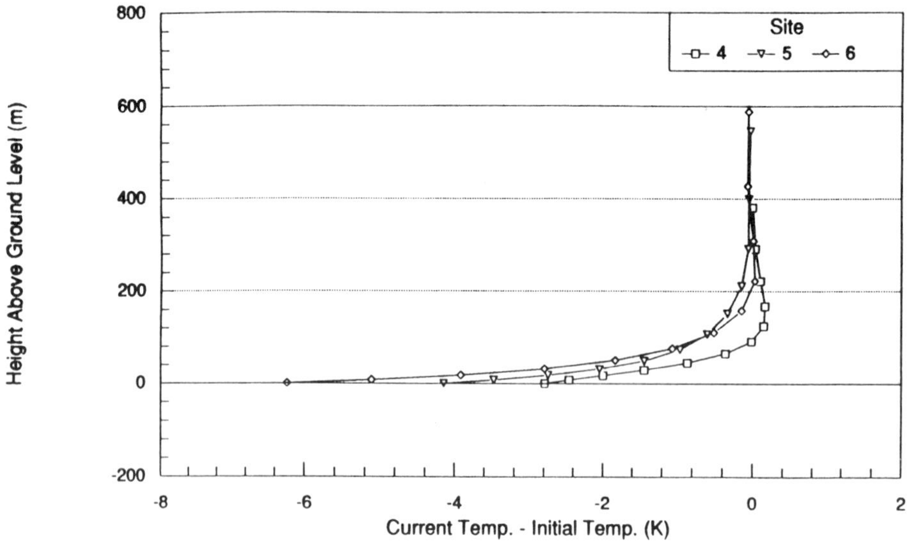

Figure 7. Vertical profiles of the total temperature change over two hours at the sites marked as 4, 5, and 6 in Fig. 1.

DISCUSSION

While a comprehensive comparison with data is not yet possible, we surveyed some preliminary observations to gain insight into the performance of the model. Because the night for which data were available had a non-zero geostrophic wind rather than the assumed calm conditions, the comparison must be treated as indicative rather than absolute. The results of the comparison demonstrated that the model is qualitatively correct except within the Plateau River Valley. Further, there are indications that the model results may be correct in some details. For example, the observations show that site 3 had the weakest drainage flow of the cross basin sites with an observed difference from the flow at site 3 of approximately 0.9 m/s. This is remarkably similar to the model result of a difference of 1.0 m/s. Also, the downslope wind speed at site 2 was observed to be 3.2 m/s which is very similar to the 2.6 m/s wind produced by the model. However, not all comparisons were as favorable. The model did not produce a significant difference between sites 1 and 2 while the observations showed that site 3 was 2 m/s stronger than at site 2. The model seems to overpredict the depth of the drainage flow by a factor of 2 or more. While the available data were not sufficient to determine if the model was correct in predicting that site 5 would be located east of the drainage flow maximum, the observation that the flow at site 1 was stronger than at site 5 could imply that the model was correct. In order to determine exactly how much of the agreement is serendipitous, we will have to wait until the full data set is available.

As the data become available, we plan to make in depth comparisons with the model results with the two-fold objective of improving the modeling capabilities and increasing our understanding of the phenomena lurking behind the relatively sparse data. We believe that a cooperative exchange between modeling and observational efforts will yield the maximum benefit for the study of terrain driven flows.

REFERENCES

King, C. W., P. H. Gudiksen, and C. A. Russell (1990): "Sodar-Derived Nocturnal Drainage Flow Classifications," to be presented at the Fifth Conference on Mountain Meteorology, Boulder, Colorado, June 25-29, 1990.

Leone, J.M., Jr. and R.L. Lee (1989): "Numerical Simulation of Drainage Flow in Brush Creek, Colorado," *J. Appl. Met.* **28,** pp 530-542.

McNider, R.T. and R.A. Pielke (1981): "Diurnal Boundary Layer Development Over Sloping Terrain," *J. Atmos. Sci.,* **38,** pp 2198-2212.

Sun, W. (1980): "A Forward-Backward Time Integration Scheme to Treat Internal Gravity Waves," *Mon. Wea. Rev.,* **108,** pp 402-407.

DISCUSSION

M.E. BERLYAND

How do you take into account the exchange coefficients?

J.M. LEONE

The vertical exchange coefficients were modeled using the formulation of McNider and Pielke wherein the coefficients are a function of the local Richardson number. The horizontal exchange coefficients were constant with a value of 100 $m^2 s^{-1}$.

C.J. NAPPO

Please say a few words on your lateral boundary conditions. Also, do you see instances of reflections at the lateral boundaries?

J.M. LEONE

The lateral boundary conditions used are the so-called finite element natural boundary conditions which arise from the integration by parts of the total stress terms. We have used these conditions quite successfully in a number of simulations, e.g. isothermal flow past a cylinder and forced flow in a bottom heated channel. Each of these simulations have time-dependent eddies which travel across the domain and exit at the outflow boundary. We see very few deleterious effects from the outflow boundary conditions. In the simulation reported here, we do not see evidence of reflection at the lateral boundaries, but they do not mimic the larger scale forcing properly.

S. SAKIYAMA

Have you turned "the heat on" in your model, where you examine the upslope flow situation?

J.M. LEONE

No, we have not yet begun any examinations of the upslope flow problem, but we have concentrated on a drainage flow case.

REGIONAL FLOW FIELDS IN NORTHRHINE WESTFALIA -

A CASE STUDY COMPARING FLOW MODELS OF DIFFERENT COMPLEXITY

Klaus Maßmeyer

Gesellschaft für Reaktorsicherheit mbH
Schwertnergasse 1, D 5000 Köln 1

Kai Born

Meteorologisches Institut der Universität Bonn
Auf dem Hügel 20, D 5300 Bonn

Bernd Erbshäußer and Carola Hauk

Meteorologisches Institut der Universität zu Köln
Kerpener Strasse 13, D 5000 Köln 41

Thomas Flassak

Institut für Technische Thermodynamik der Universität Karlsruhe
Kaiserstrasse 12, D 7500 Karlsruhe

INTRODUCTION

In air quality studies the knowledge of the spatially inhomogeneous wind fields is often essential. In order to gain information about the structure of flow fields, monitoring networks have been built up or mesoscale models of different complexity have been used. Due to a couple of reasons both methods will not yield identical results. Measurements in a nonhomogeneous topography are exposed to individual influences, which in general cannot be resolved by the numerical grid of a mesoscale model on a regional scale. Also different model formulations (i.e. approximations of the basic equations, turbulence parameterizations or numerical schemes) will come up with different model results.

As a consequence of the comprehensive work in model development during the last decade, some models will be incorporated in technical guidelines to calculate regional dispersion of pollutants over complex terrain. Especially the models REWIMET (Heimann, 1986) and RKM (Regionales Klima-Modell) (Frenzen et al., 1987) will be part of the guideline system of the Verein Deutscher Ingenieure (VDI), where climatological applications as well as special scenarios are envisaged. In order to derive some benchmark values concerning the accuracy of the models, measurements carried out in a stable nighttime atmosphere will be used for a verification study in the terrain of Northrhine Westfalia, which comprises typical topographic patterns of large portions of FRG.

Air Pollution Modeling and Its Application VIII, Edited by H. van Dop and D. G. Steyn, Plenum Press, New York, 1991

Northrhine Westfalia is a federal state of the Federal Republic of Germany. The nonhomogeneous topography is characterized by an industrial area in flat terrain in the Northwest - approximately the region between Cologne and the Ruhrgebiet - and a rural area with partly subalpine mountain ranges in the East and South. The dense populatet industrial region of the Ruhrgebiet forms an area source of different pollutants. In order to monitor air quality the Landesanstalt für Immissionsschutz (LIS-NW) has established a network of sampling sites partly equipped with meteorological instruments. Together with the sites of the Deutscher Wetter-Dienst (DWD) wind informations from 50 measuring sites should be available on an hourly basis. It was anticipated that models of different comlexity will show typical differences to those measurements.

In order to evaluate those differences the nighttime measurements of both networks of October 3rd, 1987 have been compared to the results of the following models:

- RKM (Frenzen et al., 1987)
- REWIMET (Heimann, 1986)
- FOOT (based on REWIMET) (Born, 1989)
- NOABL (Traci et al., 1978)
- CONDOR (Moussiopoulos et al., 1988)

TOPOGRAPHICAL DATA

The topography of Northrhine Westfalia has been digitized onto a 5km squared grid (46 x 43 grid points), the region around Cologne/Bonn is mapped on a 2km grid with 20 x 29 grid points. The contour lines are shown in Fig. 1. The position of the 2km grid is indicated. Minimum orographic heights of 5m above N.N. are found in the north-western

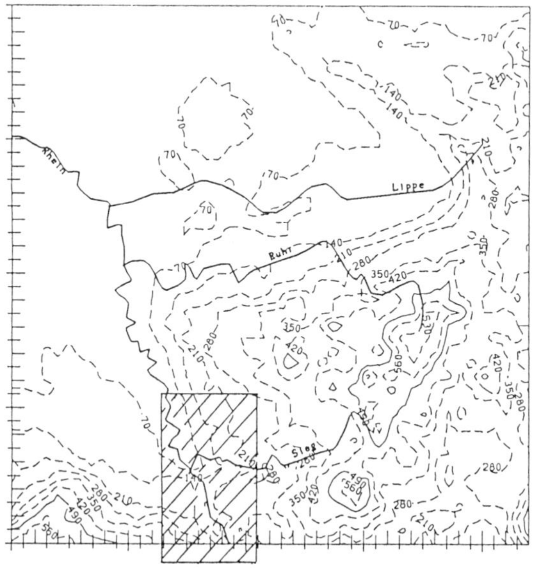

Fig. 1. Orography of Northrhine Westfalia in a 5km grid. The position of the 2km grid of the bight of Cologne is indicated by a rectangle.

parts, the maximum height of 660m above N.N. is located south-east of the center position. Roughness lengths, which are not shown, vary between 0.1m for grass land up to 2m for urban areas.

MODEL COMPARISON

The models, which have been used in this verification study, have extensively been described in the literature. In particular the calculations have been established with:

- the nonhydrostatic model RKM (Frenzen et al., 1987) based on the model FITNAH developed by Gross (see Gross (1986) for example)

- the hydrostatic three layer models REWIMET (Heimann, 1986) and FOOT (Born, 1989). FOOT is based on the same concept as REWIMET, but differs in the turbulence parameterization, the numerical solution scheme and the handling of the upper boundary.

- the mass consistent models CONDOR (Moussiopoulos et al., 1988) and NOABL (Traci et al., 1978). The main differences between these models consist in the handling of the transmission coefficients, which vary spatially in CONDOR, but are constant in NOABL. The Poisson equation in NOABL is solved by a line over-relaxation scheme in contrast to a direct elliptic solver using FFT in conjunction with a conjugate gradient method in CONDOR.

Because of the stable stratification of the atmosphere a marked influence of the orography on the flow in the surface layer can be expected. Additionally all wind measurements have been carried out between 10m and 30m above ground. Therefore the evaluations have been focussed on the surface layer wind field.

Following Pielke (1984) the first step in a model intercomparison is the subjective evaluation, i.e. a qualitative discussion of modelled against modelled or observed phenomena. The second step, point quantitative validation, is much more complicated, because all models calculate grid averaged values of the dependent variables. Despite those difficulties, Keyser and Anthes (1977) utilized a comparison technique for individual calculations and observations at identical grid points (see also Pielke, 1984). Following this method a model shows skill if the criteria

- standard deviation of measured and calculated values are nearly identical

- root mean square error (E) is smaller than the standard deviation of the measurements

- root mean square error after a constant bias is removed (E UnBiased - EUB) is smaller than the standard deviation of the measurements are fulfilled.

We adopt this technique to study model performance for the nighttime situation of October 3rd, 1987. This date was chosen because of mainly two reasons: some airborne measurements within the framework of another R&D project (Ehhalt and Poppe, 1990) have been carried out in parallel and the synoptic situation was nearly stationary.

SYNOPTIC SITUATION - SURFACE DATA

During the 3rd of October 1987 a high pressure system centered over the Baltic-Sea caused south-easterly geostrophic wind in 850 hPa and at the surface, connected with wind speeds of about 8 m/s. The temperature sounding near the radiosonde Essen - which was used to initialize the prognostic models - shows a stable atmosphere throughout the whole PBL with a temperature gradient in the lowest 100m of about 1K/100m. The diagnostic models were initialized with the measurements of the DWD set. The measured surface wind field is shown in Fig.2 for the LIS and DWD nets. It must be stressed that the LIS data have a resolution of 1 degree in the wind direction in contrast to a 10 degree resolution of DWD data. This must be kept in mind during the model evaluation. The quality of wind measurements in both nets is thought to be nearly identical because the LIS stations have been positioned in agreement with DWD.

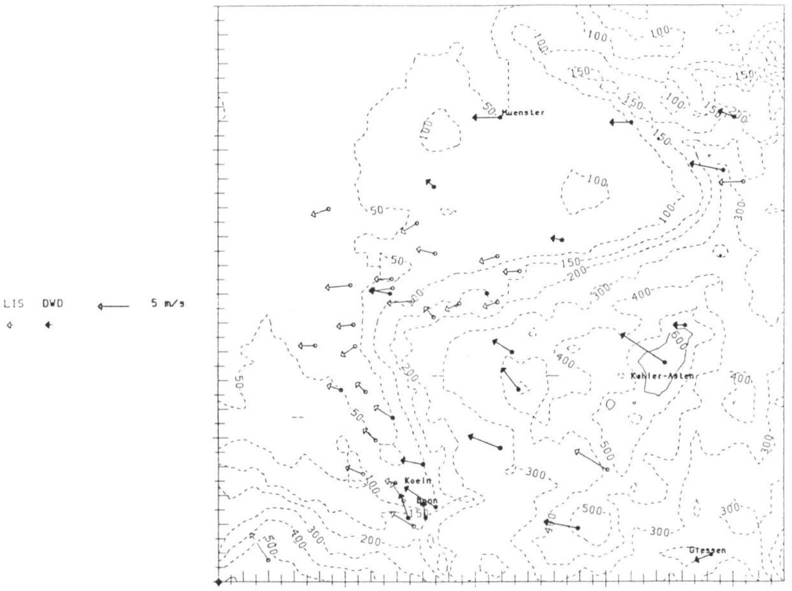

Fig.2. Mean values of measured surface winds for 2.00 UTC on October 3rd, 1987 at 26 LIS and 18 DWD sites. The geostrophic wind in this situation is about 8 m/s from 135 degree.

The surface wind field of Fig.2 shows two significant features. In the region of the Rhine valley a channeling of the flow is observed whereas in the northern parts of the simulation area the expected deviation from the geostrophic wind is detected in both nets. Stations at higher altitudes measure higher wind speeds than those at lower altitudes.

As expected, the wind field shows an overall horizontal variability, which is characterized by a standard deviation of 25 deg.. An analysis of the time series measured at all sites (2 hours before and after the time of interest) reveals a nearly stationary wind field with typical standard deviations of 5 deg. in wind direction and 0.5 m/s in wind speed. No trend can be detected. Only a very few sites, which measure low wind speeds of about 1 m/s (3.5 m/s is the mean value of all sites) show standard deviations in wind direction up to 20 deg.. No essential difference can be detected between the LIS and DWD data.

MODEL CALCULATIONS

The wind speed as modelled with RKM for heights of 20m above ground is shown in Fig.3. Fig. 4 gives the differences between RKM and REWIMET coded as follows. The number displayed at each second grid point in x- and y-direction is the difference in wind direction between RKM and REWIMET, the arrow points to north, if this difference is zero it turns clockwise in case of a positive difference. The length of the arrow is proportional to the difference in wind speed, normalized to the arrow at the upper right corner.

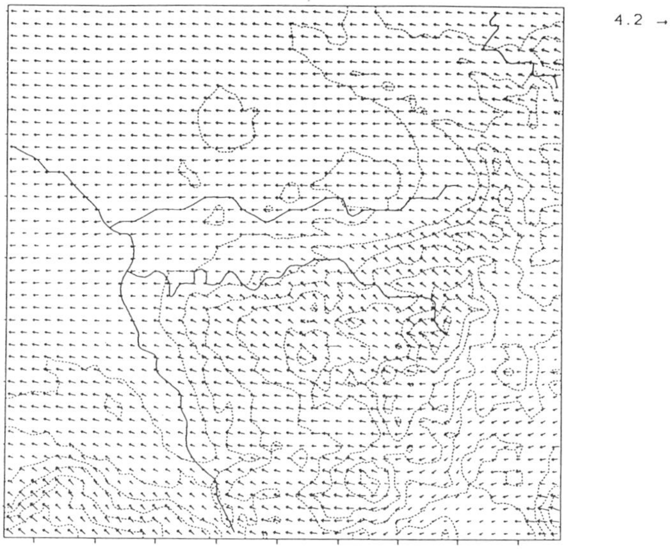

Fig.3. Calculated wind field with RKM for October 3rd, 1987, 2.00 UTC in the region of Northrhine Westfalia.

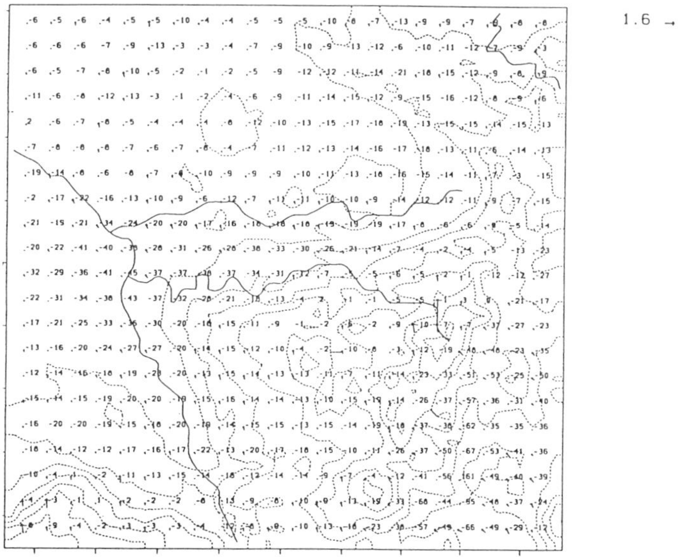

Fig.4. Difference in wind direction (number at every second grid point) and wind speed (proportional to the length of the arrow at the upper right corner) between RKM and REWIMET for October 3rd, 1987, 2.00 UTC.

All models reflect the gross structure of the observed flow field, namely the channeling of the flow in the Rhine valley and the more easterly flow in the northern part of the simulation area.

The difference between RKM and REWIMET is most pronounced in the regions of the Rhine valley (about 20 deg.), the downslope flow of the Sauerland (30 deg. in the middle of Fig.3 and 4) and the south eastern part of the calculated area (40 deg. and more), where RKM is the only model, which reproduced the east-north-easterly flow observed at Giessen upwind of the Sauerland (lower right portion of Figs. 3 and 4).

FOOT and REWIMET produced similar results with differences at their grid points normally smaller than 10 deg.. REWIMET produces maximum wind speeds in the upwind region of mountain ridges, FOOT in downwind regions. This can be explained by the different numerical diffusion in both models. FOOT results are much closer to REWIMET than to RKM.

CONDOR and NOABL also show good overall agreement except in the region of the measuring sites. CONDOR retained the wind information at these positions whereas NOABL produced a smoothed field in these regions. Although CONDOR theoretically should give different results to NOABL - due to the modeling of variable transmission coefficients in space - our comparisons reveal only small deviations among the models.

QUANTITATIVE COMPARISON WITH MEASUREMENTS

Tables 1-4 give a summary of the model evaluation following Keyser and Anthes (1977). The same abbreviations as in the chapter "Model Comparison" have been used. Tables1 and 2 contain information of all measurements. The calculations leading to the data in Tables3 and 4 were done omitting the sites at which the difference between the RKM result and the measurement exceeds 25 deg. (6 sites in the LIS network, 5 sites in the DWD network). In those cases we anticipated marked subscale effects on the measurements. The comparisons of Tables 1 and 3 as well as Tables 2 and 4 allow a rough assessment of the models overall performance and their ability to reproduce regional flow patterns.

For both networks the RKM results agree better with the measurements, if sites with extreme wind direction deviations were excluded from the analysis, a result which also holds for NOABL and CONDOR concerning the LIS network.

The results of the three layer models REWIMET and FOOT also show large deviations from the measurements at those sites, which are not incorporated in Tables 3 and 4. REWIMET and FOOT show only minor improvements comparing Tables 1 and 3, a fact, which is due to the lower overall agreement between measurement and calculation compared to the other models. If a systematic deviation is removed, REWIMET and FOOT show some skill in predicting the wind direction in 10 degree wide sectors as can be seen comparing Tables 2 and 4.

CONCLUSIONS

The straightforward application of different prognostic and diagnostic mesoscale models in a stable nighttime atmosphere revealed

Table 1. Comparison of calculated and measured values at the 26 LIS sites
Mean value of measurements : 98deg.
Standard deviation of measurements : 25deg.

Model	RKM	REWIMET	FOOT	NOABL	CONDOR
Mean dir.	109	131	130	114	117
Sigma	14	10	8	10	11
E(RMS)	24	35	42	25	26
EUB	21	23	32	21	21

Table 2. Comparison of calculated and measured values at the 18 DWD sites. The calculated values have been rounded to ten degrees.
Mean value of measurements : 112deg.
Standard deviation of measurements: 25deg.

Model	RKM	REWIMET	FOOT
Mean dir.	107	123	124
Sigma	16	14	8
E(RMS)	22	27	29
EUB	22	25	26

Table 3. Comparison of calculated and measured values at the 20 LIS sites
Mean value of measurements : 105deg.
Standard deviation of measurements : 23deg.

Model	RKM	REWIMET	FOOT	NOABL	CONDOR
Mean dir.	106	130	129	115	116
Sigma	13	8	8	10	12
E(RMS)	13	31	34	21	20
EUB	13	25	25	18	20

Table 4. Comparison of calculated and measured values at the 13 DWD sites. The calculated values have been rounded to ten degrees.
Mean value of measurements : 102deg.
Standard deviation of measurements: 14deg.

Model	RKM	REWIMET	FOOT
Mean dir.	105	123	124
Sigma	16	12	8
E(RMS)	8	24	25
EUB	7	13	14

Mean dir: mean wind direction calculated for the model results at the measuring sites (deg.)
Sigma: corresponding standard deviation (deg.)
E(RMS): root mean square error (deg.)
EUB: root mean square error after a constant bias is removed (deg.)

that all models are able to reproduce the gross features of the observed surface layer wind field for this special case. However, the quantitative analysis showed that only the nonhydrostatic model nearly always fulfills the criteria given by Keyser and Anthes (1977). The performance of the mass consistent models is between RKM and the three layer models, which show some skill if a constant bias is removed and the comparison is done with data classified in 10 degree wide sectors. As expected, the models generally produced smoother data than those measured. The largest variability is given by RKM.

The superior performance of RKM does not come without costs. The CPU-time needed to produce the results was about 60000s on a Cyber 962. The mass consistent models CONDOR and NOABL are much faster, but need representative measurements as input information, which is seldom available. Using the routine measurements of the DWD however, their results are superior to those of the three layer models REWIMET and FOOT, which need about 600s CPU-time to solve the problem.

The results, so far, support the experience that detailed modeling of a specific flow field is possible using a nonhydrostatic fully threedimensional model. Wether similar results can be achieved using a hydrostatic model is a question still to be answered. Mass consistent models can be used on a regional scale if representative measurements are available. The application of three layer models should be restricted to climatological studies, where they may give sufficient results.

REFERENCES

Born, K., 1989, FOOT - Ein Dreischichtenmodell auf der Basis von REWIMET zur Simulation der dynamischen Grenzschicht, Diploma Thesis, Met. Inst. University of Bonn.

Ehhalt, D.H. and Poppe, D., 1990, Bestimmung der Deposition von Luftverunreinigungen der Region Sauerland, Abschlußbericht des Forschungsvorhabens FP32, Forschungsschwerpunkt des Landes Nordrhein-Westfalen "Luftverunreinigung und Waldschäden".

Frenzen, G., Heimann, D. and Wamser, M., 1987, Dokumentation des regionalen Klimamodells (RKM) auf der Basis von FITNAH, Deutsche Forschungs- und Versuchsanstalt für Luft- und Raumfahrt, DFVLR-Mitt. 87-07.

Gross, G., 1986, A numerical study of the land and sea breeze including cloud formation, Beitr. Phys. Atmosph., 59,p.97-114.

Heimann, D., 1986, Estimation of regional surface layer wind field characteristics using a three layer mesoscale model, Beitr. Phys. Atmosph., 59, p. 518-537.

Keyser, D. and Anthes, R.A., 1977, The applicability of a mixed layer model of the planetary boundary layer to real data forecasting, Mon. Weather Rev., 105, p.1351-1371.

Moussiopoulos, N., Flassak, Th. and Knittel, G., 1988, A refined diagnostic wind model, Environmental Software, 3(2), p. 85-94.

Pielke, R.A., 1984, Mesoscale meteorological modeling, Academic Press, Orlando, Florida.

Traci, R.M., Phillips, G.T. and Patnaik, P.C., 1978, Developing a site selection methodology for wind energy conversion systems, DOE/ET/20280-3, N.T.I.S, Springfield, Virginia.

DISCUSSION

A. VENKATRAM

Why do you allow your prognostic non-hydrostatic model to go to steady state? How do you justify this assumption of quasi-steadiness?

K. MASSMEYER

The wind field calculations have been carried out to be compared to the measurements made during one specific hour. This hour was embedded in a period of nearly stationary meteorological conditions. Because the time series of the wind measurements showed only a small variation in time we prescribed constant boundary conditions to the model and stopped the calculations after the variation of the wind-field between two time steps dropped below a prescribed value.

J.C.R. HUNT

How do your two models (3-layer and RKM) account for changes in surface roughness? Does the 3-layer model have an internal boundary layer as the flow adjusts from one area to another?

K. MASSMEYER

The parametrization of the turbulent fluxes in the model equations is done either by transfer coefficients (3-layer model) or calculation of turbulent kinetic energy (RKM). In both methods the profile functions of Businger are used which incorporate the roughness length via a term ln z/zo which is calulated for each grid area. As a representative value for the roughness length a weighted mean of typical values for different land use classes has been used for those areas.

The 3-layer model does not have an internal boundary layer.

G. SCHAYES

1. How do the CPU-time of the different models compare?

2. What are the lower boundary conditions of the models, particularly for the temperature?

K. MASSMEYER

1. A comparison of the CPU-time is a little bit difficult because the models came to nearly stationary wind fields after different periods of integration time. Having in mind this factor of 4 difference the nonhydrostatic model needs roughly 20 times longer to come to a wind field of similar stationarity compared to the one produced by the hydrostatic 3-layer model. If you do the comparison on the basis of the same amount of model time - i.e. integration time of the models equation - the difference is a factor of 80 which is nearly the value we present in our conclusions.

These values are only valid for the specific case I talked about. Variation of field dimensions or resolution of orography will lead to other differences between the models.

2. The lower boundary conditions of the models were: i) no slip condition concerning the wind field, and ii) prescribing the temperature of the earth surface (3-layer model) or the lowest model layer (RKM).

SPATIAL VARIATIONS OF AERODYNAMIC RESISTANCE OVER A HILL

J. Padro and J.L. Walmsley

Atmospheric Environment Service
4905 Dufferin St., Downsview, Ont., M3H 5T4
CANADA

INTRODUCTION

In Padro and Walmsley (1990), vertical profiles of wind speed, shear stress, pollutant concentration, and pollutant flux were shown as solutions of a linearized set of perturbation equations, appropriate to a small hill with a gentle slope. Results over the summit were presented for idealized terrain features: isolated two- and three-dimensional hills and ridges and an infinite series of ridges.

In the present study we describe results of vertical profiles of the above variables and their influence upon aerodynamic resistance at various locations on a hill of cosine-squared cross-section, height 100m and radius 500m. Such studies are important for understanding the influence of subgrid-scale terrain features upon larger scale flows. Horizontal variations in wind, stress, concentration, and pollutant flux have important impacts upon the profiles and horizontal variations of deposition flux rates on surfaces. Wesely and Lesht (1989) attempted to incorporate changes in surface conditions in order to adjust large-scale, grid-averaged, flows to site-specific conditions and calculate their influence on dry deposition fluxes. They ignored the hill effect on vertical mixing.

In order to capture the salient features of the hill effect, we shall discuss results for a species such as HNO_3 or SO_2 over a wet surface. For these species, the aerodynamic resistance becomes a large term in comparison with the laminar boundary layer and surface resistances. This implies a surface concentration of zero (c_s=0) both upstream and on the hill. These conditions can be relaxed, resulting in a change in magnitude of the variables, but little additional insight would be gained.

THEORY

In Padro and Walmsley (1990), the hill perturbation solutions of wind speed, pollutant concentration, and pollutant flux were obtained. To these we now add aerodynamic resistance and write the upstream profiles, respectively, as:

$$u_0=(u_*/\kappa)\ln[(Z+z_0)/z_0] \quad (1)$$
$$c_0=c_{s0}+(c_*/\kappa)\ln[(Z+z_0)/z_0] \quad (2)$$
$$F_0=u_*c_* \quad (3)$$

Air Pollution Modeling and Its Application VIII, Edited by H. van Dop and D. G. Steyn, Plenum Press, New York, 1991

$$r_{a0}=(1/\kappa u_*)\ln[(Z+z_0)/z_0], \tag{4}$$

where Z is height above ground, z_0 is roughness length, u_* is friction velocity, c_* is the scaling concentration, $\kappa=0.4$ is Von Karman's constant, and c_{s0} is the upstream surface concentration, taken to be zero. Note that flux is defined to be positive downward. From Wesely and Hicks (1977) we obtain

$$c_0=F_0(r_{a0}+r_{b0}+r_{s0}), \tag{5}$$

where r_{a0}, r_{b0}, r_{s0} are the upstream aerodynamic, laminar boundary layer and surface resistances, respectively. In our study the upstream surface roughness length is $z_0=0.02$m, typical for short grass. The term F_0r_{s0} in (5) is the surface concentration c_{s0} which, although neglected in the present case, would have a perturbation described by

$$c_{s1}=F_{s1}r_{s0}+F_0r_{s1}, \tag{6}$$

where $r_s=r_{s0}+r_{s1}$ is the surface resistance on the hill, $F_s=F_0+F_{s1}$ is the surface flux on the hill and can be obtained from Padro and Walmsley (1990). Here, we take r_s, r_{s0}, c_{s0}, and c_{s1} to be zero. If we assume $r_a \gg r_b$ and $r_{a0} \gg r_{b0}$, (5) reduces to $c_0=F_0r_{a0}$ and on the hill,

$$r_a = c/F = (c_0+c_1)/(F_0+F_1). \tag{7}$$

We shall illustrate vertical profiles of r_a at three locations over the cosine-squared hill described in the introduction: the summit of the hill, 250m upstream and 250m downstream of the summit. The corresponding profiles of concentration, wind speed, stress and flux will also be shown, non-dimensionalized with respect to c_*, u_*, u_*^2, and u_*c_*, respectively.

DISCUSSION OF RESULTS

Figure 1(a) shows vertical profiles above the summit of the hill of wind speed, concentration, and aerodynamic resistance. The wind speed, starting from zero at the roughness length reaches a maximum at 10m and then remains fairly constant in approaching the upstream value at 200m. The wind speed is larger than the upstream at all heights. Up to a height of 5m, the vertical gradient of wind speed is larger than in the upstream area. This is reflected in a larger shear stress in Figure 1(b), indicating increased turbulence and a smaller aerodynamic resistance. The summit concentration perturbation, on the other hand, is nearly zero up to 1m, increases to a maximum within the 5-10m layer and then gradually approaches its upstream value. The increase in concentration at 5-10m is associated with flux convergence in the layer above and positive advection within the layer, as described in Padro and Walmsley (1990).

Figure 1(b) illustrates the variation with height above the summit of the stress and pollutant flux. Up to 1m both are approximately independent of height, showing the dominance of the mixing-length theory. Above 1m, the stress decreases to reach a minimum at 10m and then approaches the upstream value at 20m and above. The pollutant flux, however, reaches a local minimum at 5m, remains relatively constant to 20m and then approaches the upstream value, reducing the perturbation to near zero at 100m.

In order to get a better understanding of the hill effect, it would be useful to consider locations away from the summit of the hill. In particular, our attention is focused on the perturbation of the aerodynamic resistance because of its important implications in air quality problems.

Figure 2(a) illustrates the vertical profiles of wind speed at

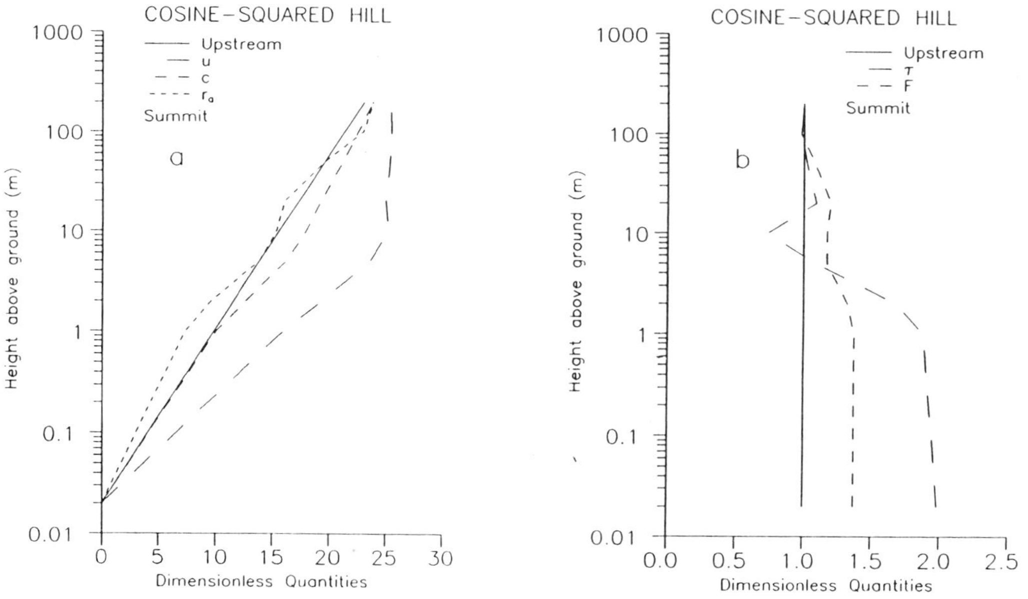

Fig.1 Vertical profiles of dimensionless quantities upstream and at the summit of the cosine–squared hill (a) Wind speed, u; Pollutant concentration, c; Aerodynamic resistance, r_a. (b) Shear stress, τ; Pollutant flux, F.

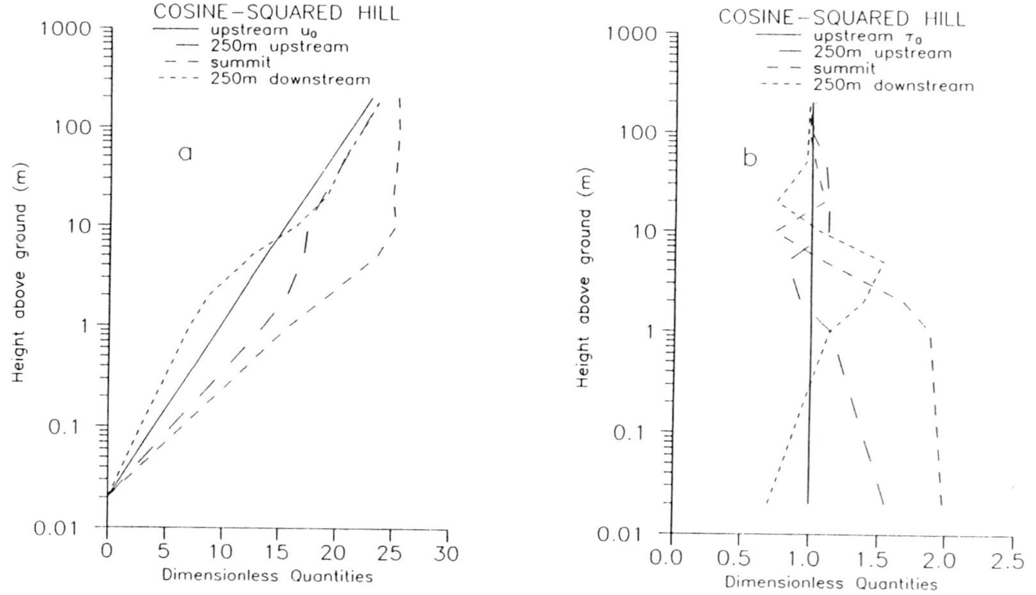

Fig.2 Vertical profiles of (a) Wind speed, u, at: upstream; 250m upstream of summit, summit, 250m downstream of summit. (b) Shear stress, τ, at same locations as (a).

locations 250m upstream and downstream of the summit and includes the upstream (u_0) and summit profiles for reference. We recall that the radius of the hill is 500m. The profile 250m upwind of the summit (i.e. half-way from the leading edge of the hill to the summit) shows that the wind speed below about 10m has increase significantly but has not yet attained its summit values. At the location 250m downwind of the summit, the wind speed perturbation below 10m is negative. At both locations, vertical gradients of wind speed are smaller below 10m than at the summit. These features are important for understanding the behaviour of the aerodynamic resistance as described below.

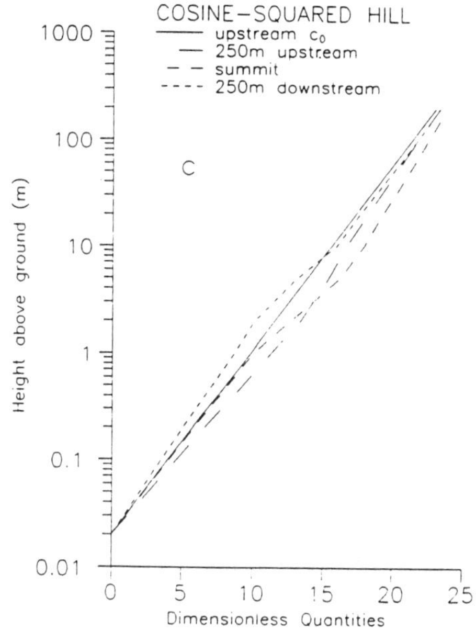

Fig.2 (c) Pollutant concentration, c, at same locations as (a).

Figure 2(b) illustrates the variation of the vertical profiles between the summit and the other two locations. Below about 3m the stresses are smaller on the upwind and downwind slopes than at the summit. This is in agreement with the smaller wind shear at those locations. Moreover, the stress varies with height on the upwind and downwind slopes, whereas it is nearly independent of height at the summit. Such variations may be attributed to contributions from advective, production or dissipation terms. Calculations in Walmsley and Padro (1990), for example, illustrate the importance of the "production due to wind shear perturbation" term in determining the shape of the vertical profile of stress.

Figure 2(c) illustrates the profiles of the concentration for the same locations as for the wind speed and the stress. Again, the upstream and the

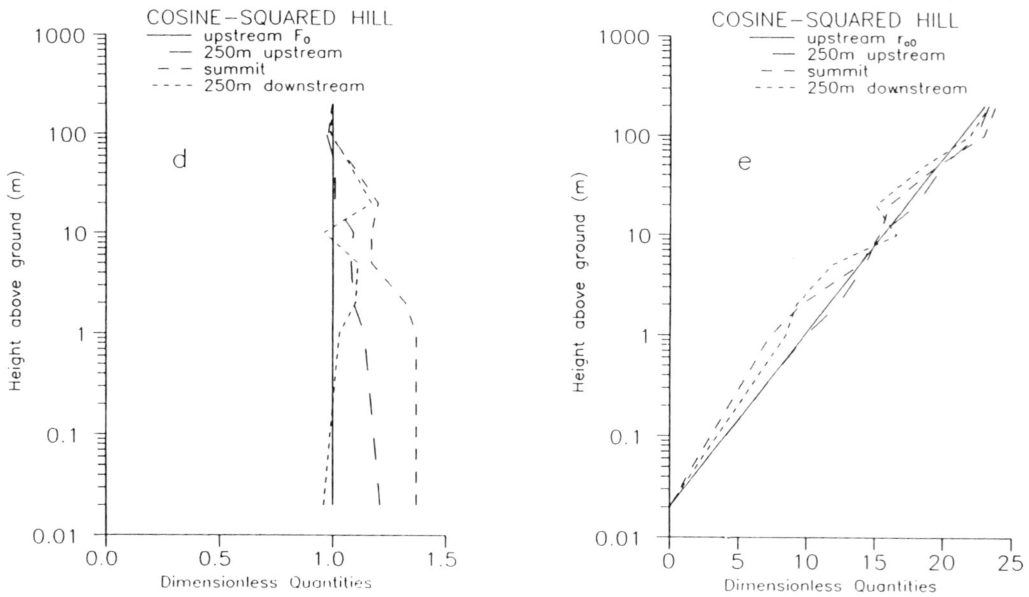

Fig.2 (d) Pollutant flux, F, at same locations as (a). (e) Aerodynamic resistance, r_a, at same locations as (a).

summit profiles are included for the purpose of comparison. Below about 2m, concentration perturbations on the upwind and downwind slopes are positive and negative, respectively, compared with near zero values at the summit. At levels above 10m, concentration perturbations are smaller than at the summit.

Figure 2(d) illustrates vertical profiles of pollutant flux at the same three locations. The surface flux on the upwind slope is roughly half that at the summit, whereas on the downwind slope it is almost the same as in the undisturbed flow. The fact that all three profiles are nearly independent of height below 1m, suggests that mixing-length theory works quite well at low levels which, in general, is not the case for the corresponding stress profiles in Figure 2(b).

Figure 2(e) shows the vertical profiles of aerodynamic resistance for the same locations on the hill in comparison with the upstream values. Below about 1m, the aerodynamic resistance on the upwind slope is almost unchanged from the values in the undisturbed flow. At the summit, however, r_a shows a small negative perturbation that becomes larger on the downwind slope. Variations in r_a can be related directly to variations in concentration and flux, as shown in (7). These, in turn, are related to variations in the terms in the governing equations given in Padro and Walmsley (1990). Those terms involve perturbations of vertical velocity, turbulent kinetic energy and dissipation, all of which depend, directly or indirectly, on perturbations of the wind flow and turbulence (Beljaars et al. (1987)).

When calculations are made at a location 250m to the right of the summit (with respect to the flow direction), we find that the results are similar to the summit profiles of Figure 1, with the exception that perturbations are reduced.

CONCLUSIONS

Although wind profiles perturbed by a small hill of a gentle slope have been studied by a number of authors in the past (see Padro and Walmsley, 1990), perturbations of pollutant concentration, pollutant flux, and aerodynamic resistance have received less attention in the literature. We have demonstrated that the hill-perturbed wind field and its associated perturbed stress affect the horizontal as well as the vertical structure of the concentration, pollutant flux, and aerodynamic resistance.

We would emphasize that our model is capable of producing results at high spatial resolution at low computing cost when compared with 3D finite-difference models. We can learn a great deal from our model about spatial variations in wind speed, concentration, flux, and aerodynamic resistance due to idealized terrain features and changes in surface roughness and surface resistance. In order to apply such knowledge in regional models that estimate dry deposition over large grid areas, however, we need to parameterize the effects of complex terrain in some convenient way. This will be a challenge, but should contribute to improved simulations.

ACKNOWLEDGEMENTS

We would like to thank Dale Woolridge for an excellent job of preparing the text and diagrams.

REFERENCES

Beljaars, A.C.M., Walmsley, J.L. and Taylor, P.A.: 1987, 'A Mixed Spectral Finite-Difference Model for Neutrally Stratified Boundary-Layer Flow Over Roughness Changes and Topography', Boundary-Layer Meteorol., 38, 273-303.

Padro, J. and Walmsley, J.L.: 1990, 'A Mixed Spectral Finite-Difference Model for Pollutant Concentrations over a Hill', Boundary-Layer Meteorol., 51, 343-363.

Walmsley, J.L. and Padro, J.: 1990, 'Shear Stress Results from a Mixed Spectral Finite-Difference Model: Application to the Askervein Hill Project Data', Boundary-Layer Meteorol., 51, 163-172.

Wesely, M.L. and Hicks, B.B.: 1977, 'Some Factors that affect the Deposition Rates of Sulphur Dioxide and Similar Gases in Vegetation', J. Air Pollut. Control Assoc., 27, 1110-1116.

Wesely, M.L. and Lesht, B.M.: 1989, 'Comparison of RADM Dry Deposition Algorithms with a site-specific method for inferring Dry Deposition', Water, Air Soil Pollut., 44, 273-293.

DISCUSSION

C.J. NAPPO

Your model is a linear model and hence it cannot simulate flow separation. Can you comment on the differences between your model results and a real flow situation when flow separation is possible?

J.L. WALMSLEY

Referring to Figures 9 and 10 of Beljaars et al. (1987), the results of the Askervein experiment and associated modelling efforts showed that the MSFD model agreed quite well with field measurements along the 'AA' line of towers running through the centre point, 'CP', of the hill. The agreement was, in fact, excellent on the windward side of the hill and at the summit. On the lee side, model results were about 80% of the upstream undisturbed flow, whereas measurements were about 60%.

For the same situation but along the 'A' line through the hill top, 'HT', the terrain was steeper and the measured wind speed on the lee side of the hill was reduced to about 30% of the upstream undisturbed flow. In this case, although MSFD gave excellent comparison with measurements on the windward side and at the summit, results on the lee side were only reduced to about 80% of the undisturbed flow.

A. VENKATRAM

The downward flux of material will evolve with distance even in the absence of the hill. How does the hill affect this evolution?

J.L. WALMSLEY

The premise is incorrect. The downward flux does not evolve with distance in the absence of the hill. We are implying a perfectly absorbing surface by assuming zero surface concentration. The upstream pollutant concentration and flux profiles given by Equations (2) and (3) apply at all (x, y) locations. In the absence of the hill, there is no mechanism to produce a non-zero perturbation in either concentration or flux, assuming the surface conditions are unchanged. Pollutants are brought downward and are absorbed at the surface. Thus, there is a downward flux, but its perturbation would be zero everywhere in flat, uniform terrain.

The hill has an effect on the downward pollutant flux, as shown for the summit and two other locations in Figure 2(d) where the dimensionless flux upstream (or in the absence of the hill) is given by the straight line with value 1.0.

JOSEPH C. CHANG — What kind of vertical coordinate did the model use?

J.L. WALMSLEY — The model used terrain-following coordinates. The governing equations are first transformed to this coordinate system. Then we perform a Fourier transformation in both horizontal equations which changes the governing equations from three-dimensional partial differential equations to a set of ordinary differential equations (one for each wavenumber pair) in the vertical coordinate. A wavenumber-dependent log-linear vertical grid is used for the calculations in Fourier mode. This computational grid has higher resolution near the surface for high wavenumbers than for low wavenumbers. Details are given in Beljaars et al. (1987) and Karpik (1988).

J.L. WALMSLEY — One additional note ..... Walmsley and Padro (1990) show the effect of inserting a parameter, missing in the original MSFD flow model, on the shear stress results. This correction, which also affects the pollutant flux, appears in the work of Padro and Walmsley (1990) and in the present paper.

REFERENCE

Karpik, S.R.: 1988, 'Improved Method for Integrating the Mixed Spectral Finite Difference (MSFD) Model Equations', Boundary-Layer Meteorol., 43, 273-286.

SIMULATION OF DISPERSION IN MODERATELY COMPLEX TERRAIN

Leif Enger

Department of meteorology
University of Uppsala
Uppsala, Sweden

INTRODUCTION

In operational applications of dispersion calculations in areas on the meso-γ-scale, which covers roughly the interval 2-20 km, one often uses models that presumes the wind- and turbulence fields to be horizontally uniform over the area concerned. This assumption could be very incorrect in an area with complex terrain. The present model works as follows:
1. With the aid of a dynamic model a lot of wind and turbulence fields have been simulated for the actual area. In this way we have produced a data base with some thousand typical wind and turbulence fields, each of which represents a certain defined meteorological situation.
2. In simulating dispersion at a special occasion wind speed, wind direction, vertical standard deviation and acoustic backscatter from a doppler sodar together with a temperature profile and wind speed and wind direction in a 10 m high mast are used to determine which of the several thousand wind fields in the data base that agree best with the actual situation. This wind and turbulence field is used for the dispersion calculations.
3. To simulate the climatological concentration pattern one uses a statistical material obtained from the sodar and the mast measurements. Every hour mean values from the sodar and mast measurements are used to calculate how many times a certain wind and turbulence field in the data base has occurred during the actual time period studied. The dispersion calculations are done for all occurring situations to get the climatological concentration pattern.

THE DYNAMIC MODEL

The dynamic model is hydrostatic and it is based on Yamada and Mellor's (1975) Level 2.5 model but with the change that the pressure-temperature interaction term is parameterised according to Lumley's (1979) suggestions.This means

$$\frac{1}{\rho}\overline{\theta\frac{\partial p}{\partial z}} = \frac{q}{A_2\lambda}\,\overline{w\theta} + \alpha_2\frac{g}{\Theta_o}\,\overline{\theta^2} \tag{1}$$

where α_2 is a constant equal to 1/3 according to the theory (Wyngaard, 1982).
The topography is introduced to the model by application of a terrain following coordinate system (Pielke, 1984). The new vertical coordinate, η, is defined as

$$\eta = s\,\frac{z - z_g}{s - z_g}$$

In this coordinate system the mean wind components, the mean temperature and the turbulent energy equations are solved prognostically.

The turbulence time scales appearing in the parameterisations have been taken as proportional to the ratio between a master length scale and a turbulent velocity scale, taken to be the square root of twice the turbulent kinetic energy. Thus all turbulence time scales have the same vertical structure in the model. The closure problem is thus reduced to the design of a master length scale formulation sufficiently general to handle the various types of flow that appear in the boundary layer of the atmosphere.
With time-dependency, diffusion and advection kept only for the turbulent energy it is possible to extract diagnostic expressions for the turbulent transport terms in the mean momentum equation and in the heat equation. The general forms obtained are

$$K_{M,H} = f_{1,2}\left(q^2, \lambda, \frac{\partial U}{\partial \eta}, \frac{\partial V}{\partial \eta}, \frac{\partial \Theta}{\partial \eta}\right) \tag{2}$$

where $f_{1,2}$ are complex diagnostic functions derived from balance equations from all the second order moments.

NUMERICAL METHODS

The numerical methods have, on purpose, been kept simple to reduce the CPU-time needed by the model as we will run the model for about 150 24-hours simulations. One may say that some sacrifice has been made in numerical accuracy, in order to have a more detailed description of the turbulence properties.
All advection terms are approximated by 'up-stream in space' - 'forward-in-time' finite differences. For the diffusion terms semi-implicit 'centered-in-space' - 'forward-in-time' finite differences is used. The dissipation terms in the turbulent energy equation and the Coriolis term in the momentum equations are evaluated implicity. The model grid is telescopic both horizontally and vertically, to achieve the highest resolution close to the surface in the centre of the model and still to have a large enough domain to reduce the influence of lateral boundary errors propagating in to the area of interest.

THE EXPERIMENTAL AREA

The experimental area is situated in the south-western part of Sweden around the towns Trollhättan and Vänersborg. The area is characterized by a relatively flat agricultural area in the

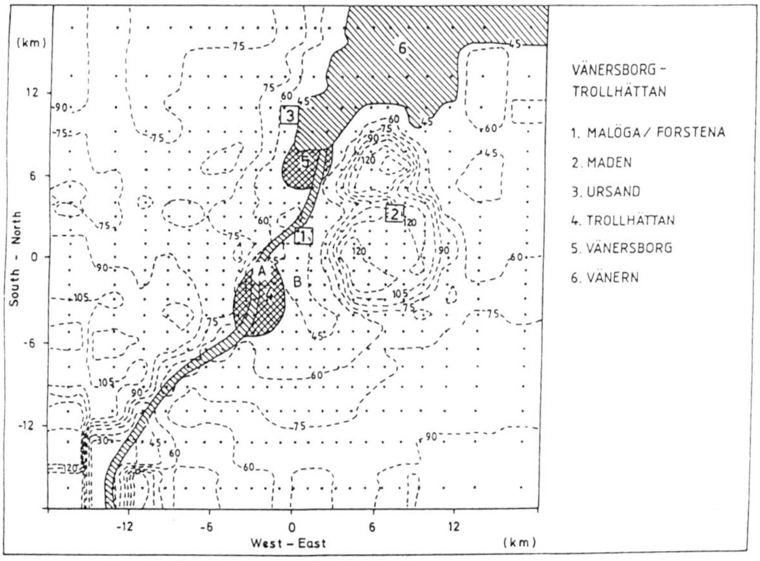

Fig. 1. Map of the geographical are of the simulations showing the terrain (dashed line), the distribution of gridpoints (dots), locations where meteorological measurements were performed (figures) and the release sites during the tracer experiments (letters). The hatched areas are lake Vänern and the river Göta älv and the crossed areas are the towns of Trollhättan and Vänersborg.

centre, about 5 by 10 km in size surrounded by higher terrain, Fig. 1. East of the farmland there are two hills, Halleberg and Hunneberg, rising some 80-100 m above the centre, the highest part being in the south-west corner of the area. North-east of the farmland the lake Vänern is situated,Sweden's largest lake, and from this the river Göta älv is flowing across the model area to its south-west corner. The area surrounding the flat farmland in the middle is covered with (mainly) pine forest.

The results from the model simulations have been compared with measurements performed in the experimental area during three campaigns: November 1983, September-October 1984, and September 1985. In the area two Doppler sodar systems where operated, as well as two 25-30 m high masts for profile measurements of wind, temperature and humidity. During some occasions there were also turbulence measurements in the masts. Furthermore measurements of wind, temperature and humidity profiles were performed with help of a tethered balloon and with radiosonde measurements. During the field experiment September 1989 dispersion was investigated by carrying out SF_6 (sulphur hexaflouride) tracer experiments. The tracer experiments were performed by the Luftfororenings laboratory, Risö, Denmark. For a more detailed description of the tracer experiment see Lyck, 1987. Six tracer experiments were carried out during the campaign, two of the experiments during convective conditions and four during stable conditions.

VALIDATION OF THE DYNAMIC MODEL

Several episodes during the field experiments in this area featured significant flow distortion phenomena, generated by the topography, specially during stable conditions. One such event was observed during the night between November 2 and 3, 1983. Model verifications are presented in Figs 2a-2c for temperature, wind direction and wind speed respectively. In the figure are given, in addition to the measurements, the results from simulations with a K-model and with the higher-order closure formulation with and without terrain included in the model. As seen from the Fig. 2b the wind direction varied rapidly with height during the experimental period. Apparently the higher order closure formulation simulate this feature quite well, whereas the K-formulation gives a too rapid wind direction change with height. This is also true when looking at the wind profile, Fig 2c, where again the higher order closure formulation gives a much better agreement with the measurements than simulations with a K- model. The influence of the topography on the simulated wind direction is shown by the result obtained with the higher order closure model with terrain excluded in the model formulation. It is seen from this result that almost all of the change in wind direction is caused by the terrain. The measurements were performed at the very flat agricultural area with a distance of two to three km to the nearest terrain relief.

During the last campaign, in September 1985, turbulence instruments where located at two places in the area. At site 1, Fig 1, in the middle of the agricultural area, turbulence measurements were performed at three levels in the mast, and at site 3 at one level above a 18 m high

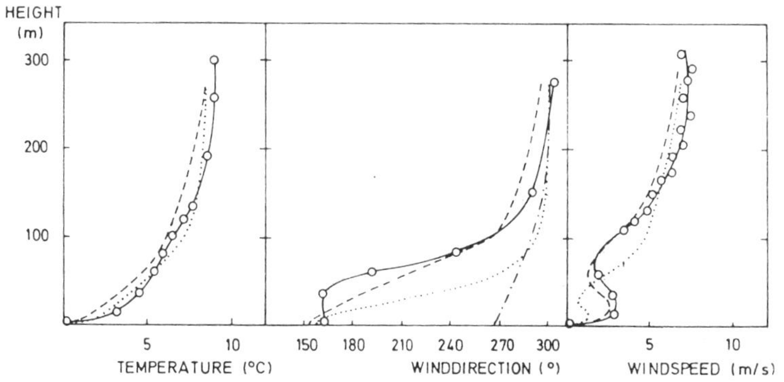

Fig. 2. Measured and simulated profiles of potential temperature, wind direction and wind speed for 83-11-02 at site 1 in Fig. 1. Full lines with circles are measured data. Dashed lines: profiles obtained from simulations with the present model. Dotted lines: profiles obtained from simulations with a K-model. Dot-dashed line: profile obtained from simulation with the present model with terrain excluded.

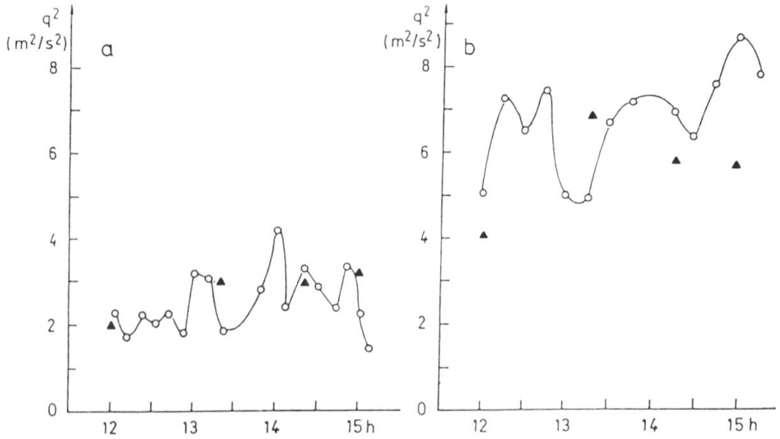

Fig. 3. Measured (o) and simulated (▲) turbulent energy as a function of time 85-09-11. a) Site 1 at 10 m, b) Site 3 at 10 m above the zero displacement height of the forest.

forest. In these comparisons the different types of mean values obtained from different sources have to be kept in mind, i.e. model volume averages versus measured time averages in one point. This difference in representation is likely to give less good agreement between measured and simulated data close to the surface than higher up in the boundary layer in an area with horizontally inhomogeneous terrain. Furthermore the agreement should be better during unstable situations than during stable situations as the measurements at a certain height are influenced by a larger local area in the first case than in the latter case.
During convective situations the model seems to simulate the turbulent energy quite realistically both over flat farmland, site 1, and over the forest, site 3. Figure 3 show the measured and simulated turbulent energy at these two sites as function of time for September 11, 1985. At site 1 both the measured and the simulated values are from the 10 m level. At Site 3 the turbulence measurements were performed at 26 m height above the ground. The zero displacement for this forest is supposed to be about 15 m. The measured turbulence data is therefore compared with the simulated values at 10 m. We see from the figure that the turbulence above the forest is much higher than over the flat farmland. This difference between the two sites is very well simulated by the model. The agreement between the simulated and the measured turbulent energy is satisfactory for most of the cases studied during stable stratification. As an example the results for September 8, 1985, between 00.00 LST and 03.00 LST is shown. During this case the Monin-Obukhov length varied from L= 23 m at 0.00 LST to L= 5 m at 02.30 LST according to the turbulent measurements above the flat farmland, site 1. The

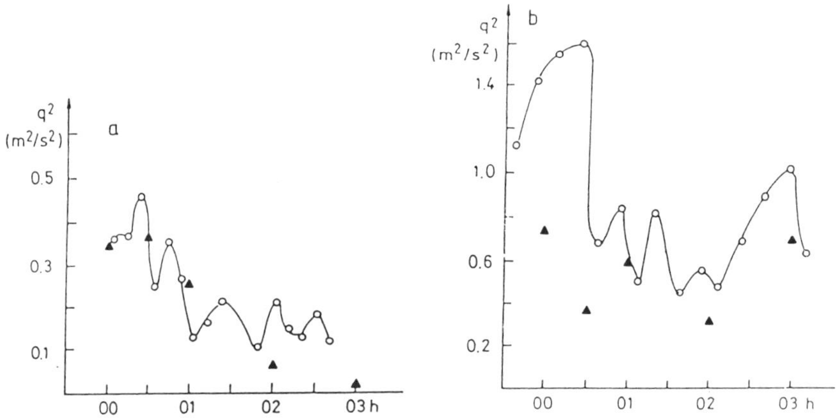

Fig. 4. As fig. 3 but for 85-09-8.

simulated turbulent energy agrees reasonable well with the measured one at this site, Fig. 4a. Over the forest the turbulent energy is much higher, Fig 4b, both according to measurements and simulations.

THE DISPERSION MODEL

The dispersion model uses mean and turbulence quantities derived from the dynamic model. A large number of wind- and turbulence fields are simulated with the fluid dynamic model. These simulations are put into a data base and can be used in the calculations of dispersion. To get relevant meteorological indata to the model a doppler sodar and a ten meter high mast with a temperature profile and wind speed and direction at one level are used. The model calculates a trajectory for the plume centerline from the simulated wind field, and approximates the concentration field with a bi-Gaussian distribution. The horizontal and vertical standard deviations for the plume are determined by using the simulated turbulence quantities from the dynamic model and Eulerian velocity spectra.

DETERMINATION OF THE STANDARD DEVIATIONS

Both the vertical and cross-wind standard deviations are derived from Eulerian velocity spectra together with the simulated turbulence quantities from the dynamic model. An equation for the standard deviation, σ_y, of the cross wind spread of material which is released continuously from a point source, was derived by Taylor in 1921. Especially in a stably stratified boundary layer the effect of wind shear on the horizontal spread can be considerable. According to Smith (1965) the cross wind spread at a given level after time t can be expressed in three terms:

$$\sigma_y^2(t,z) = \sigma_v^2 t^2 \int_0^\infty F_L(n)\left(\frac{\sin\pi nt}{\pi nt}\right)^2 dn - \left(\frac{\overline{vw}}{\sigma_w}\right)^2 t^2 \int_0^\infty G_L(n)\left(\frac{\sin\pi nt}{\pi nt}\right)^2 dn +$$

$$+\frac{1}{4}\sigma_w^2\left(\frac{\partial V}{\partial z}\right)^2 t^4 \int_0^\infty G_L(n)\left(\frac{\sin\pi nt - \pi nt\cdot\cos\pi nt}{(\pi nt)^2}\right)^2 dn \qquad (3)$$

The first term on the right hand side is the direct effect of the horizontal component of turbulence. The third term is the effect of the systematic wind shear and the second term on the right hand side is the effect of that part of the horizontal component which is correlated with the vertical component. The second term acts in opposition to the third term.
Pasquill (1974) shows that the standard deviation of the cross wind spread can be obtained from Eulerian spectrum instead by using numerical filters that effectively block out turbulent fluctuations with time scales less than the travel time t and greater than the sampling time T_s.
These filters also includes the ratio, β, between the Lagrangian and the Eulerian time scales. According to Pasquill (1974) various theoretical estimates of the ratio $\beta = t_L/t_E$ all lead to $\beta = c/(\sigma_v/U)$, where c is a constant ranging from 0.35 to 0.8. We have chosen the value c = 0.6.
We will use these equations to calculate the vertical and cross wind standard deviation of spread of material from a point source. First we want to get expressions for the Eulerian velocity spectra as function of height and stability and therefore we fit the experimental normalized spectra with analytical expressions. The Kansas measurements, Kaimal et al (1972), indicate similarity in shape for the stable spectra if normalized by their respective variance and expressed as functions of f/f_o, where f_o is the intercept of the inertial subrange spectrum with the line $nE_\alpha(n)/\alpha^2 = 1$. The velocity spectra for the stable boundary layer could be expressed as:

$$\frac{nE_\alpha(n)}{\alpha^2} = \frac{A\, f/f_0}{1 + (f/f_0)^{5/3}} \qquad (4)$$

where α = u,v or w. As the variance is equal to the integral of the spectra over all frequencies the constant A must be 0.181. To get the spectra for different stabilities we use expressions of f_o as functions of z/L according to the Kansas measurements. According to Enger (1986) the experimental normalized spectra for the convective boundary layer could be expressed as

$$\frac{nE_\alpha}{w_*^2 \psi^{2/3}} = \frac{0.2\ f_i^{-2/3}}{1 + (f_{max}/\ f_i)^b/(1.5b - 1)} \quad (5)$$

where b and f_{max} are functions of z/z_i. The dimensionless velocity variance, $\sigma^2_\alpha/\ w_*$, can be obtained from the equation above by integrating over all frequencies.
As we get the horizontal and vertical components of the turbulent energy from the dynamic model we can normalize the absolute spectrum function, E_α with σ^2. This also means that we get rid of the problem of determining the Ψ-function.
When looking at the cross wind standard deviation we also have to take into consideration the spread caused by the low frequency part of the wind spectra. Concentration data are often presented as one hour mean values. This means that in the calculations of the cross wind standard deviation the low frequency part of the spectra has the most pronounced influence during stable situations. Olesen et al (1984) start with a form for the scalar energy spectrum of the 3-dimensional turbulence obtained by Lumley (1964) and they end up with an expression for the low frequency part of the u- and v-spectra:

$$nS(n) = \gamma_\beta\ N^2\kappa^{-2} \quad (6)$$

where N is the Brunt-Väisälä frequency, which seems to be a logical choice of scaling parameter if the low frequency part of the spectra is due to mainly gravity waves. We have to use a filter that effectively block out turbulent fluctuations with time scales greater than the sampling time T_s also for the low frequency part of the spectra.
The total cross wind standard deviation we get from

$$\sigma_{tot}^2 = \sigma_y^2 + \sigma_c^2 \quad (7)$$

To be able to take into consideration the change of the meteorological situation with distance from the source we solve the standard deviation as follows:
At a certain distance from the source, or at a certain time, t_n, from the source, $\sigma^2_{t_n}$ is calculated according to the equations given above and by using the meteorological parameters at that distance. At the same time we calculate a standard deviation up to the previous grid distance at time t_{n-1} by using the same meteorological indata, giving us a standard deviation $\sigma_{t_{n-1}}$. The standard deviation, σ_n, used for the grid point n is finally calculated from

$$\sigma^2_n = \sigma^2_{n-1} + (\sigma^2_{t_n} - \sigma^2_{t_{n-1}}) \quad (8)$$

In an area with complex terrain there can be very large wind direction changes both horizontally and with height. Especially during stable conditions these changes are very pronounced. To take into consideration these wind changes with height and horizontally we have supposed that the plume follows a trajectory calculated for the height of the centre of the gravity of the plume, z_m.

VALIDATION OF THE DISPERSION MODEL

In the experiment performed in the area sulphurhexaflouride was used as a tracer. Samples were taken at two semi-circles about 3 and 7 km from the source. Three consecutive 20 minutes samples were taken at each sample point.
The source height during convective conditions were 100 m. The agreement between the simulations and measured concentration is very good during unstable situations. This shows that the wind model gives very satisfactory results both for wind speed and turbulent energy during convective conditions and that the Gaussian form is quite adequate to use in a convective boundary layer in this type of area. The use of integrated spectra together with simulated turbulent energy seems also to give the right standard deviations for the plume.
During stable conditions we used near ground level release (2.5 m above surface) in the tracer

experiments. Figure 5a shows the result from the tracer experiment performed at 02.00 - 03.00 1985-09-03. In this experiment the tracer release rate was 2.93 g/s and the release point was at site B, see figure 1. The meteorological situation was characterized by a geostrophic wind from about south-west and very stable conditions (Obukhov's length L = 15 m). The wind direction at 12 m was from south over the experimental area. The figure shows the concentration at a distance of 7.5 km from the source if the low frequency part of the spectra is excluded in determining the lateral standard deviation. We therefore compare the simulations with the measured concentration for a 20 minutes sampling period. The same wind and turbulent fields are used in the simulations of dispersion with the higher order dispersion model and the operational model presented here, which means that one can compare these two methods directly against each other. In the figure a comparison of surface concentration with the present model, a higher order closure dispersion model (Enger, 1990) and measurements are made. The higher order closure model (dashed line) and the present model (full line) both agree quite well with measured concentration values.
The next case we will study was performed the 5th Sept 1985 at 00.00h - 01.00 h. The wind field at 12 m over the tracer experimental area was very similar to the wind field during the previous experiment just discussed. The geostrophic wind during this experiment was however from WSW instead of from SW as in the previous simulation. This means that the wind direction shear is much larger during this occasion. The release was performed at the same place and the release rate was 2.20 g/s. Figure 5b shows the measured and simulated ground level concentration for a 20 minutes period at a distance of 7.5 km from the source. We can notice that the plume is much wider during this case than during the previous case. In the figure the results from simulations with a higher order closure dispersion model are also presented. The present simple model gives about the same width of the plume and about the

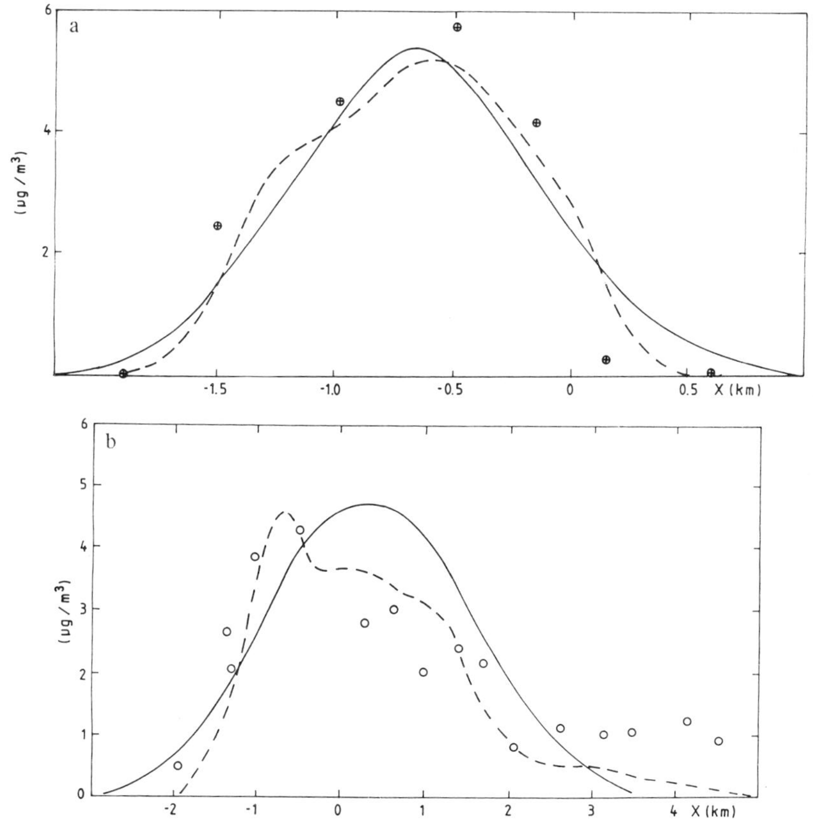

Fig. 5. 20-minutes mean SF_6-concentration at ground level at a distance of 7.5 km from the source. Full line: simulations with the present model. Dashed line: simulations with a higher order closure dispersion model. Circles: measured 20-minutes mean concentrations. a) 85-09-03 b) 85-09-05

same maximum value as the more advanced higher order closure model. This supports the method of determining the standard deviation from spectra. In this case when there is a very large directional shear the third term in Equ (3) is dominating.

OPERATIONAL USE OF THE MODEL

To obtain a model that can be used for operational use we proceed as follows:
We generate a data base with some thousand wind and turbulence fields with the higher order closure model. For areas in Sweden that have implemented the operational model simulations have been performed for three geostrophic wind speeds: 5, 9 and 15 m/s. For each of these wind speeds simulations have been done for 36 wind directions, i.e. every 10th degree. For each combination of geostrophic wind speed and wind direction simulations have been performed for a whole day to get different stabilities by changing the surface temperature according to a sine function and by starting the simulations with a weakly stable stratification through the whole boundary layer. The simulated wind-, temperature- and turbulence fields have been stored in a data base, one per simulated hour.
To simulate the dispersion on a certain occasion we need to know which of these wind and turbulence fields in the data base we shall use. Within the area we therefore have doppler sodar and a 10 m high mast with a temperature profile and wind speed and wind direction at one level. The model search for that case in the data base that agree best with the sodar measurements and with the mast measurements. The 10 m high mast is situated in an area where the topography has a very marked influence on the flow, which makes it easier to find that case in the data base that agree quite well with the actual situation. As the wind shear has a very pronounced influence on the dispersion it is important to find a wind field that has a correct wind shear.
To simulate dispersion for a certain time period, for example a month, one build up statistics on how many times each of the several wind and turbulence fields in the data base has occurred. These statistics together with the data base are used for the calculations of the mean concentrations and percentiles.

REFERENCES

Enger, L., 1986, A higher order closure model applied to dispersion in a convective PBL, Atmospheric Environment, 20, 879-894.

Enger, L., 1990, Simulation of dispersion in moderately complex terrain, Part B. The higher order closure dispersion model, Submitted to Atmospheric Environment.

Kaimal, J. C., Wyngaard, J. C., Isumi, Y. and Coté, O. R., 1972, Spectral characteristics of surface layer turbulence, Q. Jl. R. Met. Soc., 98, 563-589.

Lumley, J. L., 1964, The spectrum of nearly inertial turbulence in a stably stratified fluid, J. Atmos. Sci., 21, 99-102.

Lyck, E., 1987, SF_6-measurements performed in Vänersborg-Trollhättan region, Sweden, MST-LUFT A106, Risö National Laboratory, Denmark.

Pasquill, F., 1974, "Atmospheric Diffusion," John Wiley, New York.

Pielke, R. A., 1984, "Mesoscale Meteorological Modeling," Academic Press.

Smith, F. B., 1965, The role of wind shear in the horizontal diffusion of ambient particles, Q. Jl. R. Met. Soc., 91, 318-329.

Wyngaard, J. C., 1982, Boundary layer modeling. in: "Atmospheric Turbulence and Air Pollution Modelling," Nieuwstadt F.T.M. and van Dop H., ed, Reidel D., Dodrecht.

Yamada, T. and Mellor, G., 1975, A simulation of the Wangara atmospheric boundary layer data, J. Atmos. Sci., 31, 2309-2329

DIAGNOSTIC DERIVATION OF BOUNDARY LAYER PARAMETERS FROM THE OUTPUTS OF ATMOSPHERIC MODELS

A. A. M. Holtslag*, and R. M. van Westrhenen

Royal Netherlands Meteorological Institute (KNMI)
De Bilt, The Netherlands

ABSTRACT

In this paper we discuss the derivation of boundary-layer similarity parameters on the basis of information provided by atmospheric models. The quantities which are considered are the surface fluxes, the Obukhov length, the near-surface wind speed and the boundary-layer height. The necessary inputs are the mean values for wind, temperature, and humidity on grid points of the atmospheric model. We compare observations with calculations for a receptor point over land, using forecasted and analyzed outputs of the limited area model at KNMI. The quantities of the scheme are intended for use in air-pollution dispersion studies, ocean-wave and storm-surge models, and for direct use by forecasters.

INTRODUCTION

At present there is an increasing need for detailed description of the atmospheric boundary layer (ABL) for regional and global air-pollution dispersion simulations. This need is also felt for ocean-wave prediction modelling, and for direct use by weather forecasters. In principle, weather forecast (or climate) models may provide wind, temperature, and humidity with the resolution of the model's grid. Often, however, requests are made for quantities not directly provided by the atmospheric model, such as near-surface wind, and the height of the ABL.

The ABL parameters are directly related to the surface fluxes of momentum, heat, and moisture. Generally, the surface fluxes will depend on the specified characteristics of the model surface, such as roughness length and moisture availability. The latter model quantities may differ from the actual values at a receptor point for which a more detailed forecast is needed. In addition, the horizontal resolution of an atmospheric model will typically not be sufficient to represent the land/sea transition near a coastline. In such cases one may consider refining the calculation of the surface fluxes using the output of the atmospheric model.

In the next section we will describe the formulation of the surface fluxes, the Obukhov length, and the near surface wind speed. Subsequently, a routine for the derivation of the boundary layer height is presented. Finally, some results with the proposed routines are given. A comparison is made with point observations of the Cabauw observational facility, in the Netherlands.

---

*On leave at the National Center for Atmospheric Research, Boulder, Colorado USA.

SURFACE LAYER PARAMETERS

In atmospheric models, the surface fluxes of momentum (τ), sensible heat (H), and latent heat (λE), are usually expressed as (e.g., Louis, 1979)

$$\frac{\tau}{\rho} \equiv u_*^2 = C_M U_1^2 , \tag{1}$$

$$\frac{-H}{\rho c_p} \equiv u_* \theta_* = C_H U_1 (\theta_1 - \theta_s) , \tag{2}$$

$$\frac{-\lambda E}{\rho \lambda} \equiv u_* q_* = C_Q U_1 (q_1 - q_s) , \tag{3}$$

where ρ is the density, c_p is the specific heat and λ is the latent heat of vaporization. U_1, θ_1, and q_1 are wind speed, potential temperature, and specific humidity at the first model level in the atmosphere; θ_s and q_s are the corresponding surface values, and C_M, C_H, C_Q are exchange coefficients for momentum, heat and moisture. With (1) to (3), the friction velocity (u_*), the temperature scale (θ_*), and the moisture scale (q_*) are defined.

The exchange coefficients for momentum, heat, and moisture can be derived by applying surface-layer similarity theory, as has been discussed by e.g., Louis (1979) and Holtslag and Beljaars (1989). This leads to expressions as

$$C_M = \frac{k^2}{\left(\ln\left(\frac{z_1 + z_{0M}}{z_{0M}}\right)\right)^2} F_M (\mathrm{Ri}_0, z_1/z_{0M}) , \tag{4}$$

where k is Von Kármán's constant ($k = 0.4$), z_1 is the height of the first model level, and z_{0M} is the surface roughness length for momentum. F_M denotes a functional dependency of z_1/z_{oM} and the bulk Richardson number Ri_0 between the surface and the first model level. For neutral conditions $F_M = 1$. Similar equations as (4) for momentum, apply for heat and specific humidity. The exchange coefficients for momentum, heat, and moisture have been the subject of many studies (e.g. Carson and Richards, 1978; Louis, 1979; Louis et al, 1982; Delage, 1988). In particular for stable conditions, there have been several proposals (see Holtslag and Beljaars, 1989). We use the expressions by Louis et al. (1982) for unstable conditions, and we utilize the profile function of Holtslag and De Bruin (1988) for stable conditions. The latter is consistent with the proposals by Delage (1988).

The Obukhov length is given as (Busch, 1973)

$$L = \frac{u_*^2}{k\frac{g}{T}\left(\theta_* + 0.61\overline{T}q_*\right)} , \tag{5}$$

where $\overline{T}$ is the mean temperature across the lowest model layer. Finally, the wind speed $U(z)$ at height z in the surface layer is calculated with

$$U(z) = \frac{u_*}{k}\left[\ln\left(\frac{z}{z_{oM}}\right) - \Psi\left(\frac{z}{L}\right) + \Psi\left(\frac{z_{0M}}{L}\right)\right] , \tag{6}$$

where $\Psi(z/L)$ is a function of stability. The latter is chosen in such a way that it is consistent with the exchange coefficients in Eqs. (1)-(3), as discussed by Geleyn (1988). In a similar way the potential temperature, specific humidity, and other passive scalars can be calculated.

THE BOUNDARY LAYER HEIGHT

The height of the atmospheric boundary layer defines the thickness of the turbulent layer and is therefore a crucial parameter for air pollution studies (Verver and Scheele, 1988). The boundary-layer height is also important to characterize the structure of the boundary layer (Holtslag and Nieuwstadt, 1986). Atmospheric models usually do not provide the ABL height directly. Therefore, a diagnostic routine is developed which provides the height on the basis of the outputs of atmospheric models. Since, normally the vertical resolution of atmospheric models is not very large, the routine may be relatively simple.

In neutral conditions, the boundary-layer height h_N can be expressed as (Deardorff, 1972)

$$h_N = c_1 \frac{u_*}{f}, \tag{7}$$

where f is the Coriolis parameter and c_1 is a constant. Duynkerke (1988) shows that the value of c_1 is sensitive to temperature gradients aloft, which inhibit the growth of the ABL. As such he obtains c_1 in the range of 0.12 to 0.20 for small temperature gradients of 2 K/km and 1 K/km, respectively. In the light of these results we take $c_1 = 0.15$. In stationary stable conditions ($L > 0$), the ABL-height h_s is often described with

$$h_s = c_2 \left(\frac{u_*}{f} L \right)^{1/2}, \tag{8}$$

where c_2 is a constant (Zilitinkevich, 1972).

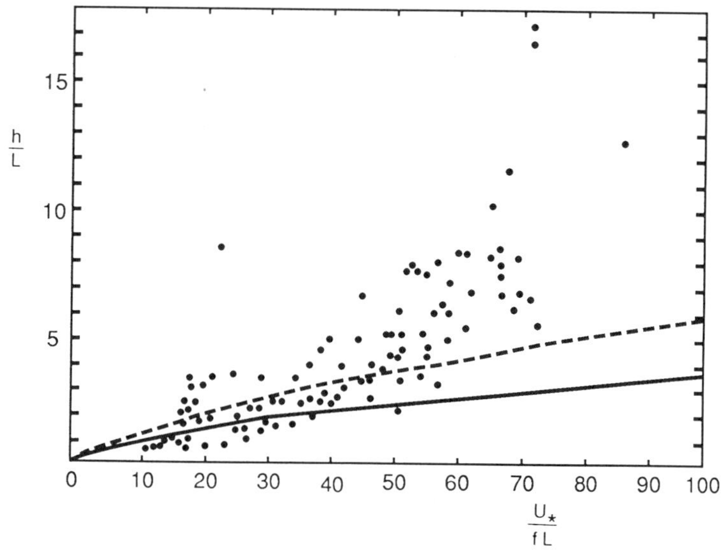

Fig. 1. A comparison of Cabauw data by Nieuwstadt (1981) with Eq. (9), using $c_1 = 0.3$ and $c_2 = 0.4$ (full line), and $c_1 = 0.15$ and $c_2 = 0.7$ (dashed line).

An interpolation formula between (7) and (8) is given by Nieuwstadt (1981), which reads in dimensionless form as

$$\frac{h}{L} = \frac{c_1 u_*/(fL)}{1 + c_3 h/L}, \tag{9}$$

where $c_3 = c_1/c_2^2$. Fig. 1 shows a comparison of (9) with the data of Nieuwstadt (1981). The solid line represents (9) with $c_1 = 0.3$ and $c_2 = 0.4$ as used by Nieuwstadt, and the dashed line represents $c_1 = 0.15$ and $c_2 = 0.7$. The latter value has been proposed by Businger and Arya (1974). From Fig. 1 it is seen that the dashed line is in better agreement with the data, although the differences with the actual data are not small for both parameter sets. Note that a linear relationship between h/L and $u_*/(fL)$ is also in reasonable agreement (see Koracin and Berkowicz, 1988). However, here we prefer (9) because this equation has a better theoretical foundation in stable conditions (Nieuwstadt, 1981).

In well-developed unstable conditions ($H > 0$; $L < 0$), a simple determination of the ABL height (h_c) is given by the dry parcel intersection method (Holzworth, 1964). This method is illustrated in Fig. 2. For a given virtual temperature profile within the atmospheric model, a parcel with virtual temperature θ_p will be lifted up to h_c.

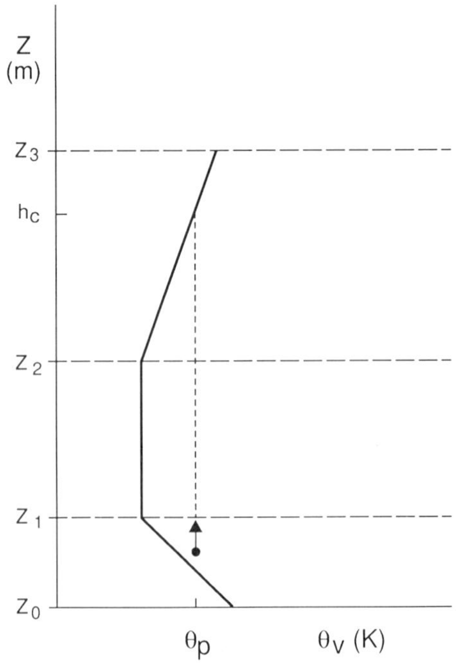

Fig. 2. Determination of the convective boundary layer height h_c by lifting a parcel with virtual temperature θ_p.

Troen and Mahrt (1986) propose to derive θ_p from

$$\theta_p = \theta_s + c_4\, \overline{w\theta_0}/w_* \,, \tag{10}$$

where θ_s is a surface-layer virtual temperature, c_4 is a constant taken as 6.5, $\overline{w\theta_0}$ is the kinematic surface heat flux related to H of (2) by $H = \rho c_p\, \overline{w\theta_0}$, and w_* is the convective velocity scale defined by

$$w_* = \left(\frac{g}{T}\overline{w\theta_0}h_c\right)^{1/3} . \tag{11}$$

With (10) and (11), h_c can be determined by iteration. A first guess is obtained by using $\theta_p = \theta_s$, which then defines w_* and results in an improved estimate for θ_p with Eq.(10). This procedure normally converges within three steps. For θ_s we use the virtual temperature at the height of 10 m. The latter is calculated with a similar expression as (6). If during unstable conditions h_c drops below h_N of Eq. (7), we use the latter as the best estimate for the boundary-layer height. In all cases a lower limit of 50 m is used for h.

SOME RESULTS

In this section we will illustrate the outputs of the present scheme for a grid point above land. The KNMI limited area model (LAM) has been used to produce the inputs of the present scheme. The LAM is in many aspects similar to the formulation of the ECMWF global model, as has been summarized by Cats et al (1987). The horizontal resolution, however, is about 60 km. Table 1 contains a list of the parameters obtained from the LAM with their vertical resolution.

In Figs. 3 until 5, we have given some of the outputs of the present scheme for a grid point near Cabauw, the Netherlands (51°58′N, 4°56′E). The roughness length z_0 is taken as $z_0 = 0.2$ m (Wieringa, 1981). The scheme has been applied both to forecasted and analysed outputs of the LAM. The initial time of the forecasts is April 1, 1989, 00 h. The model has been integrated for 30 h. Every three-hours output is generated, which can be compared with the actual analyses. The analyses are obtained from meteorological data observed around the analysis time and a first-guess field. The latter field results form a 3 h forecast with the model from the previous analysis. This means that the analyses are influenced by the quality of the numerical model (Cats et al, 1987).

Table 1

Vertical resolution of the limited area model (LAM) for the input parameters of the present scheme. Here $\sigma = P/P_o$ (P is pressure, P_o is surface pressure), z is height in standard atmosphere, T_i is air temperature at level i, q_i is specific humidity at level i; U_1 and V_1 are the wind components at the first model level in the atmosphere.

σ-level	Height z (m)	Quantity
1.0	0	T_0, q_0
0.9926	60	T_1, q_1, U_1, V_1
0.9405	490	T_2, q_2
0.8522	1280	T_3, q_3
0.7419	2390	T_4, q_4
0.6214	3800	T_5, q_5

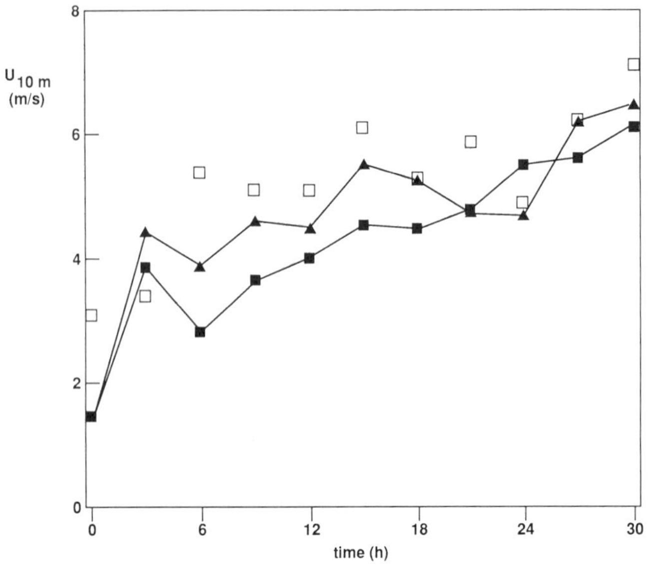

Fig. 3. Forecasts of 10 m wind speed ($U_{10\ m}$) up to 30 h ahead (solid squares), in comparison with analyzed values (triangles) and observations (open squares) at Cabauw for the period of April 1, 00 h - April 2, 06 h, 1989.

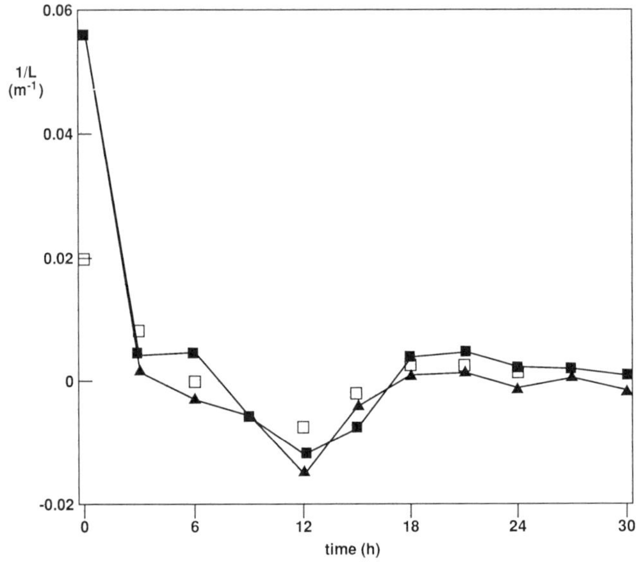

Fig. 4. As Fig. 3, but for the reciprocal Obukhov length ($1/L$).

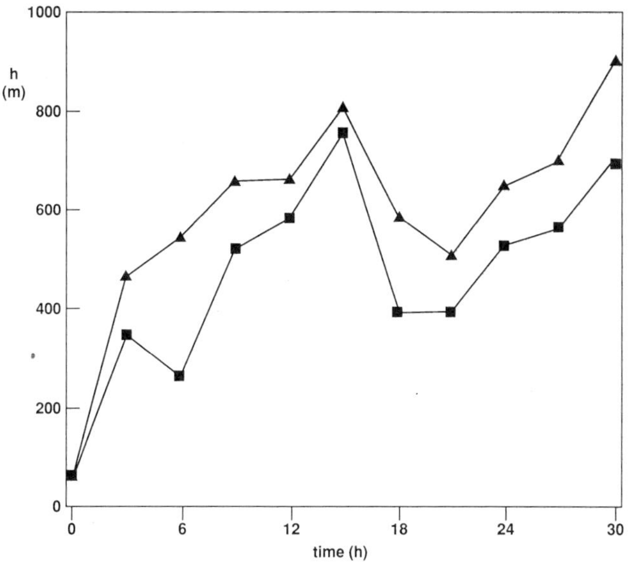

Fig. 5. As Fig. 3, but for the boundary-layer height (h).

In Fig. 3 for the wind speed at a height of 10 m, also the point observations (10-minute averages) at Cabauw are given. Despite of the fact that the forecasts and analyses are representative of a grid square of about 60 × 60 km, the agreement with the point observations is relatively good. The observed wind speed at Cabauw is somewhat larger than the results of the LAM and the scheme for this case. In Fig. 4 the calculated reciprocal Obukhov length is given and compared with the value obtained directly from observations at Cabauw, utilizing the profile method (see Van Ulden and Holtslag, 1985). The agreement with the model predictions and analyses is good. Fig. 5 shows the results of the scheme of section 3 for the boundary-layer height (h). Unfortunately, no direct observations of h were available for comparison. It is seen that h increases at the end of the period, which is caused by an increasing wind speed.

SUMMARY AND DISCUSSION

In this paper we have discussed the derivation of boundary-layer parameters from the outputs of atmospheric models. For a grid point above land, a comparison is given for the outputs of the scheme by using forecasts and analyzed outputs of a limited area model. Also a comparison with point measurements has been presented for the 10 m wind speed and the Obukhov length. Similar findings as presented here, can be obtained for the other surface quantities in Cabauw (Holtslag and Van Westrhenen, 1989). The later report also contains a comparison of calculated boundary layer quantities with observations over the North-Sea. Before definitive conclusions can be drawn, however, it will be necessary to run the scheme for a longer period using the outputs of an atmospheric model. The present scheme may be adjusted based on the outcome of such a study. Also the atmospheric model itself may benefit from such a study, because more insight will be given on the quality of its outputs.

The surface fluxes and the Obukhov length as calculated with the proposed routines may differ from the ones in the atmospheric model for several reasons. First, we may use a more detailed horizontal specification of surface characteristics than is possible and useful within the atmospheric model. In that case, the required model outputs may be interpolated to the specific locations of interest before applying the present scheme. Second, the formulation of the exchange coefficients may be different to correct for errors made in the original model due to low vertical resolution (Delage, 1988).

REFERENCES

Businger, J.A. and S.P.S. Arya, 1974: Height of the mixed layer in the stable stratified planetary boundary layer. Adv. Geophysics, 18A, Academic Press, 73–92.

Carson, D.J. and P.J.R. Richards, 1978: Modelling surface turbulent fluxes in stable conditions. Bound.-Layer Meteor., 14, 67–81.

Cats, G.J., B.J. de Haan, L.M. Hafkenscheid, 1987: Verification of an objective analysis scheme. Sci. Rep. KNMI 87-4, 15 p.

Delage, Y., 1988: A parameterization of the stable atmospheric boundary layer. Boundary-Layer Meteor., 43, 365–381.

Duynkerke, P., 1988: An application of the $E-\epsilon$ turbulence closure model to the neutral and stable atmospheric boundary layer. J. Atmos. Sci., 45, 865–880.

Geleyn, J.F., 1988: Interpolation of wind, temperature and humidity values from model levels to the height of measurement. Tellus, 40A, 347–351.

Holtslag, A.A.M. and F.T.M. Nieuwstadt, 1986: Scaling the atmospheric boundary layer. Bound.-Layer Meteor., 36, 201–209.

Holtslag, A.A.M. and H.A.R. de Bruin, 1988: Applied modelling of the nighttime surface energy balance over land. J. Appl. Meteor., 22, 689–704.

Holtslag, A.A.M. and A.C.M. Beljaars, 1989: Surface flux parameterization schemes; Developments and experiences at KNMI. Proceedings: "Parameterization of fluxes over land surface", ECMWF Reading 24-26 Oct. 1988. Also available as KNMI Sci. Rep. 88-6, 1988, 27 p.

Holtslag, A. A. M. and R. M. van Westrhenen, 1989: Diagnostic derivation of boundary layer parameters from the outputs of atmospheric models. KNMI Sci. Rep. 89–04, 26p.

Holzworth, G.C., 1964: Estimates of mean maximum mixing depths in the contiguous United States, Mon. Wea. Rev., 92, 235–242.

Koracin, D. and R. Berkowicz, 1988: Nocturnal boundary-layer height: Observations by acoustic sounders and predictions in term of surface-layer parameters. Bound.-Layer Meteor., 43, 65–83.

Louis, J.F., 1979: A parametric model of vertical eddy fluxes in the atmosphere. Bound.-Layer Meteor., 17, 187–202.

Louis, J.F., M. Tiedtke and J.F. Geleyn, 1982: A short history of the PBL parameterization at ECMWF. Workshop on Boundary Layer Parameterization, 25–27 Nov. 1981, ECMWF, Reading.

Nieuwstadt, F.T.M., 1981: The steady-state height and resistance laws of the nocturnal boundary layer: Theory compared with Cabauw observations. Bound.-Layer Meteor., 20, 3–17.

Troen, I. and L. Mahrt, 1986: A simple model of the atmospheric boundary layer: Sensitivity to surface evaporation. Bound.-Layer Meteor., 37, 129–148.

Van Ulden, A.P. and A.A.M. Holtslag, 1985: Estimation of atmospheric boundary layer parameters for diffusion applications. J. Clim. Appl. Meteor., 24, 1196–1207.

Verver, G.H.L. and M.P. Scheele, 1988: Influence of non-uniform mixing heights on dispersion simulations following the Chernobyl accident. Proceedings: 17th Intern. Techn. Meet. Air Pollution Modelling, Sept. 19–22, 1988, Cambridge (UK).

Wieringa, J., 1981: Estimation of mesoscale and local-scale roughness for atmospheric transport modelling. 11th Intern. Tech. Meet. Air Pollution Modelling. Plenum, p. 279-295.

Zilitinkevich, S.S., 1972: On the determination of the height of the Ekman boundary layer. Bound.-Layer Meteor., 3, 141-145.

DISCUSSION

C.J. NAPPO

How do you define the boundary layer depth during stable conditions? The boundary layer depth means different things to different people. Many air pollution studies assume uniform mixing through the boundary layer depth, but this is clearly not correct during stable conditions. Instead, turbulence and diffusion decrease with height alone the surface. Thus it is not clear that $h = c\left(\frac{u_* L}{f}\right)^{1/2}$ is the correct depth for air diffusion models, i.e. the Gaussian plume.

A.A.M. HOLTSLAG

The boundary layer depth is defined here as the height of the turbulent layer. Obviously, we do not assume that mixing is uniform through this depth in stable conditions. In fact, the height is only used to separate the diffusion in and above the atmospheric boundary layer (ABL).

J.C.R. HUNT

Is there any evidence for the variation of h with Coriolis parameter f? As Deardorff (1969) showed in his L.E.S. the upper layer stratification (N_u) is important (as you mentioned) and also conditions need to be rather steady (on a time scale of order f^{-1}).

A.A.M. HOLTSLAG

Evidence for variation of boundary layer height h with Coriolis parameter f is given by the well-known Ekman solutions for the neutral ABL, and also for the stable ABL (Nieuwstadt, 1984, J. Atmos. Sci., 41, p. 2202-2216). Anyhow, we plan to incorporate the effect of the upper layer stratification in further improvements of the scheme.

A. VENKATRAM

Nieuwstadt and Tennekes (1981, J. Atmos. Sci. 38, p. 1418-1428) showed that the stable boundary layer height is almost completely determined by the initial boundary layer height when the convective boundary layer height collapses. Can you justify the use of a model for the height that neglects this observation?

A.A.M. HOLTSLAG

In the paper by Nieuwstadt and Tennekes it is shown that the boundary layer height is governed by rate equations. As you mention, it appears that the solution for h is very sensitive to the initial condition. The above authors use an observation value for the latter, while for our purpose a reasonable *estimate* is needed. Unfortunately, it is not known how to obtain such an estimate for the initial condition. Moreover, in a more recent paper Nieuwstadt (1984, J. Atmos. Sci., 41, p. 2202-2216) showed that our Eq. (8) did well in stationary conditions for the location of Cabauw, the Netherlands.

R. BROST

Using a steady state equation for the height of the stable boundary layer (equation 8) should only perform well above flat, uncomplicated surfaces, well away from a change in surface conditions. Although equation 8 may verify well over the ocean and the Netherlands, I would be very surprised if C_2 was not location dependent, varying up and down by at least a factor of 2 for even modest terrain (Brost and Wyngaard, JAS, 1978).

A.A.M. HOLTSLAG

As stated above, Eq. (8) has been shown to work well for the Cabauw cases in relatively homogeneous terrain. I agree that the coefficient is uncertain (as discussed in the paper), and that it also might be site dependent. However, at this moment we do not have an acceptable alternative for our purposes.

LONG RANGE TRANSPORT OF HEAVY METALS FROM POLAND COMPUTED BY AN EULERIAN MODEL

Jerzy Bartnicki*

International Institute for Applied Systems Analysis
Laxenburg, Austria

INTRODUCTION

The modeling activity on the long range transport of air pollutants has been mainly focused on sulfur and nitrogen oxides for the past several years. Compared to sulfur and nitrogen, much less development can be noticed in modeling long range transport of heavy metals. The existing models were designed either for computing transport of heavy metals into the Baltic and the North Sea (Grasl et al., 1989; Krell and Roeckner, 1988; Jaarsveld et al., 1986), or to the Arctic (Pacyna and Ottar, 1985; Pacyna et al., 1985, 1989). Except Jaarsveld's climatologic approach and the stochastic model of Krell and Roeckner, the above models are of Lagrangian type based on the trajectory computations. Another climatologic model (Alcamo et al., 1990) was designed to compute long range transport of heavy metals in Europe but because of the resolution of the climatologic data, there might be some difficulties to run it for the transport period shorter than one year.

The main purpose of developing the model presented in this paper is the computation of the transboundary atmospheric transport of heavy metals in Europe. This model can be applied both for the episodes and annual transport of metals. Also, because of the Eulerian approach and positive definite pseudo-spectral (PDPS) numerical algorithm which does not produce negative values, it can be potentially used for modeling atmospheric transport of mercury, the case when re-emission becomes a serious limitation for the applications of the trajectory models.

In this study, the model was used to compute long range atmospheric transport of As and Pb emitted from the sources located in Poland. It has been run for four months of 1985: January, April, July and October which represented four seasons of 1985. The main goals of these runs were: (1) organize a computer tool for estimating atmospheric transport of heavy metals from Poland, and (2) to investigate the influence of seasonal variation of the meteorological conditions on the transport pattern.

MODEL DESCRIPTION

Atmospheric long range transport of As and Pb from Poland is simulated by the two-dimensional Eulerian model. Basic transport equation for As and Pb takes into account emission, advection, and dry plus wet deposition of the metals. It has the same form for As and Pb:

$$\frac{\partial c_i}{\partial t} + u\frac{\partial c_i}{\partial x} + v\frac{\partial c_i}{\partial y} = -(\frac{v_{di}}{h} + \frac{P\,W_i}{h})\,c_i + (1-\alpha_i)\,Q_i \qquad (1)$$

*On leave from the Institute of Meteorology and Water Management in Warsaw, Poland.

Air Pollution Modeling and Its Application VIII, Edited by H. van Dop and D. G. Steyn, Plenum Press, New York, 1991

where subscripts i=1,2 refer to As and Pb variables, respectively. In equation (1): $c_i = c_i(x,y,t)$ is the concentration; $u(x,y,t)$ and $v(x,y,t)$ are the components of the 925 hPa wind field; v_{di} is the dry velocity deposition; h=$h(x,y,t)$ is the mixing height; P is the precipitation amount for the last 6 hours; W_i is the scavenging ratio; Q_i is the emission and α_i is a part of the emission which is deposited at the same grid where emission occurs. The values of the dry deposition velocity, scavenging ratio and α are given in Table 1.

Table 1. Parameter values for As and Pb.

Metal	Parameter		
	v_d $(m\,s^{-1})$	W (*unitless*)	α (*unitless*)
As	0.002	500000	0.10
Pb	0.002	500000	0.15

The velocity and precipitation fields are updated every 6 hours during the model runs. Mixing height is kept constant for 24 hours and updated at 1200 GMT.

The advective part of the transport equation (1) is solved by means of the PDPS method (Bartnicki, 1986; 1989). The advantage of PDPS method over the pseudospectral approach is the elimination of artificial negative values from the solution. This method is also more accurate compared to another positive definite schemes, like for example, Spline, Smolarkiewicz, Finite Element and Flux Corrected algorithms (Bartnicki et al.,1990).

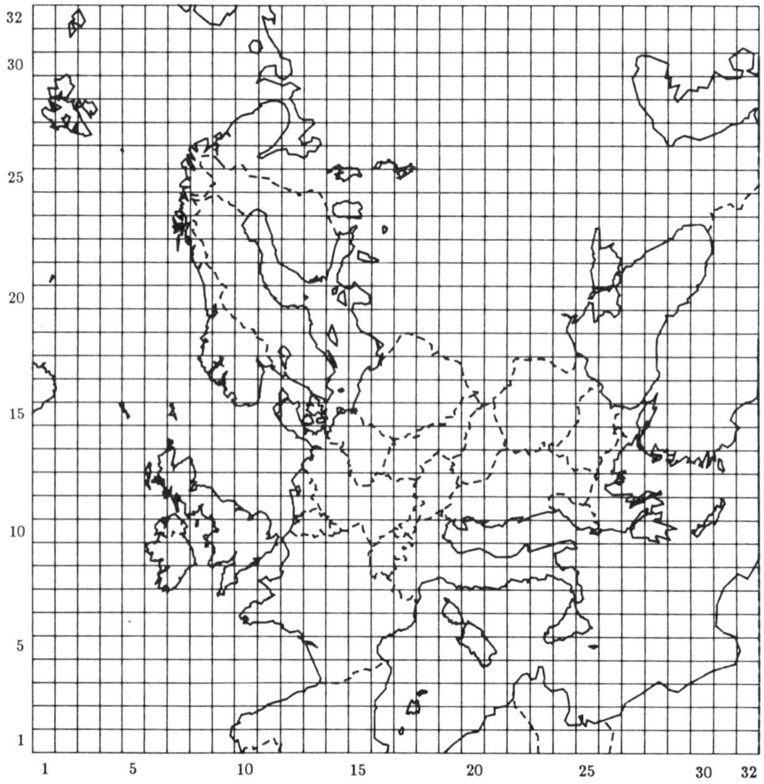

Fig. 1. Area of the model computations.

The transport equation is solved numerically in the grid shown in Figure 1 which is a part of the EMEP mesh for Europe. Grid size is 150 × 150 km and there are 32 points in each direction. Because of the complex form of Fast Fourier Transforms (FFT) concentrations of As and Pb are calculated simultaneously in the model runs.

Two types of the boundary conditions were compared in the computations. The first type proposed by Christensen and Prahm (1976) results in dumping concentrations in the outflow points with the efficiency dependent on the outflow wind speed. The second type assumes a simple constant 90% dumping of the boundary concentrations in the last two rows and columns of the model grid. An optimal factor of 90% has been selected after a series of the numerical experiments. Such a dumping may be interpreted as a continuous precipitation in the boundary region. The second type of the boundary conditions was implemented in the model because compared to the first type : (1) it was more effective in dumping the outflow mass and (2) reflections from the boundaries were much smaller.

The FORTRAN code of the model is comprehensive and does not require any special computer. It can be easily executed on a standard PC, in which case, simulation of As and Pb transport for one month period requires approximately one hour for 12 MHz processor.

METEOROLOGICAL AND EMISSIONS DATA

There are three meteorological input fields for the model: precipitation, mixing height and 925 hPa wind. These fields averaged over winter (January) and summer (July) months are presented in Figure 2. This picture clearly shows the variability of the meteorological conditions in the model computations. The maxima and minima for each field occur at different geographical regions in January and July, and the shape of the isolines is also different for these months. Seasonal variation of the meteorological conditions is also confirmed in Table 2 where the averages over each month and the entire model area are given. Especially, periodic changes over the year can be noticed for average values of the mixing height and wind speed. However, maximum and minimum of the mixing height corresponds to minimum and maximum of the wind speed, respectively.

Table 2. Meteorological fields averaged over each month and model area.

Meteorological Field	Month			
	January	April	July	October
Precipitation (mm)	45.1	34.3	37.2	36.3
Mixing height (m)	755.3	1304.5	1400.4	1081.7
Wind speed (m)	11.5	10.8	7.8	10.7

Emissions of As and Pb from Poland in 1985 were calculated based on the inventory of 75 large stationary point sources in Poland and in addition mobile sources distributed according to the population density (Pacyna and Munch, 1988; Pacyna, 1988). The total emission distributions of As and Pb in Poland are presented in Figure 3. Both for As and Pb, there is a steep maximum of the emissions near to the Katowice region in the south of Poland.

RESULTS OF THE MODEL RUNS

Before reporting on the results of the model runs, it is necessary to mention that the concentrations and depositions computed by the model were not compared to the observations in this study. The reason, in the case of As, was a lack of measurement for the computed transport periods. In addition, both for As and Pb, there was a problem with the spatial distribution of emissions. However, estimation of the total As and Pb emissions should not be too far from the real values, the spatial distribution of these emissions is still preliminary. Some additional As and Pb deposition values in Scandinavia can be calculated from

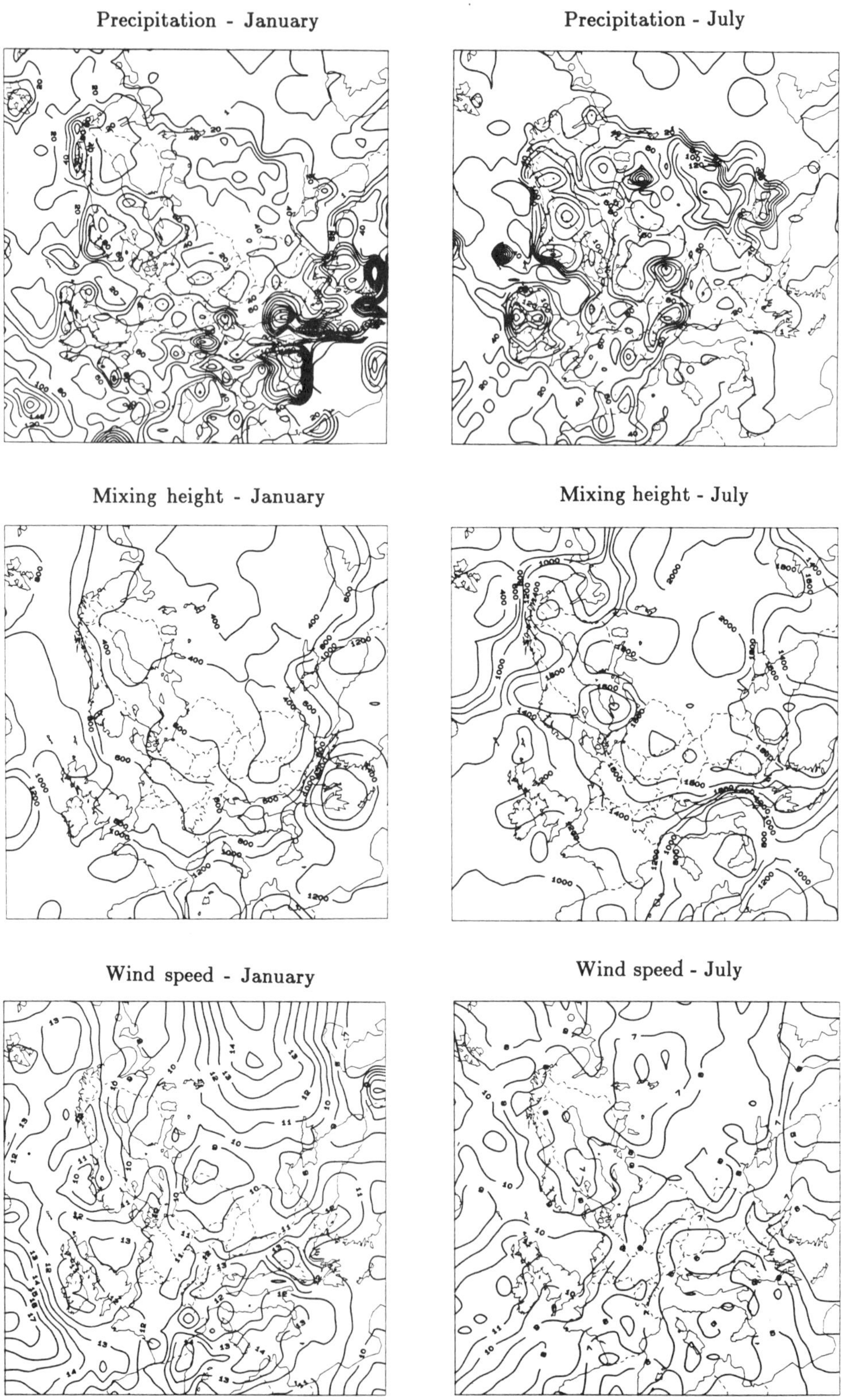

Fig. 2. Isolines of input meteorological fields averaged over January and July.

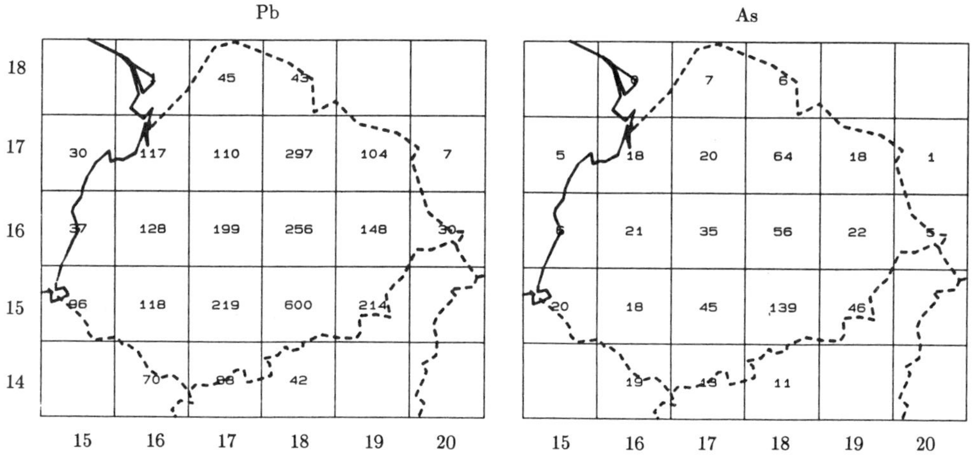

Fig. 3. Emissions of As and Pb from Poland in tonnes per year.

the metals content in the moss (Rühling et al., 1987). Comparison of the calculated and measured depositions will be the first task after the improvement of emission inventories for Europe and completion of the moss data base.

Total depositions of As in Europe due to emissions in Poland are shown in Figure 4 for each month of the computations. In January, two maxima of the depositions can be observed in Poland, the stronger of them exceeding 90 $\mu g\ m^{-2}$ as a monthly averaged value. This maximum is associated with relatively low value of the monthly average mixing height for January: 500–600 m, visible in Figure 2. For other months (seasons), the maxima of the total As deposition also occur in Poland but values are much lower – 20–30 $\mu g\ m^{-2}$. In addition to the strong maximum in Poland, there are many local deposition maxima for each month caused by the irregular pattern of precipitation (Figure 2).

Since two out of three model parameters are identical for As and Pb, the similar results can be expected for these metals. This is confirmed by Table 3 which gives monthly average concentrations and depositions for As and Pb. Concentrations and dry depositions of both metals show very similar pattern of the seasonal variation with maximum in January, and minimum in April. Wet deposition of As and Pb has also maximum in January but minimum in October. Difference between maximum and minimum are much higher for concentrations and dry depositions than for wet and total depositions.

Table 3. Mean values (over month and entire model grid) for concentrations and depositions of arsenic.

Metal	Computed variable	Month			
		January	April	July	October
As	Concentration (ngm^{-3})	0.22	0.08	0.09	0.19
	Dry deposition ($\mu g\ m^{-2}$)	1.15	0.39	0.49	1.02
	Wet deposition ($\mu g\ m^{-2}$)	1.48	1.12	1.40	0.93
	Total deposition ($\mu g\ m^{-2}$)	2.63	1.51	1.89	1.95
Pb	Concentration (ngm^{-3})	1.04	0.38	0.44	0.92
	Dry deposition ($\mu g\ m^{-2}$)	5.46	1.90	2.28	4.86
	Wet deposition ($\mu g\ m^{-2}$)	7.15	5.26	6.66	4.51
	Total deposition ($\mu g\ m^{-2}$)	12.61	7.16	8.94	9.37

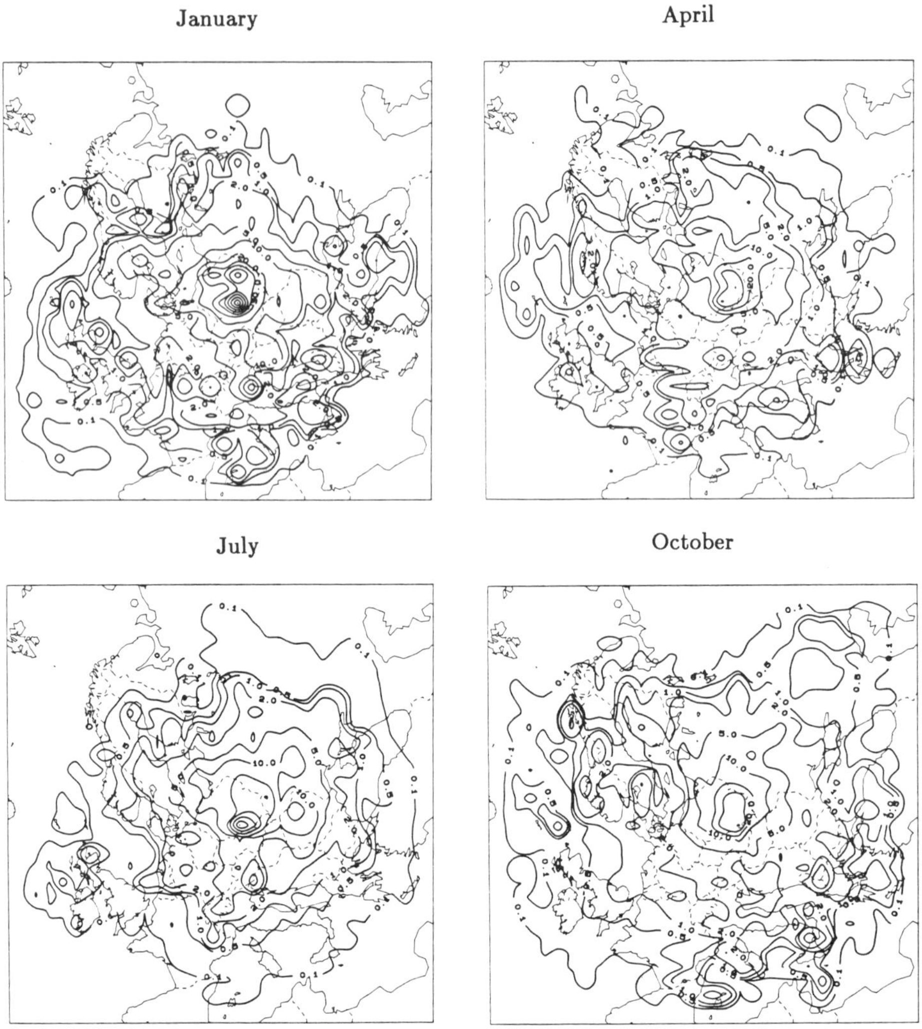

Fig. 4. Total (dry + wet) monthly deposition of As in Europe due to emissions in Poland. Units – $\mu g\, m^{-2}$.

The ratio of maximum to minimum in Table 3 is the highest for the dry deposition: 2.95 - As, 2.87 - Pb, and the lowest for the wet deposition: 1.32 - As, 1.36 - Pb. The isolines of total Pb deposition presented in Figure 5 show also similar pattern to the isolines of As total deposition (Figure 4).

As and Pb emitted in Poland are mainly deposited outside the country. Therefore, the model was used to estimate transport of heavy metals to the neighboring countries and the Baltic Sea. The results of these computations for each of the four months are presented in Table 4. Most of the As and Pb exported to USSR in July is 22.8% and 22.2%, respectively. The smallest amount of Polish emissions is deposited in the GDR, also in July in the case of Pb - 0.6%, and in October for As - 0.6%. negative correlation of the transport to USSR and GDR is consistent with geographical locations of these countries and meteorological conditions during transport. The export of As and Pb from Poland is strongly dependent on the season. Minimum occurs in January when 73.1% of As and 70.4% of Pb emitted in Poland travels outside the country. In April, atmospheric transport of metals from Poland is the most efficient - 78.1% of As and 74.5% of Pb deposited outside country. In average for four months, the export of As - 73% and Pb - 70% is approximately the same as annual average export of sulfur in 1985 - 70% (Eliassen et al., 1988).

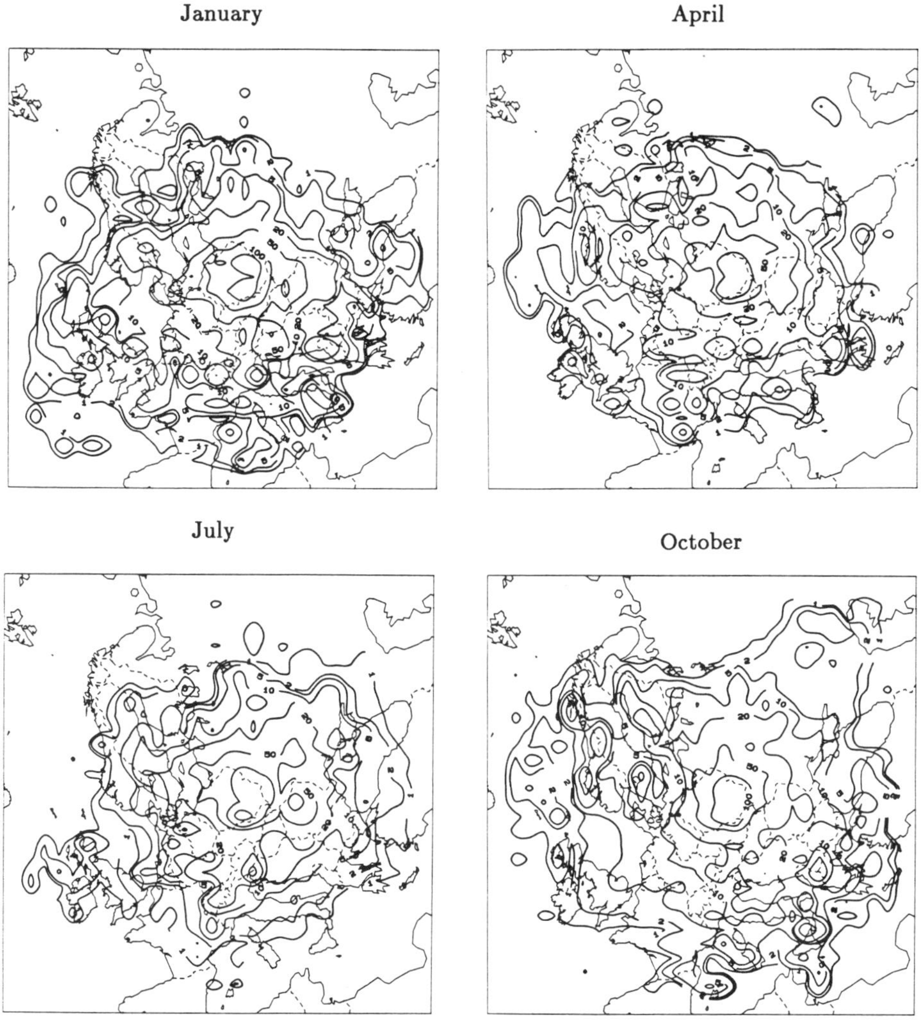

Fig. 5. Total (dry + wet) monthly deposition of Pb in Europe due to emissions in Poland. Units - $\mu g\, m^{-2}$.

Table 4. Transport of As and Pb from Poland to selected regions in Europe (in % of the total Polish emissions).

Deposition region	Month							
	January		April		July		October	
	As	Pb	As	Pb	As	Pb	As	Pb
Poland	36.9	39.6	21.9	25.5	24.4	28.1	24.1	27.8
Czechoslovakia	5.9	5.3	2.7	2.5	3.6	3.2	3.1	2.7
GDR	1.8	1.8	0.7	0.7	0.6	0.6	0.7	0.7
Baltic Sea	4.7	4.6	2.6	2.6	3.1	3.1	1.7	1.7
USSR	14.0	13.6	16.3	15.7	22.8	22.2	20.2	19.8

SUMMARY AND CONCLUSIONS

The Eulerian long range transport model of As and Pb has been described in the paper. This model was applied to the computations of the atmospheric transport of As and Pb emitted from Poland. Calculations have been performed for four months of 1985 representing four seasons of this year. The main purpose of the study was: (1) to develop a computer tool for the estimation of transboundary transport of heavy metals emitted in Poland, and (2) to evaluate the variation of the concentrations and depositions of these metals due to different meteorological conditions in each season of 1985. The following conclusions can be drawn from the model runs:

1. All meteorological fields vary significantly from one season to another. This variability is visible both in the spatial distribution and averaged values over each month and the entire model grid.
2. Variability of the meteorological fields is also reflected in the concentration and deposition fields for As and Pb which are different for each season. Maximum of the As and Pb concentrations and depositions occurs in January, minimum in April.
3. There is a strong negative correlation between monthly mean mixing height and monthly mean concentration. Maximum of average As and Pb concentrations in January corresponds to minimum of the average mixing height in this month. The reverse situation occurs in April and July when average mixing height is close to maximum value and concentrations of As and Pb are at the lowest level.
4. Dry deposition of As and Pb may exceed the wet deposition (for example, in the case of October mean values). Therefore, for further studies it is important to implement a better parametrization of the dry deposition.
5. Meteorological variability has a clear impact on the long range transport of heavy metals emitted from Polish sources to selected regions in Europe. In the case of Polish neighbouring countries and the Baltic Sea, most of the As and Pb is exported to USSR (22.8%, 22.2%) and the smallest amount to GDR (1.8%, 0.6%). Seasonal pattern of the transport to USSR and GDR is negatively correlated which is consistent with geographical and meteorological conditions.
6. On average, long range transport of As - 73% and Pb 70% from Poland is of the same order as the transport of sulfur - 70% in 1985.

The Eulerian model described here is relatively efficient and easy to use from the numerical point of view. However, the major drawback of the present study is a lack of the systematic comparison between the model results and measurements. This will be the first priority task for the future work.

REFERENCES

Alcamo, J., Bartnicki, J., Olendrzynski, K., 1990, Modeling heavy metals in Europe's atmosphere: A combined trajectory-climatologic approach. This volume.

Bartnicki, J., 1986, An efficient positive definite method for the numerical solution of the advection equation. IIASA Working Paper WP-86-35. International Institute for Applied Systems Analysis, Laxenburg, Austria.

Bartnicki, J., 1989, A simple filtering procedure for removing negative values from numerical solutions of the advection equation. Environmental Software, 34:187–207.

Bartnicki, J., Olendrzynski, K., Abert, K., Seibert, P., and Morariu, B., 1990, Numerical approximation of the transport equation: Comparison of five positive definite algorithms. IIASA Working Paper WP-90-10. International Institute for Applied Systems Analysis, Laxenburg, Austria.

Christensen, O., and Prahm, L.P., 1976, A pseudospectral model for dispersion of atmospheric pollutants. Journal of Applied Meteorology, 15:1284–1294.

Eliassen, A., Hov, O., Iversen, T., Saltbones, J., and Simpson, D., 1988, Estimates of airborne transboundary transport of sulphur and nitrogen over Europe, EMEP/MSC-W Report 1/88:1–79.

Grasl, H., Eppel, D., Petersen, G., Schneider, B., Weber, H., Gandras, J., Reinhardt, K.H., Wodarg, D., and Flies, J., 1989, Stoffeintrag in Nord- und Ostsee über die Atmosphäre. GKSS Report 89/E/8. GKSS-Forschungszentrum Geesthacht GmbH, Hamburg, FRG.

Jaarsveld, J.A., van Aalst, R.M., and Onderdelinden, D., 1986, Deposition of metals from the atmosphere to the North Sea:model calculations. National Institute for Public Health and Environmental Protection (RIVM), Bilthoven, The Netherlands. Report 842015002.

Krell, U., and Roeckner, C., 1988, Model simulation of the atmospheric input into the North Sea. Atmospheric Environment, 22:375–381.

Pacyna, J.M., 1988, Atmospheric lead emissions in Europe in 1985. NILU Report OR 18/88. Norwegian Institute for Air Research. Lillestrøm, Norway.

Pacyna, J.M., Ottar, B., Tomza, U., and Maenhaut, W., 1985, Long-range transport of trace metals to Ny Alesund, Spitsbergen. Atmospheric Environment, 19:857–856.

Pacyna, J.M. and Ottar, B., 1985, Transport and chemical composition of the summer aerosol in the Norwegian Arctic. Atmospheric Environment, 19:2109–2120.

Pacyna, J.M. and Munch, J., 1988, Atmospheric emissions of Arsenic, Cadmium, Mercury and Zinc in Europe in 1985. NILU Report OR 17/88. Norwegian Institute for Air Research. Lillestrøm, Norway.

Pacyna, J.M., Bartonova, A., Cornille, P., and Maenhaut, W., 1989, Modelling of long-range transport of trace elements. A case study, Atmospheric Environment 23:1:107–114.

Rühling, A., Rasmussen, L., Pilegaard, K., Mäkinen, A., and Steinnes, E., 1987, Survey of atmospheric heavy metal deposition, Report prepared for The Steering Body for Environmental Monitoring, The Nordic Council of Ministers. NORD 1987:21:1–44.

ACKNOWLEDGEMENTS

The author wish to acknowledge Jørgen Saltbones from the Norwegian Meteorological Institute for his help with the meteorological data and useful discussions. I am also indebted to Josef Pacyna Norwegian Institute for Air Research, who discussed with me emission data for Poland and Joseph Alcamo from International Institute for Applied Systems Analysis who reviewed this manuscript. A part of this study has been done at the Norwegian Institute for Air Research (NILU) under grant from Norwegian Ministry of Environment.

DISCUSSION

A. VENKATRAM

I noticed that the average values of wind speed (> 10 m/s) and mixed layer height (> 1200 m) were large compared to those we observe in the United States. Can you comment?

J. BARTNICKI

The 950 hPa wind data used in the model runs were prepared by EMEP MSC-W Center in Norway based on the observed wind speed over the land and the results of the Norwegian weather prediction model over the Ocean. Since the wind speed values over the sea are significantly higher than over the land, and approximately one third of the model area is covered by water, the average wind speed over the model grid is higher than the speed over the land. This is the reason why the computed average wind speed is higher than the observed in the United States over the land.

Regarding mixing height, the average value is strongly dependent on the season. It is higher than 1200 m during the summer and significantly lower - 755 m - during the winter.

H.J.A. VAN JAARSVELD

Is your calculated dry deposition flux not sensitive to the value of the α-factor as chosen, expecially in areas with high emissions?

Generally, dry deposition flux is sensitive to the α factor close to the emission sources. For the longer distances, wet deposition processes become more important and the dry deposition flux is less sensitive to the local deposition factor α.

TIME EVOLUTION OF BREEZE CIRCULATION IN ALPINE VALLEYS

G. Clerici (**), R. Salerno (**) and S. Sandroni (*)

(**) ARS S.p.A. (ENI group)
Via Medici del Vascello 26
I-20138 Milan
(*) Institute of Environment
Joint Research Centre
I-21020 Ispra

1. Introduction

Within the framework of EUROTRAC, dispersion and transport of air pollutant over complex terrain have been recognized as important research topics. As a part of TRACT (TRAnsport over Complex Terrain), a Eurotrac subproject, a research proposal named TRANSALP is included; it concerns the possible transport of atmospheric trace constituents across the Italian-Swiss Alps from the Po valley to the Swiss plateau and viceversa. The Italian-Swiss Alps include some of the highest peaks in the Alps and its average elevation is 3500 m. Actually little is known on the dynamics of the planetary boundary layer over such a barrier. Some episodes, however demonstrated the feasibility of a transalpine transport: the Saharian dust found in the Rhein valley and the radioactive fall-out after the Chernobyl accident on both slopes of the Alps are two examples. On a smaller scale the foehn wind shows that north-south oriented valleys play an important role in the exchange of trans-alpine air masses. Basically three mechanisms of transport can be postulated, namely (1) the injection from the PBL into high-level geostrophic winds, (2) the channeling along north-south oriented valleys and (3) the convective flow over the mountain ridges connected with the PBL coupling between adjacent valleys. To approach the TRANSALP project preliminary experiments based on meteorological and chemical measurements and tracer releases have been performed in south-Alpine valleys.

A first experiment was carried out on July 1988: a tracer (C_7F_{14},perfluoromethylcyclohexane) was released under the condition of a cold prealpine lake and a well developed anabatic wind on the sunny south-oriented slope of a mountain (Campo dei Fiori, 1226 m). The height difference between the lake surface and the mountain ridge is about 800 m (Gaglione et al.,1988). The atmospheric conditions were very unstable and winds were very light (few $m \cdot s^{-1}$). Air samples were taken at several points downwind around the lake, on the mountain ridges and in the valley (Valganna) behind the mountain. Measurements of tracer concentration in air samples taken at half-an-hour intervals at ground level and at altitude by an helicopter showed that the effect of the cold lake surface delays the arrival of the tracer plume at the downwind lake-shore; a strong channeling effect on the plume is created by the mountain and the tracer concentration on the mountain ridge and leewards is affected by the strong vertical wind component before the ridge. An attempt to simulate the dispersion by a segmented plume model failed completely (Gaglione et al., *loc. cit.*).

This paper describes the development and application of a time-dependent Eulerian model for this experiments and the comparison between computed and measured ground level tracer concentration.

2. Wind field reconstruction

Air masses circulation in breeze wind conditions, frequent in this area (Giuliacci,1985; Ambrosetti et al.,1987), has been simulated by a tridimensional time-dependent Eulerian model based on Navier-Stokes equations completed with the equation of pollutant transport. A stationary model is used for the initialization of the set of equations. Starting from available meteorological data, a steady tridimensional wind field has been reconstructed by solving the continuity equation in which the effects of orography, surface thermal gradients, atmospheric stability end energy of air masses have been introduced (Clerici and Sandroni.,1986). This set of values, together with the initial distribution of temperature and pollutant concentration, allow us to solve the time-dependent equations if the boundary conditions are provided. Particular care must be given to temperature in the energy equation responsible for a correct description of breeze rotation. Katabatic and anabatic winds are well described by this model, which can also be applied to the evaluation of the deposition of chemical species.

2.1 The model

The initial model describes the static tridimensional mesoscale wind fields over orographically complex regions. It takes into account two phenomena typical for the local and regional scale, i.e. the orographic forcing and the heath islands (such as lakes and urban areas), and two fundamental characteristics of air masses flow, namely the stability of air masses in the flow and their energy. Orography, temperature gradients, stability and energy parameters are inserted into the balance equations which constitute the physical basis of the model (Clerici et al.,1988). The value of the temperature field and a first coarse representation of the horizontal wind field are obtained in each grid cell by an orography-considering interpolation.

A next step is the calculation of vertical velocities, considering the vertical deflections of the horizontal flow due to orography and thermal gradients. The extension of the vertical motion to the upper air layers is linked to the atmosphere stability. This term is provided by Richardson number as follows

$$w(z_0 + dz) = w(z_0) \cdot e^{-|Ri|^{1/2} dx}$$

where w is the vertical velocity, dz is the vertical raising and Ri is the average value of the Richardson number in the layers up to dz.

The vertical wind field has to be corrected in order to minimize the global mass flux through the external boundary of the considered region. The final horizontal wind field is computed solving the following approximation of the not-evolutive continuity equation:

$$\mathrm{div}(\vec{v}) + \frac{w}{\rho_0}\frac{\partial \rho_0}{\partial z} = 0$$

where ρ_0 is the average horizontal density. If we define

$$u = \frac{\partial \Phi}{\partial x} \qquad v = \frac{\partial \Phi}{\partial y}$$

we obtain a Poisson equation:

$$\nabla^2 \Phi = -\frac{\partial w}{\partial z} - \frac{w}{\rho_0}\frac{\partial \rho_0}{\partial z}$$

The average density is computed from the perfect gas state equation.

2.2 Wind fields

Tracer experiments were performed on three days (12, 19 and 21 July 1988). Wind fields and ground concentrations of tracer were reconstructed for the three days; we limit here the analysis to the 21st July experiment.

21st July was a sunny day (29 C at ground level at the release time) with a typical haze for summertime in the Po valley. Heat flux and convection were normal for the season. From noon to 6 p.m. Pasquill's stability class varied from A to B (Hasenjäger,1989). Wind direction over area varied from SW in the upper layers to SE-SW near the ground. The reconstructed wind fields were obtained by the following procedure.

A *first step* was the reconstruction of the wind field over a large area includes the western Alps. The chosen area (Fig. 1) was limited by Bern at the North, Geneva at the West, Turin at the South and Milan at the SE. The area was subdivided by a grid with a horizontal mesh of 5 Km, corresponding to 56x42 cells units at the ground; 6 layers of variable thickness were chosen in the vertical direction. The meteorological data were available from the three WMO sounding stations and from ca. 40 ground stations operating in the region, distribuited at altitudes ranging from 100 to 3500 m a.s.l.. The model reconstructed the wind field by interpolation of available data.

In a *second step* the tridimensional wind fields obtained from the first step were applied to a smaller area included in the previous one. This new area is limited to the Ticino basin (Fig. 1). It was subdivided by a finer grid with a horizontal mesh size of 1 Km. In this area a relatively large numbers of Swiss meteorological stations were present, which allowed us a better spatial resolution of the reconstructed wind field in that area.

In the *final step* we reconstructed the wind fields over the *Campo dei Fiori* area (Figs. 1 and 2). The final horizontal grid mesh used was 500 m for a total area of 20x20 square kilometers. Six vertical layers were always considered. The reconstructions were possible by the use of wind and temperature fields previously obtained in the second step and used as boundary conditions and by introducing the supplementary meteorological measurements in the test area and at the JRC Ispra.

By this procedure wind and temperature fields for the *Campo dei Fiori* area were generated. Figs 3,4,5,6,7,8,9,10 show the wind fields for some layers reconstructed for 21st July at 12 and at 15 GMT.

3. Evaluation of chemical species deposition

The reconstructed wind fields are a necessary input for the evaluation of trace gases deposition at ground level. At each node of the grid mesh we assigned altitude, roughness, wind speed and direction, stability and all other parameters necessary for the computation. We employed a gaussian-type model (Hanna,1986;Pasquill,1983) which computes concentrations by taking into account all parameters mentioned above (Manzi and Salerno,1990). In particular orography and vertical transport are considered (Salerno et al.,1988), so it is able to describe the effects of deflection or climbing of air masses over a mountain crest. Fig. 11 gives an example of computed tracer deposition on 21st July 1988 at 14 GMT in the *Campo dei Fiori* area.

The model is static and in its original form is unable to solve the tridimensional transport equations. In order to simulate the transport of the tracer, we moved the source point along the streamlines, taking into account the depletion of the tracer plume due to deposition at each step. As a result of these calculations, the evidence of two concurrent effects is obtained, i.e. the tracer is partly transported around the mountain and partly climbs over it, both processes contributing to a deposition in the nearby Valganna valley.

From a qualitative point of view, both mechanisms are well described. A quantitative evaluation and a comparison between observed and computed tracer deposition is difficult, due to the intrinsic limitation of the model and to the low

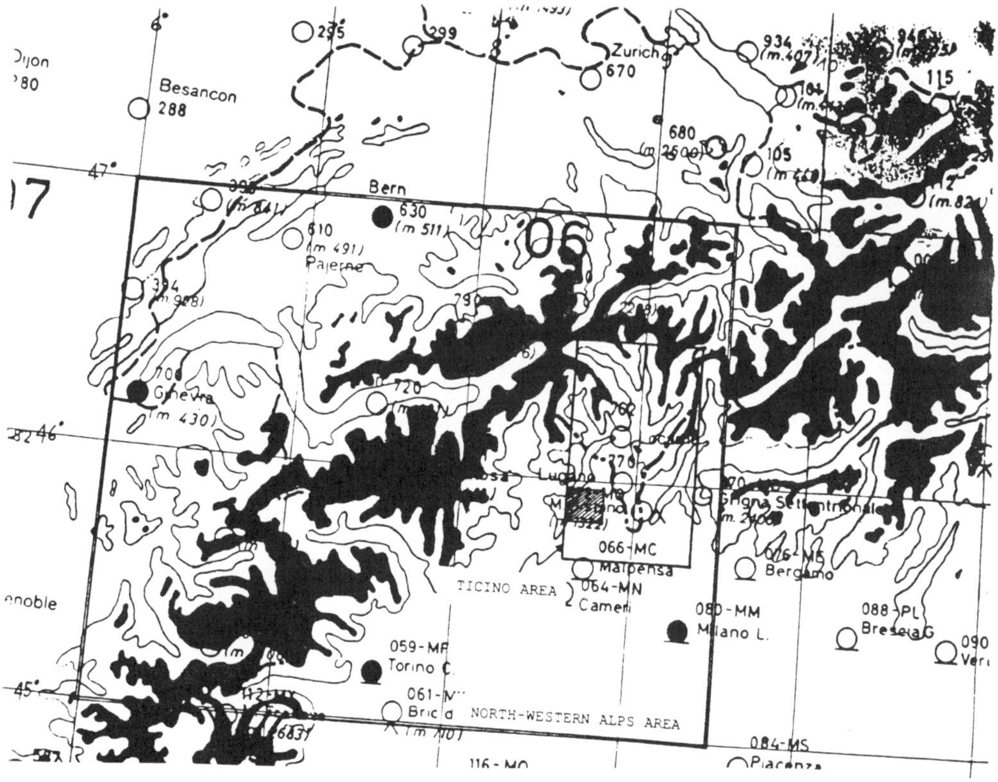

Fig. 1. North-Western Alps and Ticino area;the small square is Campo dei Fiori region.

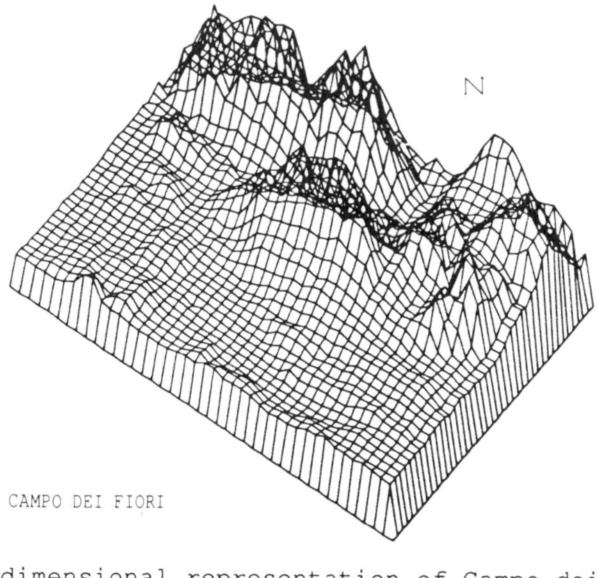

Fig. 2. Tridimensional representation of Campo dei Fiori area.

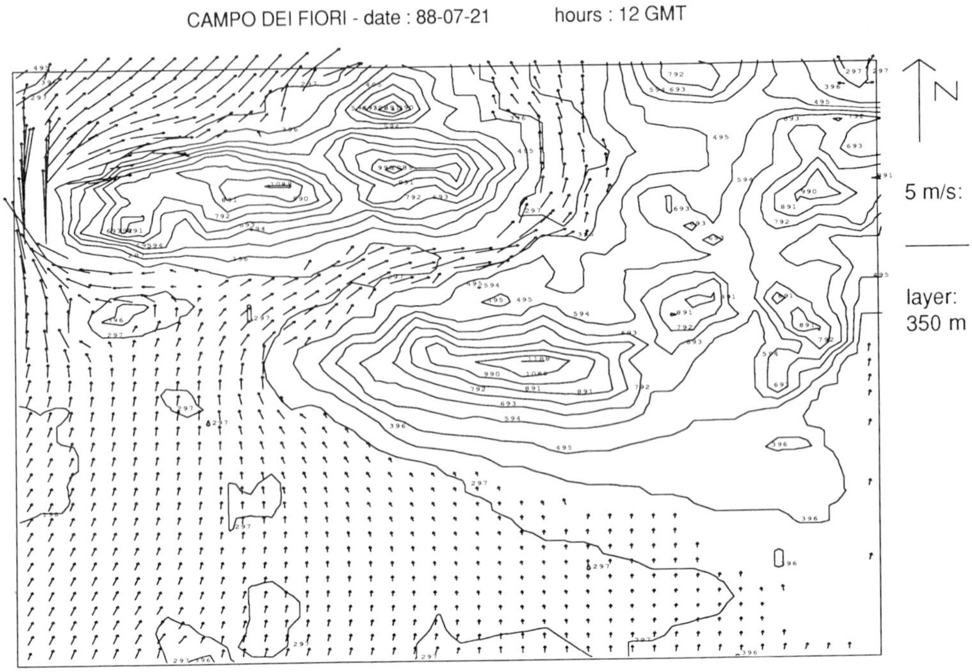

Fig. 3. Wind field at 12 GMT 21st July 1988 at 350 m. a.s.l.

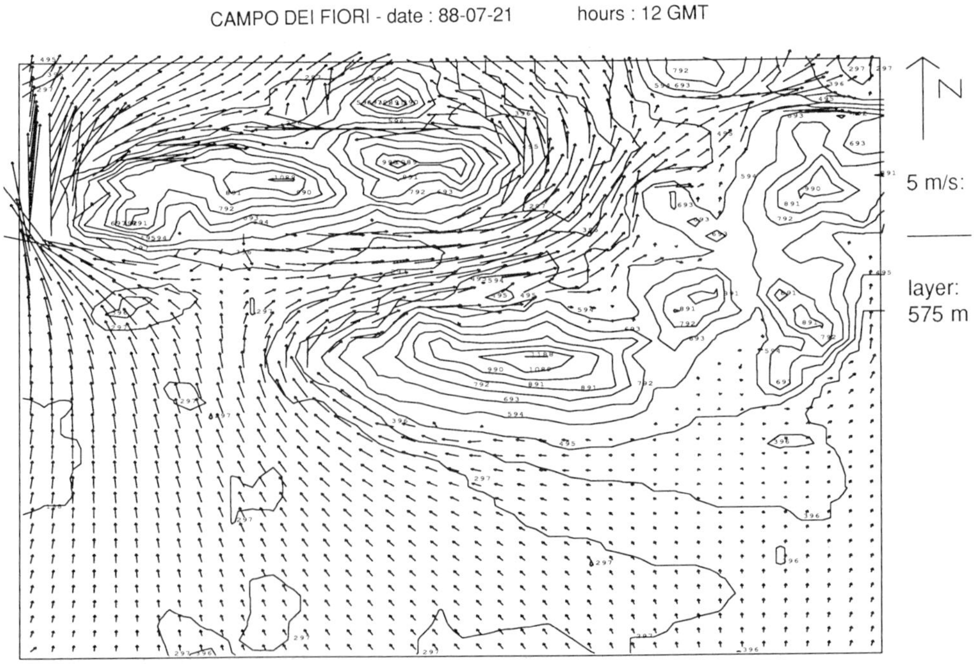

Fig. 4. Wind field at 12 GMT 21st July 1988 at 575 m. a.s.l.

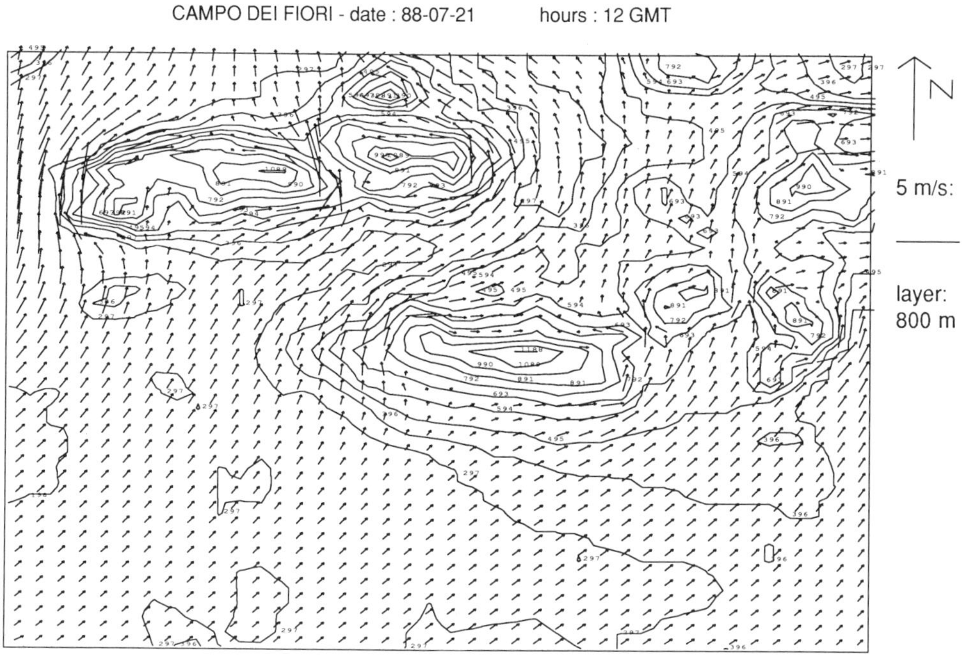

Fig. 5. Wind field at 12 GMT 21st July 1988 at 800 m. a.s.l.

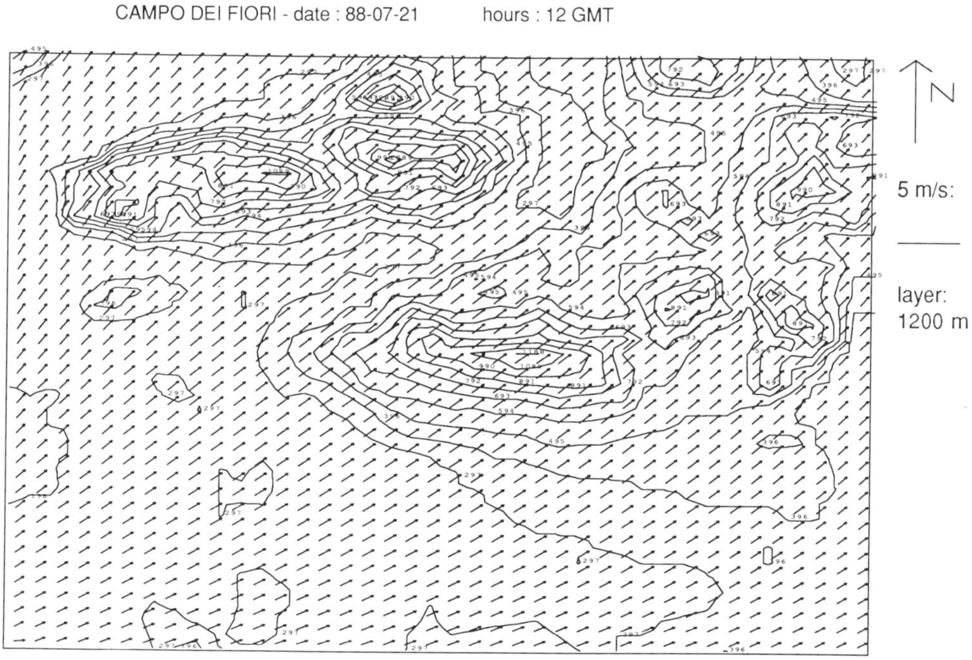

Fig. 6. Wind field at 12 GMT 21st July 1988 at 1200 m. a.s.l.

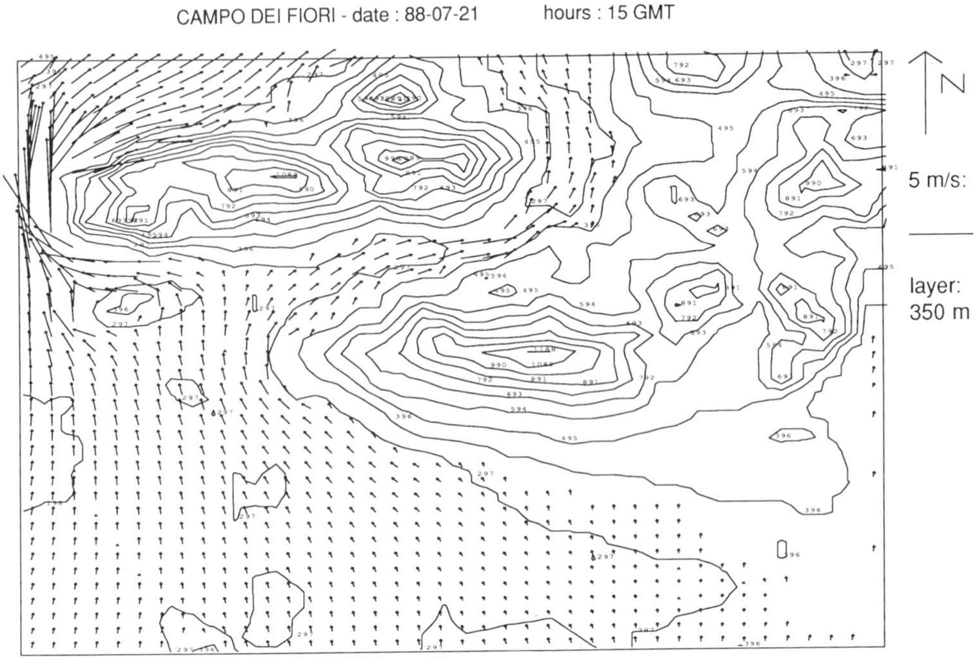

Fig. 7. Wind field at 15 GMT 21st July 1988 at 350 m. a.s.l.

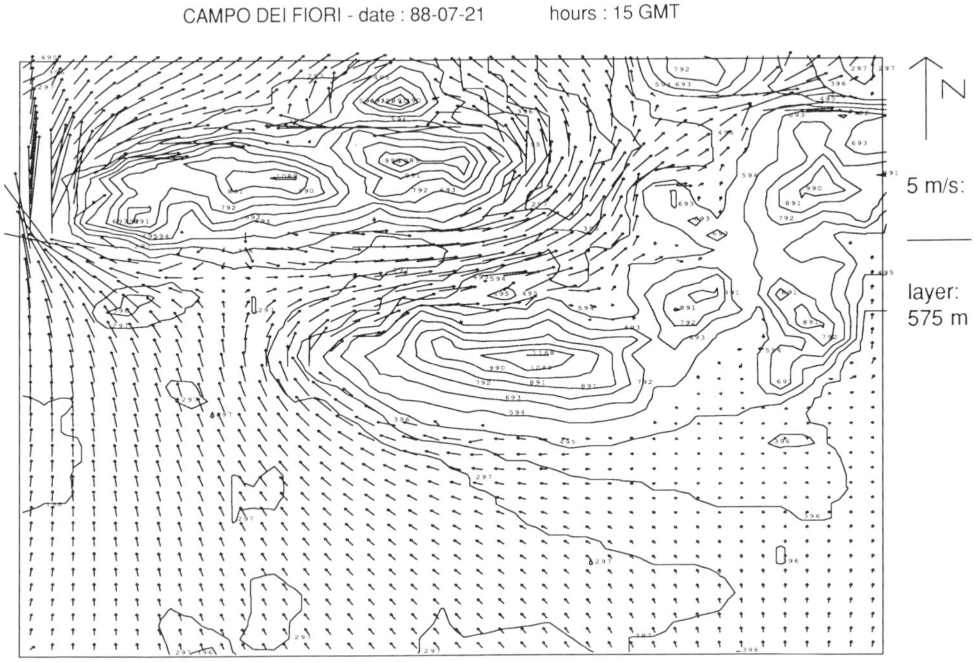

Fig. 8. Wind field at 15 GMT 21st July 1988 at 575 m. a.s.l.

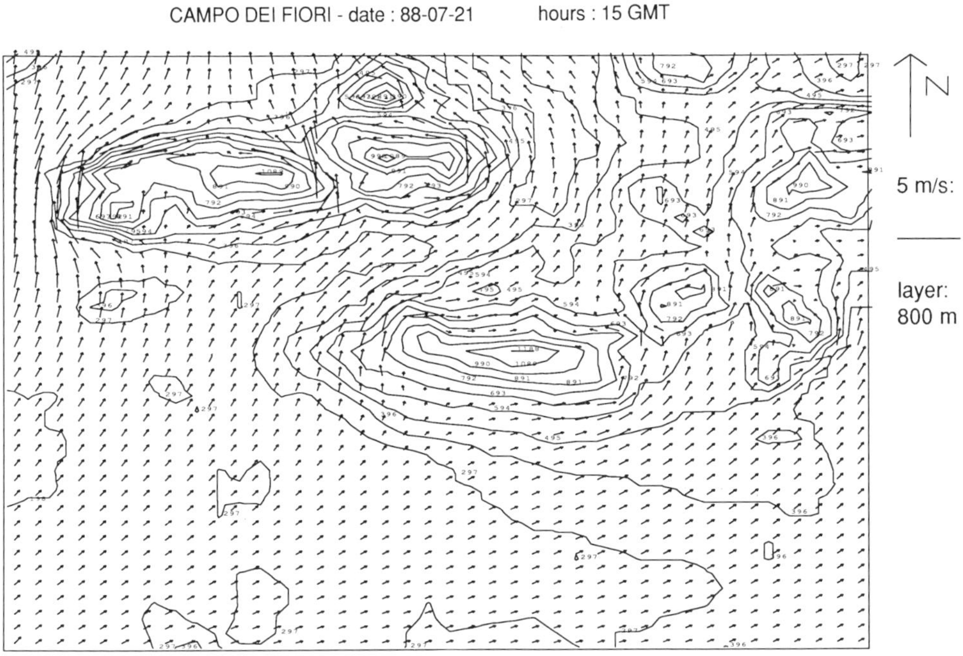

Fig. 9. Wind field at 15 GMT 21st July 1988 at 800 m. a.s.l.

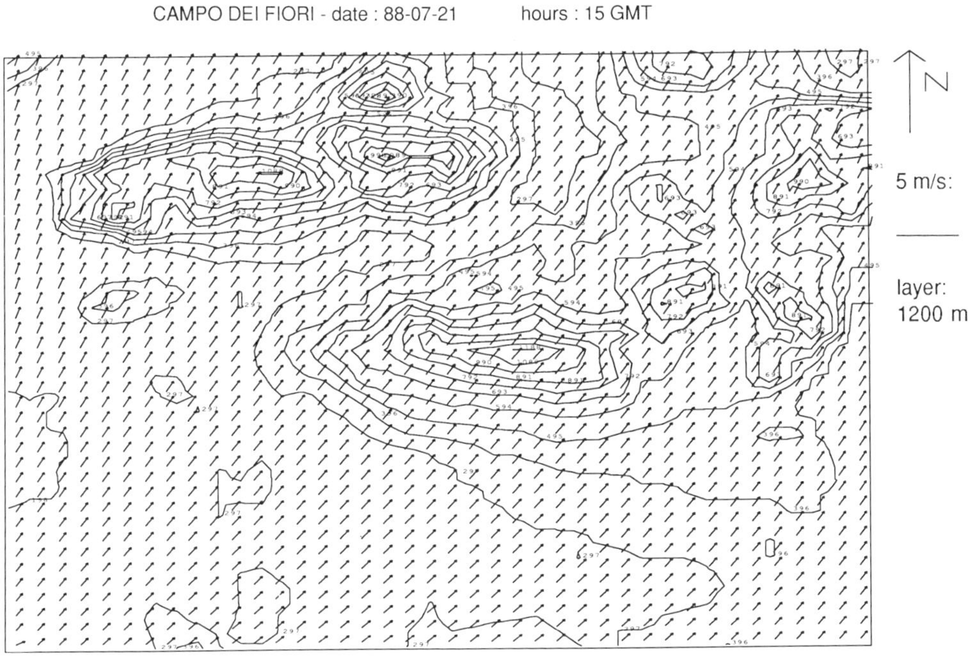

Fig. 10. Wind field at 15 GMT 21st July 1988 at 1200 m. a.s.l.

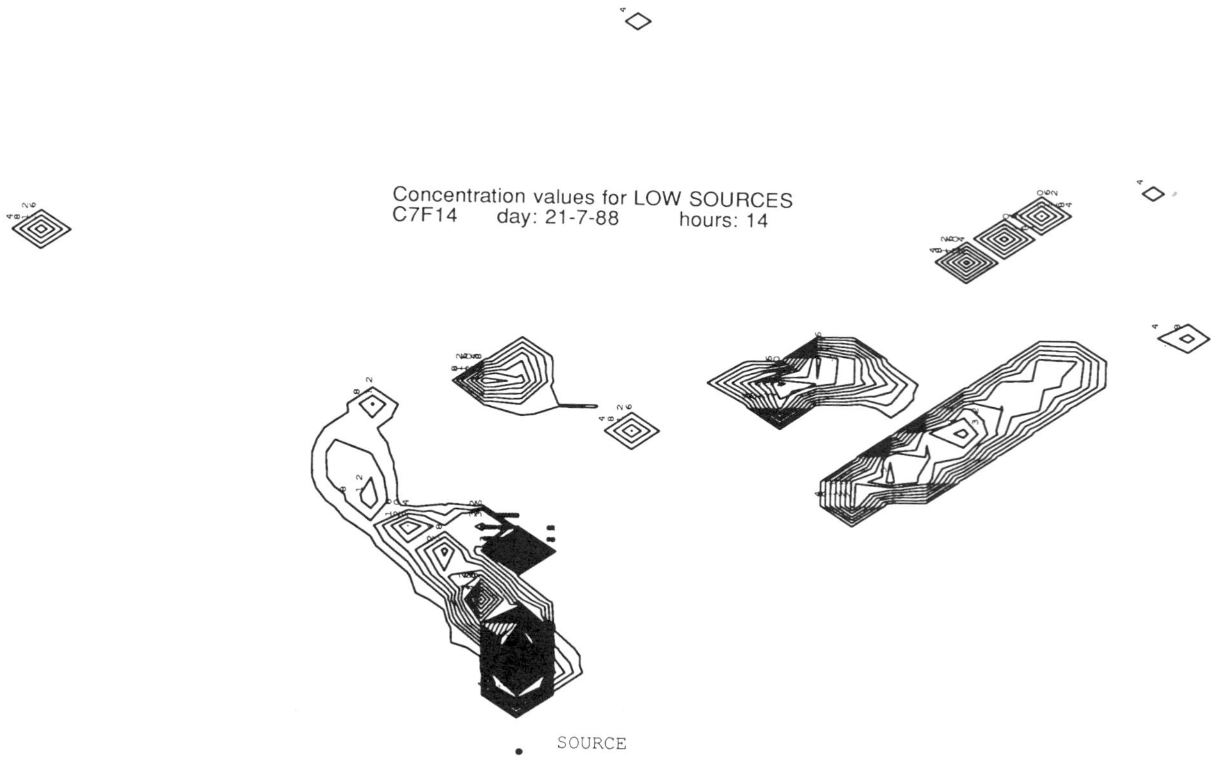

Fig. 11. Isoconcentration lines at 14 GMT.

Table 1. Observed and computed values at 13.00

Observable Points	Observated values	Calculated values
101	302	250
102	52	80
103	98	130
104	40	65
105	18	18
106	8	-
201	40	130
202	15	8
203	13	25
204	28	60
205	34	40
206	23	20
305	150	180
306	16	60
307	27	8
312	35	63
401	3	10
402	2	-
403	5	-
404	5	-
405	4	-
501	-	-
503	-	-
504	2	2
506	-	-
507	2	-
508	-	-
601	2	-
602	1	-
603	3	3

concentrations measured, which are in the ppt range. A comparison between observed and calculated data at 13 GMT, however, is shown in Table 1.

4. Conclusions

It is possible to simulate air masses circulation under breeze wind conditions with an accuracy sufficient to reproduce all typical phenomena occurring over a complex terrain. By the initialization of the steady tridimensional model it is also possible to describe the time evolution of breeze circulation in complex terrain such as alpine valleys. We are now improving these models. Preliminary calculations of deposited masses seem to promise a better agreement with the measured data than hitherto available, which represents an interesting starting point for an improved time-dependent model.

Bibliography

Ambrosetti P., Kappenberger G. and Roesli H.P.,1987: La meteorologia dell'inquinamento atmosferico della Svizzera Italiana. *Boll. Soc. Tic. Sci. Nat.* LXXV, 63-74.

Clerici G.C.,Micheletti S.,Sandroni S.,1988:Modelisation of the Tridimensional Wind Field Above the Alpine Region. *Air Poll. Mod. and its Appl.* VI. Plenum Press, N.Y.

Clerici G.C., Sandroni S.,1986: A Wind Field Model for Interpretation of Remote Sensing data in a Complex Area. *Air Poll. Mod. and its Appl.* V. Plenum Press, N.Y.,383-400.

Gaglione P., Graziani G., Gryning S.E.: Perfluorocarbon Tracer Experiments in a Lake-Mountain Area (Campo Dei Fiori Experiment). *EURASAP Conf. Proc.* Riso National laboratory, Denmark, Oct. 25-27th. 1988.

Giuliacci M.,1985: *Climatologia statistica e dinamica della Val Padana.* Italian National Research Council Report AQ 3/8.

Hanna S.R.,1986: *Handbook on Atmospheric Diffusion.* J.S.Smith Pub. Edt.

Hasenjäger H.,1989: *Micrometeorological Measurements on July 12-19-21,1988.* Technical Note I.89.39,JRC,Ispra.

Manzi D., Salerno R., 1990: Modello di dispersione atmosferica su regioni ad orografia complessa. Relazione int., Milano.

Pasquill F. and Smith F.B.,1983: *Atmospheric Diffusion.* Ellis Horwood,Chichester,West Sussex.

Salerno R.,Clerici G.C.,Micheletti S.,Giuliacci M. 1988: Modello per la stima quantitativa delle precipitazioni su regioni ad orografia complessa. *Proc. Int. Conf. of Alp. Met.* Sestola (Italy),sept 13-18,1988.

MESOSCALE FLOW-FIELDS AND DISPERSION IN THE COASTAL TERRAIN OF EASTERN KORINTHIAN GULF IN GREECE DURING SUMMER

George Kallos

University of Athens
Department of Applied Physics
Ippocratous 33, Athens 10680, Greece

INTRODUCTION

In order to accurately evaluate environmental impacts from the operation of industrial installations it is necessary to identify some crucial atmospheric parameters such as the development of the Planetary Boundary Layer (PBL) and the 3-D wind field. This is especially true in complex coastal areas where a variety of phenomena including sea-land breezes, convergence zones, drainage flows from surrounding topography, and the development of thermal internal boundary layers make the dispersion processes very complicated (Hanna, 1987; McKendry, 1989). In almost all cases where standard dispersion modelling approaches are applied, some over-simplifications are made for the PBL depth and the wind fields. Such over-simplifications are the spatial homogeneity of the PBL depth and its simple diurnal variation which is usually approximated from indirect methods based on observations and empirical formulation. However, previous studies (Pielke et al., 1983; Glendering et al, 1986; Segal et al., 1988; McKendry, 1989) suggest that large spatial and temporal variations in PBL depth may occur over coastal regions when marine air is advected over land. Of course, significant variations in to the wind field are observed at the same time. These variations are more significant when the marine air incursion occurs over irregular terrain with significant topographic variations. In such cases, even if a dense observational network is available, it is almost impossible to accurately resolve the detailed spatial and temporal variations in both the wind field and PBL depth, parameters considered crucial for detailed air quality investigations. One good approach for such investigations is the use of a mesoscale numerical model capable to accurately describe the PBL depth and wind fields. In addition, with the use of a Lagrangian type dispersion - diffusion model it is possible to investigate the environmental impacts from industrial activities.

In this study, some results from a numerical investigation of the mesoscale circulation and pollution transport in the coastal area of the Eastern Korinthian Gulf in Greece are presented. The pollutant source is in the area of Thisvi near the northern coast of the Gulf from an alumina plant which will be erected. The pollutants, mostly SO_2, result from petroleum combustion. The released SO_2 is estimated up to 1.604 tn/hour for full operation. This quantity will be releasd from eight stacks with heights ranging from 30 to 50 m above ground in a position having a mean elevation of 130 m above the sea-level. The topography around the plant is very complex and the site is about 1.5 km from the nearest coast (see Figure 1). To the N of

Air Pollution Modeling and Its Application VIII, Edited by H. van Dop and D. G. Steyn, Plenum Press, New York, 1991

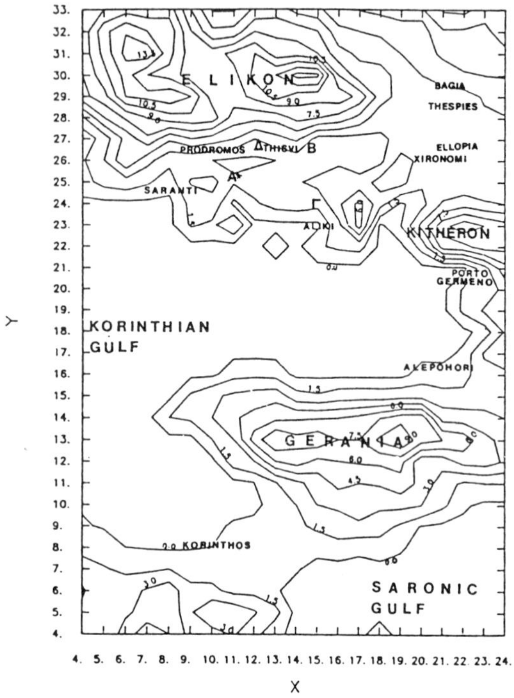

Fig. 1. Map of the modelling domain showing main topographic features and populated areas. Contours are in hundreds of meters. The numbers at the show grid numbers. Grid interval is 2 km.

the site, there is a small valley. Further to the N there is mount Elikon with its peak at approximately 1500 m. Some other hills up to 450 m are around this valley and near the coast. A gap between these hills is located at the plant's position with an elevation of 130 m above sea-level. Some other important topographic features around this area are mountain Kitheron in the E with elevation at about 1000 m and mountain Gerania in the SE with elevation at about 1100 m. The Gulf has a width of about 30 km except the eastern edge which has a width of about 12 km. At the S of Gerania there is the Isthmus of Korinthos with an average elevation of about 70 m. In a distance of 3-5 km to the N of the site there are three small populated areas where some others are in the NE in a distance of 15-30 km. Korinthos is locatet in a distance of 35 km S from the plant.

In such complicated terrain local circulations like sea-land breezes, drainage flows etc are developing in a regular basis. The interaction between these local circulations with the synoptic flow is a very complicated phenomenon and cannot be described easily. Consequently, the dispersion of the released air-pollutants is also very complicated and cannot be described with conventional methods. For these reasons a two-stage procedure was employed: In the first stage the three-dimensional version of the Colorado State University Mesoscale Model (CSUMM) was used to predict the meteorological fields over this area. In the second stage, a Lagrangian Particle Dispersion Model (LPDM) was used. A brief description of the two models is given below while the results for sea breeze simulations during summer are discussed.

MODEL DESCRIPTIONS

Mesoscale Model. The mesoscale model used in this study was originally developed by Pielke (1974) and modified later by Mahrer and Pielke (1977). The

CSUMM is a hydrostatic, prognostic model based on the primitive equations of motion, energy, continuity, and water content. It uses the terrain-following coordinate system and exchange coefficients for parameterization of the turbulence exchange processes. For the PBL a detailed parameterization is used where short and long-wave radiative contributions are included in the surface energy balance calculations. The depth of the convective boundary layer is computed from Deardorff's (1974) prognostic equation. This equation was modified according to the suggestions from Physick et al. (1989) for a more accurate prediction of the PBL depth. For the surface layer the similarity theory was used (Businger, 1973). The exchange coefficient profile in the PBL was specified by O'Brien's (1970) cubic polynomials. At the top of the model, an absorbing layer is applied to avoid reflections of the generated gravity waves.

Lagrangian Particle Dispersion Model (LPDM). This model uses as input the predicted fields from the mesoscale model described previously. It is a Lagrangian-type dispersion model on which the motions of discrete mass elements (which represent the pollutants) are tracked inside the model domain as they move with the synoptic, mesoscale and microscale (turbulent) wind components. Synoptic and mesoscale wind components are these calculated directly from the mesoscale model while the turbulent components are deduced from the mesoscale model closure scheme. A more detailed description of this technique is presented in McNider (1981), Pielke et al. (1987), and McNider et al. (1988).

DATA USED

Model simulations were performed for August 3, 1987. During this day the synoptic conditions were favorable for the formation of sea-land breeze circulation in the area. The initialization of the model was made with data from a tethered ballon system at 02:00 LST for the lowest 1.0 km and above this level, the radiosonde observations from Athens airport (located approximately 40 km ESE from the site) at 02:00 LST were used. Initial profiles and additional data used for this simulation are shown in Table 1. Experimental data collected in four different locations (shown in Figure 1 with the first four Greek capital letters) were used for comparison. More details about the experimental data for this day and in general for the experimental period are given in Kallos et al. (1989). Simulation starts at 02:00 LST and the first three hours were used for dynamic initialization, allowing the initial fields to adjust to the terrain before sunrise where the heating of the surface begins.

For the simulations with the LPDM the results from the CSUMM were used (stored every 10 minutes) while additional data such as stack parameters were also provided. Because the eight stacks are in small distances (less than 300 m between each other), one stack with representative parameters was used. Effective stack height was calculated every 10 minutes.

DISCUSSION

As it is shown in Figure 1, the modelling area is characterized by a complex terrain and frequent changes from land to sea. As it was foun from observations (Kallos et al., 1989) in the vicinity of the plant, sea breeze circulations are frequent all the months of the year when the synoptic circulation is not strong (weak anticyclonic circulation with small pressure gradients over Greece and Aegean). This happens because of the orientation of the mountain slopes, the relatively dry land and the small percentage of vegetation cover. At the site of the plant it was found that the wind direction is biased in N/S directions because of the topography near the coast.

Table 1. Model input data used for simulation of August 3, 1987.

Initial profile

Height(m)	Pot. Temp.(K)	Spec. Hum.(g/Kg)	U(m/s)	V(m/s)
5.	295.20	0.0137	68.0	-97.3
10.	295.21	0.0136	69.6	-101.3
50.	295.40	0.0104	70.3	-109.1
150.	297.00	0.0099	67.6	-113.7
300.	297.90	0.0089	65.0	-112.6
500.	298.52	0.0085	79.0	-139.6
700.	299.50	0.0084	93.0	-166.6
900.	300.50	0.0073	.107.0	-193.6
1100.	301.45	0.0070	121.0	-220.6
1300.	302.20	0.0061	135.0	-247.6
1500.	302.78	0.0059	149.0	-274.6
1800.	304.10	0.0057	170.0	-315.1
2100.	305.90	0.0053	191.0	-355.6
2500.	308.25	0.0045	219.0	-409.6
3000.	310.84	0.0029	230.7	-410.3
3500.	314.75	0.0027	242.3	-411.0
4000.	317.10	0.0024	254.0	-412.0
4500.	320.05	0.0022	265.7	-412.3
5000.	322.96	0.0019	277.3	-412.9
6000.	326.94	0.0009	300.7	-414.3
7000.	328.99	0.0008	324.0	-415.7
8000.	331.89	0.0007	324.0	-415.7
9000.	332.99	0.0007	324.0	-415.7

Additional data

Albedo	0.20
Latitude	38 N
Roughness length	15. cm
Surface pressure	1012. mb
Sea-surface temperature	298.5 K
Initial PBL height	300. m
Horizontal grid interval	2000. m
Time step	12. s
Soil density	1.5 g/cm3
Soil specific heat	0.3 cal/gK
Wetness of the soil	0.08
Soil thermal conductivity	0.003 cal/cmKs
Time start	02:00 LST
Grid points	X:27, Y:36, Z:23

Figures 2a-c show the horizontal wind fields at the first model level (5 m above ground) for three different hours. During the morning, after sunrise, the sea breeze starts near the coast in both sides of the Gulf while at the slopes facing toward S or SE upslope winds are observed (Figure 2a). The synoptic circulation was light from NW directions. Therefore, over the Gulf light westerly winds are observed veering toward N near the northern coast and toward S near the coast of Peloponnese. E of mountain Elikon the flow is moving around the mountain and consequently is from N or NE directions because the sea breeze did not penetrate yet through the gap between Elikon and Kitheron. In the area of Isthmus of Korinthos the winds are from NW veering toward N after passing over Saronic Gulf because of the beginning of the sea breeze and the upslopes over the land S of Gerania. The sea breeze during

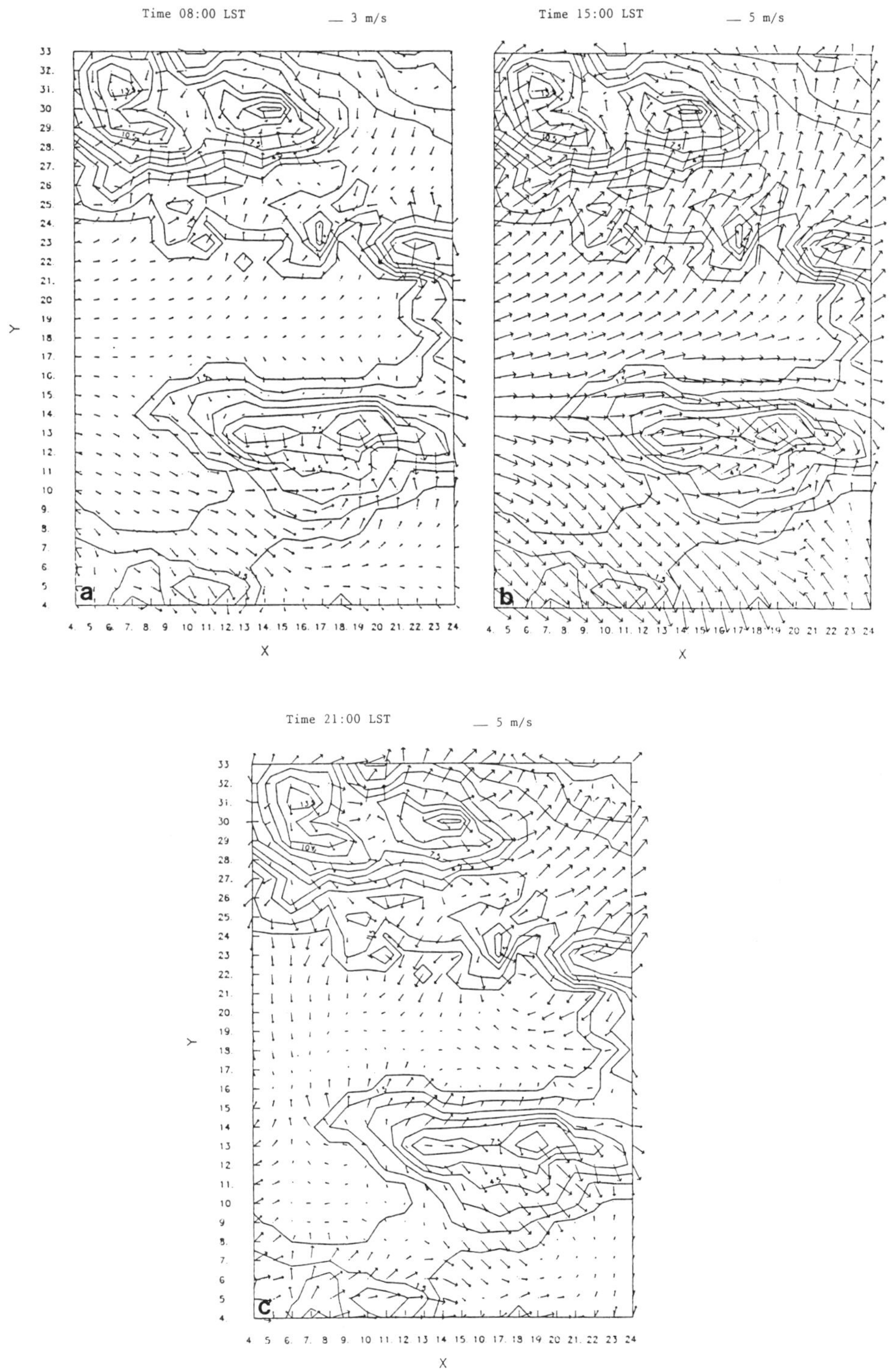

Fig. 2. Horizontal wind field at (a) 08:00 LST, (b) at 15:00 LST, and (c) at 21:00 LST, 5m above ground.

these hours has a small vertical extension (a few hundreds of meters) growing with time. Later during the day, when the sea breeze is almost at full development, stronger winds are observed over the entire area (Figure 2b). During this time, the sea breeze becomes strong enough to penetrate at the gap between Elikon and Kitheron. The vertical extension of the inland flow is approximately 900 m. The return flow is merging with the synoptic one and therefore the winds over the northern coast and over the sea are from N veering to NW over the southern coast and the Isthmus of Korinthos (not shown here). A relatively shallow stable layer is developing over the sea because of the subsidense. During the evening hours the sea breeze ceases over the plant's position and downslope winds appear over the slopes of the mountains. Very light winds are shown over the sea (Figure 2c). E of Elikon and N of Kitheron a relatively strong flow is observed because of the remnants of the sea breeze and the katabatic flows. This wind pattern continues to appear most of the night-hours until next day in the morning when the sea breeze starts again. Comparison between the model results and the surface observations showed a very good agreement. Small differences in wind direction were found in the area of the plant which must be due to the local topographic effects.

Some results from the LPDM are shown in Figures 3a-c. In these figures the position of each particle is shown for different hours while the position of the plant is shown with an asterisk. The release of particles started at 05:00 LST and Figure 3a shows the positions at 09:00 LST at the begining of sea breeze and the rise of PBL height. Particles are moving toward N or NE over the small valley near the plant and the populated areas of Thisvi and Domvraina. The widespread of the particles shows good dispersion conditions. During the afternoon hours, when the sea breeze is at full development, the released particles are moving initially toward NE and with the return flow toward SE, passing over the eastern edge of the Gulf and Gerania to reach Saronic Gulf (see Figure 3b for 17:00 LST). This situation occurs for a few hours in the afternoon and therefore it is not expected to have significant increase of concentrations of air pollutants over Saronic Gulf. During the night-hours the atmosphere is quite stable and the released particles are moving toward east in a narrow strip (see Figure 3c for 03:00 LST of the next day). The movement of the particles toward east is due to the flow above the PBL. Drainage flows from Elikon have a very small vertical extent smaller than the effective stack height.

Having known the spatial and temporal distribution of the released particles and the emission rates, is easy to estimate time averages of non reacting pollutants near the ground. The main pollutant released from the stacks is SO_2. Figure 3d shows the average concentrations of SO_2 for 24-hour time period (05:20 to 05:10 LST of the next day) for the layer near ground with a depth of 25 m. Maximum concentrations are observed N and near the released position and at the gap between Elikon and Kitheron. The maximum near the release point is due to the releases during the morning-hours (during inversion break-up) while the other at the gap between Elikon and Kitheron is due to the day and night releases. The second maximum is mainly due to the day releases which were initially over Elikon and in the evening moved to this area with the aid of drainage flows. Finally, small concentrations appear over the northern part of the Korinthian Gulf. This is due to the secondary transport of released particles during day-hours which moved over sea during night-hours with the aid of drainage flows. Other simulations for different days of the year were performed for synoptic conditions similar to these described previously. The results showed a similar flow pattern but the sea breeze starts later and stops earlier while the PBL height is significantly smaller during winter compared to the summer simulation.

CONCLUSIONS

In this paper, some results from a combined use of a mesoscale atmospheric

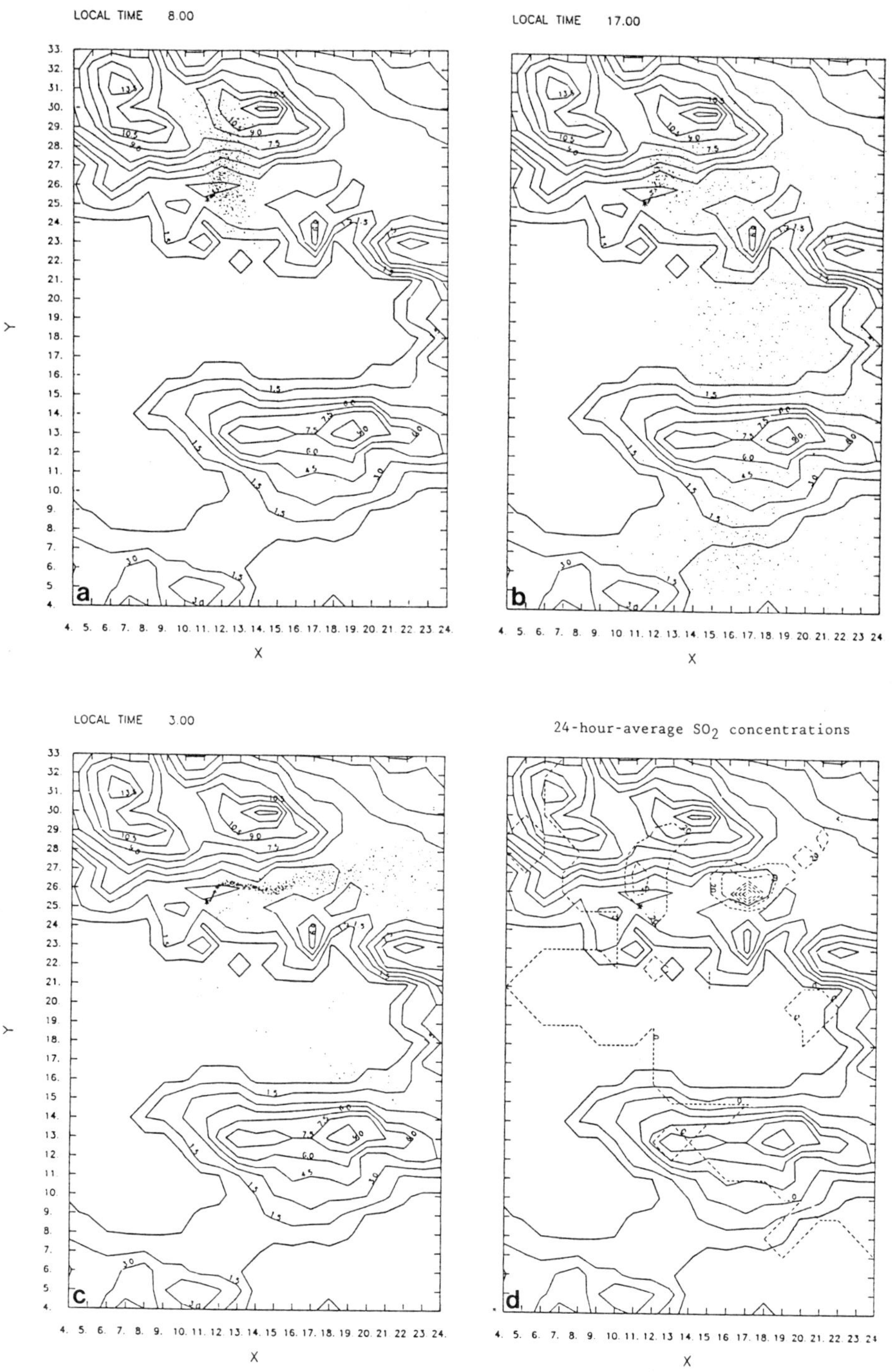

Fig. 3. Plan view of particle plume at (a) 08:00 LST, (b) at 17:00 LST, and (c) at 03:00 LST of the next day. (d) 24-hour-average near surface SO_2 concentrations in $\mu g/m^3$ (dotted lines). The release point is shown by an asterisk. Contour interval 20 $\mu g/m^3$.

model with a Lagrangian particle dispersion were presented. In such complex terrain the flow fields and the PBL height show temporal and spatial variations and therefore the dispersion characteristics are varying significantly. The results from the mesoscale model were in good agreement with observations and therefore the method could be considered quite reliable. The conclusions derived from these simulations could be used in order to minimize the environmental impacts from the operation of this industrial installation.

ACKNOWLEDGEMENTS

Support for the work reported in this paper was provided by the Hellenic Alumina Industry Inc. I would like to thank R. Pielke, M. Moran, M. Segal and R. Arritt for their assistance and useful suggestions for the model simulations.

REFERENCES

Businger, J.A., 1973: 'Turbulent transfer in the atmosphere surface layer'. Workshop in Micrometeorology, Amer. Meteor. Soc., Chap.2.

Deardorff, J., 1974: Three dimensional numerical study of the height and mean structure of a heated planetary boundary layer. Boundary-Layer Meteorol., 7, 81-106.

Glendering, J.W., B.L. Ulrickson, and J.A. Businger, 1986: Mesoscale variability of boundary layer properties in the Los Angeles basin. Mon. Wea. Rev., 114, 2537-2549.

Hanna, S.R., 1987: An empirical formula for the height of the coastal internal boundary layer. Boundar-Layer Meteorol., 40, 205-207.

Kallos,G., A. Stathi, D. Asimakopoulos, C. Helmis, C. Varotsos, and D. Deligiorgi, 1989: Micrometeorological study of the area of Thisvi, Viotia. Final Report prepared for the Hellenic Alumina Industry Inc. 380 pp.

Mahrer, Y. and R.A. Pielke, 1977: A numerical study of the airflow over irregular terrain. Beit. zur Phys der Atm., 50, 98-113.

McKendry,, I.G., 1989: Numerical simulation of sea breezes over the Auckland region, New Zeland - Air quality implications. Boundary-Layer Meteorol., 49, 7-22.

McNider, R.T., 1981: Investigation of the impact of topographic circulations on the transport and dispersion of air pollutants. Ph.D. Dissertation, University of Virginia, Charlottesville, Virginia, 210 pp.

McNider, R.T., M. Moran, and R.A. Pielke, 1988: Influence of diurnal and inertial boundary - layer oscillations on long-range dispersion. Atmos. Environ., 22, No. 11, 2445-2462.

O'Brien, J.J., 1970: A note on the vertical structure of the eddy exchange coefficient in the planetary boundary layer. J. Atmos. Sci., 27, 1213-1215.

Physick, W.L., D.J. Abbs, and R.A. Pielke, 1989: Formulation of the thermal internal boundary layer in a mesoscale model. Boundary-Layer Meteorol., 49, 99-111.

Pielke, R.A. (1974), A three-dimensional numerical model of the sea breezes over South Florida. Mon. Wea. Rev., 102, 115-139.

Pielke, R.A., R.T. McNider, M. Segal, and Y. Mahrer, 1983: The use of a mesoscale numerical model for evaluations of pollutant transport and diffusion in coastal regions and over ittegular terrain. Bull. American Meteor. Soc., 64, pp. 243-249.

Pielke, R.A., R.W. Arritt, M. Segal, M.D. Moran, and R.T. McNider, 1987: Mesoscale numerical modelling of pollutant transport in complex terrain. Boundary-Layer Meteorol., 41, pp. 59-74.

Segal, M., C-H. Yu, and R.A. Pielke, 1988: Model evaluation of the impact of thermally induced valley circulation in the Lake Powel area on long range transport. J. Air Poll. Control Ass., 38(2), 163-170.

DISCUSSION

S. LEWELLEN

You reported good agreement between your ensemble mean model results and the single realization observations. Were any model iterations used?

G. KALLOS

No model iterations were used. Comparisons were made between hourly model outputs and hourly surface observations and tethered balloon observations when available.

J.C.R. HUNT

In your model you are using a 1-particle model in a mesoscale model. This leads to ensemble average concentrations over many realisations of the <u>same</u> unsteady large scale velocity field (such similar realisations may occur rather seldom perhaps?). Surely relative (2-particle) diffusion effects need to be considered when calculating typical concentrations for one realisation at a particular location? Have you or R. Pielke compared concentrations obtained from 1 and 2 particle computations for continuous releases in flows with highly variable trajectories? It would be interesting to know.

G. KALLOS & R.A. PIELKE

One-particle models should differ little from two-particle computations with respect to mesoscale and larger scale transport and dispersion. This is a result of the dominance of dispersion due to differential mesoscale and synoptic advection (which is a resolved effect) in the Lagrangian particle model. To the extent that stochastic motions on the subgrid scale influence the dispersion (i.e. through diffusion), we do produce an ensemble plume result. It would be an appropriate evaluation, of course, to generalize the undependent one-particle model to determine if significant differences in mesoscale and larger scale dispersion results.

MESO-METEOROLOGICAL STUDIES IN THE IBERIAN PENINSULA AND WESTERN MEDITERRANEAN AREA

M. Millán, B. Artíñano, I. Palomino, J.C. Bezares
M. Martín and R. Salvador

CIEMAT-ipryma
Av. Complutense 22
Madrid, E-28040, Spain

INTRODUCTION

This work introduces the project "Meso-Meteorological Cycles of Air Pollution in the Iberian Peninsula" (**MECAPIP**). It reviews some aspects which influence the behavior of pollutants in Southwestern Europe, presents an overview of the experimental programs now under way and identifies some of the modelling needs. The project leads the CEC efforts to develop a comprehensive program to characterize the Physico-Chemical behavior of polluted airmasses in the Mediterranean Basin and to provide the necessary inputs for more realistic modelling of local episodic conditions and the long-range transmission of pollutants over the whole of Europe.

BACKGROUND: GEOGRAPHICAL AND METEOROLOGICAL SETTINGS

The Mediterranean basin is situated in the subtropical latitudes, is surrounded by high mountains, clear days are frequent and the ground properties (vegetation cover, wetness, albedo, etc.) undergo very large annual variations. The topography is also complicated, and there are several peninsulas with extensive coastal areas surrounding an interior sea which is quite deep and much warmer than the oceanic seas, i.e. abyssal temperatures of approximately 13 degrees vs. 4 for the oceans.

Important effects on the atmospheric flows are caused by the major mountain ranges. The Pyrenees, Alps and Balkans can act as constraining barriers to the North-South movement of air masses to generate either blocking effects, channeling or both, depending on the general synoptic situation and specific place. Their southern slopes can also be strongly heated during the day in the summer. Finally, semi-arid terrains with high plateaus also prevail in many areas. For the purpose of understanding and modelling the behavior of atmospheric pollutants in this area, the following general situations must be considered:

Summer Conditions

In this period, the large scale meteorological systems which affect the rest of Europe have a very weak interaction with the processes which take place in this area (Figure 1). The western Mediterranean remains at lower temperatures than the surrounding regions, acting as a large energy

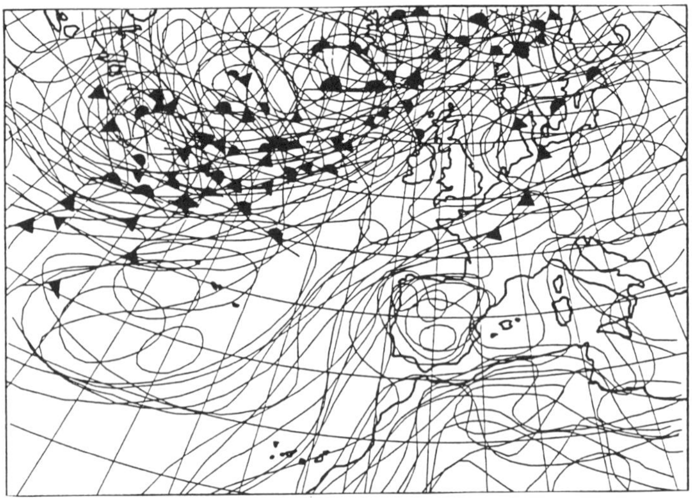

Figure 1. Overlay of the 12:00 Z synoptic charts for July 2 to 7, 1987. The representation is intended to show that: (1) the frontal systems which travel via the North of the British Isles, and influence the central and northern regions of Europe, do not reach the Western Mediterranean area, (2) at this time of day there is a semi-permanent low over the Iberian peninsula. This is the Iberian thermal low which undergoes a pronounced diurnal cycle (Figure 2), and (3) there is another semi-permanent higher pressure area over cooler waters of the Gulf of Lyon.

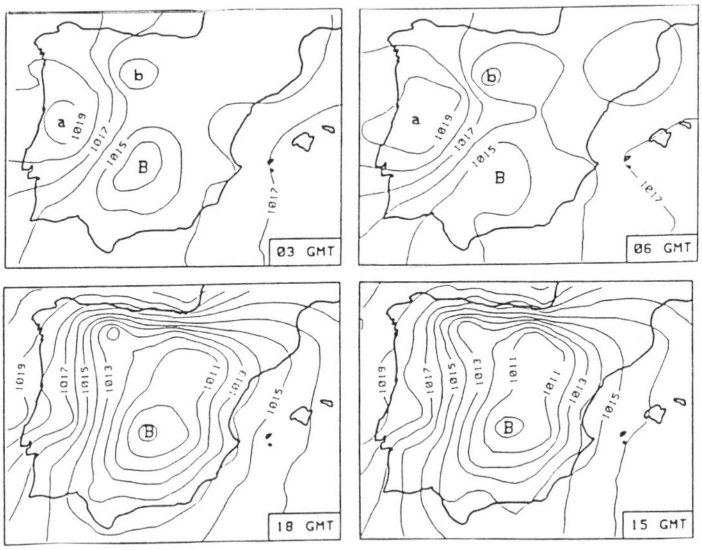

Figure 2. Stages in the evolution of the Iberian thermal low on high resolution maps for July 20, 1984. The weakest stage is reached just about sunrise (approx. 0600 h GMT), and the deepest extent between 1500 and 1800 h GMT. To get an insight about the gradients involved, it can be noticed that the vertical distance from Gibraltar to the North coast is approx. 830 km. This means that the gradients in the North coast of Spain can reach values as high as 8 hPa in approx. 100 km. A small cyclone indeed!.

sink which favors the subsidence of air over the cooler water. Continuity, however, requires that a complementary outflow must take place at, or very near, the surface, and this is strongly constrained to flow parallel to the coastal mountain barriers, eg. along the East coast of Spain.

Air pollution undergoes marked diurnal cycles governed by land-sea and mountain-valley wind systems. At the sub-synoptic scale, large thermally driven convective systems can develop during the day over Spain (Figure 2), Italy (Cantú and Gandino, 1977) and the major islands. The circulations are strong and the wind-travelled distance can easily reach 50 to 100 or more km during the day. During the night the travel in different or opposite direction is much shorter. This situation is typical of convergence at ground level and it simply indicates that, in balance, the air and pollutants, must go up over the center of the heat source.

While the full cycle of convergence across the axis of the Italian peninsula may take only one day and result in the formation of storms over the Apenine mountains in the late afternoon, it can take two or three days over Iberia. Through these mechanisms, some of the pollutants can be aged and injected into the middle troposphere: as high as 5 km over Iberia and higher over Italy, i.e. as much as 10 km, if storms develop. The result of these processes is to gather a large mass of aged pollutants into a "blob" at the top of the convective system(s) during the afternoon.

There is already enough experimental evidence in the Iberian case (Millán et al.,1990; Martín et al.,1990) that, during the diurnal part of the cycle, oxidants are intaken along the Eastern coast and added to the locally generated pollutants as they are carried inland. Finally, as soon as the convective processes decay during the evening and night, and atmospheric decoupling from the ground takes place, the "blobs" of aged pollutants are free to move and participate in long range transport out of these regions. On the basis of the 750 and 500 mb maps, a substantial fraction of the pollutants thus injected over the Mediterranean area may end up over Central Europe within the reach of the frontal rain systems, which still operate at those latitudes during the summer.

Another important mechanism for the mid-tropospheric injection of pollutants occurs as a result of the up-valley or mountain winds. These develop along the South-East to South-West facing slopes during the period of strong solar heating. In the summer, the insolation period is long and the winds can carry pollutants from the bottom of their airshed to the top of the mountain and act, essentially, as orographic chimneys. This can result in up-slope injection of pollutants from the Po valley and other industrial areas of Southern Europe.

Late Autumn and Winter Situations

During this period the Mediterranean Sea acts mainly as a large source of energy. Warm core anticyclones migrate to, and stagnate over Central Europe and can favor the development of episodic conditions over extensive areas. In the Mediterranean basin, however, a number of processes can also occur to make the pollutant cycles different from those occurring over the rest of Europe.

For example, the advection of cold continental air over the warmer sea can induce the formation of convective instability and lead to the formation of Genoa-type depressions over the Western Mediterranean. Similar processes, albeit not as frequent, can also occur over the Central and Eastern Mediterranean (Barry and Chorley, 1987). The most important differences between these Lows and those forming over Spain or Italy in the summer are as follows:

* Once they are formed, they persist for a few days without the diurnal cycle of growth and decay associated with the thermal (sensible heat) lows, which are also land-locked.
* To form they need a positive temperature contrast between the sea-air system, i.e. their frequency of occurrence is higher during the Fall, with

the first invasions of cold continental air, and decrease as soon as the Mediterranean cools off during winter.

* As their source of energy is the latent heat released from water vapor, they are not land-locked. They are mobile and can act, essentially, as travelling vacuum cleaners, i.e. pollutants will be sucked in near the surface and rained out and/or transported along the storm track.

* By this mechanism, pollutants emitted from industries located in the Western Mediterranean coastal areas may end up deposited over the Apennines, the Italian Alps region or transported into the Central Mediterranean.

* While these depressions form and evolve over the Western Mediterranean, neighboring coastal areas may experience heavy rains and strong winds. The situation over Central Europe or other Mediterranean areas, however, may remain anticyclonic and episodic.

During the Autumn-Winter conditions the locally driven thermal circulations become increasingly weaker. The wind travelled distance in either direction becomes nearly equal for day and night and, in most situations, rather short. In places where topography may limit or block weak synoptic flows (eg., under anticyclonic conditions), and the available directions for the flow are constrained (eg., valleys), serious pollution episodes can develop as a consequence of the sloshing of pollutants within confined or semi-confined airsheds. This situation can occur in the Po Valley, Athens (Lalas et al.,1983), Bilbao (Millán et al.,1984), Madrid, and other areas of Southern Europe. In these conditions, an important aspect relating to atmospheric chemistry is that airmass ageing can go on for days in the bottom of the valleys, as a result of the forementioned local oscillatory cycles, while new emissions are continuously added to it.

THE MECAPIP PROJECT

Within the Mediterranean general setting, the Iberian Peninsula also has some unique characteristics in that it is:

(1) A high plateau barrier between the Atlantic and Mediterranean (cold vs. warm seas).

(2) Mountainous and semi-arid.

(3) Sub-continental in size.

The project "Mesometeorological Cycles of Air Pollution in the Iberian Peninsula" was proposed to the CEC in 1986, as accumulating experimental evidence showed that: (a) the formation of the Iberian thermal low in the summer, was intimately associated with complex dispersion behavior of atmospheric pollutants emitted within the whole of Spain and in the coastal areas, in particular, and (b) other important differences existed between the observed behavior of polluted airmasses in the Mediterranean area and in more northerly regions (Millán et al. 1987). The General Objectives are:

* To characterize the meteorological conditions and transport mechanisms which can result in a sub-continental circulation covering most of the Iberian peninsula during the summer.
* To determine natural pathways for the inflow of pollutants emitted in the coastal areas during the convergence cycle of the circulation forced by the formation of the Iberian thermal low.
* To increase the knowledge of meso-scale processes in the interphase zone between Atlantic and Continental-Mediterranean air masses.
* To characterize wet deposition processes under chosen urban plumes for conditions of episode breakup by precipitation.

The project has four phases. The first one includes the gathering and interpretation of available historical data on plume dispersion and supporting meteorological data. The second and third phases will cover experimental work at the preliminary and intensive field campaign level, respectively, which are intended to document the inflow for the periods identified as a result of the first phase and, finally, the last phase will address the data interpretation.

The overall approach considered will contemplate the characterization of the flows as the first priority. The second level will address the identification of the cycles of the chemical constituents associated with the flows, while the third step, modelling of both the observed flows and chemical cycles, will proceed as soon as the data base and its interpretation have progressed to the point where the model development and verification can be performed judiciously.

MAIN RESULTS TO DATE

To characterize the processes involved, tall stack plumes from four sites in Spain have been tracked with COSPEC remote sensors and used as tracers of opportunity of the flow at their level. This information is being combined with available historical meteorological data to compose a mosaic of atmospheric circulation patterns for the whole of the Iberian peninsula and surrounding regions. The analysis of the data obtained during the first phase of this program (1987 and 1988) shows that:

* The formation of the thermal low forces a convergence of the atmospheric flows towards the center of the peninsula. Coastal emissions can be followed for more than 100 km inland. These processes have been documented using plume dispersion data (Figure 3) and tracers.
* The convergence reaches a maximum in the early afternoon and relaxes during the night. The observed inflow is mostly ageostrophic. It is also limited to, and strongly channelled along, natural mountain passes lower than $\approx$ 1200 m in height.
* The analysis of ground level concentrations and associated meteorology shows a direct relationship between the meso-meteorological cycles and the concentrations of primary and secondary pollutants in this area.
* The marine airmass pulled inland along the Spanish East coast is highly enriched with O_3 (Figure 4). The O_3 levels could be explained as a result of regional transport of precursors from source regions in this general area (Millán et al. 1989). The verification of this hypothesis forms part of the actual experimental program.

PRESENT WORK AND MODELLING NEEDS

Early in 1989 the ISPRA JRC Environmental Research Institute proceeded to define a research program with emphasis in Southern European Pollution problems, and the study of the effect of thermal lows in the Mediterranean area was chosen as one of their priority areas. As a result of this, ISPRA supported the MECAPIP project with a tracer study and the use of the Fraunhofer Institute for Atmospheric Environmental Research instrumented aircraft. With this support, the experimental program was modified to place most emphasis in characterizing the inflow conditions during the formation of the thermal low. Figure 5 shows the experimental deployment area covered in the July, 1989 field campaign. The data are presently being evaluated, and Figure 6 shows the O_3 concentration evolution for the stations located along the Mijares River (one of the identified natural channelling paths). The stations cover a distance of approximately 120 km from the sea shore to the Mora de Rubielos site.

For atmospheric modeling purposes, some aspects to consider are the following:

(1) The most relevant processes are all time dependent, with very few near stationary conditions. The dominant period is diurnal and involves flow reversal and/or coupling-decoupling cycles between upper and surface winds.

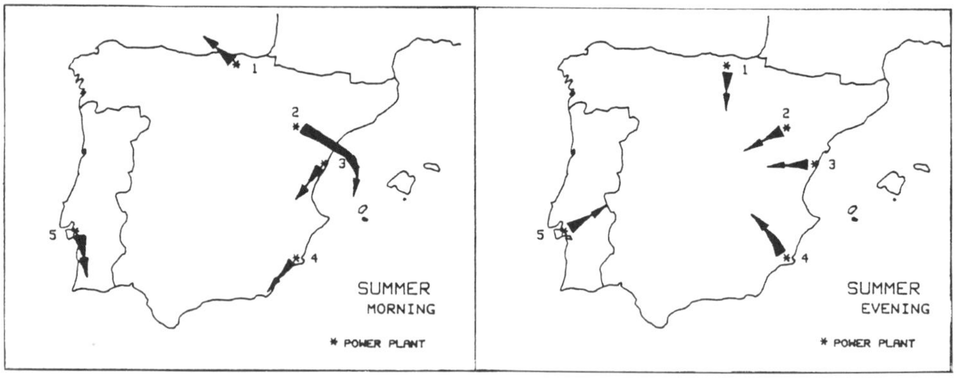

Figure 3. Map of the Iberian Peninsula showing the location of the four tall stacks used to characterize the effect of thermal lows. 1 - Bilbao (Vizcaya), 2 - Andorra (Teruel), 3 - Castellón, 4 - Cartagena (Murcia). The plume sketches synthesize the observed plume behavior at about sunrise and late in the afternoon. Some circumstantial evidence available to the authors indicates that the plume from the Setubal power plant in Portugal follows a similar cycle.

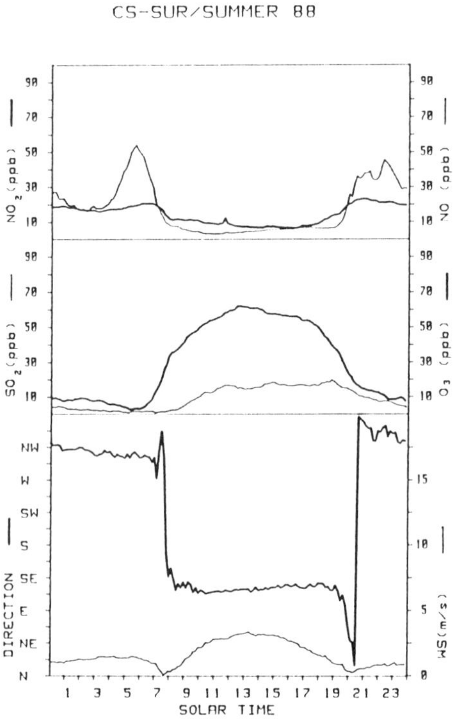

Figure 4. Average of the high resolution diurnal graphs (based on ten minute average values) for all days with a well developed sea-breeze in the July-August campaign. The upper part shows the diurnal variation of the NO and NO_2 values, the O_3 and SO_2 evolutions are in the middle, and the lower graphs show the land-sea to sea-breeze transitions that take place during the day. Land-sea and drainage flow is from the W to NW, and the sea-breeze from the E to SE.

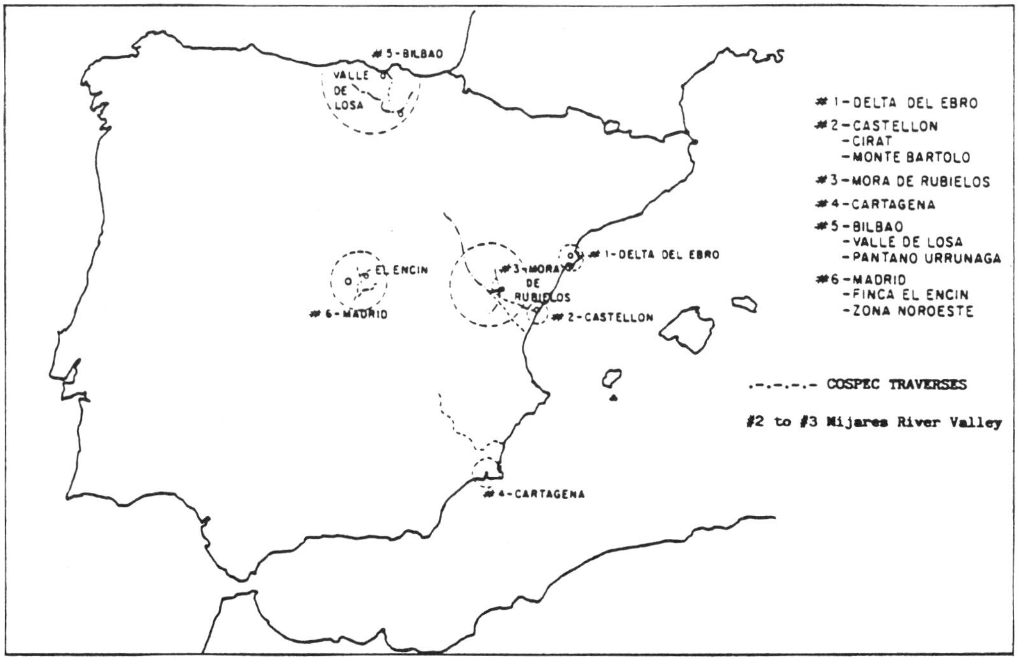

Figure 5. Map of the Iberian Peninsula showing the instrumental deployment areas for the MECAPIP, 1989 Summer campaign.

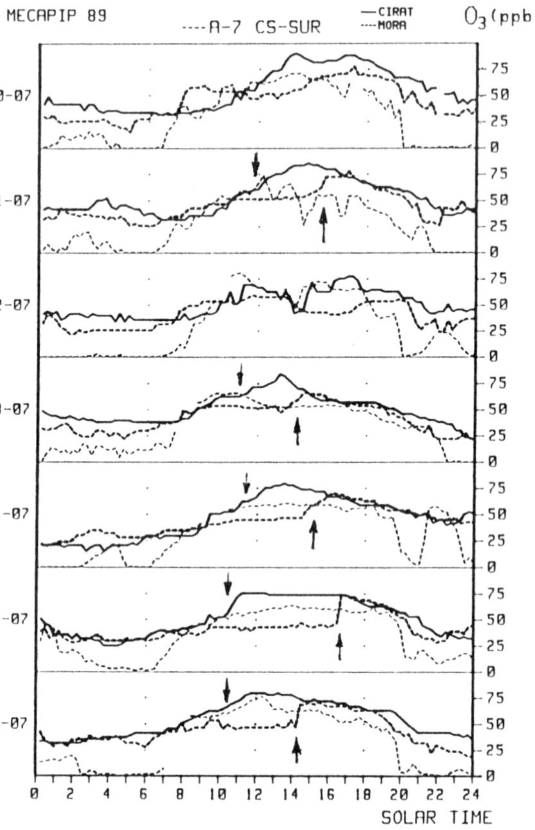

Figure 6. Diurnal evolution of Ozone concentrations for three sites along the Mijares River valley (a natural path from the East coast to the central plateau). The stations are: CS-sur at 13 km from the coast, Cirat at ≈40 km inland, and Mora Rubielos at ≈ 120 km inland. The rises in the O3, values which can be attributed to the arrival of the sea-breeze and the pull of the thermal low, are indicated with arrows.

(2) Most of the observed mechanisms involve convergence at the local to regional scale, which is followed by transport of the accumulated material at the synoptic scale.

(3) The applicable models should be able to handle recirculatory flows (including transport in different directions at two heights) and/or sloshing cycles (direct oscillation of the airmass with accumulation of pollutants within an airshed without significant vertical exchange). They would also require dealing with orographic and convective injection mechanisms, which require suitable parametrization.

(4) A great deal of effort will be required to parametrize meso-scale processes in order to link existing knowledge about local phenomena, resulting in the accumulation of pollutants over large regions, with their continuity and transport at the synoptic scale.

ACKNOWLEDGEMENTS

Part of the work presented here has been financed by the Commission of The European Communities, under project EV4V-0097-E (A). We would also like to thank the Environmental Institute of the CEC's Joint Research Centre ISPRA, and in particular to Drs. F. Geiss and A. Stingele for their support with the aircraft measurements and tracers study. Nearly eighty persons participated in the various campaigns, and we would like to express our deepest appreciation to all of them for the quality of their work. The authors would also like to thank other institutions which made this project possible: The Environmental Vicecouncelry of the Basque Government, The Township of Cartagena, ENDESA's Research Division, Hidroeléctrica Española and the Valencia Regional Centre of the Spanish Meteorological Institute.

REFERENCES

Barry R.G. and Chorley R.J.(1987) Atmosphere Weather and Climate. (5th Ed.). Methuen, New York.

Cantú V. and Gandino P.(1977) Rivista di Meteorologia Aeronautica XXXVII-N° 3, 265-267.

Lalas D.P., Asimakopoulos D.N., Deligiorgi D.G. and Helmis C.G. (1983) Sea-breeze circulation and photochemical pollution in Athens, Greece. Atmos. Environ., 17, 1621-1632.

Martín M., Plaza J., Andrés M.D., Bezares J.C. and Millán M.M. (1990) Comparative study of seasonal air pollutant behavior in a Mediterranean coastal site: Castellón (Spain). Submitted to Atmospheric Environment.

Meteorological Office (1962) Weather in the Mediterranean I & II. (2nd Ed.), MO 391 and MO 391b, HMSO, London.

Millán M.M., Alonso L.A., Legarreta J.A., Albizu M.V., Ureta I. and Egusquiaguirre C.(1984) A fumigation episode in an industrialized estuary: Bilbao, November 1981. Atmos. Environ., 18, 563-572.

Millán M.M., Otamendi J.E., Alonso L.A. and Ureta I. (1987b) Experimental characterization of atmospheric diffusion in complex terrain with land-sea interactions. JAPCA, 37, 807-811.

Millán M.M., Artíñano B., Alonso L., Navazo M. and Castro M. (1990) The effect of meso-scale flows in the regional and long-range atmospheric transport in the western Mediterranean area. Atmospheric Environment, In press.

Millán M.M., Salvador R., Artíñano B. and Palomino I.(1989) Power plant plume dispersion study in a coastal site with complex terrain. Air Pollution Modeling and its Applications VII. Plenum Press, New York.

FLOW AND PARTICLES TRAJECTORIES OVER TERRAIN THERMAL INHOMOGENEITIES

G. A. Dalu [1,4], M. Baldi [2,5], R. A. Pielke [3], and A. Guerrini [2]

1 - mailing address: G.A. Dalu, CIRA-CSU, Foothills Campus, Fort Collins, Colorado 80523
2 - Research Area Frascati, CNR, Rome, Italy
3 - Department of Atmospheric Science, Colorado State University ATM-CSU
4 - Institute for Atmospheric Physics, IFA-CNR, Rome, Italy
5 - Presentor

ABSTRACT

We present an analytical evaluation of the perturbations to mesoscale atmospheric flows induced by surface thermal inhomogeneities. We apply the theory to a sea breeze case in the presence or in the absence of an inland lake. We show also some particles trajectories for these two cases. Furthermore we show that at least to the extend of the linear response of the atmosphere terrain thermal inhomogeneities which have a separation of the order of the depth of the boundary layer need not be resolved in a numerical model, because these inhomogeneities are spatially averaged by the atmospheric response.

The problem is approached using Laplace transform theory for the time behavior and Green function theory for the spatial structure.

GOVERNING EQUATION

In order to focus on the impact of terrain thermal inhomogeneities on atmospheric flow, we assume that the intensity of the large-scale wind is negligible; then from Rotunno (1983), and Dalu and Pielke (1989) the linearized two-dimensional equations describing thermally-forced flow can be reduced to a single equation for the stream function:

$$\left[\left(\frac{\partial}{\partial t}+\lambda\right)^2+N^2\right]\frac{\partial^2}{\partial x^2}\psi+\left[\left(\frac{\partial}{\partial t}+\lambda\right)^2+f^2\right]\frac{\partial^2}{\partial z^2}\psi=-\frac{\partial}{\partial x}Q \qquad (1)$$

where f is the Coriolis parameter, b is the buoyancy force, N is the Brunt-Väisälä frequency, and λ^{-1} is the damping time due to friction, i.e., the persistency of the perturbation after the forcing is removed. Q is the diabatic buoyancy source. We recall the relation between the diabatic forcing Q and the height h of the CBL (Green and Dalu, 1980):

$$h=\frac{2}{N^2}Q_0\int_0^t dt'\,q(t')=h_0\frac{1-\cos(\omega\,t)}{2} \qquad (2)$$

and the asymptotic Rossby radius is:

$$R_0=h_0\frac{N}{\sqrt{\lambda^2+f^2}} \qquad (3)$$

THE NON-DIMENSIONAL EQUATION, ITS LAPLACE TRANSFORM AND SOLUTION IN TERMS OF GREEN FUNCTIONS

Using the Laplace transform theory and introducing non-dimensional variables, the streamfunction equation (1) can be written as:

$$\frac{\partial^2}{\partial\hat{\xi}^2}\hat{\psi}+\frac{\partial^2}{\partial\hat{\eta}^2}\hat{\psi}= -\hat{\beta}(p)\frac{\partial}{\partial\hat{\xi}}\hat{Q} \tag{4}$$

We assume that the terrain is flat and that the vertical velocity vanishes at the ground:

$$w(x,z=0,t)=\tilde{w}(\hat{\xi},\hat{\eta}=0,\tau)=\hat{w}(\hat{\xi},\hat{\eta}=0,s)=0 \tag{5}$$

The Green function for the Poisson equation (4) in the upper semi-plane, i.e. which satisfies the boundary condition $\hat{w}(\hat{\xi},\hat{\eta}=0,\tau)=0$ (5), is:

$$\hat{g}_\psi=-\frac{1}{2\pi}\ln\left[\frac{(\hat{\xi}-\hat{\xi}')^2+(\hat{\eta}+\hat{\eta}')^2}{(\hat{\xi}-\hat{\xi}')^2+(\hat{\eta}-\hat{\eta}')^2}\right]^{1/2} \tag{6}$$

from which we derive the Green functions for the horizontal and the vertical velocities:

$$\hat{g}_u=\frac{\partial}{\partial\hat{\eta}}\hat{g}_\psi=\frac{1}{2\pi}\left[\frac{(\hat{\eta}-\hat{\eta}')}{(\hat{\xi}-\hat{\xi}')^2+(\hat{\eta}-\hat{\eta}')^2}-\frac{(\hat{\eta}+\hat{\eta}')}{(\hat{\xi}-\hat{\xi}')^2+(\hat{\eta}+\hat{\eta}')^2}\right] \tag{7}$$

$$\hat{g}_w=-\frac{\partial}{\partial\hat{\xi}}\hat{g}_\psi=\frac{1}{2\pi}\left[\frac{(\hat{\xi}-\hat{\xi}')}{(\hat{\xi}-\hat{\xi}')^2+(\hat{\eta}+\hat{\eta}')^2}-\frac{(\hat{\xi}-\hat{\xi}')}{(\hat{\xi}-\hat{\xi}')^2+(\hat{\eta}-\hat{\eta}')^2}\right] \tag{8}$$

The variables with the tilde are non-dimensional; the variables with the hat are their Laplace transforms.

ATMOSPHERIC RESPONSE

The atmospheric response is given by :

$$\hat{\psi}=\tilde{Q}_0\hat{q}(s)\hat{\beta}(p)\hat{H}_\psi(\hat{\xi},\hat{\eta});\quad \hat{u}=\tilde{Q}_0\hat{q}(s)\hat{\beta}(p)\hat{H}_u(\hat{\xi},\hat{\eta});\quad \hat{w}=\tilde{Q}_0\hat{q}(s)\hat{\beta}(p)\hat{H}_w(\hat{\xi},\hat{\eta});$$

$$\hat{v}=-\frac{\tilde{f}}{p}\hat{u};\qquad \hat{b}=\frac{1}{s}\hat{Q}-\frac{1}{p}N^2\hat{w} \tag{9}$$

$$\text{where}\quad \hat{H}(\hat{\xi},\hat{\eta})=\int_D d\hat{\xi}'\int_0^{\hat{h}}d\hat{\eta}'\hat{g}(\hat{\xi}-\hat{\xi}',\hat{\eta}-\hat{\eta}')\frac{\partial\hat{r}(\hat{\xi}')}{\partial\hat{\xi}'}$$

$\hat{r}(\hat{\xi})$ is the horizontal structure of the forcing, non-zero over a region D; and $\hat{h}$ is the depth of the convective boundary layer (CBL). In physical units, through the use of the Faltung theorem (Fodor, 1965), we have:

$$\psi(x,z,t)=\frac{Q_0h_0}{TN}\{\sin(\omega t)*\{\exp(-\lambda t)\cdot\{J_0(ft)*H_\psi(x,z,t)\}\}\} \tag{10}$$

where $\{\ \ *\ \ \}$ denotes the convolution product in time.

The intensity of the flow evolves through a series of damped inertial-gravity waves. However, when $\lambda\geq O(\omega)$, i.e. when friction is the same order as the rate of change of the forcing or larger, then we can make the following approximations:

$$\psi(x,z,t)\approx\psi_0\,H_\psi(x,z,h(t))\quad\text{with}\quad\psi_0=\frac{Q_0h_0}{\omega N}=\frac{Nh_0^2}{4} \tag{11}$$

$$u(x,z,t) \approx u_0\, H_u(x,z,h(t)) \quad \text{with} \quad u_0 = \frac{\psi_0}{h_0} = \frac{Nh_0}{4} \tag{12}$$

$$w(x,z,t) \approx w_0\, H_w(x,z,h(t)) \quad \text{with} \quad w_0 = u_0 \frac{h_0}{R_0} = \frac{Nh_0^2}{4R_0} \tag{13}$$

$$v(x,z,t) \approx -\frac{f}{\lambda} u(x,z,t) \tag{14}$$

$$b(x,z,t) \approx N^2 \left\{ (h-z)\, r(x)\, He(h-z) - \frac{w(x,z,t)}{\lambda} \right\} \tag{15}$$

For $r(x) = He(x)$, the structure functions $H(x,z,h(t))$ are

$$H_u(x,z,h(t)) = \frac{1}{2\pi} \ln \left\{ \frac{\left[\left(\frac{x}{R_0}\right)^2 + \left(\frac{z+h(t)}{h_0}\right)^2\right]^{1/2} \left[\left(\frac{x}{R_0}\right)^2 + \left(\frac{z-h(t)}{h_0}\right)^2\right]^{1/2}}{\left[\left(\frac{x}{R_0}\right)^2 + \left(\frac{z}{h_0}\right)^2\right]} \right\}$$

$$H_w(x,z,h(t)) = -\frac{1}{2\pi} \left[\tan^{-1}\left(\frac{R_0}{h_0}\frac{z+h(t)}{x}\right) - 2\tan^{-1}\left(\frac{R_0}{h_0}\frac{z}{x}\right) + \tan^{-1}\left(\frac{R_0}{h_0}\frac{z-h(t)}{x}\right) \right]$$

$$H_\psi(x,z,h(t)) = \frac{1}{2\pi} \left\{ \frac{x}{R_0} \left[\tan^{-1}\left(\frac{R_0}{h_0}\frac{z+h(t)}{x}\right) - 2\tan^{-1}\left(\frac{R_0}{h_0}\frac{z}{x}\right) + \tan^{-1}\left(\frac{R_0}{h_0}\frac{z-h(t)}{x}\right) \right] \right.$$

$$+ \frac{z+h(t)}{h_0} \ln \left[\left(\frac{x}{R_0}\right)^2 + \left(\frac{z+h(t)}{h_0}\right)^2 \right]^{1/2} - \frac{2z}{h_0} \ln \left[\left(\frac{x}{R_0}\right)^2 + \left(\frac{z}{h_0}\right)^2 \right]^{1/2}$$

$$\left. + \frac{z-h(t)}{h_0} \ln \left[\left(\frac{x}{R_0}\right)^2 + \left(\frac{z-h(t)}{h_0}\right)^2 \right]^{1/2} \right\} \tag{16}$$

TRAJECTORIES

The along-x displacement of a particle, in buoyant equilibrium with the environment, is:

$$\Delta_x(x_0, y_0, z_0, h(t)) = \int_0^t dt\, u(x_0, z_0, t) = \int_0^h \frac{dh}{\dot{h}} u(x_0, z_0, t) = \frac{2u_0}{\omega} \int_0^h \frac{dh}{h_0} H_u(x_0, z_0, h(t))$$

$$= \frac{2u_0}{\omega} \frac{1}{2\pi} \left\{ \frac{x_0}{R_0} \left[\tan^{-1}\left(\frac{R_0}{h_0}\frac{z_0+h(t)}{x_0}\right) \right] - \tan^{-1}\left(\frac{R_0}{h_0}\frac{z_0-h(t)}{x_0}\right) \right.$$

$$\frac{z_0+h(t)}{h_0} \left[\ln\left(\left(\frac{x_0}{R_0}\right)^2 + \left(\frac{z_0+h(t)}{h_0}\right)^2 \right)^{\frac{1}{2}} - \ln\left(\left(\frac{x_0}{R_0}\right)^2 + \left(\frac{z_0}{h_0}\right)^2 \right)^{\frac{1}{2}} - 1 \right]$$

$$\left. - \frac{z_0-h(t)}{h_0} \left[\ln\left(\left(\frac{x_0}{R_0}\right)^2 + \left(\frac{z_0-h(t)}{h_0}\right)^2 \right)^{\frac{1}{2}} - \ln\left(\left(\frac{x_0}{R_0}\right)^2 + \left(\frac{z_0}{h_0}\right)^2 \right)^{\frac{1}{2}} - 1 \right] \right\} \tag{17}$$

The along-y displacement is:

$$\Delta_y(x_0, y_0, z_0, h(t)) = -\frac{f}{\lambda}\frac{2u_0}{\omega}\int_0^h \frac{dh}{h_0} H_u(x_0, z_0, h(t)) = -\frac{f}{\lambda}\Delta_x(x_0, y_0, z_0, h(t)) \tag{18}$$

The along-z displacement is:

$$\Delta_z(x_0, y_0, z_0, h(t)) = -\frac{2w_0}{\omega}\frac{1}{2\pi}\left\{\frac{z_0 + h(t)}{h_0}\tan^{-1}\left(\frac{R_0}{h_0}\frac{z_0 + h(t)}{x_0}\right)\right.$$

$$-2\frac{h(t)}{h_0}\tan^{-1}\left(\frac{R_0}{h_0}\frac{z_0}{x_0}\right) - \frac{z_0 - h(t)}{h_0}\tan^{-1}\left(\frac{R_0}{h_0}\frac{z_0 - h(t)}{x_0}\right)$$

$$\left. -\frac{x_0}{R_0}\left[\ln\left(\left(\frac{x_0}{R_0}\right)^2 + \left(\frac{z_0 + h(t)}{h_0}\right)^2\right)^{\frac{1}{2}} - \ln\left(\left(\frac{x_0}{R_0}\right)^2 + \left(\frac{z_0 - h(t)}{h_0}\right)^2\right)^{\frac{1}{2}}\right]\right\} \tag{19}$$

$\vec{S} \equiv (x_0 + \Delta_x, y_0 + \Delta_y, z_0 + \Delta_z)$ is the particle trajectory from the initial position (x_0, y_0, z_0) as a function of time trough the growing boundary layer $h(t)$.

DISPERSION

The diffusion equation is:

$$\frac{\partial P}{\partial t} = K\frac{\partial^2 P}{\partial z^2}$$

which, assuming K proportional to h^2, can be written as:

$$\frac{h_0\,\omega}{2}\frac{\partial P}{\partial h} = \alpha h^2\frac{\partial^2 P}{\partial z^2}; \quad \frac{\partial}{\partial t} = \dot{h}\frac{\partial}{\partial h} = \frac{h_0\,\omega}{2}\frac{\partial}{\partial h} \quad \text{and} \quad K = \alpha h^2$$

Then the particle diffusion around the average trajectory is given by:

$$P(\vec{S}, h) = \left(\frac{h_0}{4\pi h}\right)^{\frac{1}{2}}\left\{\exp\left[-\beta\left(\frac{h_0}{h}\right)^3\left(\frac{z - (z_0 + \Delta_z)}{h_0}\right)^2\right] - \exp\left[-\beta\left(\frac{h_0}{h}\right)^3\left(\frac{z + (z_0 + \Delta_z)}{h_0}\right)^2\right]\right.$$

$$\left. - \exp\left[-\beta\left(\frac{h_0}{h}\right)^3\left(\frac{z - (2h - z_0 - \Delta_z)}{h_0}\right)^2\right]\right\}; \quad \beta = \frac{\omega}{8\alpha}$$

the second term is the reflection from the ground, the third term is the reflection from the top of the CBL.

RESULTS

In Fig.1a we show the sea breeze streamfunction at $t = 6$ and 12 hours after sunrise. In Fig.1b we present the time evolution of the sea breeze when there is an inland lake, placed between the 40 km and 120 km. Comparison between the pure sea breeze case (shown in Fig.1a) and the sea breeze (Fig.1b) with a lake within a Rossby radius from seashore, shows that the extension of flow is about the same but the overall intensity is weaker, because the sensible heat over the lake is negligible. Fig.2a shows the deformation in time of a material surface initially placed at an altitude of 500 meters and the particles trajectories. Air subsides over the sea and rises over the land. When there is an inland lake, Fig.2b, the air subsides over the lake, and the air rises faster over the strip of land between the sea and the lake.

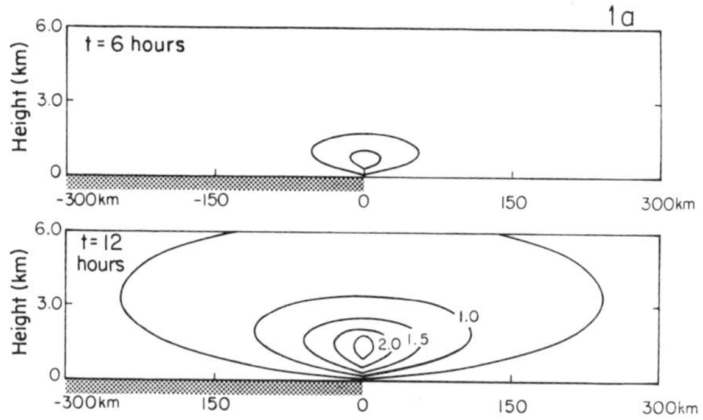

Fig.1a. Evolution of the flow for a sea breeze case at 40° of latitude ($\lambda = \omega$); $h_0 = 2000\ m$; $\Delta\psi = 5 \cdot 10^3\ m^2/s$. The sea is shown as a shaded area.

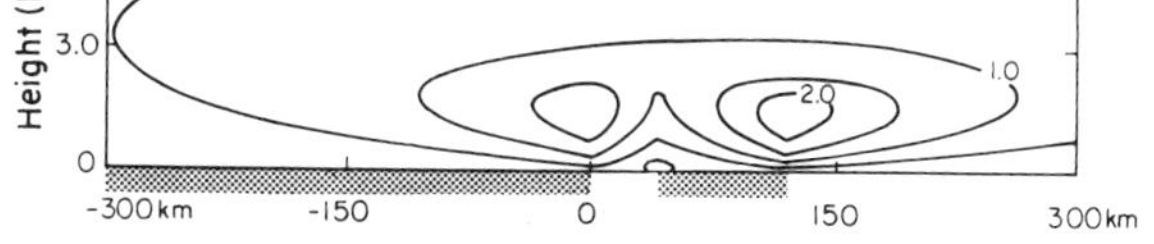

Fig.1b. Evolution of the flow for a sea breeze case like in Fig.1a but with a lake placed between 40 and 120 km from the coastline. The sea and the lake are shown as shaded areas.

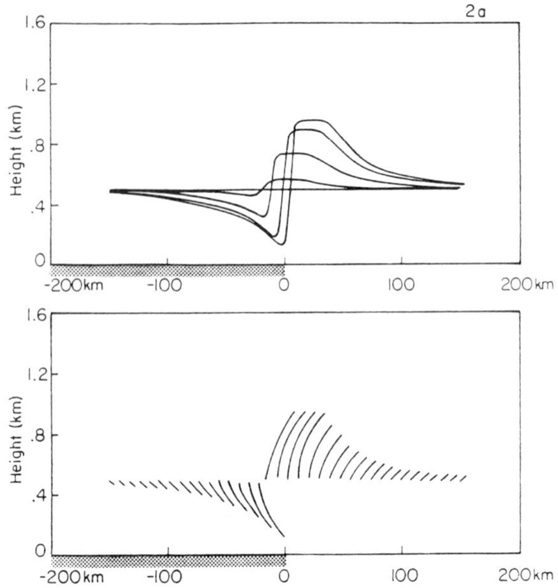

Fig.2a. Twelve hours evolution of a material surface, initially placed at $z = 500\ m$ (top); and particles trajectories (bottom). The sea is shown as a shaded area.

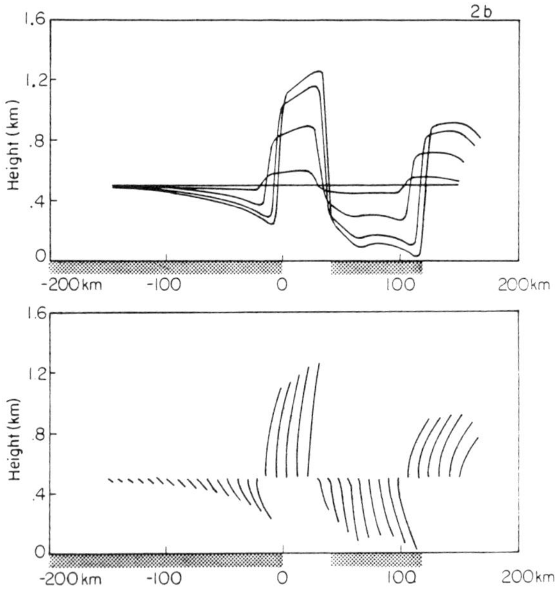

Fig.2b. Twelve hours evolution of a material surface, initially placed at $z = 500\ m$ (top); and particles trajectories (bottom). The sea and the lake are shown as shaded areas.

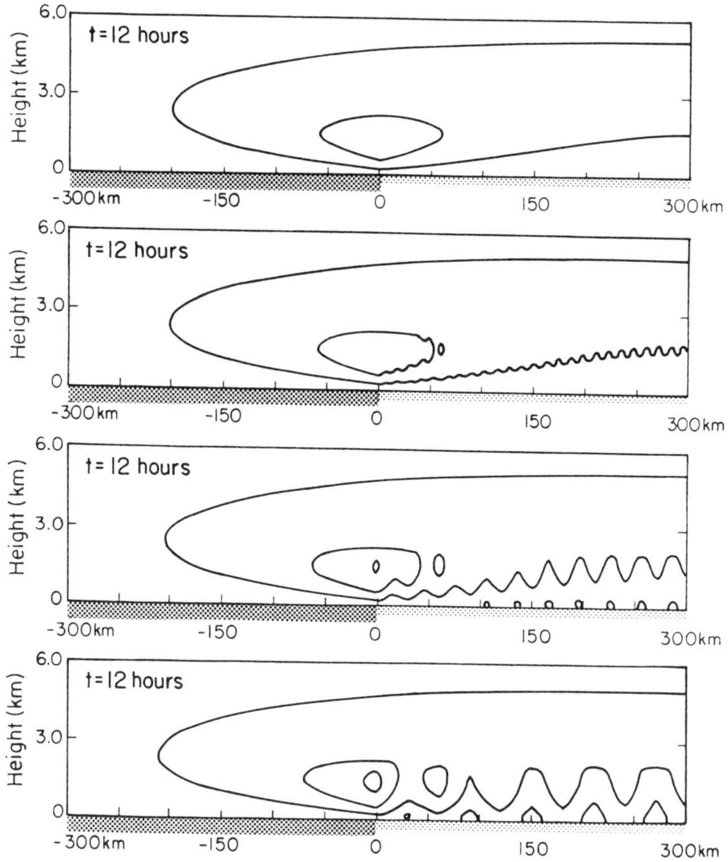

Fig.3. Streamfunction time evolution of a sea breeze when there are periodic terrain thermal inhomogeneities on the land side (heated patches alternated with non-heated patches of equal width). The figure shows the cases when the width of the small lakes is 3 *km*, 6 *km*, 15 *km*, 30 *km*. The sea is shown as a shaded area, the small lakes alternated with land patches are shown as a light shaded area.

EVOLUTION OF A SEA BREEZE WITH A PERIODIC THERMAL INHOMOGENEITIES ON THE LAND SIDE

In Fig3 we show a sea breeze case with periodic thermal inhomogeneities on the land side (i.e. heated patches of l width alternated with non-heated patches of equal width). The general result is that the sea breeze intensity and spatial extent as are about half of the sea breeze intensity and spatial extent without periodic thermal inhomogeneities on the land side. The intensity is reduced by the same factor as the horizontal scale. More specifically, when $l << R_0$, the horizontal periodic pattern of the forcing almost does not show in the sea breeze streamfunction except through a reduction of the surface heat flux. Therefore, based on this analytical theory, small scale thermal inhomogeneities can be averaged beforehand in a numerical model. As l becomes larger and a larger fraction of the Rossby radius, the periodic structure becomes more and more evident as a modulation of the streamfunction.

ACKNOWLEDGMENTS

We acknowledge the support of the Cooperative Institute for Research in the Atmosphere (CIRA), of the National Science Foundation (NSF), under grant # ATM-8616662, and of the Office of Naval Research, under contract # N00014-88-K-0029. G.A. Dalu, M. Baldi and A. Guerrini acknowledge the support of the Italian project *Clima, Territorio e Ambiente nel Mezzogiorno.*

REFERENCES

Dalu G.A. and R.A. Pielke, 1989: An Analytical Study of the Sea Breeze. J. Atmos. Sci., 46, 1815-1825.

Dalu G.A. and R.A. Pielke, 1989: Impact of Terrain Thermal Inhomogeneities on Mesoscale Atmospheric Flow: Theory of Subgrid-Scale Parameterization. Submitted to J. Atmos. Sci.

Fodor G., 1965: Laplace Transform in Engineering. Hungarian Academy of Science. Budapest.

Green J.A.S. and G.A. Dalu, 1980: Mesoscale Energy Generated in the Boundary Layer. Quart. J. R. Met. Soc., 106,721-726.

Rotunno R., 1983: On the linear Theory of Land and Sea Breeze. J. Atmos. Sci., 40, 1999-2009.

MODELING HEAVY METALS IN EUROPE'S ATMOSPHERE: A COMBINED TRAJECTORY-CLIMATOLOGIC APPROACH

Joseph Alcamo, Jerzy Bartnicki, Krzysztof Olendrzynski

International Institute for Applied Systems Analysis
Laxenburg, Austria

Introduction

The long range transport of heavy metals in the atmosphere leads to low but steady deposition of heavy metals into soils, lakes and forests in Europe. Although their atmospheric fluxes are low, these metals can accumulate in the detritus of soils. This accumulated metal may be mobilized by acidification and may then disturb soil organisms, which in turn can lead to disturbance of organic matter decomposition and nutrient cycling (Ottar et al., 1989). Circumstantial evidence from wind sector analysis and trajectory modeling (Pacyna et al., 1984), and receptor modeling (Stevens et al., 1984) indicates that the deposition of metals may originate from sources many hundreds of kilometers away. It is thought that large particles ($> 10\ \mu$) settle out in the vicinity of the stacks but stack gases containing metals condense into small particles which then coalesce within hours into particles with diameters of 0.1 to 1.0 μ; particles of this size are too coarse to be efficiently removed by brownian diffusion yet too small to settle out by gravity. Hence these metal particles may travel long distances before being removed by precipitation or dry deposition.

A set of two mathematical models have been developed to quantify the long range transport and deposition of heavy metals in Europe. The models compute the air concentration and dry and wet deposition of As, Cd, Pb, and Zn. These metals were selected because they are recommended for monitoring by the Paris and/or Helsinki Commissions which are concerned with pollution of the North and Baltic Seas respectively. Also As and Cd are known carcinogens (WMO, 1989), Pb has a variety of possible health impacts (WMO, 1989), and Zn at high concentration levels is moderately toxic to plants (Rühling, et al., 1987). The models are intended to be tools for computing source-receptor relationships, country balances of emissions and deposition, and for evaluating control strategies. The purpose of this paper is to describe the development and preliminary testing and application of these models.

Modeling Strategy

Two models have been developed in parallel. The basic tool used to compute source-receptor relationships throughout Europe is a climatologic model called the TRACE model (**Tr**ace Metal **A**tmospheric **C**oncentrations in **E**urope). This model uses annual average meteorological and deposition data from 9 years. Model calculations are checked against measurements made during this 9 year period.

Although climatologic models give good results for long term deposition and are compatible with the lack of detailed emissions and observational data of heavy metals in

Europe, they nevertheless rely on highly aggregated deposition and transport parameters that are difficult to justify scientifically. To improve the scientific credibility of the climatologic model, a more detailed trajectory model is used to estimate the parameters of the climatologic model; these include parameters for wet and dry deposition processes and the time of travel between sources and receptors. The trajectory model is a version of the EMEP model of sulfur and nitrogen in Europe which has been modified to reflect theoretical understanding of the behavior of metals in the atmosphere.

The advantage of the combined approach over using a single model is that the climatologic model is relatively easy to run compared to a trajectory model (requiring an order of magnitude less time) because it requires less meteorological data. Hence the model can be run for all Europe and for several years, and can make use of the very scattered measurements available for metals. At the same time the trajectory model provides a scientific basis for estimating aggregated climatologic model parameters.

The Trajectory Model - Modified EMEP

A description of the meteorological assumptions of the trajectory model is outside the scope of this paper and the reader is referred to Eliassen and Saltbones (1983) or Eliassen et al. (1988) for this description. In the following paragraphs we only present the method we use in the modified model to compute dry and wet deposition because this is especially relevant to how we use the trajectory model to generate parameters for the climatologic model.

Dry Deposition. Dry deposition d_d is computed along trajectories between sources and receptors at six hour intervals by:

$$d_d = v_d \cdot C \tag{1}$$

where C is the metals air concentration and v_d is the deposition velocity. The dry deposition velocity is computed from the semi-empirical model of Sehmel (1980). This deposition model is based on wind-tunnel experiments and also takes into account theoretical removal rates by brownian diffusion and gravitational settling. Dry deposition velocities are computed as a function of particle size, surface roughness and friction velocity. The advantage of using this model over assigning deposition velocities *a priori* is that it provides an independent basis for estimating the spatially-varying dry deposition velocity throughout Europe as a function of local meteorological conditions. Also, it requires data that are available for all Europe – namely, surface roughness and friction velocity. When the appropriate surface roughness and friction velocity values are used, Sehmel's model gives results consistent with Dulac et al.'s (1989) computation of dry deposition velocity of Cd and Pb into the Mediterranean, of Georgii et al.'s (1982) estimation of Cd and Pb deposition velocity into a German forested area, and the range of literature values for different receptor surfaces reported by Semb and Pacyna (1988). The disadvantages of this model are that it is based partly on empirical wind tunnel measurements and that it requires an estimation of the mass size distribution of metal particles.

In practice we compute the mass-weighted dry deposition velocity at different locations from

$$v_d = \sum v_d \left(D_p, u_*, z_o\right) f\left(D_p\right) dp \tag{2}$$

where $v_d(D_p,u_*,z_o)$ is taken from the curves in Sehmel (1980). Data for u_* and z_o comes from the EMEP Synthesizing Center West. The variable $f(D_p)$ is the mass size distribution of metal particles. Provisionally, we have used the mass size distributions measured by Dulac et al. (1989) in the Mediterranean for Cd and Pb because they are one-year averages collected at sites not affected by local sources but are nevertheless influenced by anthropogenic sources on the European continent. Dulac et al. (1989) however, did not measure either the As or the Zn distribution. Consequently, as a first assumption, we use the measured Cd mass size distribution also for As and Zn. This is consistent with the observation that As, Cd, and Zn have roughly the same source profile in Europe (Pacyna and Münch, 1987) and may therefore on the average have similar mass size distributions in the lower atmosphere. The sensitivity of model calculations to the assumption of mass size distribution is discussed later in this paper.

Wet Deposition. Wet deposition d_w is also computed at six hour intervals using the conventional scavenging coefficient approach:

$$d_w = w_q \cdot C \cdot P \tag{3}$$

where w_q is the scavenging ratio of the mass of pollutant in rain versus its mass in air, C is the air concentration of metal, and P is the 6 hour precipitation. This is the same approach used in the original EMEP model.

Despite the simplicity of this approach, it does take into account the amount of precipitation at a particular location, which accounts for much of the spatial variability of wet deposition. Consistent with our assumption that As, Cd, and Zn have similar mass size distributions on the average, we assign them the same $W_q = 500{,}000$. For Pb we assign $W_q = 200{,}000$. These are close to the average values measured by Chan et al. (1982) who performed one of the most thorough (spatial and temporal) studies of the scavenging ratio of metals available in the literature. In addition, they are among the very few researchers who provide data for Cd, Pb and Zn for the same locations and time periods. A disadvantage of using their data is that it comes from Ontario, Canada rather than Europe.

The use of a larger scavenging ratio for As, Cd, and Zn than for Pb is also consistent with experimental evidence (e.g. Gatz, 1975) and the theoretical argument (Slinn, 1984) that the washout ratio increases with increasing particle diameter. In our model we assume that As, Cd, and Zn have substantially larger mass median diameters than Pb. But in general, all assumptions regarding the wet scavenging ratio are highly uncertain. Therefore, the sensitivity of model calculations to a range of values will be reported in future publications.

An evaluation of the trajectory model is outside the scope of this paper but will be reported in later publications. In this paper we will concentrate only on evaluation of the climatologic model.

The Climatologic Model - TRACE

The basic idea of the TRACE model is to track the concentration and deposition of metals in an air parcel as it travels from a source to a receptor. The contribution of all sources to a receptor are weighted by the percent of time that winds come from the direction of different sources in one year.

The calculation of concentration and deposition is divided into two steps. In the first step, the model computes the loss of metals from the air parcel along the way from the source to the receptor by the simple loss function:

$$C(x) = c_o \cdot \frac{f}{x} \cdot F(s)e^{-(\bar{k}_d + \bar{k}_w)t} \tag{4}$$

where c_o is the initial air concentration, f is an integrating factor obtained by assuming that emissions originate from the center of each grid element, function $F(s)$ describes the probability of transport from one of eight geographical sectors, and the coefficients $\bar{k}_w$ and $\bar{k}_d$ are the spatially and temporally-aggregated wet and dry removal rates computed from the trajectory model. The variable t is the time of travel between source and receptor.

Table 1. Measurement data base and average model results.

Metal	Constituent	N[1]	Countries with Observations	Years with Observations	Computed Avg.[2]	Observed Avg.[2]
As	Air Conc.	6	Bel, Ger, Nld, Nor	1978, 79, 81, 85	1.87	1.78
As	Wet Depos.	9	Nld, Nor	1978, 80, 82	0.55	0.45
Cd	Air Conc.	41	Bel, Ger, Nld, Nor	1978, 79, 80, 81, 82, 83, 85	0.70	0.82
Cd	Wet Depos.	31	Bel, Ger, Hun, Nld, Nor, Swe	1978, 79, 80, 81, 82, 83, 84, 85	0.12	0.38
Pb	Air Conc.	41	Bel, Fra, Ger, Nld, Nor	1978, 79, 80, 81, 82, 83, 85	85.77	68.76
Pb	Wet Depos.	15	Ger, Nld, Nor, Swe	1979, 80, 81, 83, 84, 85	9.53	11.84
Zn	Air Conc.	8	Bel, Fra, Ger, Nor	1978, 80, 81, 85	32.28	126
Zn	Wet Depos.	12	Dnk, Nld, Nor, Swe	1978, 82, 83, 84, 85	6.22	12.80

[1]Number of measurements.

[2]Average of all stations. Units: Concentration = ng/m^3, Wet deposition = $mg/m^2 \cdot a$.

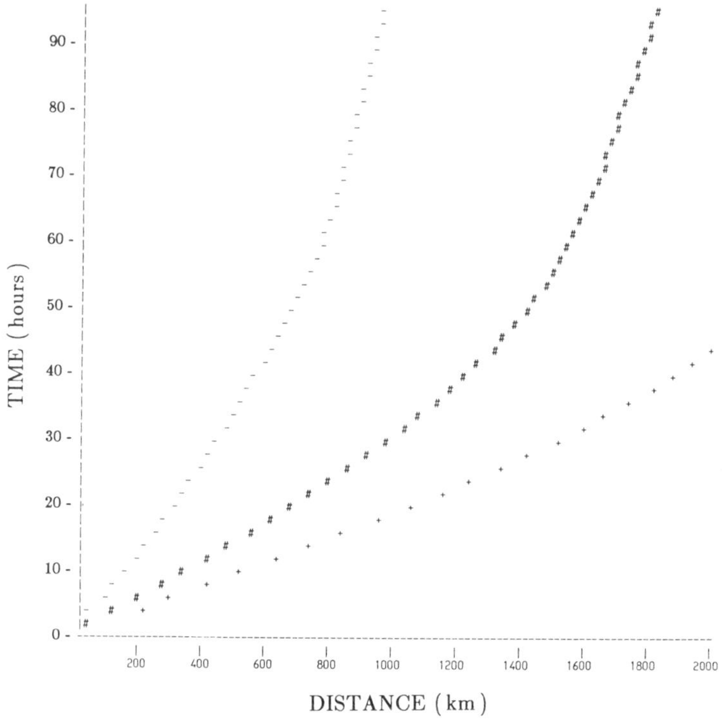

Fig. 1. Computed travel time vs distance between sources and receptors.

In the second step, the air concentration of metal remaining in the air parcel when it arrives at the receptor is subject to wet and dry deposition. This deposition is calculated using expressions (1) and (3) except that all variables and coefficients are annual averages rather than 6-h averages as in the trajectory model. The concentration C comes from equation (4)*, and the remaining coefficients are derived from the trajectory model.

Meteorological Assumptions. The usual way of estimating the transport time between a source and receptor in a climatologic model is to assume a constant ***a priori*** wind velocity. As an alternative, we compute a more realistic transport time using a relationship between distance and transport time derived from the trajectory model. This relationship (Figure 1) was computed from the travel times and transport distances of trajectories arriving at three receptors in Europe over the course of one year (1985), and accounts for the fact that air parcels seldom travel in a straight line at a constant speed between sources and receptors. The two light lines in this figure indicate the standard deviation. The main relationship is approximated by:

$$t = 6.67 \times 10^{-2}\, x + 2.59 \times 10^{-5}\, x^2 \tag{5}$$

where t and x have units of hours and kilometers, respectively. Equation (5) is used to compute t in Equation (4).

After obtaining the transport time between a source and receptor, we must still estimate F(s) in Equation (4), the percent of time in one year that winds come from a given source to a receptor. This is computed from wind trajectory sector analyses which give the percentage of 96-h back trajectories (based on the 925 hPa wind) that arrive at a receptor from different compass directions. Sector analyses were computed by the authors for each year from 1978 to 1986 based on data from the EMEP Synthesizing Center West. These analyses are available at 84 to 107 stations in Europe (the number of stations with data varied from year to year.)

Other Model Inputs. Emissions on the 150 $\times$ 150 km EMEP grid covering all Europe were taken from Pacyna (1988) and Pacyna and Münch (1988). Estimates are available for 1982 for As, Cd, Pb, and Zn and for 1985 for Pb. These estimates are preliminary. A "local deposition coefficient" of 0.10 to 0.15 was assigned to the emissions to account for the fraction of metals deposited in the vicinity of the emission source.

Testing the TRACE Model Against Measurements

Results for the TRACE climatologic model were close to results from the trajectory model. Details of this comparison are outside the scope of the present paper but will be reported in later publications. In this paper we will concentrate on the comparison of TRACE model results with measurements.

Because there is no coordinated monitoring program for heavy metals in Europe, it was difficult to select measurements suitable for comparing to model calculations. Our measurement data were drawn from a wide range of published and unpublished sources. A large error is expected in these data because they largely lack quality control, intercomparison analysis, or any agreed-upon standard sampling or analytical method. In addition, measurements have been taken at different times and at widely scattered locations throughout Europe and therefore lack any semblance to a statistically-valid network. Also, concentration measurements were rarely taken at the same time as deposition measurements; and most of the wet deposition measurements were taken with bulk samplers which are contaminated by an unspecified amount of dry deposition. Even under ideal conditions aerosol sampling has a large uncertainty. The data base is summarized in Table 1.

Results for computed air concentration and wet deposition of As, Cd, and Pb are shown in Figure 2. Included in this figure are measurements from 9 years and 8 countries in

*This assumption may lead to a technical inconsistency between the trajectory and climatologic models because the C in eqs (1) and (3) are not the same for the climatologic and trajectory models. This point will be taken up in later publications.

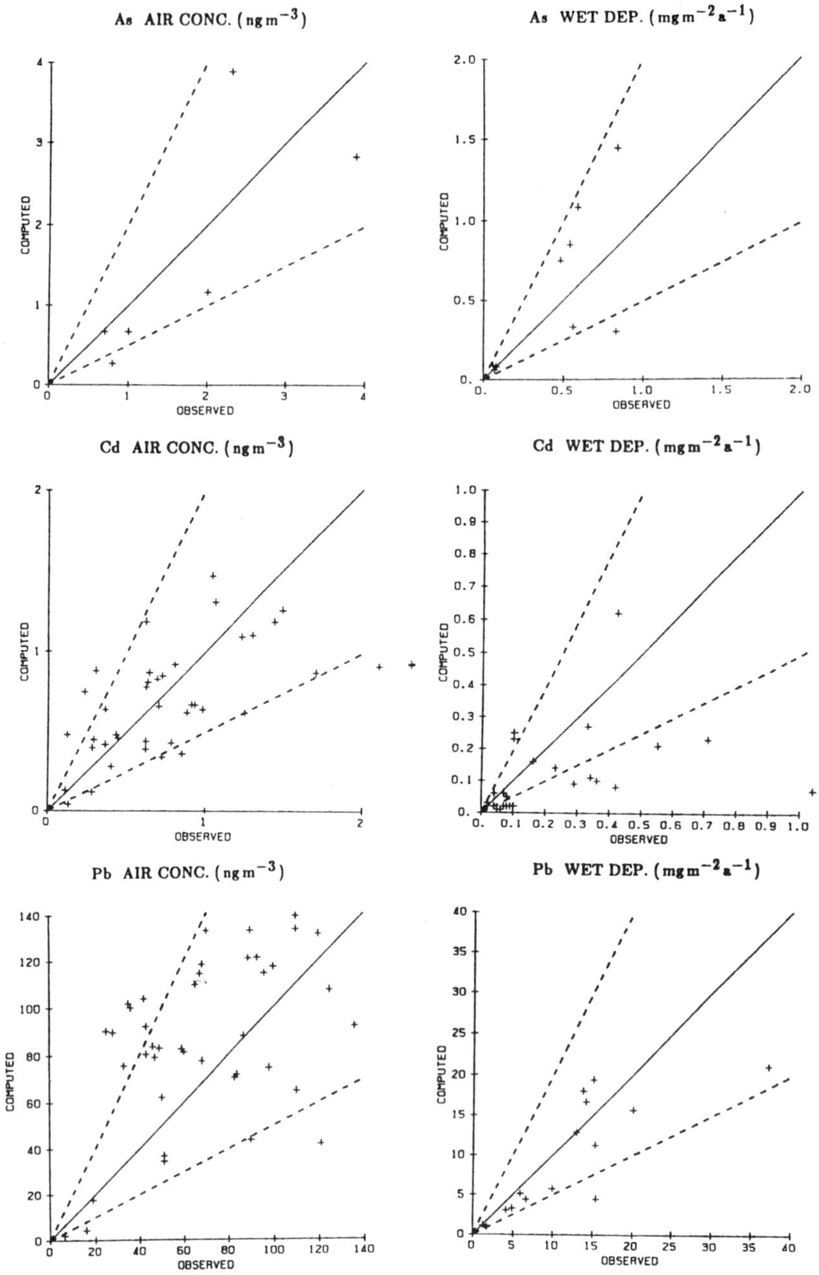

Fig. 2. Computed vs observed annual air concentration and wet deposition for As, Cd, and Pb. Solid line indicates perfect agreement, dashed lines indicate agreement within a factor of two.

Europe. Average model results for all metals are summarized in Table 1. These results are preliminary because the model is still under development.

Computations of annual air concentration of As, Cd and Pb compare favorably with measurements; most calculations are within a factor of two of measurements (Figure 1). Average computed concentrations are also close to average measurements, indicating no strong bias in calculations (Table 1). No statistics for the comparisons other than averages are presented in Table 1 or Figure 2 because of the preliminary nature of model results. Agreement of model and observations is also good for As and Pb wet deposition, but not as good for Cd. Nevertheless, the model computes within the correct order of magnitude of Cd measurements, computing very low deposition values where low values are observed, and higher values where high values are observed. As pointed out above, part of the disagreement of model calculations and deposition measurements may be attributed to the collection of an unspecified amount of dry deposition by bulk precipitation samplers. As expected, when we compare the Cd deposition measurements in Figure 1 with wet plus dry deposition calculations, the model does not underestimate measurements as much.

In all of the plots in Figure 2, the station Birkenes is an outlier. The model consistently underestimates concentration and deposition of all metals at this station. Perhaps this station is influenced by local emission sources. Evidence of local sources is suggested by maps of concentration of these metals in moss which show strong spatial gradients in the vicinity of Birkenes (Rühling et al., 1987).

Calculations of Zn concentration and deposition underestimate the measurements (Table 1), although average calculated wet deposition is within an order of magnitude of the average of measurements. It is possible that Zn emissions have been underestimated.

Results

Atmospheric Residence Time of Metals. It can be shown (Seinfeld, 1986) that the residence time in the regional atmosphere of a typical particle, assuming first order removal processes, is the sum of the inverse of its wet and dry deposition rates (not accounting for the fraction of metals deposited in the vicinity of emission sources). For As, Cd, and Zn, $k_w \approx 7.81 \times 10^{-6}\ s^{-1}$ and $k_d \approx 2.49 \times 10^{-6}\ s^{-1}$ averaged over one year and all Europe. This yields a residence time of 27 hours. For Pb, $k_w \approx 3.12 \times 10^{-6}\ s^{-1}$ and $k_d \approx 1.25 \times 10^{-6}\ s^{-1}$ which yields a much longer residence time of 64 h. This is due to the larger number of smaller-diameter particles in the Pb mass size distribution compared to the other metals which results in a lower dry deposition velocity for Pb. Lead was also assigned a lower wet scavenging ratio. We can compute the transport distance of the typical metal particles by using the relationship between distance and transport time in Figure 1. This yields a transport distance of about 900 km for As, Cd, and Zn, and 1600 km for Pb. There is, of course, great variability in the distance particles travel. We can compute the distance particles travel 90% of the time by multiplying the standard deviations indicated in Figure 1 by 1.64. Hence we estimate that the transport distance of the typical particle of As, Cd and Zn is between 100 and 1700 km, 90% of the time. These calculations assume that 10–15% of the metals from emission sources settle out in the vicinity of sources.

Deposition Maps. Figure 3 presents deposition (wet plus dry) maps for Europe computed by the TRACE model. Meteorological inputs from 1985 are used for the calculations. The As deposition pattern shows two peaks – one in the FRG and one in southern Poland. The maximum deposition is above 4 mg $m^{-2}\ a^{-1}$ in southern Poland and declines to 0.1 mg $m^{-2}\ a^{-1}$ on the edge of the continent.

The Cd pattern shows several peaks – Donetsk in the USSR, northern Spain, southern Poland-northern Czechoslovakia, and the Benelux countries. The maximum deposition is 1.4 mg $m^{-2}\ a^{-1}$, and decreases to less than 0.1 mg $m^{-2}\ a^{-1}$ in the Nordic countries.

The pattern of Pb deposition does not show the strong gradients of the other metals. There are two likely factors; the first is that the predominant sources of Pb in Europe are widely distributed mobile sources, as compared to isolated point sources for the other metals. The second is that the estimated average transport distance is longer for Pb than for the other metals. Hence, a unit of Pb emissions is deposited over a wider area than a unit of emissions from the other metals. A large area of Central Europe has Pb deposition

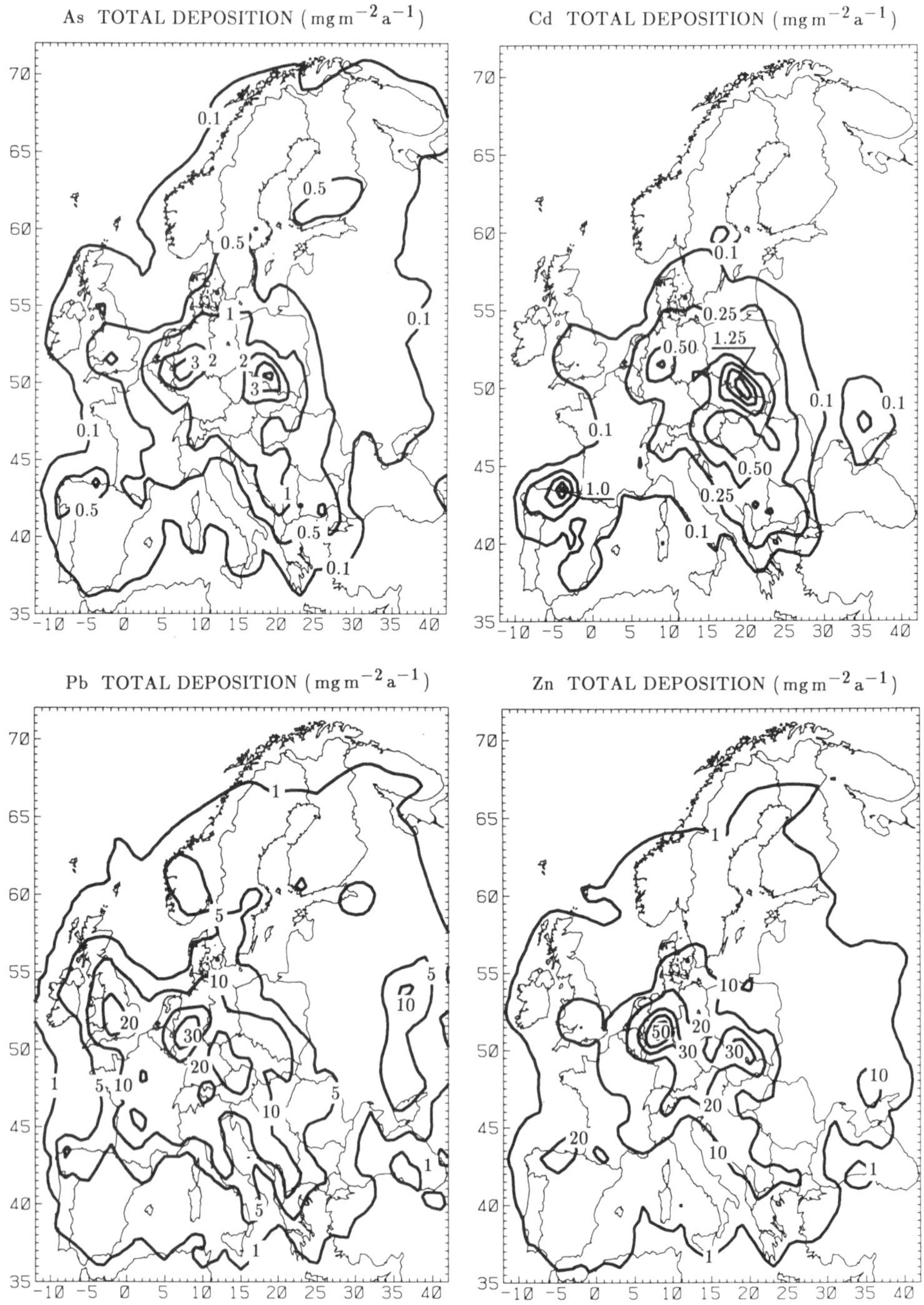

Fig. 3. Computed total (wet plus dry) deposition of As, Cd, Pb and Zn. Meteorology from 1985 is used for calculations.

greater than 10 mg m^{-2} a^{-1}. Deposition decreases to 1 mg m^{-2} a^{-1} in the Nordic countries.

Zinc has a deposition pattern similar to Arsenic which reflects the fact that a major source of As is Zn smelting. Deposition levels of Zn, however, are much higher than As, peaking at 57 mg m^{-2} a^{-1} in the FRG and only declining to 1 mg m^{-2} a^{-1} in the Nordic countries. These are probably underestimates of actual total deposition because, as pointed out previously, the TRACE model underestimates Zn observations.

All calculations presented in this paper should be considered preliminary because of improvements currently being made to the TRACE model. For example, calculations for 1979–1986 were made with trajectory sector analyses from those years but with precipitation, mixing height, and friction velocity data from only 1985. In the future we will use improved meteorological data from each year between 1979 and 1986. Also emissions for As, Cd and Zn were only available for 1982; we expect to use updated improved data in the future. Future publications will also deal with sensitivity and uncertainty analyses of the model, which were outside the scope of this paper.

Summary and Conclusions

A combined trajectory-climatologic approach is used to compute the long range transport and air concentration/deposition of heavy metals in Europe. A modified version of the EMEP trajectory model is used for generating time- and space-average inputs to the TRACE climatologic model. The advantage of the combined approach is that the climatologic model can be conveniently run for all Europe and several years, and thereby can make use of the very scattered measurements available for metals. At the same time the trajectory model provides a scientific basis for estimating the aggregated parameters of the climatologic model.

Most TRACE model calculations of As, Cd, and Pb agree within a factor of two of observed annual air concentrations and deposition. Agreement for Zn is within an order of magnitude. Using the model, a typical residence time of 27 h in Europe's atmosphere was computed for As, Cd, and Zn and 64 h for Pb (not accounting for the fraction of metals deposited in the vicinity of emissions sources).

The computed total deposition pattern of As, Cd and Zn shows strong gradients in Europe whereas the Pb pattern is more smooth. This is probably due to the predominance of mobile sources for Pb, and by lead's longer transport distance, as compared to the other metals.

Acknowledgements

The authors are indebted to Jozef Pacyna for providing emissions data for our calculations, and to Jørgen Saltbones for his help with the meteorological data.

References

Chan, W.H., Tang, A.J.S., Chung, D.H.S., and Lusis, M.A., 1982, Concentration and deposition of trace metals in Ontario-1982, Water, Air, and Soil Pollution 29:373–389.

Dulac, F., Buart-Menard, P., Ezat, U., Melki, S., and Bergametti, G., 1989, Atmospheric input of trace metals to the western Mediterranean: uncertainties in modelling dry deposition from cascade impactor data, Tellus, 41B:362–378.

Eliassen, A. and Saltbones, J., 1983, Modelling of long-range transport of sulphur over Europe: a two-year model run and some model experiments. Atmospheric Environment, 17:1457–1473.

Eliassen, A., Hov, O., Iversen, T., Saltbones, J., and Simpson, D., 1988, Estimates of airborne transboundary transport of sulphur and nitrogen over Europe, EMEP/MSC-W Report 1/88:1–79.

Georgii, H.W., Perseke, C., and Rohbock, E., 1982, Feststellung der deposition von sauren und langzeitwirksamen luftverunreinigungen aus belastungsgebieten, Umweltforschungsplan des Bundesminiters des Innern, Luftreinhaltung Forschungsprojekt 104 02 600.

Gatz, D.F., 1975, Pollutant aerosol deposition into southern lake Michigan, Water, Air and Soil Pollution 5:239–251.

Ottar, B., Lindberg, S.E., Voldner, E., Lingqvist, O., Mayer, R., Semb, A., Steinnes, E., and Watt, J., 1989, Special topics concerning interactions of heavy metals with the environment, in: Control and Fate of Atmospheric Trace Metals: 365–372, J.M. Pacyna and B. Ottar, ed., Kluwer Academic Publishers.

Pacyna, J.M., Semb, A., and Hanssen, J.E., 1984, Emission and long-range transport of trace elements in Europe, Tellus 36B:163–178.

Pacyna, J.M., 1988, Atmospheric lead emissions in Europe in 1985, NILU OR:19/88:1–10.

Pacyna, J.M., and Münch, J., 1987, Atmospheric emissions of As, Cd, Pb and Zn from industrial sources in Europe, in: S.E. Lindberg, and T.C. Hutchinson, ed., Heavy Metals in the Environment, Vol.1, New Orleans, CEP Consultants, Edinburgh.

Pacyna, J.M., and Münch, J., 1988, Atmospheric emissions of arsenic, cadmium, mercury and zinc in Europe in 1982, NILU OR:17/83:1–33.

Rühling, A., Rasmussen, L., Pilegaard, K., Mäkinen, A., and Steinnes, E., 1987, Survey of atmospheric heavy metal deposition, Report prepared for The Steering Body for Environmental Monitoring, The Nordic Council of Ministers. NORD 1987:21:1–44.

Sehmel, G.A., 1980, Particle and gas dry deposition: a review, Atmospheric Environment, 14:983–1011.

Seinfeld, J.H., 1986, Atmospheric chemistry and physics of air pollution, John Wiley & Sons, New York.

Semb, A., and Pacyna, J.M., 1988, Toxic trace elements and chlorinated hydrocarbons: sources, atmospheric transport and deposition, Miljørapport, NORD 1988:74. pp. 1–96.

Slinn, W.G.N., 1984, Precipitation scavenging, in: Atmospheric Science and Power Production, D. Randerson, ed., Technical Information Center, Office of Scientific and Technical Information, United States Department of Energy.

Stevens, R.K., Dzubay, T.G., Lewis, C.W., and Shaw Jr., R.W., 1984, Source apportionment methods applied to the determination of the origin of ambient aerosols that affect visibility in forested areas, Atmospheric Environment Vol. 18, No. 2:261–272.

WMO, 1989, Air Quality Guidelines, Review Draft, Vol. 11, World Health Organization, Regional Office for Europe, Copenhagen, Denmark.

APPLICATION OF THE RECEPTOR ORIENTED APPROACH IN MESOSCALE DISPERSION MODELING

Marek Uliasz[1] and Roger A. Pielke

Colorado State University, Department of Atmospheric Science
Fort Collins, CO 80523, USA

SOURCE ORIENTED AND RECEPTOR ORIENTED MODELING

A traditional source oriented approach in dispersion modeling consists in solving model equations forward in time for given sources of pollutant (Figure 1). As a result time and space dependent concentration fields **C** are obtained. In order to investigate another variant of emission **Q**, the solution of the model equations must be repeated. In many practical applications, air pollution at a given receptor is of primary interest and an alternative receptor oriented modeling should be considered as a more effective approach. In this case, air quality at the receptor is characterized by an integral of pollution concentration over the modeling domain and time of simulation $\Phi[\mathbf{C}]$. The integral can be defined in accordance to the aim of the study using a receptor function **R**. Instead of concentration, an influence function $\mathbf{C}^*$ is calculated which provides information on contributions from different sources to air pollution at the receptor.

The influence function can be calculated from backward trajectories of particles if a Lagrangian dispersion model is used. Eulerian models governed by partial differential equations are formulated in a variational framework and then the influence function is obtained as a solution of adjoint equations backward in time with the receptor function **R** as a source term. The influence function calculated for a given receptor depends on meteorology, deposition, and transformations of pollutant in the atmosphere but is independent of emission sources. The air quality integral can be calculated in the alternative way by averaging emission field **Q** with the influence function as a weight function. It is possible to simply calculate $\Phi[\mathbf{C}]$ for any distribution of pollutant sources (e.g., to investigate different heights of a planned stack) without repeating the solution of model equations. The receptor oriented modeling is computationally more effective than the source oriented one if the number of considered receptors is less than the number of emission distributions to be investigated.

The aim of this paper is to formulate the receptor oriented approach for Eulerian dispersion models. Examples of mesoscale dispersion calculations for a coastal environment of the Baltic Sea are also presented.

VARIATIONAL FORMULATION OF A DISPERSION MODEL

Let us consider an Eulerian dispersion model given by a set of linear advection-diffusion equations using K-theory which describe the dispersion of a number of pollutants in the spatial domain $\{0 \leq x \leq L_x, 0 \leq y \leq L_y, 0 \leq z \leq H\}$. The model equations are written in a vector form, where **C** represents chemical species:

[1] on leave from Technical University of Warsaw, Poland

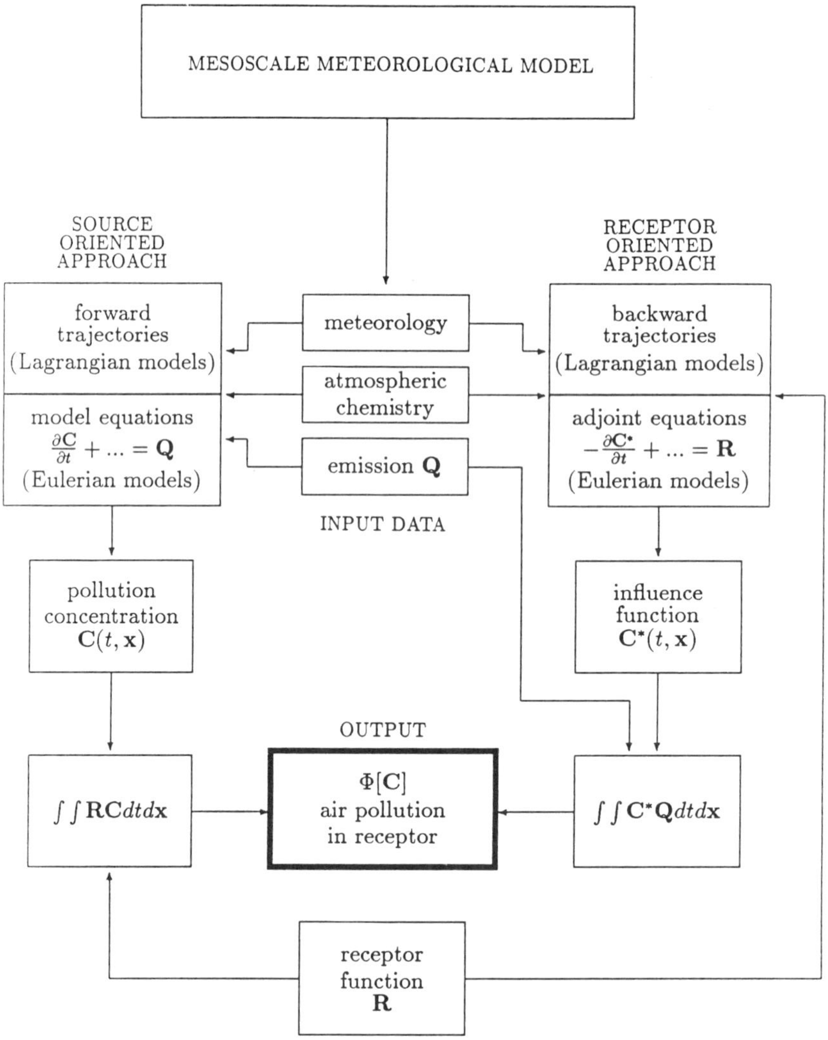

Fig.1. Source oriented and receptor oriented approaches in air pollution modeling

$$\frac{d\mathbf{C}}{dt} = \Delta_H \mathbf{C} + \frac{\partial}{\partial z} K_z \frac{\partial \mathbf{C}}{\partial z} + \mathbf{AC} + \mathbf{Q} \tag{1}$$

with the initial condition

$$at \quad t = 0, \qquad \mathbf{C} = \mathbf{C}_0 \tag{2}$$

and boundary conditions

$$x = 0, \quad K_H \frac{\partial \mathbf{C}}{\partial x} = \tilde{u}(\mathbf{C} - \mathbf{C}_E), \qquad x = L_x, \quad K_H \frac{\partial \mathbf{C}}{\partial x} = \hat{u}(\mathbf{C} - \mathbf{C}_W) \tag{3}$$

$$y = 0, \quad K_H \frac{\partial \mathbf{C}}{\partial y} = \tilde{v}(\mathbf{C} - \mathbf{C}_S), \qquad y = L_y, \quad K_H \frac{\partial \mathbf{C}}{\partial y} = \hat{v}(\mathbf{C} - \mathbf{C}_N) \tag{4}$$

$$z = 0, \quad K_z \frac{\partial \mathbf{C}}{\partial z} = \mathbf{v}_d \mathbf{C} - \mathbf{q}, \qquad z = H, \quad K_z \frac{\partial \mathbf{C}}{\partial z} = \hat{w}\mathbf{C} \tag{5}$$

where

$$\frac{d}{dt} = \frac{\partial}{\partial t} + u\frac{\partial}{\partial x} + v\frac{\partial}{\partial y} + w\frac{\partial}{\partial z}, \qquad \Delta_H = \frac{\partial}{\partial x} K_H \frac{\partial}{\partial x} + \frac{\partial}{\partial y} K_H \frac{\partial}{\partial y},$$

$\tilde{a} = max(a,0)$, $\hat{a} = min(a,0)$, $a = u,v,w$ - wind velocity components, K_H, K_z - eddy diffusivities, $\mathbf{A}$ - matrix describing chemical reactions between pollutants as well as removal processes such as washout, $\mathbf{Q}$ - emission, $\mathbf{q}$ - surface area emission sources, $\mathbf{v}_d$ - deposition velocity, $\mathbf{C}_E$, $\mathbf{C}_W$, $\mathbf{C}_S$ and $\mathbf{C}_N$ - background concentrations in the flow from the east, west, south and north, respectively which characterize the effect of distant sources located outside the modeling domain.

Air quality at the receptor is characterized by an integral of the weighted concentration field over the time and space modeling domain:

$$\Phi[\mathbf{c}] = \int_{t_0}^{T}\int_{0}^{L_x}\int_{0}^{L_y}\int_{0}^{H} \mathbf{R}\mathbf{C}\,dxdydzdt \qquad (6)$$

The receptor function can be written in a general form $\mathbf{R} = \eta\mathbf{B}$. The function η defines a location and form of the receptor (point, area, volume), and a required averaging time of concentration. It is equal zero outside the receptor. The diagonal matrix $\mathbf{B}$ takes into account contributions of particular pollutants. The integral (6) can be interpreted as an average of concentration fields over time and space with the receptor function as a weight function. Other forms of the integral (6) can be also considered (e.g., average deposition flux of pollutant for a selected area).

The influence function $\mathbf{C}^*$ should be determined in order to express the integral (6) explicitly in terms of pollution sources $\mathbf{Q}$, $\mathbf{q}$, $\mathbf{C}_E$, $\mathbf{C}_W$, $\mathbf{C}_S$ and $\mathbf{C}_N$. For this purpose, a variational formulation of the model equations is useful (Marchuk, 1982, 1985, Uliasz, 1983). The model equation (1) is premultiplied by $\mathbf{C}^{*T}$ (superscript T denotes a transposed matrix) and integrated over the spatial modeling domain and the time of simulation $\{t_0, T\}$. The time, advection, and diffusion terms are then integrated by parts (diffusion terms twice). It allow us to introduce the initial and boundary conditions (2-5) into the integral expression:

$$\begin{aligned}
&\int_{t_0}^{T}\int_{0}^{L_x}\int_{0}^{L_y}\int_{0}^{H}\left(-\frac{d\mathbf{C}^{*T}}{dt} - \Delta_H K_H \mathbf{C}^{*T} - \frac{\partial}{\partial z}K_z\frac{\partial \mathbf{C}^{*T}}{\partial z} - \mathbf{C}^{*T}\mathbf{A}\right)\mathbf{C}\,dtdxdydz\, +\\
&- \int_{t_0}^{T}\int_{0}^{L_x}\int_{0}^{L_y}\int_{0}^{H}\mathbf{C}^{*T}\mathbf{Q}\,dtdxdydz + \int_{0}^{L_x}\int_{0}^{L_y}\int_{0}^{H}\left\{\left[\mathbf{C}^{*T}\right]_{t=t_0}\mathbf{C}_0 - \left[\mathbf{C}^{*T}\mathbf{C}\right]_{t=T}\right\}dxdydz\\
&+ \int_{t_0}^{T}\int_{0}^{L_y}\int_{0}^{H}\left\{\left[\left((u-\hat{u})\mathbf{C}^{*T} + K_H\frac{\partial \mathbf{C}^{*T}}{\partial x}\right)\mathbf{C} + \hat{u}\mathbf{C}^{*T}\mathbf{C}_W\right]_{x=L_x} -\right.\\
&\qquad\left. - \left[\left((u-\tilde{u})\mathbf{C}^{*T} + K_H\frac{\partial \mathbf{C}^{*T}}{\partial x}\right)\mathbf{C} + \tilde{u}\mathbf{C}^{*T}\mathbf{C}_E\right]_{x=0}\right\}dtdydz\\
&+ \int_{t_0}^{T}\int_{0}^{L_x}\int_{0}^{H}\left\{\left[\left((v-\hat{v})\mathbf{C}^{*T} + K_H\frac{\partial \mathbf{C}^{*T}}{\partial y}\right)\mathbf{C} + \hat{v}\mathbf{C}^{*T}\mathbf{C}_N\right]_{y=L_y} -\right.\\
&\qquad\left. - \left[\left((v-\tilde{v})\mathbf{C}^{*T} + K_H\frac{\partial \mathbf{C}^{*T}}{\partial y}\right)\mathbf{C} + \tilde{v}\mathbf{C}^{*T}\mathbf{C}_S\right]_{y=0}\right\}dtdxdz\\
&+ \int_{t_0}^{T}\int_{0}^{L_x}\int_{0}^{L_y}\left\{\left[\left((w-\hat{w})\mathbf{C}^{*T} + K_z\frac{\partial \mathbf{C}^{*T}}{\partial z}\right)\mathbf{C}\right]_{z=H} -\right.\\
&\qquad\left. - \left[\left(-\mathbf{C}^{*T}\mathbf{v}_d + K_z\frac{\partial \mathbf{C}^{*T}}{\partial z}\right)\mathbf{C} - \mathbf{C}^{*T}\mathbf{q}\right]_{z=0}\right\}dtdxdy = 0 \qquad (7)
\end{aligned}$$

The equation for the influence function, $\mathbf{C}^*$, can now be determined in order to reduce the expression to the relationship between the integral $\Phi[\mathbf{C}]$ and the source terms. All terms with concentration, $\mathbf{C}$, are required to be equal to the receptor function $\mathbf{R}$ or zero. These conditions form after transposing the following equation for $\mathbf{C}^*$

$$-\frac{d\mathbf{C}^*}{dt} = \Delta_H\mathbf{C}^* + \frac{\partial}{\partial z}K_z\frac{\partial\mathbf{C}^*}{\partial z} + \mathbf{A}^T\mathbf{C}^* + \mathbf{R} \tag{8}$$

with the initial condition

$$t = T, \quad \mathbf{C}^* = 0 \tag{9}$$

and boundary conditions

$$x = 0, \quad K_H\frac{\partial\mathbf{C}^*}{\partial x} = -\hat{u}\mathbf{C}^*, \qquad x = L_x, \quad K_H\frac{\partial\mathbf{C}^*}{\partial x} = -\tilde{u}\mathbf{C}^* \tag{10}$$

$$y = 0, \quad K_H\frac{\partial\mathbf{C}^*}{\partial y} = -\hat{v}\mathbf{C}^*, \qquad y = L_y, \quad K_H\frac{\partial\mathbf{C}^*}{\partial y} = -\tilde{v}\mathbf{C}^* \tag{11}$$

$$z = 0, \quad K_z\frac{\partial\mathbf{C}^*}{\partial z} = \mathbf{v}_d\mathbf{C}^*, \qquad z = H, \quad K_z\frac{\partial\mathbf{C}^*}{\partial z} = -\tilde{w}\mathbf{C}^* \tag{12}$$

The final form of the integral equation (7) allows one to express air pollution at the receptor defined by (6) as a sum of contributions from different sources:

$$\begin{aligned}
\Phi[\mathbf{C}] = & \underbrace{\int_{t_0}^{T}\int_0^{L_x}\int_0^{L_y}\int_0^{H}\mathbf{C}^{*T}\mathbf{Q}\,dt\,dx\,dy\,dz}_{I} + \underbrace{\int_{t_0}^{T}\int_0^{L_x}\int_0^{L_y}\left[\mathbf{C}^{*T}\right]_{z=0}\mathbf{q}\,dt\,dx\,dy}_{II} + \\
+ & \underbrace{\int_0^{L_x}\int_0^{L_y}\int_0^{H}\left[\mathbf{C}^{*T}\right]_{t=t_0}\mathbf{C}_0\,dx\,dy\,dz}_{III} + \\
+ & \underbrace{\int_{t_0}^{T}\int_0^{L_y}\int_0^{H}\left(\tilde{u}\left[\mathbf{C}^{*T}\right]_{x=0}\mathbf{C}_E + \hat{u}\left[\mathbf{C}^{*T}\right]_{x=L_x}\mathbf{C}_W\right)dt\,dy\,dz}_{IV} + \\
+ & \underbrace{\int_{t_0}^{T}\int_0^{L_x}\int_0^{H}\left(\tilde{v}\left[\mathbf{C}^{*T}\right]_{y=0}\mathbf{C}_S + \hat{v}\left[\mathbf{C}^{*T}\right]_{y=L_y}\mathbf{C}_N\right)dt\,dx\,dz}_{IV}
\end{aligned} \tag{13}$$

where

- I - contribution from sources inside the modeling domain described by $\mathbf{Q}$;
- II - contribution from area surface sources described by $\mathbf{q}$;
- III - contribution from the initial pollution, $\mathbf{C}_0$, at time $t = t_0$;
- IV - contributions from distant sources outside the modeling domain characterized by background concentrations $\mathbf{C}_E$, $\mathbf{C}_W$, $\mathbf{C}_S$ and $\mathbf{C}_N$.

COMPUTER IMPLEMENTATION

The dispersion model in the variational formulation has been implemented for a single pollutant and included in Mesoscale Dispersion Modeling System (MDMS) designed for personal computers and workstations (Uliasz, 1990a,b). The main part of this system is a 3-D hydrostatic mesoscale meteorological model with a second-order turbulence closure. The same numerical algorithm based on the chapeau function method is used to solve the dispersion model equation and the adjoint equation. In the receptor oriented option, meteorological fields created by the mesoscale model are read backward in time.

The K-theory dispersion model is applicable for large area or volume emission sources (or receptors). To overcome this limitation, a hybrid Lagrangian-Eulerian model is under development, where the Lagrangian particle model available in the MDMS is applied to simulate subgrid scale dispersion close to point and/or line emission sources (or receptors). After sufficiently long travel time, particles are assumed to contribute to a volume emission field in the Eulerian grid model and then disappear. The hybrid dispersion model is expected to be more efficient than the pure Lagrangian particle model which requires tracking of several thousand particles.

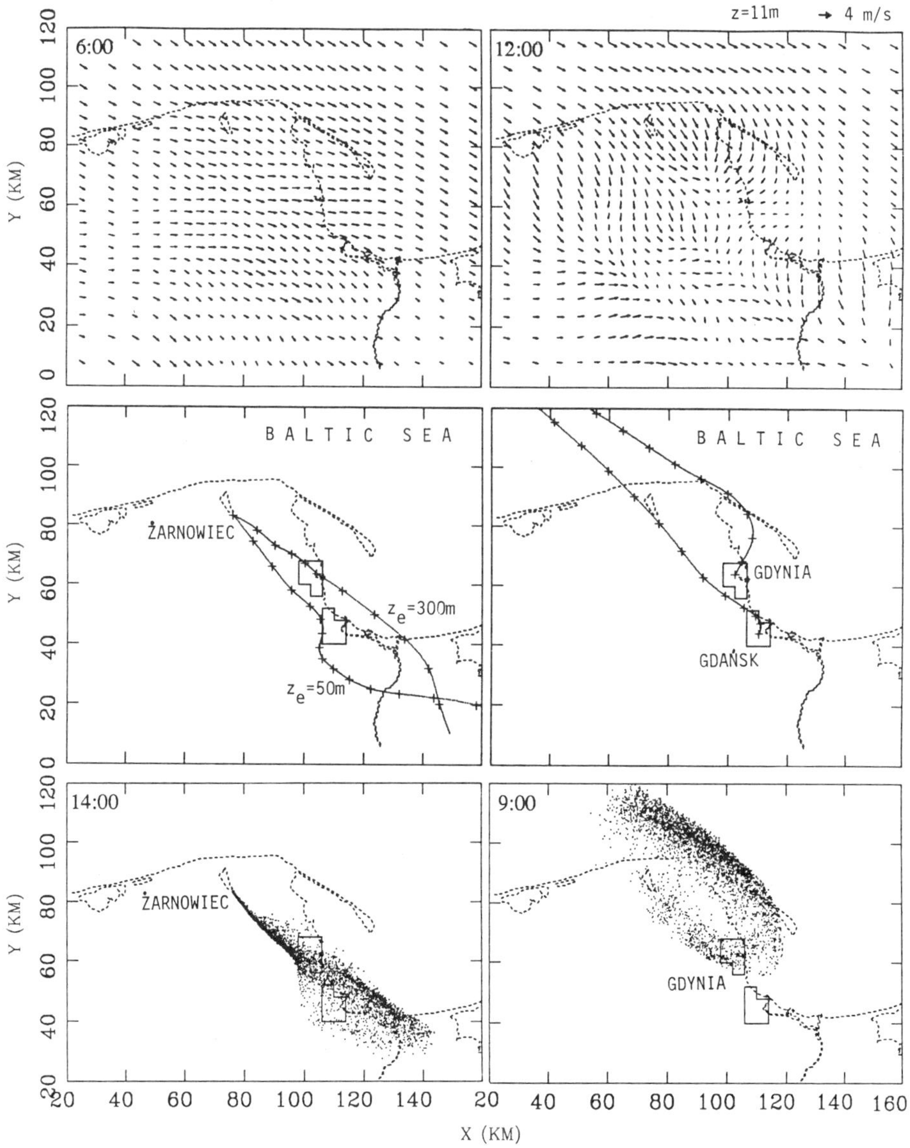

Fig.2. Simulated surface wind fields for the 2nd day of simulation (top). Forward trajectories starting at 6:00 at Żarnowiec (z_e = 50 m, z_e = 300 m), and backward trajectories starting at 12:00 over two area receptors at z = 50 m; particle positions are marked by cross every hour (middle). Particle distribution calculated forward in time for continuous release from the Żarnowiec point source (z_e = 50 m) starting at 6:00 and particle distribution calculated backward in time for the Gdynia receptor with release time 12:00 - 15:00 (bottom).

SAMPLE RESULTS

Sample calculations of influence functions have been performed with the MDMS for a hilly coastal zone of the Baltic Sea in northern Poland. Results from a 48-hour 3-D meteorological simulation (June cloudless conditions, a synoptic wind of 5 m/s from NW, $30 * 25 * 20$ gridpoints, $\Delta x = 4$ km in the central part of the $170 * 120 * 3$ km modeling domain) are used as input for the dispersion model. Figure 2 shows simulated surface wind fields and some results of the source oriented and receptor oriented modeling using the Lagrangian particle dispersion model. The model has been run forward in time for a point emission source (Żarnowiec nuclear power plant under construction) and backward in time for two area receptors - cities of Gdynia and Gdańsk. The backward particle trajectories calculated without atmospheric turbulence are an example of the simplest receptor modeling within a Lagrangian framework.

The influence functions are computed for these two area receptors and two averaging times of concentration (6:00 - 9:00 and 12:00 - 15:00 for the 2nd day of simulation) using the Eulerian dispersion model. For a single point emission source fixed in time and space $Q = e\delta(x - x_e)\delta(y - y_e)\delta(z - z_e)$ with constant emission rate e, the integral expression (13) reduces to:

$$\Phi[C] = e \int_0^T C^*(x_e, y_e, z_e, t)dt \tag{14}$$

Therefore, the time integrated field of influence function is well suited to characterize air pollution meteorology for a given receptor (Figure 3). $\int C^*(x, y, z, t)dt$ provides information on the contribution of the point emission source located at x, y, z to air pollution at the receptor.

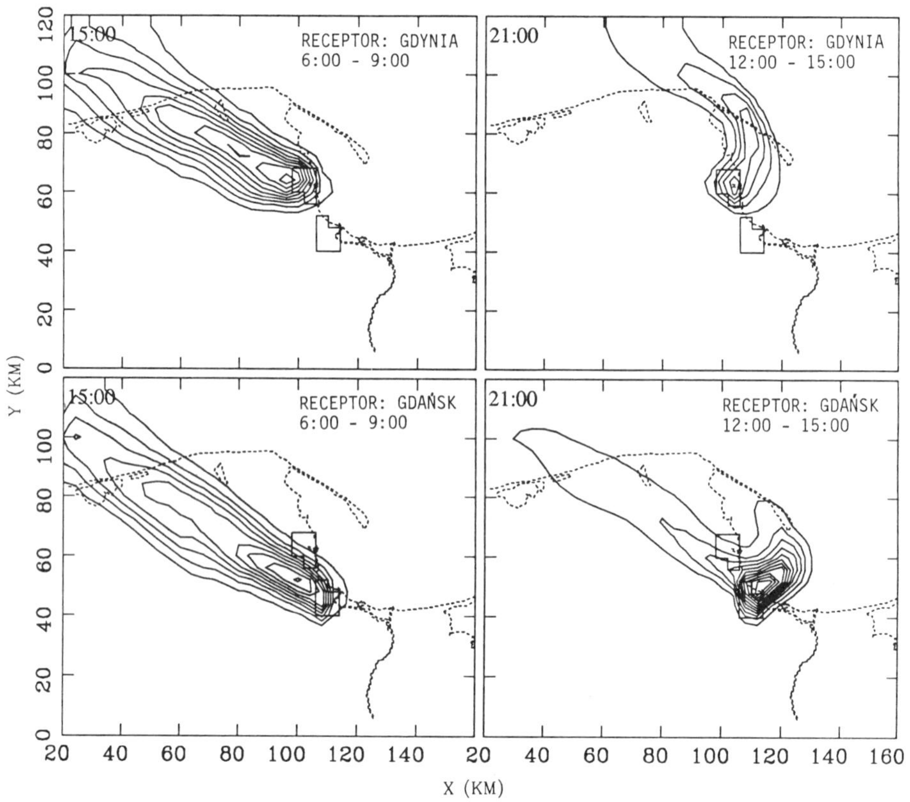

Fig.3. Time integrated fields of influence functions C^* ($z = 100$ m) calculated for two receptors and two averaging times. Contour step is $10^{-6}\ sm^{-3}$ and the outer most contour has a value $10^{-6}\ sm^{-3}$.

Figure 4 presents time runs of the influence functions at Żarnowiec at different heights. The period of time when the influence function is nonzero describes an effective travel time of pollutant between an emission source and a receptor. It indicates as well how long the model is influenced by its initial conditions since the value of $C^*(x, y, z, t)$ characterizes contribution from the initial concentration $C_0(x, y, z, t)$ for the start of the simulation at time t.

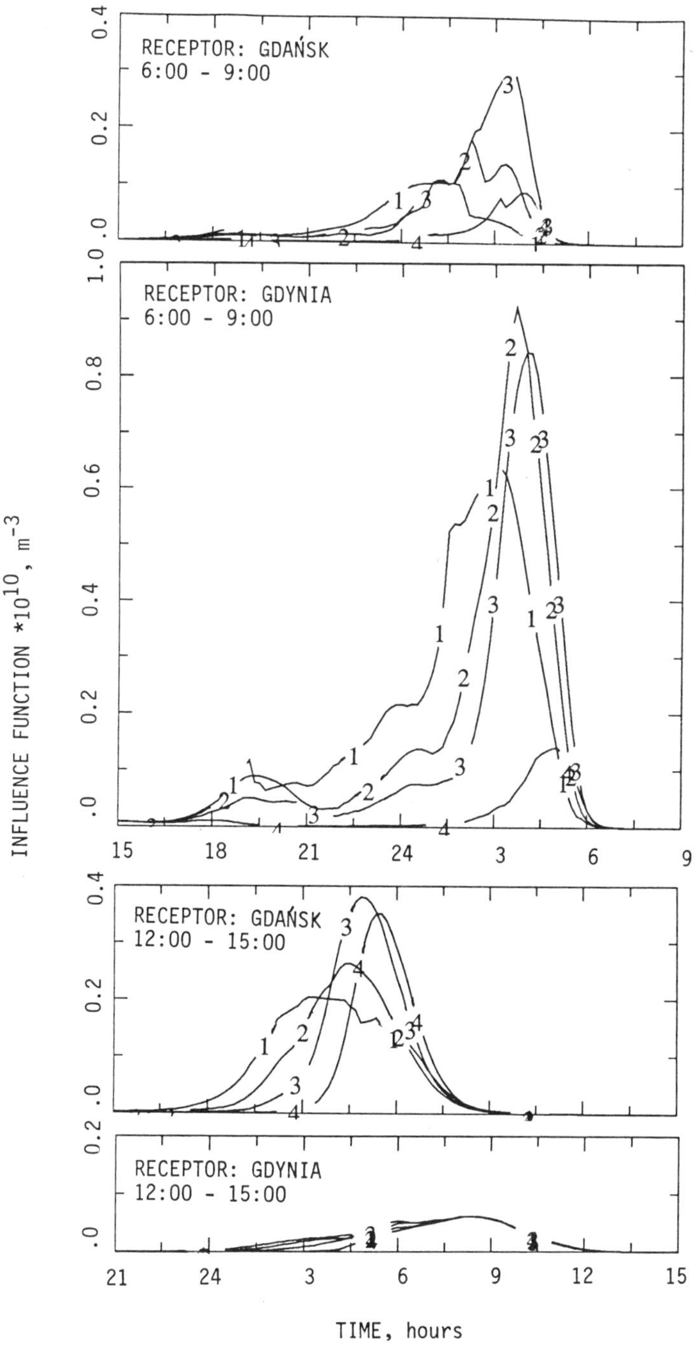

Fig.4. Time run of influence functions at Żarnowiec at different heights z_e: 1 → 0 m, 2 → 100 m, 3 → 200 m, and 4 → 300 m

CONCLUSIONS

The variational formulation of the Eulerian dispersion model allows us to apply both source oriented and receptor oriented modeling as complementary tools in air quality studies. The receptor oriented modeling is especially relevant to such applications as emission control, planning locations of new emission sources and assessing contributions from different sources to air pollution in a given region.

Applicability of the receptor oriented option is limited to linear dispersion models and integrals describing air quality at the receptor. It is necessary to assume that all chemical reactions of pollutants are linear and that pollutants do not effect the atmospheric dynamics. However, the variational approach may be still useful to perform sensitivity analysis of dispersion models with nonlinear chemistry (Uliasz, 1983, 1985). In this case, the solution of the adjoint equation depends on pollution concentration and the integral equation (13) has a more general form of relationship between a variation of the air quality integral $\delta\Phi[\mathbf{C}]$ and variations of model input parameters distributed in time and space. All model parameters can be considered (e.g., emission, wind velocity, coefficients of chemical reactions), but only small changes in parameters are allowed.

The dispersion code is currently being linked to Regional Atmospheric Modeling System (RAMS) developed at CSU (e.g., Pielke et al., 1990) and will be applied to investigate air pollution in selected national parks of northern Arizona and southern Utah.

ACKNOWLEDGMENTS

This research was supported by the National Park Service under contract NA81RAH00001, Amendment 17, Item 15.

REFERENCES

Marchuk,G.I., 1975: Formulation of the theory of perturbation for complicated models. Appl. Math. Optim., **2**, 1-33.

Marchuk,G.I., 1982: Mathematical issues of industrial effluent optimization. J.Meteor.Soc.Jap. **60**, 481-485.

Pielke,R.A., Lyons,W.A., McNider,R.T., Moran,M.D., Stocker,R.A., Walko,R.L., Uliasz,M., 1990: Regional and mesoscale meteorological modeling as applied to air quality studies. *Proc., 18th ITM on Air Pollution Modelling and Its Application*, May 14-17, Vancouver, Canada.

Uliasz, M., 1983: Application of the perturbation theory to sensitivity analysis of an air pollution model. Zeitschrift für Meteorologie, **33**, 355-362.

Uliasz, M., 1985: Comparison of the sensitivity analysis methods for meteorological models. Zeitschrift für Meteorologie, **35**, 340-348.

Uliasz,M., 1990a: Development of the mesoscale dispersion modeling system using personal computers. Part I: Models and computer implementation Zeitschrift für Meteorologie, **40**, No 2, 104-114.

Uliasz,M., 1990b: Development of the mesoscale dispersion modeling system using personal computers. Part II: Numerical simulations. Zeitschrift für Meteorologie, **40**, (In press).

DISCUSSION

A. VENKATRAM

Can you comment on the physics of reversing the diffusion operator in time?

M. ULIASZ

The diffusion operator works exactly in the same way in the model equations and the adjoint equations. Time and advection terms are only reversed in time in the adjoint equations. The proposed influence function for linear dispersion models can be considered as an extension of Green function approach.

A. ANDRÉN In using your variational approach to study sensitivity of non-linear chemistry models, are you then restricted to small variations of input parameters?

M. ULIASZ Yes, it is true. The permitted sizes of variations in model input parameters may depend on the problem under consideration, i.e. meteorology, chemistry, and a configuration of emission sources and the receptor. It will be a subject of the further study.

EULERIAN MODELING OF ATMOSPHERIC DISPERSION OVER PORTUGAL: BACKGROUND CONCENTRATIONS AND EMISSION DATA PREPARATION

C.Borrego, M.Coutinho and J.Rua

Department of Environment
University of Aveiro, 3800 Aveiro - Portugal

INTRODUCTION

A project for the implementation of an air pollution interregional model to Portugal is being develloped (Coutinho et al., 1989). An integrated study with the application of two dispersion models is being undertaken. In the first stage of the project, a puff-gaussian dispersion model has been applied. The final aim of this project is the adaptation of an Eulerian dispersion model to Portugal.

This work will use an episodic interregional model of medium complexity framed over the Airshed model of Systems Applications Inc. (Ames et al., 1985).This paper describes the input preparation work for the implementation of this photochemical air pollution model to Portugal. The Airshed model, as most of air pollution dispersion models of this level, requires a great amount of input data. The preparation of the input data has already been described by other authors (Stern et al., 1982; Builtjes at al.,1984). The application of the model to a non-heavily industrialized region, where natural emissions play a major role, rises new problems.

The main purpose of this project is the study of atmospheric dispersion over Portugal for the design of the National Air Quality Network and the determination of optimal control strategies to minimize atmospheric pollutants concentrations. Moreover, in spite of the relatively high hourly ozone concentrations that are expected, no regular photochemical modeling activities at a medium range (200-300 km) have taken place in southern Europe.

DESCRIPTION OF THE MODEL

The SAI-Airhed Model (Ames et al., 1985) was designed to calculate the concentrations of both reactive and inert air pollutants through simulation of the numerous physical and chemical processes that take place in the atmosphere. It is an Eulerian long-range transport model aiming the calculation of

Air Pollution Modeling and Its Application VIII, Edited by H. van Dop and D. G. Steyn, Plenum Press, New York, 1991

ozone concentrations through a non-linear chemical mechanism.

This model can simulate the dynamic behaviour of up to 20 pollutants. When photochemical simulations are carried out, 11 species must be included: NO, NO_2, O_3, single-bonded carbon atoms (PAR), double-bonded carbon atoms except ethylene (OLE), aromatic-bonded carbon atoms (ARO), carbonyl-bonded carbon atoms (CARB), ethylene, benzaldehyde (BZA), peroxyacetyl nitrate (PAN) and CO. In addition, the model can provide predictions of SO_2 and selected aerosol parameters.

The basis of the model is the resolution of the continuity equation, which expresses the conservation of mass of each pollutant in a turbulent fluid in which chemical reactions occur. Turbulent diffusion is parameterized using gradient transfer theory. Simulations are performed on a three-dimensional grid selected by the user to cover the region of interest.

The first application of the model to the portuguese situation will be performed over the South part of the country. The southern region of Portugal, with a modeling area of 350 x 225 km^2 (Figure 1), is divided into 25 x 40 grid cells of 7.5 x 7.5 km^2 in the horizontal and 4 cells in the vertical. This region does not suffer steep orographic gradients and offers the best meteorological data availability.

EMISSIONS

An air pollution emission inventory was developed for the application of interregional models. At this stage, the inventory considers the emission from the main industrial producers of the region:

- 4 thermal-power plants
- 5 chemical plants
- 3 pulp and paper mill plants
- 1 oil refinery
- 1 steel plant

Those emissions are introduced as elevated sources into the correspondent vertical cell. Emissions from mobile sources in the Lisbon region are introduced as area sources at ground level, and suffer a diurnal time dependence. The emission data-base including emissions of NO_x, SO_2, CO and non-methane hydrocarbons (NMHC) speciated as carbonyls, paraffins, olefins, aromatics and ethylenes is summarized in Table 1.

Portuguese industry is located near the Atlantic coast and concentrated in quite dense industrial complexes. Anyway, most of the modeling area is under rural or forest exploration. In order to develop feasible modeling studies, it is necessary to quantify, in a reliable way, natural emissions resulting from these activities. Naturally emitted volatile organic compounds (VOC) might outweigh those from natural sources, contributing to the formation of photochemical oxidants and playing a role in the acidification of the environment. A recent report (Lubkert et al., 1989) suggests that the total amount of VOC from forests in Portugal may be

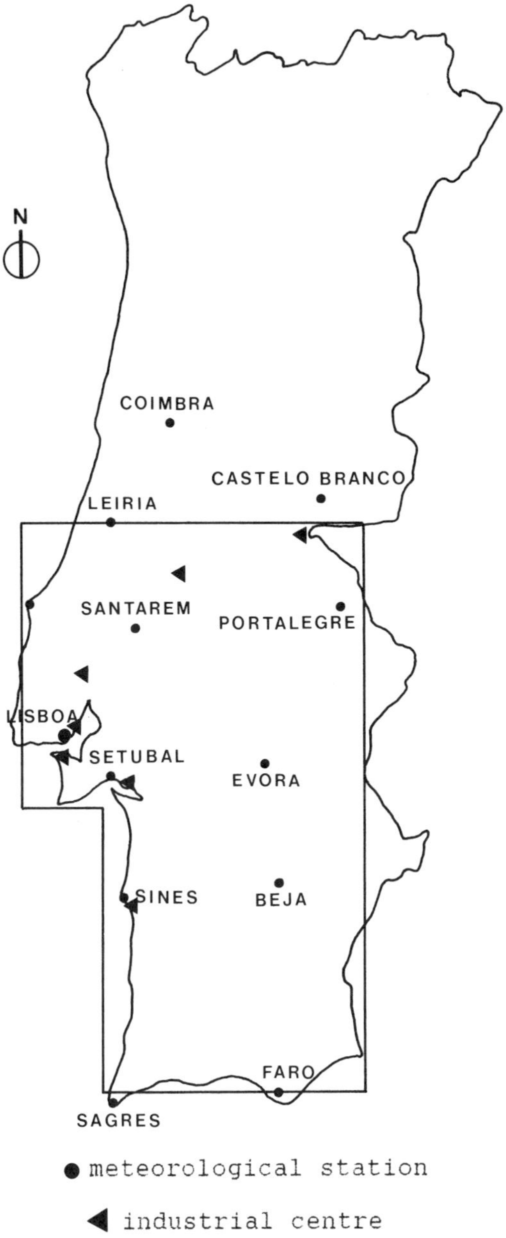

Fig. 1. Modeling region in Southern Portugal, indicating meteorological monitoring network and the location of industrial centres

Table 1. Total emissions in the modeling region and percentage of mobile sources.

	Emissions (kg/hr)	% traffic
NO_x	4039.32	33.26
SO_2	14238.73	1.19
CO	15981.74	100.00
NMHC	2869.91	66.41

greater than that of anthropogenic emissions. This problem is particularly significant during summer when strong insolation and high temperatures occur. The next stage of the emission inventory must consider forest emissions, as well as, pollutant emissions from small industrial units scattered over the region.

BOUNDARY CONCENTRATIONS

In order to specify the boundary concentrations, a careful literature review on atmospheric constituents concentrations measurements was performed. In spite of lack of air quality data, a clean air approach was assumed, framed over a strong qualitatively knowledge of the region. The strong influence of the Atlantic Ocean over the Portuguese airshed was also considered in the definition of typical trace constituents concentrations. Values of boundary concentrations defined during this study are displayed in Table 2. A short summary of the literature review is presented next.

Ozone

Ozone is an important product of photochemical processes and might be used as an index for the extent of these processes. Ozone concentrations exceeding 20-40 ppb result from photochemical production in the atmospheric boundary layer (Guicherit et al., 1977). Background levels of ozone measured in rural areas are clearly distinguished from the urban input. An evaluation of the natural contribution to photochemical smog performed in a rural area in Italy (Ciccioli et al.,1986), showed that maximum values of O_3 never exceeded 30-40 ppb.

Non-methane hydrocarbons

Non-methane hydrocarbons play a major role in the generation of photochemical oxidants, through chemical relationships much more complicated than initially supposed. Analysis of data from ambient air studies in rural areas in the USA, showed values of total NMHC between 2 and 24 ppbC (Sexton et al., 1984). Speciation factors for lightly polluted air (NMHC < 50 ppbC) were used (Rao, 1987):

PAR = 51.% NMHC   ARO = 25.% NMHC   CARB = 2.% NMHC
ETH = 3.% NMHC   OLE = 0.% NMHC

Table 2. Background concentrations used as boundary conditions

	Background conc.
NO	1.0 ppb
NO_2	2.0 ppb
O_3	20.0 ppb
CO	0.5 ppm
SO_2	300.0 ppt
PAN	0.2 ppb
NMHC	6.0 ppb

Peroxyacetil nitrate (PAN)

Peroxyacetil nitrate is generally considered as a very convenient indicator of polluted air masses, as it appears to have no large natural sources. Nevertheless, PAN has also been measured in rural areas, probably implicating long-range transport of polluted air masses. Substantial amounts of PAN could be transported from northern european countries and released in southern Europe (Ciccioli et al., 1988).

PAN measurements in a forested area in Donon, France, show monthly mean values of 0.1-0.5 ppb (Perros et al., 1988). Similar values were reported in a rural area at Harwell, UK (Brice et al., 1984).

Sulfur dioxide

Even if a lot of research has been dedicated to the physico-chemical behaviour of SO_2 and sulfate-aerosols, the collection of data on the distribution of the different sulfur components in unpolluted air has started more recently (Georgii, 1978). SO_2 measurements carried out on research ships and by aircraft above the Atlantic Ocean showed background levels between 100 and 300 ppt (Ockelmann et al., 1986).

INITIAL CONCENTRATIONS

Initial concentrations of pollutants over the entire region were assumed to be equal to the boundary concentrations, previously defined as clean air. Moreover, former modeling results were also used as a complementary tool for the establishment of the initial ground-level concentrations.

A puff-gaussian model has been used for the simulation of atmospheric dispersion over southern Portugal during 4 different episodes (Borrego et al., 1989). The analysis of its result allowed the definition of regions where initial ground level will be considered the double and quadruple of clean air concentrations. The first episode that will be studied is set over a 5-day data-base built for August 1988.

Figure 2 shows the SO_2 dispersion pattern for 3-Aug-88 and 4-Aug-88. Daily average isopleths are drawn with concentration peaks of approximately 30 $\mu gr.m^{-3}$. It should be noticed that no background concentrations or area sources are considered in this study. The regional effect of the conjugation of different point source plumes is clearly demonstrated.

CONCLUSIONS

The input data preparation for the application of a relatively complex dispersion model to Portugal is described in this paper. The model requires a great deal of detailed information, difficult to obtain in countries with incipient environmental networks.

The definition of the boundary and initial concentrations was framed over a "clean air" approach. Results of the application of a puff-gaussian model were also used. The

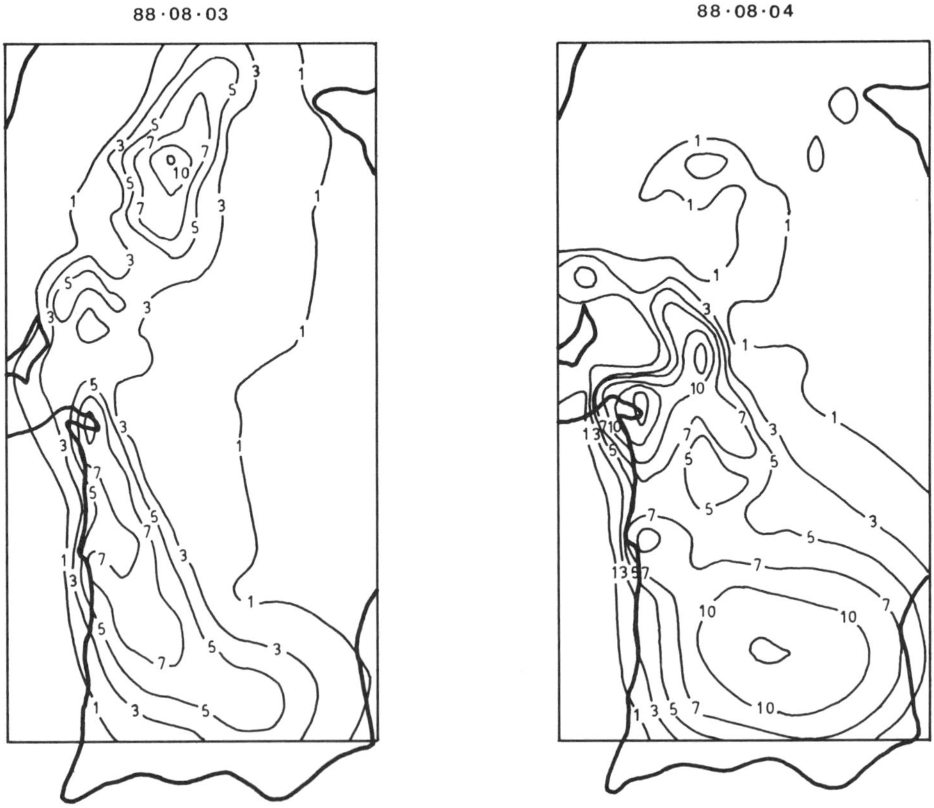

Fig. 2. Daily average SO_2 ground-level isopleths ($\mu gr.m^{-3}$) for 3-Aug-88 and 4-Aug-88 in southern Portugal. Isolines are drawn over the portuguese coast and boarder lines.

implementation of this relatively simple model was extremely important to understand the problems of input data availability in Portugal.

The study of air pollution in southern Europe should follow the methodologies developed in more industrialized countries. The know-how concentrated in existing dispersion models must be used for the development of control strategies. Nevertheless, the adaptation of dispersion models must be done with special care. In fact, most models were developed for the analysis of urban air pollution, underestimating the effects of natural emissions and the existence of a rural chemistry mechanism. In non-industrialized areas at moderate latitudes, natural emissions might have a regional impact in the production of photochemical oxidants. Moreover, the importance of meso-meteorological flows in the dispersion patterns in southern Europe has already been demonstrated (Millan et al., 1988; Flassak et al., 1989). These specific features are not yet described in available dispersion models and might represent new topics of research.

REFERENCES

Ames J., Myers T., Reid L., Whitney D., Golding S., Hayes S., Reynolds S., 1985, SAI Airshed Model operations manuals, EPA/600/8-85/007, U. S. Environmental Protection Agency.

Borrego C., Coutinho M. and Rua J., 1989, Interregional modeling of air pollution in Portugal: preliminary results, 5th Eur. Symp. on Physico-Chemical Behaviour of Atmospheric Pollutants, COST 611, Varese.

Brice K., Penkett S., Atkins D., Sandalls F., Bamber D., Tuck F. and Vaughan G., 1984, Atmospheric measurements of Peroxyacetilnitrate (PAN) in rural, South-East England: seasonal variations Winter photochemistry and long range transport, Atm. Env., 12:2691.

Builtjes P., van den Hout K. and Reynolds S., 1984, Evaluation of the performance of a photochemical dispersion model in pratical applications, in: Air Pollution modeling and its Applications III, Wispelaere C., Ed., Plenum Press.

Ciccioli P., Brancaleoni E., di Palo C., Brachetti A. and Cecinato A., 1986, Daily trends of photochemical oxidants and their precursors in a suburban forested area. A useful approach for evaluating the relative contributions of natural and anthropogenic hydrocarbons to the photochemical smog formation in rural areas in Italy, 4th Eur. Symp. on Physico-Chemical Behaviour of Atmospheric Pollutants, Stresa.

Ciccioli P., Brancaleoni E., Cecinato A., Brachetti A., Sciortino M., Carillo A. and Bargagli A., 1988, Seasonal and daily trends of PAN in a suburban area of central Italy, in "Field Measurements and their Interpretation", EUR 11960, Beilke S., Morelli J. and Angeletti G., Eds, Commission of the European Communities.

Coutinho M., Borrego C., Rua J. and Costa M.J., 1989, Application and implementation of an atmospheric pollution interregional model to Portugal, in: "Man and his Ecosystem", L.J. Brasser and W.C. Mulder, Eds.,

Publ. Elsevier, Amsterdam.
Flassak T. and Moussiopoulos N., 1989, Simulation of the sea breeze in Athens with an efficient non-hydrostatic mesoscale model, in: "Man and his Ecosystem", L.J. Brasser and W.C. Mulder, Eds., Publ. Elsevier, Amsterdam.
Georgii H., 1978, Large scale spatial and temporal distribution of Sulfur compounds, Atm. Env., 12:681.
Guicherit R. and van Dop H., 1977, Photochemical production of Ozone in Western Europe (1971-1975) and its relation to meteorology, Atm. Env., 11:145.
Lubkert B. and Schopp W., 1989, A model to calculate natural VOC emissions from forests in Europe, WP-89-082, IIASA, A-2361 Laxenburg.
Millan M., Palomino I., Artinano B., Pujadas M., Inclan R., Castro M., Alonso L., Gangoti G. and Ilardia J., 1988, The effect of thermal lows on the dispersion of tall stacks plumes in Spain, in "Field Measurements and their Interpretation", EUR 11960, Beilke S., Morelli J. and Angeletti G., Eds, Commission of the European Communities.
Ockelmann G., Burgermeister S., Ciompa R. and Georgii H., 1986, Aircraft measurements of various Sulfur compounds in a marine and continental environment, 4th Eur. Symp. on Physico-Chemical Behaviour of Atmospheric Pollutants, Stresa.
Perros P., Tsalkani N. and Toupance G., 1988, Continous measurements of Peroxyacetil Nitrate (PAN) in a forested area of the Vosges mountain, in "Field Measurements and their Interpretation", EUR 11960, Beilke S., Morelli J. and Angeletti G., Eds, Commission of the European Communities.
Rao S ., 1987, Application of the Urban Airshed Model to the New York Metropolitan Area, EPA 450/4-87-011,U. S. Environmental Protection Agency.
Sexton K. and Westberg H., 1984, Nonmethane hydrocarbon composition of urban and rural atmospheres, Atm. Env., 18:1125.
Stern R. and Scherer B., 1982, Simulation of a photochemical smog episode in the Rhine-Rhur area with a three dimensional grid model, in: Proc. 13th ITM on Air Pollution modeling and its Applications.

DISCUSSION

A. VENKATRAM

How do you justify applying the Urban Airshed Model, with three (physical not computational) vertical layers, to a scale of 300 km? I suspect that mesoscale flows with complex features govern such spatial scales. Can the UAM handle such features?

M. COUTINHO

The UAM has been regularly applied to regions of this size (e.g. the Netherlands, Rhine-Ruhr area) even if complex mesoscale flow may play a major role. In fact, as a result of this study, we found the interest of detailing our research over the meso-meteorological features of the atmospheric flow over Portugal.

J. BARTNICKI

What is the origin of the background concentrations in the south of Europe?

M. COUTINHO

Background concentrations in Portugal are mostly affected by forest and rural emissions, as well as the input form ocean air masses.

A. VENKATRAM

How can the model handle complex flow with only 3 vertical levels?

M. COUTINHO

The number of vertical levels is not fixed at 3.

P.K. KARAMCHANDANI

How is the VOC speciation performed for the VOC classes considered by the model?

M. COUTINHO

VOC speciation was framed over the results from former UAM applications. N.Y. application by S. Rao was particularly useful.

DRY AEROSOL DEPOSITION OVER THE NORTH SEA ESTIMATED FROM AIRCRAFT MEASUREMENTS

Carlos M. Rojas, Philippe M. Otten and René E. van Grieken

Department of Chemistry, University of Antwerp (UIA)
Universiteitsplein 1, B-2610 Antwerp-Wilrijk
Belgium

and

Remi Laane

Tidal Waters Division, Rijkswaterstaat
P.O. Box 20907
The Hague, The Netherlands

ABSTRACT

A mathematical approach based on the Monin-Obukhov similarity theory is used to predict the wind speed, friction velocity and drag coefficient, which are then introduced in the well-known deposition model of Slinn and Slinn (1980), to calculate the dry deposition of heavy metals into the North Sea. This model is perfectly suitable for aircraft sampling considering the fact that usually due to safety reasons, flights at the reference height used in deposition models (10 m), are not possible. To check this approach, deposition velocities were calculated based on the airborne concentrations of Cu, Cd, Zn and Pb obtained by sampling with the aid of an aircraft over the Dutch continental shelf of the North Sea. Results are in agreement with those found in the literature. A rough estimation of the atmospheric input for these heavy metals and comparison with riverine inputs and direct discharges is also included.

INTRODUCTION

There has been an increasing interest in studying the fate and removal mechanisms of gases and particulate matter over the oceans both in remote areas (Arimoto et al., 1985; Buat-Ménard and Duce, 1986) and polluted areas such as the Western Mediterranean (Dulac et al., 1987; Gomes et al., 1988). In the coastal and shelf systems of the North Sea there has been a growing concern about large quantities of heavy metals being deposited from the atmosphere. Several authors, such as Van Aalst et al. (1983), concluded that the atmosphere plays an important role in the pollution of the North Sea and estimated that the input of heavy metals is comparable to that from river and direct discharges. Due to the natural limitations imposed on sample collection at sea, it has been necessary to develop mathematical models in order to estimate such heavy metal inputs. These theoretical approaches began more than 15 years ago (Sehmel and Sutter, 1974) and were reviewed by Sehmel (1980). Models for dry deposition on natural waters such as that of Slinn and Slinn (1980) have recently been used in the Western Mediterranean Sea (Dulac et al., 1987) and the North Sea (Steiger et al., 1989; Dedeurwaerder, 1988). Williams (1982) proposed another model which has been discussed by Slinn (1983) and Liss et al. (1988). In order to assess the input of heavy metals to the North Sea Van Aalst et al. (1983) and Van

Air Pollution Modeling and Its Application VIII, Edited by H. van Dop and D. G. Steyn, Plenum Press, New York, 1991

Jaarsveld et al. (1986) used a trajectory model based upon anthropogenic emisions and in-land meteorological data. Krell and Roeckner (1988) performed stochastic trajectory modeling estimates for wet and dry deposition of Pb and Cd.

In this paper we report on the use of the Monin-Obukhov (1954) similarity theory to predict the wind speed, friction velocity and drag coefficient at a certain height (usually 10 m), taking into account the air temperature, atmospheric stability, wind speed, atmospheric pressure and relative humidity, measured at higher altitude, and the sea temperature. This can be applied directly to aircraft samplings since safety reasons usually impede to fly at altitudes lower than 20 m. The calculated parameters are then introduced in Slinn and Slinn's formulation to determine the dry deposition velocity of airborne particulate matter. A preliminary estimation of dry deposition velocities and atmospheric inputs based on average concentrations of Cu, Cd, Zn and Pb for a few sampling flights on the North Sea is included.

THEORETICAL BACKGROUND

The shape of the wind profile over sea depends on the aerodynamic surface roughness length and atmospheric stability, which in turn depends on the vertical density (temperature) gradient and on wind stress. Thus, from the Monin-Obukhov (1954) similarity theory, the profiles for wind, temperature and humidity are:

$$\frac{\kappa z}{u_*}\frac{\partial u}{\partial z} = \psi_m\left(\frac{z}{L}\right) \tag{1}$$

$$\frac{z}{t_*}\frac{\partial \theta}{\partial z} = \psi_t\left(\frac{z}{L}\right) \tag{2}$$

$$\frac{z}{q_*}\frac{\partial q}{\partial z} = \psi_q\left(\frac{z}{L}\right) \tag{3}$$

$$\kappa u_* t_* = -\overline{t'w'} \tag{4}$$

$$\kappa u_* q_* = -\overline{q'w'}$$

where, z is the height, $\kappa = 0.4$ von Kármán's constant, u_* = friction velocity, t_* and q_* are the scaling temperature and humidity, respectively. The subscripts m, t and q stand for momentum, temperature and humidity; t', q' and w' are the vertical temperature, humidity and wind speed variations. The mean wind speed is denoted by u, ψ represents the stability function and L the Monin-Obukhov length.

The integration of (1), (2) and (3) yields,

$$\frac{u(z)}{u_*} = \frac{1}{\kappa}\left[\ln\left(\frac{z}{z_0}\right) - \psi_m\left(\frac{z}{L}\right)\right]$$

$$\frac{\theta(z) - T_s}{t_*} = \ln\left(\frac{z}{z_{ot}}\right) - \psi_t\left(\frac{z}{L}\right) \tag{5}$$

$$\frac{q(z) - q_s}{q_*} = \ln\left(\frac{z}{z_{oq}}\right) - \psi_q\left(\frac{z}{L}\right)$$

where z_0 is the aerodynamic surface roughness length; $\theta(z)$ and T_s are the potential temperature of the air and sea surface, respectively and q_s is the absolute humidity of air saturation at sea surface temperature.

For the surface roughness height we used Charnock's (1955) formula, with a parameter $\alpha = 0.0185$ (Wu, 1982) instead of the original 0.011. This was motivated by the wide range of wind speeds used by Wu to determine this parameter. The resulting surface roughness length from the formulae of Charnock (1955) and Businger (1973) for smooth surfaces is:

$$z_0 = 0.0185 \frac{u_*^2}{g} + \frac{\nu}{9u_*} \tag{6}$$

where ν is the dynamic air viscosity and g the gravitational acceleration.

Using the following parameterizations:

$$\begin{aligned} C_d(z)u^2(z) &= u_*^2 \\ C_t(z)u(z)\left(T_s-\theta(z)\right) &= -\kappa u_* t_* \\ C_\theta(z)u(z)\left(q_s-q(z)\right) &= -\kappa u_* q_* \end{aligned} \tag{7}$$

for the momentum, heat and humidity fluxes and solving iteratively (5), (6) and (7) including the expressions of Large and Pond (1982) for ψ_m, ψ_t and ψ_q for stable and unstable atmospheres and a stability parameter (z/L) given by Williams (1982), one can predict the drag coefficient and friction velocity at a given height.

For the dry deposition model we used Slinn and Slinn's (1980) formulation, including a slip correction factor for the diffusion sub-layer. Brownian diffusivity was calculated using Seinfeld's (1986) approach. The effect on particle deposition due to particle growth by condensation has been taken into account assuming that particles are either hydrophobic or hygroscopic. In the latter case particles were considered as having the same hygroscopic properties as ammonium sulphate particles and their wet diameter was calculated using the results of Fitzgerald (1975).

EXPERIMENTAL

From September 1988 to August 1989 several aircraft samplings have been carried out over the Dutch continental shelf of the North Sea, using a twin engine Piper Chieftain PA 31-350 aircraft belonging to the company Geosens BV (Rotterdam, The Netherlands). During the flights SO_2 and O_3 were monitored. Wind speed was determined with the aid of a Doppler radar, together with height above sea level. Also dew point temperature and air temperature were measured.

Each flight was scheduled based on 36-h backward trajectories provided by the Royal Dutch Meteorological Institute (KNMI), and consisted of 6 tracks equally spaced between the inversion height and the sea level. The normal flight height for the lowest track was 20 m.

Samples were collected on 0.4 μm pore-size Nuclepore filters using an isokinetic inlet located on the top of the fuselage. The flow rate was of 50 $l.min^{-1}$ and the average sampling time per track was 25 min. The collected airborne matter was first analyzed by X-ray fluorescence (XRF) using a Tracor Northern 5000 X-ray spectrometer equipped with a Rh target X-ray tube and a Si(Li) solid state detector; more details on the analysis protocol can be found elsewhere (Rojas et al., 1990). Then, samples were analyzed by Particle-Induced X-ray Emission (PIXE) using a 2.4 MeV proton beam supplied by the isochronous cyclotron of the Institute of Nuclear Sciences at the University of Ghent, Belgium. Details on this analytical technique are given by

Maenhaut (1989). The sea temperature data were supplied by Management Unit for the North Sea and Scheldt Estuary Mathematical Model, belonging to the Belgian Ministry of Public Health and Environment.

RESULTS AND DISCUSSION

In all calculations an atmospheric pressure of 1013 hPa, relative humidity of 75% for the turbulent layer and a particle density of 2.5 g/cm^3 were assumed. Deposition velocities were computed for 0.5 μm particle aerodynamic diameters, which correspond to the centroid of the size distribution of Pb particles in the North Sea reported by Steiger et al. (1989).

Table 1 summarizes the results of the model calculation for 5 different flights, with altitudes ranging from 20 to 50 m above sea level. Here only the second row corresponded to an unstable atmosphere. For the 4 cases of stable atmosphere the average difference in wind speed with respect to the reference height is 20%, with a minimum of 6% corresponding to an air-sea temperature difference of 0.4 °C. The surface roughness length values are of the same order of magnitude as those reported by Krauss (1972) for oceanic conditions. Drag coefficients varied from 0.18 x 10^{-3} for stable to 1.33 x 10^{-3} for unstable conditions, respectively. Dry deposition velocities are, as expected, a function of wind speed. The average difference in dry deposition velocity for hydrophobic and hygroscopic particles was 32%. The smallest difference (16%) was observed for the unstble case with a relatively warm sea.

Preliminary results of 4 size segregated samplings of airborne particulate matter collected with a cascade impactor (Dierck et al., 1990) and the average bulk concentration for the first 8 flights for Cu and Cd taken from Otten et al. (1989) and the average of PIXE and XRF for Zn and Pb from Dierck et al. (1990) were used to determine the average weighted dry deposition velocities for Cu, Cd, Zn and Pb, shown in Table 2. The dry deposition velocities of Cu, Cd and Pb are very close to those used by Van Jaarsveld et al. (1986) and Krell and Roeckner (1988) (0.2 cm/s). The deposition velocity for Zn is about 40% higher. Table 3 shows some rough estimates of the atmospheric input calculated for both particle categories and computed from the average concentration for tracks 1-5 (first column), for tracks 1-6 (second), while the third corresponds to track 6 (nearest to the sea). These heavy metals inputs are within the variation range of those reported by Kersten et al. (1988) for Cd, Zn and Pb in both particle categories and somewhat lower for Cu. If we assume that particles are indeed hydrophobic then, the atmospheric

Table 1. Results obtained using data corresponding to 5 flights. Here (1) and (2) stand for hydrophobic and hygroscopic particles, respectively.

Given Data				Calculated Data				
Z (m)	T°C Sea	T°C Air	u m/s	u at 10 m	z_0 (m) x 10^{-4}	C_d x 10^3	V_d (1) cm/s	V_d (2) cm/s
20	13	16	4.1	2.9	0.4	0.18	0.003	0.006
20	13	10	8.9	8.5	2	1.33	0.008	0.01
20	10.4	10.8	9.3	8.7	1.8	1.17	0.008	0.009
40	10	12	5.9	4.3	0.4	0.47	0.004	0.007
50	6.8	9.2	14	11	3.6	0.98	0.008	0.01

Table 2. Calculated dry deposition velocities (cm/s) for both particles categories.

Element	Hydrophobic particles	Hygroscopic particles
Cu	0.18	0.30
Cd	0.18	0.27
Zn	0.36	0.52
Pb	0.27	0.38

Table 3. Rough estimation of the atmospheric input (Tonnes/Y) of these heavy metals, based upon the average airborne concentration for 8 flights. Here 1-5 stand for the average from track 1-5, 1-6 for the average from track 1-6 and 6 is the average for the last track.

	Hydrophobic particles			Hygroscopic particles		
	1-5	1-6	6	1-5	1-6	6
Cu	190	180	175	320	310	290
Cd	21	21	21	31	31	31
Zn	3900	4200	5400	5700	6000	7800
Pb	2000	2300	2700	3000	3200	3900

inputs of Cu and Cd are lower compared to riverine inputs and direct discharges, whereas that of Zn lays between them. The input of Pb is higher than the other two, probably due to leaded fuel combustion.

FINAL REMARKS

A mathematical approach using similarity theory has been used to predict the wind speed, drag coefficient and friction velocity to be introduced in the well-known deposition model of Slinn and Slinn (1980). In this way, it is possible to determine the dry flux of heavy metals when the sampling conditions (aircraft) do not reach the required reference height of 10 m. Dry deposition velocities were similar to those of Van Jaarsveld et al. (1986) and Krell and Roeckner (1988). Even though atmospheric input determinations in this work have a preliminary nature, since they are based on a reduced data base, results seem to be promising.

ACKNOWLEDGEMENTS

We are grateful to W. Maenhaut from the Institute of Nuclear Sciences, University of Ghent, Belgium, for the PIXE analyses, and A. Pollentier of the Management Unit for the North Sea and Scheldt Estuary Mathematical Model for providing sea temperature data. We acknowledge the financial support provided by Rijkswaterstaat, Dienst Getijdewateren, The Netherlands, through project NOMIVE*5 task n° DGW-920.

REFERENCES

Arimoto, R., Duce, R.A., Ray, J., and Unni, C.K. (1985) J. Geophys. Res. 90, D1, 2391-2408.
Buat-Ménard, P., and Duce, R.A. (1986) Nature 321, 508-510.
Businger, J.A. (1973) Amer. Meteorol. Soc. 67-100.
Charnock, H. (1955) Quart. J. Roy. Meteorol. Soc. 81, 639-640
Dedeurwaerder, H.L. (1988) Ph.D. Dissertation, University of Brussels (VUB), Belgium.
Dierck, I., Injuk, J., Otten, Ph., Rojas, C., and Van Grieken, R. (1990) Report 3 on project NOMIVE*5 to Rijkswaterstaat, The Netherlands.
Dulac, F., Buat-Ménard, P., Arnold, M., Ezat, U., and Martin, D. (1987) J. Geophys. Res. 92, 8437-8453.
Fitzgerald, J.W. (1975) J. Appl. Meteorol. 14, 1044-1049.
Gomes. L., Bergametti, G., and Chatenet, B. (1988) Environ. Tech. Lett. 9, 1295-1302.
Kersten, M., Dicke, M., Kriews, M., Naumann, K., Schmidt, D., Schulz, M., Schwikowski, M., and Steiger, M. (1988) In: Pollution of the North Sea: An Assesment. Salomons et al. (eds.) Springer Verlag.
Krauss, E.B. (1972) Atmosphere-ocean Interaction. Oxford University Press, Oxford, 275.
Krell, U., and Roeckner, E. (1988) Atmos. Environ. 22, 375-381.
Large, W.G., and Pond, S. (1982) J. Phys. Oceanog. 12, 464-482.
Liss, P.S., Jickells, T.D., and Buat-Ménard, P. (1988) In: Pollution of the North Sea: An Assessment. Salomons et al. (eds.) Springer-Verlag, Berlin.
Maenhaut, W. (1989) In: Control and Fate of Atmospheric Trace Metals. Pacyna, J. and Ottar, B. (eds). Kluwer Academic Publisher, Dordrecht, The Netherlands.
Monin, A.S., and Obukhov, A.M. (1954) Trudy Geophys. Inst. Ansssr 24, 163 USSR.
Otten, Ph., Rojas, C., Wouters, L., and Van Grieken, R. (1989) Report 2 on project NOMIVE*5 to Rijkswaterstaat, The Netherlands.
Rojas, C.M., Artaxo, P., and Van Grieken, R.E. (1990) Atmos. Environ. 24B, 2, 227-241.
Sehmel, G.A., and Sutter, S.L. (1974) J. Res. Atmos. 8, 911-920.
Sehmel, G.A. (1980) Atmos. Environ. 14, 983-1011.
Seinfeld, J. (1986) Atmospheric Chemistry and Physics of Air Pollution. Wiley Interscience, New York, 740.
Slinn, S.A., and Slinn, W.G.N. (1980) Atmos. Environ. 14, 1013-1016.
Slinn, W.G.N. (1983) In: Air-Sea Exchange of Gases and Particles. Liss, P.S., and Slinn, W.G.N. (eds.) NATO ASI series, Reidel publisher, Dordrecht, The Netherlands.
Steiger, M., Schulz, M., Schwikowski, M., Naumann, K., and Dannecker, W. (1989) J. Aerosol Sci. 20, 8, 1229-1232.
Van Aalst, R.M., Van Duyzer, J.H., and Voldt, C. (1983) Report R83/222 TNO Delft, The Netherlands.
Van Jaarsveld, J.A., Van Aalst, R.M., and Onderdelinden, D. (1986) Report RIVM 842015002, Bilthoven, The Netherlands.
Williams, R.M. (1982) Atmos. Environ. 16, 1933-1938.
Wu, J. (1982) J. Geophys. Res. 87, C12, 9704-9706.

DISCUSSION

G. SCHAYES Up to which height did you apply surface layer formulation on your data?

C.M. ROJAS Since this approach is based on turbulent features of air parcels, it can be successfully used provided certain boundary conditions are met. For example, problems may arise in trying to estimate the wind speed at a certain height when the wind speed at a higher altitude (which in our case did not exceed 50 m) is low (a few m/sec), and atmospheric conditions are stable. Then, the two heights will be separated by a layer of undisturbed air, which makes any calculation difficult.

J. ALCAMO I noticed that the literature values for V_d that you quote in your paper (Table 1) are a factor of ≈50 lower than your calculations (Table 2). What is the reason for this?

C.M. ROJAS The values you mention have not been cited in the literature; they are the result of this model applied to the centroid of the size distribution reported in the literature. The main difference is that values shown in Table 2 correspond to a weighted average taking into account the whole size distribution (and a set of size distributions), which might be in our case shifted towards larger particles.

C.J. WALCEK Do you see evidence that particles you collect are affected by evaporation or growth due to condensation or evaporation of water vapor during collection? If so, how might this artifact affect your calculated deposition fluxes?

C.M. ROJAS We have not yet taken into account such effects. However, if particle growth due to condensation of water vapor really takes place during collection, I would expect higher concentrations on the first impactor stages, which could lead to a biased particle size distribution and therefore an overestimation of the dry flux.

TEMPORAL AND SPATIAL RESOLUTION REQUIREMENTS FOR REGIONAL-SCALE DISPERSION MODELS

Michael D. Moran[1], Roger A. Pielke[1], and Richard T. McNider[2]

[1]Dept. of Atmospheric Science, Colorado State University, Fort Collins Colorado, U.S.A. 80523
[2]Dept. of Mathematics and Statistics, Univ. of Alabama at Huntsville Huntsville, Alabama, U.S.A. 35899

INTRODUCTION

Today the atmospheric regional scale is the meeting place of two very different groups of numerical modelers: atmospheric dispersion modelers and meteorological modelers (plus a third group — atmospheric chemistry modelers). This is a relatively recent development, and the two groups have arrived by different paths. Twenty years ago, with the notable exception of studies of the continental- and planetary-scale transport of fallout-derived radionuclides, most studies of atmospheric dispersion focussed on the local dispersion of air pollutants over travel distances and travel times of at most 10 to 20 km and several hours, respectively. Such spatial and temporal scales correspond to Orlanski's micro-α-scale (0.2–2 km) and meso-γ-scale (2–20 km) (Orlanski, 1975; Fujita, 1986). The development of *larger-scale* dispersion models, variously termed regional-scale dispersion models (RSDMs), 'long-range' transport models, and mesoscale dispersion models, to study the transport and diffusion of air pollutants over travel distances of 20 to 2000 km and travel times of several hours to several days (meso-β-scale and meso-α-scale) was spurred by the surge of concern and interest in the late 1970s in transboundary acidic deposition and regional oxidant episodes.

Regional-scale meteorological models were also introduced in the 1970s, first as *smaller-scale* research adjuncts to, and then as operational extensions of, the synoptic-scale numerical weather prediction (NWP) models of the 1950s and 1960s. Regional or limited-area 'short-range' (0-48 h) forecast models generally have smaller spatial domains and smaller horizontal grid sizes than synoptic NWP models; horizontal grid resolution for current operational regional-scale meteorological models is 50–250 km compared to a typical synoptic-scale model horizontal resolution of 400 km (Anthes, 1983). Such resolution is fine enough to resolve many meso-α-scale meteorological features (e.g., midlatitude cyclones, jet streaks, and mesoscale convective systems). Smaller-scale (and smaller domain) mesoscale meteorological models with horizontal grid resolutions of 1–20 km have also been developed (Pielke, 1984). This class of meteorological research model is capable of resolving many meso-β-scale meteorological phenomena such as frontal circulations, land-sea breezes, drainage winds, and lake-effect storms.

Despite different goals (and different perceptions of scale), both of these groups of modelers are interested in describing regional-scale wind fields and subgrid-scale (SGS) mixing and diffusion. However, the two groups have in general taken different approaches to generate wind and turbulence fields. Dispersion modelers have tended to use diagnostic wind-field models or meteorological 'preprocessors' that are based on the objective analysis of available meteorological observations while meteorological modelers have relied on prognostic dynamical models. This difference is not really surprising since the goal of meteorological modelers is to predict weather, including the atmospheric flow field itself, whereas dispersion modelers wish to predict the dispersive behaviour of a pollutant in a prescribed atmospheric flow. It is also consistent with the different developmental paths taken by the two groups. Many RSDMs are an *upscale* extension of local dispersion models in which turbulent diffusion is assumed to occur within a stationary, homogeneous mean flow field. Regional-scale meteorological models, on the other hand, are a *downscale* extension of synoptic-scale prognostic NWP models.

Each approach has its own advantages and disadvantages. The diagnostic approach is much simpler but is limited by the essentially synoptic-scale resolution of the standard upper-air observing network, that is, 12 h temporal resolution and 400 km spatial resolution (Kuo et al., 1985; Kahl and Samson, 1986), and may produce wind fields that are not dynamically consistent. The prognostic approach is much more demanding conceptually and computationally but can provide "gridded datasets possessing a degree of dynamical consistency, spatial coverage, and spatial and temporal resolution exceeding that obtainable from contemporary observing systems" (Keyser and Uccellini, 1987). Ignoring questions of complexity and cost for the moment, the fundamental question which regional-scale dispersion modelers must answer is whether either approach provides a sufficiently detailed flow-field description with which to model regional-scale dispersion, especially when pollutant sources are almost always subgrid-scale and initial dispersion is dominated by local SGS flow properties. In particular, by using only synoptic observations, mesoscale flow scales and processes are not resolved by the diagnostic approach so that their contribution to regional-scale dispersion is neglected.

Consider briefly the concepts of model scale and model resolution. Numerical modelers have found that the smallest temporal or spatial features which can be realistically treated in a numerical simulation are those $4\Delta t$ or $4\Delta x$ in size (Avissar et al., 1990). $2\Delta t$ or $2\Delta x$ features are impossible to handle accurately with any available numerical solution techniques. For temporal resolution, then, Δt can be no larger than 6 h if the 24 h diurnal cycle is to be resolved. As for spatial resolution, the smallest resolvable feature for a horizontal grid increment Δx of 50 km would be 200 km in size; if the horizontal grid increment is increased to 400 km, only flow features larger than 1600 km could be resolved. Thus, even a regional-scale meteorological model with 50 km horizontal grid spacing will be unable to resolve such common meso-β-scale circulations as sea and land breezes and mountain-valley winds. Note when considering the stated resolution of some diagnostic wind field models that simple numerical interpolation from a coarse grid to a finer grid does not increase resolution since no new information is provided; the resolution of the input field is the limiting factor.

In the rest of this article, we will refer to the transport and diffusion of an air pollutant, tracer, or other atmospheric constituent over horizontal distances of 20 to 2000 km and time periods ranging from several hours to several days as mesoscale atmospheric dispersion. Observations of mesoscale atmospheric dispersion from both controlled mesoscale tracer field experiments and tracers of opportunity suggest that mesoscale dispersion is an even more complex phenomenon than turbulent dispersion over smaller scales. The additional complexity arises because mesoscale processes and circulations can also contribute to atmospheric transport and diffusion over longer travel times, especially within the planetary boundary layer (PBL). To quote Pasquill and Smith (1983), "... at longer scale and over longer release durations, the mean wind itself becomes variable and is part of the spectrum of turbulence." Mesoscale terrain inhomogeneities often generate surface-forced mesoscale circulations and introduce mesoscale spatial structure in the flow field. Downscale nonlinear flow interactions and organized mesoscale convection are additional sources of mesoscale flow features. In addition, even for horizontally homogeneous flows over flat, uniform terrain, mesoscale frequencies will usually be present. These mesoscale space-time flow scales should be considered when selecting the spatial and temporal resolution of an RSDM or, sometimes, when selecting the model itself. Otherwise, significant components of mesoscale transport and diffusion may be neglected as discussed below.

OBSERVATIONS OF MESOSCALE DISPERSION

In what ways does mesoscale dispersion differ from smaller-scale atmospheric dispersion? One important consequence of Taylor's statistical theory of diffusion is that in stationary, homogeneous turbulence, the crosswind standard deviation σ_y of the ensemble-mean plume from a continuous point source will initially grow linearly with travel time τ, but then as τ becomes large compared to the Lagrangian integral time scale T_L, σ_y will asymptotically approach a square-root dependence on τ (Pasquill and Smith, 1983). Hanna (1981) reported average values for T_L of 60-90 s in the daytime boundary layer. Thus the parabolic $\tau^{1/2}$ regime for lateral plume spread should be approached after travel over several kilometres if the appropriate T_L is the PBL convective time scale. However, observations of a number of mesoscale plumes suggest that the linear time regime may persist out to mesoscale distances of 500 km or more (e.g., Gifford, 1983, 1986; Pasquill and Smith, 1983), and many Lagrangian-puff RSDMs use Heffter's (1965) formula for lateral plume spread, $\sigma_H = 0.5\tau$ where σ_H is in meters and τ is in seconds, for all mesoscale travel times (Carhart et al., 1989). As discussed by Csanady (1973) and Taylor (1982), vertical speed and direction shear is largely responsible for this linear dependence in the first few hours. At longer travel times when the flow can no longer be considered steady, Gifford (1983) has argued that the inertial period may be the appropriate T_L for mesoscale dispersion and McNider et al. (1988) have suggested that the combined influence of two mesoscale time scales, the inertial period and the diurnal period, is important for mesoscale lateral dispersion in the PBL.

A second difference between mesoscale and smaller-scale dispersion is the non-Gaussian, multimodal cross-plume concentration distributions often observed at large travel distances. Carras and Williams (1988) have presented plots of five multimodal plume cross-sections for the plumes from the Mt. Isa smelter in northern Australia and the Kalgoorlie smelter in southwestern Australia at downwind distances ranging from 230 km to 1200 km. A common feature of all five cross-sections is that in each case the plume has experienced at least one morning or evening PBL transition prior to the measurement. What is especially interesting about the Mt. Isa measurements is that transport has taken place across fairly homogeneous terrain, suggesting that terrain-forced mesoscale circulations due to mesoscale physiographic inhomogeneities are not the responsible agent. Another example may be found in the set of three-hour-average measurements of perfluorocarbon surface concentration made along an arc of samplers 600 km from the release site during the 1980 Oklahoma mesoscale tracer experiment (Ferber et al., 1981). Measurements along this sampler arc for each of the first two 3-hour observing periods showed three relative concentration maxima across the tracer puff. The westernmost peak was approximately two orders of magnitude smaller than the other two peaks but was still at least five times greater than background levels.

A third difference is the occurrence of along-wind multimodal surface concentrations. A simple Gaussian dispersion model would predict a single ground-level concentration maximum and then monotonically decreasing concentrations at greater downwind distances. A Lagrangian-puff RSDM with a Heffter-type parameterization of horizontal diffusion would predict similar behaviour. However, in the mesoscale Cross-Appalachian Tracer Experiment (CAPTEX) in 1983, some six-hour-average surface concentration fields and composite maximum concentration fields showed two or more concentration relative maxima (Ferber et al., 1986, Figs. 2, 4, 11). Multimodal surface concentration patterns were also measured in the Across North America Tracer Experiment (ANATEX) in 1987 (e.g., Draxler, 1989, Figs. 2–5).

MESOSCALE TEMPORAL SCALES

McNider et al. (1988) carried out idealized simulations of the mesoscale behaviour of the Mt. Isa plume using a 1D, 22-level meteorological model coupled with a Lagrangian particle dispersion model (see article by Pielke et al. in this volume). Despite the use of horizontally homogeneous meteorological fields, their simulated Mt. Isa plume displayed both linear horizontal growth over mesoscale travel times and multimodal cross-plume concentration patterns. They attributed this behaviour to the combination of time-dependent vertical shear with time-dependent vertical mixing as follows.

Even during the day, vertical shear is present in the PBL due to surface friction, baroclinicity, and other sources (Kahl and Samson, 1988a). As a result turbulent diffusion will be shear-enhanced. At night, the formation of a nocturnal surface inversion due to longwave radiational cooling decouples the nocturnal boundary layer from the remainder of the well-mixed daytime PBL above (termed the residual layer by Stull [1988]). The wind speed in the nocturnal boundary layer increases and the wind direction veers due to the new force balance between the pressure gradient force, Coriolis force, and weakened friction force. The residual layer begins to sweep through an inertial oscillation, clockwise in the northern hemisphere and counterclockwise in the southern hemisphere, resulting in the well known nocturnal low-level jet (Blackadar, 1957; Bonner, 1968; Thorpe and Guymer, 1977). Increased stratification reduces turbulent mixing but vertical variations in horizontal advection due to the wind-direction and wind-speed shear caused by the inertial oscillation distort the pollutant cloud. When morning arrives and the daytime PBL begins to grow anew, the shear-distorted cloud again mixes in the vertical, resulting in a wider ground-level plume 'footprint'. Unlike traditional Taylor shear dispersion, which is due to mixing caused by the *simultaneous* interaction of vertical shear and vertical diffusion, this shear-driven lateral plume growth is due to *delayed* mixing, that is, to vertical diffusion *following* differential advection. Pasquill may have been the first to describe this mesoscale diffusion mechanism (Pasquill, 1962, p. 79).

Although the timing of the formation and subsequent erosion of the nocturnal jet is controlled by the diurnal cycle, the jet itself is an inertial oscillation. Thus *two* natural mesoscale time scales may need to be considered for pollutant travel times of more than one or two hours: the 24-hour diurnal period and the latitude-dependent inertial period $2\pi/f$, where f is the Coriolis parameter. The inertial period may be longer or shorter than the diurnal period; for instance, the inertial period is 35 h at 20 degrees of latitude, 18.6 h at 40 degrees, and 13.8 h at 60 degrees. The shape of the diurnal forcing is also dependent upon latitude and upon season.

The natural time scale for horizontally-bounded mesoscale circulations such as land-sea breezes that are forced by differential surface heating (or cooling) is the forcing time scale, that is, the duration of heating (or cooling). Depending upon latitude and season, this time scale may vary from 1 to 24 hours. Short-duration forcing is less likely to generate a mesoscale atmospheric response, but some shallow circulations such as drainage winds can form in one or two hours. Mesoscale circulations due to the mechanical forcing of stably-stratified flows by topographic features (see next section) will have the time scale of the larger-scale flow even though they are mesoscale in spatial extent. Still other mesoscale time scales will be associated with flow-forced mesoscale phenomena such as fronts and jet streaks and latent-heat-driven phenomena such as mesoscale convective complexes or mesoscale rainbands, but these time scales will usually be comparable to or larger than the inertial or diurnal periods.

MESOSCALE SPATIAL SCALES

If atmospheric flows are not steady over mesoscale time periods, neither are they usually horizontally homogeneous over mesoscale distances. The presence of such topographic features as mountains, plateaus, and valleys can result in mountain waves, mesoscale gyres, and channelled flows due to *mechanical forcing* (e.g., Durran, 1986; Wilczak and Glendening, 1988) and to slope winds and mountain-valley circulations due to *thermal* forcing (Pielke and Segal, 1986). Horizontal inhomogeneities in roughness length can result in the generation of internal boundary layers and horizontal inhomogeneities in such surface properties as water cover, soil moisture, vegetation type, vegetation cover, and snow cover can have a pronounced effect on dispersion by forcing mesoscale atmospheric circulations such as land-sea breezes through differential heating (Moran et al., 1986; Pielke and Segal, 1986; Segal et al., 1990). Differences in surface heating due to surface inhomogeneities may also trigger convection; convective clouds in turn may vent pollutants up out of the planetary boundary layer or deposit them on the ground through precipitation scavenging and washout.

These various *surface-forced* mesoscale circulations possess a number of natural spatial scales. In the case of mechanically-forced circulations, the horizontal and vertical scales will depend on obstacle height h, obstacle width L, flow speed U, stratification (represented by the Brunt-Väisälä frequency N), and density scale height H. Relevant combinations of these parameters include the Rossby number ($Ro = U/fL$), the internal Froude number ($Fr = U/Nh$), the nondimensional scale height (NH/fL), the Scorer parameter ($\ell^2 = N^2/U^2 - U^{-1}d^2U/dz^2$), the Rossby radius of deformation based on scale height (NH/f), and the Rossby radius of deformation based on obstacle height (Nh/f). For instance, for $Ro > 1$

and $Fr < 0.5$, upstream flow below the height h of a 2D ridge will be strongly blocked for a horizontal distance on the order of the Rossby radius of deformation Nh/f (Pierrehumbert, 1986). For larger Froude numbers, a deep, vertically propagating, upstream-tilting hydrostatic mountain wave may form if the ridge width L is much greater than ℓ^{-1}, where ℓ^2 is the Scorer parameter, but only a weak shallow flow perturbation will occur if L is much less than ℓ^{-1} (Durran, 1986). In general, the vertical scale of mechanically-forced mesoscale circulations will range between the height of the obstacle h and the density scale height H while the horizontal scale will lie between the horizontal scale of the obstacle L and the Rossby radius of deformation based on obstacle height Nh/f. In the case of thermal forcing, thermally-forced mesoscale circulations tend to be shallow so that the PBL height h_{BL} is a reasonable vertical scale while the Rossby radius of deformation Nh_{BL}/f is a limiting horizontal scale (e.g., Dalu and Pielke, 1989). Convection will, of course, tend to be cumulus-scale although organized mesoscale convective systems can be much larger.

Horizontal variations in atmospheric fields may also arise due to larger-scale atmospheric dynamics. Mesoscale circulations can be generated by synoptic-scale dynamic forcing and/or mesoscale variations in atmospheric properties; examples of such circulation systems include frontal and dryline circulations, squall lines, and jet streaks (e.g., AMS, 1986). Synoptic-scale and mesoscale baroclinicity can also affect vertical profiles of mean wind and turbulence within the PBL (McNider et al., 1988) and other interactions may occur. For instance, over the Great Plains the nocturnal low-level jet is strengthened by the contribution of daytime ageostrophy due to upslope flow over the heated sloping terrain.

IMPACT OF MODEL RESOLUTION ON MESOSCALE DISPERSION SIMULATIONS

The concept of a trajectory as the path followed by an air parcel or by the center of mass of a pollutant cloud has been widely used in RSDMs, especially Lagrangian puff and plume models. A number of studies have attempted to evaluate RSDM performance based on the comparison of calculated trajectories. However, the use of trajectories in this manner must be interpreted very carefully since it is impossible with current technology to actually measure a 3D air-parcel or puff center-of-mass trajectory. Tetroons do not make good surrogate air parcels since they follow constant-density surfaces rather than constant-entropy surfaces. Moreover, the path of one tetroon is only a single realization. A number of tetroons should be released simultaneously to permit the estimation of an ensemble-mean trajectory. On the other hand, after travelling for some hours, a dispersing mass of tracer gas is likely to be an elongated, possibly noncontiguous, 3D object extending from the surface to some height in the free troposphere. Surface concentration measurements may give only an approximate measure (or 'footprint') of cloud size, especially at night when differential advection may be occurring, and there is no guarantee that the surface density-weighted concentration maximum will lie beneath the tracer center-of-mass location though this may be the case for a well-mixed PBL; conceivably there might not be any tracer at the cloud center-of-mass location if the original tracer cloud has been fragmented. The dispersion of a tracer release is also only a single realization, and the length of the release, the travel time, and the sampling time must all be considered in evaluating the inherent uncertainty in the concentration measurements. Despite these caveats, surface measurements from mesoscale tracer experiments are still the best available means to evaluate the performance of an RSDM. However, it must be remembered that in such comparisons it is the *combined* treatment of transport and diffusion that is being evaluated and not transport alone (Draxler, 1987).

Nevertheless, Kuo et al. (1985) were able to evaluate just the treatment of transport in an extensive set of observing system simulation experiments which used detailed 3D wind fields generated by a regional-scale meteorological model. They explored the sensitivity of trajectory model calculations to the spatial and temporal density of the input meteorological fields by treating the air-parcel trajectories calculated from the highest-resolution wind fields as 'truth' and then systematically degrading the meteorological fields in 13 different experiments. Kuo et al. (1985) concluded that (1) restricting the trajectories to a 2D surface through the use of an isobaric, isosigma, or isentropic assumption resulted in significant degradations in trajectory accuracy, and (2) the major limitation of the current operational upper-air network is its 12 h observational frequency. Several other purely meteorological evaluations of trajectory sensitivity to rawinsonde temporal and spatial density have been carried out by Kahl and Samson (1986, 1988b) based on the CAPTEX experiment. Because only isobaric trajectories were calculated in these two studies and no 'true' trajectories were available, their usefulness is somewhat limited. However, they did conclude from six CAPTEX cases that spatial resolution was more important than temporal resolution in approving trajectory accuracy and suggested that the difference between their conclusion and that of Kuo et al. (1985) might be at least partly explained by the different meteorological cases considered.

Draxler (1987) ran a multi-level Lagrangian puff model for five different versions of the CAPTEX meteorological data set and found that enhanced temporal resolution resulted in slightly more accurate surface tracer trajectories than did enhanced spatial coverage. Brost et al. (1988) estimated CAPTEX surface tracer trajectories using an Eulerian advection-diffusion model and a variety of versions of the meteorological input data set. They concluded that increasing the spatial or temporal resolution of the observations made comparable improvements in the tracer simulation. They also found that the Eulerian model was able to calculate more accurate surface tracer trajectories than an isosigma trajectory model for the same meteorological input.

The 1980 Oklahoma mesoscale tracer experiment illustrates the significant impact one mesoscale flow feature, the nocturnal low-level jet, can have on mesoscale transport and diffusion and also shows the errors which can result due to RSDM inability to resolve this scale. Two arcs of samplers were deployed in this experiment, one 100 km N of the release site at Norman, Oklahoma and the other 600 km NE of the release site. The decision was made during the experiment to start the first three-hour sampling period along the 600-km-arc stations 13 h after the start of the three-hour perfluorocarbon tracer release at 1900 GMT (1400 CDT). Given the moderate 4 ms^{-1} southerly flow over the experiment domain on the backside of a stagnant high-pressure system, a trajectory calculation based on the NMC forecast indicated that it would take 18 h for the leading edge of the tracer cloud to reach the 600-km-arc samplers, allowing a five-hour margin of safety. However, some of the first set of samples contained significant tracer concentrations; in fact, these were the highest values measured and subsequent concentration levels decreased monotonically with time. Since the synoptic pattern changed only slightly over this period, the likeliest explanation for the accelerated tracer transport is a nighttime wind speed increase not predicted by the synoptic-scale forecast model, for example, the nocturnal low-level jet. Synoptic conditions of clear skies and light pressure gradient were ideal for the formation of a nocturnal jet (e.g., Bonner, 1968; McNider and Pielke, 1981). Comparison of two special soundings taken at Tinker Air Force Base north of Norman at 0000 GMT and 0300 GMT on July 9, 1980, five and eight hours after the start of the tracer release, shows a marked increase in wind speed between 200 m and 1100 m AGL (Table 5 of Ferber et al., 1981). Wind-speed measurements from a very tall, instrumented TV broadcast tower near Norman made up until 0200 GMT on July 9 also showed increasing wind speeds at the 177 m, 266 m, and 444 m levels as night fell.

An EPA-sponsored study has compared the predictions of seven regional-scale dispersion models against measurements taken at the two sampler arcs during the Oklahoma experiment (Policastro et al., 1986; Carhart et al., 1989). Figure 1 shows a comparison of model-predicted tracer puff center-of-mass positions and approximate puff widths with observations for the first three-hour sampling period at the 600 km sampler arc. Three model deficiencies are evident from this figure: (1) too slow transport of tracer; (2) underestimation of horizontal puff spread; and (3) transport direction too far to the east in some models. Each of these deficiencies may be explained based on neglect of the nocturnal jet and differential advection. All of the models tested except the ARRPA model relied on diagnostic wind fields generated from 12-hourly upper-air soundings and all models except RADM employed the well-mixed assumption. The ARRPA model uses winds at 10 levels from the Boundary Layer Model of the National Weather Service and RADM employs a random-walk approach for dispersion. As shown in Figure 2, 12-h temporal resolution is not adequate to resolve the inertial oscillation and supergeostrophic low-level jet. Linear interpolation between the two synoptic sampling times would give too low a transport wind speed and a transport direction biased to the east, both problems with these model predictions. Moreover, differential advection due to the inertial oscillation and subsequent vertical mixing will give a wider puff than predicted by the models.

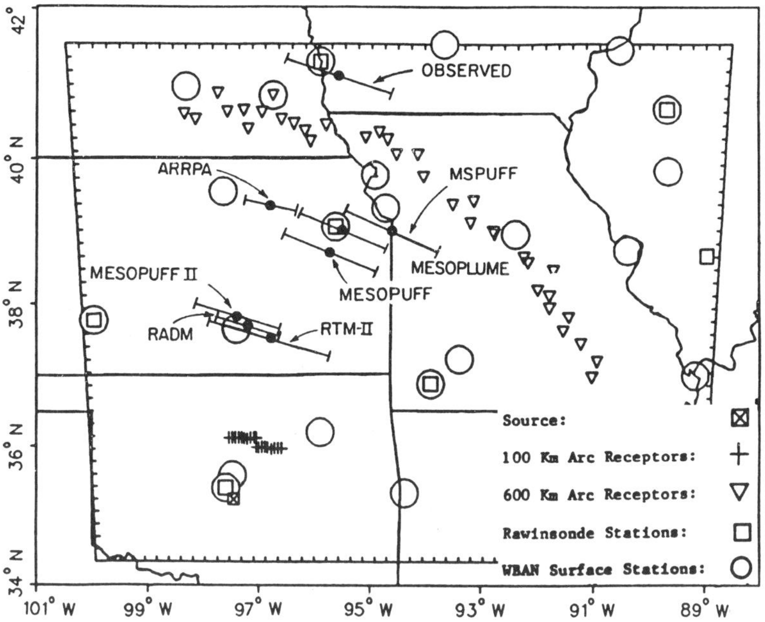

Figure 1. Schematic showing the Oklahoma mesoscale tracer experiment domain and the tracer puff centers (indicated by solid black dots) and approximate puff widths (indicated by bars passing through puff centers) predicted by six RSDMs for the first 3-hour sampling period beginning at 0800 GMT, 9 July 1980. An estimate of the actual tracer puff position is also shown based on an extrapolation of 600-km-arc sampler measurements since it is impossible to tell from the measurements when the tracer material first arrived at this arc. The 3-hour release of perfluorocarbon tracer took place at Norman, Oklahoma (lower lefthand corner) beginning at 1900 GMT, 8 July 1980 (based on Policastro et al., 1986).

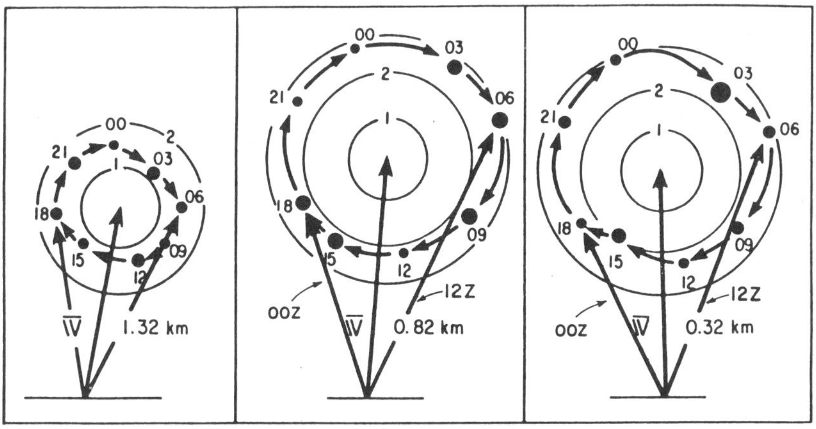

Figure 2. Hodographs of the summertime diurnal wind variation at selected levels for Fort Worth, Texas including the mean wind vector and wind vectors at the two standard synoptic observing times (0000 GMT/1800 CST, 1200 GMT/0600 CST). Speeds are in meters per second. The size of the black dots indicates the magnitude of probable errors in determining the deviation vector [adapted from Bonner and Paegle, 1970].

CONCLUSIONS

Although neglected by the majority of current RSDMs, atmospheric flows over land commonly possess mesoscale time and space scales arising from the diurnal heating cycle, the atmospheric inertial mode, thermal and mechanical forcing due to mesoscale terrain inhomogeneities, downscale energy transfer from the synoptic-scale to mesoscale due to nonlinear flow interactions, and mesoscale precipitation systems. While unresolved SGS motions are often thought of as 'turbulence', many mesoscale circulations are organized, long-lived, and geographically fixed, that is, nonrandom. Regional-scale pollutant transport and diffusion are affected by these flow scales through variations in the mean transport wind, differential advection due to vertical and horizontal wind shear, and both simultaneous and delayed vertical mixing.

Even for horizontally homogeneous flow over flat, uniform terrain, the diurnal heating cycle and frequent formation of a nocturnal low-level jet suggest the need for RSDM time resolution of at least six hours and vertical resolution of at least three layers (nighttime boundary layer, nighttime residual layer, and free troposphere). Resolution of terrain-forced mesoscale circulations such as land-sea breezes will require even finer temporal and vertical resolution. The minimum horizontal resolution required depends on the nature of the atmospheric flow field and physiographic features, but several studies have suggested that improved spatial resolution will produce better RSDM performance (Kuo et al., 1985; Kahl and Samson, 1986; Kahl and Samson, 1988b; Draxler, 1987; Brost et al., 1988).

There does appear to be increasing recognition of the need for mesoscale temporal and spatial resolution in RSDMs. Scholtz et al. (1988) have used an operational synoptic-scale NWP model as a dynamical interpolator together with a quasi-diagnostic 1D PBL model to diagnose physically-consistent hourly wind fields, including vertical motion, as well as such other meteorological fields as temperature, humidity, cloud cover, precipitation rate and type, and PBL depth. Much improved temporal and vertical resolution is obtained although horizontal resolution is still effectively synoptic-scale. As mentioned earlier, a number of other investigators have employed prognostic dynamical models to predict meteorological fields for regional-scale transport or dispersion studies. One other indication of this trend is the increase in the size of the meteorological component of mesoscale tracer experiment data archives. The meteorological data in the CAPTEX (1983) data archive consisted of rawinsonde and instrumented tower observations (Ferber et al., 1986); for ANATEX (1987), on the other hand, surface station observations and gridded meteorological fields from the U.S. operational regional-scale Nested Grid Model (NGM) were included along with rawinsonde and tower data (Draxler and Heffter, 1989). The archived NGM fields have two-hour temporal resolution, roughly 91 km horizontal resolution, and fair vertical resolution (ten sigma-p levels, including winds at σ_p =0.982, 0.943, 0.897, 0.844, and 0.784). It should be pointed out that prognostic dynamical models also do not provide perfect wind and turbulence fields and that the magnitude of error in the prognosed fields increases with forecast time; however, four-dimensional data assimilation techniques can be used to incorporate post-model-initialization meteorological observations and reduce the size of this error (e.g., Kao and Yamada, 1988).

ACKNOWLEDGEMENTS

The authors appreciate the help of Dallas McDonald in preparing the manuscript and that of Judy Sorbie in preparing the figures. The work described in this paper was supported by the Electric Power Research Institute under Contract No. RP-1630-53, by the National Park Service under Contract NA81RAH00001, Amendment 17, Item 15, by the National Aeronautical and Space Administration under Grant No. NAS8-36479, and by Cray Research, Inc. through its University Research Program.

REFERENCES

AMS, 1986: *Mesoscale Meteorology and Forecasting.* American Meteorological Society, Boston, 793 pp.

Anthes, R. A., 1983: Regional models of the atmosphere in middle latitudes. *Mon. Wea. Rev.*, **111**, 1306–1335.

Avissar, R., M. D. Moran, G. Wu, R. N. Meroney, and R. A. Pielke, 1990: Operating ranges of mesoscale numerical models and meteorological wind tunnels for the simulation of sea and land breezes. *Boundary-Layer Meteor.* (In press).

Blackadar, A. K., 1957: Boundary layer wind maxima and their significance for the growth of nocturnal inversions. *Bull. Amer. Meteor. Soc*, **38**, 283–290.

Bonner, W. D., 1968: Climatology of the low level jet. *Mon. Wea. Rev.*, **96**, 833–850.

Bonner, W. D. and J. Paegle, 1970: Diurnal variations in boundary layer winds over the south central United States in summer. *Mon. Wea. Rev.*, **98**, 735–744.

Brost, R. A., P. L. Haagenson, and Y.-H. Kuo, 1988: Eulerian simulation of tracer distribution during CAPTEX. *J. Appl. Meteor.*, **27**, 579–593.

Carhart, R. A., A. J. Policastro, M. Wastag, and L. Coke, 1989: Evaluation of eight short-term long-range transport models using field data. *Atmos. Environ.*, **23**, 85–105.

Carras, J. N. and D. J. Williams, 1988: Measurements of relative σ_y up to 1800 km from a single source. *Atmos. Environ.*, **22**, 1061–1069.

Csanady, G. T., 1973: *Turbulent Diffusion in the Environment.* D. Reidel Publishing Co., Dordrecht, 248 pp.

Dalu, G. A. and R. A. Pielke, 1989: An analytical study of the sea breeze. *J. Atmos. Sci.*, **46**, 1815–1825.

Draxler, R. R., 1987: Sensitivity of a trajectory model to the spatial and temporal resolution of the meteorological data during CAPTEX. *J. Climate Appl. Meteor.*, **26**, 1577–1588.

Draxler, R. R., 1989: Overview and preliminary results from the Across North America Tracer Experiment (ANATEX). In *Proc. Sixth AMS/APCA Conf. on Applications of Air Pollution Meteorology*, American Meteorological Society, Boston, 82–85.

Draxler, R. R. and J. L. Heffter ,Editors, 1989: Across North America Tracer Experiment (ANATEX). Volume I: Description, ground-level sampling at primary sites, and meteorology. NOAA Tech. Memo. ERL ARL-167, Air Resources Laboratory, National Oceanic and Atmospheric Administration, Silver Spring, Maryland, 83 pp.

Durran, D. R., 1986: Mountain waves. In *Mesoscale Meteorology and Forecasting*, Ray, P. S., editor, American Meteorological Society, Boston, 472–492.

Ferber, G. J., J. L. Heffter, R. R. Draxler, R. J. Lagomarsino, F. L. Thomas, R. N. Dietz, and C. M. Benkovitz, 1986: Cross-Appalachian Tracer Experiment (CAPTEX '83) final report. NOAA Tech. Memo. ERL ARL-142, Air Resources Laboratory, National Oceanic and Atmospheric Administration, Silver Spring, Maryland, 60 pp.

Ferber, G. J., K. Telegadas, J. L. Heffter, C. R. Dickson, R. N. Dietz, and P. W. Krey, 1981: Demonstration of a long-range atmospheric tracer system using perfluorocarbons. NOAA Tech. Memo. ERL ARL-101, Air Resources Laboratory, National Oceanic and Atmospheric Administration, Silver Spring, Maryland, 74 pp.

Fujita, T. T., 1986: Mesoscale classifications: their history and their application to forecasting. In *Mesoscale Meteorology and Forecasting*, Ray, P. S., editor, American Meteorological Society, Boston, 18–35.

Gifford, F. A., 1983: Atmospheric diffusion in the mesoscale range: the evidence of recent plume width observations. In *Proc. 6th AMS Symp. on Turbulence and Diffusion*, American Meteorological Society, Boston, 300–304.

Gifford, F. A., 1986: Atmospheric diffusion in the range 20 to 2000 kilometers. In *Air Pollution Modeling and Its Application V*, De Wispelaere, C., Schiermeier, F. A., and Gillani, N. V., editors, Plenum Press, New York, 247–252.

Hanna, S. R., 1981: Lagrangian and Eulerian time-scale relations in the daytime boundary layer. *J. Appl. Meteor.*, **20**, 242–249.

Heffter, J. L., 1965: The variation of horizontal diffusion parameters with time for travel periods of one hour or longer. *J. Appl. Meteor.*, **4**, 153–156.

Kahl, J. D. and P. J. Samson, 1986: Uncertainty in trajectory calculations due to low resolution meteorological data. *J. Climate Appl. Meteor.*, **25**, 1816–1831.

Kahl, J. D. and P. J. Samson, 1988a: Shear effects on wind interpolation accuracy. *J. Appl. Meteor.*, **27**, 1299–1301.

Kahl, J. D. and P. J. Samson, 1988b: Uncertainty in estimating boundary-layer transport during highly convective conditions. *J. Appl. Meteor.*, **27**, 1024–1035.

Kao, C.-Y. J. and T. Yamada, 1988: Use of the CAPTEX data for evaluations of a long-range transport numerical model with a four-dimensional data assimilation technique. *Mon. Wea. Rev.*, **116**, 293–306.

Keyser, D. and L. W. Uccellini, 1987: Regional models: emerging research tools for synoptic meteorologists. *Bull. Amer. Meteor. Soc*, **68**, 306–320.

Kuo, Y.-H., M. Skumanich, P. L. Haagenson, and J. S. Chang, 1985: The accuracy of trajectory models as revealed by the observing system simulation experiments. *Mon. Wea. Rev.*, **113**, 1852–1867.

McNider, R. T., M. D. Moran, and R. A. Pielke, 1988: Influence of diurnal and inertial boundary-layer oscillations in long-range dispersion. *Atmos. Environ.*, **22**, 2445–2462.

McNider, R. T. and R. A. Pielke, 1981: Diurnal boundary-layer development over sloping terrain. *J. Atmos. Sci.*, **38**, 2198–2212.

Moran, M. D., R. W. Arritt, M. Segal, and R. A. Pielke, 1986: Modification of regional-scale

pollutant dispersion by terrain-forced mesoscale circulations. In *Trans. Second APCA Int'l. Spec. Conf. on Meteorology of Acidic Deposition*, Laznow, J. and Stensland, G. J., editors, Air Pollution Control Association, Pittsburgh, 136–157.
Orlanski, I., 1975: A rational subdivision of scales for atmospheric processes. *Bull. Amer. Meteor. Soc*, **56**, 527–530.
Pasquill, F., 1962: Some observed properties of medium-scale diffusion in the atmosphere. *Quart. J. Roy. Meteor. Soc.*, **88**, 70–79.
Pasquill, F. and F. B. Smith, 1983: *Atmospheric Diffusion*, 3rd ed. John Wiley & Sons, New York, 437 pp.
Pielke, R. A., 1984: *Mesoscale Meteorological Modeling.* Academic Press, Orlando, Florida, 612 pp.
Pielke, R. A., W. A. Lyons, R. T. McNider, M. D. Moran, D. A. Moon, R. A. Stocker, R. L. Walko, and M. Uliasz, 1990: Regional and mesoscale meteorological modeling as applied to air quality studies. In *Proc. 18th NATO/CCMS Int'l. Tech. Mtg on Air Pollution Modelling and Its Application.* This volume.
Pielke, R. A. and M. Segal, 1986: Mesoscale circulations forced by differential terrain heating. In *Mesoscale Meteorology and Forecasting*, Ray, P. S., editor, American Meteorological Society, Boston, 516–548.
Pierrehumbert, R. T., 1986: Lee cyclogenesis. In *Mesoscale Meteorology and Forecasting*, Ray, P. S., editor, American Meteorological Society, Boston, 493–515.
Policastro, A. J., M. Wastag, L. Coke, R. A. Carhart, and W. E. Dunn, 1986: Evaluation of short-term long-range transport models. Volume I: Analysis procedures and results. Technical Report EPA-450/4-86-016a, U.S. Environmental Protection Agency, Research Triangle Park, North Carolina, 215 pp.
Scholtz, M. T., B. Weisman, L. Mahrt, and A. D. Christie, 1988: Generation of meteorological data fields for the ADOM Eulerian regional model. In *Air Pollution Modeling and Its Applications VI*, van Dop, H., editor, Plenum Press, New York, 391–402.
Segal, M., J. H. Cramer, R. A. Pielke, J. R. Garratt, and P. Hildebrand, 1990: Observational evaluation of snow-cover effects on the generation and modification of mesoscale circulations. *Mon. Wea. Rev.* (Submitted).
Stull, R. B., 1988: *An Introduction to Boundary Layer Meteorology.* Kluwer Academic Publishers, Dordrecht, 666 pp.
Taylor, A. D., 1982: Puff growth in an Ekman layer. *J. Atmos. Sci.*, **39**, 837–850.
Thorpe, A. J. and T. H. Guymer, 1977: The nocturnal jet. *Quart. J. Roy. Meteor. Soc.*, **103**, 633–653.
Wilczak, J. M. and J. W. Glendening, 1988: Observations and mixed-layer modeling of a terrain-induced mesoscale gyre: the Denver cyclone. *Mon. Wea. Rev.*, **116**, 2688–2711.

DISCUSSION

R. BORNSTEIN — What is the relationship between relative diffusion and differential advection?

M.D. MORAN — We have used the term "differential advection" in the sense of the straining motion of a resolved flow field, that is, grid-scale "stirring". On this scale the motions of adjacent particles will be correlated. Turbulent diffusion corresponds to subgrid-scale (SGS) advective and mixing processes. As currently parameterized in our model, particle movement due to these SGS processes is not spatially correlated. Thus the model includes relative diffusion due to grid-scale differential advection but not to SGS motions.

F.L. LUDWIG — Can you reduce the data requirements by being "smart" about where and when those data are collected?

M.D. MORAN — Not if you are using wind fields obtained solely through the objective analysis of observed winds. However, the use of a prognostic dynamical model such as a three-dimensional mesoscale meteorological model or a one-dimensional boundary-layer model as part of a data assimilation scheme could reduce the required temporal and spatial resolution of the input data, especially under quasi-stationary conditions with strong surface forcing.

C.J. NAPPO

You have not mentioned relative diffusion, yet we know that it is this process and not single-particle diffusion that controls dispersion on the mesoscale. Also, a real plume cannot be sampled or simulated using serially-released particles. All plumes begin with a finite width, not at a point. Can you comment on these points.

M.D. MORAN

To the contrary, it is single-particle diffusion that controls dispersion on the mesoscale. After long travel times, typical spatial separations between members of particle pairs will be larger than the scale of energy-containing eddies so that the particles travel independently. Concerning the second point, serially-released independent particles cannot be used to model an instantaneous plume but they can be used to model an ensemble-average plume. However, at increasingly larger scales, the difference between the two becomes increasingly smaller. Finally, at mesoscale distances, small differences in sopurce size are undetectable, but if source size effects are a concern, particles can easily be released from a line, area, or volume source.

M.M. MILLAN

Our experience on tracking the Sudbury INCO plume over 400-600 km in Southern Ontario, and other tall stacks plumes in Spain, indicates that the observed diffusion at any one point is greatly influenced by advected properties (i.e., what has happened before, up to 6 or more hours) and the local terrain (e.g., long-range transport of the INCO stack depended on night releases and subsequent travel over the cold water of Georgian Bay to keep the plume sufficiently coherent to reach Southern Ontario the following morning) [see article by Millán in *Regional and Long-range Transport of Air Pollution*, 1987, S. Sandroni, Editor, Elsevier, Amsterdam, pp. 249-280].

M.D. MORAN

This is consistent with our numerical simulations of mesoscale plume dispersion. Plume width at mesoscale downwind distances is significantly influenced by atmospheric stability at the time of release and subsequent variations during transport. Release of pollutants into a stable atmosphere followed by transport over cold water will minimize vertical mixing and subsequent lateral dispersion by differential advection.

A PRELIMINARY EVALUATION OF ADPIC MODEL PERFORMANCE ON SELECTED ANATEX RELEASES USING OBSERVED, ANALYZED, AND DYNAMICALLY PREDICTED WINDS

D. J. Rodriguez and R. T. Cederwall

Lawrence Livermore National Laboratory
Livermore, CA 94550 USA

INTRODUCTION

The Chernobyl nuclear power plant accident demonstrated the need for long-range transport and diffusion models to assist in the real-time evaluation of the consequences of such events on man and his environment. Based on the efforts of the Atmospheric Release Advisory Capability (ARAC) to estimate the dispersion of radioactivity from Chernobyl (Dickerson and Sullivan, 1986; Gudiksen and Lange, 1986), work was undertaken to expand the domain of application of the mesoscale atmospheric diffusion particle-in-cell (ADPIC) model to the hemispheric scale.

The mesoscale and hemispheric versions of the ADPIC model are essentially identical. They both use Lagrangian markers to represent the movement of pollutant mass or activity within an Eulerian framework (Lange, 1978). They differ only in their utilization of either raw or processed wind velocity data. The mesoscale ADPIC uses surface and upper air wind observations, which are interpolated and extrapolated to a regular matrix of grid points using well-tested routines (Rodriguez, et al, 1982). However, it was decided that the acquisition and software manipulation of large volumes of wind observations over continental or hemispheric scales would cause unacceptable delays in ARAC's response to an emergency situation. Consequently, arrangements were made to obtain processed fields of winds, which the Air Force Global Weather Central (GWC) produce on a routine basis.

Until the recent release of the Across North America Tracer Experiment (ANATEX) database, there have been few opportunities for testing whether the GWC wind fields provide an accurate depiction of large-scale flow features between sources and samplers. In one study, Rodriguez and Peterson (1989) used GWC wind fields in the hemispheric ADPIC model to demonstrate the probable linkage between fresh fission products, detected by samplers across Europe in March 1987, and the venting of radioactiviy from an underground nuclear test in the U.S.S.R. A crucial finding of more general interest was that the hemispheric ADPIC model is capable of resolving a plume over continental distances provided, of course, that a sufficient number of particles (10,000 in this case) are injected into the modeling domain.

The ANATEX database provides a fertile ground for testing the sensitivity and uncertainty of long-range models. The primary purpose of this paper is to focus on a single aspect of modeling uncertainty—the wind fields that describe the advection component of a transport and diffusion calculation. The ANATEX database has been used to evaluate the performance of ADPIC based on the use of wind-field products from three different sources: 1) the GWC's Whole Mesh Grid covering the northern hemisphere (Stobie, 1986), 2) a more refined grid of advection velocities generated by in-house interpolative procedures applied to rawinsonde data and supplementary soundings in the source-receptor region, and 3) dynamically consistent u, v and ω fields from the National Oceanic and Atmospheric Administration's (NOAA's) Nested Grid Model (Hoke, et al, 1989).

WIND FIELD ATTRIBUTES

The GWC winds are based on a polar stereographic projection of High Resolution Analysis System (HIRAS) data. This system, which produces global analyses on a 2.5 × 2.5 degree latitude-longitude grid, combines observations from land stations, ships, buoys, aircraft, balloons and satellites. Its two main components are a "first-guess" model and an analysis model, with the first-guess field produced by a 40-wave, 12-layer global spectral model (GSM) developed by the National Meteorological Center (Sela, 1982). Interpolation and projections of the resulting analyses onto polar stereographic planes produce a grid having a 381 km horizontal resolution and a vertical resolution corresponding to mandatory pressure levels. These data are mapped to evenly-spaced grid points in the vertical (Δ z = 100 m and grid top = 1400 m) by linear interpolation before their insertion into hemispheric ADPIC. Although these wind fields are produced every 6 h, the 00 and 12 UTC products were the only ones used in this study because of the dubious quality of the data at intervening times (Norquist, 1989).

Upper air soundings for all of North America, including supplemental soundings (06 and 18 UTC) at Glasgow, St. Cloud and four rawinsonde stations near these sources, were archived during the course of ANATEX. Observed wind vectors were scaled and rotated so as to conform to a Lambert Conformal projection and then interpolated to a regular grid having 100 and 120 km resolutions in the north-south and east-west directions, respectively, and a vertical resolution of 100 m. Data interpolations were performed using inverse-distance squared weighting in the horizontal dimensions and linear interpolation in the vertical dimension. Since upper air soundings provide discrete samples of winds over limited volumes of the atmosphere, they may not represent conditions over the temporal and spatial scales in the model. Nevertheless, they offer the potential for an enhanced description of the synoptic wind structure since, unlike the GWC product, they contain information at both mandatory and significant pressure levels.

Outputs from the Regional Analysis and Forecast System (RAFS) of the National Meteorological Center represent the most promising product with regard to wind field characterization. The RAFS has three primary components—the regional optimum interpolation analysis, the nonlinear normal mode initialization and the forecast component, also known as the Nested Grid Model (NGM). Using initialized fields based on processed observational data, the NGM solves the primitive equations in terrain-following coordinates, therein producing forecast data at 16 levels every 2 h; 6 h forecasts were used in this study to maintain a measure of comparability with the other products under evaluation. In the NGM, vertical motion is represented by ω. In this study, the ω's were converted to w-wind components and the pressure levels were transformed to pressure heights using the hypsometric formula. Wind speeds and wind directions were assumed to vary linearly between data levels. The resolution of the final grid was 94 km × 94 km × 100 m.

ANATEX TRACER STUDY

ANATEX was designed to provide a database for establishing levels of performance of long-range air pollution models (Draxler and Heffter, 1989). The experiment

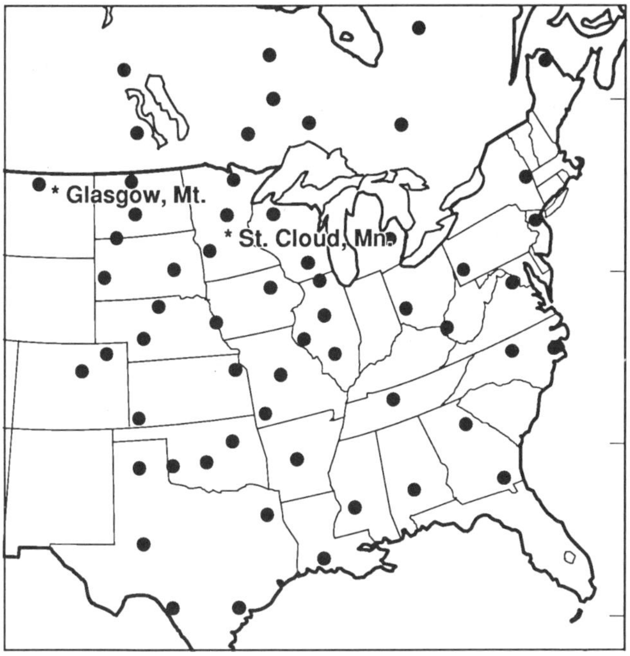

Fig. 1. ANATEX source locations and ground-based sampling sites (filled circles).

consisted of simultaneous, 3 h releases of perfluorocarbon tracers—PTCH at Glasgow, Montana, and PDCH at St. Cloud, Minnesota—every two and one-half days over an 84 day period beginning on 5 January 1987. Releasing these gaseous tracers on a cyclic (daytime/nighttime) basis provided greater opportunities for studying the process of dispersion under a variety of meteorological conditions. Generally speaking, previous long-range experiments relied on forecasts to direct tracers along narrowly-defined corridors, thereby maximizing the potential for capturing the tracer by a limited array of samplers (Ferber, et al, 1986; Heffter, et al, 1984). Unlike ANATEX, these "constrained" schedules of emission had the unfortunate effect of diminishing the challenge to long-range models due to homogeneous synoptic conditions.

The simultaneous release of unique perfluorocarbons at sources separated by a distance of roughly 1000 km and their subsequent collection provided information to researchers concerned with source attribution problems. Twenty-four hour samples taken by 77 ground-level, sequential air samplers over the central and eastern U.S. and southern Canada provided information concerning the rates of plume growth over unprecedented distances (see Figure 1). The use of airborne sampling platforms within 500 km of the release sites at altitudes ranging from 150 m to 2000 m provided information concerning the changing vertical distributions of tracer materials due to diffusive and wind shear effects. Finally, the creation of an archive consisting of rawinsonde data from the National Weather Service, supplemented by six-hourly soundings around the sources, provided a substantial body of information for model simulations and sensitivity studies.

CASE STUDY SELECTION AND MODEL PERFORMANCE MEASURES

The perfluorocarbon releases at Glasgow and St. Cloud provided data for testing the transport component of the ADPIC model. Two criteria were developed to facilitate the selection of case studies: 1) there must be two or more 24-h surface concentrations greater than or equal to 1 fl/l over four or more consecutive days, and 2) the surface concentrations must be unambiguously attributable to a specific release. Application of these rules yielded nine PDCH and 12 PTCH cases. From this reduced set, four cases were selected for initial examination, two of which involved the simultaneous releases of PTCH and PDCH and another two involving only PTCH. The transport environments for each case were characterized as being dominated by advection or weak synoptic forcing. Tracer releases into advection-dominated flows were preferred since they allowed us to study the differences between wind-field products more effectively and to better define the locations of concentration-weighted centroids at daily intervals.

The protocols used for rating the performance of the ADPIC model were devised by the U.S. Environmental Protection Agency (Clark, et al., 1986). These protocols are based on concentration-weighted centroids of the daily tracer "footprints". The protocols are based on three distance measures: D_{ms}, the distance between the source and the concentration-weighted centroid of the measured surface concentrations, D_{cs}, the distance between the source and the center of mass of the calculated concentrations, and D_{cm}, the distance of separation between the calculated and measured centroids. The ratio D_{cs}/D_{ms} provides a measure of how well the model matches the speed of the observed plume; a value of one would be ideal. The ratio D_{cm}/D_{ms} provides a measure of how well the model matches the speed and direction of the plume; if the locations of the measured and calculated concentration centroids were coincidental, thereby signifying perfect agreement, the ratio would be zero.

MODEL RESULTS

Before discussing the model results in global terms, it may be insightful to consider the bevavior of the ADPIC model when observed, analyzed and dynamically-predicted winds were used during the course of selected releases. Two separate cases, representing the "best" and the "worst" sets of model results in a composite sense, are highlighted in Figures 2 and 3. The various symbols, and the numbers within these symbols, represent the daily positions of the concentration-weighted centroids of the observed and calculated plumes following a 3-h morning release of PTCH at Glasgow on 10 January 1987 (see Figure 2). Two semi-permanent synoptic features of importance during the four days in which the tracer cloud was detected were the Great Basin High over northern Utah and a developing trough in the Hudson Bay region. Additionally, a transient center of low pressure, which formed over the Great Lakes, moved off the Canadian coastline approximately two days after the Glasgow release. When this occurred, the isobaric pattern flattened out, providing optimal conditions for producing good model results, regardless of the wind-field product being used. As can be seen in Figure 2, no plume observations were recorded on the first day after release, suggesting the possibility of lofting. On the first, second and third days, the calculated centroids formed reasonably tight groupings, the one exception being the NGM plume, which underwent an unexplained acceleration between the second and third days. Overall, the direct use of upper air profiles (UAP) seemed to provide the best characterization of transport conditions during this case study.

Results of the worst-case simulation are presented in Figure 3. As PDCH was being released at St. Cloud on 24 March 1987, a low pressure center was developing over northeastern Kansas. It maintained its position for a day and a half before moving rapidly into the Great Lakes region over the next 12 h. This shift in low

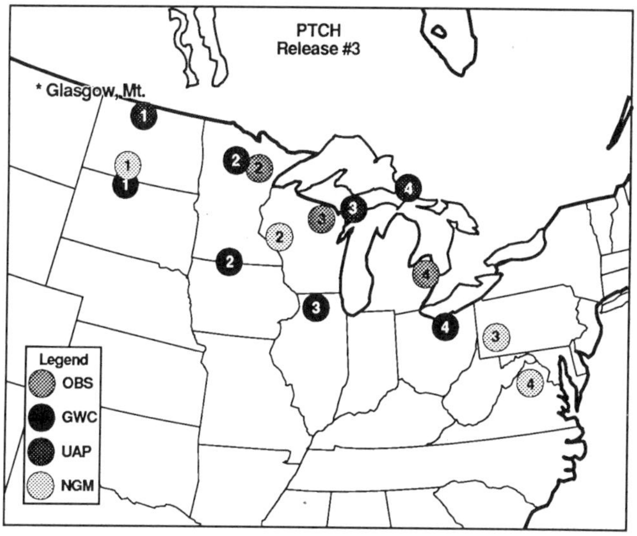

Fig. 2. Daily locations of the concentration-weighted centroids of observed and calculated plumes for the "best-case" simulation.

pressure was accompanied by a build-up of high pressure over northern Texas. Regardless of the wind-field product used, the ADPIC model was unable to replicate the transport environment with any degree of success. As suggested in Figure 3, all calculated plumes remained near the source of emissions on the first day. On the second day, circulation around the backside of the low pressure center occurred as the center began its migration towards the Great Lakes. A significant divergence in the centroid locations is clearly evident in the final two days—the observed plume finds itself in a regime of accelerated flow, the UAP plume attempts, but fails to keep pace with the observed plume, and the NGM and GWC plumes are trapped in a completely different regime dominated by high pressure. In all three simulations, the model's inability to match the observed plume trajectory was largely the result of persistence; six- and twelve-hourly updates in the transport fields were too infrequent to accurately portray the rapidly changing meteorology.

The results for all case studies are presented in Figure 4. The top panel indicates the errors in wind speed, as depicted by the ratios D_{cs}/D_{ms}; the ideal value is one. Each symbol represents the median of anywhere from three to five values. The bottom panel provides a quantitative measure of the total spatial (trajectory) error, as depicted by the ratios D_{cm}/D_{ms}; here, the ideal value is zero. Interpretation of the results on a case-by-case basis requires a coupling of the information in the top and bottom panels. Using the PTCH results for Case A as an example, the NGM fields clearly produce the best overall results, both in terms of wind speed and the combination of wind speed and wind direction. However, if we shift our attention to the PDCH results for Case B, we see that the model has been able to accurately match the speed, but not the trajectory, of the observed plume when it uses the GWC fields. Finally, horizontal lines, representing the averages of the medians for each symbol group, have been added to indicate the overall performance of the ADPIC model. The model seems to perform best when using the NGM fields, but the results appear to be comparable when it uses the UAP and GWC winds. However, considering the amount of scatter which is present in Figure 4, the significance of these numbers is perhaps questionable.

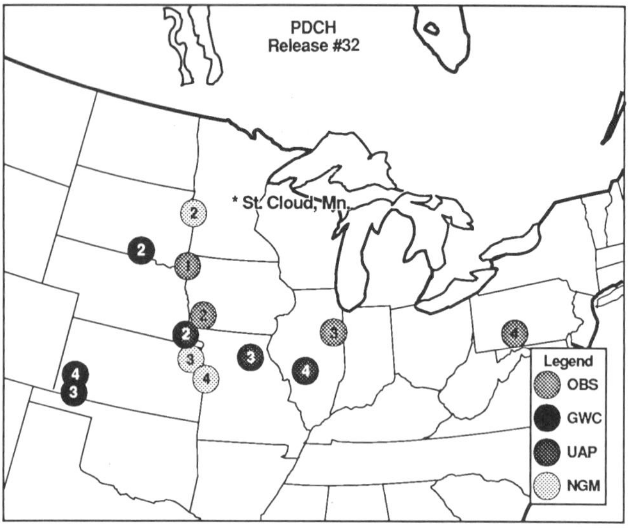

Fig. 3. Daily locations of the concentration-weighted centroids of observed and calculated plumes for the "worst-case" simulation.

CONCLUDING REMARKS

Our research on the comparative value of the analyzed, observed and dynamic wind fields is far from being completed. We have, however, come to the realization that more stable statistics than distance ratios based on center of mass calculations are needed. In spite of this conclusion, our comparison of the medians of the ratios for all case studies suggests that the NGM and "upper air" wind fields provide a better representation of synoptic-scale circulation than the one offered by the GWC wind field. We also found that the inclusions of space- and time-varying mixing height surfaces in ADPIC enhances performance, regardless of the wind-field product. This improvement to ADPIC has not yet been formally developed. Instead, we calculated the mass-weighted centroids of the observed plumes for each 24-h sampling period. The radiosondes nearest the centroids provided the information necessary to estimate the mixing heights, which were then input to the model as a temporal array.

Several areas of research relating to wind-field representation are still outstanding. They include: 1) a more thorough examination of the role of wind-field persistence vis-a-vis model performance, 2) a classification of results on the basis of the release environment, thereby helping us to better understand model limitations under different meteorological conditions, and 3) an expanded set of case studies and the incorporation of aircraft measurements.

ACKNOWLEGMENTS

This research is supported by the Air Force Technical Applications Center. Work was performed under the auspices of the U.S. Department of Energy by the Lawrence Livermore National Laboratory under contract number W-7405-ENG-48.

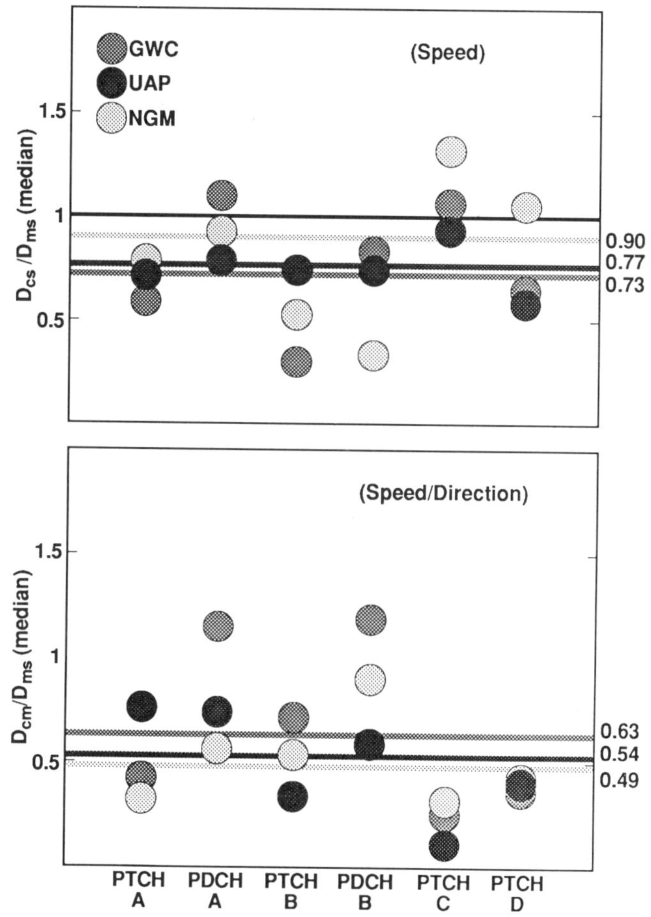

Fig. 4. Medians of distance ratios, representing errors in plume speed (top) and plume trajectory (bottom) associated with perfluorocarbon releases at Glasgow, MT (PTCH) and St. Cloud, MN (PDCH) for selected case studies (A–D).

REFERENCES

Clark, T.L., R.D. Cohn, S.K. Seilkop, R.R. Draxler and J.L. Heffter, "Comparison of Modelled and Measured Tracer Gas Concentrations During the Across North America Tracer Experiment (ANATEX)," Proc. NATO/CCMS ITM on Air Pollution Modeling and Its Application, Sept. 19-22, 1988, Cambridge, UK.

Dickerson, M.H. and T.J. Sullivan, 1986, "ARAC Response to the Chernobyl Reactor Accident," Lawrence Livermore National Laboratory Report UCID-20834, Livermore, CA.

Draxler, R.R. and J.L. Heffter (Eds.), 1989, "Across North America Tracer Experiment (ANATEX)," Vol. I: Description, Ground-Level Sampling at Primary Sites, and Meteorology. NOAA Tech. Memo. ERL ARL-167, Silver Spring, MD.

Ferber, G.J., J.L. Heffter, R.R. Draxler, R.J. Lagomarsino, F.L. Thomas, R.N. Dietz and C.M. Benkovitz, 1986, "Cross-Appalachian Tracer Experiment (CAPTEX-83)", NOAA Tech. Memo. ERL ARL-142, Silver Spring, MD.

Gudiksen, P.H. and R. Lange, 1986, "Atmospheric Dispersion Modeling of Radioactivity Releases from the Chernobyl Event," Lawrence Livermore National Laboratory Report UCRL-95363, Livermore, CA.

Heffter, J.L., J.F. Schubert and G.A. Mead, 1984, "Atlantic Coast Unique Regional Atmospheric Tracer Experiment (ACURATE)", NOAA Tech. Memo. ERL ARL-130, Silver Spring, MD.

Hoke, J.E., N.A. Phillips, G.J. DiMego, J.J. Tuccillo and J.G. Sela, 1989, "The Regional Analysis and Forecast System of the National Meteorological Center," *Weather and Forecasting*, **4**, 323-334.

Lange, R., 1978, "ADPIC—A Three-Dimensional Particle-In-Cell Model for the Dispersal of Atmospheric Pollutants and Its Comparison to Regional Tracer Studies," *J. Appl. Meteor.*, **17**, 320-329.

Norquist, D.C, *private communication.*

Rodriguez, D.J., G.D. Greenly, P.M. Gresho, R. Lange, B.S. Lawver, L.A. Lawson and H. Walker, 1982, "User's Guide to the MATHEW/ADPIC Models," Lawrence Livermore National Laboratory Report UASG-82-16, Livermore, CA.

Rodriguez, D.J. and K.R. Peterson, 1989, "Simulating the Venting of Radioactivity from a Soviet Nuclear Test," *Atmos. Environ.*,**23**, 5, 953-958.

Sela, J.G., 1982, "The NMC Spectral Model," NOAA Tech. Report NWS 30, Silver Spring, MD.

Stobie, J.G., 1986, "AFGWC's Advanced Weather Analysis and Prediction System," Air Weather Service Report AWS/TN-86/001, Scott Air Force Base, IL.

THE INFLUENCE OF BIOGENIC EMISSIONS ESTIMATES ON OZONE PRECURSOR CONTROL REQUIREMENTS FOR ATLANTA

Richard D. Scheffe and Gerald L. Gipson

U.S. Environmental Protection Agency
Research Triangle Park, N.C.

Ralph E. Morris

Systems Application, Inc.
San Rafael, CA

INTRODUCTION

Over the past 15 years photochemical air quality simulation models have been used to simulate observed spatial and temporal patterns of ambient ozone, and evaluate the merits of different emissions control approaches for reducing ozone. A coincident increase in our knowledge of atmospheric chemistry responsible for ozone formation has accompanied the use of source driven models which couple atmospheric transport processes with chemistry to provide concentration estimates. Most modeling applications for urban areas included only anthropogenic emissions sources and, typically, control of reactive organic gases (ROG) was shown to be an effective measure for reducing peak ozone. In contrast, the recent modeling work of Chameides et al. (1988) on the Atlanta area suggested that when natural emissions (principally isoprene and alpha-pinene) are included as sources, the effectiveness of anthropogenic ROG control is substantially reduced (relative to the case where natural emissions are not included). The Chameides work using the OZIPM4 (EPA, 1989) suggested that the level of anthropogenic ROG control needed to attain the ozone NAAQS (0.12 ppm) increased from roughly 30% to 70% when natural emissions were included for the modeling day, June 4, 1984. The OZIPM4 is a relatively simple trajectory model, however, and a natural question to be raised is whether similar results would be obtained with a more physically realistic model.

We have produced a similar modeling exercise (expanded to include NOx emissions controls) based on the Urban Airshed Model (UAM) (EPA, 1990a) with carbon bond-4 (CB4) chemistry, EPA's preferred regulatory tool for ozone modeling. The UAM uses a grid based, three dimensional structure which offers more complete characterization of the transport processes, source and concentration distributions relative to OZIPM4. The Atlanta modeling domain (Figure 1) covered an area 160 by 160 km (40x40 4 km cells) and included 5 vertical layers (3 layers above the nighttime inversion height). Simulations commenced at 1200 LST, June 3 and proceeded to 2000 LST, June 4.

MODEL SETUP

The Atlanta modeling was performed as part of EPA's 5-City UAM study (EPA, 1990b), a demonstration project broad in scope which looked at the feasibility of applying the UAM without conducting a detailed field study as well as various emission control strategy approaches. Accordingly, results reported are intended to represent relative effects (rather than absolutes) incurred as a result of following certain control pathways. Cursory procedural descriptions for developing primary modeling inputs are provided below.

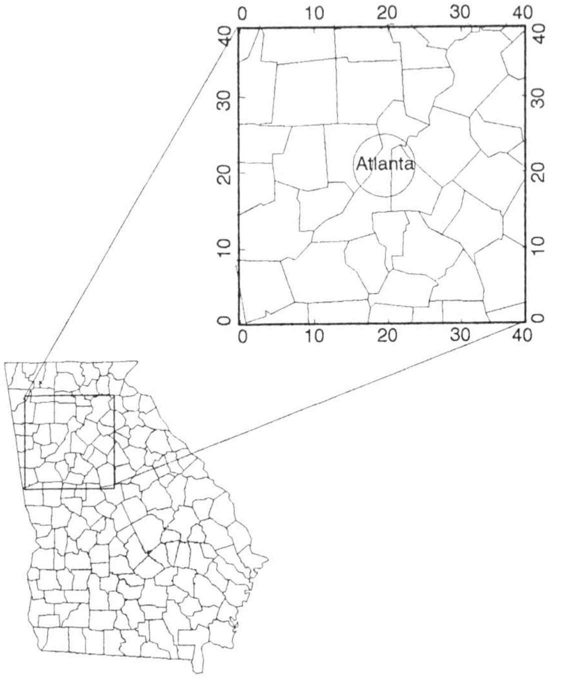

Fig. 1. UAM modeling domain consisting of a 40 by 40 array of 4 km grid cells.

Emissions

Anthropogenic emissions were based on the 1985 National Acid Precipitation Assessment Program (NAPAP) (EPA, 1988). Emissions data were translated to model inputs through a multi-step procedure which performed the necessary gridding, temporalization and speciation. Day specific temperature adjustments were made for mobile source evaporative and running loss emissions. Natural ROG emissions (biogenics) were developed as part of EPA's Regional Oxidant Model (ROM) program (Novak and Regan, 1986). On a domain-wide basis, biogenics accounted for 59% of all ROG emissions. An acknowledged high level of uncertainty (accuracy on the order of +/- 200%) has been associated with biogenic emissions.

Initial and Boundary conditions

Sensitivity to boundary and initial concentrations was mitigated by the large domain and early (June 3) initiation period. The following "clean" initial concentrations were used:

ROG	25 ppbC
ISOPRENE	0.001 ppb
NOx	1 ppb (split 3/4 NO_2, 1/4 NO)
O_3	40 ppb
CO	200 ppb

Boundary concentrations were identical to initial concentrations, except for the following changes: ISOPRENE, 1ppb; ROG, 40ppbC.

Windfield and Mixing Heights

Hourly wind fields (x and y components for each cell) were produced using SAI's diagnostic wind model (DWM) (EPA, 1990a) which is driven by a domain mean wind (based on upper air observations) adjusted for terrain effects and weighted by surface observations (5 NWS/FAA stations). Mixing heights were calculated with the RAMMET meteorological processor (EPA, 1977) using temperature soundings taken from the Athens, Georgia upper-air site. Maximum daily mixing heights reached 1350 m above ground.

RESULTS AND DISCUSSION

A limited performance evaluation (EPA, 1990b) of the model's ability to simulate observed ozone concentrations showed a bias towards underprediction, for three operational ozone monitors (peak observed ozone, 147 ppb; peak simulated, 132 ppb - unmatched in time and space). Model performance was enhanced with a simulation assuming a 50% reduction in wind speed. The available surface wind monitoring network reported wind velocities sampled over short (1-3 minute) time scales. The UAM requires hourly average values; thus, the short averaging period might overstate hourly average velocities. This wind speed simulation was viewed as a sensitivity analysis illustrating the importance of data collecting and reporting practices. The control strategies presented below were based on unadjusted wind speeds, unless otherwise noted.

A series of anthropogenic ROG reduction scenarios across all grid cells and all hours (across-the-board) were simulated for cases with and without inclusion of biogenics. Simulated peak ozone reached a value of only 123 ppb when biogenics were not included (relative to a simulated value of 132 ppb with biogenics and an observed peak of 147 ppb). Given these differences in base case zero control scenarios, all peak ozone model results reported in Figures 2-4 were normalized to the observed ozone value of 147 ppb. These adjustments allowed visual comparisons of peak ozone responses emanating from identical starting points (147 ppb), and also provided a potential calculation technique which accounts for model bias to determine emissions controls needed to demonstrate attainment of the ozone NAAQS.

Peak ozone sensitivity to anthropogenic ROG control was influenced strongly by biogenic emissions (Figure 2). The results clearly show the same general effect observed in the Chameides study - inclusion of biogenic emissions increases the predicted level of anthropogenic VOC control required to reduce peak ozone for Atlanta on June 4, 1984. To reach predicted attainment based on normalized percentage reductions of peak ozone, the biogenics and nonbiogenics cases require 100% and 62% anthropogenic ROG control, respectively.

A comparison of the relative efficiencies of across-the-board ROG only and NOx only reductions (biogenics included for both cases) show peak ozone to be more responsive to NOx emissions changes (Figure 3). Clearly, the degree of ROG loading, due largely to the biogenics fraction, is creating an atmosphere (for Atlanta on the day modeled) which is NOx limited with respect to ozone sensitivity.

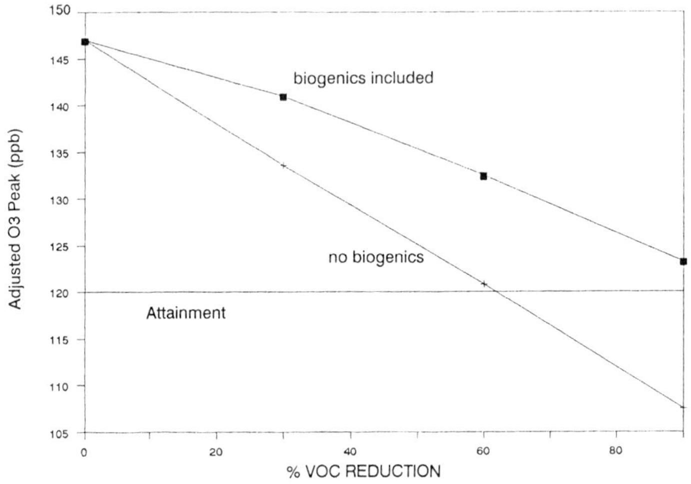

Fig. 2. Comparison of peak ozone sensitivity to VOC emissions reductions (man made) with and without biogenics.

Given the uncertainty in characterizing prevailing wind fields, the modeling runs shown in Figure 2 were repeated with the adjusted, reduced speed wind field. The base case simulation with zero control resulted in peak ozone values of 147 ppb and 129 ppb (unmatched in space and time) for the biogenics and nonbiogenics cases, respectively. Results from identical VOC control strategy simulations with the new wind field (Figure 4) suggest a reduced impact, relative to the original windfield, on control calculations due to biogenic emissions. Simulations based on the new windfield produced VOC control levels of 62% and 53% to reach attainment for biogenic and nonbiogenic cases, respectively (compared to 100% and 62% with the original windfield). Rather than deciding which wind field is more representative, the new wind field might be viewed as an additional modeling day with a slower mean trajectory. In this scenario anthropogenic emissions now comprise a greater fraction of total VOC emissions because the slower wind speed increases the residence time ratios of anthropogenic to biogenic emissions, relative to the original windfield with higher wind speeds. Thus, it is evident that meteorological variability influences ozone precursor control calculations. Certainly, multiple episodes will need to be examined when control calculations are performed.

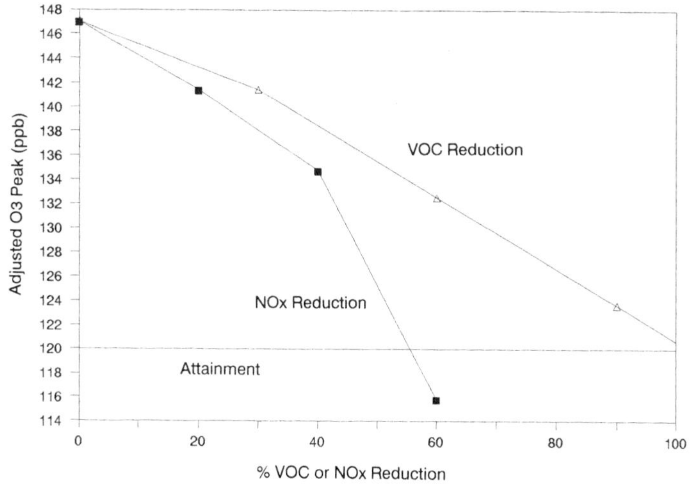

Fig. 3. Comparison of peak ozone sensitivity to changes in VOC and NOx emissions reductions (biogenics included).

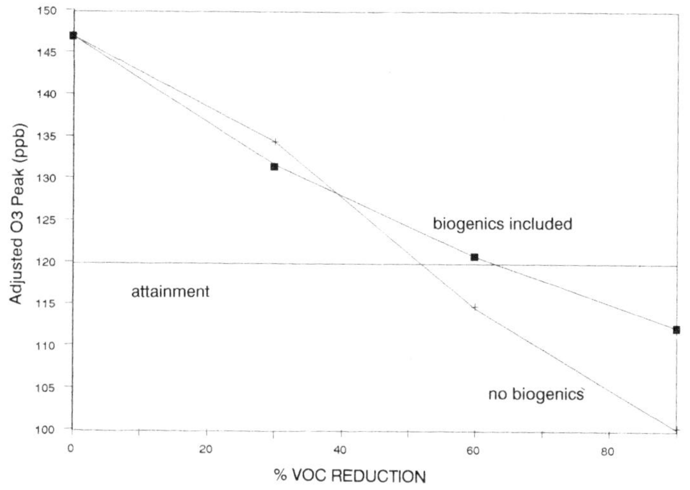

Fig. 4. Comparison of peak ozone sensitivity to anthropogenic VOC reductions with and without biogenics. These simulations were based on the adjusted (50% speed reduction) wind field.

CONCLUSIONS

Grid based modeling results presented here support Chameides earlier findings - inclusion of biogenic emissions increases the level of anthropogenic ROG control required to reduce peak ozone in Atlanta for the modeled day, June 4, 1984. Biogenic emissions must be considered as part of the total emissions loading when the effectiveness of anthropogenic control strategies is evaluated through modeling. The results indicate that NOx based control approaches may be more effective in reducing peak ozone in certain areas, using Atlanta as an example.

How effective would NOx strategies be in other locations? It must be emphasized that these results may not be representative of all situations. Atlanta is dissimilar from other major urban areas in several ways. First, biogenic emissions may be higher in the southern U.S. due to the relatively dense distribution of high emitting vegetation types. Second, biogenic emission models exhibit strong temperature sensitivity, and southern U.S. areas are characterized by high summer temperatures that lead to high emission rates. Third, Atlanta is a predominantly residential/commercial city with a limited urban core and more diffuse residential areas relative to other major urban areas. Although NOx control was shown to be effective for Atlanta, recent similar UAM simulations performed on Dallas-Fort Worth and St. Louis (EPA, 1990b) have shown that aggressive NOx control can potentially offset benefits resulting from ROG control.

Additional analyses, modeling as well as observational, need to be conducted on other cities and combined with regional scale analyses (to better characterize background hydrocarbon levels) to determine the most effective control pathways for reducing ozone. Before a good deal of confidence can be associated with the quantitative capability of those analyses, the uncertainty of natural (as well as anthropogenic) emissions must be reduced. The importance of adequate meteorological data and its attendant influences on control strategy results and the model's ability to depict proper response behavior can not be understated.

REFERENCES

Chameides, W.L., R.W Lindsay, J. Richardson and C.S. Kiang, 1990, The role of biogenic hydrocarbons in urban photochemical smog: Atlanta as a case study, Science, 241:1473.

Novak, J.H. and J.A. Reagan, 1986, A compilation of natural and man made hydrocarbon emission inventories necessary for regional acid and oxidant modeling, presented at the 79th APCA conference, Minneapolis, MN.

EPA, 1977, "User's manual for single source (CRSTER) model," U.S. Environmental Protection Agency," RTP, NC, EPA-450/2-77-013.

EPA, 1988, "Anthropogenic Emissions for the 1985 NAPAP Inventory," U.S. Environmental Protection Agency, RTP, NC, EPA-600/7-88-022.

EPA, 1989, "User's manual for OZIPM4 (Volume 1)," U.S. Environmental Protection Agency, RTP, NC, EPA-450/4-89-009a.

EPA, 1990a, "User's guide for the Urban Airshed Model," U.S. Environmental Protection Agency," RTP, NC, EPA-450/4-90-007a-d.

EPA, 1990b, "5-City Urban Airshed Model Report Series," U.S. Environmental Protection Agency," RTP, NC, EPA-450/4-90-006a-g.

Specification of Source Characteristics for Ozone Prediction in a Complex Airshed

Shinji Wakamatsu*, Itsushi Uno* and Richard A. Wadden**

*National Institute for Environmental Studies
Tsukuba, Ibaraki, 305, Japan
**School of Public Health, University of Illinois
Chicago, Illinois, 60680, USA

Introduction

Photochemical O_3 is formed in the atmosphere as a result of complex non-liner interaction of meteorology, emission distribution and chemical reactions. To investigate the urban O_3 reduction strategy a three-dimensional grid model including meteorology, chemistry and emissions based on the real conditions should be needed.

The National Institute for Environmental Studies, Tsukuba, Japan, has carried out a series of summer-time field studies in urban and rural settings which include detailed measurements of meteorology and pollution concentrations both at the surface and aloft[1,2,3,4]. The purpose of such studies is to better understand the medium and long range interaction of weather with pollutant emissions. And from this understanding develop appropriate models to predict pollutant levels, particularly those such as O_3 which are formed in the atmosphere.

One such model is the three-dimensional Urban Air-Shed Model which has been adapted to calculate O_3 concentrations in the Tokyo area. However, accurate calculations require proper specification of boundary conditions and of emission characteristics. This study is directed towards determining the sensitivity of the O_3 prediction procedure to variation in boundary values; and to variation in the source emission of organics and nitrogen oxides. The effects of using different boundary assumptions are compared. All calculated results are also compared, using actual meteorology, with a large data set of ambient concentrations (NO, NO_2, O_3, and 17 Hydrocarbons) collected both aloft and at ground-level in the Tokyo region in the summer of 1981.

O_3 prediction model

The Urban Air Shed Model (UAM) is a three-dimensional Eulerian, photochemical air quality simulation model[6]. In the application of the model to the Tokyo data base[7], the area modeled was 170km wide and 150km long. Each individual cell was 6.8km wide and 5.5km long.

Air Pollution Modeling and Its Application VIII, Edited by H. van Dop and D. G. Steyn, Plenum Press, New York, 1991

Schematic drawing of horizontal section of the three-dimensional Urban Air-Shed Model and surface monitoring points are shown in Figure 1.

In a previous study[5] we specified five vertical layers of cells, the bottom three layers simulating the mixed layer and the top two representing the volume above the mixed layer. The maximum height of the model volume was 2000m. However, in this study, since our major interest was in a sensitivity analysis, 20 layers were specified, each with a height of 100m. The maximum height was maintained at 2000m.

Model simulations were conducted for 20 successive one-hour intervals covering the period which began at 0400JST, July 16, 1981. Data from sixty-two selected ground monitoring stations, 4 upper-air soundings, and 23 pilot balloons were used to determine values for the boundary conditions and meteorological parameters. The carbon bond model (version CBM-II)[8] was used to estimate conversion of precursor materials to O_3.

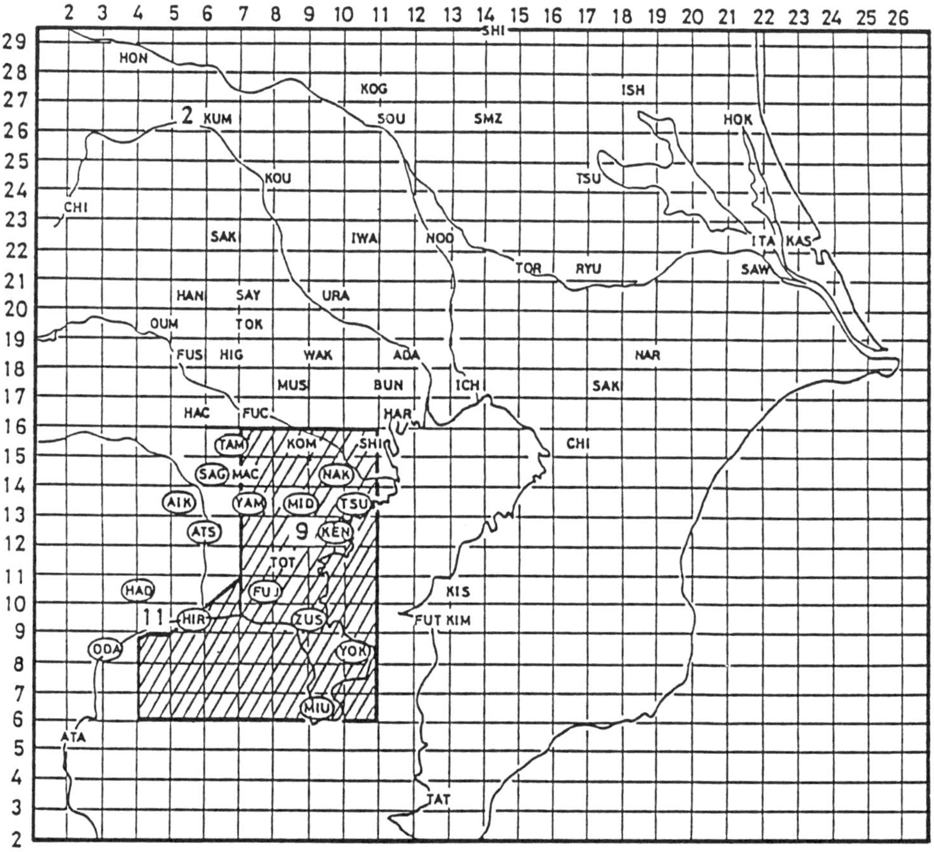

Figure 1. Schematic drawing of horizontal section of the three-dimensional Urban Air-Shed Model and typical aircraft flight area on 16 and 17 July 1981. The circled 16 points are used to calculate the surface pollutants concentration showed in Figure 2. Mesh shows 6.8km wide and 5.5km long

Emission

The industrial complexes located in the Tokyo bay area and Tokyo Metropolis is the largest contributor of pollutants from mobile sources. Almost 26% and 14% of the total NO_x and SO_x emissions from all of Japan come from this modeled region. Annual emission rates of NO_x, SO_x and nonmethane hydrocarbons(NMHC) of this area are shown in Table 1. Hourly emission rates of NO_x and NMHC from area and point sources were calculated for each grid location.

NMHC emissions were determined in bond-structure classes based on the particular organic fingerprint which was appropriate for each source category. To accomplish this the ratio of 18 NMHC species from the five major NMHC source categories were determined. The source fingerprints of hydrocarbons for vehicle exhaust, gasoline vapor, petroleum refineries, petro-chemical production, and paint solvents were used. The development of those has been previously described[9,10]. Annual emission rates for each bond-structure classes from mobile and fixed sources are shown in Table 2.

Table 1. Annual emission rates of NO_x, SO_x and NMHC (nonmethane hydrocarbons) from the modeled region in Figure 1

	NO_x	SO_x	NMHC
Mobil Source	143027.05 (41.4%)	15783.12 (8.4%)	57232.88 (12.8%)
Fixed Source	202613.81 (58.6%)	171760.12 (91.6%)	388843.12 (87.2%)
Point(40m<)	155950.36	121084.87	
Area (40m>)	46663.45	50675.25	
Total	345650.37	187543.69	446076.00

(Ton/Year)

Table 2. Annual emission rates for each bond-structure classes of olefins(OLE), paraffins(PAR), aromatics(ARO), carbonyls(CAR) and ethylene(ETH) from mobil and fixed sources

	OLE	PAR	ARO	CAR	ETH	Total
Mobil Source	1170 (2.6%)	20867 (45.7%)	11800 (25.8%)	1717 (3.8%)	10094 (22.1%)	45648
Fixed Source	559 (0.1%)	113539 (29.7%)	260715 (68.2%)	-	7686 (2.0%)	382499
Total	1729	134406	272515	1717	17780	428147

(Ton/Year)

Field observation

Aircraft measurements were conducted six times per day on 16 and 17 July 1981. O_3 and NO_x measurements were conducted continuously, but the values listed in Figure 2 are the averaged values while air was being sampled for HC analysis. Non-methane HC concentration and the ratio of NMHC to NO_x concentrations are shown in Figure 2 respectively. The ratio between NMHC and NO_x is important in determining O_3 abatement strategies. This observational results showed that NMHC/NO_x exceeded 20 in units of ppbC/ppb. Surface values shown in Figure 2 are the averaged concentrations of 16 ground-monitoring stations which were located near or inside of the aircraft flight areas. The flight areas of these 2 day are shown in Figure 1.

During these 2/s the me meteorological conditions were quite similar and the polluted air masses circulated over the Tokyo area. On 16 and 17 July 1981 the polluted air mass formed in the southern Tokyo Metropolitan area were trapped due to the easterly wind from Kashima sea. Under these circumstances pollutant concentration aloft showed the same diurnal variation. NO concentrations were higher in the morning, NO_2 showed a peak at noontime, and O_3 showed maximum peak at noontime aloft on both days.

Simulation conditions

The effect on O_3 prediction of three parameters was determined. These were the concentrations of O_3, NO_x, CO, and organics at the study area boundaries and at time =0; the composition of NMHC for each source category: and the NMHC/NO_x ratio for emissions. The O_3, NO_x, CO and organic concentrations were either set to zero or to observed values. In addition calculation runs were also made with O_3 set to 75% and 50% of the observed value. The NMHC/NO_x emission ratio was varied on a mass basis using multiples of the emission inventory values of 2:1, 1:1, 0.5:1, 1:0.5, and 1:2. NMHC/NO_x =2:1 means the NMHC emissions were specified at twice the inventory value for this pollutant, while emissions for NO_x were set equal to the inventory values.

Simulation results

The choice of boundary conditions (either observed concentrations of O_3, NO_x, CO and NMHC, or all concentrations = 0) was important depending on how close the monitoring station was to the study area boundaries. Figure 3,4 and 5 are for Kumagaya (Station 2 on Figure 1), Kencho (Station 9 on Figure 1) and Hiratsuka (Station 11 on Figure 1), respectively. Kumagaya (Figure 3), for instance, is close to the study boundaries. With observed initial and boundary concentrations, the predicted O_3 values for this station are much higher (and better match actual observation) than when the starting and boundary conditions are zero. On the other hand, Figure 3 for the Kencho site (Station 9), which is in the middle of the study area, indicates that the model predictions are the same, regardless of initial and boundary conditions. There is also poor agreement between model predictions and observations for this monitoring station.

The Hiratsuka station (11 on Figure 1) shows a somewhat different pattern than either of the above stations. While geographically close to the study area boundaries, model predictions for the morning hours

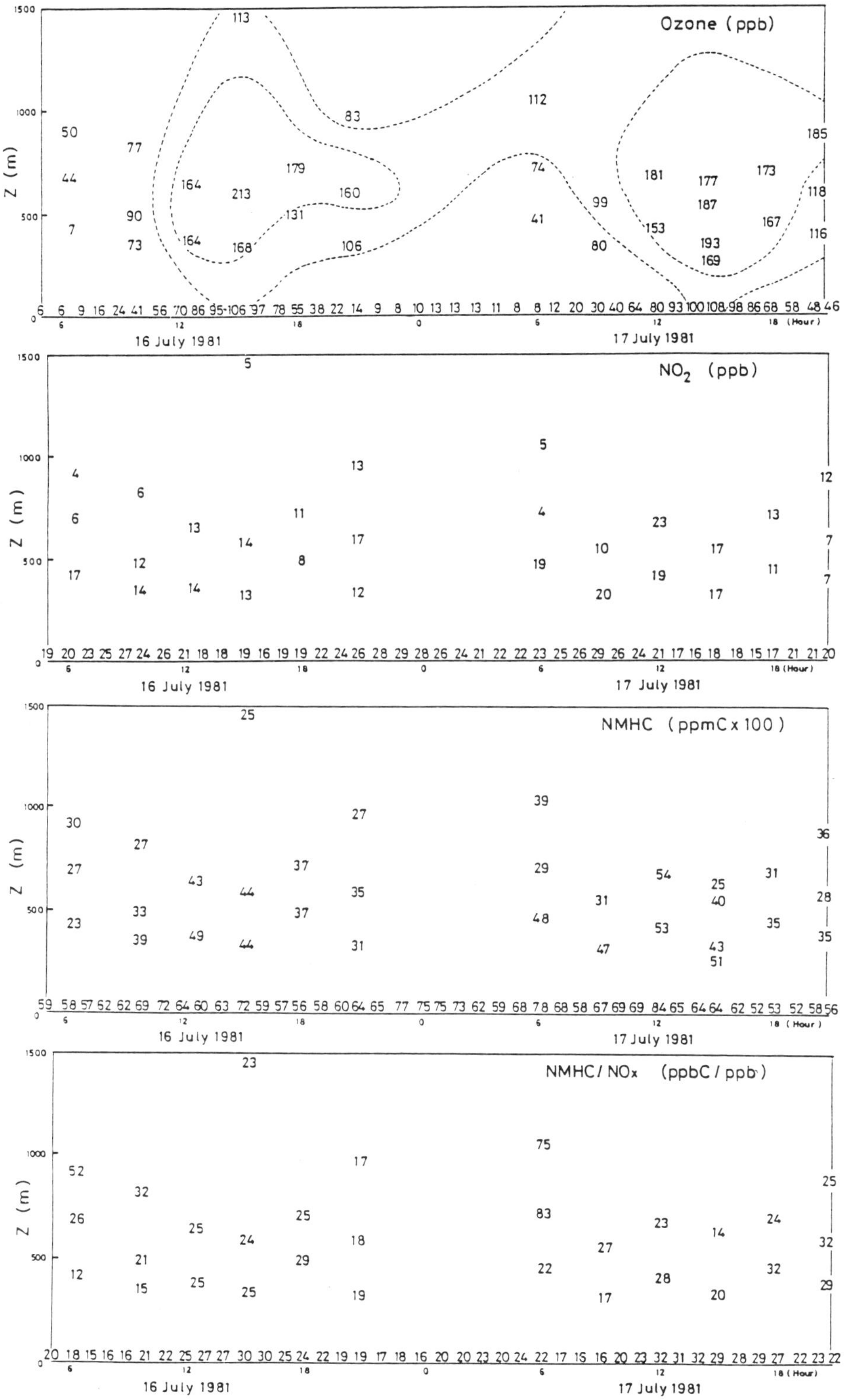

Figure 2. Averaged O_3, NO_2 NMHC concentration and ratio between NMHC and NO_x during 16-17 July 1981 at the surface and aloft. Surface pollutants concentration is the averaged value of 16 monitoring points showed in Figure 1

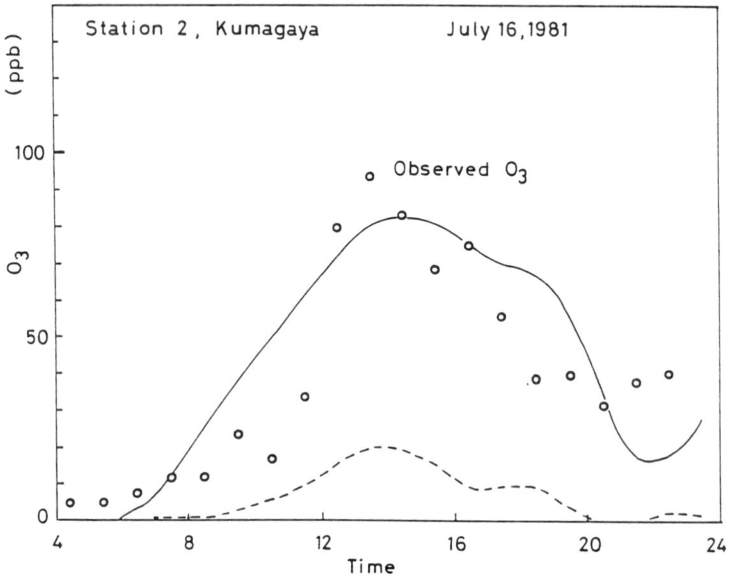

Figure 3. Effect of boundary value specification at Kumagaya station.
Open circle (○) ; Observations for this monitoring station
Solid line (——); Boundary condition =observed values
Dotted line(-----); Boundary condition = 0

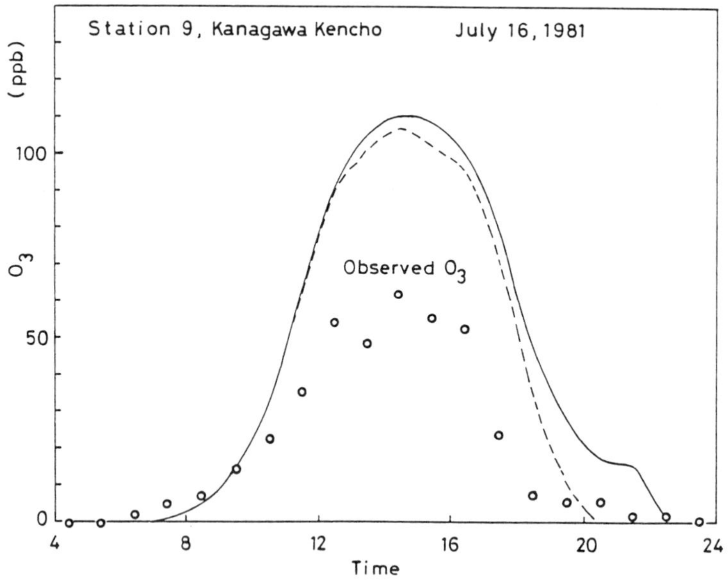

Figure 4. Effect of boundary values specification at Kencho station.
Open circle (○) ; Observations for this monitoring station
Solid line (——); Boundary condition =observed values
Dotted line(------); Boundary condition = 0

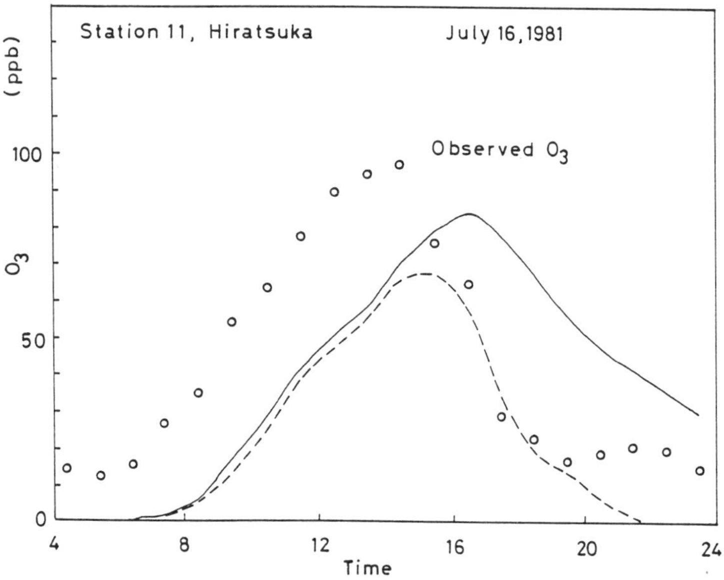

Figure 5. Effect of boundary values specification at Hiratsuka station.
Open circle (○) ; Observations for this monitoring station,
Solid line (——); Boundary condition =observed values,
Dotted line(------); Boundary condition = 0

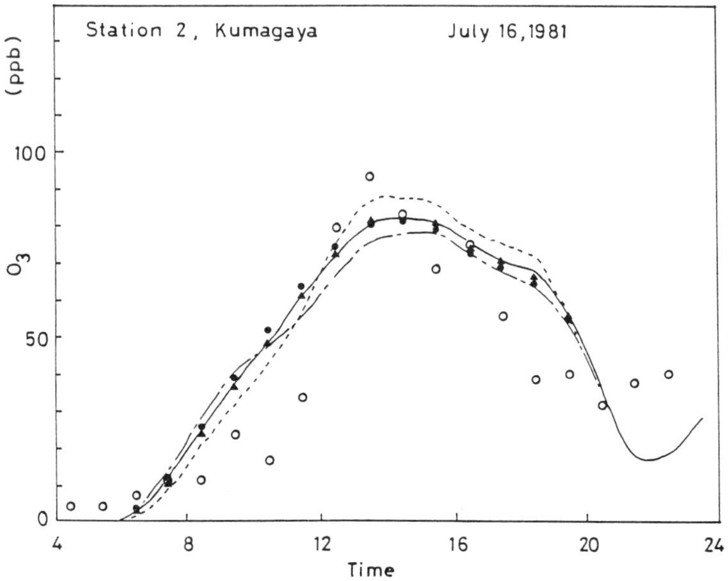

Figure 6. Effect of $NMHC/NO_x$ emission ratio for Kumagaya station.
Boundary condition = Observed values
Open circle (○) ; Observations for this monitoring station
Solid line (——) ; $NMHC/NO_x$=1:1
Filled circle (●) ; $NMHC/NO_x$=2:1
Filled triangle(▲) ; $NMHC/NO_x$=0.5:1
Dotted line (------) ; $NMHC/NO_x$=1:2,
Broken line (—·—) ; $NMHC/NO_x$=1:0.5

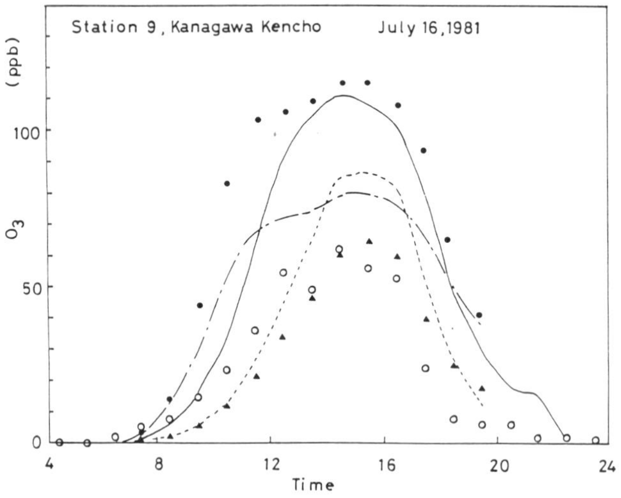

Figure 7. Effect of NMHC/NO_x emission ratio for Kencho station.
Boundary condition = Observed values
Open circle (○) ; Observations for this monitoring station
Solid line (———) ; NMHC/NO_x=1:1
Filled circle (●) ; NMHC/NO_x=2:1
Filled triangle(▲) ; NMHC/NO_x=0.5:1
Dotted line (--------) ; NMHC/NO_x=1:2
Broken line (—·—) ; NMHC/NO_x=1:0.5

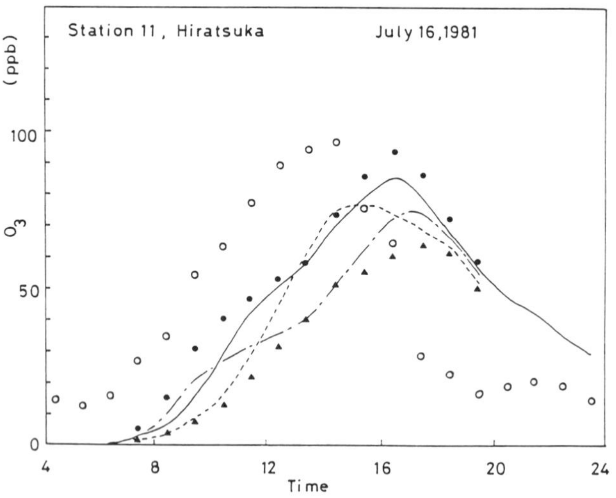

Figure 8. Effect of NMHC/NO_x emission ratio for Hiratsuka station.
Boundary condition = Observed values
Open circle (○) ; Observations for this monitoring station
Solid line (———) ; NMHC/NO_x=1:1
Filled circle (●) ; NMHC/NO_x=2:1
Filled triangle(▲) ; NMHC/NO_x=0.5:1
Dotted line (--------) ; NMHC/NO_x=1:2
Broken line (—·—) ; NMHC/NO_x=1:0.5

do not differ because of different boundary specifications. This is probably because a land breeze was in effect during the morning and so the boundary conditions which would affect the Hiratsuka prediction are located far to the north. On the other hand, the O_3 predictions for the afternoon begin to diverge quite noticeably depending on boundary conditions. This corresponds to the occurrence of an afternoon sea breeze, movement of air from the south, and the return of pollutants swept out to sea in the early morning. The Hiratsuka station is now much closer to the boundary (southern) of the study area and the predictions are accordingly influenced. Compared with observations, The morning predictions are too low, but in the afternoon the two predictions bracket the observed values.

Figures 6,7 and 8 show the effect caused by variation of the NMHC/NO_x emission ratio at the three stations. Figure 6 for Kumagaya indicates little difference in O_3 levels due to different NMHC/NO_x assumptions. The UAM also gives a reasonable fit of the observed data. At Kencho (Figure 7) the model consistently overpredicts the maximum O_3, more so for NMHC/NO_x = 1:1, NO concentrations at this station are considerably higher, relative to the other sites. The model overproduction may occur because of the quenching effect high NO concentrations have on measured O_3 at the Kencho site. The predicted concentration patterns for Hiratsuka (Figure 8) had the same shape as the other stations, but with significant underprediction of O_3 in the morning and overprediction in the afternoon.

Conclusions

UAM predictions have been compared with observed concentration values for the Tokyo area. The results indicate that the patterns of simulated and observed O_3 concentrations are in qualitative agreement. However, a sensitivity analysis has indicated that choice of boundary conditions for initial concentrations has an important influence on model prediction if the monitoring station is close to the region boundaries. It is significant to note that determining "how close" a station is to the boundaries is a function of the wind direction.

Model predictions are also quite sensitive to the mass ratio of NMHC/NO_x emission. Consequently, mis-specification of the emission inventory can lead to major to differences between observed and calculated O_3 concentrations.

Acknowledgments

The authors wish to express their appreciation to Keenth L. Schere of Atmospheric Research and Exposure Assessment Laboratory of US-EPA for his help to conduct a three-dimensional grid model study.

References

1. Wakamatsu S., Y. Ogawa, K. Murano, K. Goi and Y. Abramoto,(1983) Aircraft survey of the secondary photochemical pollutants covering the Tokyo Metropolitan Area. Atmos. Environ. 17, 827-835.

2. Uno I., S. Wakamatsu, M. Suzuki and Y. Ogawa, (1984) Three-dimensional behavior of photochemical pollutants covering the Tokyo Metropolitan Area. Atmos. Environ. **18**,751-761.

3. Uno I., S. Wakamatsu, R.A. Wadden, S. Konno and H. Koshio, (1985) Evaluation of hydrocarbon reactivity in urban air. Atmos. Environ. **19**, 1283-1293.

4. Wakamatsu S., I. Uno, and M. Suzuki, (1990) A field study of photochemical smog formation under stagnant meteorological condition. Atmos. Environ. **24**, 1037-1050.

5. Wakamatsu S., I. Uno and K.L. Schere, (1988) Air pollution modeling and it's application, IV, 259-270 Plenum Publishing Corporation.

6. Ames J., T.C. Myers, L.E. Reid, D.C. Whitney, S.H. Golding, S.R. Hayes and S.D. Reynolds, (1985) Users' Manual for the SAI Airshed Model. EPA Report- 600/8-85-007a,b.

7. Wakamatsu S., K.L. Schere, J.H. Shreffler and I. Uno, (Feb. 27-28, 1987) A study using a three-dimensional photochemical smog formation model. presented at the 9th US-Japan Meeting on Air Pollution-Related Meteorology, Tokyo Japan.

8. Killus J.P. and G.A. Whitten, (1982) A New Carbon-bond Mechanism for Air Quality Simulation Modeling. EPA Report-600/3-82-041.

9. Wadden R.A., I. Uno and S. Wakamatsu, (1986) Source discrimination of short-term hydrocarbon samples measured aloft, Environ. Sci. Technol. **20**, 473-483.

10. Wakamatsu S., I. Uno and R.A. Wadden, (Dec 3-4, 1984) A study of NMHC source fingerprints and their photochemical reactivity. In Proc. of the Eighth US-Japan Conference on Photochemical Air Pollution, U.S. EPA, Research Triangle Park, NC, pp.45-76.

DISCUSSION

A. VENKATRAM

Did you compare model predictions of the precursors such as NO_x with corresponding observations?

S. WAKAMATSU

Simulated NO and NO_2 concentrations were also compared with the observation. Simulated results showed good qualitative agreement with observations but with underestimation of the ground data. This discrepancy between the simulation and ground data can be explained by the vertical mixing effect of the model calculation.

III: GLOBAL SCALE DISPERSION STUDIES INCLUDING SUB-GRID SCALE PARAMETERIZATION OF TRANSPORT AND EXCHANGE PROCESSES

chairman: G. McBean

rapporteur: D.G. Steyn

PARAMETERIZATION OF CLOUD TRANSPORT OF TRACE SPECIES IN GLOBAL 3-D MODELS

Richard A. Brost

Max Planck Institute for Chemistry
P.O. Box 3060, D-6500 Mainz, FRG

Martin Heimann

Max Planck Institute for Meteorology
Bundesstr. 55, D-2000 Hamburg 13, FRG

1. INTRODUCTION

It is important to understand the effects of warming of the earth by trace gases and polluting the environment by acid rain. For these and other reasons we need to be able to simulate atmospheric chemistry in the global troposphere. Perhaps the biggest meteorological uncertainty in such modeling is determining vertical transport of trace species by clouds. There has been considerable work on the parameterization of vertical transport of heat and moisture by clouds in meteorological models (see reviews by Anthes et al., 1986 and Frank, 1983); such meteorological parameterizations, however, do not imply a unique tracer transport, and additional assumptions are required. Cloud transfer in global three-dimensional (3-D) tracer models has sometimes been partly resolved but usually is parameterized using a variety of methods, including eddy diffusivities (K), transfer matrices, and convective mass fluxes.

The dependence of subgrid-scale motions on grid-scale resolved variables is complex and poorly understood, so that parameterizations of subgrid-scale motions are controversial. In a region of a few hundred kilometers on its side (a single grid cell in a global model), perhaps 99.9% of the horizontal area has vertical motions that require days to move air from the surface to the tropopause, whereas the remaining 0.1% of the area (a thunderstorm updraft) needs only 1 h to travel the same distance.

The first parameterizations used gradient diffusion, which has been properly criticized as being inappropriate when the eddies doing the transporting are large compared to the scale of the flow, but misleadingly criticized as being too slow for a species with life time τ less than a month (Chatfield and Crutzen, 1984). It is probably true that when one uses a single K to represent vertical motion in the entire grid cell, this effective K must increase as τ decreases from infinity (a stable species) to a few hours; when τ becomes less than 1 h, K must finally decrease sharply. This means that a single K is not appropriate for every species. Any parameterization that handles the separation in time scales between thunderstorm updrafts and larger-scale motions, however, has either explicitly or implicitly separated the grid cell into at least these two regions. To be fair to gradient transport, the transfer in the updraft could be represented with a $K = 10^5$ m^2 s^{-1} and that in the non-updraft region with $K = 10$ m^2 s^{-1}, assuming that there is a parameterization of the fractional area of updraft. The same K's could be used with any species with τ greater than a few hours. Even this procedure, however, could not produce isolated detrainment levels, which are important for single clouds, but less so when averaged over large areas (Levy and Moxim, 1989). Moreover, the basic asymmetry of up and down motions in the troposphere makes K's at least inconvenient to use, if not inappropriate.

The asymmetry of vertical transport in the troposphere can be more easily handled by another method of parameterization, a transfer matrix, which can be multiplied with a vertical column of grid-scale concentrations at the current time to give the column of grid-scale concentrations at the new time, that is, the matrix tells what fraction of the mass in any level moves to every other level during one convective time step. This may not seem to have explicit consideration of subgrid-scale variations, but the matrix could only have been developed with some more detailed model of the subgrid-scale behavior. But by not maintaining the distinction between regions that are cloudy and clear, the matrix approach may not be amenable to the addition of aqueous chemistry.

Parameterizations that explicitly determine the fraction of the grid cell that contains clouds, the liquid water content, and the updraft velocity would be the most versatile for aqueous chemical calculations and vertical transport descriptions. Because the rate at which air is processed by clouds will prdbably be emphasized in the future due to the importance of aqueous chemistry (Lelieveld and Crutzen, 1990), and because the flux of air mass at least partly determines the flux of trace species, we will review vertical transport of air mass by clouds in three-dimensional, global tracer and chemical models, with also some references to regional scale models. This comparison of convective fluxes should be an interesting test of the ability of meteorological models to supply such detailed information to chemical models. Because

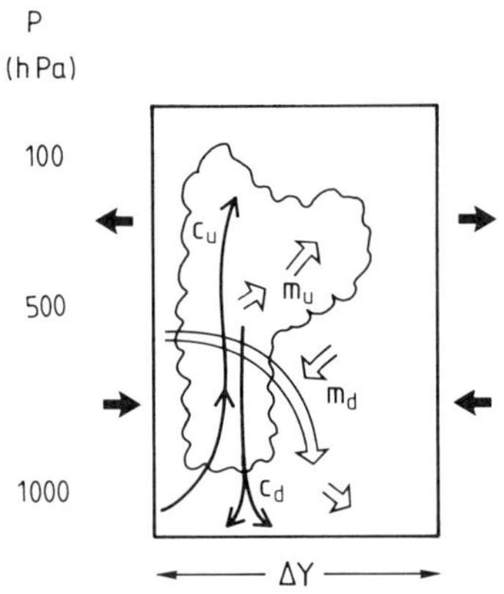

Fig. 1. Schematic of vertical motions in a tropical cloud cluster. We show the resolved motion into the grid box (solid arrows), the cumulus scale updrafts and downdrafts (c_u and c_d), and the mesoscale updrafts and downdrafts (m_u and m_d). Adapted from Zipser (1977). The figure is not drawn to scale because the mesoscale drafts on the right side occupy 10 times the area of the convective drafts on the left side.

there has been so much work on cloud transport and there are limitations on the size of this paper, we do not pretend to be comprehensive. For example, we do not have space to discuss related issues such as the fractional area of cloudiness and the cycling time between cloudy and clear air in the grid cell, i.e., the frequency of cloud transport events. We ask the reader's forgiveness for omissions we may make.

Figure 1, adapted from Zipser (1977), schematically shows a horizontal grid cell (and several cells in the vertical), indicating the up and down motions possible in cloud clusters, using a tropical squall line as an example. The solid, horizontal arrows outside the box represent the resolved-scale convergence in the lower troposphere and divergence in the upper troposphere. (Of course what is resolved and what is subgrid scale, which must be parameterized, depends on the spatial and temporal scales of the numerical model.) The sum of all the subgrid-scale effects (open and thin arrows in the box) must provide the upward motion required in order to have mass continuity. Single arrows represent cloud-scale updrafts and downdrafts, which travel at m s^{-1} and occur on horizontal scales of 1-10 km. Double arrows represent mesoscale updrafts and downdrafts of 10's of cm s^{-1} that occur on scales of 10's-100's of km. In the cloud-free part of the grid cell, environmental-scale rising motion or subsidence can occur with velocities of cm s^{-1} and scales of the size of the grid cell (not shown).

2. OBSERVED AIR MASS FLUXES IN A CLOUD CLUSTER

Figure 2 shows the mass fluxes in a tropical cloud cluster (Johnson, 1980). Some of the general characteristics of such a cluster may also found in midlatitude cloud clusters and squall lines (Rotunno et al., 1988; Johnson, 1984).

The quasi-parabolic shaped curve $\overline{M}$ is the net upward mass flux in the grid cell, as well as the mass flux required to balance the resolved horizontal convergence in the lower troposphere and divergence in the upper troposphere. Typically $\overline{M}$ maximizes between 400 and 600 mb, although particular cloud clusters have been analyzed or

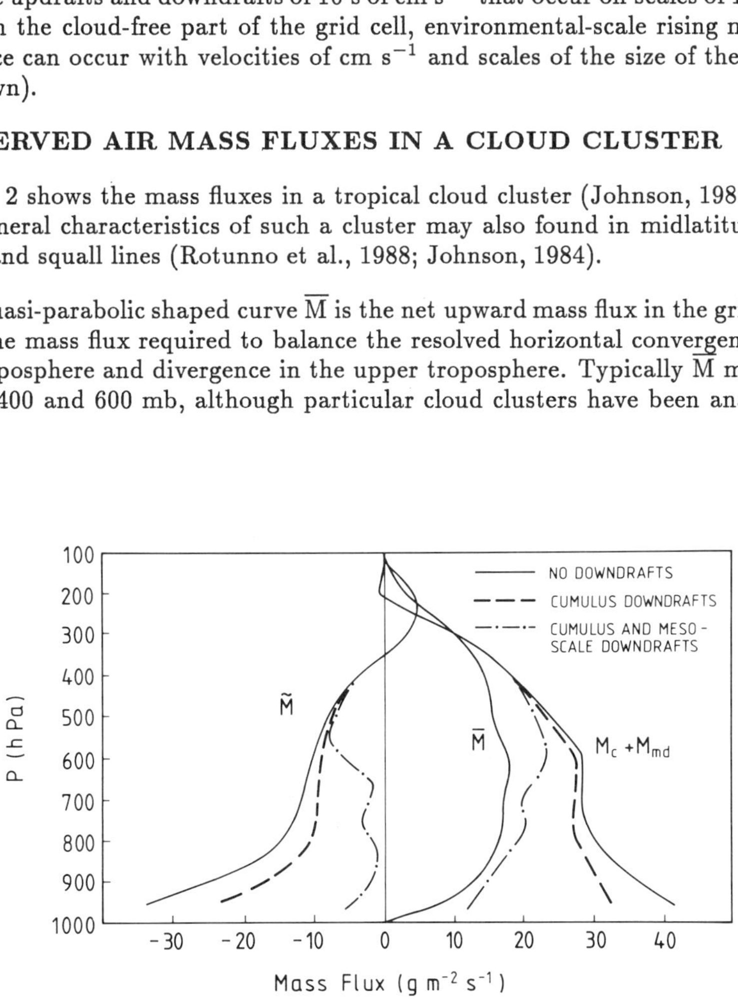

Fig. 2. Mass fluxes for a tropical cloud cluster, showing the average mass flux $\overline{M}$ for the region (i.e., grid cell), the induced environmental subsidence $\tilde{M}$, and the cumulus updraft, with and without downdrafts (from Johnson, 1980).

modeled with peaks much higher or much lower. A reasonable generalization is that shallow convection would tend to produce a maximum flux at a lower height, whereas more vigorous deep convection would produce a maximum at a higher level (Houze and Betts, 1981). There must be net entrainment into updrafts and downdrafts below (and net detrainment above) the level of the maximum net flux. $\overline{M}$ is the sum of the convective and mesoscale fluxes on the right-hand side and the matching (line pattern) environmental subsidence curve on the left-hand side: a complicated combination of slow and fast, and up and down motions.

The solid line on the right-hand side of Fig. 2 is the cloud updraft, which normally decreases with height above PBL and almost always exceeds the net flux. Because the cloud updraft moves at a few to 10's of m s^{-1}, air from the PBL can travel across most or all of troposphere in a time step of a global model (typically, 0.5-4 h); in effect, air from the PBL jumps to all higher levels in the troposphere. The sum of the (slow-moving) environmental subsidence (solid curve on the left) and the convective updraft could account for the net flux $\overline{M}$. In general, the air that goes up does not immediately turn around and come back down as fast, so that we may have to consider the detailed up and down motions, and not just the net flux.

In the convective downdraft, the mass flux is the difference between the solid line on the right and the dashed line next to it, and averages from 0 to 50% of the mass flux in the convective updraft. The downdraft usually starts near 500 mb and becomes more important approaching the surface. The speed in a downdraft is comparable to that in an updraft, so downdraft air can go from $z = 5$ km to the surface in one time step. Generally the air in the downdraft is not recycled updraft air from the same cloud event. Again, the sum of the dashed curve on the left, the environmental subsidence, and the convective up and down mass fluxes equals $\overline{M}$.

The difference between the dashed and dash-dot curves on the right of the figure show the mass flux in the mesoscale downdraft. As precipitation falls from the mesoscale anvil, the precipitation melts as it crosses the level at which temperature $T = 0°$ C. Below that level some precipitation evaporates and cools the air, driving the mesoscale downdraft. Thus, the mesoscale downdraft starts around the the melting level (600 mb) and increases in importance approaching the surface. This downward flux in the lowest kilometers is sufficient to suppress small cumulus (Houze and Betts, 1981). Traveling at 10-50 cm s^{-1} (Rutledge et al., 1988), air in the mesoscale downdraft may not go more than one model level in one time step. In this example the mass flux in the mesoscale downdraft exceeds the mass flux in the convective downdraft. Here the sum of these three drafts (convective up and down, and mesoscale down) is almost $\overline{M}$, so that relatively little additional environmental subsidence is required.

Latent heat release in the mesoscale anvil can drive upward motions at 20-40 cm s^{-1} above about 600 mb (Rutledge et al., 1988). Including a mesoscale updraft in Fig. 2 could increase the net upward flux by 20-30% above the 600 mb level (Houze and Cheng, 1981), requiring an increased environmental subsidence above 600 mb. Although the mesoscale updraft occupies 10 times the area of the convective updraft, the mesoscale updraft moves only about one-tenth as fast as the convective updraft (Johnson, 1984), so that the air in the mesoscale updraft may or may not go more than one layer in one step, but the mass flux in the mesoscale updraft can be comparable to that in the convective updraft.

Perhaps the parameterized fluxes should depend on the life time of the species transported. For a long-lived species, only the net flux matters, and the details of canceling up and down motions are unimportant. For a short-lived species (with a surface source) individual components may matter, especially the fast-moving cloud updraft and downdraft, although the single most important draft is the cloud updraft from the PBL.

3. SIMPLE ESTIMATES OF UPDRAFT FLUXES

Cloud mass fluxes can vary by many orders of magnitude in space and time. Although presenting large-scale averages is potentially misleading, because parameterizations with the same large-scale averages may differ substantially in detailed simulations, the large-scale averages will show whether substantial differences exist among parameterizations. Here we only consider annual averages; also we only consider zonal averages for 10° latitudinal belts or global averages.

When we need to assume an air density ρ_0 to calculate the fluxes, we use $\rho_0 = 1$ kg m^{-3}, which is approximately appropriate for the top of the planetary boundary layer (PBL); such flux estimates then decrease with height as air density decreases: for example, $\rho = 0.5$ kg m^{-3} at $z \approx 7$ km or $P \approx 400$ mb, and $\rho_0 = 0.2$ kg m^{-3} at $z \approx 15$ km or $P \approx 100$ mb.

3.1. The Equatorial Trough Zone

Thunderstorms, the "hot towers" of Riehl and Malkus (1958) transport energy vertically in the intertropical convergence zone (ITCZ), where the zonal average vertical mass flux may be the largest on earth. Synoptic disturbances occupy 10%, convective clouds 1%, and active updrafts 0.1% of the horizontal area of the ITCZ region (Riehl and Simpson, 1979). Updrafts of 1-3 m s^{-1} (justified later) imply a flux of 1-3 g m^{-2} s^{-1} for 10° trough zone.

In developing a model of cloud clusters in the ITCZ, Chatfield and Crutzen (1984) surveyed data, such as Fig. 2, and gave 20 g m^{-2} s^{-1} as a best estimate of the flux out of the PBL by updrafts in the active thunderstorm region. Zonally averaging around world would dilute this estimate. For example, Gray (1973) suggested that the active thunderstorm region was 20% of the belt, whereas Riehl and Simpson (1979) suggested 10%, giving a range of 2-4 g m^{-2} s^{-1} for the zonal averaged flux.

Taking a different approach, vertical motion can be estimated from the energy balance of the atmosphere. The net effect of the vertical gradient of the absorption of solar radiation and the emission of terrestrial radiation is a cooling of about 1 K d^{-1} in the free troposphere (FT). This cooling is ultimately supplied by solar energy absorbed at the surface and carried to the FT by convective motions, particularly clouds. In the tropics, clouds carry 10 times as much latent heat as sensible heat, so we ignore the latter. As the updraft air moves upward and stays saturated by condensing more liquid water, latent heat is being converted to potential energy and sensible heat. The precipitation efficiency (PE) is the ratio of rainout to water vapor inflow to the cloud. When the PE$<$ 1, condensate may be left behind by the cloud, and must subsequently evaporate and locally cool the environment. In the compensating subsidence and adiabatic compression of clear air, the potential energy stored by the cloud is converted to sensible heat. Thus, clouds can locally cool but globally warm the environment, and latent heat release in clouds appears on the larger scales through subsidence warming between clouds (Yanai et al., 1973).

We can estimate the subsidence required to balance the net radiative cooling of the FT. In the tropics, the tropospheric potential temperature increases about 3.3 K per kilometer of increase in height, so that each 1 K d^{-1} of radiative cooling must be balanced by a subsidence of 3.5 10^{-3} m s^{-1}, or an average flux of 3.5 ρ/ρ_0 g m^{-2}

s^{-1} downward over 99.9% of the area. (Of course the same total mass moves upward in the approximately 0.1% of the area occupied by thunderstorm updrafts.) Thus a plausible estimate of the upward mass flux in thunderstorm updrafts, averaged over the entire ITCZ belt, would be provided by assuming PE=100% and 1 K d^{-1} radiative cooling, giving 5.5 g m^{-2} s^{-1} fast upward in thunderstorms, 3.5 slow downward in clear air, for net of 2 (upward), which is about the annual and zonal average strength of the resolved upward Hadley cell flux (shown later). Because a 2 m y^{-1} annual precipitation corresponds to about 1.6 K d^{-1} latent heating in the FT, and because the PE can approach 1 for thunderstorms in the ITCZ, we do not believe that the upward flux should be much more than this estimate.

For the ITCZ, these simple estimates for the annual and zonal average updraft flux from the PBL ranged from 1-5 g m^{-2} s^{-1}, decreasing with height to 1/2 that flux at 400 mb and 1/5 at 100 mb. Now we must consider whether detailed observations of thunderstorm updrafts and counts of thunderstorm occurrence can explain this range of fluxes.

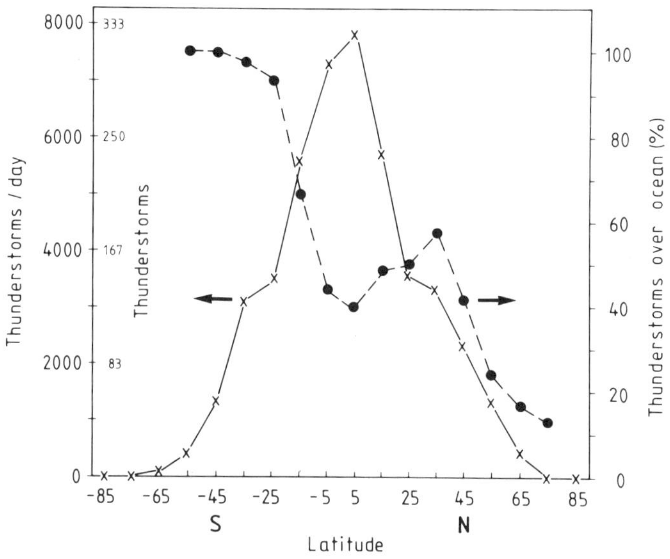

Fig. 3. Annual average number of thunderstorms in 10° latitude belts (solid curve). Percentage of thunderstorms over the oceans (dashed curve).

3.2. Global Flux Estimates From Counting Thunderstorms

3.2.1. Fluxes using the distribution of thunderstorm days. We take the simple-minded approach of estimating the flux $F = \sum_i N_i t_i$, where N_i is the number of thunderstorms in class i and t_i is the transport or rate of processing air per thunderstorm in class i, where we may assign a different rate per thunderstorm over land and

Table 1. 50th and 90th Percentiles of Updraft Core Characteristics

	Velocity ($m\ s^{-1}$)		Diameter (km)		Mass Processed ($kg\ s^{-1}$)	
Data	50%	90%	50%	90%	50%	90%
Midlatitude Continents[1]	6.3	11.8	4.0	7.0	$1.0\ 10^8$	$5.8\ 10^8$
Hurricanes[2]	3.8	6.7	3.2	5.0	$3.9\ 10^7$	$1.7\ 10^8$
Tropical Atlantic[3]	2.9	5.0	1.0	1.8	$2.9\ 10^6$	$1.6\ 10^7$

[1] US Thunderstorm Project, see LeMone and Zipser (1980), Zipser and LeMone (1980)

[2] Gray (1965)

[3] LeMone and Zipser (1980), Zipser and LeMone (1980)

over water, and in midlatitudes (poleward of 30°) and tropical regions (equatorward of 30°). The transport $t_i = \rho v_i A_i$, where v_i is the updraft velocity, A_i the horizontal area, and again we use air density $\rho = \rho_0 = 1$ kg m^{-3}. There will necessarily be large uncertainties in these estimates.

Figure 3 shows the latitudinal variation of annual average thunderstorm activity, as well as the percentage of thunderstorms over water. This figure was constructed in the following way. On a grid of 10° latitude and 15° longitude, we digitized a global map of thunderstorm days found in Court and Griffiths (1986). (A thunderstorm day is a day on which thunder is heard, regardless of the actual, hard to determine, number of thunderstorms.) The thunderstorm-day data provided a relative measure of the horizontal variation of thunderstorm activity. For the globe, the number of thunderstorms at any instant averages about 2000, with each one lasting about 1 h, and a total of about 46,000 per day (Pierce, 1986). These estimates of thunderstorm activity based on atmospheric electricity have remained unchanged for several decades. Thus, we scaled the thunderstorm-day data to give a global total of 2000 thunderstorms. This resulting scaled data showed that 80% of the thunderstorms are in the tropics, and 20% in midlatitudes, mostly the northern hemisphere with 13%. About 1/2 of the thunderstorms are over land and 1/2 over the ocean. A separate data set showed that thunderstorms are present over the ocean about 0.1% of time, but there are peak values of occurrence over the ocean of about 1 % (Sanders and Freeman, 1986).

Only a few experiments (Project Thunderstorm over the United States, GATE over the tropical Atlantic Ocean, and some hurricane data) sampled enough thunderstorms to provide statistics on updraft sizes and velocities. (In Table 1, 50% of the events are smaller than the the 50th percentile and 90% of the events are smaller than 90th percentile.) In general Table 1 shows that the air mass taken from the PBL per second per thunderstorm is greater over land than over the ocean, and greater over midlatitudes than over the tropics. (These differences in the rate of processing air mass by a storm seem reasonable and can be qualitatively explained in terms of vertical profiles of temperature, moisture, and wind velocity, as well as CCN

Table 2. Rate of Processing Updraft Air Mass per Thunderstorm (kg s^{-1}) From Data

Source	Mass Processed (kg s^{-1})
Average Cumulus Congestus or Growing Thunderstorm (US)[1]	$4.5\ 10^{7}$
10 Exceptionally Strong or Isolated Thunderstorms (US)[2]	$1\text{-}10\ 10^{8}$
19 Hailstorms (US)[1]	$2.2\ 10^{8}$
Several Hailstorms (US)[3]	$2\text{-}5\ 10^{8}$
2 Squall Lines (Many Cells) (US)[1]	$4.1\ 10^{8}$
Extremely Strong Hailstorm (US)[1]	$4.3\ 10^{9}$
Supercell (US)[4]	$3.6\ 10^{9}$
South China Sea[5]	Like GATE
Tropical Cloud Cluster (10's of cells)[6]	$2\text{-}4\ 10^{9}$

[1] Newton (1968) and references therein
[2] Sikdar et al. (1970)
[3] Frank and Foote (1982)
[4] Miller et al. (1982)
[5] Churchill and Houze (1984)
[6] Zipser and LeMone (1980) and references therein

concentration.) For the tropical Atlantic, all sizes of updraft cores made important contributions to the total flux. In other cases, however, the fluxes may be dominated by the few large storms (LeMone and Zipser, 1980).

There is a discrepancy between our interpretation of the GATE data and that of Zipser and LeMone (1980). We have taken the core draft statistics and assumed that each thunderstorm has one core; this procedure derives trivial rates of processing air for GATE thunderstorms. To derive a vertical transport by updrafts comparable to that inferred from the 300-km scale convergence in the lower troposphere, Zipser and LeMone (1980) noted that aircraft flights through squall lines encountered 4 cores, on average, and they assumed that these 4 cores ran unbroken for the 300 km length of the squall line, but this seems implausible, because there must be breaks in a tropical squall line (Rotunno et al., 1988). (A reasonable alternative explanation is that at least 1/2 of the required upward mass flux was in slower motions on a larger scale than the convective updrafts.) In a 300 by 300 km region in GATE, a cloud cluster vertically transports $2 - 4 \times 10^{9}$ kg s^{-1} of air mass (Table 2). The clusters average more than 10 mature thunderstorms and many more growing and dying clouds (Houze and Betts, 1981). Cloud clusters are separated by 3000-4000 km, so that there are 10-13 clusters circling the globe at any instant in a 10° latitudinal belt. Our annual average number of thunderstorms between 0° and 10°N is about 330, or about 30 per cluster, so that each thunderstorm would need to process about 10^{8} kg s^{-1}, which is larger than the median rate but less than the 90th percentile rate for hurricanes. We conclude that our GATE estimates are artificially low because we only included one core per thunderstorm, so we will use the hurricane data for all thunderstorms over the ocean.

Table 3. Rate of Processing Updraft Air Mass per Thunderstorm (kg s^{-1}) Assumed in Simple Models

	Ocean		Land	
Model	Tropics	Midlatitudes	Tropics	Midlatitudes
HMD	$3.9\ 10^7$	$3.9\ 10^7$	$1.0\ 10^8$	$1.0\ 10^8$
H90	$1.7\ 10^8$	$1.7\ 10^8$	$1.7\ 10^8$	$5.8\ 10^8$
H13	$1.3\ 10^9$	$1.3\ 10^9$	$1.3\ 10^9$	$1.3\ 10^9$

Table 2 shows additional, less complete observations of air mass processed per second by thunderstorms. (A danger in surveying such data is that the most studied storms are generally abnormally strong.) Over North America, the rate of processing air is one hundred times greater in supercells than in average thunderstorms. From Tables 1 and 2 we estimate that median cells everywhere process less than 10^8 kg s^{-1}, the 90th percentile cells over North America and in hurricanes process $10^8 - 10^9$ kg s^{-1}, and what are perhaps 99th percentile storms (supercells) over North America process more than 10^9 kg s^{-1}.

Based on Tables 1 and 2, we devised the three simple models shown in Table 3. The median model, HMD, used the hurricane median flux over the ocean and the midlatitude median flux over all land. The 90th percentile model, H90, used the 90th percentile over midlatitude land and the hurricane 90th percentile everywhere else. Finally, the third model, H13, used a constant 1.3×10^9 kg s^{-1}, just slightly weaker than the strongest super cells. This was meant to be an upper bound, and the actual flux could only be this large if there were more than 2000 thunderstorms. These simple models are compared with other results presently.

3.2.2. Fluxes from cloud coverage statistics. Olofsson (1988) estimated the mass flux out of the PBL by getting the areal coverage of clouds from a climatological cloud atlas and assigning updraft speeds to each type of clouds. The zonal average total flux for the sum of convective clouds and frontal clouds is the solid curve in Fig. 4a. The dashed curve shows that the H13 model gives only slightly less flux. We believe that these fluxes are probably too large.

3.2.3. Fluxes adjusted to simulate long-lived tracers. Langner and Olofsson (1990) took the spatial pattern of mass fluxes from Olofsson (1988) but tuned the magnitude of the mass flux (Fig. 4b) to improve their two-dimensional simulations of a relatively long-lived species (CO_2). Their best simulations used mass fluxes about one-tenth of Olofsson's original values. Our median model HMD gives about 1/3 the flux of Langner and Olofsson (Fig. 4b), but we believe that fluxes in the HMD model are too small because relatively infrequent large events carry a disproportionate flux. Our 90th percentile model H90 gives a flux almost indentical to that of Langner and Olofsson, except for the midlatitudes of the northern hemisphere where H90 has a peak (i.e., in the H90 model the relative vigor of the assumed updrafts compensates for their relatively small number).

Based on assigning updraft strengths to each class of thunderstorm and estimating the thunderstorms in each class, it seems likely that the maximum zonal average flux out of the PBL is much less than 10 g m^{-2} s^{-1}, and probably is as small as 1 g m^{-2} s^{-1}, unless the global total of thunderstorms is far more than 2000.

4. CLOUD PARAMETERIZATIONS IN 3-D MODELS

Most tracer models have normally been run offline, that is, separate from the meteorological model. We will only note the exceptions where the tracer is simulated at the same time as (online with) the meteorological model.

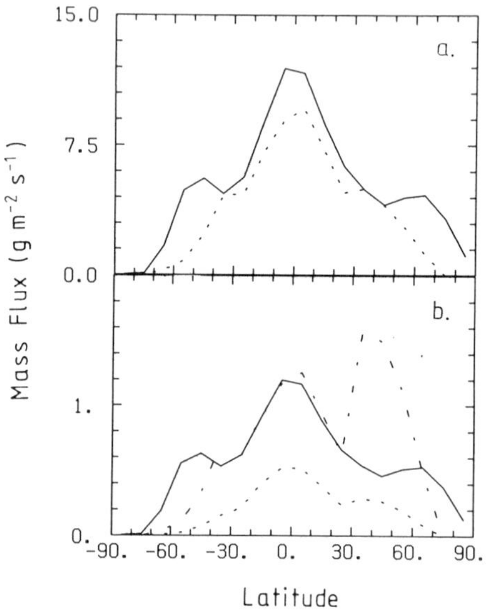

Fig. 4. Zonal and annual average of the air mass fluxes from the PBL produced by thunderstorm convective updrafts. Fluxes averaged over 10° latitudinal belts. Fig. 5a: solid line shows Olofsson (1988); dotted line, the current study using the thunderstorm distribution of Fig. 4 and a transport of 1.3×10^9 kg air s^{-1} for each of 2000 thunderstorms (H13 in Table 3). Fig. 5b: solid is Langner and Olofsson (1990); dash-dot, H90 model (Table 3); dotted, HMD model (Table 3).

Generally parameterizations assume that the subgrid-scale flux by clouds is much greater than that by the resolved motions. (Remember that the flux by the convective updraft is much greater than the resolved flux in the lower troposphere, but the fluxes are comparable in the upper troposphere.) Thus the convection parameterization is treated as an "add on" feature that must have zero net vertical flux. The resolved scale motions may be consulted to see if there is precipitation and resolved convergence of water vapor into the column, in which case the cloud parameterization is turned on, but the net vertical mass flux in the grid cell is not considered as being produced by the clouds! Thus, for the case of Fig. 2, some of the following parameterizations would double count the upward fluxes (as both resolved and subgrid scale motions), and would also have too much downward subgrid-scale flux.

4.1. *Local and Nonlocal Parameterizations*

If air mass can go more than one grid level in the vertical during one convection time step (typically 30 min-4 h), then a nonlocal cloud parameterization is appropriate. Because layers are often a few kilometers thick in the middle troposphere, let us assume that if the air moves 5 km or more it needs a nonlocal scheme. The time required to move 5 km is 500, 5000, and 5×10^4 s for a convective updraft velocity of $w = 10$ m s^{-1} over a midlatitude continent, for a convective velocity of $w = 1$ m s^{-1} over a tropical ocean, and for a mesoscale motion of $w = 0.1$ m s^{-1}, respectively. Thus, in one model time step, a midlatitude convective updraft normally moves 5 km or more, a mesoscale updraft or downdraft almost never does, and a tropical convective updraft only moves 5 km for longer convective time steps. With these values in mind, we will compare the patterns and lengths of the jumps used in convection parameterizations.

Mahlman and associates at GFDL were the first to simulate global tracer evolution in three-dimensions (Mahlman and Moxim, 1978; Mahlman et al., 1980; Levy et al., 1980, 1982, 1985). Their effort has continued to be an extreme in several senses. They have used the same well-documented meteorology from their meteorological model for more than 13 years. They have consistently used the highest spatial and temporal resolutions (265 km horizontally, 11 vertical levels, six-hour average winds) and hence

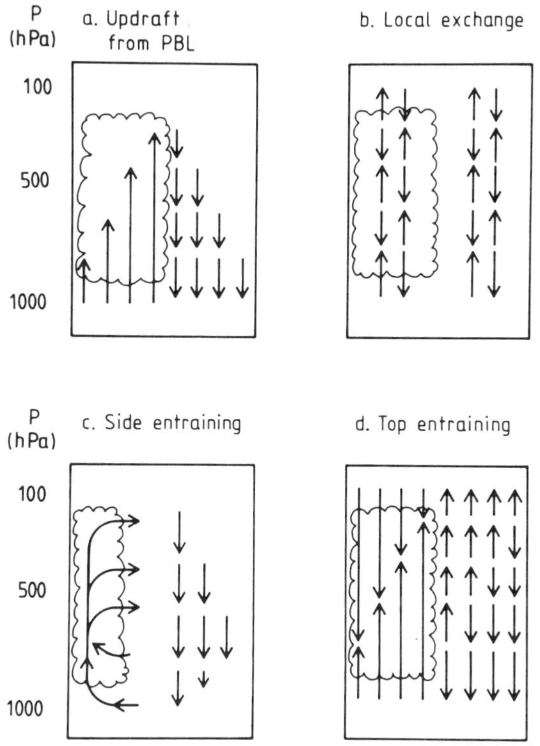

Fig. 5. Schematic of cloud models used in some global and regional models: (a) Updraft from the PBL from GISS (Prather et al, 1987); (b) Local exchange, either by eddy diffusivities (GFDL) or local convective fluxes (Walton et al., 1988) (c) Side entraining (Feichter and Crutzen, 1990; Feichter et al. in this session); and (d) top entraining from RADM (Walcek and Taylor, 1986).

have arguably the best meteorology on the resolved scales. Because they resolve the meteorology down to the mesoscale, they have thus far succeeded in using the simplest cloud model: they simply increase the K in a grid cell that experiences a convection event, and this increased K provides enhanced local exchange throughout the grid cell (Fig. 5b). The zonal average subgrid-scale K has maxima in the middle troposphere of midlatitudes in both hemispheres, but the K in the ITCZ region is surprisingly small (Plumb and Mahlman, 1987). Apparently the resolved vertical motion in the GFDL model simulates well the precipitation in the tropics–a result in agreement with the analysis of Gray (1973). There are perhaps indications, however, of too much vertical mixing in the tropics by the resolved vertical wind (Hiram Levy II, private communication). Their model has been used for climatological simulations of longer-lived species, and will not transport species from the surface to the tropopause in less than a few days, so that insoluble species with lifetimes of less than one day might not be well simulated, although Levy and Moxim (1989) argue that this is unclear.

Although the other models described here are Eulerian, GRANTOUR is a Lagrangian model using local convective fluxes (Walton et al., 1988).

The most widely used global model is the one developed at Harvard and GISS (Fung et al., 1983; Prather et al, 1987; Jacob et al., 1987; Jacob et al., 1990). This model has generally used low to moderate horizontal resolution of 500-1000 km, 9 levels, model episodic winds, and model quasi-episodic clouds. The GISS model resolves less than the GFDL model, and thus the cloud parameterization is much more important in the GISS model than in the GFDL model. The convection parameterization was developed by Prather et al. (1987) and is a matrix that depends on horizontal position and time and that gives the fraction of the mass that moves during one time step (4 h) from each height in the model to every other height. The updraft from the PBL is parameterized as a jump from the PBL to other heights in the troposphere as well as the compensating subsidence in which mass can normally move down only one layer in one time step (Fig. 5a).

The climatological global model MOGUNTIA (Zimmermann, 1988; Zimmermann et al., 1989) uses the cloud parameterization of Feichter and Crutzen (1990). MOGUNTIA has 10 vertical levels, low horizontal spatial resolution (10°), and low temporal resolution meteorology (mean monthly winds and K's based on temporal variations of winds on time scales less than a month). Air flowing into clouds in the lower troposphere can jump to all detraining levels in the middle and upper troposphere during one convective time step (2 h; see Fig. 5c). In this model the updraft air is taken from the PBL and heights above the PBL, so that the upward jumps are not necessarily as large as in the Prather scheme. Again there is a compensating (bump-down) subsidence. Of the 3-D models discussed here MOGUNTIA uses meteorology that resolves the least and therefore needs a subgrid scale parameterization (whether called a cloud model or not) to parameterize the most vertical transport.

The ECMWF model has developed and added new parameterizations faster than some other global meteorological models, such as the GFDL model. The ECMWF model has been taken by Erich Roeckner and associates at the University of Hamburg and adapted for climate studies and online simulations of trace species transport. This new Hamburg model has been named ECHAM. Its changes useful for atmospheric chemistry include the calculation of explicit liquid water content, important for simulations with aqueous chemistry, and the inclusion of online tracer transport (Feichter et al., this session). ECHAM uses moderate to high spatial and temporal resolution (19 vertical levels, 5.6° horizontal resolution, 40 minute time step). ECHAM resolves about as much as the GFDL model, but also parameterizes about as much as GISS. The cloud model is a Kuo-type scheme based on net horizontal moisture convergence into a vertical column (Fig. 5c).

The primary source of cloud air comes comes from below cloud, but the secondary source is controversial and may be from the side or from the top. Some evidence exists that nonprecipitating clouds undergo top entrainment, and this was used in RADM by Walcek and Taylor (1986) (Fig. 5d). Downdrafts can only originate from 500 mb or lower, so that mature thunderstorms can only have side entrainment. There is perhaps some agreement that precipitating clouds have lateral entrainment and nonprecipitating clouds have top entrainment.

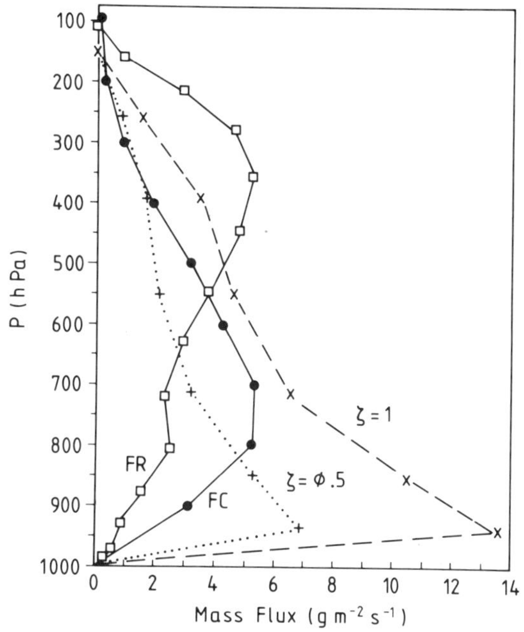

Fig. 6. Global average vertical profiles of mass fluxes. FC is Feichter and Crutzen (1990). Two curves are the GISS convection statistics (including shallow clouds and clear air PBL turbulence) as used by Prather et al. (1987, curve $\zeta = 1$) and by Heimann and Keeling (1989; $\zeta = 0.5$). FR is Feichter et al. (this session)

4.2. 3-D model parameterized cloud fluxes

Figure 6 shows annual and global average vertical mass fluxes used in some three-dimensional models. The curves are not directly comparable because the Prather et al. (1987) fluxes include PBL turbulence, as well as shallow and deep clouds, the Feichter and Crutzen (1990; hereafter FC) parameterization must account for everything not represented by a K for PBL turbulence and another K for mixing near the polar front, and the ECHAM parameterization (Feichter et al., this session; FR) only includes deep convection.

The original Prather fluxes given to Heimann are labelled $\zeta = 1$ in Fig. 6; these are an older version of the fluxes used by Jacob et al. (1990) to simulate radon. Heimann and Keeling (1989), however, improved their simulations of the longer-lived CO_2 by multiplying Prather's mass flux by 0.5 ($\zeta = 0.5$). The two Prather curves have profiles appropriate for the convective updraft alone, whereas FC and FR have profiles more appropriate for the net flux, the sum of the convective up and down and mesoscale down motions. Although the observed net fluxes normally have peaks between 400 and 600 mb, profiles exist in the literature for individual storms or individual cloud clusters that have peaks located as high as the FR curve or as low as the FC curve. For example, in McBride (1981) the western Pacific result resembles FR and the GATE curve looks like FC.

The various cloud models have their peak fluxes at different latitudes as well as different heights. For comparison with the cloud fluxes, Figure 7a shows the annual and zonal average mean flux from ECMWF analysis using the grid resolution of Heimann and Keeling (1989). In the middle troposphere, the Hadley cell displays an upward flux of 2 g m^{-2} s^{-1} near the equator, and downward fluxes of 2 at 25°N and 1.6 at 35°S. The GISS results in Fig. 7b (for $\zeta = 1$) show maximum fluxes coming out of PBL (900 mb) in midlatitudes (±35°). Although PBL turbulence somewhat masks the cloud convection, the 4 isoline extends up to 300 mb in the tropics. In the middle troposphere near both poles, there are also anomalous fluxes that are common in global meteorological models. FC (Fig. 7c) show a primary peak from deep convection in the tropics (5°N, 700 mb), as well as secondary peaks in midlatitudes, with the southern hemisphere (−45°N, 800 mb), having stronger convection than the northern hemisphere, due to stronger baroclinic slope convection in the southern hemisphere. FR (Fig. 7d) has an extremely large maximum in the upper tropical troposphere (−5°, 300 mb), and hardly an indication of maxima in midlatitudes.

Although 3-D model parameterizations also have detailed variations in longitude and in season or month, the results differ enough even when heavily averaged over space and time, so that a more detailed comparison is not yet appropriate.

Many models have compared simulations of ^{222}Rn with observations in order to calibrate or validate cloud parameterizations (Brost and Chatfield, 1989; Feichter and Crutzen, 1990; Heimann et al., 1990; Jacob et al., 1990). There are, however, insufficient vertical profiles of radon, and these tend to be located over continents in the middle and high latitudes (35° − 55°N), so that simulating radon only tests fluxes in these areas. Simulating radon provides a good test for the Prather model because the parameterization has a vertical profile appropriate for a short-lived species and has a maximum flux near the latitudes where the radon data was taken. Radon provides only a moderately good test for the FC model, because the mass fluxes at 35°-55°N are only about 1/2 the peak fluxes, and because the vertical mass flux profile of the FC parameterization is appropriate for a long-lived species. Radon provides a rather poor test of the FR parameterization because significant fluxes only occur south of 45°N. For any scheme, fluxes tuned to reproduce radon data may have to be decreased a factor of two or more to simulate long-lived species, but this adjustment may be smaller than the adjustment that a K model may require to handle different species. More radon data is needed over continents in the tropics in order to test the fluxes there. Also, radon will never provide a good test of vertical transport over the oceans; other tracers are needed there.

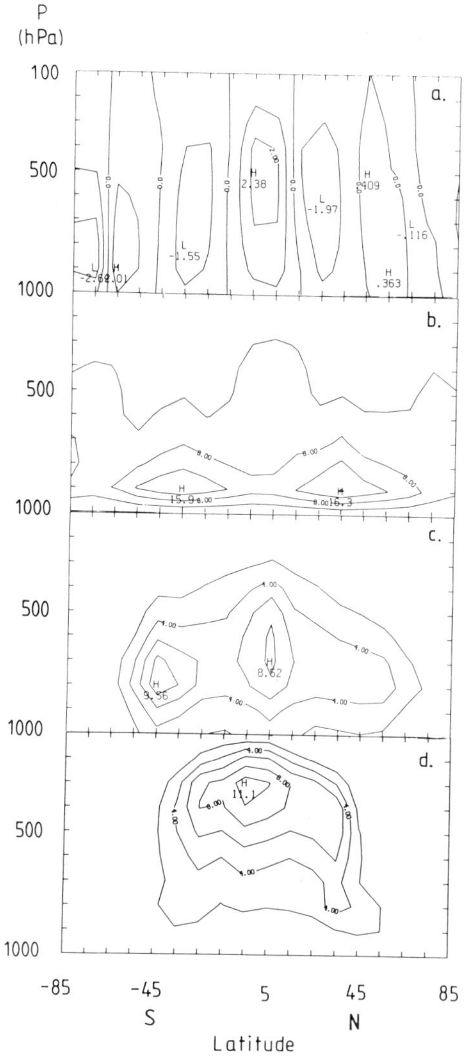

Fig. 7. Zonal average vertical mass fluxes: Fig. 7a, resolved motions; Fig. 7b, convective parameterization of Prather et al. (1987; clouds plus PBL turbulence; uses $\zeta = 1$); Fig. 7c, Feichter and Crutzen (1990); Fig. 7d, Feichter et al. (this session)

5. CONCLUSIONS

There are large variations in parameterized vertical fluxes of mass by cumulus convection in the 3-D models compared here. The heights and latitudes of the maxima vary widely, but so do the observational results, so that, based on the patterns, we have difficulty picking which parameterization is best.

We can offer, however, some generalizations. First, the Prather parameterization describes the updraft only and seems preferable for species with a surface source and with a τ less than a few days. Second, the vertical profile of a net flux parameterization (such as FR and FC) seems preferable for species with $\tau > 7$ d.

For the ITCZ, simple analysis estimated zonal average fluxes from the PBL of 1-5 g m^{-2} s^{-1}. From counting thunderstorms, it seems hard to justify maximum zonal average fluxes out of the PBL of much more than 1 g m^{-2} s^{-1} unless there are far more than 2000 thunderstorms globally at any instant. If we assume that the updraft flux profile monotonically decreases with height and must always be greater than the net flux, then for the FC and FR parameterizations the extrapolated convective updraft flux leaving the PBL would have to be more than 10 g m^{-2} s^{-1}, which seems large based on our simple estimates, but not impossible considering the uncertainties in these estimates. Although the total Prather flux leaving the PBL (for $\zeta = 1$) exceeds 8 g m^{-2} s^{-1} at the equator, we cannot say that it is too large because it also includes the effects of shallow clouds and PBL turbulence. On the other hand, reducing the Prather flux for longer-lived species is consistent with this analysis.

From the Johnson (1980) analysis of a particular cloud cluster, we can conclude that there may be no significant downward mass fluxes that move so fast they go more than one level in one time step. Thus, the customary bump-down scheme seems adequate to represent the subsidence, although care should be taken with the finite difference scheme used (Tiedtke, 1989). Also from Johnson's analysis, the net subgrid-scale flux should supply the resolved-scale flux, rather than having a net subgrid-scale flux that sums to zero. Thus, as horizontal grid resolutions decrease toward 100-300 km, some parameterizations seem to run the risk of double counting the same upward net flux as both resolved and subgrid-scale, and then also having too much parameterized downward flux for a given net resolved upward flux.

Chemical models have started to use parameterizations that require subgrid-scale information that is hard for meteorological models to supply because of the complexity of convection processes in the atmosphere, as the large variations in convective fluxes from different parameterizations demonstrate in this paper. The GFDL approach of not introducing a nonlocal cloud parameterization can be considered a conservative and defensible approach considering the uncertainty surrounding cloud parameterizations, although the approach is best for long-lived species.

Acknowledgments. We thank Hans Feichter for the fluxes from the ECHAM model and Michael Prather for the fluxes from the GISS model; we also thank Paul Crutzen, Jim Fankhauser, Peggy LeMone, Hiram Levy II, Michael Tiedtke, and Morris Weisman for useful discussions and references.

References

Anthes, R. A., Orville, H. D., and Raymond, D. J., 1986, Mathematical modeling of convection, in: "Thunderstorm Morphology and Dynamics," E. Kessler, ed., University of Oklahoma Press, Norman and London.

Brost, R. A., and Chatfield, R. B., 1989, Transport of radon in a three-dimensional synoptic scale model, *J. Geophys. Res.*, 94:5095.

Chatfield, R. B., and Crutzen, P. J., 1984, Sulfur dioxide in remote oceanic air: Cloud transport of reactive precursors, *J. Geophys. Res.*, 89:7111.

Churchill, D. D., and Houze, R. A. Jr., 1984, Mesoscale updraft magnitude and cloud-ice content deduced from the ice budget of the stratiform region of a tropical cloud cluster, *J. Atmos. Sci.*, 41:1717.

Court, A., and Griffiths, J. F., 1986, Thunderstorm climatology, in: "Thunderstorm Morphology and Dynamics," E. Kessler, ed., University of Oklahoma Press, Norman and London.

Feichter, J., and Crutzen, P. J., 1990, Parameterization of deep cumulus convection in a global tracer transport model and its verification with ^{222}Radon, *Tellus B*, in press.

Frank, H. W., and Foote, G. B., 1982, The 22 July 1976 case study: Storm airflow, updraft structure, and mass flux from triple-doppler measurements, in: "Hailstorms of the Central High Plains," Charles A. Knight and Patrick Squires, ed., Colorado Associated University Press, Boulder, 131.

Frank, W. M., 1983, The cumulus parameterization problem, *Mon. Wea. Rev.*, 111:1859.

Fung, I., Prentice, K., Matthews, E., Lerner, J., Russell, G., 1983, Three-dimensional tracer model study of atmospheric CO_2: response to seasonal exchanges with the terrestrial biosphere, *J. Geophys. Res.*, 88:1281.

Gray, W. M., 1973, Cumulus convection and larger scale circulations 1. Broadscale and mesoscale considerations, *Mon. Wea. Rev.*, 101:839.

Gray, W. M., 1965, Calculations of cumulus vertical draft velocities in hurricanes from aircraft observations, *J. Appl. Meteor.*, 4:463.

Heimann, M., and Keeling, C. D., 1989, A three-dimensional model of atmospheric CO_2 transport based on observed winds: 2. Model description and simulated tracer experiments, in: "Aspects of Climate Variability in the Pacific and the Western Americas," D. H. Peterson, ed., American Geophysical Union, Washington, D. C.

Heimann, M., Monfray, P., and Polian, G., 1990, Modeling the long range transport of ^{222}Rn to subantarctic and antarctic areas, *Tellus*, in press.

Houze, Jr., R. A., and Betts, A. K., 1981, Convection in GATE, *Rev. of Geophys. and Space Phys.*, 19:541.

Houze, R. A. Jr., and Cheng. C.-P., 1981, Inclusion of mesoscale updrafts and downdrafts in computations of vertical fluxes by ensembles of tropical clouds, *J. Atmos. Sci.*, 38:1751.

Jacob, D. J., Prather, M. J., Wofsy, S. C., McElroy, M. B., 1987, Atmospheric distribution of ^{85}Kr simulated with a general circulation model, *J. Geophys. Res.*, 92:6614.

Jacob, D. J., and Prather, M. J., 1990, Radon-222 as a test of convection in a general circulation model, *Tellus*, in press.

Johnson, R. H., 1980, Diagnosis of convective and mesoscale motions during Phase III of GATE, *J. Atmos. Sci.*, 37:733.

Johnson, R. H., 1984, Partitioning tropical heat and moisture budgets into cumulus and mesoscale components: Implications for cumulus parameterizations, *Mon. Wea. Rev.*, 112:1590.

Langner, J., and Olofsson, M., 1990, Parameterization of subgrid scale vertical tracer transport in a global 2D-model of the troposphere, submitted to *J. Geophys. Res.*.

Lelieveld, J., and Crutzen, P. J., Influences of cloud photochemical processes on tropospheric ozone, *Nature*, 343:227.

LeMone, M. A., and E. J. Zipser, 1980, Cumulus vertical velocity events in GATE. Part I: Diameter, intensity and mass flux, *J. Atmos. Sci.*, 37:2444.

Levy II, H., Mahlman, J. D., and Moxim, W. J., 1980, A stratospheric source of reactive nitrogen in the unpolluted troposphere, *Geophys. Res. Lett.*, 7:441.

Levy II, H., Mahlman, J. D., and Moxim, W. J., 1982, Tropospheric N2O variability, *J. Geophys. Res.*, 87:3061.

Levy II, H., Mahlman, J. D., Moxim, W. J., and Liu, S. C., 1985, Tropospheric ozone: The role of transport, *J. Geophys. Res.*, 90:3753.

Levy II, H., and Moxim, W. J., 1989, Examining the global impact of local/regional air pollution: The role of global chemical transport models, in: "Air Pollution Modeling and Its Application VII," H. van Dop, ed., Plenum, New York.

Mahlman, J. D., and Moxim, W. J., 1978: Tracer simulation using a global general circulation model: Results from a mid-latitude instantaneous source experiment, *J. Atmos. Sci.*, 35:1340.

Mahlman, J. D., Levy II, H., and Moxim, W. J., 1980: Three-dimensional tracer structure and behavior as simulated in two ozone precursor experiments, *J. Atmos. Sci.*, 37:655.

McBride, J. L., 1981, An analysis of diagnostic cloud mass flux models, *J. Atmos. Sci.*, 38:1977.

Miller, L. J., Harris, F. I., and Fankhauser, J. C., 1982, The 22 June 1976 Case Study: Structure and evolution of internal airflow, in: "Hailstorms of the Central High Plains," Charles A. Knight and Patrick Squires, ed., Colorado Associated University Press, Boulder, 35.

Newton, C. W., 1968, Convective cloud dynamics–a synopsis, in: Proceeding of International Conference on Cloud Physics, Toronto, AMS, Boston, 487.

Olofsson, M., 1988, Global vertical mass transport by clouds–a two-dimensional model study, Report CM-74, ISSN 0280-445X, Dept. of Meteor., Univ. of Stockholm, Sweden.

Pierce, E. T., 1986, Storm electricity and lightning, in: "Thunderstorm Morphology and Dynamics," E. Kessler, ed., University of Oklahoma Press, Norman and London.

Plumb, R. A., and Mahlman, J. D., 1987, The zonally averaged transport characteristics of the GFDL general circulation/transport model, *J. Atmos. Sci.*, 44:298.

Prather, M., McElroy, M., Wofsy, S., Russell, G., and Rind, D., 1987, Chemistry of the global troposphere: Fluorocarbons as tracers of air motion, *J. Geophy. Res.*, 92:6579.

Riehl, H., and Malkus, J. S., 1958, On the heat balance in the equatorial trough zone, *Geophysica (Helsinki)*, 6:503.

Riehl, H., and Simpson, J. M., 1979, The heat balance of the equatorial trough zone, revisited, *Contributions to Atmospheric Physics*, 52:287.

Rutledge, S. A., Houze, R. A. Jr., Biggerstaff, M. I., and Matejka, T., 1988, The Oklahoma-Kansas mesoscale convective system of 10-11 June 1985: Precipitation structure and single-doppler radar analysis, *Mon. Wea. Rev.*, 116:1409.

Rotunno, R., Klemp, J. B., and Weisman, M. L., 1988, A theory for strong, long-lived squall lines, *J. Atmos. Sci.*, 45:463.

Sanders, F., and Freeman, J. C., 1986, Thunderstorms at sea, in: "Thunderstorm Morphology and Dynamics," E. Kessler, ed., University of Oklahoma Press, Norman and London.

Sikdar, D. N., Soumi, V. E., and Anderson, C. E., Convective transport of mass and energy in severe storms over the United States–an estimate from a geostationary altitude, *Tellus*, 22:521.

Tiedtke, M., 1989, A comprehensive mass flux scheme for cumulus parameterization in large-scale models, *Mon. Wea. Rev.*, 117:1779.

Walton, J. J., MacCracken, M. C., Ghan, S. J., 1988: A global-scale Lagrangian trace species model of transport, transformation, and removal processes, *J. Geophys. Res.*, 93:8339.

Yanai, M., Esbensen, S., and Chu, J.-H., 1973, Determination of bulk properties of tropical cloud clusters from large-scale heat and moisture budgets, *J. Atmos. Sci.*, 30:611.

Zimmermann, P. H., Feichter, J., and Crutzen, P. J., 1989, A global three-dimensional source-receptor model investigation using ^{85}Kr, *Atmos. Environ.*, 23:25.

Zimmermann, P. H., 1988, MOGUNTIA: A handy global tracer model, in: "Air Pollution Modelling and its Application," H. van Dop, ed., Plenum, New York.

Zipser, E. J., 1977, Mesoscale and convective-scale downdrafts as distinct components of squall-line circulations, *Mon. Weather Rev.*, 105:1568.

Zipser, E. J., and LeMone, M. A., 1980, Cumulonimbus vertical velocity events in GATE. Part II: Synthesis and model core structure, *J. Atmos. Sci.*, 37:2458.

DISCUSSION

M.D. MORAN

Do any current global 3D tracer models include a stratospheric sink for tracer due to injection of tracer by overshooting thunderstorm tops?

R.A. BROST

Yes. For example, in modeling the tropospheric distribution of chlorofluorocarbons, Prather et al. (1987) included a stratospheric sink for the species in order to prevent the chemicals from re-entering the troposphere. However, no global 3D model has a good parameterization of thunderstorm injection into the stratosphere.

R.A. PIELKE

Your analysis of the importance of non-zero net cumulus mass flux has important implications in GCM trace gas and meteorological simulations. However, due to the course resolution scales of these models (e.g. even the ECMWF result you mentioned had $\Delta y = 1°$ which corresponds to 440 km for a 4 Δy feature), only 1 grid point often will have deep cumulus convection. How do you propose the loss of mass flux due to insufficient resolution (e.g. because of the numerical solution algorithm used including any explicit horizontal diffusion) be handled so as to avoid this loss?

R.A. BROST

Important issues are the magnitude and vertical distribution of upward mass fluxes due to convection, and the fraction of the compensating subsidence that occurs within the grid cell experiencing the convection, and the connection between resolved and sub-grid scale motions. For most practical spatial grid resolutions in global 3D models, the shallow fair-weather cumulus motions may have about zero net vertical flux, but the motions associated with (precipitating) cumulonimbus clouds, particularly squall lines, cloud clusters, and associated mesoscale anvils, should have a net upward flux that is closely related to the resolved-scale vertical flux. (Upper tropospheric horizontal divergence can precede the occurrence of deep convection by 12 hours, and may help release the convective instability and allow the convection to start, but once the deep convection begins the resolved net upward mass flux is then largely due to convection.) Current parameterizations invariably make the "coarse spatial resolution approximation" that assumes that cloud fluxes are independent of resolved fluxes, and that assumes that vertical mass fluxes in clouds are much, much larger than resolved vertical fluxes. Regional chemical models and increasingly global ones should start to make the "fine resolution approximation" that recognizes the connection between the resolved and convective fluxes. Tiedtke (1989) recently connected the vertical mass flux in precipitating clouds to the resolved horizontal mass convergence. (The horizontal eddy diffusion produces a horizontal flux of tracer but not a net horizontal air mass flux and thus is not important for air mass convergence.) Tiedtke's parameterization should work well for long-lived species, and may be used in the ECHAM model, but upward motion from the surface will still be under-represented for short-lived insoluble species.

THE GLOBAL NUMERICAL SIMULATION OF THE DISTRIBUTION OF CO IN THE TROPOSPHERE

Rick D. Saylor (1) and Leonard K. Peters (2)

(1) Center for Applied Energy Research
University of Kentucky
Lexington, KY 40511-8433

(2) Department of Chemical Engineering
University of Kentucky
Lexington, KY 40506-0046

INTRODUCTION

In recent years, much scientific effort has been focussed on achieving a better understanding of the effect of human activities on the environment. More than thirty years ago, the atmosphere was considered to be an infinite capacity sink in which one could dump any sort of gaseous waste stream. Air pollution was viewed largely as a local problem which in many cases could be alleviated by simply building taller stacks, thus dispersing the pollutants at higher altitudes and relieving a locally noticed nuisance. Twenty years ago, with the advent of the acid rain phenomenon, air pollution began to be perceived also as a regional problem both in Europe and North America. Only in the past few years, with the discovery of the Antarctic ozone hole and discussion of a possibly imminent greenhouse effect, has a realization emerged that human activity can indeed significantly affect the global environment and possibly cause irreversible damage to worldwide ecosystems. In this context, it is vital that we gain a deeper understanding of global biogeochemical cycles so that more effective control measures can be designed.

One way to achieve this understanding of atmospheric chemical cycles is through the use of numerical models. The results of model simulations can provide valuable information about the relative importance of both anthropogenic and biogenic sources of pollutants and their precursors. Since the global distributions of many trace atmospheric species are largely source dominated, it is desirable to model the atmospheric chemistry of these species in three dimensions. The present work reports on results from simulations of such a three-dimensional atmospheric chemistry model designed to investigate the methane/carbon monoxide tropospheric cycle.

MODEL DESCRIPTION

Governing Equations

GLOBAL is an Eulerian three-dimensional numerical model that simulates the emission, advective and turbulent transport, gas-phase chemical transformation, and surface interactions of trace species in the global troposphere (Peters and Jouvanis, 1979; Kitada and Peters, 1982). The current version of GLOBAL models the methane/carbon monoxide atmospheric chemical cycle and provides detailed spatial and temporal distributions of CH_4, CO, OH, and HO_2 in the troposphere.

Air Pollution Modeling and Its Application VIII, Edited by H. van Dop and D. G. Steyn, Plenum Press, New York, 1991

The governing equations of the GLOBAL model are:

$$\frac{\partial x_i}{\partial t} + (\mathbf{v}\cdot\nabla x_i) = \frac{1}{c}\nabla\cdot c(\,\mathbf{K}\cdot\nabla x_i\,) + \frac{1}{c}R_i \tag{1}$$

with boundary conditions at the Earth's surface,

$$-\,c(\mathbf{K}\cdot\nabla x_i)\cdot\vec{n}_h = \begin{cases} \dfrac{k_l}{H_i}(cx^{*}_{i,l} - cx_i) & \text{for } \textit{ocean}, \\[2ex] k_s(cx^{*}_{i,s} - cx_i) + Q_i & \text{for } \textit{land}, \end{cases} \tag{2}$$

and at the tropopause,

$$(\mathbf{v}cx_i - c(\mathbf{K}\cdot\nabla x_i)\cdot\vec{n}_h)^{n+1} = \begin{cases} (\mathbf{v}cx_{i,St} - c(\mathbf{K}\cdot\nabla x_i)\cdot\vec{n}_h)^{n} & \text{for } \textit{inflow}, \\[2ex] (\mathbf{v}cx_i - c(\mathbf{K}\cdot\nabla x_i)\cdot\vec{n}_h)^{n} & \text{for } \textit{outflow}, \end{cases} \tag{3}$$

where,

x_i = the mixing ratio of species i (mole i/mole air),
$\mathbf{v}$ = the wind velocity vector (m/s),
c = the air molar density (moles air/ m^3),
$\mathbf{K}$ = the turbulent eddy diffusivity tensor (m^2/s),
R_i = the molar rate of production (destruction) of species i by chemical reaction,
n = the nth time step,
$(n+1)$ = the (n+1)st time step,
$\vec{n}_h$ = the unit vector normal to the earth's surface,
k_l = the air-ocean mass transfer coefficient (m/s),
k_s = the air-soil mass transfer coefficient (m/s),
H_i = the Henry's law coefficient for species i (moles i aqueous/moles i),
$x^{*}_{i,l}$ = the mixing ratio of species i in equilibrium with the ocean (moles i/mole air),
$x^{*}_{i,s}$ = the mixing ratio of species i in equilibrium with the soil (moles i/mole air),
Q_i = the emission rate of species i (moles i/m^2-s),and
$x_{i,St}$ = the mixing ratio of species i in the lower stratosphere.

Chemistry

The CO/CH_4 chemical reaction mechanism that has been incorporated into the model is adapted from that of Kitada and Peters (1982) and is shown in Table (1). It consists of 36 reactions and 26 species. Using this mechanism, chemical production/destruction of CO and CH_4 are calculated explicitly based on prescribed three-dimensional distributions of O_3 and NO_x. The steady state approximation is used at each grid point to calculate the concentrations of the intermediate species, mainly free radicals. The photolysis reaction rate constants are calculated as a function of the solar zenith angle and altitude, employing absorption cross-section and quantum yield data.

The prescribed global O_3 distribution is based on the annually averaged latitudinal distribution of Fishman and Crutzen (1978) which was derived from various worldwide observations. NO_x distributions are prescribed so that their diurnal variations correspond to the variations of anthropogenic CO emissions. The data of Platt et al. (1979) and Garnett (1979) were used to set the maximum NO_x concentration to 10 ppb near highly polluted areas. At less polluted locations, the NO_x concentration was prescribed in proportion to the CO source emission rate, while in unpolluted regions the profiles were set according to the data of Hales and Warneck (1981) and Kley et al. (1981).

Table 1. Methane/Carbon Monoxide Chemical Reaction Mechanism

Reaction	Rate Constant[b]
[1] $O_3 + h\nu \rightarrow O(^1D) + O_2$	$1.89 \times 10^{-4} \exp(-1.93/\cos\theta)$[c]
[2] $O(^1D) + H_2O \rightarrow 2OH$	2.3×10^{-10}
[3] $O(^1D) + M \rightarrow O + M$	$2.0 \times 10^{-11} \exp(107/T)$[d]
[4] $O + O_2 + M \rightarrow O_3 + M$	$1.1 \times 10^{-34} \exp(510/T)$
[5] $NO_2 + h\nu \rightarrow NO + O$	$1.55 \times 10^{-2} \exp(-0.48/\cos\theta)$
[6] $NO + O_3 \rightarrow NO_2 + O_2$	$9.0 \times 10^{-13} \exp(-1200/T)$
[7] $NO + HO_2 \rightarrow NO_2 + OH$	8.1×10^{-12}
[8] $NO + CH_3O_2 \rightarrow NO_2 + CH_3O$	$3.3 \times 10^{-12} \exp(-500/T)$
[9] $NO + OH \rightarrow HNO_2$	2.0×10^{-12}
[10] $NO + NO_2 + H_2O \rightarrow 2HNO_2$	6.0×10^{-38}
[11] $HNO_2 + h\nu \rightarrow NO + OH$	$2.75 \times 10^{-3} (\cos\theta)$
[12] $NO_2 + OH \rightarrow HNO_3$	2.0×10^{-12}
[13] $NO_2 + O_3 \rightarrow NO_3 + O_2$	$1.2 \times 10^{-13} \exp(-2450/T)$
[14] $NO_2 + NO_3 \rightarrow N_2O_5$	$1.48 \times 10^{-13} \exp(861/T)$
[15] $N_2O_5 \rightarrow NO_2 + NO_3$	$1.24 \times 10^{14} \exp(-10317/T)$
[16] $N_2O_5 + H_2O \rightarrow 2HNO_3$	1.0×10^{-20}
[17] $HNO_3 + h\nu \rightarrow OH + NO_2$	$9.88 \times 10^{-7} (\cos\theta)$
[18] $NO + NO_3 \rightarrow 2NO_2$	1.9×10^{-11}
[19] $CH_4 + OH \rightarrow CH_3 + H_2O$	$2.36 \times 10^{-12} \exp(-1710/T)$
[20] $CH_3 + O_2 \rightarrow CH_3O_2$	2.2×10^{-12}
[21] $2CH_3O_2 \rightarrow 2CH_3O + O_2$	1.6×10^{-13}
[22] $CH_3O + O_2 \rightarrow CH_2O + HO_2$	$1.6 \times 10^{-13} \exp(-3300/T)$
[23] $CH_2O + h\nu \rightarrow HCO + H$	$3.92 \times 10^{-5} \exp(-0.825/\cos\theta)$
[24] $CH_2O + h\nu \rightarrow H_2 + CO$	$1.23 \times 10^{-4} \exp(-0.652/\cos\theta)$
[25] $CH_2O + OH \rightarrow HCO + H_2O$	$3.0 \times 10^{-11} \exp(-250/T)$
[26] $HCO + O_2 \rightarrow HO_2 + CO$	5.7×10^{-12}
[27] $H + O_2 + M \rightarrow HO_2 + M$	$2.08 \times 10^{-32} \exp(290/T)$
[28] $CO + OH \rightarrow CO_2 + H$	$2.1 \times 10^{-13} \exp(-115/T) + 7.3 \times 10^{-33} n_m$[e]
[29] $HO_2 + OH \rightarrow H_2O + O_2$	3.0×10^{-11}
[30] $2HO_2 \rightarrow H_2O_2 + O_2$	2.5×10^{-12}
[31] $2OH + M \rightarrow H_2O_2 + M$	$1.25 \times 10^{-32} \exp(900/T)$
[32] $HO_2 + O_3 \rightarrow OH + 2O_2$	$1.4 \times 10^{-14} \exp(-580/T)$
[33] $H_2O_2 + h\nu \rightarrow 2OH$	$1.81 \times 10^{-5} \exp(-0.3297/\cos\theta)$
[34] $H_2O_2 + OH \rightarrow HO_2 + H_2O$	$1.0 \times 10^{-11} \exp(-750/T)$
[35] $HO_2 + NO_2 \rightarrow HO_2NO_2$	1.78×10^{-12}
[36] $HO_2NO_2 \rightarrow HO_2 + NO_2$	$1.26 \times 10^{16} \exp(-11700/T)$

[a] After Kitada and Peters (1980).
[b] Units are cubic centimeters, molecules, and seconds.
[c] θ represents the solar zenith angle.
[d] T represents the absolute temperature, K.
[e] n_m is the density of air expressed as molecules per cubic centimeter.

Emissions of CO and CH_4

An essential input into the GLOBAL model is emissions data for CO and CH_4. Emissions of CO included in the present analysis include anthropogenic CO from the combustion of fossil fuel and CO from biomass burning. The anthropogenic source strength was estimated from fuel consumption data and CO emission factors by DeHaven (1980), and include a diurnal variation simulating human activity. CO emissions from biomass burning were based on land use data and the CO flux information of Seiler and Crutzen (1980).

Another important source of CO is that produced from the oxidation of hydrocarbons emitted from vegetation. This production of CO through hydrocarbon oxidation has been incorporated into the GLOBAL model as an equivalent CO source. The equivalent CO flux from forests was estimated based on CO yields from the isoprene and terpene oxidation data of Hanst et al. (1980), the hydrocarbon emission factors given by Zimmerman et al. (1978), and the leaf biomass density data in Whittaker and Marks (1975) and Tadaki (1977). The total CO emission source strength from both anthropogenic and natural sources is 1200 Tg/yr.

The major sources of CH_4 emissions are enteric fermentation in animals, rice paddy fields, swamps, bogs, marshes, and other wetlands, and from coal and lignite mining. In the current analysis, CH_4 emissions from these sources total 200 Tg/yr and are based on the work of DeHaven (1980), Cicerone and Shetter (1981), and Harriss and Sebacher (1981).

Numerics

The GLOBAL governing equations are integrated numerically using the time-splitting finite element method of Mitchell (1969) and Yanenko (1971). This requires the splitting of the three-dimensional partial differential equation, Equation (1), into three time dependent one-dimensional problems which are solved sequentially. This scheme has been shown by various workers (Carmichael et al., 1980; Kasibhatla, 1988; McRae et al. 1982) to be an efficient method for the solution of the reactive advection-diffusion equation. To suppress numerical oscillations associated with locally large Peclet numbers, the asymmetric weighting functions of Christie et al. (1976) and Heinreich et al. (1977) have been incorporated into the finite element formulation.

The modeling region encompasses the entire global troposphere bounded by the Earth's surface and the tropopause. This region is discretized into $5° \times 5°$ horizontal grids and nine non-uniform vertical levels with finer vertical resolution in the boundary layer. A 1-week GLOBAL simulation requires 84 CPU minutes on an IBM 3090-300E with 3 vector processors.

Meteorological Data

Meteorological data from worldwide observations were obtained from the National Climatic Data Center for November 1981 and processed through an objective analysis module and a one-dimensional boundary layer model based on the work of Yamada and Mellor (1975). In this manner, vertical velocities w, vertical eddy diffusivities $\mathbf{K}$, temperature T, air density c, water vapor mixing ratio r, and horizontal wind components u and v, are generated for input into the GLOBAL model. Figure (1) presents an example of a typical horizontal wind field used in the current simulation.

MAPS MEASUREMENTS OF THE GLOBAL DISTRIBUTION OF CO

The Measurement of Air Pollution by Satellites (MAPS) experiment was designed to measure the global distribution of CO in the middle and upper troposphere. A gas filter radiometer was carried aboard the second Space Transport System engineering test flight (STS-2) of the U. S. space shuttle during November 12-15, 1981. The radiometer was tuned to remotely detect middle and upper tropospheric CO using a gas filter correlation technique to maintain high radiation throughput and high effective spectral resolution in the 4.67 μm fundamental band of CO. Details of the instrumentation and the observed measurements can be found in Reichle et al. (1986).

The November 1981 MAPS experiment measured CO concentrations in a band between 38° N latitude and 38° S latitude over all longitudes. In these measurements, lower mixing ratios were found in the southern hemisphere with the lowest readings (40-50 ppbv) occurring over the southeastern Pacific and southern South America. CO concentrations greater than 100 ppbv were measured in a band extending from northern South America across the Atlantic to central Africa. Values as large as 70 ppbv were observed over the north Atlantic and the eastern Mediterranean as well as off the western coast of Australia. A large peak greater than 100 ppbv was found over China with values as great as 70 ppbv extending from the mainland past Japan and over the western Pacific.

SIMULATION RESULTS

The GLOBAL model has been used to simulate the November 1981 distribution of CO in the troposphere as observed by the MAPS experiment. Figure (2) presents the ground-level CO distribution as simulated by the model. The locations with the highest CO concentrations correspond to major source regions. The industrialized areas of eastern North America and Europe exhibit peaks of 196 ppbv and 142 ppbv, respectively, illustrating the influence of anthropogenic CO emissions there.

In the tropics, the influence of CO emissions from biomass burning is very evident. Northern South America, central and southern Africa, southern Asia, and the East Indies all exhibit CO peaks greater than 100 ppbv. In locations remote from major emission sources, at the poles for example, the background CO concentration is predicted to be about 50-60 ppbv.

The MAPS experiment CO measurements from the middle troposphere can be compared to the GLOBAL simulation results at $\rho = 0.40$ and $\rho = 0.60$ shown in Figures (3) and (4), where ρ is the transformed vertical coordinate defined to be 0.0 at the Earth's surface and 1.0 at the tropopause. Several features predicted by the GLOBAL model cannot be compared to the observed data due to the lack of MAPS measurements at these locations. Specifically, no MAPS data were collected over North America, northern Europe, or southern Africa. However, in areas where the measurements were obtained, the model results exhibit many similarities to the observed MAPS data. The observed elevated concentration band extending from northern South America into central Africa is quite evident in the simulation result, as is the concentration peak that occurs over eastern China, although it is shifted northward in the simulation result.

Dissimilarities in the observed data and the simulation results occur over the northern Atlantic and western Pacific oceans where model results are 10-15 ppbv lower than MAPS measurements. This may be the result of the abbreviated 1-week simulation in which CO plumes emanating from high source regions have not had a long enough period to develop downwind. Longer simulations (a month or more) may allow these extended plumes to develop over the northern Atlantic and western Pacific.

Figure (5) presents the latitudinally-averaged CO concentrations simulated by the GLOBAL model. A strong asymmetry is noted with higher concentrations occurring in the more heavily polluted Northern Hemisphere. This result agrees with measurements made by several workers including Seiler (1974), Heidt et al. (1980), and Seiler and Fishman (1981). It must be noted, however, that the highest latitudinally averaged CO values occur in the Southern Hemisphere near the equator. This is a result of CO emissions from biomass burning, the sources of which are mostly located in the equatorial tropics.

CONCLUSIONS

A three-dimensional atmospheric chemistry model has been used to simulate the global distribution of CO in the troposphere. A simulation has been carried out using observed meteorological data for November 12-15, 1981 in an attempt to compare the model's global CO distributions with those measured in the MAPS experiment.

The model results exhibit several features in common with the MAPS measurements including a band of elevated CO concentrations extending from northern South America to central Africa. Although the GLOBAL model predicts other significant CO peaks in eastern North America, northern Europe, and Southern Africa, these results cannot be compared with MAPS observations since no measurements were obtained over these areas. These peaks do seem reasonable, however, since they occur near large CO source regions.

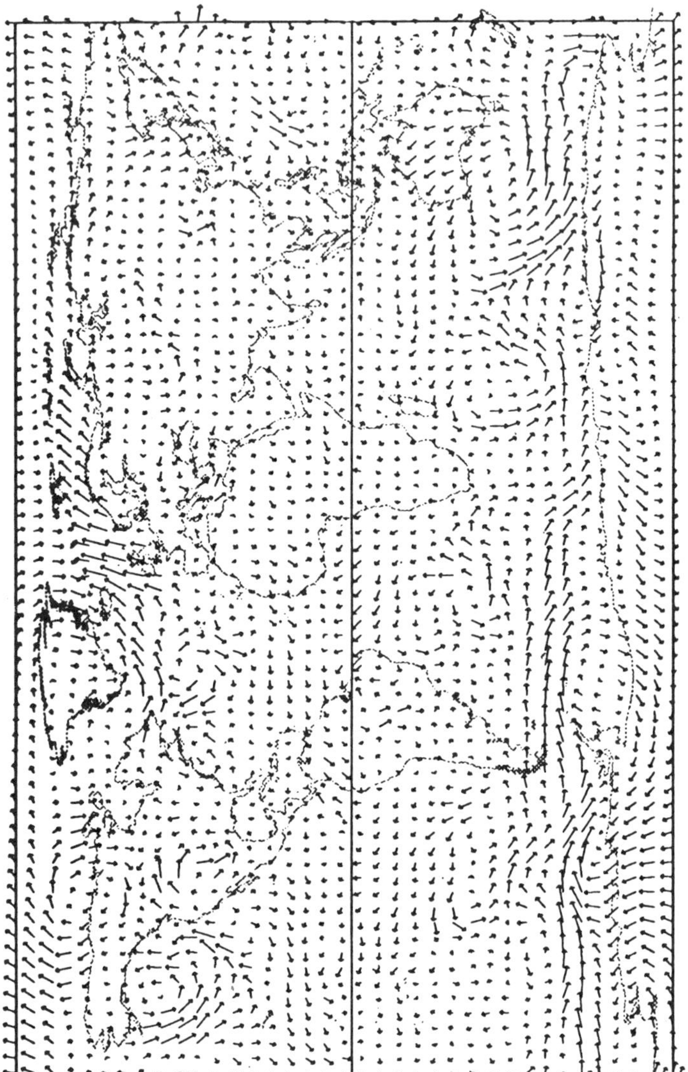

Figure 1. Horizontal wind field at ground-level on November 15, 1981 at 12:00 GMT

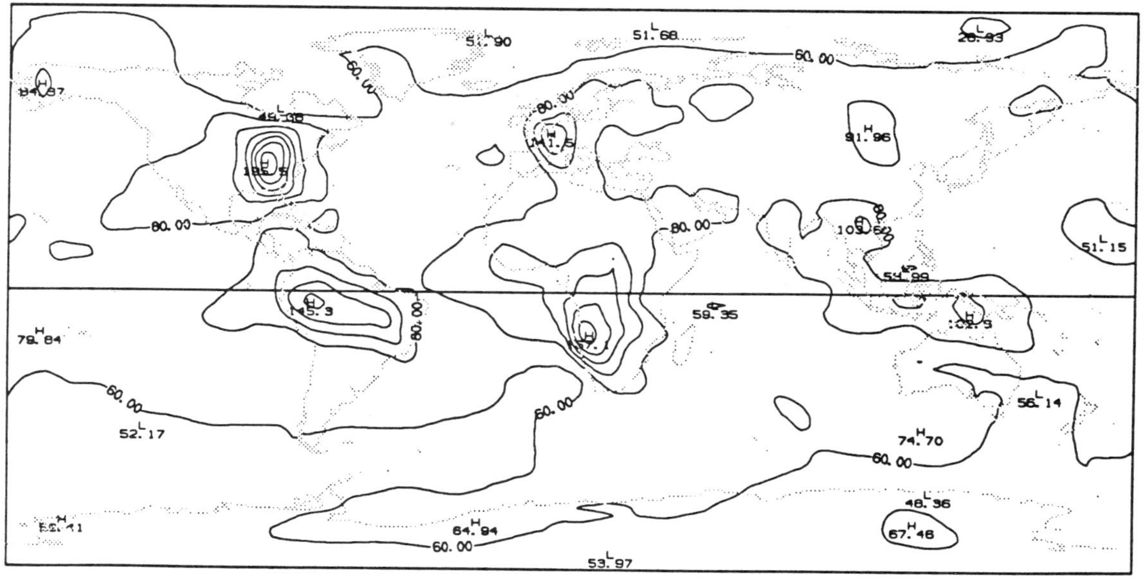

Figure 2. Simulated carbon monoxide mixing ratios at $\rho = 0.0$ on November 15, 1981 at 12:00 GMT (ppb)

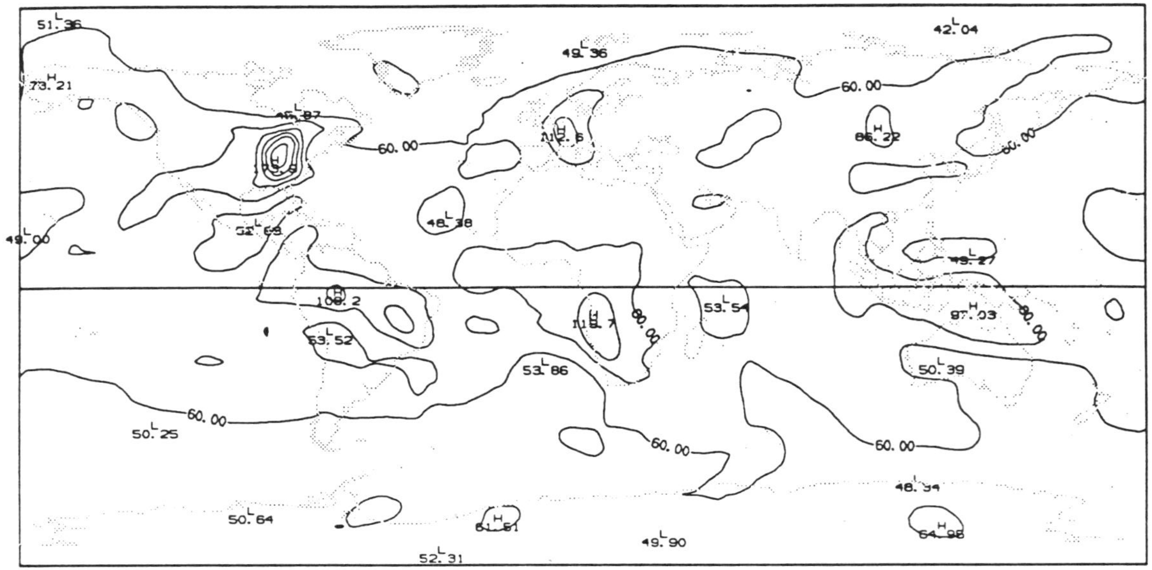

Figure 3. Simulated carbon monoxide mixing ratios at $\rho = 0.4$ on November 15, 1981 at 12:00 GMT (ppb)

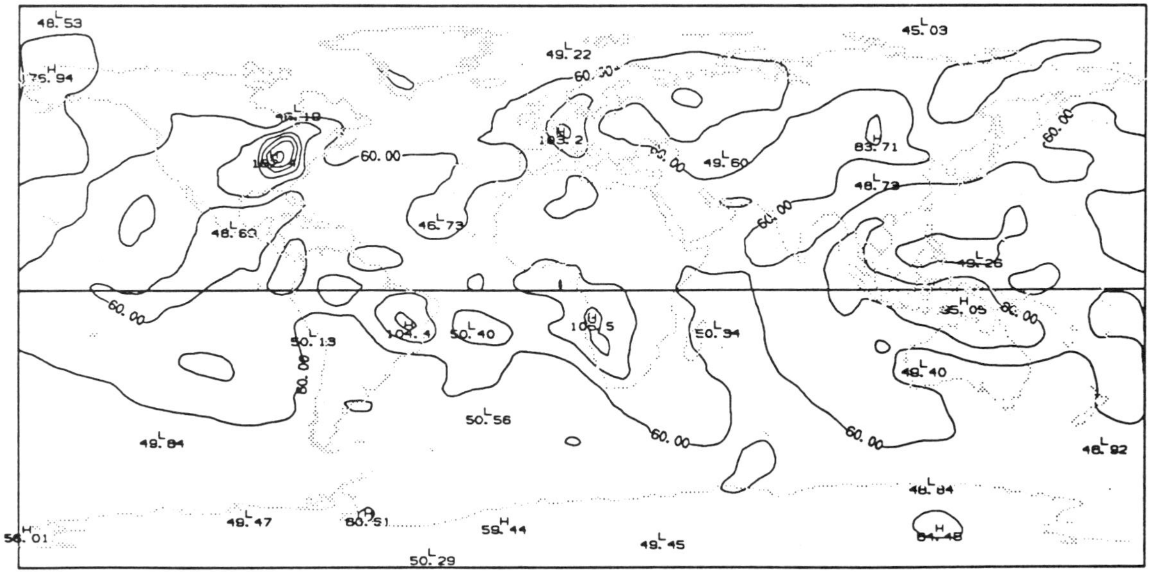

Figure 4. Simulated carbon monoxide mixing ratios at ρ = 0.6 on November 15, 1981 at 12:00 GMT (ppb)

The utility of the GLOBAL model in simulating the tropospheric *CO* distribution has been demonstrated. Future simulations for the October 1984 time period, during which another MAPS experiment was undertaken, will further explore the capabilities of the model. Future plans for GLOBAL include expansion of the chemical scheme with the explicit consideration of more tropospheric trace species and the investigation of anthropogenic influences on global tropospheric chemistry.

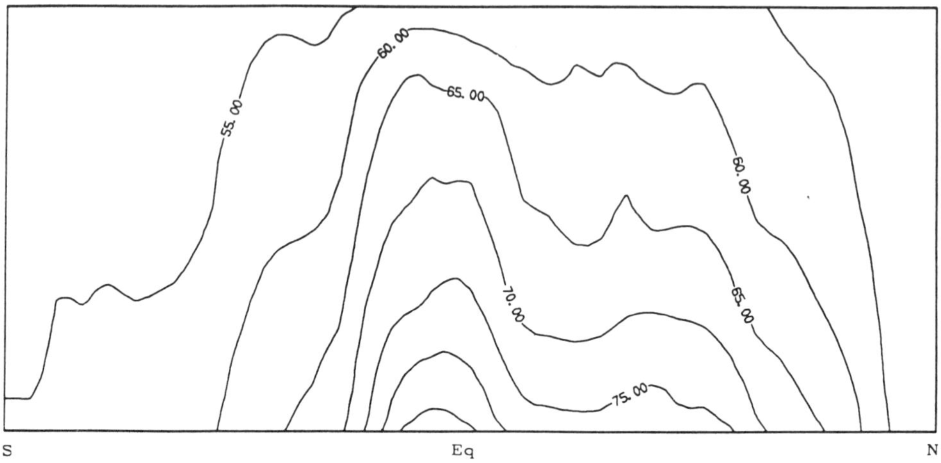

Figure 5. Simulated latitudinally averaged carbon monoxide mixing ratios on November 15, 1981 at 12:00 GMT (ppb)

ACKNOWLEDGMENTS

The authors acknowledge the University of Kentucky Center for Applied Energy Research for supporting this work and the University of Kentucky Center for Computational Sciences for providing computer time and office facilities.

REFERENCES

Carmichael, G. R., T. Kitada, and L. K. Peters, 1980, Application of a Galerkin finite element method to atmospheric transport problems, *Computers & Fluids 8*, 155-176.

Christie, I., D. F. Griffiths, A. R. Mitchell, and O. C. Zienkiewicz, 1976, Finite element methods for second-order differential equations with significant first derivatives, *Int. J. Num. Meth. Engng. 10*, 1389-1396.

Cicerone, R. J. and J. D. Shetter, 1981, Sources of atmospheric methane: Measurements in rice paddies and a discussion, *J. Geophys. Res. 86*, 7203-7209.

DeHaven, D. A., 1980, *CO and CH_4 sources and their effects on the CO-CH_4 budgets*, M. S. Thesis, Department of Chemical Engineering, University of Kentucky, Lexington, KY.

Fishman, J. and P. J. Crutzen, 1978, The distribution of the hydroxyl radical in the troposphere, Atmos. Sci. Paper No. 284, Department of Atmospheric Science, Colorado State University, Fort Collins, CO.

Garnett, A., 1979, Nitrogen oxides and carbon monoxide air pollution in the city of Sheffield, *Atmos. Environ. 13*, 845-852.

Hales, G. and P. Warneck, 1981, Background NO_x mixing ratios in air masses over the North Atlantic ocean, *J. Geophys. Res. 86*, 7283-7290.

Hanst, P. L., J. W. Spence, and E. O. Edney, 1980, Carbon monoxide production in photooxidation of organic molecules, *Atmos. Environ. 14*, 1077-1088.

Harriss, R. C. and D. I. Sebacher, 1981, Methane flux in forested freshwater swamps of the southeastern United States, *Geophys. Res. Lett. 8*, 1002-1004.

Heidt, L. E., J. P. Krasnec, R. A. Lueb, W. H. Pollock, B. E. Henry, and P. J. Crutzen, 1980, Latitudinal distributions of *CO* and CH_4 over the Pacific, *J. Geophys. Res. 85*, 7329-7336.

Heinrich, J. C., P. S. Huyakorn, O. C. Zienkiewicz, and A. R. Mitchell, 1977, An upwind finite element scheme for two-dimensional convective transport equation, *Int. J. Num. Meth. Engng. 11*, 131-143.

Kasibhatla, P. S., 1984, *Some aspects of regional scale air pollution modeling*, M. S. Thesis, Department of Chemical Engineering, University of Kentucky, Lexington, KY.

Kitada, T. and L. K. Peters, 1982, A three-dimensional transport-chemistry analysis of *CO* and CH_4 in the troposphere, Proceedings of the 2nd Symposium on the Composition of the Nonurban Troposphere, May 25-28, Williamsburg, VA.

Kley, D., J. W. Drummond, M. McFarland, and S. C. Liu, 1981, Tropospheric profiles of NO_x, *J. Geophys. Res. 86*, 3153-3161.

McRae, G. J., W. R. Goodin, and J. H. Seinfeld, 1982, Numerical solution of the atmospheric diffusion equation for chemically reacting flows, *J. Comp. Physics. 45*, 1-42.

Mitchell, A. R., 1969, *Computational Methods in Partial Differential Equations*, John Wiley & Sons, New York.

Peters, L. K. and A. A. Jouvanis, 1979, Numerical simulation of the transport and chemistry of CH_4 and *CO* in the troposphere, *Atmos. Environ. 13*, 1443-1462.

Platt, U., D. Perner, and H. W. Patz, 1979, Simultaneous measurement of atmospheric CH_2O, O_3, and NO_2 by differential optical absorption, *J. Geophys. Res. 84*, 6329-6335.

Reichle, H. G., V. S. Connors, J. A. Holland, W. D. Hypes, H. A. Wallic, J. C. Casas, B. B. Gormsen, M. S. Saylor, and W. D. Hesketh, 1986, Middle and upper tropospheric carbon monoxide mixing ratios as measured by a satellite-borne remote sensor during November 1981, *J. Geophys. Res. 91*, C9, 10865-10887.

Seiler, W. and J. Fishman, 1981, The distribution of carbon monoxide and ozone in the free troposphere, *J. Geophys. Res. 86*, 7255-7265.

Seiler, W. and P. J. Crutzen, 1980, Estimation of gross and net fluxes of carbon between the biosphere and the atmosphere from biomass burning, *Climate Change 2*, 207-247.

Seiler, W., 1974, The cycle of atmospheric *CO*, *Tellus 26*, 116-135.

Tadaki, Y., 1977, Leaf biomass, *Primary Productivity of Japanese Forests - Productivity of Terrestrial Communities*, T. Shidei and T. Kira, eds., University of Tokyo Press, Tokyo.

Whittaker, R. H. and P. L. Marks, 1975, Methods of assessing terrestrial productivity, *Primary Productivity in the Biosphere*, H. Leith and R. H. Whittaker, eds., Springer-Verlag, New York.

Yamada, T. and G. Mellor, 1975, A simulation of the Wangarra atmospheric boundary layer data, *J. Atmos. Sci. 32*, 2309-2329.

Yanenko, N. N., 1971, *The Method of Fractional Steps*, Springer-Verlag, New York.

Zimmerman, P. R., R. B. Chatfield, J. Fishman, P. J. Crutzen, and P. L. Hanst, 1978, Estimates on the production of CO and H_2 from oxidation of hydrocarbon emissions from vegetation, *Geophys. Res. Lett. 5*, 679-682.

TRACER TRANSPORT IN THE HAMBURG CLIMATE MODEL

Johann Feichter, Erich Roeckner, Ulrich Schlese, and Martin Windelband

Meteorological Institute, University of Hamburg/FRG

Abstract

A tracer transport scheme was implemented in a global general circulation model (Hamburg version of the ECMWF weather prediction model) to simulate the transport and chemistry of tracer constituents in the troposphere. Transport of tracers by advection, vertical diffusion and shallow convection is treated similarly to the transport of water vapor. A new scheme was developed for vertical mixing in deep convective clouds consistent with the Kuo-scheme used in the GCM. In order to investigate the large scale transport and the interhemispheric exchange simulation runs were performed with the chlorofluorocarbon F11. The short-lived radioactive rare gas ^{222}Radon was used to demonstrate the role of subgrid scale processes in the dispersal of tracers.

Introduction

During the last decade an increasing interest in tropospheric chemistry led to the development of several 3-dimensional tracer transport models. Two different ways can be taken to simulate a realistic global tracer distribution; Either an additional variable is introduced in a GCM (on-line model), as it was done in this work, or an independent transport model (off-line model) is driven by model generated or observed meteorology. Much lower computer costs are the great advantage of off-line models and the possibility to use observed wind fields which may be more reliable than model generated ones. Therefore, the off-line approach was often employed for tracer experiments (Mahlmann and Moxim,1978; Covey et al., 1985; Malone et al., 1986; Prather et al., 1987; Zimmermann et al., 1989). On-line models were used until now mainly for tracer studies in the middle atmosphere as by Hunt and Manabe (1968), Cunnold et al. (1975), Schlesinger and Mintz (1979), and Rose and Brasseur (1985). An advantage of on-line models is the high spatial and temporal resolution important for a realistic simulation of short-lived trace gases. Furthermore they may provide information such precipitation, liquid water content, cloud cover, and the diurnally varying boundary layer exchange, while off-line models must determine these processes from sparser meteorological data than were present in the original meteorological model.

To examine the global and regional transport we performed two tracer studies with chemical inert species. Firstly we modelled the distribution of the industrial chlorofluorocarbon $CFCl_3$ (F11), which provides a test for interhemispheric transport and regional surface distribution. Emission rates for F11 are quite well known and continual surface observations (ALE data) are available. A second tracer simulation is presented with the short-lived radioactive rare gas ^{222}Radon (half-life time = 3.83 days), to visualize subgrid scale processes like the upward transport in convective clouds or the dispersal of trace gases in the planetary boundary layer. The model is run in a climatological mode so that a comparison with data can only be made on the basis of integrations of more than a month. Our long-term objective in developing this tracer model is to include the tropospheric chemistry and further to gain more insight into the radiative-dynamical feedback between chemistry and meteorology.

Model Description

ECHAM is the Hamburg low resolution version (T-21) of the European Center for Medium Range Weather Forecast (ECMWF) model adopted for climate studies (Dümenil and Schlese, 1987). Parts of the model physics in the weather prediction model are constrained to present-day climate conditions or failed to simulate a realistic climatology, therefore parts of the physical parameterization were changed to include the following (Roeckner et al., 1990):

- Radiative transfer based on the two-stream approximation, optimized by using exact line-by-line calculations in the infrared part (Rockel et al., 1986). The cloud water path predicted by the model is used to parameterize the cloud optical properties.

- Cloud formation and dissipation, including subgrid scale effects and cloud water transport equation (Sundquist, 1978) with a temperature dependent partitioning of ice and liquid water content.

- Horizontal dissipation was reformulated (Boer et al.,1984; Laursen and Eliasen, 1988), to allow a more scale selective dissipation.

- The heat transfer within the soil is simulated by numerically integrating the diffusion equation in a five-level model with total depth of 10 m and assuming zero heat flux at the bottom.

- The soil moisture scheme is based on catchment considerations and accounts also for drainage and percolation.

Presently the spectral part of the model resolves waves up to wavenumber 21 and the physics part uses a horizontal resolution of 5.625 degrees. The vertical domain is subdivided into 19 levels covering the atmosphere from the surface to 10 hPa. The time step used in this version is 40 minutes. The CPU time on a CRAY-2 to simulate one year is about 400 minutes for meteorology and an additional 40 minutes for each tracer.

The tracer transport scheme in ECHAM is coded so that it treat an arbitrary number of chemical species at the same time. The advective and diffusive transport and the transport by shallow convection is handled in the same way as the water transport. The vertical exchange in deep convective cells is based on the calculation of a mean updraft vertical velocity within a cumulus cloud according to the method by Kuo (1974):

$$a\overline{\omega_c} = -\frac{M_c}{M_a}\frac{\Theta_c - \overline{\Theta}}{\frac{\partial\Theta_c}{\partial p}}, \qquad (1)$$

where

a fractional cloud cover

Θ potential temperature, subscript c for the value in the cloud, overbar denotes large scale values

$\frac{\partial \Theta_e}{\partial p}$ vertical gradient of the equivalent potential temperature in the unstable layer

M_c net convergence of moisture in a vertical column

M_a moisture amount available to create a convective cloud

$$M_a = \frac{1}{g} \int_{p_{bottom}}^{p_{top}} [\frac{c_p}{L}(T_c - \overline{T}) + (q_{sat}(T_c) - \overline{q})]dp.$$

Loss of air mass in the lower levels of the cloud by rising motion and gain of air in the upper levels is determined by the vertical gradient of $a\overline{\omega_c}$. This step is followed by subsidence beginning at the top of the cloud which maintains the large scale mass balance.

A great disadvantage of the spectral approach used in the ECHAM model is the occurrence of negative mixing ratios in the case of large horizontal gradients. Less noticeable, but equally serious, are regions of overshoot. The over/undershoot can be attributed to three problems with the spectral representation: The incorrect representation of discontinuities by smooth global basic functions, spectral truncation, and dispersion errors. There is an obvious need for a positive advection scheme in order to simulate tracers affected by photochemical reactions. Therefore, it is planned to implement a semi-Lagrangian scheme designed by Williamson and Rasch (1987), which avoids negative mixing ratios.

F11 experiments

Emission of F11 to the atmosphere is exclusively anthropogenic. We adopted the release scenario by Prather et al. (1987), who assumed a uniform emission throughout the year. The estimate of total production is given by the Chemical Manufacturers Association and regionally distributed in proportion to the consumption of electricity. Loss rates of F11 due to photochemical reactions in the stratosphere based on 1-dimensional model calculations and varying with latitude and season were taken from Prather et al. (1987) (corresponding to a life time of 53 yr). In order to save CPU time we used model results of a coarser model version as initial fields and performed a simulation run over two model years. The second year shown in this paper represents the year 1981.

During 1981 the yearly emission amounts to 260.7 10^6 kg (NH: 245.7 10^6; SH: 15.0 10^6), the stratospheric loss to 60.2 10^6 kg (NH: 29.2 10^6; SH: 31.0 10^6), which leads to an increase of the global F11 content of 5.5%. The global mass at the begining of year 1981 is 4039 10^6 kg, comparable to values given by Cunnold et al. (1986), who estimated a global mass of 3960 10^6 kg, but higher than the estimate of Prather et al. (1987) of 3813 10^6 kg. Nevertheless, they reported an underestimation of the global F11 burden by about 5% compared to values derived from measurements. The differences of the mass content of F11 between the two hemispheres show a pronounced seasonal variation (Fig. 1) with minima in March and July and a maximum in September. The average over the year is 52.9 10^6 kg. The interhemispheric flux (F_{N-S}) is determined as the monthly increase of F11 content in the southern hemisphere reduced by the emission and adding the stratospheric loss. The seasonal variation reflects mainly the migration of the tropical convergence zone with strong interhemispheric exchange during winter, when the Hadley cell crosses the equator and takes F11 enriched air into the southern hemisphere. Analyzing the 3-dimensional wind fields in spring and fall shows a Hadley cell on each hemisphere with inflow of air mass near the surface in the ITCZ, rising motion and poleward flux in the upper troposphere. This configuration leads to weak interhemispheric exchange (see Fig. 1). Contrary in summer and winter months there is only a single but very strong Hadley cell with rising motion above the ITCZ and subsidence in the winter hemisphere. In winter we therefore find an influx of F11 enriched air near the surface from the northern hemisphere into the region with rising motion. Most of this air is transported back in higher levels, but a small fraction flows southward and enhances in the higher troposphere the F11 content of the southern hemisphere. In summer indeed the southward transport in the upper troposphere is far stronger than in winter, but most of the air converging into the ITCZ is F11 poor air from the southern hemisphere and consequently the interhemispheric exchange is weaker in June than in December and January.

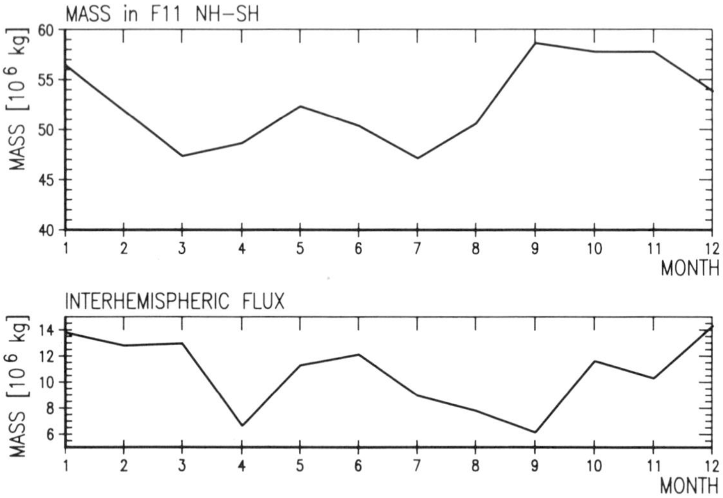

Figure 1. Monthly mean mass difference of F11 between northern and southern hemisphere (upper graph) and the monthly interhemispheric F11 net mass flux from the northern to the southern hemisphere (lower graph), both in 10^6 kg.

A mean exchange time τ between hemispheres may be defined by

$$\tau = \frac{M_N - M_S}{F_{N-S}}, \tag{2}$$

where M_N and M_S respectively denote the total mass of F11 in the northern and southern hemispheres. τ simulated by ECHAM has an annual mean value of 0.44 yr. This value is far lower than values given by other authors, e.g. 0.74 yr by Prather et al. (1987). As a consequence, a comparison of simulated surface mixing ratios with the ALE data (Fig. 2) shows a slightly stronger yearly increase in the southern hemisphere (Cape Grimm) than at locations in the northern hemisphere (Adrigole, Cape Meares) and suggests that the interhemispheric exchange simulated by ECHAM may be too fast. The ALE data have had pollution episodes removed, which we tried to simulate by eliminating values during easterly wind episodes at Adrigole and at Cape Meares and northerly wind episodes at Cape Grimm. These monthly mean values are compared with observations from the five ALE stations in Figure 2. The simulated seasonality at Adrigole and Cape Meares is too strong compared to the observations which may occur when average mixing ratios from a coarse grid are used close to the sources. The agreement between simulated and observed values is better far away from the main sources at Barbados, Samoa, and Cape Grimm.

Meridional cross sections of zonally averaged mixing ratios are shown for June and December in Figure 3. The distribution is very similar in both months with high values above the northern hemisphere sources, a well mixed atmosphere in the tropics and subtropics and a uniform distribution in the southern hemisphere with slightly higher values at upper levels. The position of the ITCZ is marked by a strong zonal gradient

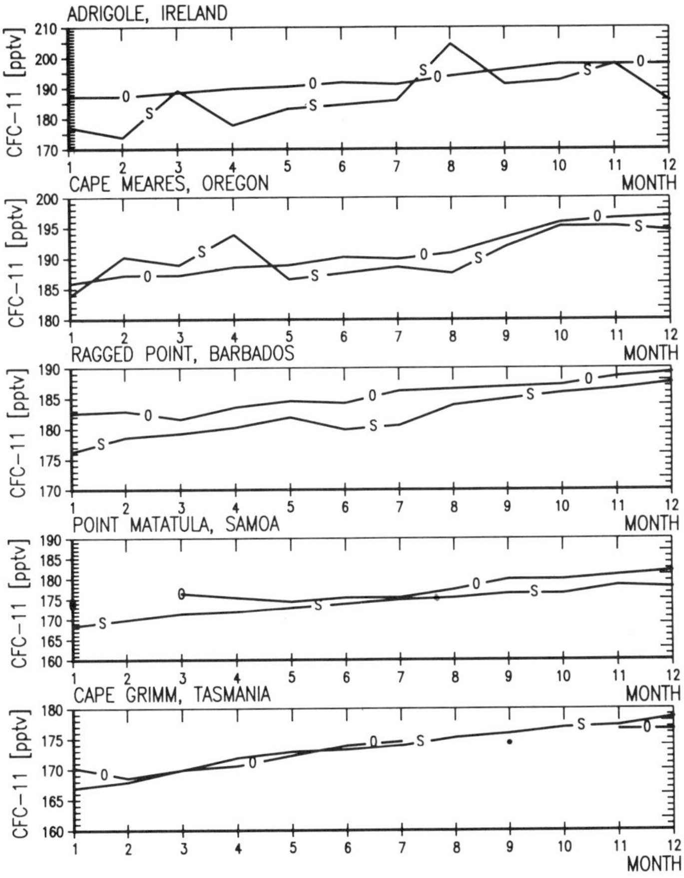

Figure 2. Comparison between observed surface values (ALE data, designated by o) and model results (s) in ppt_v.

near the surface between the equator and 5° S in December and between 15° N and 30° N in June. Below the tropopause we have a strong vertical gradient, as a result of the influence of surface sources and the weak exchange between troposphere and stratosphere. The values in the highest model level (10 hPa) are too high as compared to observations (Atmospheric Ozone, 1985) and suggest that the stratospheric loss is too weak or the exchange between troposphere and stratosphere is too strong.

^{222}Radon experiments

The radioactive rare gas ^{222}Radon is the first gas in the uranium decay chain and is mainly exhaled from land surfaces. As this process is the only source and as its half-life time is only 3.83 days, it is a suitable tracer to study the vertical mixing in the troposphere. In this tracer experiment we assumed an uniform exhalation rate of 1 atom per cm^2 and second at the surface over land.

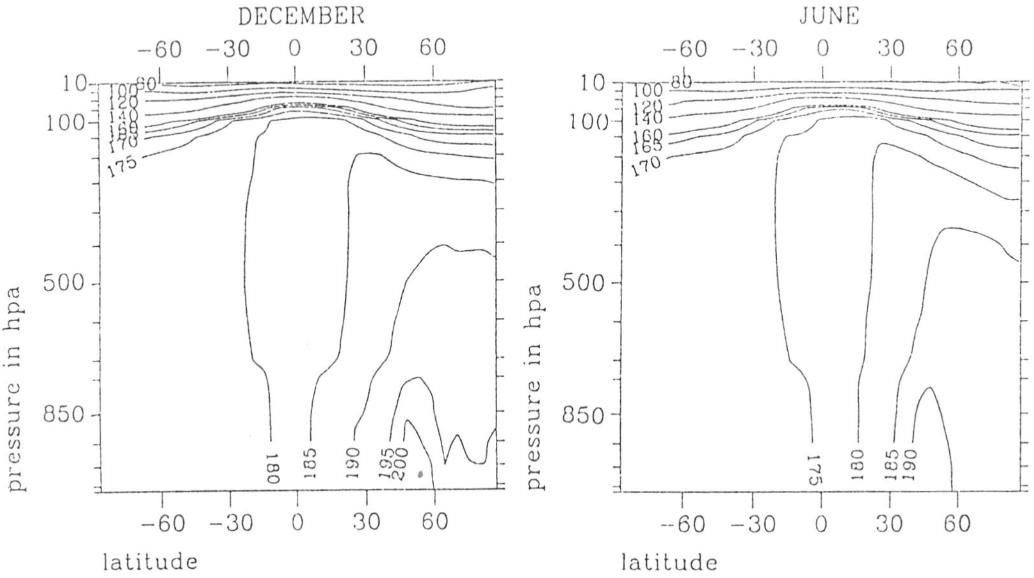

Figure 3. Meridional cross sections of zonally averaged mixing ratios (ppt_v) for December and June.

Liu et al. (1984) presented in a compilation of available vertical profiles measured at five locations in the USA and two in the USSR an averaged summer (derived from 23 soundings) and winter (7 soundings) profile, which is compared to our model simulations (Fig. 4). The winter and summer profiles are well represented by our model, but above 9 km the model overestimates the observations by about a factor of 2. The strong decrease of radon mixing ratios from the surface to about 3 km is reproduced by the model, but is less marked than in the observations. However, for a realistic test of vertical transport characteristics we would need more representative soundings of radon than are currently available.

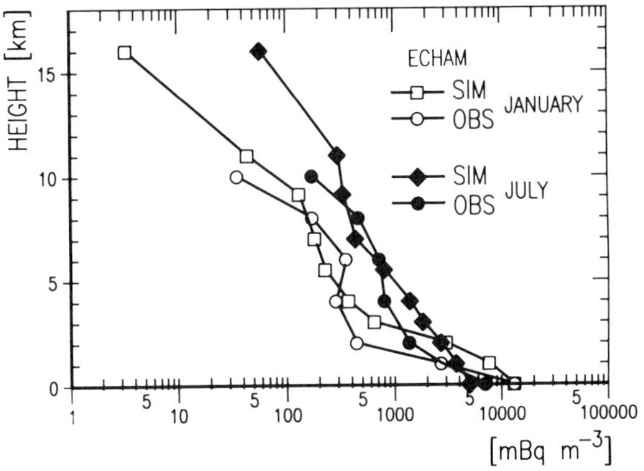

Figure 4. Comparison of observed vertical profiles of ^{222}Radon and model results in mBq m^{-3}_{STP} for summer and winter.

To visualize the vertical exchange near the surface the monthly mean diurnal variation of radon from the surface up to 500 hPa is presented in Figure 5 at Freiburg/FRG for July. During the night and in the morning we find a strong vertical gradient in the lowest 500 m and high radon values underneath. The boundary layer height is growing during the day up to 1500 m at the late afternoon and at the end of afternoon a stable layer is rapidly built up. The surface mixing ratios changes during the day by about a factor of 2, while layers above 2000 m are not affected by turbulence.

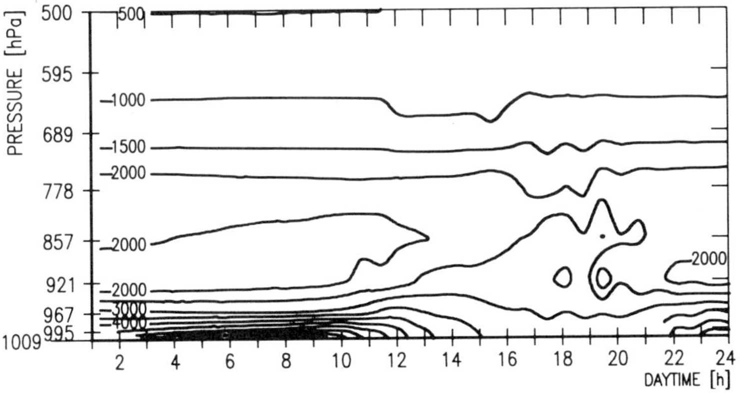

Figure 5. Monthly mean diurnal variation of ^{222}Radon in mBq m^{-3}_{STP}.

The vertical radon distribution should also be very sensitive to subgrid scale processes like the upward transport in convective clouds. Figure 5 presents the monthly and meridionally averaged radon mixing ratios by deep convective transport in January. Loss of tracer prevails in the lower atmosphere up to 600 hPa in the tropics and to 850 hPa in higher latitudes. In the tropics radon is transported mainly at altitudes between 400 and 200 hPa, but not higher than 100 hPa. The magnitude of changes of the mixing ratio is about 10 - 50 mBq m^{-3}_{STP} per day and is surprisingly high in high latitudes, although vertical mixing there is confined on the lowest 200 hPa.

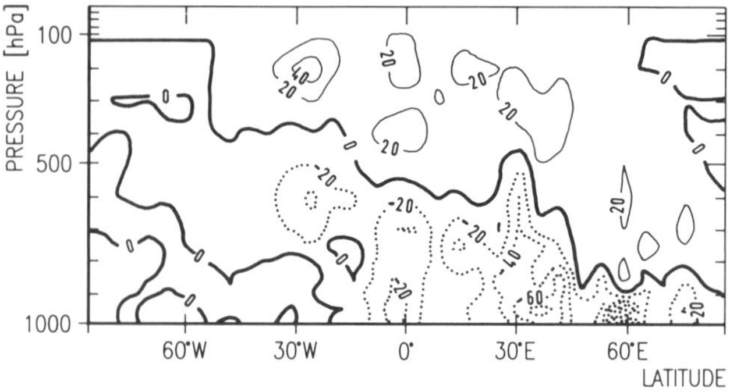

Figure 6. Monthly and meridionally averaged change of the ^{222}Radon mixing ratios in mBq m_{STP}^{-3} per day in convective clouds.

Monthly averaged mixing ratios for the 200 hPa level clearly show the regional distribution of convective activity in January with maximum mixing ratios above the Amazonas region, the southern parts of Africa and Australia. Surface values in these regions are higher only by a factor of 6, while the ratio between the winterly high surface values in Siberia and the 200 hPa values above increases to a factor of 1000.

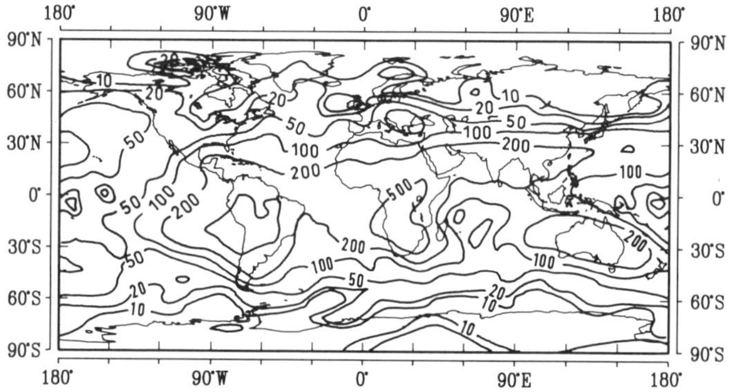

Figure 7. Monthly averaged ^{222}Radon mixing-ratios in mBq m_{STP}^{-3} in 200 hPa for January.

Conclusion

In order to demonstrate the transport characteristics of the ECHAM model tracer simulations with the long-lived species F11 and the short-lived radioactive tracer ^{222}Radon were performed. The model provides a good simulation of the F11 surface mixing ratios, although the interhemispheric exchange seems to be a little too strong. The model's performance related to vertical mixing in the planetary boundary layer and in convective clouds was illustrated in radon experiments and showed reasonable results. Vertical profiles of radon observations agree quite well with simulated ones, as far as we can expect agreement between results of a model running in a climatological mode and measurements reflecting a special meteorological situation. As the temporal resolution of the model is quite high, the model also simulates synoptic scale episodes and the diurnal variation. An examination of how realistic these variations are simulated, will be studied in further tracer experiments.

References

Atmospheric Ozone, Vol II, WMO, Global Ozone Research and Monitoring Project - Rep. NO.16, p.634.

Boer, G. J., N. A. McFarlane, R. Laprise, J. D. Henderson, and J.-P. Blanchet, 1984,The Canadian Climate Centre spectral atmospheric general circulation model, Atmosphere-Ocean 22, 397-429.

Covey, C., S. L. Thompson, and S. Schneider, 1985, Nuclear winter: a diagnosis of atmospheric general circulation simulations, J. Geophys. Res. 90, 5615-5628.

Cunnold, D., F. N. Alyea, N. Phillips, and R. G. Prinn, 1975, A three-dimensional dynamical chemical model of atmospheric ozone, J. Atmos. Sci. 32, 170-194.

Cunnold D., R. G. Prinn, R. A. Rasmussen, P. G. Simmonds, F. N. Alyea, C. A. Cardelino, A. J. Crawford, P. J. Fraser, and R. D. Rosen, 1986, Atmospheric lifetime and annual release estimates for $CFCl_3$ and CF_2Cl_2 from 5 years ALE data, J. Geophys. Res. 91, 10797-10817.

Dümenil L., and U. Schlese, 1987, Description of the general circulation model, In: Climate simulation with the ECMWF T21-model in Hamburg (ed. G. Fischer). Rep. Nr. 1, Meteorologisches Institut, University of Hamburg/FRG.

Hunt, B. G., and S. Manabe, 1968, Experiments with stratospheric general circulation model, II. Large-scale diffusion of tracers in the stratosphere, Mon. Weather Rev. 96, 503-539.

Kuo, H. L., 1974, Further studies of the parameterization of the influence of cumulus convection on large-scale flow, J. Atmos. Sci. 31, 1232-1240.

Laursen, L., and E. Eliasen, 1988, Sensitivity of a spectral general circulation model to changes in the parameterizations of the mechanical processes, Workshop on Systematic Errors in Models of the Atmosphere, Rep. No. 12, WMO/TD-No.273, 207-214.

Liu, S. C., J. R. McAffee, and R. J. Cicerone, 1984, Radon-222 and tropospheric vertical transport, J. Geophys. Res. 89, 7291-7297.

Mahlmann, J. D., and W. J. Moxim, 1978, Tracer simulation using a global general circulation model: results from a mid-latitude instantaneous source experiment, J. Atmos. Sci. 35, 1340-1378.

Malone R. C., L. H. Auer, G. A. Glatzmaier, M. C. Wood, and O. B. Toon, 1986, Nuclear winter: Three-dimensional simulations including interactive transport, scavenging, and solar heating of smoke, J. Geophys. Res. 91, 1039-1053.

Prather, M., M. McElroy, S. Wofsy, G. Russell, and D. Rind, 1987, Chemistry of the global troposphere: Fluorocarbons as tracers of air motion, J. Geophys. Res. 92, 6579-6613.

Rockel, B., B. Zhao, and E. Raschke, 1986, A flexible radiative transfer routine for GCM's: infrared part, In: G. J. Boer(ed.): Research activities in atmospheric and oceanic modelling, Rep. No. 9, WMO/TD-No. 141, 4.62-4.65.

Roeckner, E., L. Dümenil, E. Kirk, F. Lunkeit, M. Ponater, B. Rockel, R. Sausen, and U. Schlese, 1990, subm. to Research activities in atmospheric and oceanic modelling, CAS/JSC Working Group on Numerical Experimentation, WMO.

Rose, K., and G. Brasseur, 1985, Ozone during sudden stratospheric warmings: A three-dimensional simulation, In: Atmospheric Ozone, C. S. Zerefos and A. Ghazi (ed.), pp. 28-32, D. Reidel, Dordrecht.

Schlesinger, M. E., and Y. Mintz, 1979, Numerical simulation of ozone production, transport and distribution with a global general circulation model, J. Atmos. Sci. 36, 1325-1361.

Sundqvist, H., 1978, A parameterization scheme for non-convective condensation including prediction of cloud water content, Quart. J. Roy. Meteor. Soc. 104, 677-690.

Williamson, D. L., and P. J. Rasch, 1987, Shape preserving interpolators for semi-Lagrangian transport, Workshop Proceedings: Techniques for horizontal discretization in numerical weather prediction models, ECMWF, Reading/GB.

Zimmermann, P. H., J. Feichter, H. K. Rath, P. J. Crutzen, and W. Weiss, 1989, A global three-dimensional source receptor model investigation using ^{85}Kr, Atmos. Environment 23, 25-35.

DISCUSSION

JOSEPH C. CHANG: How sensitive are the results to the schemes used to parameterize deep cumulus convection?

J. FEICHTER: I have not tested different schemes. The method by Kuo for the calculation of mass fluxes in deep convective clouds was chosen in order to treat the tracer transport as similar as possible to the transport of water vapour.

LONG RANGE TRANSPORT OF TRACE GASES AND AEROSOLS IN THE NORTHEAST PACIFIC RIM

V. K. Kotamarti*, Y. Sunwoo*, G. R. Carmichael*, Hiromasa Ueda**, and Hidemi Kurita***

*Department of Chemical & Biochemical Engineering
University of Iowa
Iowa City, IA 52240

**National Institute for Environmental Studies
Tsukuba, Ibaraki 300-21, Japan

***Matsumoto Institute for Public Health
Matsumoto 390 Nagano Japan

INTRODUCTION

Chemicals released into the atmosphere can have many adverse impacts on the environment including: local air pollution problems such as smog; regional scale problems such as acid rain and reduced visibility; and global scale problems including the reduction in stratospheric ozone and global warming (due to the build-up of greenhouse gases). The regional and global scale pollution problems facing us today have taught us that neither pollutants, nor their effects, are confined to the locations where the chemicals enter the atmosphere. Rather through the atmospheric processes of long range transport the deleterious effects of pollutants can be propagated away from the urban and industrial centers and spread over widespread geographical regions. There is a clear need to improve our understanding of the atmospheric processes which distribute these chemicals throughout the troposphere, and their fate and impact throughout the "biofilm" (that layer of the earth's surface which supports all of human life; i.e., the 20-mile layer that extends from 15 miles above the earth's surface to 5 miles below it).

Understanding the long-range transport mechanisms is essential in assessing the fate and impact of anthropogenic chemicals emitted into the troposphere. The three regions of the globe with the highest emissions are the United States, Western Europe, and the Pacific Rim region of East Asia. Regional scale air pollution problems including acid deposition, regional oxidant and reduced visibility have been documented in the United States and Western Europe. The Pacific Rim region of East Asia containing China, Japan, Korea and Taiwan is characterized by high anthropogenic emissions, resulting from its high population density and energy consumption. Presently only local air quality problems have been studied in the region of Northeast Asia. However, the potential for regional scale and global problems associated with these emissions can be expected based on the large emissions and close geographical proximity between the major industrial and urban centers (e.g., Tokyo, Beijing, Taipei, Seoul, Hong Kong, etc.). The pollutants emitted from these countries can be expected to contribute in adverse ways to the background air quality of this region and to the tropospheric background composition throughout the Pacific Ocean region.

The long range transport of yellow sand is one example of a regional scale pollution problem affecting this region during the spring months. This problem has been studied (Uematsu et al., 1982; Iwasaka et al., 1983). The background aerosol measurements obtained through the SEAREX program (Prospero and Savoie, 1989) show that dust originating in the Gobi Desert in China can be transported into the central regions of the Pacific Ocean. Data at the Midway station indicates that elevated anthropogenic nitrate and sulfate concentrations accompany the transport of this dust. The springtime dust transport episodes are also often associated with periods of elevated tropospheric ozone in East Asia.

In order to quantify the impact of the continued population and economic growth in the Pacific Rim region of Northeast Asia, it is necessary to improve our understanding of the present condition of the regional troposphere and the sources, sinks, transport and transformation processes controlling the distribution of pollutants in this region. Due to the combination of stretches of sea surrounding the land masses, and because of the different lifestyles and technologies used in the various east Asian countries, the chemical nature and deposition processes may be quite different from those previously studied in long range transport episodes in the eastern United States or western Europe. We have begun to study the issues of regional scale pollution in the Northeast Pacific Rim. The modeling approach being used in this study and preliminary results are presented in this paper.

DESCRIPTION OF THE STUDY REGION AND AVAILABLE DATA

The region under study is shown in Figure 1 and includes China, and all of Korea, Japan, and a small piece of southeast USSR. This region is of particular interest because it includes major industrial and population centers of several countries; and each of these Pacific Rim countries are experiencing rapid economic growth. This region is of further interest in regards to long range air pollution transport because of the complex topography and meteorology of the region, and the long transport distances over the sea.

As mentioned in the introduction section, there is a developing data set on regional scale distribution of several tropospheric species. Aerosol data exists from the SEAREX program as well as from a series of aerosol monitoring networks within Japan (NIES report, 1989). These include three sampling sites on the northern coast of Japan and two on Oki Island located ~ 100 km north of the mainland of Japan. Five years of aerosol data are available for 1984 through 1988.

Over the last five years or so there has also developed a rich data set on ozone in this region. Data on surface O_3 at remote stations are available at the 11 sites shown in Figure 1. Records for hourly ozone over the past three to five years are available. In addition there are 7 "World Ozone Data Center Stations" located throughout this region. At these sites a variety of measurements ranging from total ozone (by Dobson spectrophotometer) and vertical profiles from Umkehr measurements and ozonesondes are available on temporal scales ranging from hours to days.

In addition total ozone (tropospheric plus stratospheric) measured from satellites on a daily basis are also available over this region. Total ozone from the TOMS instrument aboard Nimbus 7 is available from 1979. The instrument works on the principle of backscattered ultraviolet radiation, and provides measurements near local noon at spatial resolutions down to 100km.

Measurements of stratospheric ozone are also available from the SAGE I and SAGE II instruments aboard the Applications Explorer Mission satellite and the Earth Radiation Budget satellite, respectively. Both experiments use the solar occulation technique to measure the extinction of atmospheric trace gases and aerosols during local spacecraft sunrise and sunset. Ozone measurements are available at ~1 km vertical resolution from cloud top to roughly 60 km.

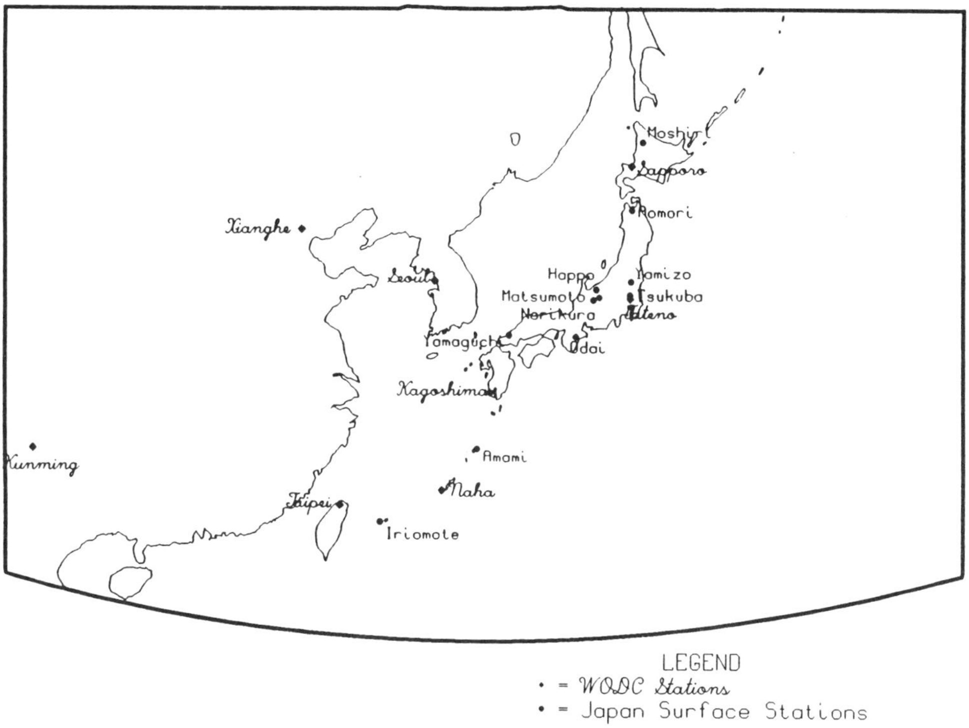

Figure 1. The Northeast Pacific Rim region under study. The locations of the surface ozone stations are identified.

Investigations of the long range transport of chemically active air pollutants in this region have only recently been initiated by investigators in Japan (Kondo et al., 1988; Ueda et al., 1988). Intensive field studies investigating long range transport of pollutants in southern Japan were recently conducted by the Japanese National Institute for Environmental Studies. Detailed chemical and meteorological observations were obtained during periods in May, 1987 and July, 1988. The meteorological conditions in May, 1987 were such that the flow was from Kyushu towards southern Korea. Concentrations of SO_2 in this plume in excess of 150 ppb were detected at Tsushima Island (located ~ 100 km north of Kyushu and ~ 80 km south of Pusan, South Korea. The SO_2 plume is due to the Sakura Jima volcano located in southern Kyushu. Prevailing winds on this day were south-easterly transporting pollutants toward the southern coastline of Korea.

Ozone in the Northeast Pacific Rim

Recently Fishman developed a method of estimating tropospheric ozone using the TOMS and SAGE data (Fishman and Larson, 1987). The residual tropospheric ozone can be determined by difference between the TOMS total ozone and stratospheric ozone measured by SAGE. The largest residual ozone values anywhere in the globe are found in the Northeast Pacific Rim region (Fishman et al., 1990).

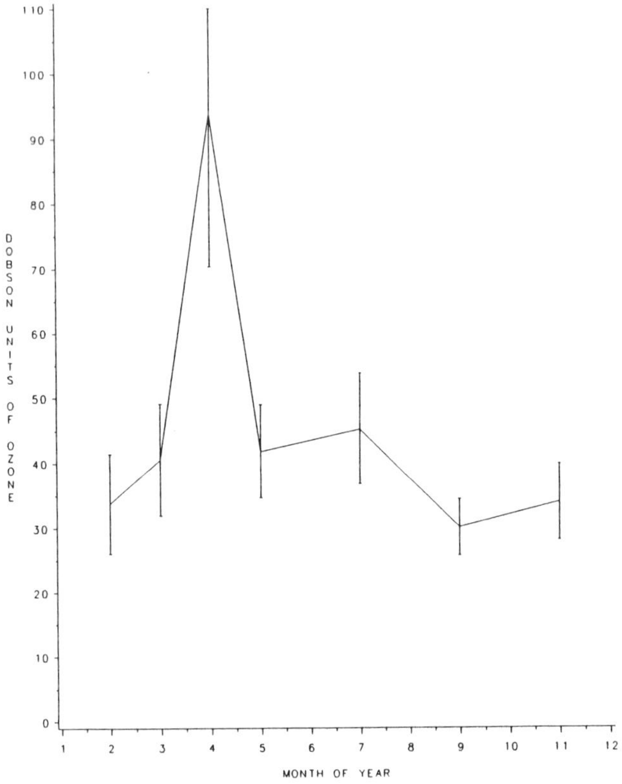

Figure 2. Residual tropospheric ozone values calculated over the Northeast Pacific Rim region for 1981.

Recently we have conducted an analysis of residual tropospheric ozone in this region using the TOMS and SAGE I data. The residual values as a function of season for 1981 are shown in Figure 2. There is a significant seasonal variation, with the maximum value occuring in the spring period. A similar seasonal variation is also observed in the surface ozone network in Japan (cf, NIES Report).

These high surface ozone values also occur on time-scales of several days at high elevation background sites. For example, observed ozone at Happo Ridge, located at an altitude of 1.8 km is shown over a two week period in May in Figure 3. Concentrations in excess of 90 ppb are observed. Furthermore this period of high ozone coincides with periods of high aerosol loadings resulting from the long range transport of dust from the deserts in central China.

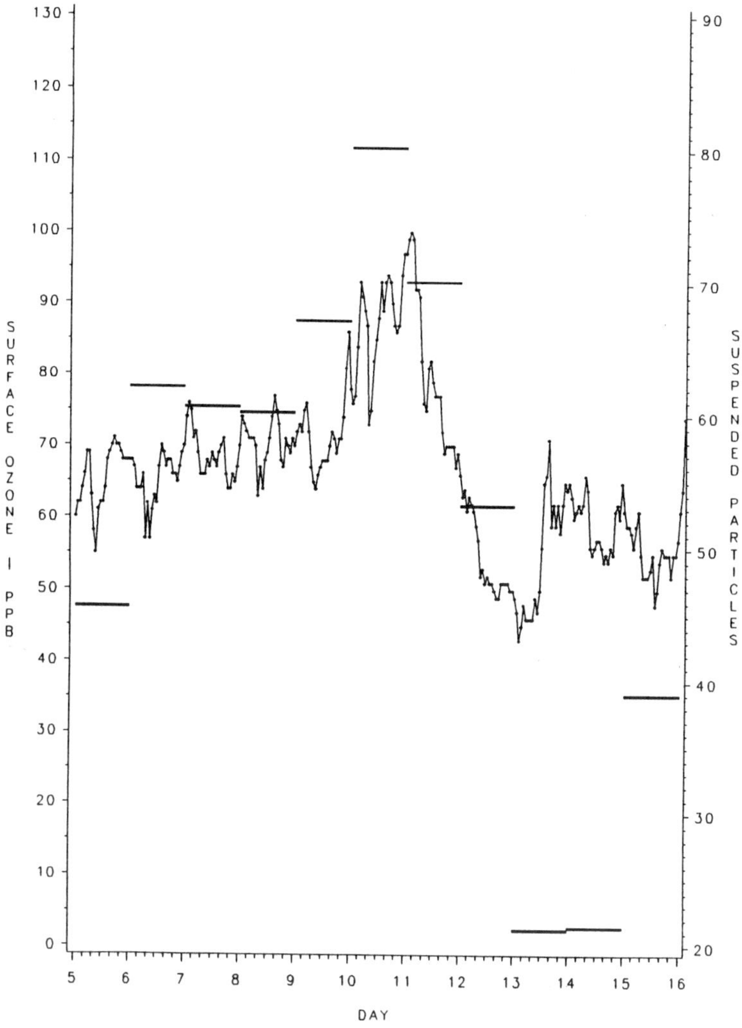

Figure 3. Surface ozone and suspended particulates measured at Happo ridge for May 5-16, 1987. The period of Kosa occurred during May 8-12, 1987.

MODELING ANALYSIS

The mechanisms causing these high ozone levels remain unquantified. Clearly large-scale subsidence is an important process (Wakamatsu, 1989). However, production of ozone in the troposphere may also be significant (Liu et al., 1987; Janach, 1989). In an attempt to quantify the processes leading to elevated ozone and to further assess the importance of long range transport processes in this region, a detailed modeling study has been initiated.

An advanced Eulerian combined transport/chemistry/removal model, referred to as STEM-II (Carmichael et al.,1986) is being used in this analysis. This model utilizes the updated gas phase reaction mechanism of Lurmann et al. (1986) which includes 112 chemical reactions and 53 chemical species. Of these species, 33 long-lived species are transported while the remaining 20 short-lived species are modeled using the psuedo steady state approximation. In addition, an equilibrium relationship for NH_4NO_3 aerosol is included in the analysis. More complete details of the STEM-II model can be found in Chang et al. (1989).

To study the behavior of the dust transport, an aerosol transport model has also been developed. Making use of a K & E model with an algebraic stress component (Uno et al., 1990) to describe the boundary layer structure, a conditioned particle trajectory analysis is imbeded into the trace gas calculations. The turbulence model is used along with the meteorological fields derived from the Japan Weather Service meteorological model to provide the input transport fields of flow and temperatures over the lowest 14 km of the atmosphere.

Emissions for the region have been compiled based on the work of Hameed and Dignon (1988). They have estimated global emissions due to energy consumption on a grid of 10° longitude by 8° latitude. We have used this data to prepare a regional scale emissions inventory for the domain shown in Figure 4 on a 80 km grid. These estimates were obtained by breaking-down the larger-scale emissions using population density as the weighting factor. The major industrial areas of Tokyo, Osaka, Seoul, Pusan, Beijing, and Shanghai are clearly depicted. The hydrocarbon (i.e., NMHC) emissions for this region were not available but were assumed to be 5-times (on a mole of carbon basis) those for NO_x.

These various model components have been formulated and initial analysis is underway. A preliminary assessment of the long range transport of pollutants in a 2-dimensional region utilizing a synthetic steady flow field has been performed. The horizontal slice chosen for analysis corresponds to a typical transport path during the spring (Merrill et al.,1985). (During the spring, the flow in this region is typically west to east.) This domain stretches from Mongolia to the west and passes through Beijing, Seoul, Pusan, and Osaka.

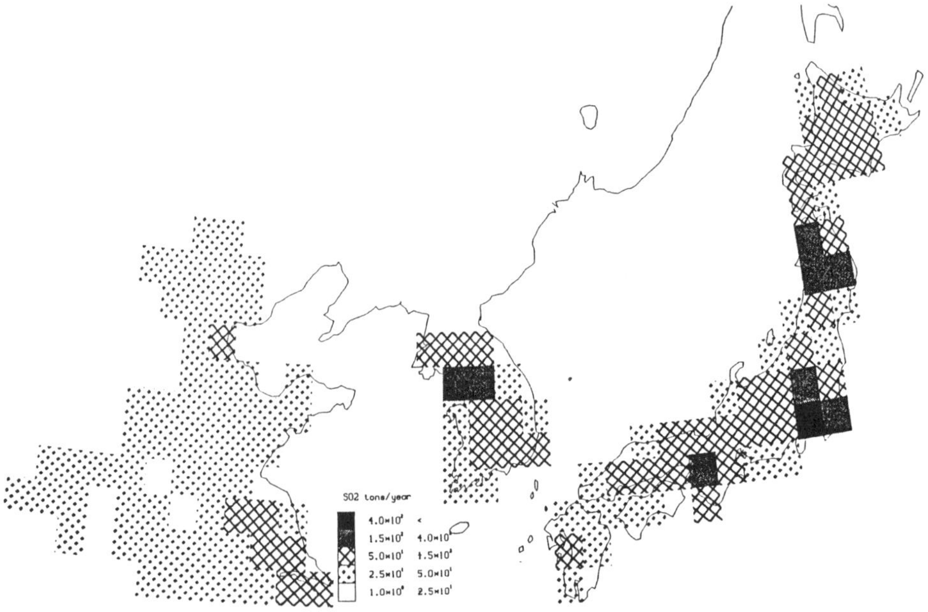

Figure 4. SO_x emissions as SO_2 for the study region estimated from Hameed and Dignon(1988).

The emissions along this path increase from China to Korea to Japan. The terrain in this region is complex showing 1.5 km plateau in Mongolia and the mountainous topography of the Korean peninsula and the main Honshu island of Japan. The eddy diffusivity profiles used in the analysis varied diurnally, and different profiles were used over land and sea surfaces. The profiles over land surfaces were derived using a boundary layer model developed by Yamada and Mellor (1975) with a maximum mixing height of 1.2 km. Over sea surfaces a mixing height of 400 m was assumed with stable stratification above the mixed layer. A uniform surface temperature of 25 ^oC throughout the domain was assumed. Relative humidity was set at 60% over land and 80% over sea with a vertical profile based on the psuedo-adiabatic chart. The dry deposition velocities used in the analysis also varied over land and sea surfaces were selected based on the detailed analysis of Kitada et al. (1986). Most species show a lower deposition velocity over the sea than over the land surfaces due to the lower surface roughness and more stable boundary layer structure over the sea. The initial conditions consisted of uniform distributions with 1 ppb of NO_x, 0.6 ppb of SO_2, 0.6 ppb of sulfate, 30 ppb of O_3, 1.4 ppb of methane, 6 ppb of NH_3, 80 ppb of CO, and 0.6 ppb of ethane.The boundary conditions were assumed equal to the initial conditions. Solar conditions for April 20th were assumed.

The predicted ozone concentrations in the air mass as it leaves Beijing at 10:00 LT on day-1 is ~30 ppb. By the time this air mass reaches the western coast of Korea the ozone levels have increased to 70 ppb. As this air mass passes over the NO_x sources of Korea the ozone levels increase to ~125 ppb (see the profiles at grid points 23 & 25). The ozone concentrations decrease as the air mass passes over the sea between Korea and Japan during the nighttime predominantly by thermal chemical reactions. This air mass as it reaches Japan has an ozone concentration of ~ 80 ppb. In contrast to ozone, the NO_2 values show little change during the nighttime transport over the sea. The NO_2 profiles under relatively low mixing conditions at 10:00 LT over a high source regions show a strong vertical gradient in the lowest 1 km; while the profiles over land during the afternoon show NO_2 distributed uniformly within a well mixed layer. The NO_2 levels above 2 km have values ranging from 20 to 100 ppt. These levels are consistent with the observations of Kondo et al. (1988) for similar flow conditions during the month of February.

CLOSURE

These results suggest that the long range transport of air pollutants may play a significant role in the air quality of this region, and may contribute to elevated spring ozone values. However these results must be considered preliminary and further studies are necessary to better quantify the importance of long range transport of pollutants in this part of the Pacific Rim. These studies are now underway. These studies in the Pacific Rim will improve our understanding of the transport and chemical characteristics of the regional Asian plume. The modeling analysis will provide an evaluation of the importance of long range transport of the Asian plume in this region, help quantify the contribution of tropospheric ozone production to the elevated tropospheric ozone levels found in spring periods in this region, provide calculated regional scale SO_x, NO_x, NO_y, and reactive hydrocarbon distributions, and provide estimates of deposition to regions within and around Korea. This information should prove valuable to future studies in this region.

REFERENCES

Carmichael, G., L. Peters, and T. Kitada, 1986, A second generation model for regional-scale transport/chemistry/deposition, Atmos. Environ., 20:173-188.

Fishman, Jack and Larsen, Jack C., 1987, Distribution of total ozone and stratospheric ozone in the tropics: Implications for the distribution of tropospheric ozone, Journal of Geophysical Research, 92:6627-6634.

Fishman, Jack, Watson, Catherine E., Larson, Jack C., and Logan, Jennifer A, 1990, Distribution of tropospheric ozone determined from satellite data, Journal of Geophysical Research, 95:3599-3617.

Hameed, S. and J. Dignon, 1988, Changes in the geographical distributions of global emissions of NO_x and SO_x from fossil fuel combustion between 1966 and 1980, Atmos. Environ., 22:441-450.

Iwasaka, Y., H. Minoura, and K. Nagaya, 1983, The transport and spatial scale of Asian dust-storm clouds: A case study of the dust storm event of April 1979, Tellus, Ser. B, 35:189-196.

Janach, Walter E., 1989, Surface Ozone: Trend details, seasonal variations, and interpretation, Journal of Geophysical Research, 94:18289-18295.

Kondo, Y., W. Matthews, A. Iwata, Y. Morita, and M. Takagi, 1987, Aircraft measurements of oxides of nitrogen along the eastern rim of the Asian Continent: winter observations, J. Atmos. Chem, 5:37-58.

Kotamarthi, V., and G. R. Carmichael, 1989 (in press),.The long range transport of pollutants in the Pacific Rim Region, Atmos. Environ,.

Liu, S.C., M. Trainer, F.C. Fehsenfeld, D.D. Parrish, E. J. Williams, D. W. Fahey, G. Hubler, and P.C. Murphy, 1987, Ozone production in the rural troposphere and the implications for regional and global ozone distribution, Journal of Geophysical Research, 92:4191-4207.

Lurmann, F.W., Lloyd, A.C., and Atkinson, R., 1986, A chemical mechanism for use in long-range transport/acid deposition computer modeling, Journal of Geophysical Research, 92:9740-9752.

Merrill, J., M. Uematsu, and R. Bleck, 1989, Meteorological analysis of long range transport of mineral aerosols over the North Pacific, J. Geophy. Res., 94:8584-8598.

NIES, 1989, Studies on the Methods for Long-Term Monitoring of Environmental Pollutants in the Background Regions and Atmospheric Pollutants on the Remote Island and Mountains: Concentrations and Variations, Research Report from the National Institute for Environmental Studies, Japan #R-123.

Prospero, J.M. and D.L. Savoie, 1989, Nitrate concentrations over the Pacific Oceans background and continental impacts, Nature,.

Ueda, H., 1988, National Institute for Environmental Studies, Japan, private communication.

Uematsu, M., R. Duce, J. Prospero, L. Chen, J. Merrill and R. McDonald, 1983, Transport of mineral aerosols from Asia over the North Pacific Ocean, J. Geophy. Res, 88:5343-5352.

Uno, I. Ueda, H., and S. Wakamatsu, 1990 (in press), Numerical modeling of the nocturnal urban boundary layer, Boundary Layer Meteorology.

Wakamatsu, S., Uno, Itsushi, Ueda, Hiromasa, and Uehara, Kiyoshi, 1989, Observational study of stratospheric ozone intrusions into the lower troposphere, Atmos. Environ., 23:8, 1815-1826.

Yamada, T. and G. Mellor, 1975, A simulation of the Wangara atmospheric boundary layer data, J. Atmos. Sci., 32:2309-2329.

VERTICAL TRANSPORT AND SCAVENGING OF POLLUTANTS BY CONVECTIVE CLOUDS

M. Laube, A. Ebel, E. Kälicke and P. Scheidgen

Universität zu Köln
Institut für Geophysik und Meteorologie
Albertus-Magnus-Platz
D-5000 Köln 41
Federal Republic of Germany

Introduction

The redistribution and transport of chemical substances in the atmosphere by convective systems are poorly understood problems in atmospheric motion. The most important chemical species originate in the atmospheric boundary layer or at the earth's surface eg. sulfur dioxide, carbon monoxide, carbon dioxide, methane, nitrogene dioxide and various complex hydrocarbons. The residence time of sulfur and nitrogen compounds are the shortest in the planetary boundary layer where they are deposited after oxidation. As discussed by Dickerson et al. (1987) their residence time increases significantly, if they are transported into the upper troposphere and spread over large distances by the faster windspeeds therein. Deep convection in the troposphere can also cause rapid downward transport of ozone for example by convective downdrafts originating in the middle or upper troposphere. The quantification of those meteorological processes has implications for the sub-grid scale representation of convective motions and transport of chemical species in large-scale or meso-scale models.

Convective clouds can perform both upward and downward transport of trace gases. The effect of vertical mixing in regional-scale acid deposition models has been parameterized by one-dimensional cloud-models of the troposphere, e.g. Walcek and Taylor, 1986. Studies suggest, that we need to know increasing details about redistribution by convective clouds. In deep convective clouds air is pumped into and out of the cloud at low and intermediate levels as well as at the top level of the cloud. We need to know, where a given parcel of air originates, and how long it spends in the cloud environment. Downdrafts may represent either upper tropospheric air as well as boundary layer air being recycled through the cloud. We attempt to address some of these questions by using a two-dimensional time dependent dynamic and microphysical model.

Air Pollution Modeling and Its Application VIII, Edited by H. van Dop and D. G. Steyn, Plenum Press, New York, 1991

The Cloud Model

The model is nonhydrostatic and prognostic in that it yields the distribution of the horizontal and vertical wind components and the fields of temperature, air density, water vapor, cloud water, rain water and ice. The initial vertical profiles of temperature, water vapor and the horizontal wind component must be specified. Full details of the prognostic and diagnostic equations and the cloud microphysics have been given by Scheidgen and Laube (1989) and Scheidgen (1989). Therefore only a brief description of the microphysical processes will be given here.

The microphysical processes include condensation, deposition, evaporation, collection, freezing, riming and melting. The model calculates cloud water, rain water and ice mixing ratios. Water vapor, cloud water and ice are assumed to be advected with the airflow, but rain water and aggregates, while moving with the wind, are assumed to fall with their terminal velocity.

The model is two-dimensional in the x-z plane, where x is the horizontal and z the vertical coordinate. The horizontal direction is perpendicular to the long axis of the storm and the coordinate-system is assumed to move with mean velocity of the storm. So the model actually applies to the strong updraft center. The model domain is 45 km long and 15 km deep with a horizontal and vertical grid resolution of 333 m.

Transport and Scavenging of Chemical Species

The concentration of soluble gases in cloud water is assumed to be in Henry's law equilibrium with the species in the gas phase. The in-cloud scavenging of soluble particles is modeled only through the nucleation process. As the drop forms, all cloud condensation nuclei contained in the air will go into solution. Other processes, that could contribute to particle scavenging, are not considered. The collection of the cloud water concentration by rain or ice respectively is represented by the collection or freezing terms of the continuous formulation of microphysics.

Each chemical species in each microphysical constituent, e.g. cloud water, is represented by a mass continuity equation. The source and sink terms in the equations describe either the transfer of a pollutant from one cloud species to another, e.g. pollutant in cloud water to pollutant in ice by freezing, or a chemical reaction in the same cloud species, e.g. oxidation of sulfur dioxide to sulfate in rain water.

The model initial values are specified by vertical profiles of the chemical species and these fields are assumed to be initially horizontally homogeneous. Point or area sources at the earth's surface can be added. The tracers are calculated using the Smolarkiewicz positive definite advection scheme. The models outputs are the time dependent two-dimensional fields of the mass mixing ratios of the chemical species in the air, in the cloud water, in the rain water or ice respectively. A single gas species or soluble aerosol, respectively, only has been considered in order to investigate the impact of microphysics on transport and scavenging. No chemical reactions have been included into the model.

Results

The simulation is based on the 22 June soundings from the COPT 81 (convection profonde tropicale in 1981) experiment in West-Africa. The mayor features of the model results are a strong updraft region and a trailing anvil of stratiform ice clouds.

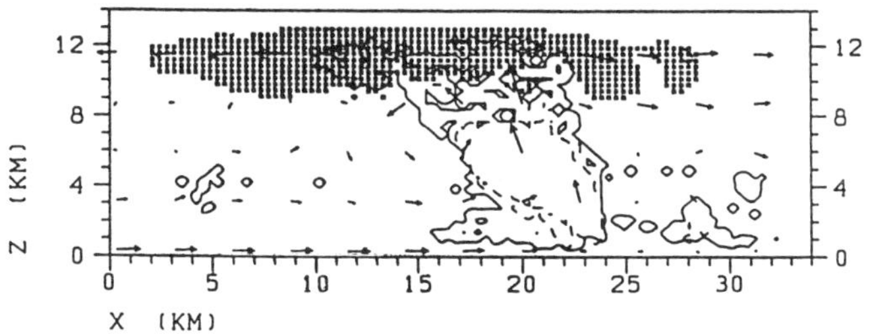

Fig. 1. Model computed cloud after 60 minutes. The arrows indicate direction and speed of wind. Full lines indicate the cloud water mixing ratio, dashed lines the rain water mixing ratio and asterisks the cloud ice mixing ratio >0.1 gr/kg. The storm is moving from left to right.

The lifetime of the cloud is about 120 minutes. The model cloud produces heavy rain in front of the region of the strongest updraft. Figure 1 shows the cloud model results at 60 minutes after the start of simulation. Most of the airflow into the cloud comes from the rear region (on the left side) within the height from 0 to 4 km. In the anvil region the air is detraining from the cloud into the front and rear parts of the cloud. The full lines, indicating cloud water content 0.1 gr/kg, are showing the contours of the cloud containing liquid water. In the upper region the ice cloud, indicated by asterisks, is spreading into both directions. Most of the ice particles originate from freezing of water drops.

In order to get more insight into the movement of air parcels, parcel trajectories have been calculated from the model predicted two- dimensional wind fields. The trajectories have been started at different locations. The trajectory paths are calculated from the evolving wind fields of the model. Each marker indicates the movement of an air parcel over a 2 minutes period. Examples of trajectories starting at different locations in 1 km height at the same time when the convection begins are shown in Figure 2. The air parcels started near the center of the cloud are entering the cloud after about 10 minutes and spend the rest of the time within the cloud. Air parcels starting 5 to 10 km away from the storm center are moving horizontally for a distance of 15 to 20 km and then are rising to the mid or upper level of the cloud. Figures 3 and 4 show examples of trajectories of air parcels starting at somewhat higher levels. These examples show that air parcels, starting close together, may have a different life history. The air parcels enter the cloud from the rear side, rise to the upper level and exit the cloud as well to the front region as to the rear region. But there are some air parcels starting in the same hight which never enter the cloud.

The concentration of a water soluble gas, e.g. CO, initially located in the boundary layer is shown in Figure 5 after 50 and 70 minutes of simulation, respectively. The gas enters the cloud in the updraft region and is transported upwards and diluted by the upper level wind. There are no gas phase concentrations outside the cloud. The reason is that all the CO in gas phase is able to enter the cloud drops. After a simulation time of about 50 minutes, a considerable amount of CO remains in the gas phase and is transported to the upper levels very rapidly.

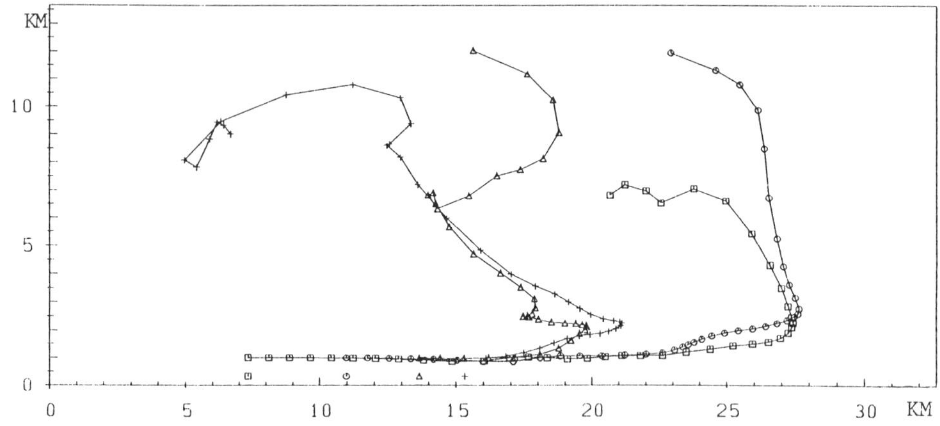

Fig. 2. Air parcel trajectory paths, calculated from the time dependent wind field, from 0 to 80 minutes. The markers indicate the movement of an air parcel over a 2 minutes period. Starting height of the air parcel is 1 km, the starting points are indicated.

Figure 6 shows the field of a passive tracer after a simulation time of 60 minutes. The tracer is emitted by an area source with constant source strength during the simulation time. Most of the tracer is transported upward through the updraft region, while a part is transported into the front region of the storm. The tracer concentration maximum is attained in the anvil region. The vertical profile of the horizontal average of the concentration shows the maximum at an altitude of about 10 to 11 km.

In Figure 7 the concentration field of a water-soluble aerosol, initially located in the layer between 1 and 1.66 km, after a simulation time of 60 minutes is quite different from the concentration of gaseous tracers. In the updraft region most of the aerosol has entered the cloud water, only a small amount remains in the air. Since due to the modelling approach the aerosol is captured by the nucleation process only, the aerosol

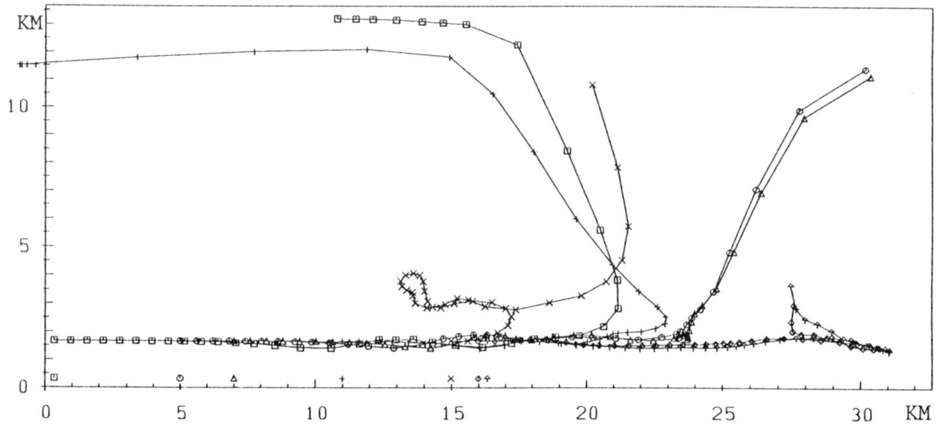

Fig. 3. Same as Fig.2 except the starting height is 1.66 km.

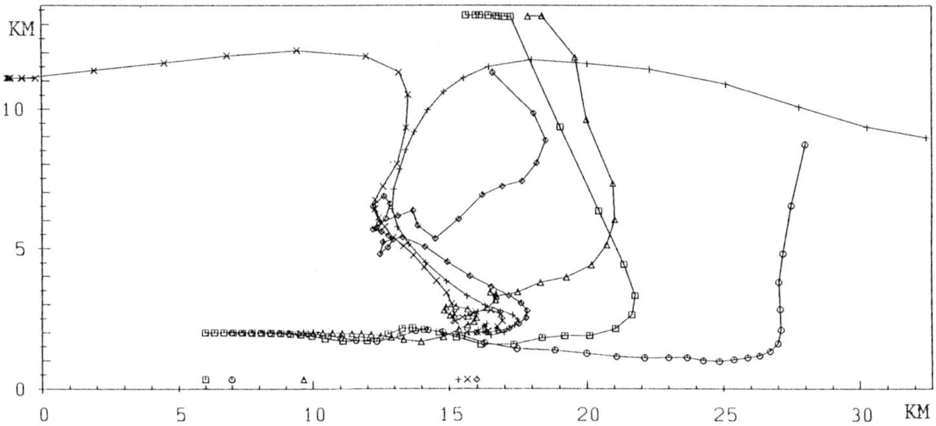

Fig. 4. Same as Fig.2 except the starting height is 2 km.

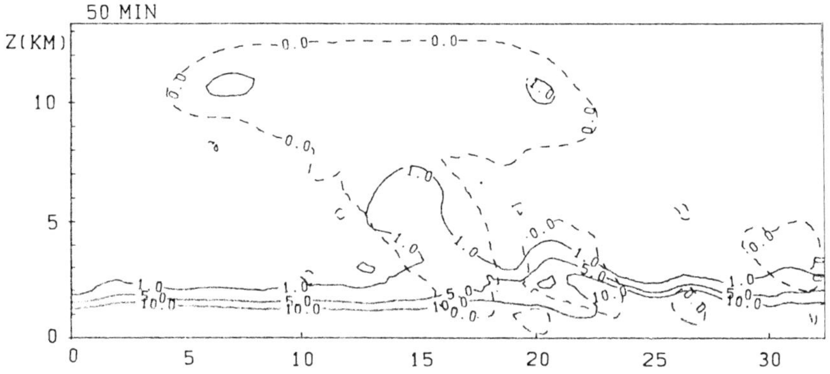

Fig. 5. Field of a soluble gas (CO) after 50 minutes of simulation. Full lines: concentration in the gas phase, dashed lines: concentration in the aqueous phase.

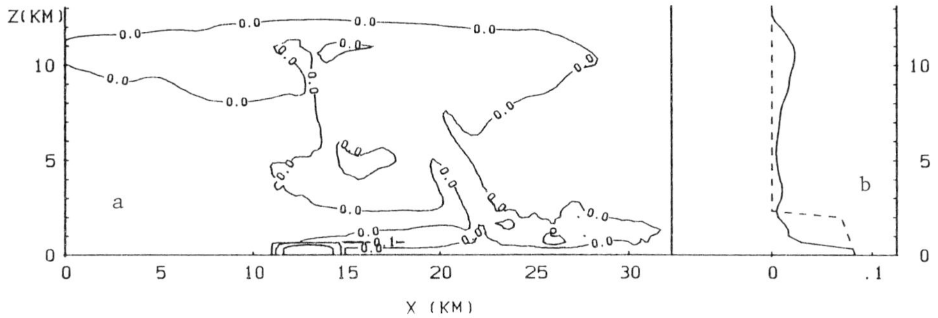

Fig. 6. a) Field of passive tracer after 50 minutes of simulation. The tracer has been emitted by an area source located between 11 and 15 km.
b) Vertical profile of the averaged concentration in each layer of the model domain.

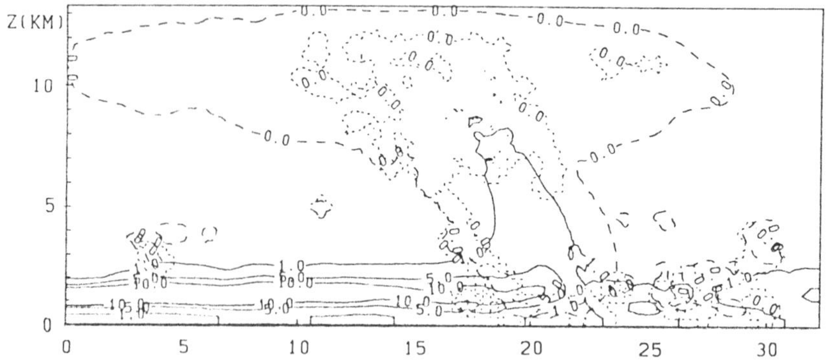

Fig. 7. Field of soluble aerosol after 60 minutes of simulation. The aerosol originates from the level 1 to 1.66 km.
Full lines: concentration in air, dotted lines: concentration in water or ice, dashed lines: contours of the cloud.

is captured neither by rain water below the cloud nor when entering the cloud from the front or rear side. To take into account these processes the parameterization of microphysics must be improved.

Conclusions

In this paper we have demonstrated that the tracer transport by a mesoscale convective system can be obtained by a numerical simulation approach. The cloud model is able to simulate vertical transport of chemical species.

The cloud model can provide consirable details of the cloud transport mechanisms. A residence time of an air parcel in the cloud environment of 10 to 20 minutes seems to be realistic.

The fully developped stage of the cloud is most important for vertical transport processes. To simulate scavenging processes and downward transport correctly all microphysical processes must be included into the model.

References

Dickerson, R.R., G.J. Huffman, W.T. Luke, L.J. Nunnermacker, K.E. Pickering, A.C.D. Leslie, C.G. Lindsey, W.G.N. Slinn, T.J. Kelly, P.H. Daum, A.C. Delany, J.P. Greenberg, P.R. Zimmerman, J.F. Boatman, J.D. Ray and D.H. Stedman, 1987: Thunderstorms: An important mechanism in the transport of air pollutants. Science, Vol. 235, p. 460-465

Scheidgen, P., 1989: Numerische Simulation einer Böenlinie. Mitteilungen aus dem Institut für Geophysik und Meteorologie der Universität zu Köln, Heft 64

Scheidgen, P. and M. Laube, 1988: The Simulation of a West African Squall Line. Report of the Second International Cloud Modelling Workshop. WMO Technical Document No. 268

Walcek, C.J. and G.R. Taylor, 1986: A Theoretical Model for Computing Vertical Distributions of Acidity and Sulfate Production within Cumulus Clouds. Journal of Atmospheric Sciences, Vol. 43, p. 339-355

NUMERICAL STUDY OF THE FORMATION OF ACIDIC SPECIES IN THE CONVECTIVE-CLOUD STREETS OVER THE JAPAN SEA IN WINTER

Toshihiro Kitada*, Pius C.S. Lee*
Eiichi Kitagawa*, and Hiromasa Ueda**

* Dept. of Regional Planning, Toyohashi Univ. of Technology Tempakucho, Toyohashi 440, Japan. ** National Institute for Environmental Studies, Onogawa, Tsukuba 305, Japan

INTRODUCTION

Long range transport of acidic pollutants from Asian continental countries, such as mainland China and Korea, to Japan begins to be recognized among the scientists in Japan and the other countries. Especially, measurements of acid deposition in Japanese cities along the Japan Sea rim suggest that such a long range transport likely occurs in winter, when cold air mass formed over the continent flows towards Japanese Islands. During the travel of the cold air mass over the warm Japan Sea, convective clouds develop and finally result in severe snowfall over Japan.

Thus, in this study we have numerically investigated the formation of acidic pollutants during the transformation of the cold and polluted continental air mass over the Japan Sea. The study has been performed in two steps. First, using an adaptation of Rutledge and Hobbs cloud model (1984), we have constructed sample fields of the precipitation and other cloud-microphysical variables associated with the convective cloud streets due to the roll vortices over the Japan Sea. Secondly, using these cloud and precipitation fields as inputs, the generation of acidic species in polluted continental air mass was simulated. The framework of the transport/transformation model of air pollutants used in this study is described in Carmichael et al. (1986). The model, however, incorporates updated chemistry systems. Recent information on the uptake process of air pollutants into ice and snow particles are also included in the model.

MODELING OF ACID SNOW AND RAIN

Simulation of the long range transport of acidified substances needs the tranport/transformation model of chemical species in gas- and hydrometeor-phases. In addition, the model has to be coupled with an appropriate cloud physics, which can provide both dynamics of hydrometeor fields (such as water vapor, cloud water, cloud ice, snow, graupel, and rain) and interphasal mass transfer rates. As far as gas phase species, the present study uses an adaptation of the model developed by the authors' group (Carmichael et al., 1986). In the following subsections, the models are described briefly.

Model For Gaseous Species

The governing equation takes two forms: Eq. (1) for advected speices; whereas, Eq. (2) for the steady state species.

$$\rho \frac{\partial C_i}{\partial t} + \rho V \cdot \nabla C_i = \nabla \cdot K \cdot \rho \nabla C_i + R_i + S_i - G_i{}^j \tag{1}$$

$$R_i - G_i{}^j = 0 \tag{2}$$

where ρ being the air density, C_i is the non-dimensional concentration of the ith species, R_i is the reaction term, and S_i denotes the source term. Moreover, the term $G_i{}^j$ describes the rate of transfer of the ith species between gas- and the jth hydrometeor-phases. The present study utilized a system of 71 chemical reactions, with the equilibrium relation among HNO_3, NH_3 and NH_4NO_3 by Stelson and Seinfeld (1982). Principal part of the mechanisms is described in Carmichael et al. (1986), except for updated SO_2 and NO_3 reactions after Lurmann et al. (1986). Eq. (1) governs the mass budget of 21 species of NO, NO_2, HNO_3, PAN, O_3, H_2O_2, SO_2, $SO_4{}^{2-}$, NH_3, NH_4NO_3, HNO_2, $RONO_2$, RO_2NO_2, ROOH, CH_4, HC_1 (alkanes), C_2H_4, HC_2 (alkenes), HC_3 (aromatics), CH_2O, and RCHO, while Eq. (2) applies to species such as OH and HO_2.

Model For Aqueous Phase Species

Besides water vapor, there are five hydrometeor phases in the model: cloud water, cloud ice, rain, snow, and graupel. The following continuity equation governs the mass transfers of the aqueous phase species:

$$\frac{\partial C_i{}^j q_j}{\partial t} + u\frac{\partial C_i{}^j q_j}{\partial x} + v\frac{\partial C_i{}^j q_j}{\partial y} + W_j\frac{\partial C_i{}^j q_j}{\partial z} - \frac{q_j \partial \rho V_j}{\rho \partial z} C_i{}^j = (R_i{}^j + {}_kT_i{}^j + G_i{}^j)/\rho \tag{3}$$

where $C_i{}^j$ denotes the non-dimensional concentration of the ith species in the jth hydrometeor, in kg/kg-water; q_i being the nondimensional concentration of the jth hydrometeor, in kg-water/kg-air; W_j denotes the net vertical velocity: $W_j = w - V_j$ is used, where V_j denotes gravitational terminal settling velocity averaged over pre-determined drop-size distribution for the jth hydrometeor, e.g., Marshall and Palmer (1948). $R_i{}^j$ represents the reaction term of the ith species in the jth hydrometeor. Furthermore, ${}_kT_i{}^j$ denotes the transfer rate of the ith species from the jth hydrometeor to the kth hydrometeor.

Concerning aqueous phase chemical reactions, the present study utilizes a reaction system largely based upon a recent sensitivity study (Pandis and Seinfeld, 1989). It includes oxidation reactions of S(IV) by H_2O_2, O_3 and O_2 with Fe^{3+} and Mn^{2+} as catalysts together with equilibrium reactions, totaling a system of 21 reactions (Table 1 and 2).

Gaseous absorption has been handled by accounting for the effective Henry's Law constant with a mono-dispersed assumption for drop size. The absorption rate of the ith species can be calculated for unit volume of air as the follows:

$$4\pi r_j D_{g,i} \left(1 + \frac{4\iota}{3 r_j \alpha_{w,i}}\right)^{-1} N_j \left(\rho C_i - \frac{C_i{}^j}{H_{eff,i} RT}\right) \tag{4}$$

where r_j is the representative droplet radius of cloud water or rain; $D_{g,i}$ is the molecular diffusivity of the ith species in air; ι denotes the mean free path of gaseous molecules in air; $\alpha_{w,i}$ stands for sticking probability; N_j is the number density of the jth hydrometeor; $H_{eff,i}$ represents the effective Henry's Law constant of the ith species; and R is the universal gas constant. $D_{g,i}$ can be expressed by $D_{g,i} = \iota (8RT/M_i \pi)^{1/2}/3$, where M_i is the molecular weight of the species. We assumed $\iota = 10^{-5}$ cm and $\alpha_{w,i} = 0.01$.

Recent experiments by Valdez et al. (1989) suggested that the solid hydrometeors of cloud ice, snow and graupel, which growing due to water vapor deposition, can take in gaseous SO_2. Their observations confirmed that at the subzero temperature of -15°C, the solid ice, which is under growth due to vapor deposition, incorporates gas phase SO_2 into it as if it were liquid water at 0°C. They assumed that: adsorbed water vapor forms fluid-water-like film encompassing the ice surface; and the film absorbs gas phase SO_2; then very rapid freezing of the film does not allow complete desorption of the SO_2. The present study has accounted for these phenomena. We believe that the inclusion of these absorption mechanisms may bring rise to sizable differences in simulation results.

Lamb and Blumenstein (1987) had conducted experiments through which they quantified the amount of S(IV) remained in the ice phase when a supercooled cloud droplet freezed on the surfaces of solid hydrometeor. They found that the fraction $f_{S(IV)}$ of SO_2 entrapped by the riming process to be proportional to the drop supercooling ΔT and $f_{S(IV)}$ can be expressed with an empirical relation:

$$f_{S(IV)} = a\Delta T + b$$

where $a = (5.8 \pm 0.5) \times 10^{-3} K^{-1}$, $b = (1.2 \pm 0.6) \times 10^{-2}$, $\Delta T = T_0 - T_{air}$, $T_0 = 273$ K This relation implies that only 7% of S(IV) in a supercooled cloud droplet at -10°C can remain in its ice phase in the course of the riming process; i.e., 93% of S(IV) is returned to gas phase SO_2.

These aforementioned laboratorial findings give important information about the controlling mechanisms of air pollutant transfer for cases when a solid hydrometeor grows through vapor deposition or through riming of supercooled cloud droplets. We have done sensitivity tests on these processes.

Cloud Physics Model

The ${}_kT_i^j$, G_i^j terms in the right hand side of Eq. (3) represent mass transfer of the ith chemical species in association with the jth and the kth hydrometeors. The rates of air pollutant transfer are functions of the rates of inter-hydrometeor water transfers. The present study had tailored the cloud physics model, presented by Rutledge and Hobbs (1984) for their narrow cold-frontal rainbands' study, for the Japan Sea conditions in winter. The model can handle the hydrometeor phase of graupel, in addition to those of water vapor, cloud water, cloud ice, rain, and snow. Equations for water vapor (q_v), cloud water (q_c) and cloud ice (q_i) are given as follows: (a = vapor, cloud water and cloud ice):

$$\frac{\partial q_a}{\partial t} + u\frac{\partial q_a}{\partial x} + v\frac{\partial q_a}{\partial y} + w\frac{\partial q_a}{\partial z} = \frac{S_{oa}}{\rho} \tag{5}$$

while those for snow (q_s), rain (q_r), and graupel (q_g) are as follows (b = snow, rain and graupel):

$$\frac{\partial q_b}{\partial t} + u\frac{\partial q_b}{\partial x} + v\frac{\partial q_b}{\partial y} + (w - V_b)\frac{\partial q_b}{\partial z} - \frac{q_b \partial \rho V_b}{\rho \partial z} = \frac{S_{ob}}{\rho} \tag{6}$$

In addition, the thermodynamic energy equation can be written in terms of temperature T as:

$$\frac{\partial T}{\partial t} + u\frac{\partial T}{\partial x} + v\frac{\partial T}{\partial y} + w\left(\frac{\partial T}{\partial z} + \Gamma\right) = \frac{\partial}{\partial z}\left(K_z\left(\frac{\partial T}{\partial z} + \Gamma\right)\right) + \frac{S_h}{\rho} \tag{7}$$

In Eqs. (5) to (7), V_b is the same as V_j described in relation to Eq. (3), Γ represents the adiabatic lapse rate of air temperature ($= 9.8 \times 10^{-3}$ K/m), S_{oa}. S_{ob} are the corresponding source- and sink-terms quantifying the rate of inter-hydrometeor transfers, and S_h represents the rate of heat transfer in association with the S_{oa} and S_{ob} terms.

FORMATION OF SNOW OVER THE JAPAN SEA IN WINTER

General Description of the Phenomena

In winter, cold air masses formed over Siberia sweep east towards Japan. Within their advance over the relatively warm Japan Sea, convective cells develop in lower atmosphere, and eventually they are organized to roll vortex streets (e.g. Asai, 1983). The system accumulates water substance in the atmosphere by forming cloud streets, and which in turn give rise to snow fall over Japan.

Characteristics of the winter-snowfall over the Japan Sea can be summarized as follows (Komabayashi, 1972): (1) There exists stably stratified layer at heights around 3 to 4 km above sea level; (2) thus the cloud -top is limited below these levels. This indicates that the depth of the cloud layer is relatively thin, such as 2 km at most;(3) nevertheless, graupel and large snowflake are observed in precipitation, indicating both considerable updraft enough to provide long residence time to those particles in the atmosphere, and large condensation nuclei in the area.

2-Dimensional Simulation of the Snow-Fall

The present study considers two pairs of roll-vortices, as shown in Fig. 1, where we assumed 2-D domain moving at constant velocity in x-direction. In the simulation, physical link between flow system and cloud physics was assumed not to affect the flow; i.e., we made a kind of diagnostic analysis of the cloud formation and snow fall phenomena under given flow system. For simplicity, we chose this approach, which is, we believe, partially justified since the aforementioned stable layer at relatively low altitude suppresses the height of the convective system. The depth of the roll vortex was set at 2 km high, according to the observation (RGMM, 1988) on board the Keifu-maru, which was in the middle of the Japan Sea on 4 Feb., 1986. The width of a pair of vortices was fixed at 18 km long, which was estimated using an empirical relation between the depth and the width of the roll vortex over the Japan Sea (Asai, 1983). Flow field adopted is shown in Fig. 2. Initial vertical profiles of temperature and water vapor for the simulation were also taken from the data on the Keifu-maru.

Eqs. (5) to (7) were solved using the Locally-One-Dimensional, Finite Element Method (LOD-FEM) (Kitada et al., 1983) with the explicit Euler technique for the cloud micro-physics part (Kitada et al., 1989). Spatial grid sizes and time increment were Δy = 1.5 km, Δz = 100 m, and Δt = 4 s, respectively.

We have made three runs studying sensitivities of both hydrometeor fields and accumulated precipitation to maximum vertical wind velocity, W_{max}, and a condition of graupel formation: (1) W_{max} was either 1.5 m/s or 0.75 m/s; (2) the condition for graupel initiation through collision between snow and cloud water droplets was either (a) snow $[q_s]$> 0.1 g/kg-air and cloud water $[q_c]$ > 0.5 g/kg-air (as in Rutledge and Hobbs 1984),or (b) $[q_s]$ > 0.08 g/kg-air and $[q_c]$ > 0.18 g/kg-air (i.e. a slackened condition). Cases 1 and 2 are for W_{max} = 1.5 m/s and 0.75 m/s with condition (a) for the graupel initiation, respectively, while Case 3 represents the slackened case for (b) with W_{max} = 1.5 m/s. Fig. 2-a shows $[q_s]$ field after 6 hours simulation for Case 1. Maximum $[q_c]$ ($\simeq$ 0.32 g/kg-air) appears where updraft is strongest. The cloud base is at about 1 km high and the depth of the cloud is also around 1 km. These results agree well with the observed data observed on the Keifu-maru. Fig. 2-b is the same as in Fig. 2-a except for $[q_s]$. In Cases 1 and 2, snow was the only precipitation. In Case 3, both graupel and snow appear as shown in Fig. 3-a and -b. The main source for graupel and the main sink for snow in this case is collision of snow particle with cloud water droplet. Thus zone of high $[q_g]$ comes below that of high $[q_s]$.

For all cases, precipitation rates at surface level were maxima at y = 10.5 km, where updraft peaked: after 6 hours simulation, they were 1.60, 1.30, and 1.16 mm/hr for Cases 1, 2 and 3, respectively. The maximum precipitation rate in Case 1 was only 23% larger than that in Case 2. Accumulated amount of precipitation within the area of a pair of roll vortices was, however, roughly proportional to the vertical wind speed (see Fig. 4). Most remarkable thing is large increase in accumulated precipitation caused by slackening the criteria for graupel initiation; the amount increased from 6.4 mm in Case 1 to 10.4 mm in Case 3 (Fig. 4). This is likely because large particles, graupel, can go fast through the evaporation zone below cloud, thus effectively carrying water substance in the upper layer to the surface level.

Table 3 shows temporal change of the rate of snowfall at y = 10.5 km in Case 1. If we assume wind velocity in x-direction as $u \simeq 10$ m/s (see Fig. 1), time lapses of 6 and 9 hours correspond to 216 km and 324 km sweeps of the air mass over the Japan Sea. These distances roughly locate between the Korean peninsula and Japanese Islands. Thus we can compare our precipitation rates of 1.6 and 2.6 mm/hr at 6 and 9 hours in Table 3 with those of Estoque and Ninomiya (1975), which were 0.8 to 3.6 mm/hr along the Japan Sea rim of Japan: they were obtained by solving a coupled system of dynamical flow, heat and water transport equations but without detailed cloud physics. These two results agree qualitatively well.

Two important cloud-microphysical processes of snow in Cases 1 and 2 were the deposition and evaporation of water vapor onto- and from- snow particles, and the growth of snow particles by capturing cloud water droplets; i.e., riming.

SIMULATION OF THE FORMATION OF ACIDIC SPECIES OVER THE JAPAN SEA

To elucidate effects of the cloud and snowfall phenomena on the formation of acidic species, two series of simulations were designed: one for "without-cloud" case and the other for "with-cloud".

For both series, initial vertical profiles of pollutants were prepared by doing 1-D transport/chemistry calculation for five hours. Obtained concentrations, listed below, were generally consistent with observed data in the region near emission sources in Korea (Park, 1990): e.g., $[SO_2] = 22$, $[SO_4{}^{2-}] = 0.8$, $[NOx] = 40$, $[HNO_3] + [NH_4NO_3] = 4.3$, $[O_3] = 26$, $[H_2O_2] = 0.6$, $[NH_4] = 0.3$; these concentrations in ppb are all at surface level. These 1-D profiles were extended over entire y-domain.

For the "without-cloud" simulation, we used the "Case 1" meteorological conditions less cloud physics part: i.e., constant flow field was input as in Fig. 2, where $W_{max} = 1.5$ m/s and $V_{max} = 15$ m/s; vertical- and horizontal-eddy diffusivities were set at 10 m^2/s and 1000 m^2/s, respectively.

The simulation was started at 0800 LST, around 20 January at 40°N. Emission sources were not included in the simulation since the air mass concerned was assumed to advance over the Japan Sea.

Dry deposition process at sea surface level was also not included in this run, for simplicity. Distribution of $[SO_4{}^{2-}]$ (Fig. 5) at 2000 LST after 12 hours simulation showed uniformity in y-direction, which is common to all species. This indicates that cirulating flows in vertical plane contributed more to rapid horizontal transport of pollutants than to their vertical mixing.

Figs. 6-a and 6-b show temporal changes of $[NO_3{}^-]$ and $[SO_4{}^{2-}]$ at heights of z = 0, 1 and 1.8 km at y = 10.5 km. Findings from this figures are summarized as follows: (1) averaged increase rates at surface level are 0.12 ppb/hr (≃ 0.48 μg/m^3-hr) for [SO_4^{2-}] and 0.89 ppb/hr (≃ 2.3 μg/m^3-hr) for [NO_3^-]. Increase rate of [NO_3^-] is 7.4 times larger than that of [SO_4^{2-}], (2) nearly monotonic increase occurs at all heights.

By integrating concentrations for the region of a pair of roll vortices, we have estimated conversion rates of SO_2 to SO_4^{2-} and NO_x to [NO_3^-] for 12 hours. The results are as follows: 12% of total SO_2 and 44% of total NO_x at 0800LST were transformed to SO_4^{2-} and NO_3^- (= [HNO_3] + [NH_4NO_3]) at 2000LST, respectively.

For "with cloud" run, we have used all meteorological data from Case 1. Other conditions such as initial distributions of pollutants are the same as those for "without cloud" run. Rates of photolysis reactions for NO_2, H_2O_2, O_3 etc. were, however, reduced below cloud level and enhanced just above cloud level.

Results of this case will be discussed at the meeting.

CONCLUSIONS

In the present study, we investigated the formation of acidic species in the convective-cloud-streets over the Japan Sea in winter.

Firstly, we simulated cloud formation and snowfall in an idealized flow field in which two pairs of roll-vortices were embedded. The results agreed well with available observations in terms of the height of cloud base (1km high), the depth of cloud (1km), and the rate of snowfall (1.6 to 2.6 mm/hr).

The depositional growth and the growth due to riming of supercooled cloud-droplets were found to be the most-important mechanisms related to snow formation. This simulation utilized an adaptation of the model by Ruthledge and Hobbs (1984).

Secondly, generation of the acidic species was computed under the same meteorological conditons as in the snowfall simulation.

Two cases, i.e. "without cloud" and "with cloud", were analysed by using the transport/transformation model of gas and hydrometeor-phase species. The results of the "without cloud" showed the conversion rates of SO_2 and NO_x to SO_4^{2-} and NO_3^- within a pair of roll vortices as: during 12 hours from 0800LST to 2000LST, 12% of SO_2 and 44% of NO_x were transformed to sulfate and nitrate, respectively.

Comparison of this result with that for "with cloud" case will be presented at the meeting.

Table 1. Aqueous Phase Equilibrium Reactions

	Reactions	Equilibrium Constants, M or M/atm	Source
EQ1.	$SO_2(g) \rightleftarrows SO_2(aq)$	1.23exp[3120・f(T)]*	PS#
EQ2.	$SO_2(aq) \rightleftarrows H^+ + HSO_3^-$	1.23×10^{-2}exp[1960・f(T)]	PS
EQ3.	$HSO_3^- \rightleftarrows H^+ + SO_3^{2-}$	6.61×10^{-8}exp[1500・f(T)]	PS
EQ4.	$NH_3(g) \rightleftarrows NH_3(aq)$	75exp[3400・f(T)]	PS
EQ5.	$NH_3(aq) \rightleftarrows NH_4^+ + OH^-$	1.75×10^{-5}exp[-450・f(T)]	PS
EQ6.	$HNO_3(g) \rightleftarrows H^+ + NO_3^-$	2.6×10^{6}exp[8700・f(T)]	CH$
EQ7.	$CO_2(g) \rightleftarrows CO_2(aq)$	3.4×10^{-2}exp[2420・f(T)]	PS
EQ8.	$CO_2(aq) \rightleftarrows HCO_3^- + H^+$	4.46×10^{-7}exp[-1000・f(T)]	PS
EQ9.	$HCO_3^- \rightleftarrows CO_3^{2-} + H^+$	4.68×10^{-11}exp[-1760・f(T)]	PS
EQ10.	$O_3(g) \rightleftarrows O_3(aq)$	1.13×10^{-2}exp[2300・f(T)]	PS
EQ11.	$H_2O_2(g) \rightleftarrows H_2O_2(aq)$	7.45×10^{4}exp[6620・f(T)]	PS
EQ12.	$HO_2(g) \rightleftarrows HO_2(aq)$	2×10^{3}exp[6640・f(T)]	PS
Eq13.	$OH(g) \rightleftarrows OH(aq)$	25exp[5280・f(T)]	PS
EQ14.	$H_2O \rightleftarrows H^+ + OH^-$	1×10^{14}exp[-6710・f(T)]	PS

* $f(T) \equiv 1/T - 1/298$, where T is temperature [K].

Pandis and Seinfeld (1989)

$ Chameides (1984)

Table 2. Aqueous Phase Reactions

	Reactions	Rate Constants M^n/s	Source
R1.	$H_2O_2 \xrightarrow{h\nu} 2\cdot OH$	0.1×10^{-6}	ES[&]
R2.	$OH + HO_2 \rightarrow H_2O + O_2$	$1.1\times10^{12}\exp[-1500/T]$	PS[#]
R3.	$OH + H_2O_2 \rightarrow H_2O + HO_2$	$8.1\times10^{9}\exp[-1700/T]$	PS
R4.	$HO_2 + HO_2 \rightarrow H_2O_2 + O_2$	$2.4\times10^{9}\exp[-2365/T]$	PS
R5.	$S(IV) + O_3 \rightarrow S(VI) + O_2$	2.4×10^{4} $4.2\times10^{13}\exp[-5530/T]$ $7.4\times10^{16}\exp[-5280/T]$	PS
R6.	$S(IV) + H_2O_2 \rightarrow S(VI) + H_2O$	$3.7\times10^{12}\exp[-4430/T]$	PS
R7.	$S(IV) + 1/2\cdot O_2 \xrightarrow{Fe^{3+}, Mn^{2+}} S(VI)$	*	MA[$]

* For $S(IV)<10^{-6}$ mole/l; pH=0-4

$$\frac{d[S(IV)]}{dt} = \frac{5000.[Mn^{2+}][SO_2(aq)]\cdot EQ2}{[H^+]} + \frac{0.82[Fe^{3+}][SO_2(aq)]}{[H^+]}\cdot\left(1+\frac{EQ2}{[H^+]}\right)$$

see MA[$] for complete system of equations

PS :Pandis and Seinfeld (1989) $ Martin (1984) & Estimated for noon time in January at 40°N

Table 3. Temporal Variation of the Snowfall Rate in Case 1

Time Lapse (hr):	4	5	6	7	8	9
Snowfall Rate (mm/hr):	0.6	1.3	1.6	2.6	2.6	2.6

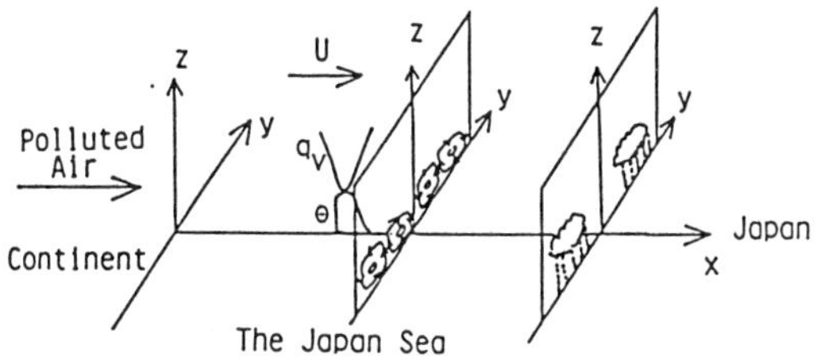

Fig. 1 Calculation domain for the convective-cloud street over the Japan Sea in winter.

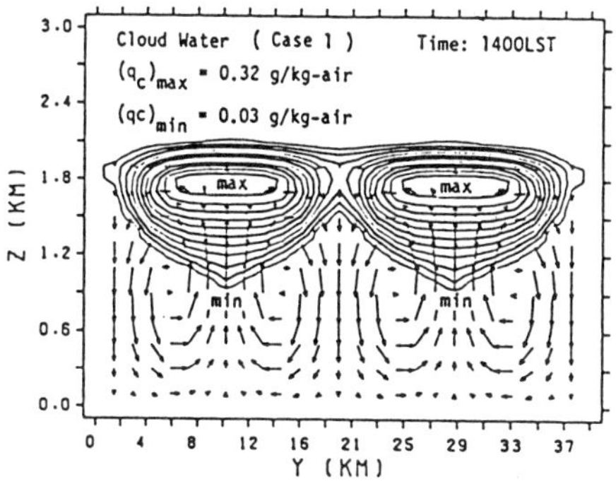

Fig. 2-a Computed cloud water field after 6 hours simulation for Case 1.

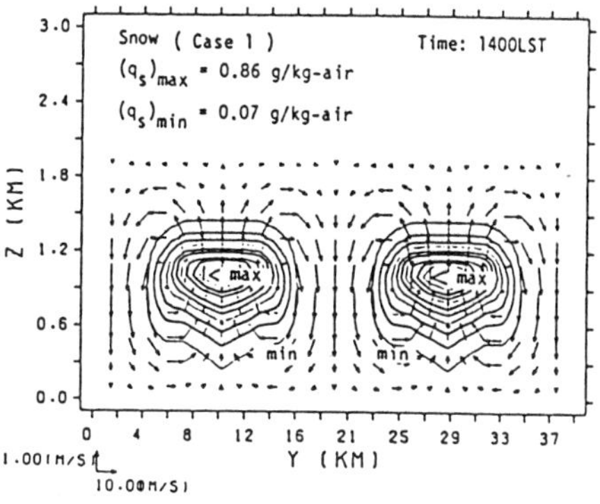

Fig. 2-b Computed snow field after 6 hours simulation for Case 1.

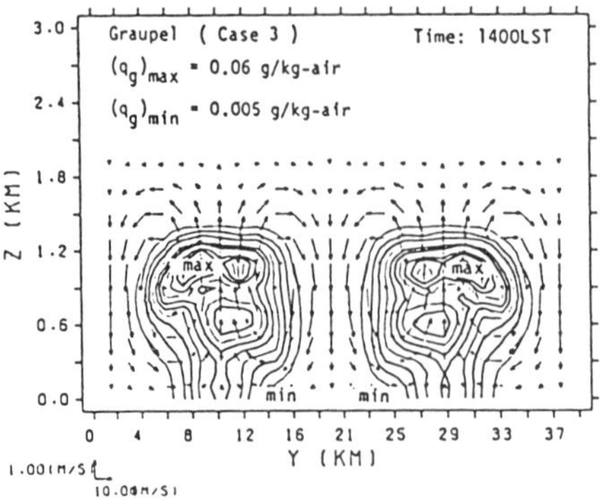

Fig. 3-a Computed graupel field after 6 hours simulation for Case 3.

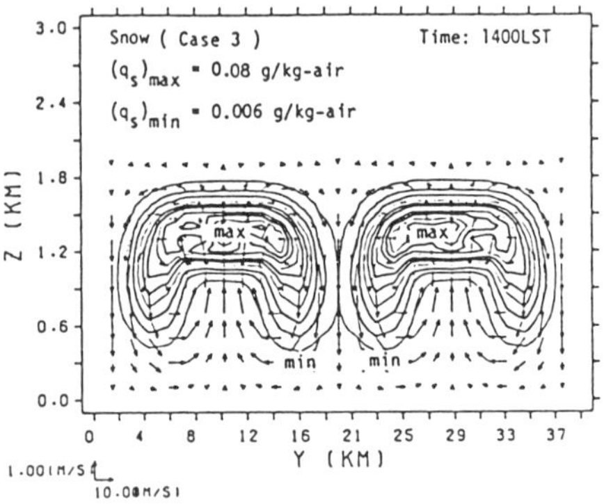

Fig. 3-b Computed snow field after 6 hours simulation for Case 3.

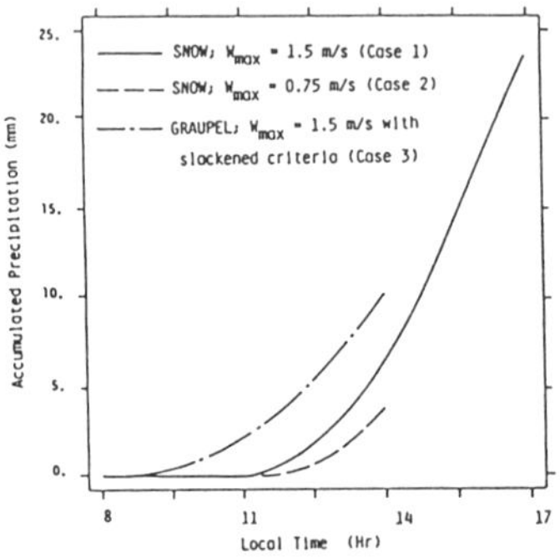

Fig. 4 Accumulated precipitation within the region of a pair of roll vortices.

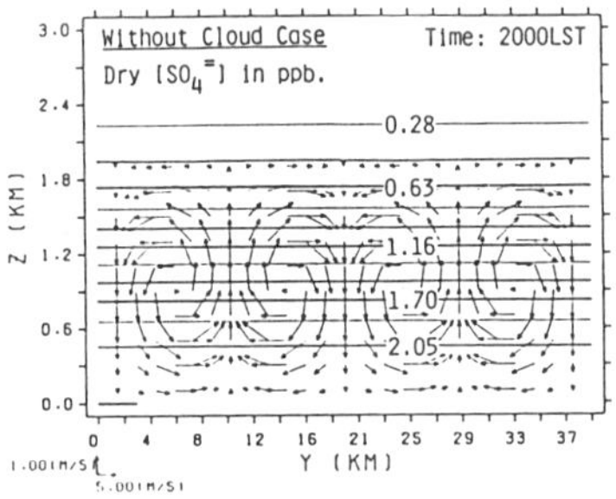

Fig. 5 Distribution of $[SO_4^{2-}]$ after 12 hours simulation.

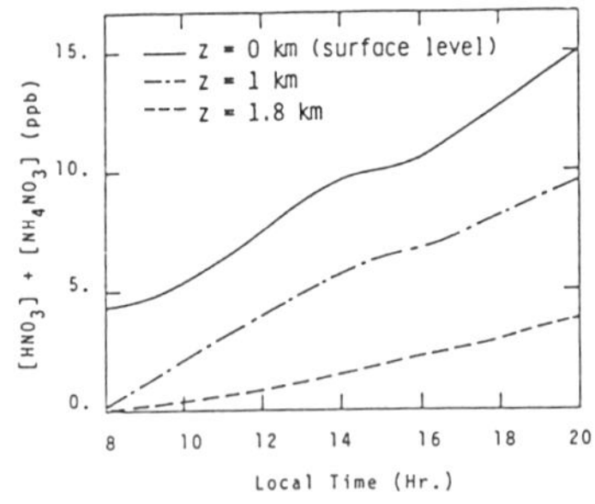

Fig. 6-a Computed $[HNO_3]$ + $[NH_4NO_3]$ concentration at y = 10.5 km

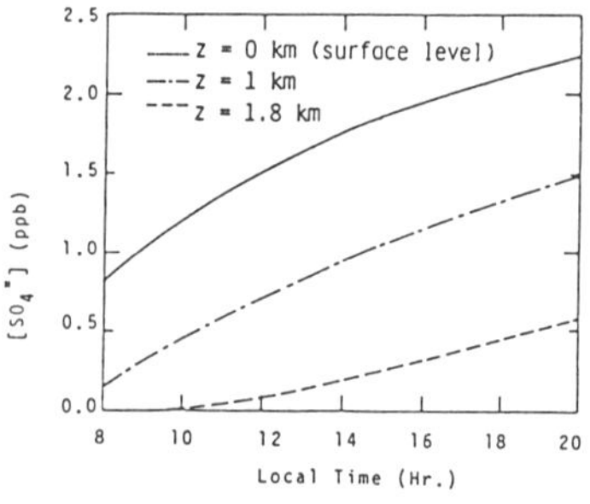

Fig. 6-b Computed $[SO_4^{2-}]$ concentration at y = 10.5 km

REFERENCES

Asai, T., 1983, "Science of atmospheric convection", Tokyo Dou, 220pp.

Carmichael, G.R., Peters, L.K., Kitada, T., 1986, A second generation model for regional-scale transport/chemistry/deposition, Atmos. Environ., 20:173-188.

Chameides, W.L., 1984, The photochemistry of a remote marine stratiform cloud, J. Geophys. Res., 89:4739-4755.

Estoque, M.A., Ninomiya, K., 1975, This data is referred to in the book "Rain and Meso-system" by Ninomiya, Tokyo Dou, 242pp.

Kitada, T., Kitagawa, E., Lee, P.C.S., Ueda, H., 1989, A FEM-LOD numerical algorithm applied for a diagnostic cloud model simulation, in: "Proc. 44th Conf. of the Japan. Soc. of Civil Eng.", 2:854.

Komabayashi, M., 1972, Meteorology and physics in the area of the Japan Sea, Kaiyo Kagaku, No. 3:77-82, in Japanese.

Lamb, D., Blumenstein, R., 1987, Measurement of the entrapment of sulfur dioxide by rimed ice, Atmos. Environ., 21:1765-1772.
Lurmann, F.W., Lloyd, A. C., Atkinson, R., 1986, A chemical mechanism for use in long-range transport/acid deposition computer modeling, J. Geophys. Res., 91:10905-10936.
Martin, L.R., 1984, Kinetic studies of sulfite oxidation in aqueous solution, in: "SO_2, NO and NO_2 Oxidation Mechanisms: Atmospheric Considerations", J.G. Calvert, ed., Ann Arbor Science, 63-100.
Marshall, J.S., Palmer, W. McK, 1948, The distribution of raindrops with size, J. Meteor., 5:165-166.
Pandis, S.N., Seinfeld, J.H., 1989, Sensitivity analysis of a chemical mechanism for aqueous phase atmospheric chemistry, J. Geophys. Res., 94:1105-1126.
Park, J.K., 1990, "Atmospheric Chemistry Research Programs in Korea", KIST, Korea.
RGMM (Research Group on Meso-scale Meteorology), 1988, On the mesoscale structure of the cloud band system over Japan Sea in winter monsoon period, Tenki, 35:237-248, in Japanese.
Rutledge, S.A., and Hobbs, P.V., 1984, The meso-scale and microscale structure and organization of clouds and precipitation in midlatitude cyclones: XII: A diagnostic model studying of precipitation development in narrow cold frontal rainbands, J. Atmos. Sci., 41:2949-2972.
Stelson, A.W., Seinfeld, J.H., 1982, Relative humidity and temperature dependence of the ammonium nitrate dissociation constant, Atmos. Environ., 16:983-992.
Valdez, M.P., Dawson, G.A., Bales, R.C., 1989, Sulfur dioxide incorporation into ice deposition from the vapor, J. Geophys. Res., 94:1095-1103.

DISCUSSION

G. DALU

What determines the aspect ratio of the cloud street?

T. KITADA

In this study, we took the aspect ratio of the longitudinal roll vortex as 9, which was estimated by giving the observed cell height of 2 km to the empirical relation derived from observations over the Japan Sea in winter monsoon (in Asai, 1983). As shown in this example, much flattened convective cells often appear in the atmosphere, although classical theoretical work by Raleigh (1916) indicates that the aspect ratio is to be 2.83 with use of diffusivities of equal magnitude for horizontal and vertical directions and slip conditions at both bottom and top boundaries. Some studies were devoted to explain the discrepancy of the aspect ratios. Priestley (1962) and Ray (1965) showed that the anisotropy of the eddy diffusivity in the convective layer, i.e. larger horizontal diffusivity compared with vertical one, can produce the flat cell. Meanwhile, Rosmond (1973) stressed the importance of the net heating due to small scale cumulus convection, thus the role of relative humidity and large scale static stability, to generate the flattened convective cell.

Though several hypotheses were proposed as above, to my knowledge, the mechanism of the large aspect ratio of the convective vortex, found in the atmosphere, has not been fully understood yet.

References

Priestley, C.H.B., 1962, The width-height ratio of large convective cells. Tellus, 14, 123-124.
Ray, D., 1965, Cellular convection with non-isotropic eddies, Tellus, 17, 433-439.
Rayleigh, 1916, On convection currents in a horizontal layer of fluid when the higher temperature is on the under side. Phil. Mag., 32, 529-546.
Rosmond, T.E., 1973, Mesoscale cellular convection. J. Atmos. Sci., 30, 1392-1409.

CLOUD CHEMISTRY CALCULATIONS USING A CHEMICAL GRID

Geert-Jan H. Roelofs

Institute of Meteorology and Oceanography
University of Utrecht, Princetonplein 5
3584 CC Utrecht, The Netherlands

INTRODUCTION

From a microphysical point of view, drops of the same size are equal. However, from a chemical point of view this need not be the case: due to processes such as entrainment, nucleation, condensation and coalescence, individual drops may have a completely different dynamical and microphysical "history". Therefore, the chemistry of the whole cloud is affected by dynamical and microphysical processes. At our institute a detailed cloud and rain chemistry model is being developed in order to study the influence of cloud dynamical and microphysical processes on cloud chemistry. At the moment a coupling of the separate chemical and microphysical modules is accomplished.

DESCRIPTION OF THE MODEL

In the model presented here, the microphysical processes that determine the evolution of a drop size distribution are taken into account when chemical processes inside the drops are computed. In order to do so the model considers two coupled drop size distributions. The first one, with a high resolution, is used for the microphysical calculations; the second is a low resolution copy of the first and forms the basis for the aqueous chemistry calculations. Every change in the first distribution (which drops collide, how much a certain drop size increases due to condensation, etc.) is also considered in the low resolution distribution and accounted for in the chemical concentration calculations.

The chemical grid results from the fact that the chemical module not only makes a distinction between drops of different sizes, but also between drops of the same size but with different chemical compositions. The parameter chosen for this purpose is the acidity (pH), since it is related to most of the occurring chemical processes. Both criteria, drop size and drop acidity, together give an indication of the microphysical and chemical history of the drops.

Microphysical processes taken into account are condensation, evaporation, coalescence and collision-induced break-up. Nucleation of drops on aerosol particles is also considered, but its treatment is still preliminary and simplified.

The chemistry of the model consists of the dissolution of gaseous SO_2, NH_3, HNO_3, O_3 and H_2O_2; dissociation of SO_2, NH_3 and HNO_3; and sulfate production resulting from the oxidation of S(IV) by O_3 and H_2O_2. The aerosol size distribution is considered to be monodisperse. All aerosols are spherical and of the same chemical composition, i.e. a user-specified combination of H_2SO_4, NH_4HSO_4 and

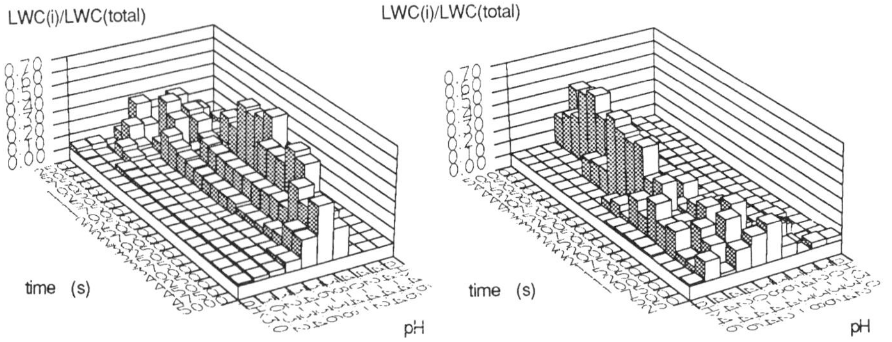

Figure 1. Distribution of the liquid water over the acidity classes during the simulation (3D-plot viewed from different angles).

$(NH_4)_2SO_4$. More information on the model and a comprehensive reference list can be found in Roelofs (1990).

ILLUSTRATION OF MODEL PERFORMANCE

In the case described here the simulated cloud is contained in one single grid cell. Initially the supersaturation is 1.25%. During the simulation with a time length of 5000 seconds it drops linearly to -1.0%. Initial gas phase concentrations of SO_2, NH_3, HNO_3, O_3 and H_2O_2 are 10, 1, 1, 40 and 0.5 ppb. There is a gas phase H_2O_2 production of 0.2 ppb per hour. The initial molar ratio NH_4^+/H^+ per aerosol particle is 3.0.

Model results are presented in figures 1, 2, 3 and 4. Figure 1 depicts the distribution of the total liquid water over the acidity classes during the simulation. It can be seen that during the first 2500 seconds the fraction of more strongly acidic water (pH $<$ 4.0) decreases while the fraction of the less acidic water (pH $>$ 4.0) increases. This is due to condensation causing the growth of drops and the dilution of their chemical contents. An opposite but less significant trend, acidification of drops as a result of chemical reactions, can be noticed as well (e.g. at times 1250 and more clearly at 2250 seconds).

In the less acidic drops the ratio of the concentrations of the sulfate resulting from S(IV) oxidation by O_3 and that by H_2O_2 is significantly higher than in more acidic drops of identical size. For example: at 2250 seconds this ratio is 1.3 in less acidic drops (pH 4.3) with a radius of 23.0 micrometers, and 0.04 in more acidic drops (pH 3.5).

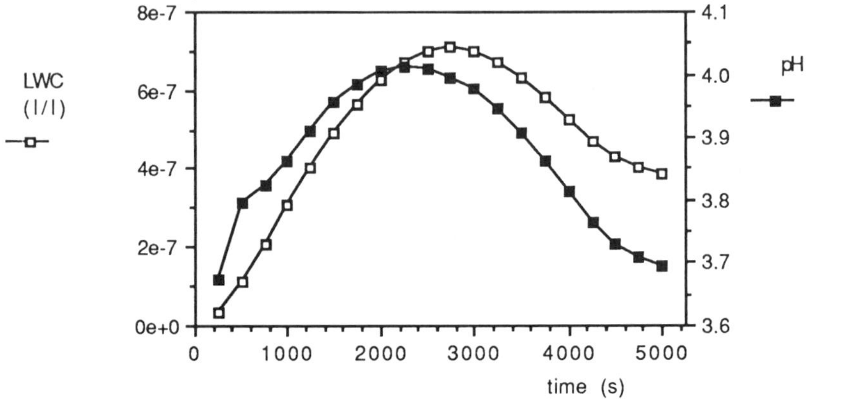

Figure 2. Total liquid water content and average pH as a function of time.

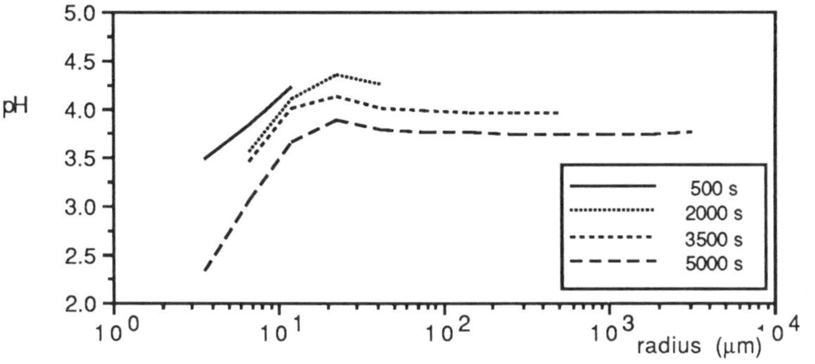

Figure 3. Average droplet pH as a function of drop size at different simulation times.

Since condensation is the predominant process in the first 2000 seconds, the average pH increases along with the liquid water content (see figure 2). After 2000 seconds the condensation rate slows down. This is caused by the decrease of the supersaturation but also by the fact that coalescence becomes more significant and causes the removal of small drops. The average pH then starts to decrease. As evaporation starts, small drops vanish causing an increase of the mean aerosol radius and a decrease of the ratio NH_4^+/H^+ of the mean aerosol composition. When most of the small drops have disappeared most of the water is comprised in the larger drops formed by coalescence, and the evaporation rate decreases.

Figure 3 shows the average pH as a function of drop size at different simulation times. From small drops to large, at first the acidity decreases, but increases again somewhat as coalescence becomes a predominant factor in the production of the drops concerned. The shape of the plot tends to remain unchanged during the simulation: the average acidity inside the large drops follows that of the smaller drops. This suggests that the large drops act mainly as reservoirs in which smaller drops are captured, and that chemical processes in these drops play a minor role compared to those in smaller drops. Figure 4 shows a "close-up" of the chemical grid after 4500 seconds of simulation time.

Other experiments, not described here, have shown that model results are influenced to a significant extent by the number of drop size and drop acidity classes used to define the resolution of the chemical grid.

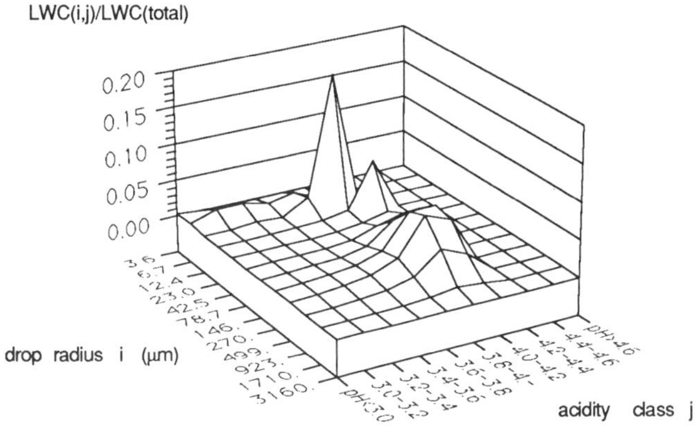

Figure 4. Distribution of the liquid water over the drop size (i) and drop acidity (j) classes after 4500 seconds simulation time.

GENERAL CONCLUSIONS AND REMARKS

The general conclusion derived from model simulations carried out so far is that microphysical processes influence cloud chemistry processes significantly. In particular the rates of pH dependent versus pH independent chemical reactions (e.g. the S(IV) oxidation by O_3 versus H_2O_2) may vary significantly between individual drops. The accuracy of cloud chemistry calculations is highly sensitive to the resolution of the chemical grid used in the model. This indicates the importance of an accurate representation of the evolution of drop size distributions in cloud chemistry models.

Presently our model is still subject to further development. Extensions to be made include explicit cloud dynamical processes, rain and ice phase microphysics and the incorporation of more detailed gas and liquid phase chemistry. We expect to have a complete and detailed model, considering dynamics, microphysics and elaborate chemistry, operational in the near future.

REFERENCES

Roelofs G.J.H. (1990) Cloud chemistry calculations with detailed microphysics. Paper submitted to Atm. Env..

DISCUSSION

C.J. WALCEK

What sensitivity do you expect if the concentration of H_2O_2 relative to SO_2 were increased?

G.J.H. ROELOFS

Thus far we have not carried out a sensitivity analysis based on the initial concentrations. I expect a relatively quick reaction of the H_2O_2 with SO_2, thereby causing a somewhat lower pH than a reaction with less H_2O_2 would have resulted in. The lower pH, in its turn, would result in a lesser yield of sulfate produced by the oxidation of O_3 in a later stadium.

A SIMPLE CHEMICAL CLOUD MODEL:

CAN SIMPLICITY APPROXIMATE REALITY?

J.P.J.M.M. de Valk

Institute of Meteorology and Oceanography, Air Chemistry Group
University of Utrecht, Princetonplein 5
3584 CC Utrecht, The Netherlands

INTRODUCTION

An important aspect of the redistribution and transformation of trace gases within the atmosphere concerns the occurrence of clouds. The flow pattern associated with clouds can transport trace gases from regions with high concentrations (normally the boundary layer) to relatively clean regions e.g. the free troposphere, (Ching 1989 and Builtjes 1988). Rain transports atmospheric constituents to the earths surface, primary those components which have an affinity towards liquid water. The introduction of an aqueous phase changes the air chemistry. Soluble species will be absorbed by the aqueous phase which prohibits them to participate in the gas phase chemistry (Lelieveld 1990). Some transformations occur at higher rates within the aqueous phase when compared to the rates in the gas phase, e.g. the oxidation of SO_2 to sulphate.

The study of cloud processes forms a focal point in the research of air chemistry. Experimental(Hegg 1982; Leaitch 1982; Isaac 1989) and theoretical studies on different cloud types have been performed. The complexity of the combined cloud dynamics, microphysics and chemistry demands a numerical approach. Most of the numerical cloud studies concentrate on one cloud type, e.g. cumulus clouds; Nair (1989), Taylor (1989), frontal rain bands; Hegg (1989) or warm stratiform clouds.; Xing (1990) Others study Flow Trough cloud models; Hong (1988) The chemistry can be limited; Hong (1988) or elaborate; Nair (1989) or even omitted when one concentrates on the redistribution of aerosol due to microphysics; Flossman (1985).

To incorporate the effect of clouds in large scale dispersion models we felt that there was a need for a cloud model which:

- would use little computing time
- could cope with different cloud types
- describes relevant physical and chemical cloud processes in an appropriately parameterized way.

A chemical cloud model was developed which describes transport and transformations in two different types, i.e. stratus and cumulus clouds. These occur frequently at midlatitudes. The model will hereafter be referred to as "simple" model. Its' simplicity is reflected in the low computing time and memory use.

PROCESS DESCRIPTION

The simple model presented here is focussed on aqueous phase chemistry. Gas phase chemistry is omitted. To gain a clear understanding of the involved processes source terms for gaseous species are not (yet) considered.

Obviously aqueous phase chemistry occurs only in the presence of liquid water. The concentrations of species depend on the quantity of liquid water. An accurate water cycle description should precede a study of concentration development of the constituents.

Liquid water in the atmosphere is either produced by melting of ice or by the condensation of water. When an air parcel, containing water vapour, is cooled, it will eventually become saturated. Further cooling supersaturates the air, and droplets will form on cloud condensation nuclei, (ccn) An upward vertical motion can induce such an adiabatic cooling. The spatial horizontal scales of these motions can vary from 100 m to 100 km. At the boundaries of the parcel, entrainment will introduce subsaturated air from the surroundings causing evaporation.

The above described large scale air motions, referred to as "cloud dynamics" provide an environment in which the small scale or microphysical processes are confined. Microphysics refers to the formation, growth and interaction of droplets with the environment. This occurs on scales of 10- 10000 μm. (Pruppacher, 1978). Here microphysics includes processes like nucleation, condensational growth, coalescence, rain formation by autoconversion and accretion. The rain falls into a subsaturated sub cloud region and partly evaporates. Eventually rain reaches the ground, where it can be collected in a rain water collector. Rain drops may evaporate completely in the sub cloud region leaving solid particles in the air which either will be scavenged by other rain drops or serve as ccn when transported in to the cloud.

The microphysical development of a droplet will influence its chemical composition (Roelofs, 1990). From a chemical view point nucleation results in small droplets with relatively high concentrations. Condensation diminishes the high concentrations. Coalescence redistributes the chemical components over the droplet spectrum, resulting in droplets with equal radii but different chemical composition. The absorption rate of gases depends on drop radius, temperature, chemical composition of the drop, and on the diffusion and Henry coefficients of the considered gases.

Here it is assumed that the internal circulation of the droplet assures a homogeneously mixed drop at all times. Evaporation causes a concentration increase due to drop volume reduction and by the same process it results in aerosol production. Aqueous chemistry finally also changes the concentrations. For different species this happens at different rates.

MODEL DESCRIPTION

To incorporate the above described physics and chemistry in the simple model, parameterizations are unavoidable. Therefore a bulk approach is chosen which is explained below. The simple model consists of two layers, of which the upper layer is fully (stratus cloud) or partially (cumulus cloud) filled with a homogeneously mixed cloud with a temperature TTC. The top of the cloud is also the rigid lid of the model. The lower layer is the sub cloud region with a temperature TTA and a relative humidity RH. See figure 1. The description of the dynamics is restricted to a simple water cycle in which neither thermodynamics nor supersaturation in the cloud is considered.

The microphysics considers three bulk hydrometeor fields, i.e. small drops with a fixed drop radius (10 μm), large drops and rain drops both with variable radii. The distinction in cloud drops is made to incorporate the different chemical development of small and large cloud drops. Nucleation occurs when an air parcel out of the sub cloud region is transported into the cloud. Due to the decreasing temperature the parcel becomes supersaturated. Nucleation generates only small drops. Autoconversion converts small drops to large drops and is proportional to the square of the small droplet number density. Large drops grow due to accretion. Here entrainment leads to the evaporation of the small drops. Rain is initiated when the large drops radii exceed a threshold value (100 μm). The evaporating rain drop contributes to the sub cloud humidity and aerosol density. The fraction of the total rain which evaporates completely is prescribed. The remaining rain scavenges the aerosol. The entrainment at the edge of the cumulus cloud is also prescribed. In the cumulus cloud model a preset cloud cover determines which part of the upper layer is filled with cloud.

The development of the concentrations is considered in the bulk of the four aqueous phases, i.e. the hydrometeor fields, the collected rain water and the gas phases, i.e. sub cloud, interstitial, and in the upper layer cloud free region.

The physics influences the concentrations of the species in gas and water phase. To facilitate the interpretation of the results, the chemistry is limited. In the gas phase it considers : SO_2, NH_3, HNO_3, H_2O_2, O_3, and $(NH_4)_{2x}H_{2y}(SO_4)_{x+y}$ aerosol, in the aqueous phase it considers the products of dissociation of SO_2, NH_3, HNO_3, H_2O and of oxidation of SO_2 by ozone and peroxide.

The development of the concentrations is considered in the bulk of the four aqueous phases, i.e. the hydrometeor fields and the collected rain water.

Both the chemistry and the physics are described by a set of first order coupled differential equations. The set is solved by a multistep Gear method which conserves mass as well as neutrality.

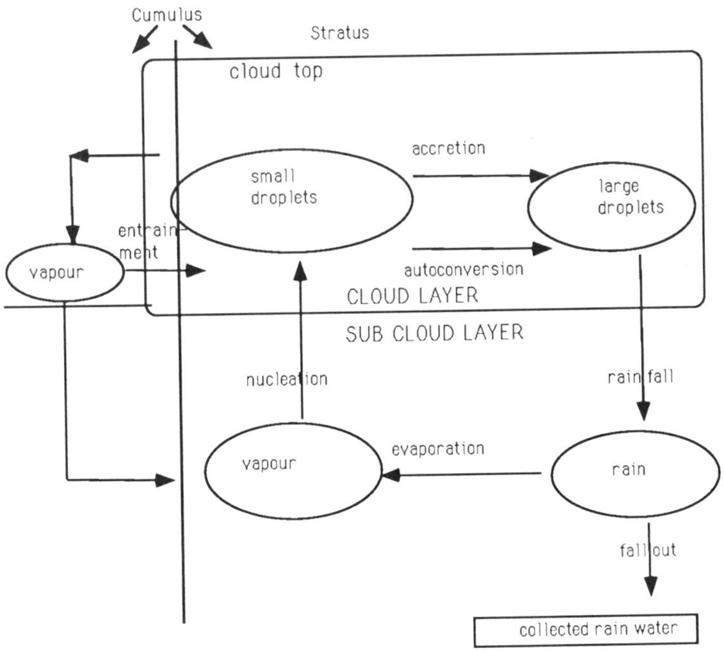

Fig. 1. The water cycle considered in the model for a stratus cloud,at the righthand side of the vertical line, and a cumulus cloud, at both sides of the vertical line. The arrows denote the conversions from one aqueous phase to another.

RESULTS

A case study for a stratus cloud is presented here to illustrate the differences in concentrations which can occur in the aqueous phases. In figure 2 the development of the liquid water content (lwc) is given. The initial lwc of small and large drops each are equal to 1.25 μl/l. The lwc of small drops decreases due to accretion and autoconversion until the decrease is compensated by nucleation. Then it rises to a value where nucleation is balanced by accretion and autoconversion.

The lwc of large drops increases due to accretion and autoconversion. The lwc decreases when it starts to rain, until the loss due to rain is compensated by accretion and autoconversion. The initiation of rain will increase the lwc of rain, which decreases due to evaporation and to the loss of rainwater reaching the ground. The fallen rain contributes to the lwc of the collected rain.

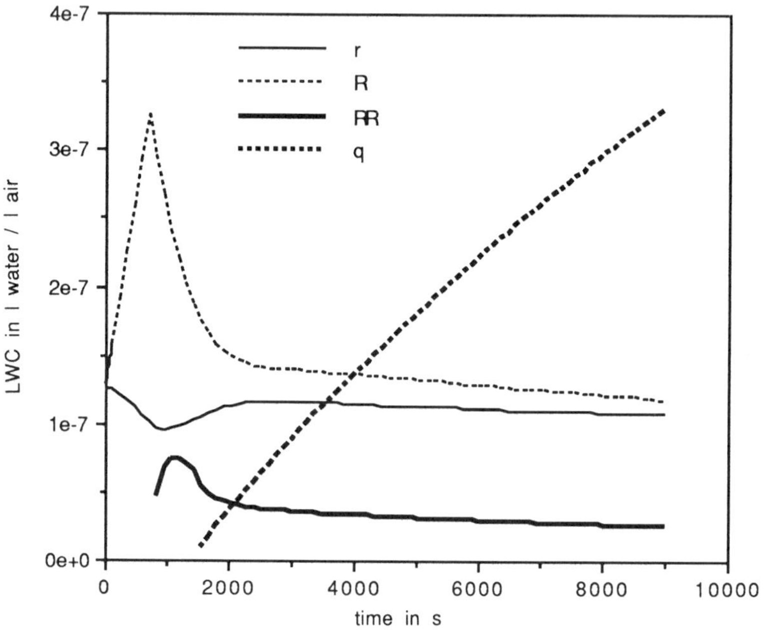

Fig. 2. The liquid water content (lwc) of the four aqueous phases, small drops (r), large drops (R), rain drops (RR), and collected rain water (q) as a function of time for a stratus cloud simulation.

The initial concentrations of SO_2, NH_3, HNO_3, H_2O_2, O_3, are 2,1,1,0.5,20 ppb respectively in the gas phases. The initial aqueous phase is unpolluted and neutral. The pH development and sulphate production are closely linked see figures 3 and 4. Minimum values of pH coincide with maximum values of sulphate concentrations. The oxidation of sulphate by peroxide is an important source of H^+. As soon as SO_2 and peroxide are absorbed in the drop they will react. Here the sulphate concentrations due to oxidation by peroxide reach a maximum when sulphate production and dilution due to introduction of relatively clean water compensate each other. In small drops dilution is due to nucleation which introduces fresh water to the small drops water bulk. This dilution eventually affects the large drops. The falling rain finds itself in a sub cloud region with high gas concentrations. This results in a high sulphate production. Also the rain becomes diluted as the peroxide becomes depleted and there is a supply of less polluted water.

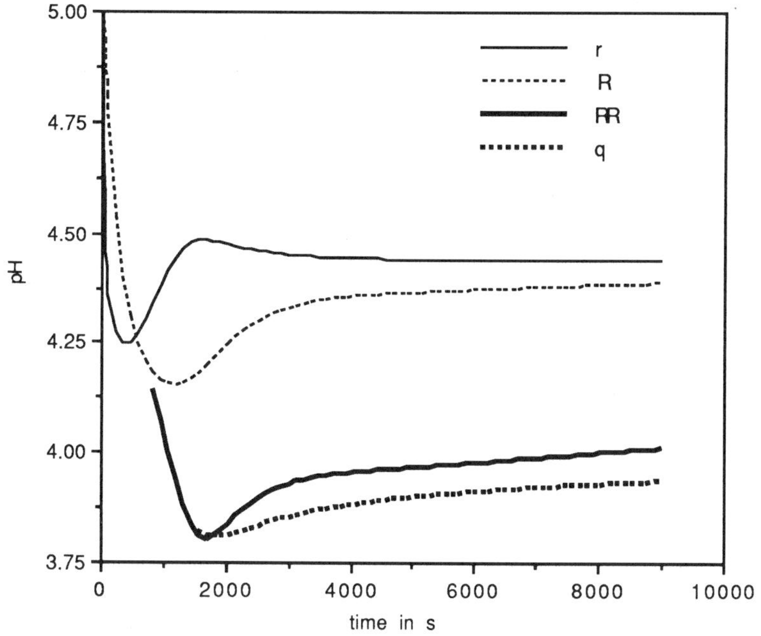

Fig. 3. pH of the small (r) and large drops (R), rain drops (RR) and collected rain water (q) as a function of time for a stratus cloud simulation.

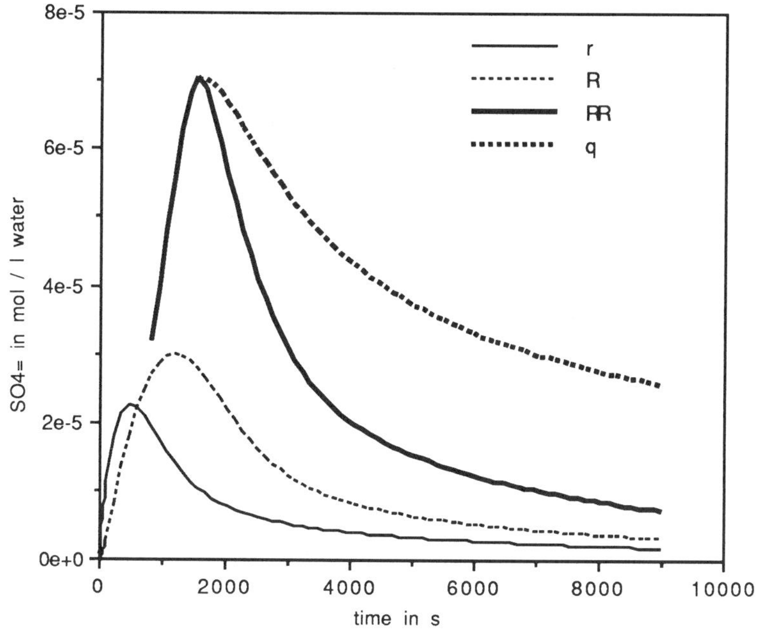

Fig. 4. Sulphate produced by peroxide oxidation of sulphur dioxide in aqueous phase of the small (r) and large drops (R), rain drops (RR) and collected rainwater (q) as a function of time for a stratus cloud simulation.

Although the sulphate production by peroxide is an important source of sulphate, it only contributes about 47% to the total sulphate in collected rain water at the end of the model simulation. Nucleation scavenging and impaction scavenging contribute 44 % and oxidation by ozone 9%, in this particular study.Those percentages are in good agreement with the results found by Taylor 1989 in his model.

The model results can be improved by comparing them to measurements. In table 1 the results are compared to flight data obtained by the Electrical Research Institute of the Netherlands (KEMA). In flight 239 on 85-06-20 three tracks were flown perpendicular on the wind direction; results of the first track were used to initiate the model. At t=5000 s the model performance is quite good. The calculated concentrations differ less than 10%, from the observed ones, except for the pH. The observed values, however, are not electroneutral. A pH of 3.5 satisfies neutrality. This implies that either there is an unconsidered cation or a failure in the measurements. At t = 8250 s the observed concentrations have increased whereas the numerical results remain approximately constant. This reversed behaviour can be explained if one realizes that the model does not consider source terms and that the observed air passes an emission area.

Table 1. The model results compared to the measurement data of flight 239. The data of the first track were used to initialize the model.

track	time in seconds	NH_4^+ in µmol/l	pH	NO_3^- in µmol/l	$SO_4^=$
1	0 s	55	4.0	70	62.5
2		80	4.0	80	145
model 2	5000	81	3.6	85	142
3		105	3.6	90	175
model 3	8250	83	3.6	86	136

CONCLUSIONS

We have developed a "simple" chemical cloud model to study the behaviour of clouds. Its computing time is kept low (i.e. <0.002 s simulation time/ s real time), for it is intended to be incorporated in large scale dispersion models

The model can cope with two cloud types, stratus and cumulus, which occur frequently above Western Europe and the United States. Model results show:
- SO_2 oxidation by peroxide is a major contributor to acidity and to sulphate of the aqueous phase.
- Next to SO_2 oxidation, nucleation and scavenging are major contributors to sulphate in the aqueous phase.
-Comparison of model results with measurements and results obtained by other models show good agreement .

Further evaluation with airborne measurement data will lead to an improvement of the model. Special attention will be given to an extension of the chemistry (the inclusion of more cation) and microphysics.

The agreement with experimental data would justifie the title : Simplicity can approximate the reality.

The author wish to acknowledge the KEMA for providing the measurement data.

REFERENCES

Builtjes, P.J.H., 1988, Interaction of planetary boundary layer and free troposphere, paper presented at the 3rd US-Dutch Symp. Atmospheric ozone research and its policy implications, Nijmegen, the Netherlands, May 8

Ching, J.K.S., 1989, Simulating vertical transport and transformation of mixed layer pollutants by non-precipitating convective cumulus clouds, paper presented at the AMS symposium on the role of clouds in atmospheric chemistry and global climate, Anaheim, Calif., January 30

Flossmann, A.I., Hall, W.D., and Pruppacher, H.R., 1985, A theoretical study of the wet removal of atmospheric pollutants, J.A.S., 42:583

Hegg, D.A, and Hobbs, P.V., 1982, Measurements of sulfate production in natural clouds, Atm. Env 16:2663

Hegg, D.A, Rutledge, S.A., Hobbs, P.V., Barth, M.C., and Hertzman, 1989, The chemistry of a mesoscale rain band, Q.J.R. Meteorol. Soc ,115:867

Hong, M.S., and Lee, I.Y., 1988, Numerical studies of acidification processes within and below clouds with a Flow-Through chemical reactor model, Atm. Env ,22:297

Isaac, G.A, Leaitch, W.R., Strapp, J.W., Anlauf, K.G., Bottenheim, J.W.,Wiebe, H.A., and Puckett, 1989, Canadian airborne measurements for model evaluation, preliminary results from the 1988 Muskoka field study, paper presented at the AMS symposium on the role of clouds in atmospheric chemistry and global climate, Anaheim, Calif., January 30

Leaitch, W.R., Strapp, J.W., Wiebe, H.A., and Isaac, G.A., 1982, Measurements of scavenging and transformation of aerosol inside cumulus, in:"Precipitation scavenging, dry deposition, and resuspension" proceedings of the fourth international conference, Santa Monica, Calif. 29 November

Lelieveld, J., and Crutzen, P.J., 1990, Influences of cloud photochemical processes on tropospheric ozone, Nature, 343:227

Nair, S.K., and Peters, I.K., 1989, Studies on non-precipitating cumulus clouds acidification, Atm. Env, 23:1399

Pruppacher, H.R., Klett, J.D., 1978,'Microphysics of clouds and precipitation', Reidel Dordrecht

Roelofs G.J.H. (1990) Cloud chemistry calculations with detailed microphysics. Paper submitted to Atm. Env..

Taylor, G.R., 1989, Sulfate production and deposition in midlatitude continental cumulus clouds. Part II: Chemistry model formulation and sensitivity analysis, J.A.S , 46:1991

Xing, L., and Chameides, W.L., 1990, Model simulations of rain out and wash out from a warm Stratiform cloud, J.A.Chem, 10:1

MESOSCALE CHARACTERIZATION OF CLOUD MICROPHYSICAL AND CHEMICAL PROCESSES: A COMPARISON OF 3 FORMULATIONS

C. J. Walcek[1], P. Karamchandani[2], and A. Venkatram[2]

[1] State University of New York at Albany
100 Fuller Rd., Albany, NY 12205 US

[2] ENSR Consulting and Engineering
1220 Avenida Acaso, Camarillo, CA 93010 US

INTRODUCTION

In order to accurately characterize the long-range transport, transformation, and removal of trace gases in the troposphere, it is necessary to account for the influence of processes occurring in clouds. These effects include:

1) vertical mixing of trace constituents by cloud-scale updrafts and downdrafts;

2) aqueous-phase chemical reactions that transform trace gases in clouds;

3) scavenging and removal of soluble constituents by precipitation;

4) scattering of shortwave radiation, inducing changes in photochemistry.

Unfortunately, the finest resolution of many long-range transport and transformation models (100's to 1000's km) is significantly larger than the typical horizontal size of many cloud updrafts (1 km or less), requiring comprehensive models of tropospheric chemistry to parametrize subgrid-scale cloud processes.

In this paper, we compare three cloud parameterizations that are used in regional-scale pollutant transport models. The three cloud models compared are: (1) the Acid Deposition and Oxidant Model (ADOM) cloud scavenging module, developed by ENSR Consulting and Engineering (Venkatram and Karamchandani, 1989); (2) the Regional Acid Deposition Model (RADM) parameterization, developed by State University of New York scientists (Chang et al., 1987); and (3) the RADM Scavenging Module (RSM) developed at Pacific Northwest Laboratories (Berkowitz et al., 1989).

These cloud scavenging models simulate the microphysics, dynamics, and chemistry associated with regional-scale cloud fields, and are nested within larger-scale regional pollution transport and transformation models. Trace species concentrations and meteorology conditions over each grid area are provided as input to the finer-resolution, multi-stream cloud models. The cloud chemical models calculate the integrated effects of cloud dynamics, scavenging, and aqueous-phase chemistry during a one hour period. Outputs of the cloud modules include the net change in all trace species concentrations at all tropospheric levels, and the wet deposition rates for soluble constituents.

Air Pollution Modeling and Its Application VIII, Edited by H. van Dop and D. G. Steyn, Plenum Press, New York, 1991

INITIALIZATION OF CLOUD MODELS

For this comparison, the three cloud models were executed for a one-hour period with the identical input meteorology and chemistry information. All three cloud models require vertical profiles of the grid-averaged temperature, humidity, and trace species concentrations. In addition, the grid-averaged precipitation rate and surface pressure are needed. The ADOM and RSM further require that the mean precipitation rate be categorized as "grid-scale" or "subgrid-scale" in nature. In addition, the ADOM cloud model must be provided with the cloud base and top elevations, as well as the fractional area of cloud coverage at each grid area. The RSM and RADM internally diagnose these cloud characteristics from the mean precipitation and temperature profiles.

Meteorology

The cloud models were compared under 18 different meteorological environments representative of warm, cold, and temperate conditions with "grid-scale" and "subgrid-scale" precipitation. Warm soundings were those with freezing levels above 3 km. Moderate soundings had freezing levels between 1 - 3 km above the surface, and cold soundings had freezing levels below 1 km. For subsequent results, reference will be made to soundings by numbers defined in Table 1.

Input soundings and precipitation rates were derived from a mesoscale meteorology model and data assimilation system simulation of a midlatitude cyclone over the northeast U. S. at 18 GMT, 23 April, 1981. This meteorology model enhances the temporal and spatial resolution of meteorology observations using a primitive equation mesoscale meteorology model that continuously incorporates observed wind fields and temperatures within the simulation domain using a 4-D data assimilation technique. The additional cloud coverage and depth information required by the ADOM cloud model were specified by the other cloud models. When ADOM was compared with RADM, the RADM-calculated cloud base and fractional area of coverage were used. When ADOM was compared with RSM, the cloud elevation and coverage information calculated by RSM were used. For subgrid-scale precipitation clouds, the ADOM calculates a cloud top from the input cloud base elevation. For grid-scale precipitation, ADOM uses the cloud top and base calculated by RADM or RSM, and assumes 100% cloud cover.

Chemistry

For each of the 18 meteorological cases discussed above, four different chemical environments were simulated. Conditions representative of "clean" and "polluted" conditions were specified for sulfate, nitrate, ammonia, ozone, and hydrogen peroxide. In

Table 1

DEFINITION OF METEOROLOGICAL CASE NUMBERS FOR COMPARISON STUDY

	Temperature classification					
	warm		moderate		cold	
relative precipitation rate ↓	precipitation type		precipitation type		precipitation type	
	grid-scale	subgrid	grid-scale	subgrid	grid-scale	subgrid
high	1	4	7	10	13	16
average	2	5	8	11	14	17
low	3	6	9	12	15	18

order to illustrate the nonlinear relationship between initial sulfur concentration and subsequent sulfur deposition, two simulations were performed under clean and polluted conditions where sulfur concentrations ($SO_2 + H_2SO_4$) were increased by a factor of 10 from a low sulfur to a high sulfur regime, and all other trace species were fixed. All trace species concentrations decreased exponentially with altitude, except for O_3 and H_2O_2, which were initially constant at all heights. Three additional inert tracers were used to assess subgrid-scale vertical transport by cloud motions. These tracers were specified as "step functions" present initially in the lower, middle, and upper troposphere. Calculations of cloud mixing were compared by tracking the displacement and dispersion of these tracers.

COMPARISON OF CLOUD MODELS

Subgrid-scale vertical mixing

All three models calculated approximately consistent displacements and dispersion of lower troposphere air when averaged over the 18 meteorological scenarios considered in this study. At grid areas where grid-scale precipitation dominates, the ADOM and RSM assume that clouds have no subgrid-scale vertical motions, and thus no vertical mixing is calculated. In contrast, RADM assumes that all clouds contain subgrid-scale motions, and thus calculates significantly greater mixing than RSM or ADOM when grid-scale precipitation is specified. For middle and upper troposphere air, the RADM calculates the most intense mixing, with average displacements and dispersions being a factor of 4 to 6 higher than the ADOM or RSM.

Scavenging of soluble species

Figure 1 shows a three-way comparison of the ADOM, RADM, and RSM calculations of nitrate deposition under 18 different meteorological conditions. On average, the three models disagree by a factor of 1.7 - 2.2, and show relatively little net bias, although scatter among the various cases is considerable. Many of the differences in calculated nitrate wet deposition shown in Fig. 1 result from differences in the condensed water content and fractional area of cloud coverage used internally by the three models. For example, at case #3, RADM disagrees with ADOM and RSM by more than a factor of 5. At this sounding, "grid-scale" precipitation dominates, and RSM and ADOM used 100% coverage by stratiform clouds. In contrast, RADM diagnosed about 3% coverage by subgrid-scale clouds, suggesting convective-like clouds at this unstable sounding. Fig. 2 shows the effect of changing the water content and fractional area of cloud coverage at this sounding in RADM. Through a combination of changing the water content and area of cloud coverage, RADM is brought into much better agreement with the other two models.

Aqueous-phase chemistry

On average, model calculations of SO_2 oxidation rate and sulfate deposition disagreed by a factor of 1.5 - 2.8, and again showed relatively little net bias, although scatter among the various cases is considerable. All three models calculate that a 10-fold increase in total sulfur concentration leads to about an 8-fold increase in sulfate deposition under polluted conditions, demonstrating nonlinear sulfur scavenging. Factors such as the fractional area of cloud coverage, cloud base and top, and condensed water content were all found to influence the aqueous-phase chemistry occurring in cloudy grid areas.

Microphysical and mesoscale cloud structure

Calculations of vertical mixing, scavenging, and aqueous-phase chemistry are all influenced by numerous microphysical and mesoscale properties of the cloud fields over each grid area. Figure 3 shows an example of the cloud microphysical structure calculated by the three models at sounding 5, where subgrid-scale precipitation dominates at a warm sounding. One can see that ADOM calculates relatively high water contents, and the RSM calculates the lowest ice contents. These differences result from different assumptions

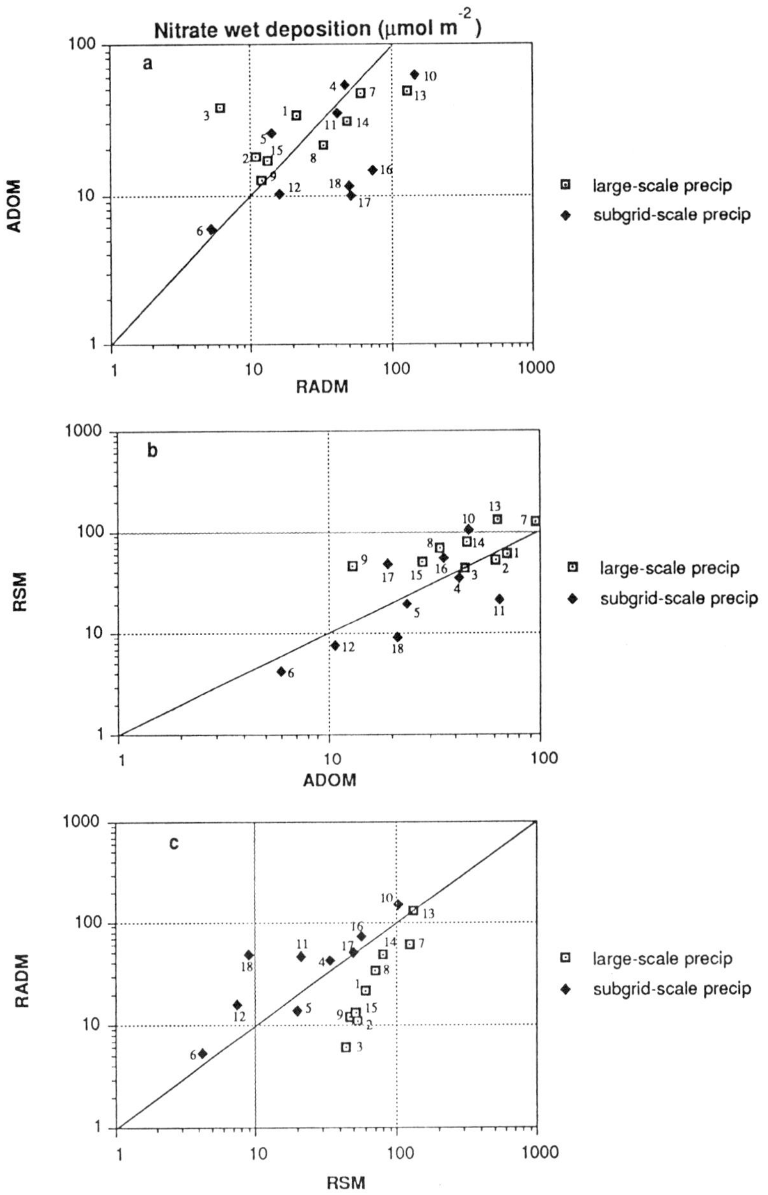

Fig. 1. Comparison of ADOM, RADM and RSM calculations of nitrate wet deposition for a sample of 18 meteorological scenarios.

about the dynamic structure of clouds producing the precipitation in the grid areas. RSM assumes a dynamical and microphysical balance between vertical motions in a two-stream convective cloud model, and area of coverage is related to the grid-averaged precipitation rate. In contrast, ADOM and RADM assume a balance between entrainment and upward mixing in these clouds, and make differing assumptions about the relationship between the area of cloud coverage and other meteorological parameters.

Fig. 4 shows the fractional area of cloud coverage used by the three models at the 18 cases studied here. Most of the time, ADOM uses the RADM or RSM area, and

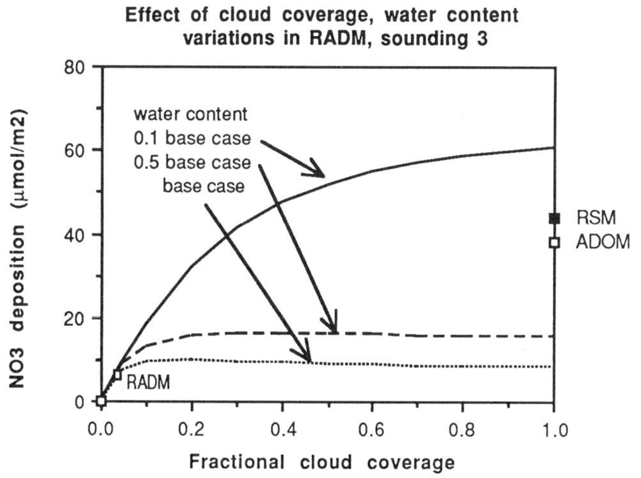

Fig. 2. Sensitivity of nitrate scavenging rate to variations in model-diagnosed fractional area of cloud coverage and condensed water content according to RADM at one sounding.

therefore the agreement between these models appears good. In contrast, RADM disagrees with ADOM and RSM when large-scale precipitation is specified. When subgrid-scale precipitation dominates, the RSM in general calculates lower areas of cloud coverage.

CONCLUSIONS

For comparison of nonzero and positive parameters calculated by various models, we define the *factor difference* between the model calculations, defined for a single comparison between two model calculations (A and B) of a given parameter as

$$\text{factor difference} \equiv \max\left(\frac{A}{B}, \frac{B}{A}\right) = \exp\left|\ln\frac{A}{B}\right| ,$$

Perfect agreement between two variables will yield a factor difference of one. As model calculations diverge, the factor difference becomes larger. Results of the comparison study are summarized in Fig. 5, showing the geometric mean of the factor differences between numerous parameters compared over the meteorological and chemical cases comnsidered in this study. Also shown on this figure are the geometric average deviation of the factor differences about the mean factor difference. When initialized with identical regional-scale meteorological and chemical information, the three model calculations of nitrate and sulfate wet deposition and SO_2 oxidation rate differ by a factor of 1.7 - 2.8. Greater differences are diagnosed for subgrid-scale vertical mixing of middle and upper tropospheric air. Difference between model calculations are associated with differences in model-diagnosed or specified fractional area of cloud coverage, water contents, cloud base elevation, entrainment assumptions, and characterization of precipitation as stratiform or convective in nature. All three models individually show correlations between these meteorological factors and model calculations of scavenging, mixing and chemistry. When these meteorological factors are artificially adjusted within models to be consistent, then the three models calculate more consistent chemical and scavenging behavior.

These comparisons suggest that further refinement of cloud effects modules in regional- or global-scale tropospheric chemistry models will require regional-scale observations of fractional cloud coverage, cloud water contents, cloud base elevations, and cloud thickness in conjunction with measurements of wet deposition and ambient concentrations of soluble and reactive trace species. Further sensitivity studies of the individual models are required to more fully quantify the reasons for the discrepancies in model calculations of scavenging, mixing and aqueous chemistry.

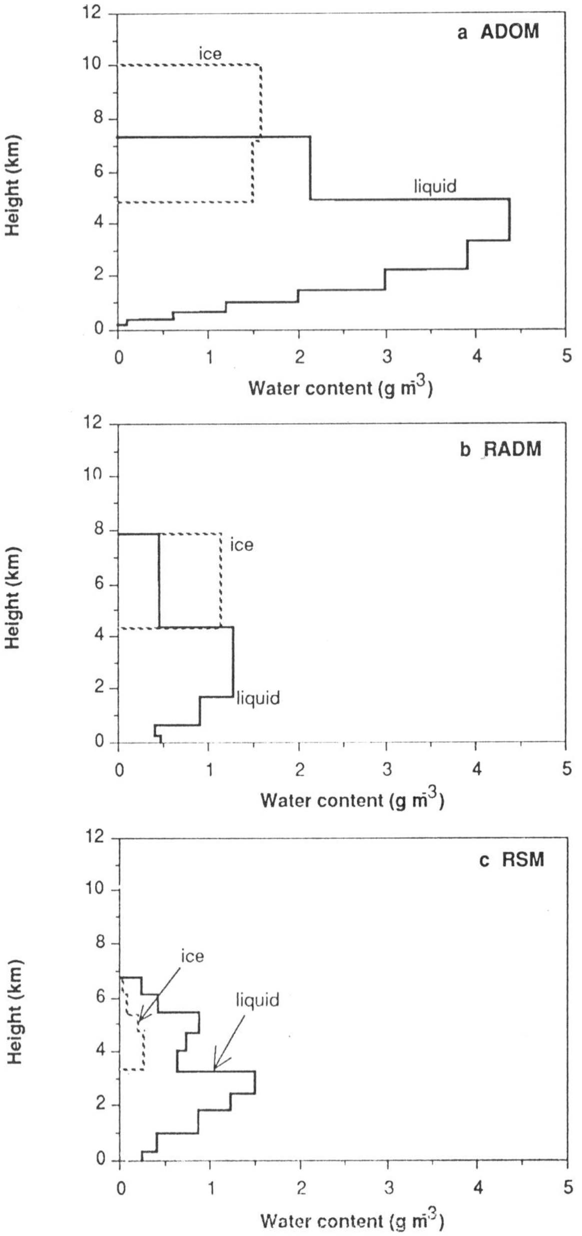

Fig. 3. Comparison of cloud microphysical structure according to three models at one sounding. Grid-averaged precipitation rate is the same for all three models.

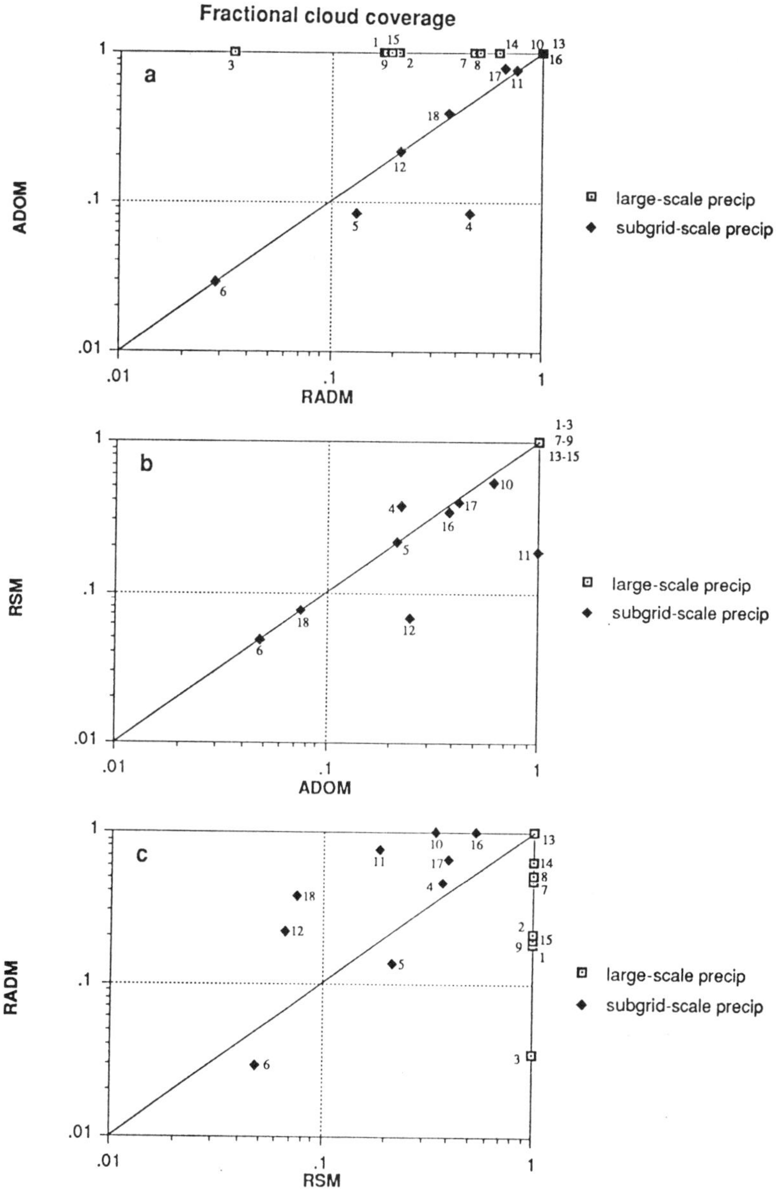

Fig. 4. Comparison of ADOM, RADM and RSM calculations of area of cloud coverage for a sample of 18 meteorological soundings.

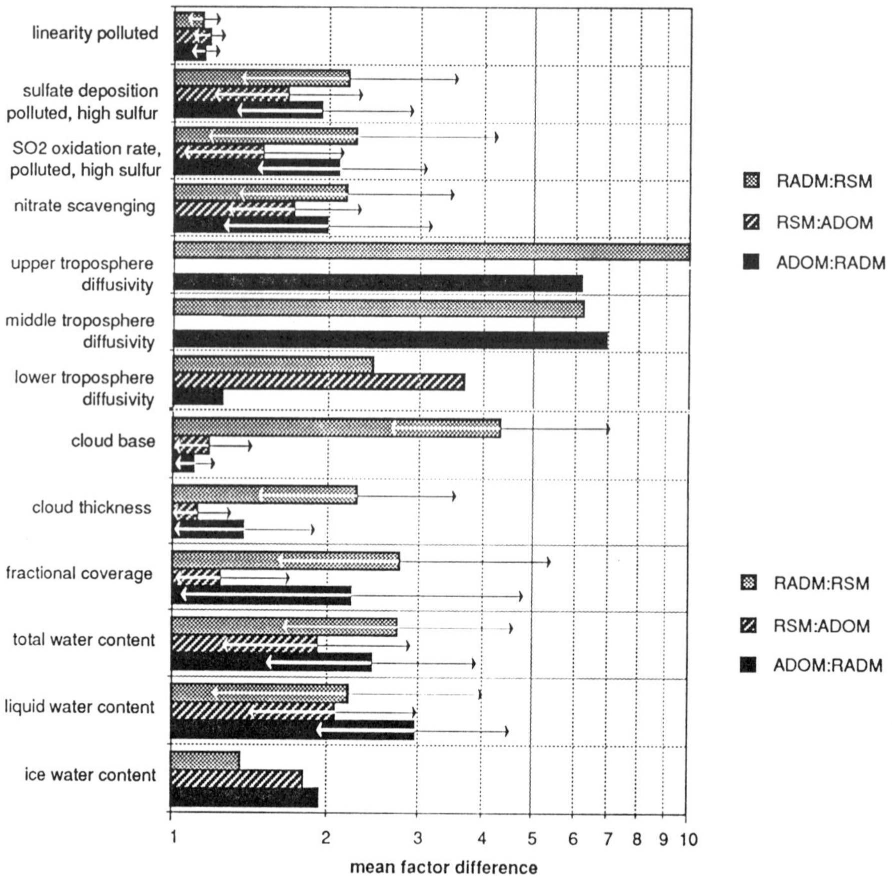

Fig. 5. Summary of the mean factor differences in model calculations of scavenging, aqueous-phase chemistry, nonlinearity, vertical mixing, and cloud mesoscale and microphysical characterization for a sample of 18 meteorological environments under numerous chemical conditions.

Acknowledgments - We are grateful to the Electric Power Research Institute for support of these research efforts.

REFERENCES

Berkowitz, C. M., Easter, R. C., and Scott, B. C., 1989, Theory and results from a quasi-steady-state precipitation scavenging model, Atmos. Environ., 23:1555.

Chang, J. S., Brost R. A., Isaksen I. S. A., Madronich S., Middleton P., Stockwell W. R., and Walcek C. J., 1987, A three dimensional Eulerian acid deposition model: physical concepts and formulation, J. Geophys. Res., 92:14681.

Venkatram A., and Karamchandani P., 1989, "The ADOM II Scavenging Module: Incorporation of an Improved Cumulus Cloud Module." ENSR Document No. 0780-004-205, ENSR Consulting and Engineering, 1220 Avenida Acaso, Camarillo, CA 93010.

DISCUSSION

R.A. BROST

You could test vertical transport in your three models by comparing model simulations with analyses of observational data, as has been done recently by M. Tiedtke (MWR, Aug., 1989).

C.J. WALCEK

I agree that certain observations of cumulus convection could be used to test vertical advection schemes to rule out obviously unrealistic mechanisms of vertical transport by clouds. Yet I feel that one cannot place a lot of faith in individual "observations" of cumulus transport. Methods of estimating vertical transport by clouds over large areas are highly uncertain, and strongly sensitive to several assumptions about the nature of cloud updrafts and cloud population distributions. The reference you mention by Tiedtke (Mon. Wea. Rev., 117, p. 1779) admits (bottom of p. 1799) that observations of cumulus convection are highly uncertain. Most "observations" are in fact extrapolations from residuals of larger-scale budgets of momentum, heat, or water vapor, and themselves rely on models to quantify subgrid-scale transport. Also, most previous observations of cumulus-scale vertical transport are performed in tropical areas, where synoptic-scale dynamic forcing mechanisms are significantly different from midlatitude conditions. Johnson (J. Atmos. Sci., 37, p. 733) explicitly states that observations of cumulus-scale vertical transport can only be used in a qualitative sense.

MESOSCALE MODELING OF POLLUTANT TRANSPORT AND DEPOSITION IN CASE OF FRONTAL RAIN

Nadine Chaumerliac[1], Robert Rosset[1] and Everett C. Nickerson[2]

1.LAMP/OPGC
Clermont-Ferrand, FRANCE
2.Forecast Systems Lab., NOAA
Boulder, CO, USA

ABSTRACT

Gases emitted in the planetary boundary layer can be transported very efficiently to the free troposphere through vertical motion along a frontal surface. Then, they can dissolve and react with other products within the cloud before being deposited on the ground. A mesoscale numerical model was used to simulate frontogenesis in a moist atmosphere (an evolving Eady wave) in order to study the relative importance of vertical transport and scavenging on the redistribution of different tracers (soluble or insoluble, reactive or nonreactive) in the atmosphere.

INTRODUCTION

Frontal systems are responsible for the transport of material between the boundary layer and the upper troposphere and can transform a local air pollution problem into a regional or global atmospheric chemistry problem. As discussed in Dickerson et al. (1987), so long as the pollutants are confined to the planetary boundary layer, they have short residence times and short transport distances. However, if those materials are transported into the free troposphere where stronger winds can lead to a much larger scale of horizontal transport, then their lifetimes increase significantly. Clouds associated with frontal systems can also alter the pollutant distribution through their feedback with radiation processes, and through the scavenging of soluble and/or reactive species in aqueous phase.

Numerical simulations of frontogenesis in a moist atmosphere have been performed using the mesoscale meteorological model of Nickerson et al., 1986, to study the relative importance of vertical transport and scavenging in the redistribution of different types of tracers (inert, partially soluble, or reactive gases). Sensitivity tests with various solubilities demonstrate the important role of microphysics in modifying gaseous concentrations in the upper troposphere.

THE FRONTAL SITUATION AND THE CHEMICAL SCENARIOS

Numerical simulations of frontogenesis in a moist atmosphere have been carried out similar to those reported on by Hsie at al.(1984). A two-dimensional version of the model (Nickerson et al., 1986) has been run for 24 h including planetary boundary layer physics (Blackadar, 1978) and diabatic effects (Berry and Reinhardt, 1973) for an Eady wave evolution. The mesoscale model is initialized with the solutions from the analytical model of Keyser and Anthes (1982).

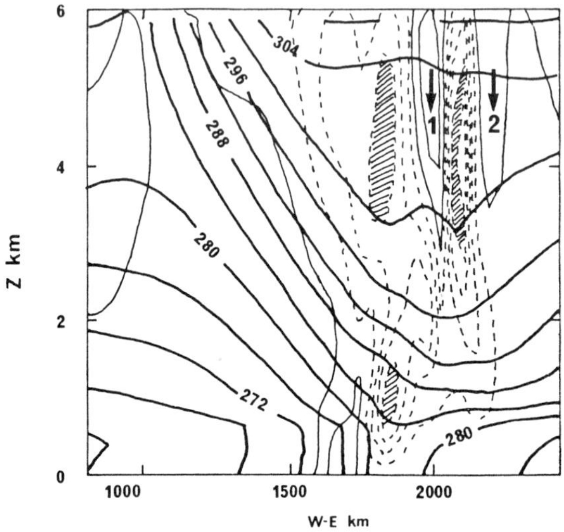

Fig. 1. Vertical cross-sections of potential temperature (heavy solid lines) and of ω in mb/h (dashed and thin lines) after 21 h of simulation time. Shaded areas indicate regions of maximum vertical ascent.

Results for the dynamical and thermodynamical fields are shown in Fig. 1 after 21 h of simulation time. They compare well with the results from Hsie at al., (1984), which show a banded structure of the vertical velocity in the warm sector. A region of maximum upward motion is found in the first kilometer of the atmospheric boundary layer ahead of the front. This low level ascent is due to horizontal convergence forced by surface friction. The other two ascending cells are in regions of moist convection, and the subsidence zones in between the regions of ascent are caused by evaporative cooling. A series of experiments were then carried out for this particular meteorological situation involving the transport of a soluble tracer, SO_2. We initialized the model with exponential profiles for SO_2(Taylor, 1989), starting with a value of 5ppb at the ground. As a first step, we consider only one tracer and assume that it is not reactive. The aim of this sensitivity study was to examine the filtering effects of clouds which scavenge the tracer and limit its vertical transport.

The soluble, nonreactive gas is described in the model by three predictive equations which represent it in three forms: its gaseous phase, its dissolved form in cloud water, and its dissolved form in rainwater. Three phase concentrations of the species X are defined: a gaseous one, a concentration in cloud water, and a concentration in rainwater, denoted respectively X_a, X_c, and X_r.

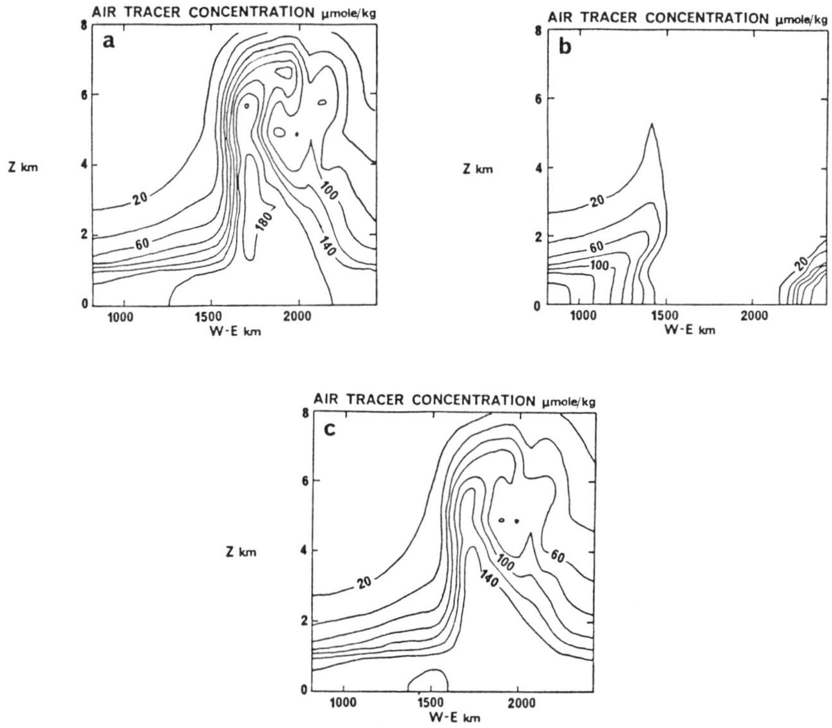

Fig. 2. Vertical cross-sections of air tracer concentration in μmole/kg for a) an insoluble one, b) a soluble one with pH=5 and $H_{eff}=10^3$M/atm, and c) a soluble one with pH=3 and $H_{eff}=10$M/atm.

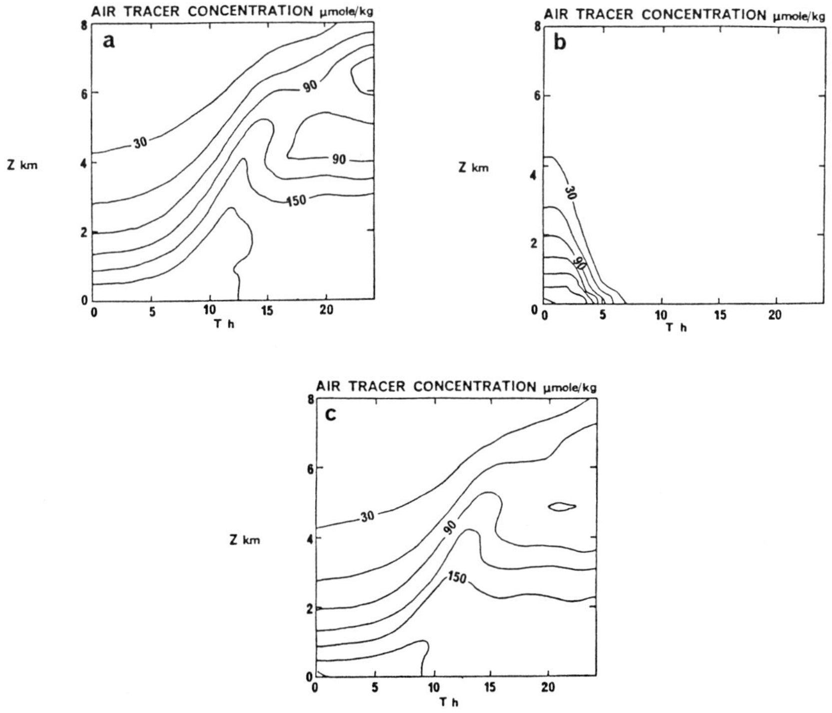

Fig. 3. Same as in Fig. 2. but for the temporal evolution of the air tracer concentration at a grid point located close to the surface front.

We thus obtain the time rate of change of each concentration for a given soluble species from the following set of equations:

$$\frac{dX_a}{dt} = -S_{dis} - S_{scav}$$

$$\frac{dX_c}{dt} = S_{dis} - S_{acc} - S_{aut}$$

$$\frac{dX_r}{dt} = S_{acc} + S_{aut} + S_{sed} + S_{scav},$$

where S stands for the various sources and sinks of the concentrations. The subscript "*aut*" refers to autoconversion of cloud droplets into raindrops and "*acc*" to accretion of cloud droplets by raindrops. The subscript "*sed*" refers to sedimentation. The subscripts "*dis*" and "*scav*" designate respectively the dissolution in aqueous phase of X and the scavenging from gas phase to rainwater through mass transfer. The left hand sides of the above equations include transport and friction. The detailed expressions for all the sources and sinks for the various phase concentrations are given in Chaumerliac et al. (1989), except for the rate of dissolution of the gaseous species into cloud water, which was treated slightly differently.

Two distinct predictive equations are considered for the concentration of the species in the air and in the cloud environment. This implies that a dissolution rate of the gaseous species into cloud water must also be considered and is expressed as

$$X_{dis} = X_a H_{eff} RT \frac{\partial q_{cw}}{\partial t}|_{cond},$$

where $\frac{\partial q_{cw}}{\partial t}|_{cond}$ is the condensation rate and H_{eff} is the effective solubility in moles per atmosphere for X in aqueous phase. It should also be noted that gas scavenging by water drops is calculated by the diffusion-controlled kinetic formula, as given in Iribarne and Cho (1989):

$$F = k_g(C - C_{eq}),$$

where F is the density of flux of gas toward the drop, C is the gas phase concentration, and C_{eq} is the value in equilibrium with the aqueous concentration in the liquid.

The coefficient $k_g = fD_g/r$ (where D_g is the diffusivity in the gas phase and r is the drop radius) contains a correction factor f that represents the mass-transfer enhancement produced by the relative motion of the drop and the air and by the drag-induced internal circulation. As discussed in Walcek et al. (1984)

for $r < 500$ μm , $f = 1$
for $500 < r < 900$ μm , $f = 2.5$
and for $r > 900$ μm , $f = 20$.

THE SOLUBLE TRACER EXPERIMENT

Three sensitivity studies were performed: in the first one, the tracer is only advected and considered to be insoluble, in the other two, the tracer is assumed to be soluble and its effective solubility is expressed as a function of the pH (as is the case for SO_2). Lowering the pH decreases the solubility. Figure 2 presents cross-sections obtained after 24 h of simulation time for the three cases. Figure 2a corresponds to the insoluble case, and Fig. 2b to the soluble case where pH=5 and the effective Henry's law constant is 10^3 M/atm. For Fig. 2c, the pH is set to 3 so that the Henry's law constant is of order 10 M/atm. This last case simulates the situation where SO_2 would be oxidized and where that oxidation would result in acidity production. In Fig. 2a, the tracer is transported from the boundary layer along the frontal surface, giving rise to concentrations of 140 μmole/kg at an altitude of 6 km. In Fig.2b, on the other hand, the tracer remains in the boundary layer. The scavenging of the tracer by clouds and precipitation causes large heterogeneous perturbations in the spatial distribution of this soluble gas which was initially uniform in the PBL. And finally, Fig. 2c shows a situation somewhere between the first two cases, where the vertical transport of the tracer is limited by wet chemistry processes. At 6 kms in altitude, we obtain a maximum of 80 μmole/kg, which is about half the value found in the non-soluble case.

To obtain further insight into the mechanisms and processes, we have drawn the temporal evolution of the vertical tracer profiles for a horizontal grid point close to the surface front. Figure 3b shows that the more soluble tracer is completely depleted in a couple of hours. For the other two cases where the solubility is zero or at least significantly reduced, we note that increased solubility limits the vertical transport of the tracer. To quantify this reduction, we have drawn in Fig. 4 the vertical profiles obtained in the different cases at 24 h as well as the initial profile (solid line). Comparing again cases a and c, we see that half of the ground tracer concentration (200 μmole/kg) is transported to the free troposphere in case c, whereas three fourths of it is transported to the upper levels in case a.

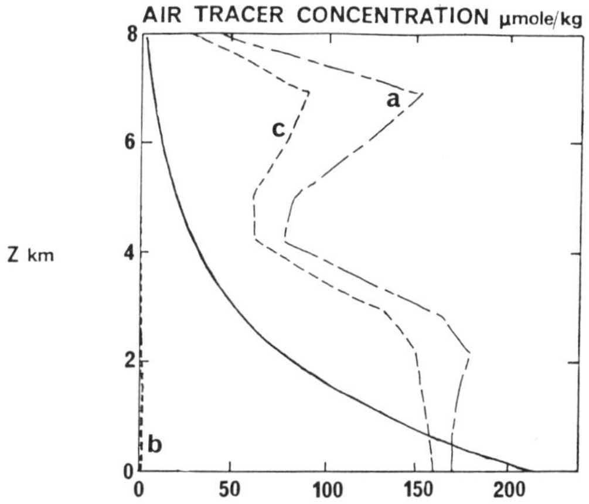

Fig. 4. Vertical profiles of air tracer concentration after 24 h for runs a, b and c (see Fig. 2.) at a horizontal grid point close to the surface front. The initial profile is indicated by the solid line.

CONCLUSION

Numerical simulations of moist frontogenesis (an evolving Eady wave) have shown that wet scavenging of gases is as important as vertical transport along the frontal surface. Even for a slightly soluble gas, scavenging by clouds and rain can limit the vertical transport to 50% of what it would be with scavenging neglected. For a more soluble gas, its lifetime is drastically reduced to several hours, whereas the vertical transport of an inert tracer can result in a residence time of several days. In such meteorological situations, scavenging processes can make an initially uniform PBL pollutant distribution quite heterogeneous and then maintain those strong concentration gradients.

This study is preliminary in nature and needs the addition of more realistic chemistry, including reactive species and pH evolution. It would be interesting to add an upper level front coupled with a surface front so that both upward and downward transport could be more fully considered. One would then be in a position to consider the mixing of two air masses with quite different chemical compositions.

ACKNOWLEDGMENTS

This work was supported by funds from A.T.P PIREN (C.N.R.S.), from the French Ministry of Environment, and from Electricité de France. Computer resources were provided by ECWMF (reading, England), CCVR (Palaiseau, France, Project n° 1717) and Météorologie Nationale (Paris).

REFERENCES

Berry, E.X., and R.L. Reinhardt, 1973: Modeling of condensation and collection within clouds. Desert Res. Inst. Phys. Sci. Publ., **16**, University of Nevada.

Blackadar A.K., 1978: High-resolution models of planetary boundary layer. Advances in Environmental Science and Engineering, **1**, J.R. Pfafflin and E.N. Ziegler, Eds., Gordon and Breach, 50-85.

Chaumerliac, N., and R. Rosset, 1989: Mesoscale modeling of acidity production in orographic clouds and rain. accepted for publication in Atmos. Env.

Dickerson R.R., G.J. Huffman, W.T Luke, L.J. Nunnermacker, K.E. Pickering, A.C.D. Leslie, C.G. Lindsey, W.G.N. Slinn, T.J. Kelly, P.H. Daum, A.C. Delany, J.P. Greenberg, P.R. Zimmerman, J.F. Boatman, J.D. Ray, and D.H. Stedman, 1987: Thunderstorms: An important mechanism in the transport of air pollutants. Science, **235**, 460-465.

Hsie E.U., R.A. Anthes, and D. Keyser, 1984: Numerical simulation of frontogenesis in a moist atmosphere. J.A.S., **41**, 2581-2594.

Iribarne, J.V., and H.R. Cho, 1989: Models of cloud chemistry. Tellus, **41B**, 2-23.

Keyser, D., and R. A. Anthes, 1982: The influence of the planetary boundary layer physics on frontal structure in the Hoskins-Bretherton horizontal shear model. J.A.S., **39**, 1783-1802.

Nickerson E.C., E. Richard, R. Rosset, and D.R. Smith, 1986: The numerical simulation of clouds, rain and airflow over the Vosges and Black Forest Mountain: A meso-β model with parameterized microphysics. Mon. Wea. Rev., **114**, 398-414.

Taylor G. R., 1989: Sulfate production and deposition in midlatitude continental cumulus clouds. Part II: Chemistry model formulation and sensitivity analysis. J.A.S., **46**, 1991-2007.

Walcek C.J., H.R. Pruppacher, J.H. Topalian, and S.K. Mitra, 1984: On the scavenging of SO_2 by clouds and raindrops: II. An experimental study of SO_2 absorption and desorption for water drops in air. J. Atm. Chem., **1**, 291-306.

IV: ACCIDENTAL RELEASES INCLUDING RADIOACTIVITY AND DENSE GAS DISPERSION

chairman: J.C.R. Hunt

ENVIRONMENTAL PREDICTION SYSTEMS: DESIGN, IMPLEMENTATION ASPECTS AND OPERATIONAL EXPERIENCE WITH APPLICATION TO ACCIDENTAL RELEASES

J. Pudykiewicz

Atmospheric Environment Service of Canada
2121 Trans Canada Highway, suite 534
Dorval, Quebec H9P 1J3, Canada

INTRODUCTION

The numerical simulation of atmospheric dispersion is used increasingly to assess the state of the environment. A well known example is the use of numerical dispersion models with complex chemistry schemes for studies of atmospheric oxidants and acid rain phenomenon. The other important application of tracer models is the simulation of accidental releases of toxic and radioactive materials. The Atmospheric Environment Service (AES) of Canada recently used such a tracer model to simulate the dispersion of nuclear debris from the Chernobyl reactor accident. This research was the beginning of the development of a sophisticated Environmental Emergency Response (EER) system which is currently operational at the Canadian Meteorological Center (CMC). Similar efforts were also undertaken by the meteorological services of other countries. The objective of this paper is to present a general description of the Canadian Nuclear Emergency Response Model (CANERM) and compare it to similar models which are being used in other meteorological centers around the world.

The development of the CANERM was initiated by the Atmospheric Environment Service of Canada in response to the Chernobyl accident. The first Canadian simulation of the Chernobyl dispersion was performed on a hemispheric grid in a pressure coordinate system using a relatively simple 2-D model. The details of the simulation are described in Pudykiewicz (1988). Following the emergency simulation, extensive development work with the three-Dimensional (3-D) tracer model was conducted. The new model was tested on the Chernobyl case using objectively analysed meteorological fields over a period of one month (Pudykiewicz 1989). The major limitation of the system employed in these studies was the use of the observed meteorological fields. This limitation was overcome when a predictive version of the tracer model was completed during the summer of 1988.

The new 3-D version of the model is implemented as an operational Emergency Response System at the Atmospheric Environment Service of Canada. The system is designed to deal with a wide variety of nuclear accidents. The operational version of the system is very flexible and could be executed either in diagnostic mode using objectively analysed meteorological fields or in predictive mode. The predictive mesoscale version of the of the system is run using output from the fine resolution Regional Meteorological Model. The meteorological fields from the hemispheric spectral model are used for the simulation of the very long range dispersion (similar to Chernobyl). Simulation of the dispersion processes on a local scale will be performed with the meteorological input provided by a nonhydrostatic mesoscale model.

The present version of CANERM is a sophisticated computer system designed to respond to various environmental emergencies related to accidental releases of toxic and radioactive materials into the atmosphere. The model is capable of treating stationary sources as well as moving sources of different origins (for example, a nuclear powered satellite decaying in the atmosphere) located within the Northern Hemisphere. Selection of the model configuration as well as the choice of the meteorological input is accomplished with the help of an interactive procedure executed on the front end computer. The menu of the procedure contains several important scenarios of nuclear accidents and some details related to the model structure. The user need specify only the time and location of the event for most accidents. If the case is beyond the scope of the automatic system, the procedure refers to the research version of the system involving a global tracer model. The special cases treated outside the fully automatic system include multiple sources, moving sources and a variety of the problems involving high atmospheric levels.

The work with the CANERM performed over the past 4 years was very important for the development of a state of the art Emergency Response System working in the environment of an operational Numerical Weather Prediction Center. We believe that this is most probably the optimal environment for this kind of computation. The tests of the recent version of the Emergency Response System on Chernobyl data indicate that the system is quite accurate and offers an accurate advisory capability for Emergency Response planners.

GENERAL DESCRIPTION OF THE TRACER MODEL

The most general form of the equations governing the transport of atmospheric tracers is given by a set of advection equations in flux form:

$$\frac{\partial n_v^i}{\partial t} = -\nabla \cdot \vec{J}^i + S^i(n_v^1, \ldots\ldots, n_v^M, X) \tag{1}$$

where: n_v^i - volumetric mixing ratio $(particles*m^{-3})$, $\vec{J}^i = n_v^i\vec{V} + n_v^i v_g^i \vec{k}$, v_g^i - settling velocity, $\vec{k} = [0,0,1]$ – unit vector indicating vertical direction, ∇ - gradient operator, S^i - sum of all external sources and sinks, X - vector of all external variables describing the state of the atmosphere.

In the analysis of problems involving radioactive tracers, it is convenient to transform equation (1) to the specific activity form:

$$\frac{\partial \Phi^i}{\partial t} = -\vec{V} \cdot \nabla \Phi^i - \frac{1}{\rho} \nabla \vec{k} v_g^i \rho \Phi^i + S^i(\Phi^1, \ldots\ldots, \Phi^M, X) \tag{2}$$

where: Φ^i - specific activity for radioactive tracers or mixing ratio for chemical species

In our study, eq. (2) was transformed to the terrain-following coordinate system σ (Phillips, 1957). After decomposing the variables in the resulting equations into grid cell average and fluctuating parts, followed by averaging, assuming small terrain slopes and the hydrostatic approximation (Pielke, 1984) we obtain the following equation:

$$\frac{\partial \overline{\Phi}^i}{\partial t} = -\vec{\overline{V}}_H^\sigma \cdot \nabla_H^\sigma \overline{\Phi}^i - \dot{\sigma}\left(\frac{\partial \overline{\Phi}^i}{\partial \sigma}\right) + v_g^i\left(\frac{\partial \Gamma \overline{\Phi}^i}{\partial \sigma}\right) - \langle V'^\sigma_H \nabla_H^\sigma \Phi'^i \rangle + \frac{\partial}{\partial \sigma}\Gamma^2 K_z \frac{\partial \overline{\Phi}^i}{\partial \sigma} + \overline{S}_\Phi^i(\overline{\Phi}^1, \ldots, \overline{\Phi}^M, X) \tag{3}$$

where: i= 1, M; M is the number of species, $\sigma = p/\pi$ is the vertical coordinate, p, π are the pressure and surface pressure respectively, $\overline{\Phi}^i$ is the grid cell average specific activity, $\vec{V}_H^\sigma$ is the grid cell average horizontal wind vector, ∇_H^σ is the horizontal nabla operator, $\dot{\sigma}$ is the vertical motion field in the σ system, v_g^i is the gravitational settling velocity, $\Gamma = g\sigma/RT$, g is the acceleration due to gravity, R is the gas constant, T is the temperature, K_z is the vertical diffusion coefficient, < > indicate an averaging operator and $\overline{S}_\Phi^i$ is the sum of the external sources and sinks.

The horizontal subgrid-scale flux term in the σ coordinate system was left in Eq.3 in symbolic form since our knowledge of the lateral turbulent mixing in the atmosphere is not based on any rational theory. Furthermore in our numerical simulation we neglected this term completely. Experience of other researchers (Kimura and Yoshikawa, 1988) indicate that the horizontal turbulent diffusion is less important than the indirect lateral mixing caused by wind shear and vertical diffusion.

Eq.3 is solved subject to the following boundary conditions:

$$K_z \Gamma^2 \frac{\partial \overline{\Phi}^i}{\partial \sigma} = \alpha^i \Gamma \overline{\Phi}^i + \beta^i \Gamma \qquad (4)$$

$$K_z \Gamma^2 \frac{\partial \overline{\Phi}^i}{\partial \sigma} = 0 \qquad (5)$$

where: σ_B, σ_T correspond to σ at the bottom and at the top of the model domain respectively, α^i is the deposition velocity for the i-th specie, $(\alpha^i = v_d^i)$, β^i is the surface emission flux for the i-th specie.

The terms describing all internal sources and sinks have the following general form :

$$\overline{S}_\Phi^i = Q^i - W(\overline{\Phi}^i) + D_{ij}\overline{\Phi}^j \qquad (6)$$

where: Q^i is the source term, $W(\overline{\Phi}^i)$ is the wet scavenging term and D_{ij} is the matrix describing the transformation of the isotopes and radioactive decay.

In the case of radioactive tracers the matrix D has block-diagonal structure with blocks corresponding to the different decay chains:

$$G_m = \begin{pmatrix} -\lambda_1^m & 0 & 0 & \dots & 0 & 0 \\ \lambda_1^m & -\lambda_2^m & 0 & \dots & 0 & 0 \\ . & . & . & & . & . \\ . & . & . & & . & . \\ . & . & . & & . & . \\ 0 & 0 & 0 & \dots & \lambda_{k-1}^m & -\lambda_k^m \end{pmatrix} \qquad (7)$$

λ_i^m - radioactive decay constant for the i-th step in the m-th decay chain, k - number of steps in the decay chain.

The form of the matrix describing the coupling between isotopes enables us to consider advection and diffusion separately for each group of radionuclides belonging to the decay chain. The situation is then much simpler in comparison to the simulation of atmospheric chemistry where the analogous matrix is not sparse.

The representation of point forcing on the grid of an Eulerian model is a numerical modelling problem common in computational fluid dynamics. The situation is particularly difficult in the simulation of atmospheric transport processes where the gap between the scale of the source and scales resolved by the model spans six orders of magnitude. There are several solutions to the source problem in Eulerian grid point models. One of them is the application of the Particle In the Cell (PIC) method by Lange (1978). This method gained a common acceptance in the context of the modelling of atmospheric tracers on a local scale. The alternative approach is based on the simulation of the initial stages of the release by a series of puffs, as discussed in the review of simple Long Range Transport of Air Pollutants (LRTAP) models given in Eliassen (1980). A different approach to the source term problem is based on the multigrid method discussed by Pudykiewicz et al. (1985). The release in this method is simulated on a series of nested grids with increasing resolution. The multigrid method is particularly attractive in the context of the simulation of local and mesoscale transport where detailed information about the deposition pattern in the vicinity of the source is important. In the case of a model with relatively coarse resolution which is adequate to trace radioactive and toxic species on the continental and hemispheric or global scale, it is convenient to use the concept of virtual source (Pudykiewicz 1988). From the point of view of the grid system the virtual source is equivalent to forcing the system with a very narrow Gaussian distribution, simulating both release and subgrid scale mixing. In the case of a 3-D model the virtual source is given by the following relation:

$$Q^i(\zeta_x, \zeta_y, \sigma, t) = \frac{E^i(t)F(\sigma)}{2\pi\sigma_H^2}\exp\left(-\frac{r^2}{2\sigma_H^2}\right) \tag{8}$$

where: $E^i(t)$ is the amount of the i-th specie released to the atmosphere, $F(\sigma)$ is the function describing vertical distribution of the release, r is the distance from the source and σ_H is the standard deviation of the horizontal mass distribution.

The function E (t) is obtained from the accident scenario. In the most general case, the accident scenario could be obtained from a separate computer model simulating the dynamics of the discharge of the radioactive or chemical species. A simpler solution is related to the use of the accident scenario from "look up tables" containing descriptions of the most typical potential accidents.

The vertical extension of the source term is given by a function F which has to be estimated from a separate model of thermal convection. The models suitable for this application are very similar to the simple models employed for the simulation of the dynamics of convective clouds. The application of the convection model to the source region will definitely improve the overall quality of the transport model by eliminating arbitrary assumptions about the vertical structure of release.

Numerical Solution of the Transport Equations

The regional scale modelling of atmospheric dispersion is most often performed on a polar stereographic projection grid. In fact, this mesh system is also convenient for large scale applications up to hemispheric scale. The model equations are transformed to the polar stereographic projection system true at 60 degree North. The corresponding equations may be written as follows:

$$\frac{\partial\overline{\Phi}^i}{\partial t} = (L_1 + L_2 + L_3^i + L_4^i + L_5^i)\overline{\Phi}^i \tag{9}$$

$$L_1 := -m^2\tilde{V}_H^\sigma\nabla_H^\sigma, \quad L_2 := -\dot{\sigma}\frac{\partial}{\partial\sigma}, \quad L_3^i := v_g^i\frac{\partial}{\partial\sigma}\Gamma, \quad L_4^i := \frac{\partial}{\partial\sigma}\Gamma^2 K_z\frac{\partial}{\partial\sigma} + Q^i, \quad L_5^i := S^i, \quad \tilde{V}_H^\sigma = \frac{1}{m}\vec{V}_H^\sigma$$

where: m is the map scale factor.

The time discretization of (9) is performed by successively applying the operators $L_1 - L_5$ to advance a time step Δt and then successively applying the operators in reverse order to advance a further time step. This particular method of operator splitting is described in detail in McRae et al. (1982); the "flip-flop" has the virtue of giving a second-order time truncation error, rather than the first order one that would otherwise be obtained. As noted by McRae et al. (1982), the precise choice of operator splitting is somewhat arbitrary and they chose to do it on the basis of coordinate direction, whereas we have chosen to do it on the basis of physical processes. The horizontal and vertical advection and gravitational settling are performed using the semi-Lagrangian algorithm developed by Robert (1982) and subsequently used for pollution problems by Pudykiewicz and Staniforth (1984). The most interesting property of the above method is that a linear stability analysis shows that the time step is not limited by a Courant condition. The time step may therefore be chosen on the basis of an acceptable time truncation error (e.g. of the same magnitude as the spatial truncation error), rather than on the basis of stability considerations (which are usually far more restrictive), provided the other terms in the equations are also handled in a computationally stable manner.

The vertical diffusion is solved using the implicit method described by Concus et al. (1976). The last substep related to scavenging and radioactive decay is accomplished using the analytical solution of the following equation:

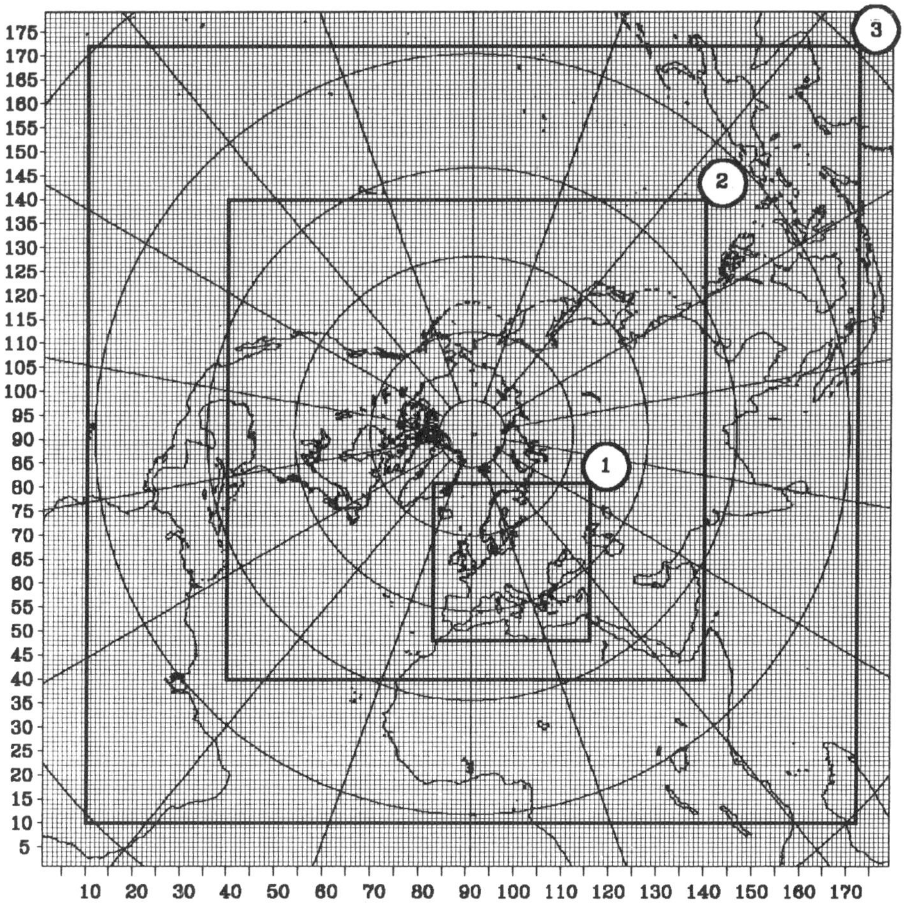

Fig. 1 The horizontal mesh over which the Hemispheric Tracer Model is integrated. The window marked 1 is used for regional scale simulations.

$$\frac{\partial \overline{\Phi}^i}{\partial t} = L_5^i \overline{\Phi}^i = -W(\overline{\Phi}^i) + D_{ij} \overline{\Phi}^j \tag{10}$$

In cases where equation (10) is nonlinear, numerical methods are employed instead of the analytical solution.

The area considered in the simulation performed with the hemispheric version of the CANERM is shown in Fig.1. The hemispheric grid of the model is uniform with a resolution of 150 km on the Polar Stereographic Projection. The corresponding grid domain is 179×179 points centered on the North Pole. The model resolution on a hemispheric scale can be increased up to 75 km. The number of levels employed in the simulation can vary between eleven and thirty. The spacing between levels depends on the nature of the source of the accidental release. Sources located close to the surface will require most of the vertical resolution to be placed within Planetary Boundary Layer (PBL) whereas the problem of high stratospheric sources will require the more uniform positioning of the model levels. The layer corresponding to $\sigma = 1$ is located at the anemometer level - z_a. The thin layer between z_a and surface is treated analytically using surface layer theory (Businger et al., 1970).

The Meteorological Input of the Diagnostic Model

The transport model run in diagnostic mode can be driven by the data stored in the archiving system of the Objective Analysis (OA) of the Canadian Meteorological Center (CMC), Rutherford (1977). The analysis system is executed in a 6 hr cycle and uses optimum interpolation of the forecast errors in order to provide analysed wind components, geopotential, temperature and dew point depression. The driving model of the OA system is a hemispheric primitive equation spectral model described originally by Daley et al. (1976). The function of the spectral model is to provide the 6 hr forecasts required in the assimilation cycle. The forecast model is initialized using Nonlinear Normal Mode Initialization. The target grid of the analysis is a gaussian latitude-longitude grid with a resolution of 128×32 points over the Northern Hemisphere. The temporal resolution of the meteorological fields from the OA system is 6 hrs. The meteorological fields for intermediate time levels are obtained by time interpolation. The vertical motion field in the σ coordinate system is obtained diagnostically from the continuity equation:

$$\dot{\sigma} = \int_{\sigma}^{1} [\vec{V}_H \cdot \nabla \ln \pi + \nabla \cdot \vec{V}_H] d\sigma - \frac{1-\sigma}{1-\sigma_T} \int_{\sigma_T}^{1} [\vec{V}_H \cdot \nabla \ln \pi + \nabla \cdot \vec{V}_H] d\sigma \tag{11}$$

where $\vec{V}_H$ is the horizontal wind vector and π is the surface pressure.

This diagnostic relation is the same as the one used internally in the spectral model of the assimilation cycle and is exact in the σ - coordinate system. Its use with an analysed velocity field is debatable but entirely justified because the assimilation model of the analysis cycle undergoes an initialization step before and after each analysis step. Consequently, the wind fields are relatively free of gravity waves and the vertical motions resulting from (11) are generally acceptable for atmospheric tracer studies on the hemispheric scale over an extended time between several days and several weeks.

The profile of the vertical diffusion coefficient required to solve eq. (3) is computed internally within the diagnostic tracer model using the analytical theory of the surface layer (Businger et Al, 1970) and the O'Brien profile above the constant flux layer Pielke (1984):

$$K_z = K_h + \left(\frac{z-h}{H-h}\right)^2 \left\{ K_h - K_H + (z-h)\left[\left(\frac{\partial K_z}{\partial z}\right)_h + 2\frac{K_h - K_H}{H-h}\right]\right\} \tag{12}$$

where $(\partial K_z/\partial z)_h$ is the derivative of the diffusion coefficient at the top of the surface layer, h, H are the height of the surface layer and the Boundary Layer respectively, K_h, K_H are the vertical diffusion coefficient at the top of the surface layer and at the top of PBL respectively.

The value of K_h and the profile of the diffusion coefficient in the surface layer were approximated using the relation provided by the analytical theory of the surface layer Businger (1970):

$$K_z = \frac{\kappa u_* z}{\Phi_n(z/L)} \tag{13}$$

$$\Phi_n(\xi) = \gamma + \alpha\xi, \quad \xi > 0; \quad \gamma(1 - \beta'\xi)^{-\frac{1}{2}}, \quad \xi < 0 \tag{14}$$

$\xi = z/L$, L is the Monin Obukhow length, z is the height, κ is the von Karman constant, u_* is the friction velocity and $\alpha = 4.7$, $\beta' = 9$, $\gamma = 0.74$.

The value of the diffusivity in the free atmosphere was assumed to be constant and equal to K_H. This approximation is sufficiently accurate for the simulation of atmospheric tracers on a hemispheric scale.

The deposition velocity appearing in the boundary condition was computed as the inverse of the sum of the resistances of the turbulent and laminar deposition layers and the resistance of the surface (Chamberlain and Chadwick, 1966):

$$v_d = \frac{1}{r_a + r_d + r_s} \tag{15}$$

$$r_a = \frac{\gamma \ln(z_a/z_0) - \psi}{\kappa u_*} \qquad r_d = \frac{\gamma \ln(z_0/z_d) + \kappa u_*/h_d}{\kappa u_*}$$

r_a, r_d, r_s are the aerodynamic resistance, deposition layer resistance and surface layer resistance respectively, ψ is the stability correction function, h_d is the transfer coefficient for the surface deposition layer, z_0, z_a, z_d are the roughness length, anemometer height and height of the top of the deposition layer respectively.

The most difficult problem in the hemispheric scale simulation of atmospheric tracers is the parameterization of wet scavenging. In general, the correct solution to this problem can be obtained only by applying a fully predictive cloud water scheme carrying explicit information about cloud and precipitating water at all model levels (Sundqvist, 1981). At this moment, the spectral model driving our assimilation cycle is not equipped with this scheme. On the other hand, the available observations of rain rates at the surface are extremely sparse and insufficient for a proper, deterministic parameterization of wet removal in the 3-D model.

Therefore, in this study we decided to employ a very simple, statistical parameterization with a wet scavenging term described by the following relation:

$$W^i = 0 \qquad for\, U < U_0 \tag{16}$$

$$W^i = \Lambda\left(\frac{U - U_0}{U_s - U_0}\right) \qquad for\, U \geq U_0$$

where: U, U_0, U_s are the relative humidity, the threshold value of the relative humidity above which the subgrid scale condensation occurs and the relative humidity for the saturation state respectively and Λ is a constant.

The parameterization employed is the rain-out type and does not take into account the difference between removal by stratiform and convective clouds. The scheme given by (16) is thus restricted to large scale application.

PREDICTIVE TRACER MODEL

The predictive models which simulate atmospheric tracers are relatively complex because of the need to represent a wide range of physical and chemical processes including atmospheric dynamics, clouds, solar radiation and boundary layer processes. To simulate the interaction between meteorology and atmospheric transport of nonconservative species one needs a set of thermohydrodynamic equations coupled with the mass conservation equations for atmospheric tracers. The system in its most general form could be written as follows:

$$\frac{dU}{dt} = F - \frac{1}{\rho}\nabla p - \Pi U \tag{17}$$

$$\frac{d\ln\rho}{dt} = -\nabla U \tag{18}$$

$$\frac{dT}{dt} = A_T + \frac{L}{c_p}Q - \frac{L}{c_p}E_0 \tag{19}$$

$$p = \rho R T \tag{20}$$

$$\frac{dq}{dt} = -Q + E_0 \tag{21}$$

$$\frac{dm}{dt} = Q - (P - E_r) - E_0 \tag{22}$$

$$\frac{dn^i}{dt} = S^i(n^1, \ldots, n^M, m, T), (i = 1, \ldots, M) \tag{23}$$

where U is the velocity field, F are the mass forces, p is the atmospheric pressure, Π is the operator representing turbulent mixing, ρ is the air density, ∇ is the gradient operator, T is the temperature, R is the gas constant, A_T is the dynamic part of tendency rate for temperature, Q is the latent heat of condensation, E_0 is the evaporation, m is the cloud water, P is the precipitation rate, E_r is the evaporation of rain, n^i is the mixing ratio for atmospheric trace species, $S^i()$ represents source and sinks for atmospheric trace species, d/dt is the material derivative, $d/dt = \partial/\partial t + \vec{U} \cdot \nabla$ and M is the number of atmospheric trace species simulated in the model.

The system (17)-(23) consists of (8+M) equations and simulates the meteorological processes and chemical reactions of atmospheric trace species. The boundary conditions for the dynamical and chemical part of the system have to be specified, and they depend on the particular geometry of the domain being considered. The first eight equations of the system are in fact a nonhydrostatic meteorological model. The model equations include the bulk representation of microphysical processes such as condensation, the formation of clouds and precipitation and the evaporation of rain. The atmospheric boundary layer processes are represented by the operator P which takes into account the momentum transfer by convective processes and by dynamically generated turbulence. The set of equations (23) represents a nonlinear atmospheric chemistry. The

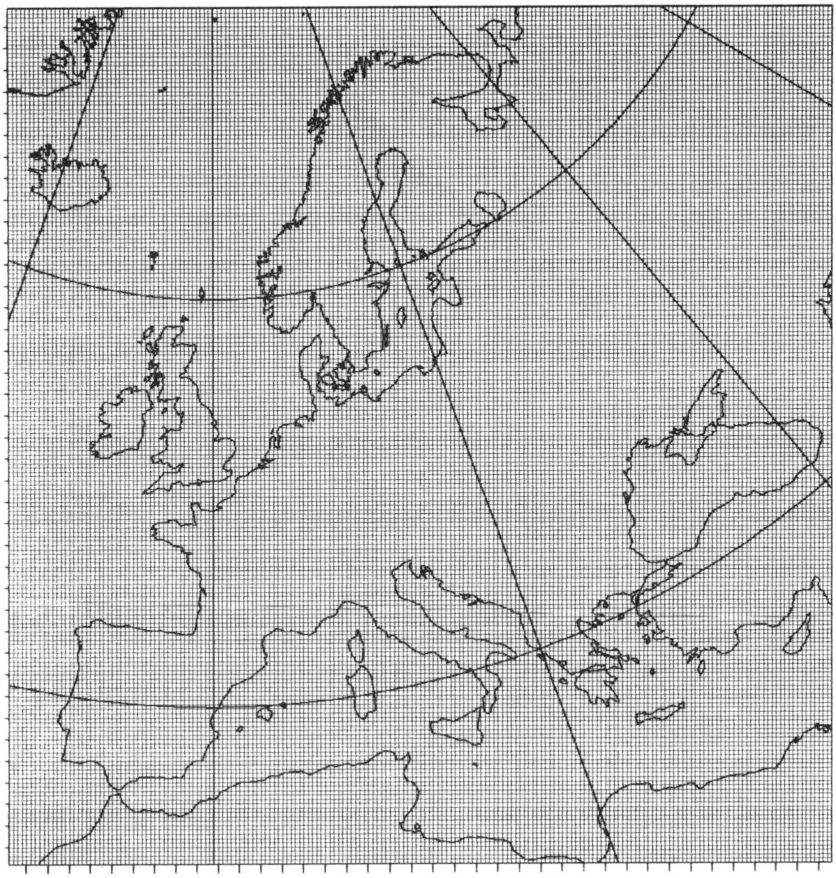

Fig. 2 The horizontal mesh used for regional scale simulation.

set given by equations (17)-(23) is designed mostly for application on a regional scale. The sample grid system for this application is shown in Fig.2.

The system for simulation of atmospheric processes on a hemispheric or global scale assumes hydrostatic equilibrium (Kasahara, 1977). This approximation leads to a tracer model driven by a primitive equations meteorological model (Pudykiewicz et al., 1985). The large scale predictive tracer model used in this study is, in fact, a set of tracer equations coupled to a spectral model (Daley et al., 1976). In general, this coupled set of models was designed to simulate tropospheric and stratospheric transport and chemistry. Formally, the equations employed by the predictive tracer model could be written as follows:

$$\frac{\partial \zeta}{\partial t} = -\nabla \cdot (\zeta + f)\vec{V} - k \cdot \nabla \times \left(RT\nabla\pi + \dot{\sigma}\frac{\partial \vec{V}}{\partial \sigma} - \vec{F} \right) \tag{24}$$

$$\frac{\partial D}{\partial t} = k \cdot \nabla \times (\zeta + f)\vec{V} - \nabla \cdot \left(RT\nabla\pi + \dot{\sigma}\frac{\partial \vec{V}}{\partial \sigma} - \vec{F} \right) - \nabla^2\left(\Phi + \frac{\vec{V} \cdot \vec{V}}{2} \right) \tag{25}$$

$$\frac{\partial T}{\partial t} = -\nabla\vec{V}T + TD + \dot{\sigma}\gamma - \frac{RT}{C_p}\left(D + \frac{\partial \dot{\sigma}}{\partial \sigma} \right) + H_T \tag{26}$$

$$\frac{\partial\pi}{\partial t}=-D-\frac{\partial\dot{\sigma}}{\partial\sigma}-\vec{V}\cdot\nabla\pi \tag{27}$$

$$\sigma\frac{\partial\Phi}{\partial\sigma}=-RT \tag{28}$$

$$\frac{\partial q}{\partial t}=-\vec{V}\cdot\nabla q-\dot{\sigma}\frac{\partial q}{\partial\sigma}-Q+E_0 \tag{29}$$

$$\frac{\partial\phi^i}{\partial t}=-\vec{V}\cdot\nabla\phi^i-\dot{\sigma}\left(\frac{\partial\phi^i}{\partial\sigma}\right)+v_g^i\left(\frac{\partial\Gamma\phi^i}{\partial\sigma}\right)+\frac{\partial}{\partial\sigma}\Gamma^2K_z\frac{\partial\phi^i}{\partial\sigma}+S^i(\phi^1,\ldots..,\phi^M,X) \tag{30}$$

where f is the Coriolis parameter, $\vec{V}$ is the horizontal wind vector, ζ is the vertical component of vorticity, $(\zeta=k\cdot\nabla\times\vec{V})$, D is the horizontal divergence $(D=\nabla\cdot\vec{V})$, T is the temperature, $\gamma=RT/C_p\sigma-\partial T/\partial\sigma$, $\dot{\sigma}$ is the vertical motion field, Φ is the geopotential height, F is the horizontal frictional force per unit mass, H_T is the diabatic heating, R is the gas constant for dry air, C_p is the specific heat of dry air at constant pressure, ϕ^i is the mixing ratio or specific activity of the i-th tracer, S^i represents the sources and sinks term for i-th specie (in general this term is described by the atmospheric chemistry scheme included in the model), v_g^i is the gravitational settling velocity for i-th specie, $\Gamma=g\sigma/RT$, g is the acceleration due to gravity and K_z is the vertical diffusion coefficient.

The model described by equations (24)-(29) is the Canadian Meteorological Centre's operational spectral model which is used to produce hemispheric weather forecasts for up to 6 days. The model uses a triangularly truncated set of spherical harmonics with the truncation at eighty-nine waves (T89). Sixteen levels are used in the terrain following coordinate system. Dynamic and thermodynamic variables are defined on staggered levels. The vertical discretization of the primitive meteorological equations is accomplished with a piecewise constant finite-element scheme. The time discretization employed in the model is semi-implicit with a typical time step of 20 minutes. The physical processes parameterized in the model include turbulent vertical fluxes, horizontal diffusion, stable and convective precipitation and radiative heating. Turbulent vertical fluxes of momentum, heat and moisture in the surface layer are calculated from similarity theory. Turbulent exchange above the constant flux layer is represented using eddy diffusion coefficients which are functions of vertical wind shear and static stability. Small scale horizontal mixing is parameterized using a weak biharmonic diffusion scheme for momentum, temperature and dry static energy. Stable and convective precipitation are included as one of the most important physical processes simulated by the model. Stable precipitation is currently calculated by removing the excess of moisture above some threshold value of relative humidity. The moist convection is parameterized following Kuo (1974). Cloud cover required in the calculation of radiative heating is obtained as a function of the local relative humidity. Surface temperature in the spectral model is calculated by solving an energy balance equation over land; over water, surface temperature is obtained from climatological data.

The tracer model consist of the set of equations (30) coupled with the primitive meteorological equations. For the purposes of this work, the set of advection-diffusion equations (30) was rewritten in a spherical coordinate system:

$$\frac{\partial\phi^i}{\partial t}=-\frac{1}{\cos^2\theta}\left[\frac{\partial}{\partial\lambda}\tilde{U}\phi^i+\cos\theta\frac{\partial}{\partial\theta}\tilde{V}\phi^i\right]-\dot{\sigma}\left(\frac{\partial\phi^i}{\partial\sigma}\right)+v_g^i\left(\frac{\partial\Gamma\phi^i}{\partial\sigma}\right)+\frac{\partial}{\partial\sigma}\Gamma^2K_z\frac{\partial\phi^i}{\partial\sigma}+S^i(\phi^1,\ldots..,\phi^M,X) \tag{31}$$

$$(\tilde{U},\tilde{V})=(U\cos\theta/r,V\cos\theta/r)$$

The time discretization of (31) is performed using a time splitting algorithm, as in the diagnostic model. The advection operator is approximated using the semi-Lagrangian technique in the spherical coordinate system. The scheme employed is described by Ritchie (1987). The vertical diffusion is solved using the implicit method described by Concus et al.(1976) subject to the boundary conditions representing the processes of deposition and surface emission. The last substep in the solution of equation (31) is accomplished using the analytical solution of the decay equation (10).

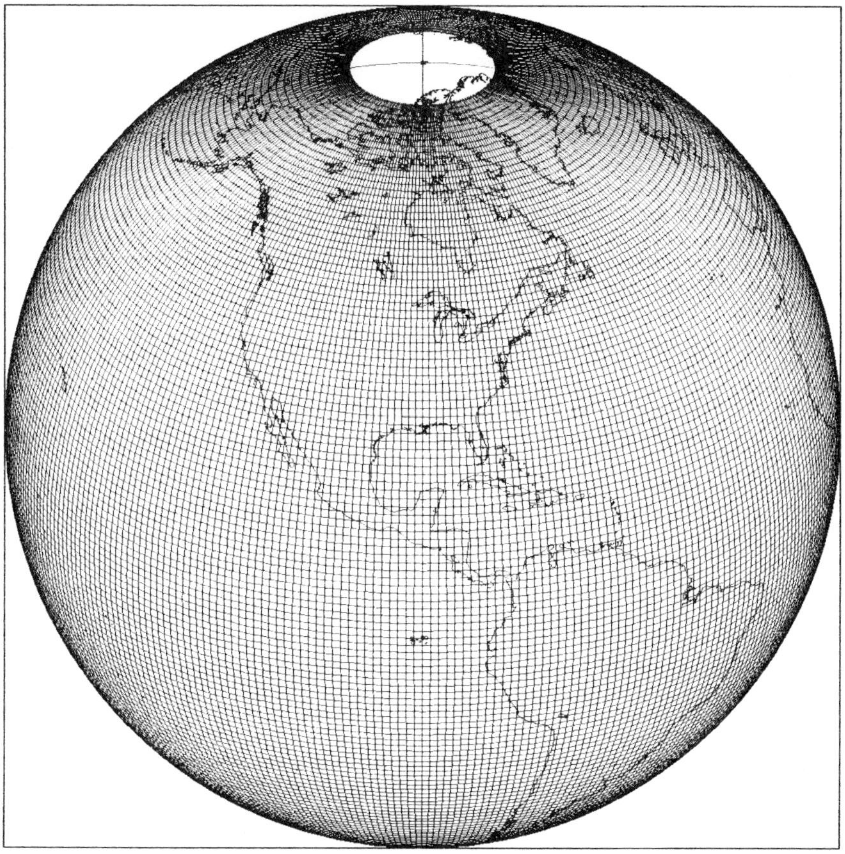

Fig.3 Grid system of the predictive tracer model

The predictive tracer model described in this chapter was obtained by coupling the spectral model with a set of advection-diffusion equations. The resulting system permits simulation of the transport processes and meteorological variables within one system. The boundary layer parameters, cloud cover and relative humidity are computed internally in the predictive tracer model. It is therefore possible to avoid expensive diagnostic calculations. This is an advantage of the fully predictive approach over a diagnostic model such as the one described in the previous section.

SIMULATION OF THE DISPERSION FROM CHERNOBYL USING THE CANERM SYSTEM

The atmospheric tracer model was executed for a period of 4 weeks, starting on April 25 1986, the first day of the release. The emission scenario employed in the computer simulation is quite complex. The release of radioactivity from the Chernobyl reactor occurred as a continuous process over a nine-day period. The initial release immediately following the explosion in the reactor accounted only for 25% of the total radionuclides injected into the atmosphere. After an extensive analysis of the observations of radioactivity in the surroundings of the Chernobyl reactor, it was concluded that there were four distinct stages in which radionuclides were released. The initial stage was a very intensive release on the first day of the accident. During the following five days, the release rates declined steadily to a minimum value six times lower that the initial release rate. This decline in the amount of material injected into the atmosphere was then followed by a four-day period of increase to reach a level of 70% of the initial release rate. The last stage of the process took place 9 days after the accident when the release was reduced to 1% of its initial value. The amount of release was estimated on the basis of radiation measurements and various technical analyses of samples taken from the vicinity of the Chernobyl reactor. It was concluded that about $1.0 \times 10^{18} - 2.0 \times 10^{18} Bq$ $(3.0 \times 10^{7} - 5.0 \times 10^{7} Ci)$ were discharged to the atmosphere from the core of the Chernobyl reactor. This estimate does not include the release of xenon and krypton, which were completely expelled from the fuel. The observation indicated that about 20% of volatile radionuclides like Iodine, Cesium and Tellurium were expelled from the damaged reactor. It is commonly accepted that the error of estimate of the Chernobyl release is 50%.

The source term in the advection equation was approximated using the concept of the 3-Dimensional virtual source. The effective height of the release was a function of time, changing from 4000m, just after the explosion, to 1500m subsequently. The initial height of 4000m was assumed to be in direct response to the initial explosion Pudykiewicz (1988). The smaller vertical extension of 1500m which was assumed during the days following the accident, mostly reflects the intensive convection over the burning reactor and subgrid-scale vertical mixing of radioactive material in the lower part of the troposphere.

The tracer equation (3) was solved initially using data stored in the archiving system of the Objective Analysis (OA) at the Canadian Meteorological Center (CMC) (Rutherford, 1977). The analysis of meteorological fields was performed every 6 hours on a hemispheric latitude-longitude grid with relatively low resolution of 128*32 points. The meteorological fields for intermediate time levels were obtained by time interpolation. The sample results from the diagnostic run are depicted in Fig.4. The most important conclusions from the results shown could be summarized as follows:

a) radioactive material at higher levels was moved along mostly with the average zonal westerly flow,
b) the transfer of the radioactivity across the Atlantic Ocean occurred because of eddy - like mixing by synoptic scale low pressure systems,
c) the influence of the synoptic scale vertical motions was manifested mostly over the eastern part of the Pacific Ocean,
d) the continent of North America was affected initially by material injected from the arctic regions.

The verification of the results from the diagnostic model was performed for a network covering the Northern Hemisphere and indicates that the simulation was quite accurate. The simulated times of arrival and times of arrival of maximum activity agree quite well with the observed values. The ratio of the time average activity simulated by the model and obtained from the measurements varies between 0.21 and 2.80 with a mean value of 1.05 (Pudykiewicz 1989). The ratio of the calculated and observed maximum values varies between 0.32 and 3.22 with an average value of 1.23. The lack of perfect agreement of specific values could be attributed mostly to the large uncertainty in the specification of the source term, errors in the meteorological fields and oversimplified physical parameterizations.

Imperfections of the numerical advection scheme also have some impact on the accuracy of the tracer simulation. A series of experiments reported by Pudykiewicz et al.(1985), Ritchie (1985) and Ritchie (1987) indicate that dispersion errors in the semi - Lagrangian scheme are smaller than those for Eulerian schemes. The accuracy of the simulation of the time of arrival presented in the figures is an additional indication of the accuracy of the dynamical part of our tracer model. We believe that the most important contribution to the error of our model is from the error of the analysed fields and from the error related to the parameterization of wet removal.

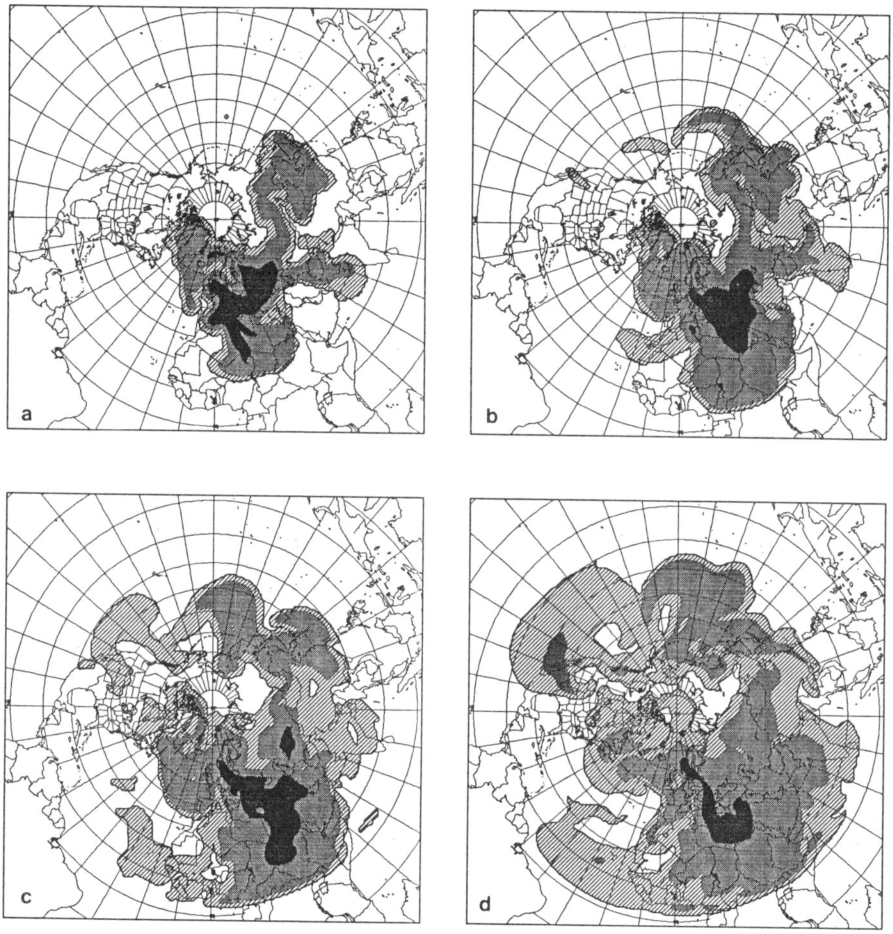

Fig.4 Activity fields (bq/ kg) for I-131 at anemometer level over window 3 of Fig.1., (a) 2 May, (b) 4 May, (c) 6 May, (d) 8 May. Activities between 0.0001 and 0.01 bq/kg are represented by the dashed pattern, activities between 0.01 and 1.0 bq/kg by the grey pattern, activities larger than 1 bq/kg by the black pattern.

Statistical verification of the model is presented in Fig.5, where the results of the simulation versus measurements are compared. The horizontal bars are equal to the model uncertainty and the vertical bars are equal to the arbitrary assumed error of the measurements. A diagram such as Fig.5 is called a scatter diagram. The analysis of the

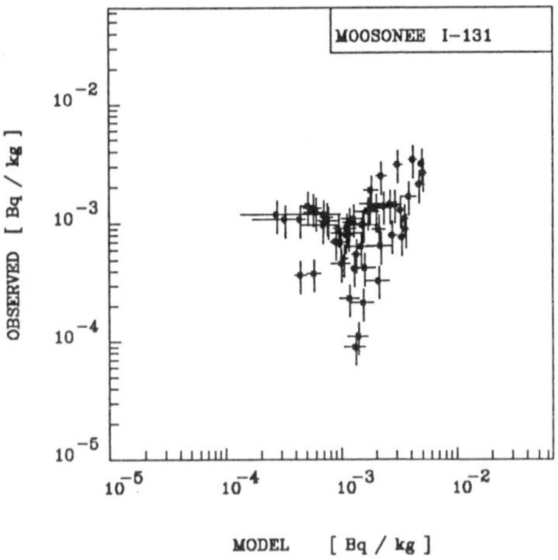

Fig.5 Scatter diagram for Moosonee (Northern Ontario) showing simulated activities of the I-131 versus measured activities. The horizontal bars indicate the uncertainty of sampling, whereas the vertical bars represent the measurement error.

scatter diagrams for the analysed receptor points indicates that in all cases the points are spread quite uniformly around the diagonal of the graphs.

The coefficients of the regression line :

$$y = \alpha_2 + \beta_2 x \tag{32}$$

fitted to the experimental points on the scatter diagram using the least squares method, are described by the following formulae:

$$\beta_2 = \frac{\mu_{11}}{\sigma_1^2} \qquad \alpha_2 = \mu_y - \beta_2 \mu_x \tag{33}$$

where:

$$\mu_{11} = \frac{1}{N}\left[\sum_{i=1}^{N}(x_i - \mu_x)(y_i - \mu_y)\right], \quad \mu_x = \frac{1}{N}\sum_{i=1}^{N} x_i, \quad \mu_y = \frac{1}{N}\sum_{i=1}^{N} y_i \tag{34}$$

$$\sigma_1^2 = \frac{1}{N}\sum_{i=1}^{N}(x_i - \mu_x)^2$$

N is the number of points in the sample, x_i are the simulated activities at anemometer level, y_i are the observed activities.

The regression line (32) could be also written as follows:

$$y - \mu_y = \beta_2(x - \mu_x) \tag{35}$$

The analogous regression line of x on y is given by:

$$x - \mu_x = \beta_1(y - \mu_y), \quad \beta_1 = \frac{\mu_{11}}{\sigma_2^2}, \quad \sigma_2^2 = \frac{1}{N}\sum_{i=1}^{N}(y_i - \mu_y)^2 \tag{36}$$

The correlation coefficient ρ defined as:

$$\rho = \frac{\mu_{11}}{\sigma_1 \sigma_2} \tag{37}$$

could be described by the following relation:

$$\rho^2 = \beta_1 \beta_2 \tag{38}$$

It is easy to prove that if and only if x and y are in a strict linear relationship, the two regression lines (32) and (35) coincide ($\rho = 1$). When x and y are uncorrelated the two regression lines are perpendicular. In the intermediate case, the angle between the regression lines is given by following formula:

$$\theta = arc\tan\left(\frac{\sigma_1\sigma_2}{\sigma_1^2+\sigma_2^2}\left(\frac{1}{\rho}-\rho\right)\right) \tag{39}$$

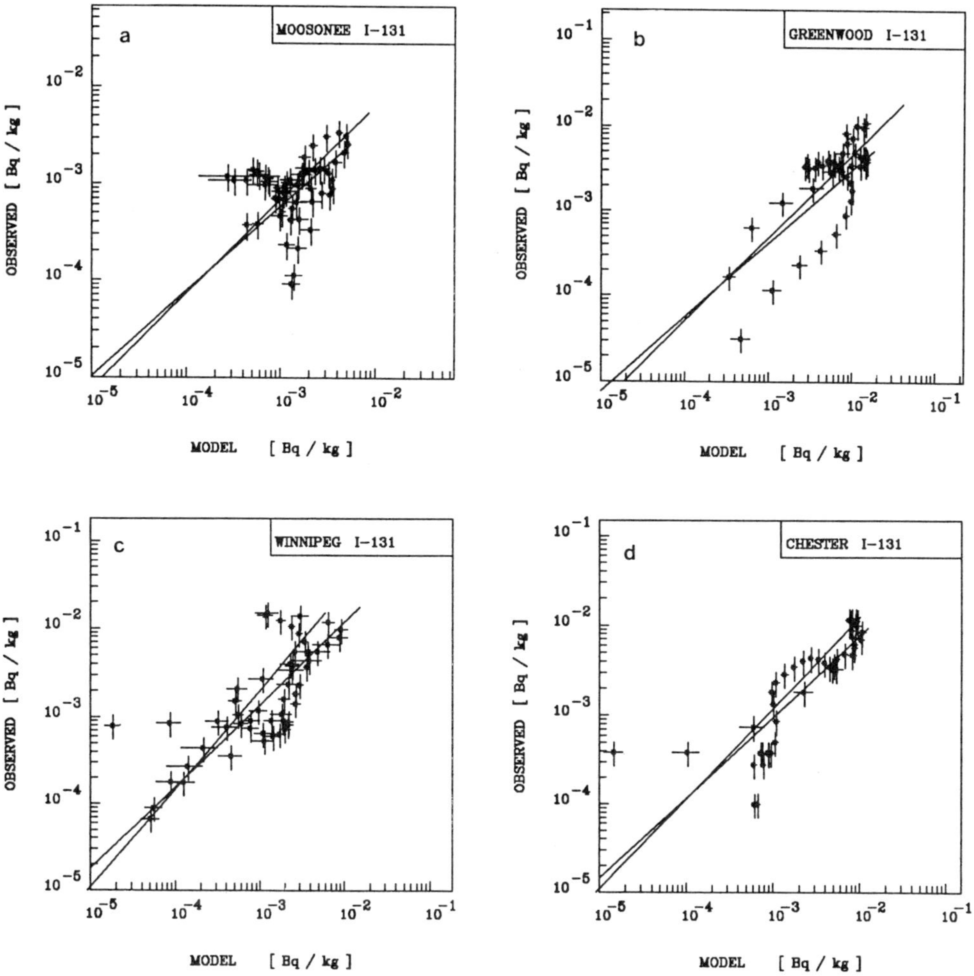

Fig.6 The scatter diagrams and regression lines for a) Moosonee (Northern Ontario), b) Greenwood (Nova Scotia), c) Winnipeg (Manitoba) and d) Chester (New Jersey). The regression lines were obtained from the least squares fitting performed for the logarithms of the activities.

The study performed shows that for the sampling points considered, the angle between two regression lines is always less then 1 degree. This indicates that the relationship between model results and observations is very nearly linear. The values of the correlation coefficient displayed in the third column of table 1 provide additional information about the coherence of the simulated and observed activities.

The activities considered in the hemispheric scale simulation are of the order of millions of Bq/kg close to the source and fractions of mBq/kg in the remote locations. Because of this extreme variability of activities the regression lines were fitted also for logarithms of the activities. The correlation coefficient for the logarithmic case is displayed in the second column of Table 1. New values of the correlation coefficient are between 0.73 and 0.95 with an average value of 0.88. The fact that these values are so large shows that the order of magnitude of the activity field predicted by the model agrees very well with observations. Fig.6 shows the results of the logarithmic least square fit for the group of four receptors The angle between two regression lines is very small which is an indication that orders of magnitude predicted by the model are related linearly to the orders of magnitude of the observations.

Table 1 Correlation coefficients for receptor points located within Northern Hemisphere.

station	ρ_{log}	ρ
Helsinki	0.73	0.70
Stockholm	0.80	0.78
Paris	0.73	0.24
Chiba	0.96	0.85
Rexburg	0.93	0.52
Moosonee	0.95	0.80
Fredericton	0.84	0.76
Greenwood	0.94	0.81
Winnipeg	0.91	0.65
Vancouver	0.90	0.63
Beaverton	0.91	0.47
Barrow	0.84	0.38
Windsor	0.86	0.72
Ottawa	0.88	0.77
Chester	0.95	0.94
Calgary	0.89	0.80
Regina	0.87	0.78

TESTING OF THE PREDICTIVE TRACER MODEL

The results of calculations performed with analysed meteorological fields represent an ideal situation which cannot possibly be achieved in a predictive mode because of inherent errors in the Numerical Weather Prediction (NWP) models. However, the good accuracy of the simulation performed with the diagnostic tracer model raises the question: "What is the maximum range of the predictability of the transport of radioactive tracers on the synoptic and hemispheric scale?".

The test of the predictive model began with an attempt to predict the transport of I-131 released from the Chernobyl reactor during the 7 days following the accident. The model employed had a triangular truncation of T89 and the corresponding hemispheric gaussian grid had a resolution of 288×72. The tracer model was started using analyzed meteorological fields for 1800 GMT, April 25, 1986, the I-131 field was initially set to zero. The source term was turned on at the moment of the accident, namely at 2100 GMT and had exactly the same form as in the diagnostic simulation. The results of the diagnostic calculations indicate that during the first two days following the Chernobyl accident, the radioactive cloud was transported mostly towards Scandinavia. A second southern segment of the cloud had spread southeastward over the Black Sea in the direction of the Middle East. The position of the radioactive cloud near the surface is depicted in Fig.7a. On April 28 the northern part of the radioactive cloud moved southward, crossing central Europe. Subsequently on April 29, a well developed westerly flow began to transport radioactive material across the USSR (Fig.7c). The results of the predictive run for 48 hrs and 96 hrs are presented in Fig.7b and 7d. It is shown that the 2 day prediction has relatively good accuracy, as the model was quite good in predicting the main directions of the transport. However, the quality of the 4 day prediction deteriorated substantially.

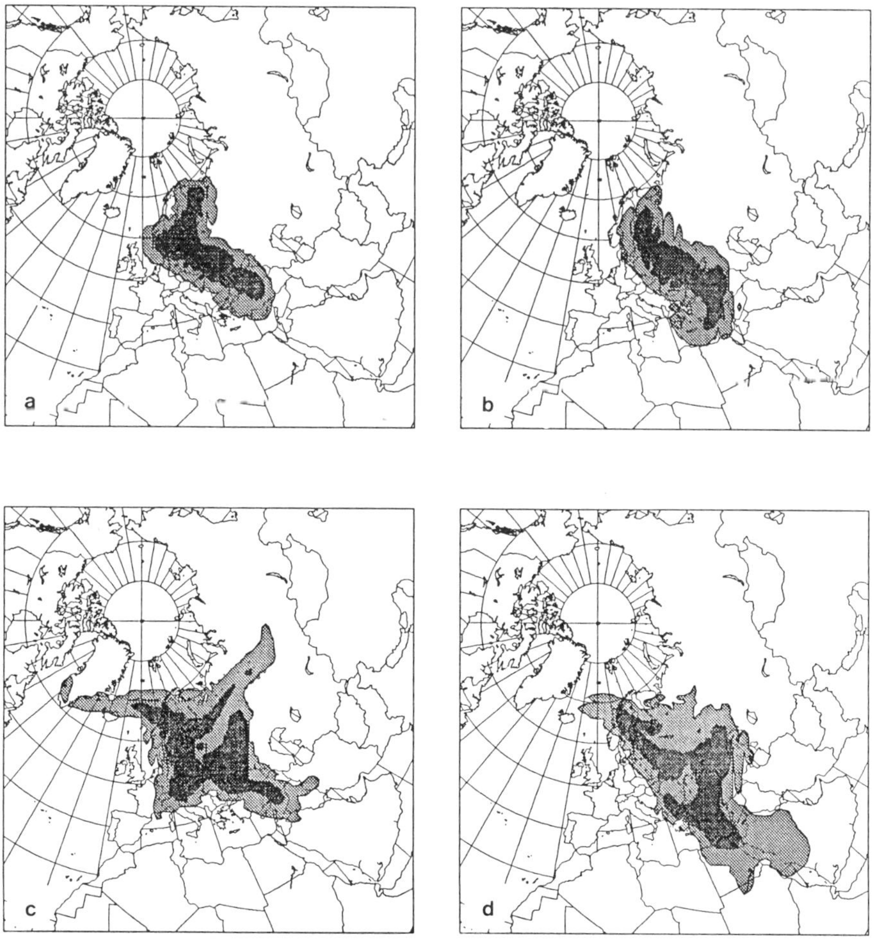

Fig.7 Predicted versus diagnosed activity fields of I-131 at $\sigma = 1$ over Europe. a) diagnosed activity field on April 28,1986 1800GMT, b) activity field predicted for 48 hrs., valid on April 28, 1986 1800 GMT. c) diagnosed activity field on April 30, 1986 1800 GMT, d) activity field predicted for 96 hrs., valid on April 30, 1986 1800 GMT.

The deficiency of the predictive model is manifested mostly by the lack of radioactive material over western Europe, Fig.7b and 7d. It is obvious that the value of the prediction beyond 4 - 5 days is very small in this case. The first run of the predictive model shows, however, that we can predict the distribution of the tracer on a regional scale for the first 2 - 3 days after the release. The development of a system capable of producing an accurate mesoscale and regional scale prediction must await future enhancements of global models and increased sophistication of regional NWP models. In order to investigate the performance of the predictive tracer model on a larger scale, we decided to continue the run of the predictive model starting from 1800 GMT May 1, 1986. The meteorological part of the initial condition was provided by the objective analysis of the Canadian Meteorological Center whereas the initial conditions for the tracer field were obtained from the diagnostic run. In other words, the procedure was to use observations to update the meteorological part of the system and a parallel diagnostic run to initialize the tracer part. In the following step, the model was integrated forward in time in order to produce a prediction of the tracer distribution. This particular form of the execution of our tracer model will be called "mixed mode" because of the use of the diagnostic model to provide the initial condition for the tracer field.

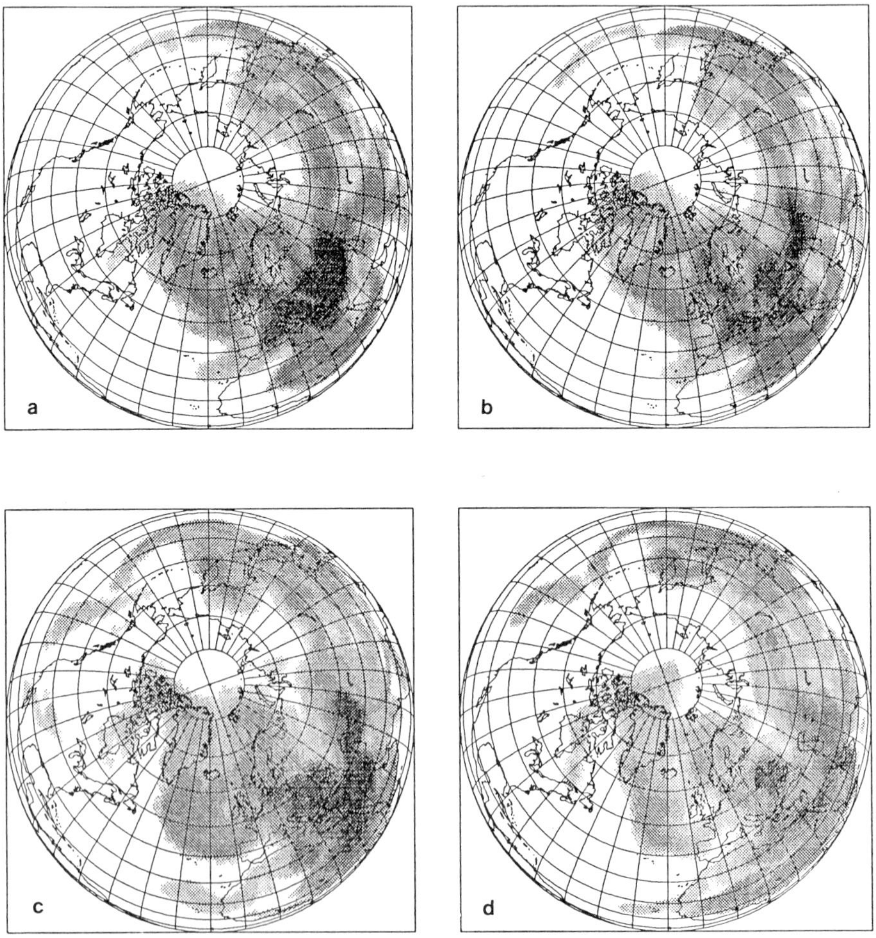

Fig.8 Predicted versus diagnosed activity fields of I-131 at anemometer level over the Northern Hemisphere a) diagnosed activity field on May 3, 1986 1800 GMT, b) activity field predicted for 48 hrs., valid on May 3, 1986 1800 GMT. c) diagnosed activity field on May 5, 1986 1800 GMT, d) activity field predicted for 96 hrs., valid on May 5, 1986 1800 GMT.

The comparison of the diagnostic and mixed mode results for May 3 and 5 at 1800 GMT is depicted in Fig.8. It is shown that the prediction for 2 days (Fig.8b) and 4 days (Fig.8d) starting from the large scale tracer distribution on May 1 is correlated quite well with the diagnostic model results. Also, one notices that the tracer model employed in the mixed mode correctly predicts the two major directions of transport of radioactive material to the Western Hemisphere. The characteristic circular distribution of mass over Greenland and the Atlantic Ocean was also predicted quite well by our tracer model.

The other main features reproduced by the predictive model are the injection of the radioactive material through the Canadian arctic to Northern Quebec and the transport of material in the westerly flow over the Pacific Ocean. On May 7, after the dissipation of the low over northern Quebec, the radioactive material over the arctic region moved rapidly southeastwards passing through Ontario, Quebec and the Atlantic provinces of Canada. The descending motions on the western side of the radioactive cloud brought down a small amount of the material from the higher levels. The radioactive material appeared along the west coast from Vancouver (British Columbia) to Alaska and over the southwestern states of the USA. The corresponding comparison of the diagnostic and predictive runs is shown in Fig.9. It is shown again that the 6 days prediction is surprisingly accurate concerning the large scale patterns.

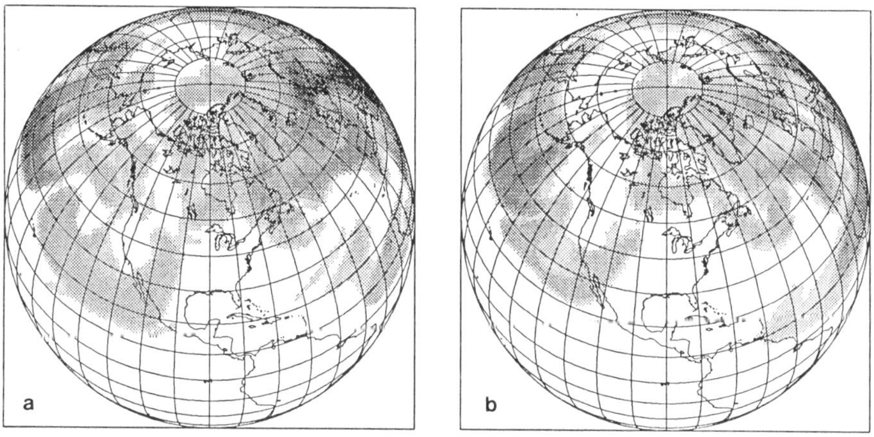

Fig.9 Predicted versus diagnosed activity fields of I-131 at anemometer level over the Northern Hemisphere a) diagnosed activity field on May 7, 1986 1800 GMT, b) activity field predicted for 144 hrs., valid on May 7, 1986 1800 GMT.

The updating procedure for the tracer field was then repeated for the next two 7 day intervals. Fig.10b and 10d show the two and four day predictions performed using the tracer field valid 1800 GMT, May 7 as the initial condition. It is shown that after May 8, the radioactive cloud attained hemispheric scale due to eddy-like mixing by the synoptic scale low pressure systems superposed with the average zonal flow. The cloud viewed on a hemispheric scale essentially shows a pattern analogous to turbulent diffusion, with low pressure systems acting as mixing eddies. It is important to note that the patterns obtained from the forecast and diagnostic modes are again in relatively good agreement (Fig.10). The analysis of the results described here shows that the prediction of the large scale distribution of the I-131 was more accurate than the initial predictions on the regional scale. This fact could be attributed partially to the poor accuracy of the initial medium range forecast on a regional scale. It is more certain however, that

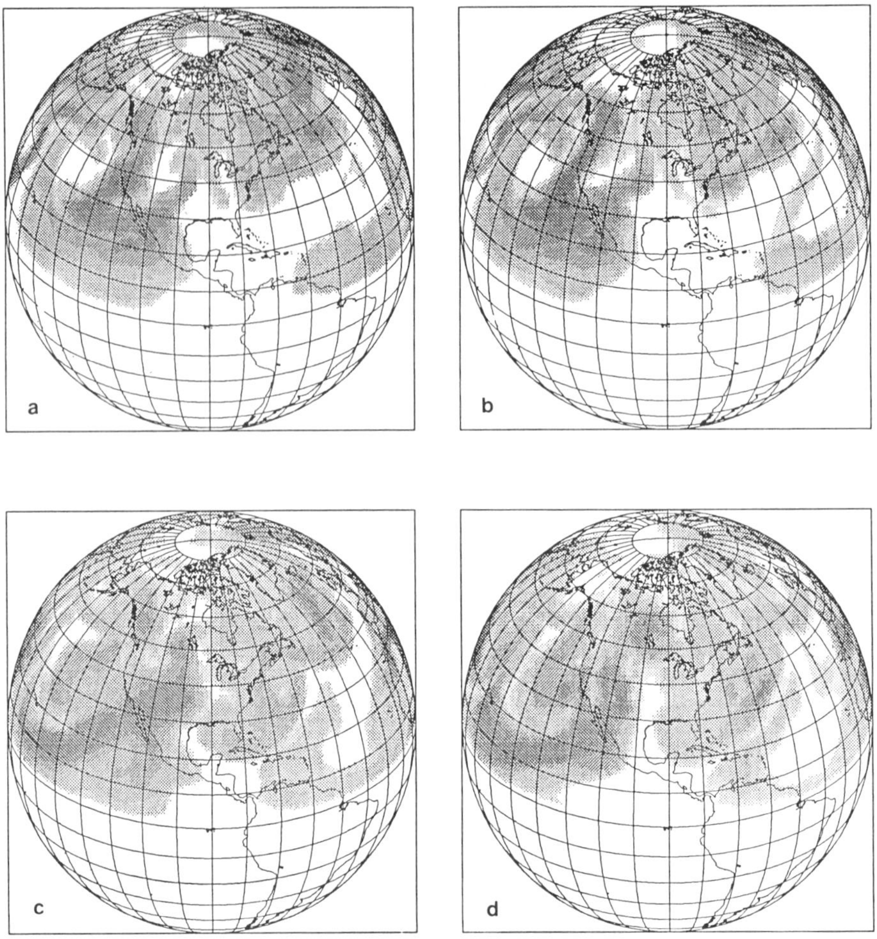

Fig.10 Predicted versus diagnosed activity fields of I-131 at anemometer level over the Northern Hemisphere a) diagnosed activity field on May 9, 1986 1800 GMT, b) activity field predicted for 48 hrs., valid on May 9, 1986 1800 GMT. c) diagnosed activity field on May 11, 1986 1800 GMT, d) activity field predicted for 96 hrs., valid on May 11, 1986 1800 GMT.

the better performance of the model on the hemispheric scale could be explained by the good accuracy of the meteorological prediction of the synoptic scale patterns.

Verification of the Predictive Model Results

The measurements of radioactivity performed following the Chernobyl accident provide a unique data base for extensive verification studies of atmospheric models. In this paper, we present the preliminary verification of the predictive tracer model against measurements from a network covering most of the Northern Hemisphere. The model verification on this scale is very important because it is difficult to estimate the predictive power of the CANERM on the basis of the regional tracer studies only. In most cases regional scale tracer experiments are limited by the size of the domain under consideration. The influence of lateral boundary conditions is often a factor limiting our ability to evaluate models over time intervals of the order of several weeks. In this study, the influence of lateral boundary conditions was not present because during the 4 week period considered, the interhemispheric exchange of radionuclides was negligible. Verification of the predictive model presented here was performed for the surface activities of I-131. The predicted activities are also compared to the results obtained from previous diagnostic calculations (Pudykiewicz, 1989). The verification results are presented in the form of graphs displaying time series of the measured and simulated activities. The predictive model verification for Stockholm is shown in Fig.11a as the black area between the two lines representing the lower and upper model bounds. The model bounds for the experimental point located within a particular grid cell are defined by the following formulas:

$$f_L(t)L = f_M(t) - f_\delta(t)$$

$$f_U(t) = f_M(t) + f_\delta(t)$$

$$f_M(t) = \sum_{k=1}^{n} \alpha_k f^{(k)} \quad , \quad n = 4 \tag{40}$$

$$f_\delta = \sqrt{\frac{\sum_{k=1}^{n} (f_M(t) - f^{(k)}(t))^2}{(n-1)}}$$

$f_L(t), f_U(t)$ are the lower and upper bounds of the model for a particular receptor point, $f^{(k)}(t)$ are the grid point values from the corner of the grid cell containing the receptor point and α_k are the weights of interpolation.

The diagnostic model results are represented by the dashed belt between the lower and upper model bounds. The measured values of the activity at the surface are represented by the vertical error bars. For display purposes, we assumed that the error is 30 %.

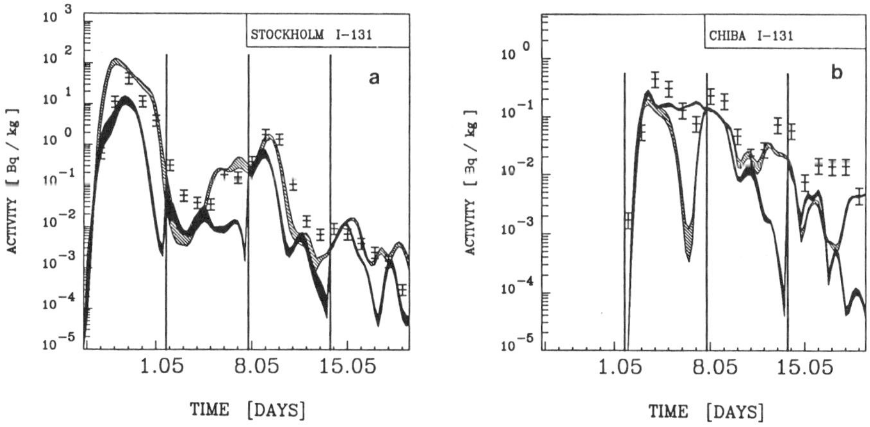

Fig.11 Comparison of the observed, predicted and diagnosed time series of I-131 activity in: a) Stockholm, b) Chiba (located on Honshu Island, Japan). The measured values are represented by vertical bars. The diagnostic model results correspond to the dashed area between the two lines, and the predictive model results are represented by the black area. The two lines depicting the model results take into account the uncertainty of the sampling process.

The tracer model described here simulates grid-cell average values of specific activity with the incorporation of wet scavenging by cloud systems resolvable by the grid system. The subgrid scale wet removal by mesoscale precipitation systems is not represented in the simulation. When comparing the observed and predicted specific activities, it was assumed that the observations are representative of the synoptic scale. The mesoscale meteorological processes certainly affected the observed values of specific activity by introducing short time scale perturbations caused by, for example, the passage of a mesofront or convective showers. The time scale of these types of perturbation is relatively short; from 0.5 hours up to a maximum of 6 hours. These high frequency perturbations were filtered out of the observational data because the observed specific activities were given as daily averages.

The analysis of Fig.11a indicates a very good coherence between specific activities observed at the surface and predicted values at the anemometer level. The predicted time variations of the specific activity are reproduced very well throughout the simulation. It is remarkable that the prediction of the time of arrival of radioactivity in Stockholm as well as the simulation of the second maximum of activity are very accurate.

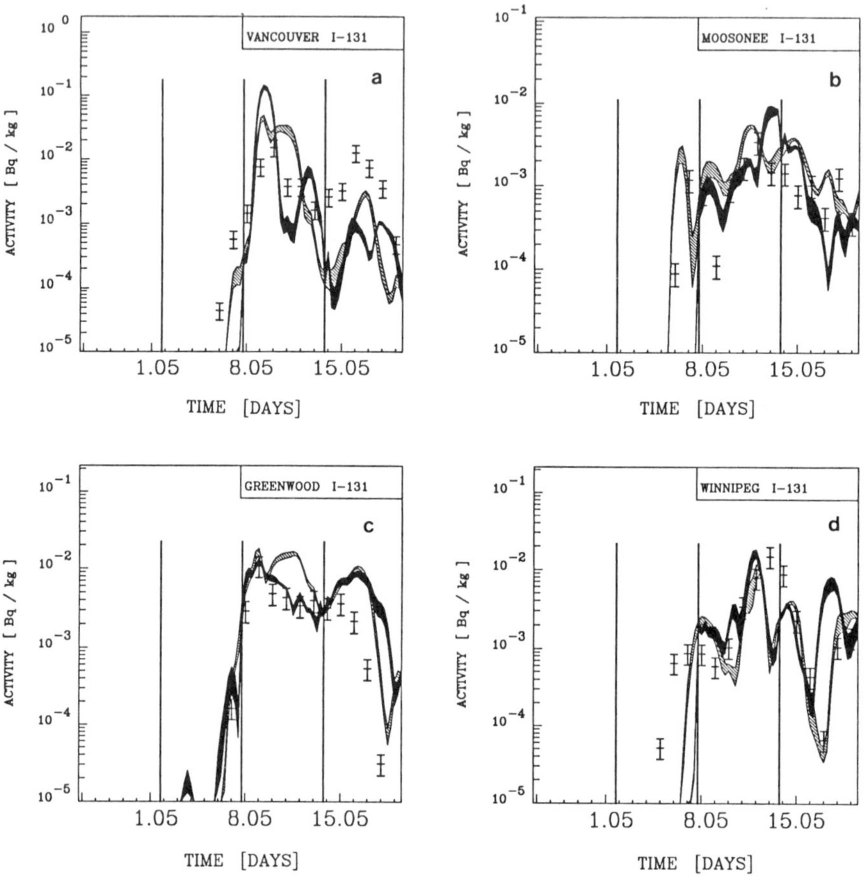

Fig.12 Comparison of the observed, diagnosed and predicted time series of I-131 activity in (a) Vancouver (British Columbia), b) Moosonee (Northern Ontario), c) Greenwood (Nova Scotia), d) Winnipeg (Manitoba).

Analysis of the model results indicates that the radionuclides from the Chernobyl accident transported by a steady westerly flow reached Japan on May 3. This conclusion, drawn from the predictive model run corresponds very well to the abrupt increase of radioactivity of the surface air at Honshu Island reported by Higuchi et al. (1988). The quantitative model verification for Japan is depicted in Fig.11b where measurements at Chiba are compared with the model results. The correspondence between the predicted and the observed times of arrival of the radionuclides is quite good.

The general pattern of transfer of the radioactive material from Chernobyl to North America is depicted in Fig.8. The results of the numerical simulation shown in Fig.9 indicate that the radioactive cloud appeared in the surface air over the southwestern states of the USA and northeastern Canada on May 5. The comparison of the activities of I-131 predicted by the model at the anemometer level to the values obtained from the measurements taken at Vancouver and Moosonee (Northern Ontario, Canada) is depicted in Fig.12 a,b. It is shown that for Vancouver the predictive model correctly simulates the general trends of specific activity, the time of arrival of the radioactive cloud and even the maximum values of activity. Concerning the model verification for Moosonee, there is a 2 day error in arrival time. However, the predicted time of arrival of the maximum values and the general trend of activity at this point are simulated correctly. The error in the prediction of the time of arrival of activity at Moosonee is a good illustration of the predictability problem for atmospheric tracers.

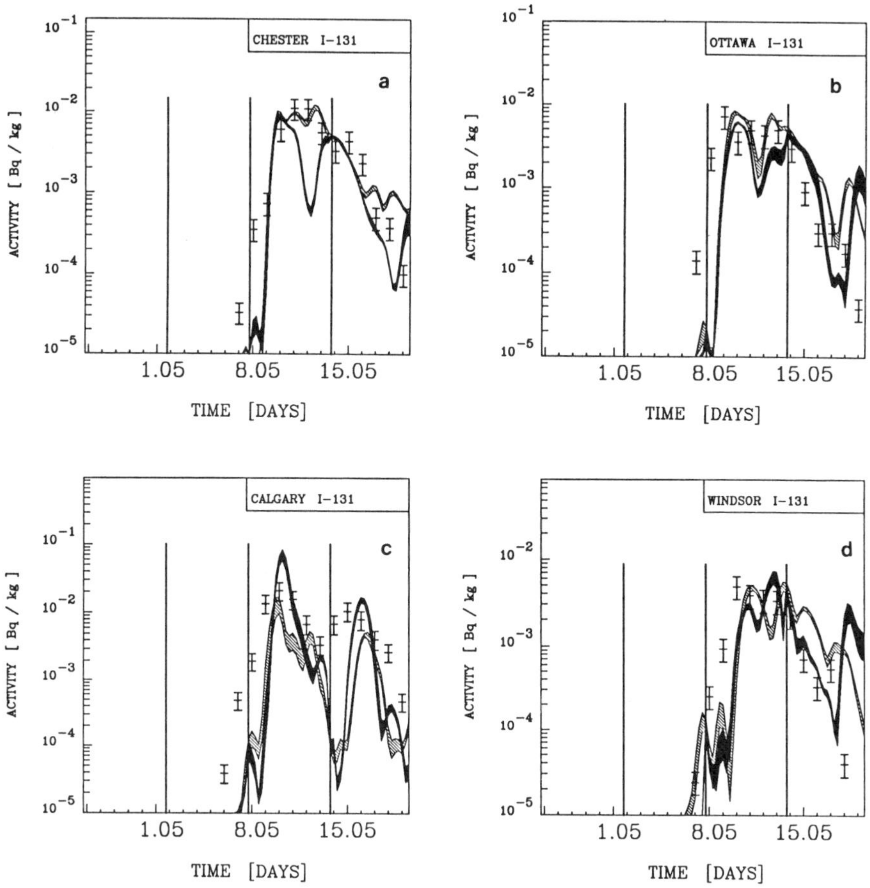

Fig.13 Comparison of the observed, diagnosed and predicted time series of I-131 activity in: a) Chester (New Jersey), b) Ottawa, c) Calgary (Alberta) and d) Windsor (Ontario)

It is interesting to note that the radioactive cloud affected the Atlantic and Central parts of Canada at the same time as it affected the West Coast, namely on May 5 and 6 (Fig 8 c). The verification of the model for Atlantic and Central Canada is shown in Fig 12 c, d. It is evident that the predictive model is able to reproduce the initial Arctic injection of the radioactive material as well as the subsequent influence of the material transferred over the Atlantic Ocean. In general the pattern of transfer of radioactive material from Chernobyl to North America was quite complex. The maximum values of the activity appeared in the surface air in different locations in North America at approximately the same time, namèly on May 9 and May 10 (Fig.10). This effect can be largely attributed to the fact that the continent was affected by material travelling from across the Pacific as well as from the arctic regions and the Atlantic Ocean. In fact, the occurrence of maximum activities of I-131 in North America coincided with the merging of the western, polar and Atlantic parts of the radioactive cloud. This process occurred as a result of horizontal advection and moreover was intensified by the subsidence of radioactive material in a high pressure system. This descent brought down to the surface material from higher levels where horizontal advection was much stronger than in the lower part of the troposphere. The large scale subsidence was therefore an important factor contributing to the observed times of arrival of the maximum activities. The time series for Chester (New Jersey), Ottawa, Calgary (Alberta) and Windsor (Ontario) depicted in Fig.13 a-d provide additional evidence of this fact. The verification of the predictive model for Canadian and American stations presented in this section clearly shows that the numerical results are quite accurate. Moreover, the predicted times of arrival and the times of arrival of maximum activity agree quite well with the observed values and the parallel run of the diagnostic model.

ENVIRONMENTAL EMERGENCY RESPONSE MODELS USED IN OTHER COUNTRIES

The comprehensive description of the Canadian Environmental Emergency Response system should provide some general information about modern capabilities in the area of numerical simulation of accidental releases. In this section some information about the American, Japanese and French system will be presented in order to provide the reader with sufficient information about the computer models used for environmental emergencies. Similar models are being used or are under development in other countries. The best known examples include the mesoscale 4 - level Eulerian model developed in the Netherlands and the most recent version of the UK Meteorological Office model based on the Lagrangian Particle in Cell method. The most common feature of all these models is their very strong dependence on the objective analysis of meteorological fields and numerical weather prediction. The development of the CANERM clearly indicates that the close link between Numerical Weather Prediction and Environmental models is very beneficial. The physical parameterization developed for the NWP models could be used directly in the Environmental models thus assuring that all the components of the atmospheric environment are simulated using a consistent approach. Similarly, the state of the art numerical algorithms developed for the Numerical Weather Prediction could be used with very minor modification for the simulation of tropospheric and stratospheric tracers.

The Accidental Release Advisory Capability (ARAC) system

The Accidental Release Advisory Capability (ARAC) is a comprehensive computer system developed in United States by researchers from the Lawrence Livermore National Laboratory. ARAC is designed to provide real time prediction of the dose levels and the extent of surface contamination resulting from a broad range of possible events including accidents, spills, extortion threats involving nuclear materials, reentry of nuclear powered satellites and atmospheric nuclear tests. The basic models of the ARAC system are the three-dimensional computer codes MATHEW and ADPIC which were developed in the mid-seventies for the simulation of local and mesoscale dispersion.

The large scale atmospheric transport processes in the ARAC system are simulated using a three-dimensional sequential puff transport and diffusion model (PATRIC) designed originally for application on a continental and hemispheric scale. This model is a simplified version of the ADPIC model which uses the Particle In Cell (PIC) method. The radioactive or chemical tracers in this method are represented by a large number of particles which reside in an Eulerian grid system. In order to simulate a release, particles are injected as a sequence of puffs at the source point and subsequently transported within the three-dimensional grid mesh. The transport velocity defined in grid points of the Eulerian grid consists of a wind velocity provided by a meteorological model and a diffusion velocity. The additional processes included in the model are gravitational settling, dry deposition and radioactive decay. The activity field is obtained by summing the resulting distribution of particles over the grid mesh volumes. The PATRIC model was applied to estimate the hemispheric transport of radioactivity from the Chernobyl accident. The model was run on a cartesian mesh consisting of 40*50*14 grid cells with 381 km horizontal resolution and constant 750 m vertical spacing. The grid system was defined on a Polar Stereographic projection true at 60 degrees north. In order to resolve the strong gradients of activity over the European continent, a fine particle sampling grid with one third of the original resolution was placed over the source region. The dispersion calculation with the PATRIC model was performed using wind fields over the Northern Hemisphere supplied by the U.S. Air Force Global Weather Central (AFGWC). The grided wind fields contained values at seven pressure levels ranging from 1000 to 250 mb. The main limitations of the model employed was the neglect of wet scavenging processes and ignoring the influence of topography on the transport of radionuclides.

Transport and dispersion models of the ARAC system

The models employed in the ARAC system are based on the Particle-in-Cell concept introduced originally by Amsden (1966). This technique involves the generation of a large number of marker particles to represent the distribution of radioactivity or a toxic chemical in the atmosphere. The most sophisticated model of the ARAC system is the Atmospheric Diffusion Particle-in-Cell (ADPIC) code. The ADPIC is a three-dimensional, Cartesian particle-diffusion model designed to simulate the time-dependent distribution of atmospheric tracers in complex topography (Lange, 1978). The model is capable of treating space and time-varying wind fields, calm conditions, space variable surface roughness, wet and dry deposition, radioactive decay and space and time-variable diffusion parameters.

The ADPIC model solves the three-dimensional advection-diffusion equation in its flux conservative form using a pseudo-velocity technique. The method employed could be described as follows. Consider the following transport equation:

$$\frac{\partial \Phi}{\partial t} + \vec{V}\nabla\phi = \nabla \mathbf{K} \nabla \Phi \tag{41}$$

where Φ, $\mathbf{K}$, $\vec{V}$ are the transported field, diffusion coefficient and nondivergent 3-Dimensional advection velocity field. Combining two divergence terms we can rewrite equation (41) in the following pseudo-velocity form:

$$\frac{\partial \Phi}{\partial t} + \nabla\left[\Phi\left(\vec{V} - \frac{K}{\Phi}\nabla\Phi\right)\right] = 0 \tag{42A}$$

$$\frac{\partial \Phi}{\partial t} + \nabla\Phi\vec{U} = 0 \tag{42B}$$

where $\vec{U}=\vec{V}-(K/\Phi)\nabla\Phi$ is defined as pseudotransport velocity. The advection field V is generally obtained from a nondivergent three-dimensional wind field model. The computational step of the ADPIC model is divided into Eulerian and Lagrangian steps. During the Eulerian step, the diffusion velocity is computed from the given distribution of particles. This information is used during the subsequent Lagrangian step for the updating of positions of the marker particles according to the following relation:

$$\vec{R}_{t+\delta t}=\vec{R}_t+\vec{U}\delta t \tag{43}$$

where $\vec{R}_{t+\delta t}, \vec{R}_t, \delta t$ are vectors indicating the position of particle and the length of the time step, respectively. The step is completed by the computation of a new concentration field based on the new positions of the particles.

The most important advantages of the particle-in-cell method for the simulation of atmospheric tracers are:

a) fictitious numerical diffusion is eliminated,

b) the particles can be tagged with their coordinates, age since release, mass, activity, chemical properties and size. This information is important for the parameterization of sinks of atmospheric tracers, and

c) the PIC algorithm is very inexpensive to run.

Despite the availability of Eulerian schemes of higher order which are able to eliminate artificial numerical diffusion, the PIC methods are still quite attractive. The most important reason for this is the positive definite characteristics of the PIC algorithms.

The Tracer Diffusion Model of the Japan Meteorological Agency

The model developed in Japan simulates movement of tracers on a global scale using particle approximation (F.Kimura and T.Yoshikawa, 1988 and T.Takasugi and H.Nakamura, 1988). The meteorological fields are supplied by a global spectral model of the Japan Meteorological Agency (Kanamitsu et Al, 1983). The global model employed in the simulation of the transport processes has triangular wavenumber truncation T42. The gaussian grid used for computation of nonlinear terms and physical processes has resolution 128 * 64 points. The tracer model is executed in a decoupled mode after the meteorological prediction is completed. The calculation of the dispersion is carried on for 48 hrs. using the wind field stored every two hours. The updating of the meteorological fields is performed every 2 days.

The dispersion model is based on the random walk approximation. The tracer in this approximation is simulated by a large number of Lagrangian particles which are advected in the three dimensional wind field perturbed by an additional random component. The random component in the present version of the model is added only to the vertical velocity in order to simulate turbulent diffusion. The equations of motion of the Lagrangian particles are solved using a simplified Runge-Kutta method having second order accuracy. A time step of two minutes is used for vertical motion terms whereas a longer ten minute time step is employed for horizontal advection. All calculations are performed in the spherical coordinate system.

The Emergency Preparedness System of Electricite de France (EDF)

Electricite de France operates over 42 nuclear reactors on 20 sites. The Emergency Preparedness System was designed to centralize the radiological data for all French nuclear plants and predict the eventual effects of accidental releases. The assessment of the effects of releases is performed using a two level structure:

a) a local scale response based on a very simple model, and

b) a centralized computer system interacting with local models. The central system is based on a complex mesoscale model and a synoptic scale model.

The local scale dispersion is simulated using a variety of gaussian type models. The model employed for mesoscale simulation is a hydrostatic code HERMES nested in the French Weather Service model PERIDOT. The function of the HERMES code is to predict the wind field and boundary layer turbulence with 5 to 10 km resolution up to 36 hrs and use this information to calculate the transport and deposition of radionuclides. The long range transport is computed using a relatively simple approach based on the segmented plume model which is executed in conjunction with a large scale weather prediction model. This model was already successfully tested by simulating Chernobyl dispersion. The system employed by EDF is under continuous development. The puff trajectory model will be introduced as an additional model for local scale calculations. Concerning the simulation of dispersion in complex terrain and under complicated meteorological conditions, a 3-D Monte Carlo code is being developed. Comprehensive discussion of this methodology is given by Pielke (1984). The Monte Carlo model will be used together with an objective analysis scheme to deal with dispersion problems in complex sites where the simple models are not applicable. The idea of introducing the specialized small scale dispersion models is quite similar to the efforts undertaken in order to develop the ADPIC model used as part of the ARAC system.

CONCLUSIONS

The simulation performed with CANERM in diagnostic and predictive mode is a good indication that with the computing power currently available and with high resolution numerical models, it is possible to treat advection dominated processes with good accuracy. We do not suggest that the model described here will perform under other circumstances as well as it has performed on the Chernobyl case. More research related to the parameterization of mesoscale processes is needed in order to further refine the technique. The important part of this process leading to the enhancement of environmental models is the exchange of information between different research centers and the intercomparison of existing models.

The results presented in this paper indicate however that the large scale distribution of an atmospheric tracer can be successfully predicted for up to 6 days. The selection of 6 days as a maximum prediction time, after which meteorological fields are updated from observation and the tracer field is updated from a parallel diagnostic run, is dictated by the observed performance of the spectral model. Prediction of the tracer distribution beyond the 6 day limit starts to deteriorate rapidly. In most practical applications with real data, this limit is quite reasonable.

We believe that in the case of future nuclear emergencies, the role of predictive meteorology will be more prominent than in the past. The numerical models available currently are sufficiently accurate for synoptic scale and hemispheric scale applications. Problems related to the simulation on the mesoscale remain to be solved. The most critical areas to be addressed in this research are related to the improvement of the representation of the complex structure of precipitation systems and the enhancement of the parameterizations of chemical processes and wet scavenging in the tracer model.

There are also other challenges in the area of development of the computer models for the analysis of environmental problems. The most important examples include increasing the horizontal and vertical resolution of weather forecast models which are employed to solve environmental problems and the creation of a coupled atmosphere-ocean model capable of treating the entire earth as a complete, interlinked system. A model of this type will be able to resolve very complex environmental issues including the interaction between meteorological processes and the biosphere.

REFERENCES

Amsden A.A., 1966: The particle-in-cell method for the calculation of the dynamics of compressible fluids. Los Alamos Scientific Laboratory, Report LA -3466

Businger J.A. , J.C.Wyngaard , Y.Izumi and E.F.Bradley 1970 : Flux - Profile Relationships in the Atmospheric Surface Layer. J. Atm. Sci 28, 181 - 189.

Chamberlain A.C. And R.C.Chadwick 1966: Transport of Iodine from Atmosphere to Ground. Tellus 18 , 226 - 237.

Concus P., G.H.Golub and D.P. O'Leary 1976: A Generalized Conjugate Gradient Method for the Numerical Solution of Eliptic Partial Differential equations. Rep. No. STAN - CS - 76 -533, Computer Sci. Dep., Stanford University.

Daley R., C.Girard, J.Henderson and I.Simmonds (1976): Short-Term Forecasting with a Multi-Level Spectral Primitive Equation model. Atmosphere 14, 98-134.

Eliassen A. 1980: A review of Long-Range modelling. J.Appl. Meteorol. 19, 231-240.

Kasahara A. (1977): Computational aspects of numerical models for weather prediction and climate simulation. Methods in Computational Physics. Vol.17, pp 2-60.

Kimura F. And T.Yoshikawa (1988): Numerical Simulation of Global Scale Dispersion of Radioactive Pollutants from the Accident at the Chernobyl Nuclear Power Plant. J. of the Met. Soc. of Japan. Vol 66, No.3, pp 489-495.

Lange.R. 1978 : ADPIC - A Three Dimensional Particle-in-Cell Model for the Dispersal of Atmospheric Pollutants and its Comparison to Regional Tracer Studies. J. Appl. Meteorol. 17 , 320 - 329.

Mc Rae G.J. , W.R.Goodin and J.H. Seinfeld 1982: Numerical Solution of the Atmospheric Diffusion Equation for Chemically Reacting flows. J. Comp. Phys. 45, 1 - 42.

Phillips N.A. 1957: A Coordinate System having some special advantages for Numerical Forecasting. J. Meteorol. 14 , 184 - 185.

Pielke A.R. 1984 : Mesoscale Meteorological Modeling. 612 pp Academic Press, London.

Pudykiewicz J. And A. Staniforth (1984): Some properties and comparative performance of the semi-Lagrangian method of Robert in the solution of the Advection-Diffusion Equation. Atmosphere-Ocean, 22, 283 - 308.

Pudykiewicz J., Benoit R. And Staniforth A. (1985): Preliminary Results from a partial LRTAP model based on existing meteorological model. Atmos. And Ocean. vol.23, 283-308.

Pudykiewicz J. (1988): Numerical Simulation of the transport of Radioactive Cloud from the Chernobyl Nuclear Accident, Tellus 40 B, 241-259.

Pudykiewicz J. (1989): Simulation of the Chernobyl dispersion with a 3 - D hemispheric tracer model. Tellus 41B.

Ritchie H. 1985 : Application of a semi-Lagrangian integration scheme to the moisture equation in a regional forecast model. Mon. Wea. Rev., 113 , 424 - 435 .

Ritchie H. 1987 : Semi-Lagrangian advection on a gaussian grid. Mon. Wea. Rev., 115 , 608 - 619.

Robert A. 1982: A semi-Lagrangian and semiimplicit numerical integration scheme for the primitive meteorological equations. J.Meteorol. Soc. Japan 60 , 319 - 325 .

Rutherford I.D. 1977: An Operational 3-Dimensional Multi- Variate Statistical Objective Analysis Scheme. Notes Scientifiques and Techniques, Division de Recherche en Prevision Numerique, Montreal, Canada.

Sundqvist H. 1981: Prediction of Stratiform Clouds: Results from a 5-day forecast with a global model. Tellus 33 , 242-253.

T.Takasugi and H.Nakamura (1988): A tracer diffusion model coupled with JMA spectral model. Japan Meteorological Agency, Numerical Prediction Division Technical Report, Tokyo, Japan.

DISCUSSION

A. VENKATRAM How do you explain the high correlation between model predictions and observations of activity? How does the model perform in explaining the magnitudes of the observed concentrations?

J. PUDYKIEWICZ The high correlation between model results and observations indicates a correct simulation of the trends of activity at all analysed receptor points as well as a relatively good simulation of the time of arrival of the radioactive cloud. These properties of the simulation of dispersion indicate that the dynamical part of the model which represents large scale meteorological processes is very realistic. An additional factor which contributed to a very good statistical evaluation of the model relates to the accurate numerical algorithms used to solve the advection diffusion equation. Finally, it is relevant to mention that the simulation described in the paper dealt with hemispheric dispersion on a time scale of several weeks. The influence of mesoscale processes, which are very difficult to simulate, was therefore much less important than in the modelling of shorter time scales of the order of hours or days. All these facts indicate that the simulation of atmospheric dispersion on a large scale could be very accurate in a statistical sense. This is possible mostly because of the advances in meteorological data assimilation techniques and the continuing improvement of the numerical algorithms in meteorological and dispersion models.

Concerning the magnitude of the observed activities the model results were also correct. In fact the ratio of the time-average activity simulated by the model and obtained from measurements varies between 0.21 and 2.80 with a mean value of 1.05 for a sample of 18 stations distributed all over the Northern Hemisphere. The lack of perfect agreement for specific values at given times could be attributed to the relatively poor accuracy of the source term and discrepancies in the parameterization of subgrid-scale meteorological processes.

H. HASS Comment to Venkatram's question: With regard to the underestimation of the first maxima after the arrival of the cloud in certain European areas we experienced the same as Dr. Pudykiewicz when modelling the Chernobyl release with the EURAD model. We suggest that this might be related to the source function used, or that the advection scheme is too diffusive.

The source term is definitely the first reason for errors in the simulated activities at receptors located relatively close to the source. Concerning the effects of numerical diffusion, its influence is probably also important on coarse grids with resolution of the order of 100 km. However, the effect of numerical diffusion could be reduced substantially when the regional scale simulation is performed on grid with resolution between 20 and 50 km. My own experience with numerical modelling of point releases indicates that the most critical factor influencing the accuracy of the results is the representation of the subgrid-scale source. When the subgrid scale source is represented by a δ function we can expect relatively large errors. The most accurate solution is to use the hybrid model where the point source release is simulated using Lagrangian particles in a manner analogous to the PIC (Particle In Cell) method.

THE WMO/IAEA/CEC MODEL INTERCOMPARISON STUDY (ATMES)

Han van Dop

World Meteorological Organisation
Case Postale 2300
1211 Geneva 2, Switzerland

The WMO is presently sponsoring a model evaluation study (ATMES), in collaboration with the International Atomic Energy Agency (IAEA) and the Commission of the European Communities (CEC). Although the primary aim is to simulate the transport and deposition of radionuclides following a nuclear accident on a continental scale, the study will also be of interest in relation to non-nuclear pollutants. Model results will be compared with the data collected from the Chernobyl accident. Best estimates of the emissions and effective source heights of Cs_{137} and I_{131} were provided to all participants. The ECMWF and the Royal Netherlands Meteorological Institute (KNMI) provide the meteorological data. For this purpose data sets were prepared consisting of wind fields at four pressure levels (500, 700, 850 and 1000 hPa), (Rubli, 1989).

Precipitation data are provided by KNMI consisting of six hourly precipitation totals over the area obtained from an interpolation procedure developed by Scheele (1989). Participants, however, are also encouraged to submit results using their own meteorological forecasts. Presented results should include:

- surface air concentration data of Cs_{137} and I_{131} for 89 prescribed locations at six hour intervals for 5 consecutive days and starting on 26 April 0300 UTC,
- high resolution time series of concentrations in air of I_{131} at two locations (Berlin and Munich),
- the wet deposition of Cs_{137} at the above 89 locations, and
- the cumulative deposition data of Cs_{137} on a grid consisting of 875 points.

More than 20 modellers have agreed to participate (see table 1). Their results will be processed in Ispra at the Euratom Joint Research Centre of the EC. In March 1991 the results will be discussed during a workshop of the participants. Final results will be presented in the summer of 1991. A timetable of the project is given in table 2.

A proposal was discussed between the above parties to support a follow-up study (starting in 1992). The study consists of two major elements:

- a well-defined and prepared dispersion experiment on a European scale consisting of three releases in Europe (see Fig. 1) of an inert, non-toxic tracer. Research organisations can use the data to evaluate and improve their models;
- a test of the ability of participating meteorological organisations to react to these releases. The meteorological organisations will be notified of the releases in the same way as would be the case in a real accidental release.

Air Pollution Modeling and Its Application VIII, Edited by H. van Dop and D. G. Steyn, Plenum Press, New York, 1991

Table 1. Participants cooperating in ATMES and their affiliation.

Name	Organisation	Country
J. Pudykiewicz	Canadian Meteorlological Centre	Canada
S. Skulec	Slovak Hydrometeorological Institute	Czechoslovakia
H. Hass and A. Ebel	Universität zu Köln	Federal Republic of Germany
I. Valkama and P.J. Airaksinen	Finnish Meteorological Institute	Finland
B. Crabol	CEA	France
F. Bompay	Direction de la Météorologie Nationale	France
A. Albergel and R.D. Wendum	EDF	France
H. Kaplan and N. Dinar	Israel Institute of Biological Research	Israel
M. Sciortino and A. Carrillo	ENEA	Italy
F. Desiato	ENEA-DISP	Italy
S. Moriuchi	Japan Atomic Energy Research Institute	Japan
S. Yamada	Japan Meteorological Agency	Japan
T. Yoshikawa	Japan Meteorological Agency	Japan
G.H.L. Verver	Royal Netherlands Meteorological Institute	The Netherlands
A. Grundling	Atomic Energy Corporation of South Africa Limited	South Africa
R. Maryon	UK Meteorological Office	United Kingdom
R. Lange	Lawrence Livermore National Laboratory	United States
M. Pendergast	Savannah River Laboratory	United States
V.N. Petrov	Institute of Applied Geophysics	USSR
M.J. Pekar	Institute of Applied Geophysics	USSR
B. Telenta	Federal Hydrometeorological Institute	Yogoslavia

Table 2. Time schedule of the ATMES project.

Submission of Results	31 May	1990
Evaluation accomplished	December	1990
Participants Workshop (ISPRA)	12-14 March	1991
Final Presentation	May	1991
Proposed Tracer Experiment	1992-1994	

Further information on the project can be obtained from the sponsoring agencies or from the project leader, W. Klug, Technische Hochschule, Darmstadt, FRG.

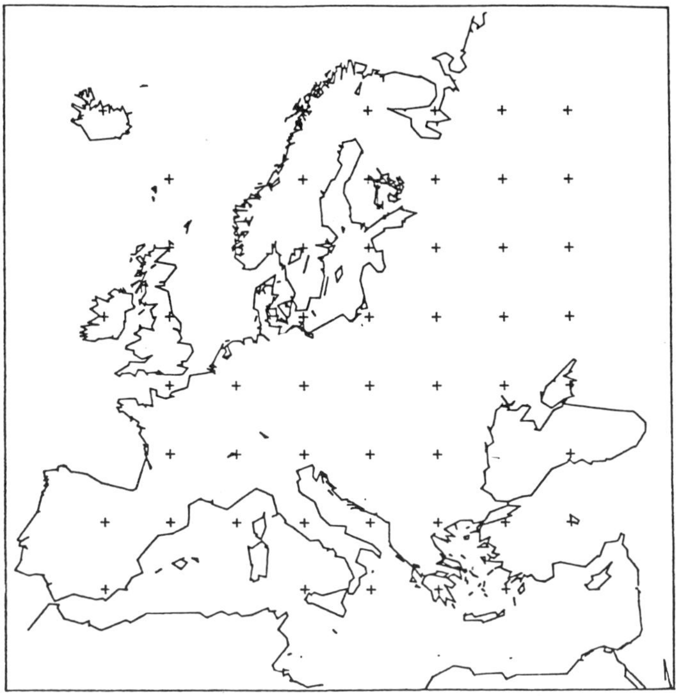

Fig. 1. Approximate region for the dispersion experiment. The crosses indicate tentative locations for 50 samplers.

REFERENCES

Rubli, A. (1989) Synoptic characteristics of the ECMWF Chernobyl data set, ECMWF Technical Memorandum No. 162 (Available on request at ECMWF, Reading, UK).

Scheele, M.P. and G.H.L. Verver (1990) Objective analysis of precipitation observations during the Chernobyl episode. Technical Report TR-122, KNMI, P.O. Box 201, 3730 AE De Bilt, The Netherlands.

MODELLING OF THE LONG-RANGE TRANSPORT OF RADIOACTIVE SUBSTANCES FROM ACCIDENTAL RELEASE

Alexander Pressman and Margarita Pekar

Meteorological Synthesizing Centre - East
of EMEP, Moscow

Serious accidents with intense release of adverse chemical and radioactive substances can affect vast ares as it was with Chernobyl accident. Eulerian three-dimensional model was developed by MSC-E for the evaluation of emergent consequencies and providing of prognostic modelling of accidents at large chemical or atomic plants. The model is based on finite-difference solution of an aquation of mass continuity allowing for dry and wet deposition, pollution transformation of the first order. The model makes integration over the EMEP grid with spatial horizontal resolution 150 km and vertical resolution 0.5 km. In addition to this main grid, grids with multiple gridsize 75 km and 37.5 km are used for the area near the source. The calculation of release of a certain duration (6-hour period is typical) is made in consecutive order, first with a small grid, then with a middle one and finally with the main one. At present meteorological data from two standard baric surfaces 1000 mb and 850 mb provided by the USSR Hydrometeorological Centre with objective analysis over the EMEP grid are used in the model. Meteorological data for small grids are obtained by interpolation.

Finite-difference approximation of advection is realized with directed difference scheme, in which wind velocity is corrected for minimization of pseudodiffusion. In each cell this correction is determined according to unlinear equation:

$$\tilde{u}_i = u_i + \frac{1}{2}(\tilde{u}_i\Delta x - \tilde{u}_i^2\tau)\left(\frac{1}{C}\frac{\partial C}{\partial x}\right)_i$$

(u - wind velocity, $\tilde{u}$ - corrected velocity, Δx - gridsize, τ - time step, C - concentration), which solution is:

$$\tilde{R}_i = \begin{cases} (-1 + \sqrt{1 + (a_i^2 - 1)R_i})/(a_i - 1), & \text{if } a_i \neq 1 \\ R_i, & \text{if } a_i = 1 \end{cases}$$

($R_i = u_i\tau/\Delta x$, $\tilde{R}_i = \tilde{u}_i\tau/\Delta x$, $a_i = C_{i\pm1}/C_i$ - signs correspond to those of velocity one).

A more thorough discussion of the model may be found in Pekar (1988).

The model was applied for calculating of concentration and deposition fields of radionuclides released to the atmosphere after Chernobyl accident.

Data of Izrael at al.(1989) were used as initial information of radioactivity release out of plant reactor. The release of each isotope was simulated by a monodisperse pollution with single dry deposition velocity (for Cs^{137} V_g = 0.15 sm/S) and washout coefficient (for Cs^{137} λ = 0.1 I h^{-1}, I - precipitation intensity - mm/h). Table 1 presents accumulated (dry + wet) deposition data of Cs^{137} in some countries on the 15th day after the accident. These data are compared with experimental and model results (Van Rheineck Leyssius at al.1988).

Table 1. Modelled and observed Cs^{137} deposition

Country	Calculation 10^{12}Bq	KBq/m²	Observed* KBq/m²	Modelled* KBq/m²
Austria	357	4.3	15	8
Belgium	62	2.0	0.8	1.4
Czechoslovakia	823	6.4	4	1.9
Denmark	228	5.3	0.8	0.3
France	626	1.1	1.3	0.9
Germany	1254	2.7	4	3.6
Italy	573	1.9	6	2.6
Holland	92	2.2	1.8	1.3
Sweden	1188	2.7	8	2
Switzerland	149	3.6	7.1	2.7
England	192	0.8	0.9	0.2
Greece	772	5.8	3.5	0.9

*Data of Van Rheineck Leyssius at al.(1988).

The agreement of measured and calculated results of deposition density is on the average within the limits of 2, though there are great discrepancies, for example, for Austria, Denmark, Sweden. Additional comparison of calculated and measured ground concentrations allowed to conclude that reason for discrepancies was improper consideration of wet deposition.

REFERENCES

Izrael Yu.A., Petrov V.N., Severov D.A., 1989, Regional model for transport and fall out of radionuclide from Chernobyl atomic plant accident, Meteorologia i Hydrologia, 6:5-14.

H.J. van Rheineck Leyssius, H.J.A. van Jaarsveld and F.A.A.M. de Leeuw, 1988, A Lagrangian model for the real time simulation of atmospheric transport and dispersion of accidently released materials, 17th Int. Tech. Meeting on Air Poll. Modell. and its Appl., 19-22 September, Cambridge, England.

Pekar M.I., 1988, Finite-difference schemes conservative in the first several moments for the long-range pollution transport, EMEP/MSC-E Scientific Report 2/88.

MEDIA: AN EULERIAN MODEL OF ATMOSPHERIC DISPERSION

VALIDATION ON THE CHERNOBYL RELEASE

J.P. Piedelièvre, L. Musson-Genon and F. Bompay

Direction de la Météorologie Nationale
SCEM/D/ES
2 avenue Rapp
75340 PARIS CEDEX 07, FRANCE

INTRODUCTION

After the Chernobyl accident, the French Weather Service had a few means to answer the questions from political authorities and press concerning the evolution of the toxic cloud. At that time, only a trajectories model was available. So, though this mean is very useful for retro trajectories, we have developed an Eulerian Model of Dispersion in the Atmosphere (Piedelièvre et al., 1990).

THE MODEL EQUATIONS

The concentration of a passive chemical passive element follows the laws of continuous medium and specially that of mass conservation.

The pollutant concentration will be advected in the wind field predicted by the operational coupling model of numerical weather prediction.

The turbulent diffusion is parameterized in terms of exchange coefficients with the same formulation as the coupling model (Louis, 1979).

The wet deposition is parameterized in terms of scavenging ratio.

The dry deposition is modelized by a coefficient which is dimensionally equal to a deposition velocity.

For a radioactive pollutant, we take into account the radioactive decay and the source term in the source mesh is described at a subgrid scale by a gaussian distribution.

THE NUMERICAL TRAITMENT

Upon the evolution of the French and ECMWF data bases of the weather numerical forecasting models, we are using a sigma coordinate with a latitude longitude grid.

The technique that we have opted for, called C.N.G. (Crank Nickolson Galerkin), combined with a splitting is generally used for the modelization of pollutant transportation which includes the physico-chemical processes (Carmichael, 1984).

The whole equation system has been applied to the simulation of a classic numerical experiment which consists in advection-diffusion of a gaussian hill in a rotational flow and gives satisfying results (the maximum error is about 5% for the top of the hill and the loss mass is reduced to less than 1%).

SIMULATION OF THE CHERNOBYL DISASTER

The simulation uses the source term given by the Soviet authorities at the conference of Vienna for the ^{137}Cs. We use the initialized data recorded at the ECMWF because we did not intend to prove the quality of the weather model forecast but the quality of the MEDIA model.

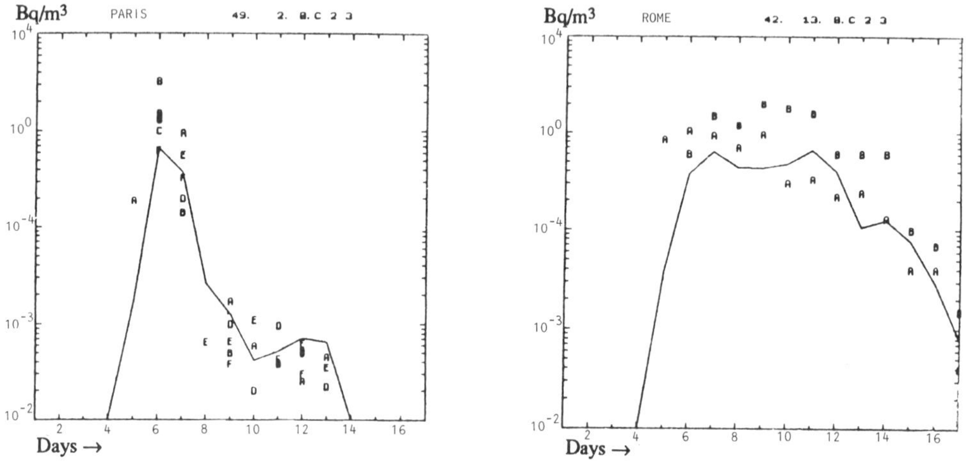

a b

Fig. 1. Daily averaged measured concentrations at Paris (a) and Rome (b).

The only way to underline the model capability of well describing the evolution of the radioactive cloud is to compare the values issued from the model to measurements taken into different towns in western Europe. The results of MEDIA are very satisfying for almost towns of western Europe. The figures 1 illustrate this quality by comparison between daily averaged measured concentration at PARIS and ROME and average concentration predicted by MEDIA in the lower layers for the nearest grid point.

CONCLUSION AND FUTURE

This MEDIA code is now operational at the French Weather Service on a fine mesh (35 km x 35 km) limited area model PERIDOT over an area covering the western Europe and on the French global model EMERAUDE for a movable window. Note that is is now possible to run PERIDOT model anyway in the world. Nevertheless, a validation alone is not sufficient and we want to use other data bases (CAPTEX and ANATEX). Besides, the physic developed in the model is rather primary and a better description of the scavenging and the dry deposition must be considered.

For regional forecast, we have at our disposal a very fine version of the PERIDOT model (5 km x 5 km) using the turbulent kinetic energy equation for vertical diffusion coupled with MEDIA model. This version allow to study, in a better way, the different parameterizations since we can get the whole information of the model.

REFERENCES

Carmichael, G.R., and L.K. Peters, 1984: An eulerian transport/transformation/removal model for SO_2 and sulfate. Atmos. Envir., 18, 5, 937-967.

Louis, J.F., 1979: A parametric model of vertical eddy fluxes in the atmosphere. Bound.-Layer Meteor., 17, 187-202.

Piedelièvre, J.P., L. Musson-Genon and F. Bompay, 1990: MEDIA, an eulerian model of atmospheric disperion. First validation on the Chernobyl release. Submitted to J. Appl. Met.

MODELLING OF THE CHERNOBYL RADIOACTIVE CLOUD DISPERSION

H. Hass, H. Geiss*, H.J. Jakobs and M. Memmesheimer

Institut für Geophysik und Meteorologie, Universität zu Köln
Projekt EURAD, Salierring 48, 5000 Köln 1, FRG

*Institut für Chemie 2, Forschungszentrum Jülich GmbH
5170 Jülich, FRG

The European Acid Deposition Model (EURAD) is used to investigate the long-range transport and deposition of radioactive material in Europe after the Chernobyl accident. First results using EURAD in a completely forecast mode were presented by Hass et al. (1990) where a detailed description of the EURAD design for modelling the Chernobyl case is given.

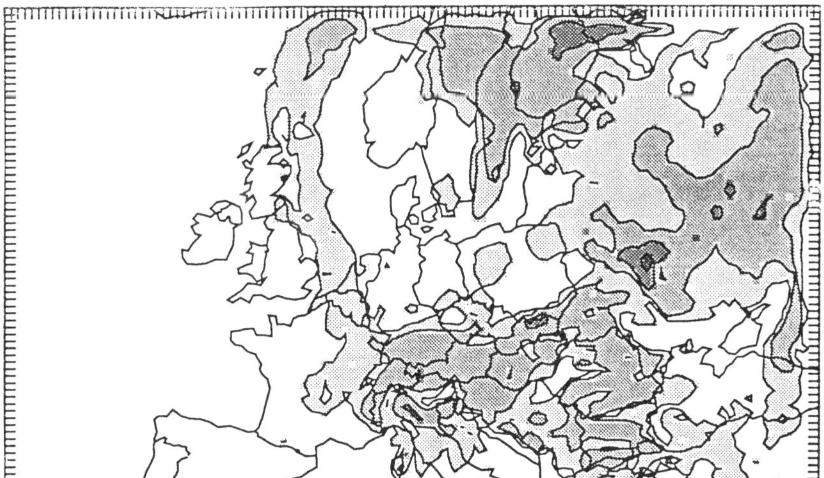

Fig. 1. Accumulated wet deposition of Cs-137 for the period 26 April - 3 May 1986. Results were achieved with a forecast mode of EURAD, i.e. meteorological fields and precipitation are predicted. Shading intervals: 1, 10, 100, 1000 kBq m^{-2}.

Air Pollution Modeling and Its Application VIII, Edited by H. van Dop and D. G. Steyn, Plenum Press, New York, 1991

The processes to be simulated are advection, mixing and deposition, both dry and wet. Hence, we use the Chernobyl event to test the modules in EURAD which numerically realize the needed processes. Together with these simulations the influence of the horizontal resolution was studied by decreasing the grid size from 80 km to 40 km. The meteorological fields used by the long-range transport model is an important factor for the quality of the results. Hass et al. (1990) used NMC data to initialize the predictions. Simulations using ECMWF data were also completed.

Fig. 1 presents the predicted wet deposition of Cs-137 accumulated for the period 26 April - 3 May 1986. These results were gained in a forecast mode, i.e. all meteorological fields and the precipitation were predicted. The initialization of the forecasts used NMC data. The grid size for the simulation was 40 km whereas Hass et al. (1990) show the result for 80 km grid size. In contrast, Fig. 2 shows the wet deposition of Cesium modelled with observed input fields. For this simulation analysed 6-hourly ECMWF meteorological fields were used together with precipitation data from the German Weather Service.

In general, the patterns of both Figures rather well coincide for the areas which were mainly affected: Scandinavia, the Alpine area and around Chernobyl. From the inspection of the time series of wet deposition both simulations show also agreement in the occurence time of the process. Compared to measured wet deposition there are of course discrepancies as both versions of the model are not able to reproduce small-scale variabilities within the precipitation systems.

Fig. 2. As in Fig. 1, but meteorological analyses are input into EURAD. The grid has been extended and slightly shifted compared to the grid of Fig. 1.

Reference

Hass H., M. Memmesheimer, H. Geiss, H.J. Jakobs, M. Laube and A. Ebel: Simulation of the Chernobyl radioactive cloud over Europe using the EURAD model, Atmos. Environm., in press, 1990.

EMERGENCY RESPONSE MODEL EVALUATION USING DIABLO CANYON NUCLEAR POWER PLANT TRACER EXPERIMENTS*

Ronald L. Baskett
John S. Nasstrom

EG&G Energy Measurements, Inc.
Pleasanton, CA

Rolf Lange

Lawrence Livermore
National Laboratory
Livermore, CA

INTRODUCTION

The Lawrence Livermore National Laboratory's (LLNL) Atmospheric Release Advisory Capability (ARAC) provides real-time emergency response support for accidental radiological releases to the atmosphere at Department of Defense (DOD) and Department of Energy (DOE) facilities throughout the U.S. ARAC uses diagnostic three-dimensional (3-D) dispersion modeling as its primary emergency response tool (Dickerson and Orphan, 1976). The regional (20 to 200 km) modeling system is built around the MATHEW (Mass-Adjusted Three-dimensional Wind field) and ADPIC (Atmospheric Dispersion Particle-In-Cell) models (Sherman, 1978; Lange, 1989). MATHEW adjusts the wind field by variational methods to be mass-conservative and to account for terrain effects. ADPIC calculates the time- and space-varying transport and diffusion (using K-theory) of source material using thousands of Lagrangian "mass" particles on a Eulerian grid. To determine the accuracy and transferability of the MATHEW/ADPIC models to a wide variety of settings and meteorological conditions, the models have been evaluated against numerous tracer experiments over the last decade (Dickerson and Ermak, 1988; Lange, 1989). This paper presents a model evaluation using a tracer experiment in the complex, coastal terrain of the Diablo Canyon Nuclear Power Plant.

DIABLO CANYON TRACER EXPERIMENT

The Diablo Canyon tracer experiment was conducted by the Pacific Gas and Electric Company (PG&E) from August 31 to September 17, 1986 (Thuillier, 1988). It was in the vicinity of the Diablo Canyon Power Plant near San Luis Obispo, California, on the central California coast. Tracer gas (SF_6 and Freon) was released from several locations in the area during the study, including from the power plant itself. Measurements of hour-averaged SF_6 and Freon air concentrations were made at 150 locations within a 40-km radius of the power plant during twelve hours of each of eight experiment days. Meteorological data were collected from 18 surface stations and 3 doppler acoustic sounders in the area.

---

* This work was supported by the U.S. Department of Energy under Contract Numbers DE-AC08-88NV10617 and W-7405-Eng-48. By acceptance of this article the publisher and/or recipient acknowledges the U.S. Government's right to retain a nonexclusive royalty-free license to any copyright covering this paper.

MATHEW/ADPIC MODELING RESULTS

As a preliminary study, the first three experiment days were modeled using MATHEW/ADPIC. A 3-D grid was used that had 40 x 40 x 14 cells covering the 40 km x 40 km x 560 m modeling domain (north-south, east-west, and vertical directions, respectively) and that included terrain elevations. Comparisons were made of modeled and observed hour-averaged air concentrations of SF_6 and Freon for the eleven sampling hours of each day. Figure 1 shows an example of southeastward transport, along the coast, from a power plant SF_6 release on Aug. 31, 1986. A divergent pattern is evident in the vicinity of San Luis Obispo Bay, with some transport inland with the sea breeze through a gap in the hills along Highway 101, and transport also continuing along the coast. A statistical analysis of the ratio of observed-to-modeled air concentrations was performed using the technique used by Lange (1989). A total of 643 observed concentrations, paired in space and time with model-calculated concentrations, were used. The results show good agreement, with 55% of the modeled concentrations within a factor of 2 of the observations and 89% within a factor of 5. For the 563 observed SF_6 concentrations, there was a tendency to overpredict with an average residual (average-observed minus average-modeled concentration) of -14% of the averaged observed concentration. For the 80 observed Freon concentrations, there was a tendency to underpredict with an average residual of 16% of the average observed concentration.

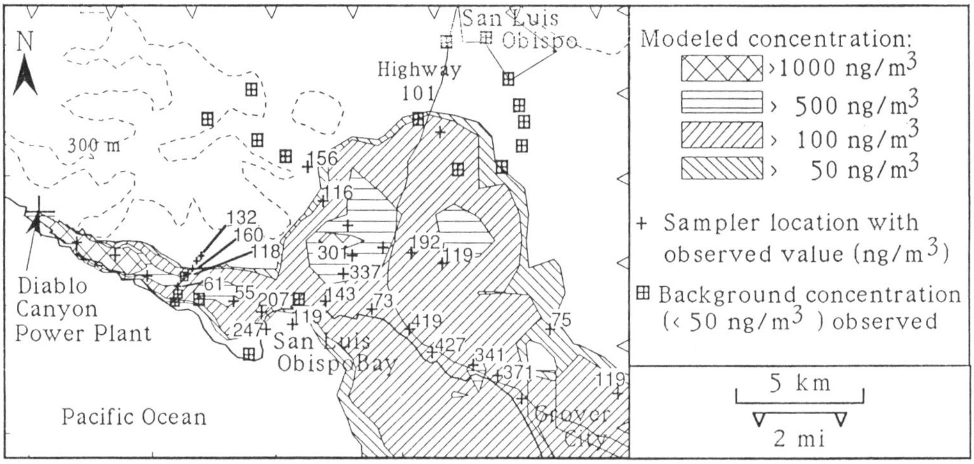

Figure 1. Modeled and observed SF_6 concentrations (ng/m^3) averaged from 1100 to 1200 PDT Aug. 31, 1986.

REFERENCES

Dickerson, M. H., and Orphan, R.C., 1976, The Atmospheric Release Advisory Capability, Nuclear Safety, 17:281.

Dickerson, M. H., and Ermak, D.L., 1988, The evaluation of emergency response trace gas and dense gas dispersion models, in: "Advanced Studies Institute Short Course on Meteorological Aspects of Emergency Response," Amer. Meteor. Soc., Boston.

Lange, R., 1989, Transferability of a three-dimensional air quality model between two different sites in complex terrain, J. Climate Appl. Meteor., 28:665.

Sherman, C. S., 1978, A mass-consistent model for wind fields over complex terrain, J. Appl. Meteor., 17:312.

Thuillier, R. H., 1988, Tracer experiments and model evaluation at Diablo Canyon Nuclear Power Plant, in: "Proc. of the ANS Topical Meeting on Emergency Response-Planning, Technologies, & Implementation," CONF-880913, Amer. Nuc. Soc., LaGrange, Ill.

FLUSHING OF DENSE GAS FROM BOTTOMLANDS BY WIND

Gary A. Briggs, Roger S. Thompson, and William H. Snyder

Atmospheric Sciences Modeling Division
Atmospheric Research and Exposure Assessment Laboratory
U.S. Environmental Protection Agency
Research Triangle Park, NC 27711, USA

ABSTRACT

Mappings of the areas of occurrence of fatalities in the aftermath of disasters such as those in Bhopal, India and at Lake Nyos, Cameroon, reveal strong influences of topography on channeling and inhibiting diffusion of dense gases. Wind tunnel experiments were carried out at the USEPA Fluid Modeling Facility to determine how fast a crosswind removes dense gas that is otherwise trapped in a topographic depression. V-shaped valleys set into the tunnel floor were used. A series of 92 steady-state experiments, with inflow of CO_2 at the valley bottom balanced by outflow due to entrainment, established relationships between outflow rate, height of the dense gas pool in the valley, reference wind speed upwind of the valley, and relative gas density difference times gravity, g'. Measurements of wind speed just above pool height, U_s, confirmed that the outflow is proportional to U_s^3/g' for Froude number <1 and large Reynolds number, Re; it does not depend on valley width, W. At low Re a weak dependence on W was found; an algebraic expression fitting both large and small Re asymptotes revealed that molecular diffusion can dominate entrainment even at real-world scales if $U_s < 2$ m/s. "Flushing" experiments using CO_2 and SF_6 with no inflow provided further support for relationships developed from the steady-state experiments.

Air Pollution Modeling and Its Application VIII, Edited by H. van Dop and D. G. Steyn, Plenum Press, New York, 1991

ATMOSPHERIC DISPERSION OF AMMONIA RELEASED IN A MAJOR CHEMICAL ACCIDENT IN IONAVA, LITHUANIA

J. Kukkonen, A.L. Savolainen, I. Valkama
S. Juntto and T. Vesala

Finnish Meteorological Institute
Air Quality Department
Sahaajankatu 22 E, SF-00810 Helsinki

INTRODUCTION

A very severe chemical accident occurred near the town of Ionava in central Lithuania, in the USSR, 20 March 1989. Approximately 1400 tons of liquefied ammonia was accidentally released into the atmosphere in a catastrophic failure of a storage container. An additional 700 tons of gaseous ammonia was released in the resulting fire.

We have estimated the transport and dispersion of pollutants released in the accident. Preliminary results were reported by Kukkonen (1990), and a more extensive paper by the present authors is in preparation.

THE COMPUTATIONAL METHODS

We have applied a long-range dispersion model based on the gradient-transfer diffusion theory for the concentration estimates. The model has been described in detail by Nordlund and Savolainen (1983) and Nordlund and Tuovinen (1986).

THE RESULTS

The meteorological situation during the atmospheric transport has been evaluated from the radio-sonding and synoptic data. Neutral and stable atmospheric stability conditions were dominant and light southern winds prevailed. No significant rainfall took place during transport.

We have computed the trajectories of air masses for the lowest atmospheric layer (height < 1000 m), starting from the accident site (Fig. 1). The time of occurrence of the accident was about 09.15 UTC and the release of ammonia vapour continued for several days. The results show that the air masses were transported quite straight to the southern coast of Finland. The computed transport times from Ionava to the southern coast of

Finland along various trajectories were about 28 - 36 hours (the corresponding distance is about 500 km).

The concentrations of gaseous ammonia were computed for various assumed release scenarios. The maximum predicted ground-level concentrations of gaseous ammonia on the southern coast of Finland were approximately from 0.5 to 20 mg/m^3. Measurements of the dry deposition of ammonia vapour and the concentration of ammonia in precipitation are being made once a day at the European Monitoring and Evaluation Programme (EMEP) stations. We have made a qualitative comparison of the predicted concentration values and experimental data.

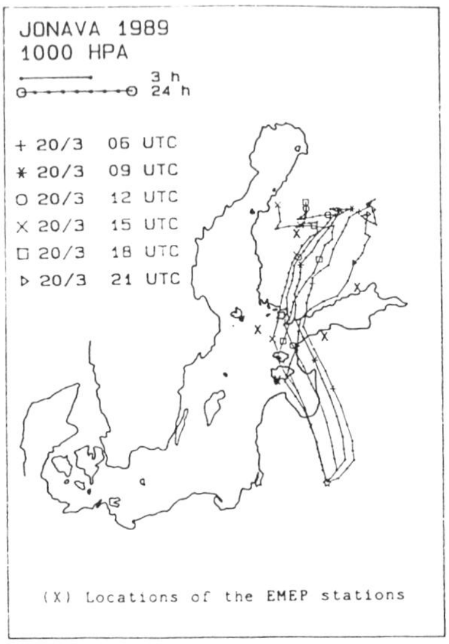

Fig. 1. The trajectories of air masses starting from the accident site on 20 March 1989. The texts in the figure show the notation for the transport times and the starting times (UTC-time) of the trajectories. The locations of EMEP-stations have also been marked.

REFERENCES

Kukkonen, J., 1990, Research at the Finnish meteorological Institute on the dispersion of hazardous materials. In: Finnish Association for Aerosol Research, Report Series in Aerosol Science 11. Helsinki, 44-58.

Nordlund, G. and Savolainen, A.L., 1983, Application of non-Gaussian calculation schemes for evaluating atmospheric transfer of radionuclides. In: International Atomic Energy Agency, IAEA-SR-85/20, Vol. I, 57-65.

Nordlund, G. and Tuovinen, J-P., 1986, Modelling long-term averages of sulphur deposition on a regional scale. In: Proceedings of the WMO Conference on air Pollution Modelling and its Application, Vol. III. WMO/TD 187, 53-61.

EVALUATION OF A SIMPLE DENSE GAS MODEL FOR INSTANTANEOUS RELEASES

Chuck Matthias

Atmospheric Environment Service
4905 Dufferin Street, Downsview, Ontario M3H 5T4
CANADA

An analytical model has been developed which describes many essential properties of a collapsing dense gas cloud upon a plane surface. The initial state is a right circular cylinder which slumps under its own weight. As it collapses, it divides into a central disk and a surrounding torus.

The central disk consists of a core and a fringe. The core concentration immediately assumes a Gaussian distribution in the vertical direction but is radially uniform. The core radius expands during the gravitational collapse but is continually eroded by edge entrainment. Edge entrainment forms a Gaussian fringe. The core concentration is controlled by top entrainment. The equations allow the cloud concentration to transform smoothly from an initial uniform slug to a Gaussian distribution. In the absence of wind, top and edge entrainment are driven by turbulence generated by the gravitational slumping. The height of the central disk decreases slowly with time.

The torus is the leading portion of the cloud during the radial expansion. It is important only during the gravitationally-dominant phase and is eventually absorbed into the central disk. A model - data comparison suggests that the torus height grows with time and is about 6% of the value of the cloud radius.

In the presence of a wind, cloud properties are simulated well by ignoring the torus and by considering the central disk alone. Acceleration of the cloud is described by considering top entrainment of momentum. The major parameter affecting the cloud concentration is the bulk Richardson number Ri. The Richardson number is a measure of the cloud stability relative to atmospheric turbulent kinetic energy. Ri decays with time but while it is large, atmospheric turbulence is ineffective in diluting the cloud. The larger the initial value Ri_0, the longer the time required to dilute the cloud. While Ri is large, concentration decays approximately as the inverse of time. When Ri drops below a critical value Ri_c, atmospheric turbulence begins to control the dilution rate. A comparison of model results and data, with initial volumes ranging over 8 orders of magnitude, and Ri_0 over 4 orders of magnitude, is shown in Figs. 1 and 2. The model entrainment coefficients were determined from a laboratory dispersion study in calm air at the University of Arkansas. The model overpredicts relative to the Thorney Island and Porton field study data. However, in field studies with fixed receptors, the path of the cloud centre usually does not coincide with the receptor location and so the receptor maximum is less than the cloud maximum.

Air Pollution Modeling and Its Application VIII, Edited by H. van Dop and D. G. Steyn, Plenum Press, New York, 1991

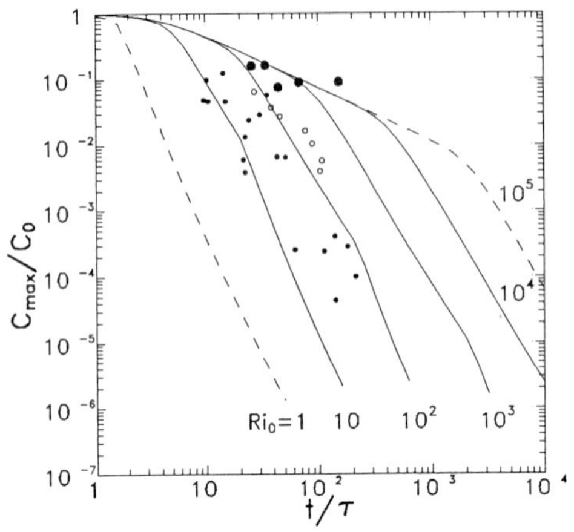

Fig. 1a. $10 < Ri_0 < 10^2$ (o,•)

$Ri_0 > 10^4$ (●)

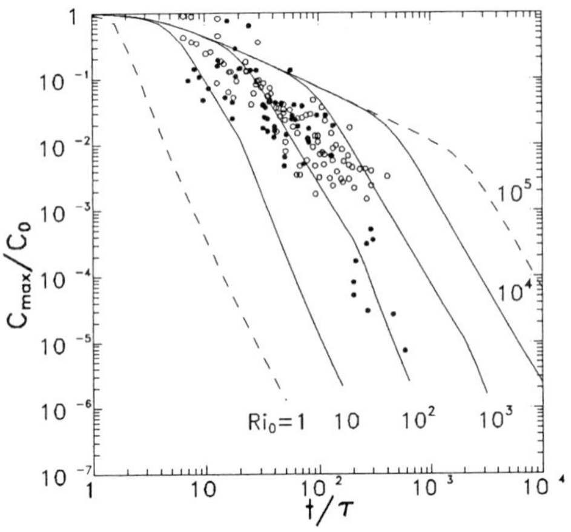

Fig. 1b. $10^2 < Ri_0 < 10^3$ (o,•)

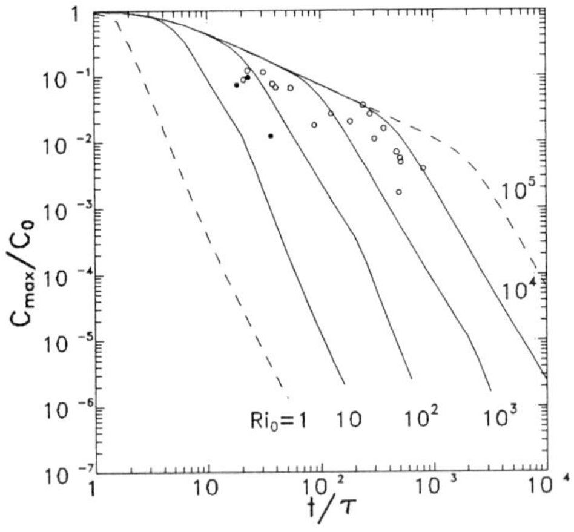

Fig. 1c. $10^3 < Ri_0 < 10^4$ (o,•)

Fig. 1. Cloud maximum concentration as a function of time. Solid and dashed lines are model predictions. Data are from Thorney Island (o) and Porton Down (•) for various initial Richardson numbers Ri_0.

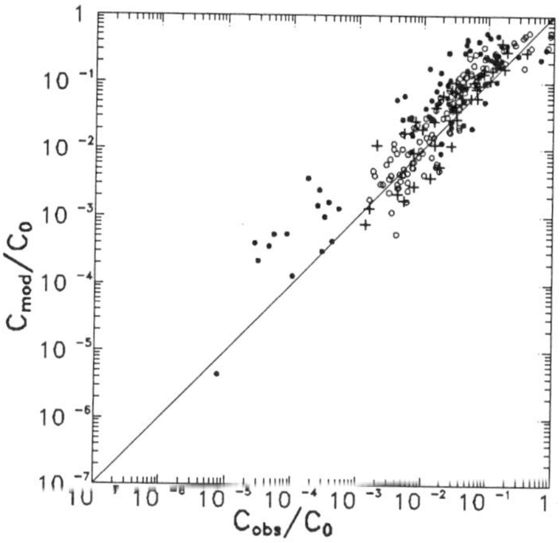

Fig. 2. Modelled vs. observed maximum concentrations. Data are from Thorney Island (o), Porton Down (•), and Colorado State University (+).

MODELLING THE DYNAMICS OF AEROSOLS IN TWO-PHASE RELEASES OF HAZARDOUS MATERIALS

T. VESALA[1] and J. KUKKONEN[2]

1. University of Helsinki, Department of Physics, Siltavuorenpenger 20 D, SF-00170 Helsinki, Finland

2. Finnish Meteorological Institute, Air Quality Department, Sahaajankatu 22 E, SF-00810, Helsinki, Finland

INTRODUCTION

The accidental releases of toxic or flammable gases often form a two-phase mixture, consisting of vapour and liquid droplets of the released species, and entrained humid air. A substantial fraction of the pollutant substance may be initially in the form of airborne liquid droplets. The aerosol phenomena may therefore have a substantial influence on the temperature and density evolution of the source term, and on the subsequent heavy gas dispersion. In particular, the deposition of substance liquid droplets may cause a substantial decrease in concentration.

THE MODEL

We have developed a model for estimating the evaporation of freely falling droplets in a gas mixture containing pollutant gas vapour and dry air (Vesala et al., 1989; Kukkonen et al., 1989). The mathematical model is based on the simultaneous numerical solution of the equations governing mass and heat transfer from the droplet surface. The model also includes the description of forced convection due to the free fall, and thermal interactions of the droplet population. Recently, the model has been generalised to allow for the condensation of water vapour on the surface of the pollutant substance liquid droplets (Vesala, 1990). The droplets are assumed to be well-mixed.

NUMERICAL RESULTS AND DISCUSSION

The model has been applied to estimate the properties of two-phase ammonia releases. In particular, the model computations give predictions for the mass fraction of species deposited on the ground. Figure 1 shows the deposited mass fraction of ammonia as a function of the initial droplet radius, for various relative humidities. The monodisperse ammonia droplet population is initially uniformly distributed in a 1 m high volume. The gas temperature is 20 °C and the total pressure is 1 atm. The ammonia vapour pressure in the gas is negligible.

Increasing relative humidity decreases the deposited mass fraction (Figure 1). This is caused by the latent heat release of condensing water vapour, which increases substantially the initial evaporation rate of ammonia. Finally, the evaporation rate of ammonia is slowed down due to the dilution of ammonia in the droplet. Clearly, the humidity of entrained air should be allowed for, when computing the evaporation rate and the deposition of the liquid phase of the pollutant species.

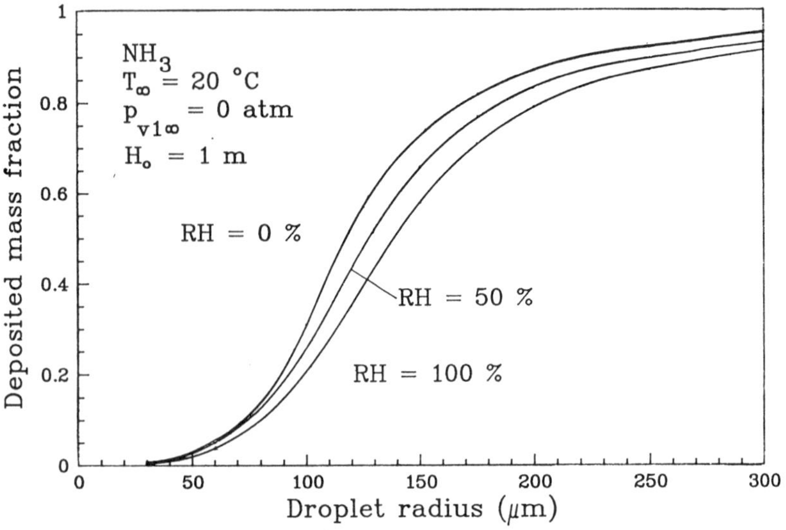

Figure 1. The deposited mass fraction of a uniform population of ammonia droplets vs. the initial droplet radius.

REFERENCES

Kukkonen, J., Vesala, T. and Kulmala M., 1989, The interdependence of evaporation and settling for airborne freely falling droplets, J. Aerosol Sci., 20:7.

Vesala, T., Kukkonen, J. and Kulmala, M., 1989, A Model for evaporation of freely falling droplets, Finnish Meteorological Institute, Publications on Air Quality 6, Helsinki.

Vesala, T., 1990, Dynamics of two-phase aerosol releases, Finnish Association for Aerosol Research, Report Series in Aerosol Science,11, Helsinki.

NUMERICAL MODELLING OF CLOUD DISPERSION IN FORESTRY PESTICIDE SPRAYING OPERATIONS

D. J. Wallace, R. E. Mickle, R. M. Hoff, F. Froude
and A. J. Arnold

Atmospheric Environment Service
4905 Dufferin St., Downsview, Ont. M3H 5T4
Canada

ABSTRACT

The dispersion of aerially released pesticides over a forest canopy is modelled. Numerical results are compared with ARAL (AES Rapid Acquisition LIDAR) measurements of cloud drift obtained during pesticide spray field trials in the province of New Brunswick. Spray cloud dispersion and trajectory predictions are generated from PKBW3 (Picot, Kristmanson, Basak-Brown and Wallace, Version 3); a numerical model developed for forestry aerial spraying. Spray cloud behaviour is determined by considering the trajectories of individual droplets released from the aircraft. Trajectories are initially calculated from a pseudo-ideal flow theory of aircraft vortex motion. After the aircraft trailing vortices have dissipated to background levels of atmospheric turbulence, droplet trajectory calculations are continued with a Monte Carlo technique for simulating droplet motion in a turbulent atmosphere.

Results for spray cloud trajectory predictions are studied for up to one minute after the droplets are released from the aircraft. Spray cloud behaviour under two different conditions of atmospheric stability (very stable and slightly unstable) is examined by comparing PKBW3 predictions with LIDAR measurements. Maxima generated from cross-sectional profiles of the spray clouds (number of drops per cross-section X total projected surface area of droplets) gave comparable decay curves in the numerical and experimental results.

INTRODUCTION

Version 1 of the PKBW model (Picot, Wallace and Kristmanson, 1987) was developed in order to understand the significance of aircraft vortex behaviour in determining the overall spray patterns in forestry spray operations. Aircraft vortex motion was simulated with a second-order closure turbulence model, VORTEX WAKE, developed by the Aeronautical Research Associates of Princeton (Bilanin, Teske and Williamson, 1977). A simplified wake model has since been installed in lieu of the WAKE code. The modified version of PKBW (PKBW3) employs a Rankine vortex model (Hallock and Eberle, 1977), with decay curves for the circulation, Γ, or total vorticity (m^2/sec). The decay curves were obtained by analyzing velocity profiles generated by the WAKE code for the specific conditions under which the spraying was conducted (e.g. aircraft type, flight speed, weather conditions, etc). This paper reveals a direct comparison in the dispersion of the spray cloud between modelled results generated from PKBW Version 3 and those measured by Atmospheric Environment Service (AES) with a LIDAR facility.

METHOD

Experimental

Two cases are presented in order to examine the performance of PKBW3 for different stability conditions; stable and unstable. The conditions are characterized in Table 1. Meteorological data for each of the cases were obtained from tethersonde profiles taken at the site at the time of spraying. Monin-Obukhov lengths and friction velocities were deduced by fitting the tethersonde velocity profiles with empirical profiles given in Wallace et al. (1989). The wind direction was within 20° of the right angle to the spray line. The LIDAR facility was placed approximately half a kilometer upwind from the spray line. Shots of the spray cloud from the LIDAR were taken at right angles to the spray line. Cloud projections ideally represent the motion of a particular cross sectional area of the cloud. LIDAR measurements are proportional to the total droplet cross-section per cross-sectional area. Scans of the spray cloud were taken every four seconds. Further details on the principles and operation of the LIDAR can be found in Hoff, Mickle and Froude (1989).

Table 1. Summary of LIDAR/Numerical Studies

LIDAR		Numerical		
DATE/TIME	AIRCRAFT/ HEIGHT	STABILITY CLASS	FRICTION VELOCITY (M/S)	MONIN-OBUKHOV (M)
08/01/87 0655	TBM AVENGER (40 M)	V. STABLE	0.05	20
08/19/89 0836	AG-TRUCK CESSNA-188 (40 M)	SL. UNSTABLE	0.36	-175

Note: Both spray trials used AU4000 Micronairs with $d_{0.5}$= 47 microns (50 percentile dropsize by mass).

Numerical

The PKBW3 code was executed with the meteorological data given in Table 1 as input in order to simulate the conditions under which the LIDAR measurements were made. Droplet size spectra data, also a required input for the code, was obtained from Picot (personal communication, 1990). Refer to Picot et al. (1987) and Wallace, Picot and Kristmanson (1989) for further details on the PKBW model.

OBSERVATIONS

Decay curves in the maxima derived from cross-sectional contour profiles of the spray clouds at different times are shown in Figures 1(a)-(b). Each maximum represents the maximum at the given time in the total number of drops per cross section X the total projected surface area of the droplets. The plots are expressed in terms of decay relative to the maximum at time zero for the unstable case, and at 9 seconds for the stable case. These results suggest a slightly greater rate of diffusion in the numerical data than in the experimental for the stable case, with the reverse being implied in the unstable case.

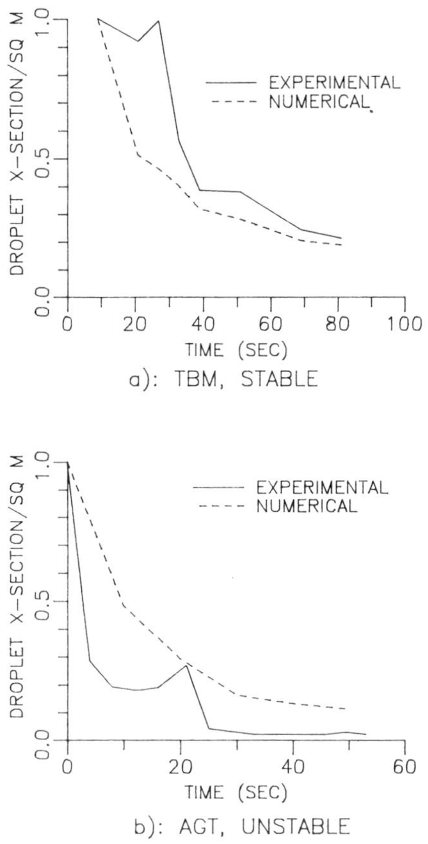

FIGURE 1. SPRAY CLOUD DISPERSION

CONCLUDING REMARKS

A direct comparison of modelled results with measured data on spray cloud transport and dispersion has been a first attempt to test some of the assumptions from which the PKBW3 code has been constructed. One phenomena not accounted for in the model is core bursting. When core bursting occurs, the vortices explode in the radial direction at the expense of continuity in the axial direction (direction of flight path). If core bursting is a significant mechanism of cloud dispersion in aerial spraying, one would expect to find numerical cloud dispersion rates to be less than those deduced from observation. This is indicated in the unstable run but not in the stable case.

Another point worth noting is the value used to normalize the experimental data. In the stable run (Figure 1(a)), the LIDAR measurements were saturated up to nine seconds after the initial time. All experimental maximum values at successive times were normalized with respect to the saturated value. If a true value were substituted as the normalizing parameter, the entire curve would be shifted downwards.

With or without the above corrections, the experimental and numerical maxima decay curves agree quite well for both of the cases studied.

REFERENCES

Bilanin, A. J., Teske, M. E. and Williamson, G. G., 1977, Vortex Interactions and Decay in Aircraft Wakes, Amer. Inst. of Aero. and Astro. Journal, 15,2:250.

Hallock, J. N. and Eberle, W. R. (ed), 1977, Aircraft Wake Vortices: A State of the Art Review of the United States R and D Program, Report No. FAA-RD-77-23 Feb.

Hoff, R. M., Mickle, R. E. and Froude, F. A., 1989, A Rapid Acquisition Lidar System for Aerial Spray Diagnostics, Amer. Soc. Agric. Eng., 32,5:1523.

Picot, J. J. C., Wallace, D. J. and Kristmanson, D. D., 1987, The PKBW Model for Prediction of Aerial Spraying of Deposition and Drift. Proc. of the Symp. on the Aerial Application of Pesticides in Forestry, Ottawa, Canada, October, 1987, ACAFA, G. W. Green, ed.

Wallace, D. J., Picot, J. J. C. and Kristmanson, D. D., 1989, A Numerical Model for Forestry Aerial Spraying Operation: PKBW2, Submitted to Crop Protection.

A PRACTICAL ON-LINE METHOD FOR ESTIMATING PASQUILL STABILITY CLASS USING THE ANALYSIS OF WIND FLUCTUATIONS

J.J. Erbrink

N.V. KEMA
Environmental Research Department
P.O Box 9035
6800 ET ARNHEM
The Netherlands

ABSTRACT

A practical and low-cost method is proposed for the on-line determination of the stability class for the application of dispersion curves in a Gaussian dispersion model. The method is based on the estimation of the intensity and the time scale of horizontal turbulent fluctuations of the wind direction measured at 10 m. Application of this method leads to a more realistic frequency distribution of the classes in which the neutral case D no longer prevails.

INTRODUCTION

Determination of the Pasquill class is often based on the observations of cloud cover and wind speed, although other schemes are also being used, based on values of e.g. w_*, u_*, Obukhov length and mixing height z_i. When these parameters are not measured, and simple wind measurements at 10 m height are available, we propose an alternative method which derives atmospheric stability from automatic and continuous measurements of wind fluctuations.

METHOD

We use direct wind measurements to calculate a fictitious plume dispersion parameter σ_z. The effects of slow fluctuations in wind direction are eliminated by substracting the moving average from the original wind direction measurements, remains the stability dependent fluctuations. From these fast fluctuations the intensity σ_v and the Eulerian time scale T_e can be calculated. Taylor's (Taylor, 1921) formula is used to calculate σ_z from these fast fluctuations, assuming the Lagrangian time scale $T_l=3*T_e$.
σ_v and σ_w represent the standard deviations of wind speed in the horizontal and the vertical plane, respectively. σ_w equals σ_v in all circumstances for $0.1\ z_i<z<0.9\ z_i$, after removing the slow fluctuations in σ_v.

Air Pollution Modeling and Its Application VIII, Edited by H. van Dop and D. G. Steyn, Plenum Press, New York, 1991

Some operational models need the stability class as model input. Assuming the validity of the Pasquill dispersion curves, the most appropriate Pasquill class is fitted into the calculated σ_z value at a distance of, for example, 3 km and a height of 100m. In light wind conditions σ_v and T_e are no longer clearly defined. At night low degrees of turbulence will be present in the atmosphere. In those cases, therefore, we select stability class F.

RESULTS

The use of our method results in a much more realistic estimation of the stability class. The frequency distribution of stability classes for summer 1989 and for winter 1989/1990 is given in figure 1. During daytime in summer, with convective conditions, our method results in higher frequencies of the instable classes, which is consistent with results of more complicated models (Weil, 1984). In the summer of 1989, the differences were not very large because of the unusually high frequency of fair weather conditions. In winter, when stable atmospheres prevail in Northern Europe, our method gives much higher frequencies of the stable classes E and F. The appearance of the neutral class D no longer prevails. Application of this method leads to a better description of not only short-time averaged dispersion processes, but also to better statistics of concentration patterns.

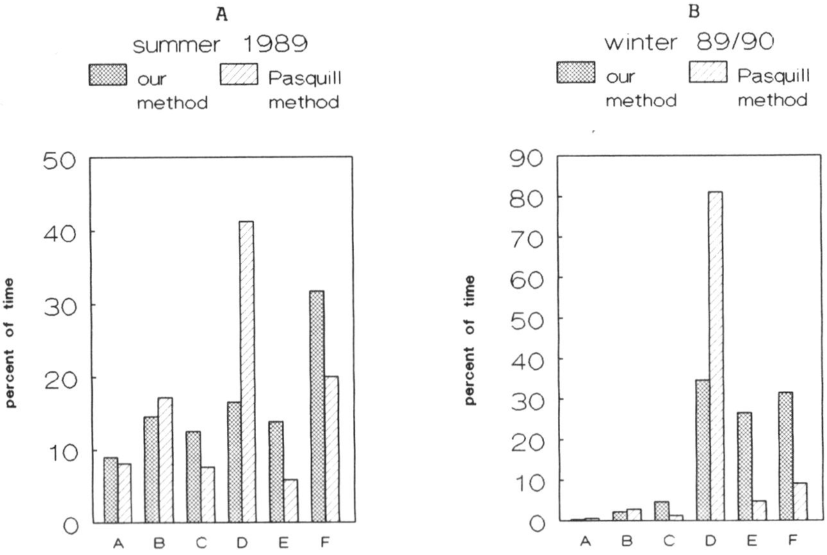

Figure 1 Frequency distributions of the observed stability classes during summer 1989 (A) and winter 1989/1990 (B). Pasquill method and our method are compared

REFERENCES

Taylor, G.I., 1921, Diffusion by continuous movements, Proc. London Meteor. Soc. 20, 196-202.

Weil, J.C. and Brower, R.P., 1984, An updated gaussian plume model for tall stacks, JAPCA 34, 818-827

WIND FIELD AND DISPERSION AT REGIONAL SCALE : THE HERMES CODE

J.Y CANEILL, S. PERDRIEL, A. ALBERGEL and Ph. GEAI

EDF/Direction des Etudes et Recherches
Département EAA
6 Quai Watier - 78400 CHATOU - FRANCE

Over the last decade a large effort has been accomplished in modelling atmospheric phenomena at mesoscale. It is now possible, taking advantage of the new computing facilities, to think about the utilization of three-dimensional atmospheric models in operational context, in order to assess the dispersion of effluents at regional scale in a predictive way. Such a task has been emphasized at EDF/DER, with the opportunity of an important research modelling program that led us recently to a practical evaluation of such tools in the framework of the ECRAN project (Estimation des Conséquences Radiologiques d'un Accident Nucléaire).
In this contribution we examine the three-dimensional model named HERMES, together with simulation results.
HERMES is a pronostic model which has been designed to adapt at regional scale (some hundred of kilometers) forecasts realized by the French Meteorological Office, in order to produce wind-fields and turbulent diffusivities fields representative of this scale. The latter are used in a second time for dispersion of radio-active effluents assessment.

The HERMES code has been developed on the basis of the work of Nickerson (1979). It consists in the numerical integration of the primitive equations of meteorology, written in the so-called sigma-coordinate. It is nested with a one-way technique, to the French Met'Office Model : PERIDOT. A description of the model, is available in L. Janvier (1987).

One of the goal of the ECRAN project, was to assess the use of this tool in an emergency preparadness system.

One of the recent applications of the model, was to study the performance of the code on different power-plant sites exhibiting typical topographical features (Rhône-River-Valley, Loire Valley,...). On the basis of twenty-one mesometeorological forecasts, results have been analysed, in a first step, to assess the improvement that one can obtain with a fine-mesh (10 km) meteorological model. Results are summarized on the Fig. 1, which presents histograms of the deviations between the calculated and the measured values for the direction of the flow at 100 meters above the ground for three sites. It is found thah a net improvement can be obtained with the mesoscale model in the lower layers of the PBL ; on the average this improvement is of the order of magnitude of ten to fifteen degrees.

Air Pollution Modeling and Its Application VIII, Edited by H. van Dop and
D. G. Steyn, Plenum Press, New York, 1991

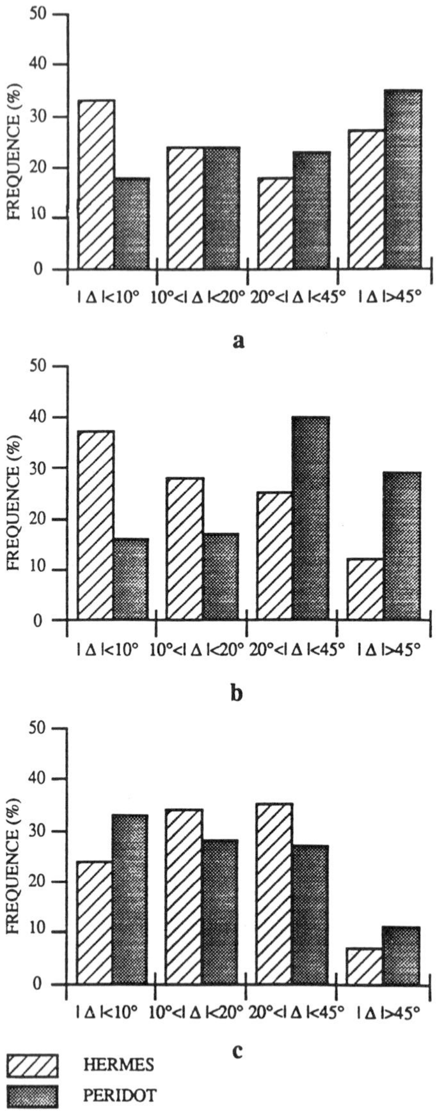

FIGURE 1 **Respective scores of the PERIDOT model (French Meteorological office) and HERMES model on the observed deviations for the wind. Distribution of the deviations at 100 m.**
(a) Cruas site
(b) Belleville site
(c) Saint-Laurent site

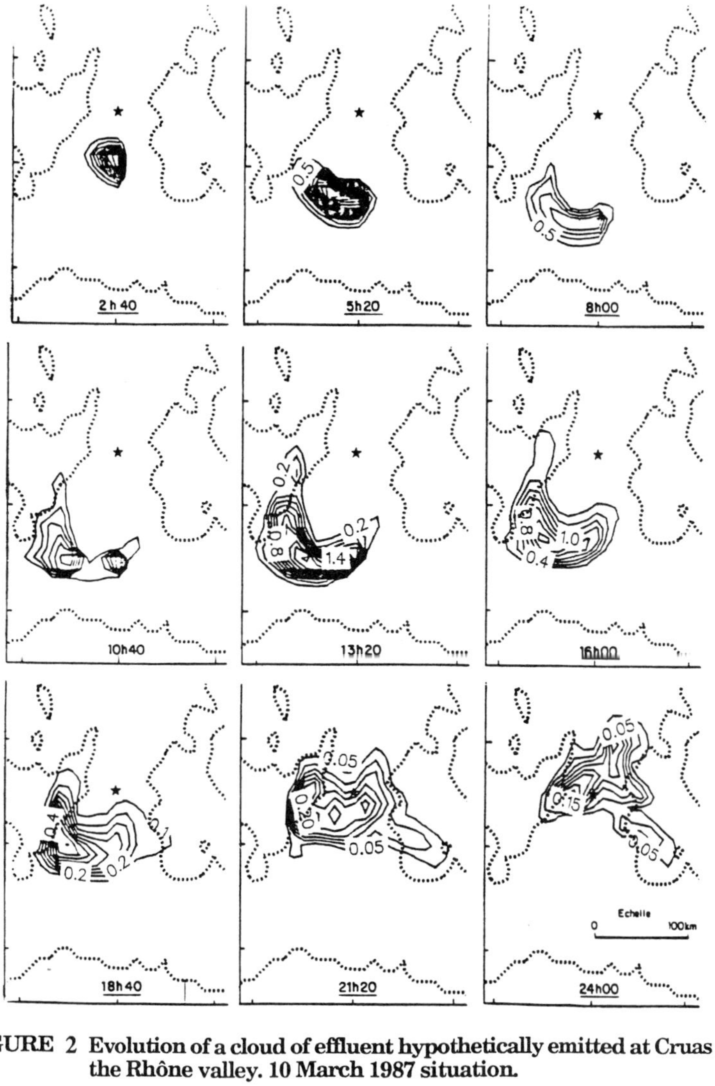

FIGURE 2 **Evolution of a cloud of effluent hypothetically emitted at Cruas in the Rhône valley. 10 March 1987 situation.**

The major improvements are found on sites which exhibit complex topography (namely Cruas in the Rhône valley, and Belleville in the Loire valley) ; on flat areas however (like the St Laurent Site), no significant improvement is depicted.
At last on different case studies, hypothetical releases were performed in the numerical simulations. The obtained results show (see fig. 2) that in very rapidly evolutive situations, mesoscale dispersion prediction results can be very useful in emergency situations. The above mentionned situation is typical of flows in the Rhône valley with occurence of reverse flow within the valley following a strong change in the sub-synoptic regime.

The results of this study, show all the interest using a mesoscale model with high spatial resolution, to get coherent winds with the real topography and consequently permitting to derive the air concentration field with a good representativity.

References

Nickerson C. 1979 — On the numerical simulation of air flow and clouds over mountainous terrain. Atmospheric Physics, vol 52, n° 3.

Janvier L. 1987 — Paramétrisations de la turbulence et de l'interface sol-atmosphère dans un modèle tridimensionnel à moyenne échelle. Thèse

Perdriel S., Geai P. 1988 — Analyse de prévisions mésométéorologiques en vallée du Rhône réalisées à l'aide du code HERMES. Rapport HE 34/88.03.

PARTICLE DIFFUSION MODEL EVALUATION AGAINST TRACER EXPERIMENT

G.Brusasca*, M.G. Morselli* and D. Anfossi+

* ENEL/DSR/CRTN Environment Department
Via Rubattino 54, Milano, ITALY
\+ C.N.R. Istituto di Cosmogeofisica
Corso Fiume 4, Torino, ITALY

INTRODUCTION

A Monte Carlo particle model for simulating pollutant dispersion in the atmosphere has been developed by our team. Its ability to predict ground level concentrations (g.l.c.) is verified against experimental data from the Karlsruhe Nuclear Research Center (F.R.G.). Tracer data were also compared to the simulations of 10 different Gaussian models in order to prove the particle model performances against simple and widely used models. A complete "model evaluation" was carried out: particle model yields satisfactory results and shows performances which are significantly better than all the Gaussian models.

MODEL EVALUATION

Numerical schemes and parametrizations of the particle model are described in Brusasca et al. (1989) where two of the above mentioned KNRC tracer experiments are examined in details; meteorological and emission data are presented in Thomas et al. (1983). The 8 exercises considered in this model evaluation are performing during daytime with a large range of meteorological conditions ranging from strong instability and low wind speed to neutrality and strong wind speed. For all the exercises (except one) two consecutively half an hour periods have been simulated; two different non buoyant tracers were simultaneously emitted at 160 and 195 m. thus giving rise to 30 tests. To estimate the goodness of the fit between observed and predicted g.l.c., the following indexes were computed:

$$\mathrm{NBIAS} = \overline{(C_p - C_o)} \,/\, \overline{C_o} \qquad \mathrm{NMSE} = \overline{(C_p - C_o)^2} \,/\, (\overline{C_p} \cdot \overline{C_o})$$

$$\mathrm{RMSFD} = \exp\left\{ \left[\overline{(\ln (C_o + b)/(C_p + b))^2} \right]^{\frac{1}{2}} \right\}$$

and the Correlation Coefficient CORRE, where C_o and C_p refer to the observed and predicted concentrations at each sampling point and b is the experimental background. NBIAS (Normalized BIAS) indicates underprediction or overprediction of the models, NMSE (Norm. Mean Square Err.) and RMSFD (Root Mean Square of the Fractional Deviation) refer the scatter, respectively, of the differences and quotients between observed and predicted g.l.c. The model evaluation was carried out on 11 models: the first five refer the Gaussian models with vertically averaged wind speed and direction and with the dispersion sigmas given according to, respectively, Pasquill-Gifford (1), Briggs open-country (2), Briggs urban (3), Brookhaven (4) and Karlsruhe (5); the following five (from 6 to 10) still refer to the Gaussian model with the above mentioned dispersion sigmas but with wind speed and direction measured at emission height; the last one (11) refers to our particle model.

RESULTS

Statistical indexes for the 11 models computed on the 8 KNRC exercises are shown in Table 1. Particle model exhibits the best performances for all the indexes; among the Gaussian models, model 8 is the best for NBIAS and NMSE, whereas model 5 is the best for RMSFD and CORRE; the worst is always model 6. Using a "bootstrap resampling" procedure suggested by Hanna (1988), we obtain that the particle model indexes are "significantly" (better than 99 percent confidence) different from all the indexes obtained by Gaussian models.

Table 1

Model	NBIAS	NMSE	RMSFD	CORRE
1	-0.6439	18.2229	5.2207	0.2900
2	-0.5555	13.7091	4.7825	0.3612
3	0.3112	5.5202	5.1186	0.4548
4	-0.3997	9.6593	4.3075	0.4379
5	-0.2647	6.5074	3.7359	0.5428
6	-0.6898	21.3650	5.4411	0.2583
7	-0.6144	16.2684	5.0324	0.3252
8	0.1178	5.3107	5.0128	0.4696
9	-0.4852	11.1991	4.5239	0.4148
10	-0.2457	6.8665	3.9330	0.5076
11	-0.0083	3.9158	3.0513	0.6749

REFERENCES

Brusasca G., Tinarelli G. and Anfossi D., 1989, Comparison between results of a Monte Carlo atmospheric diffusion model and tracer experiments, Atmospheric Environment, Vol.23, No. 6, 1263:1280

Thomas P., Hubschmann W., Schuttelkopf H. and Vogt S., 1983, in: "Experimental determination of the atmospheric dispersion parameters at the Karlsruhe Nuclear Research Center for 160 m and 195 m emission heights. Part 2: evaluation of measurements," KfK Report 3457

Hanna S.R., 1988, Air quality model evaluation and uncertainty, JAPCA 38, 406:412

MANAGEMENT OF ATMOSPHERIC POLLUTANT VENTING - APPLICATIONS AND DEVELOPMENT OF OPERATIONAL MANAGEMENT OF AIR RESOURCES (OMAR)

W.B. Goddard [1] and C.B. Goddard [2]

(1) Chief Research Engineer and Principal
(2) Manager and Environmental Planner, both of
Goddard & Goddard Engineering - Environmental Studies
6870 Frontage Road, Lucerne, CA 95458, (707) 274-2171

INTRODUCTION

An automated computerized telecommunication system using telephone and telemetry to obtain real-time air pollution alarms in case of elevated concentrations, real-time aerometric monitoring data for air dispersion assessments, and to compile and report archived data sets is described. The system, termed OMAR, is in operations in California, USA, at The Geysers, the worlds largest geothermal power plant complex (1900 MWe), and at Coso, the worlds 4th largest (240 MWe), which is within the USN China Lake Naval Weapon Center. The OMAR system allows users, who include State and Federal agencies and industrial operators, to manage and assess air quality impacts of necessary geothermal air pollutant venting using real-time aerometric compliance monitoring data and complex terrain air dispersion models.

The OMAR research and development project was funded by the California Energy Commission's (CEC) Geothermal grant program and managed by the Lake County Air Quality Management District (LCAQMD) who administer State and Federal rules pertaining to air quality (Goddard, 1989). OMAR at Coso was funded by industry (California Energy Company, Inc.)

Previously industry carried the financial costs of mandated compliance monitoring programs, but aerometric data reports were only available on a quarterly basis after months of data auditing and passing thorough quality assurance standards (QA). While QA procedures are necessary and desirable, this has not allowed, until OMAR, real-time access and use of these data.

An organizational diagram of OMAR is shown in Figure 1. Aerometric monitoring includes wind speed and direction, air temperature and gradient, humidity, precipitation and hydrogen sulfide (H2S) concentrations. The Campbell Scientific Inc. (CSI) data loggers calculate and store means, 3 minute peaks, Sigma, hourly averages, running totals, stability, and report out-of-range readings. The loggers can store up to 5 months of data unattended, will report on demand archived data, and are solar powered. The data loggers are accessed via modems through telephone lines and radio telemetry using an error correcting packet protocol. Access is initiated automatically from the remote OMAR computer at a short interval (nominally 1 hour), or upon receipt of an out-of range alarm report from one CSI monitoring site or on request from a local OMAR user. The OMAR hardware and software components are shown in Figure 2, and include IBM 386 compatible

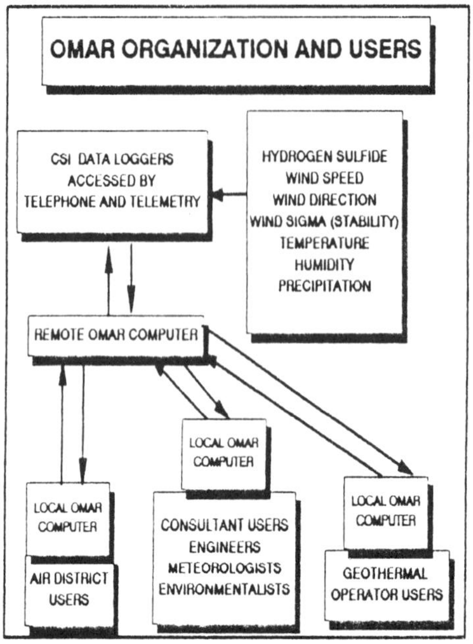

FIGURE 1. OMAR ORGANIZATION AND USERS

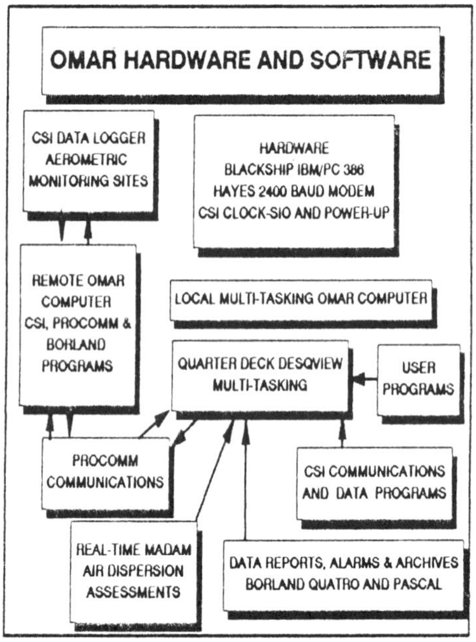

FIGURE 2. OMAR HARDWARE AND SOFTWARE

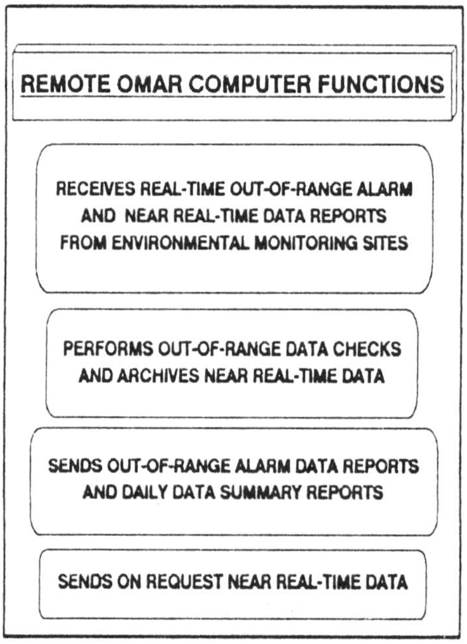

FIGURE 3. OMAR REMOTE COMPUTER FUNCTIONS

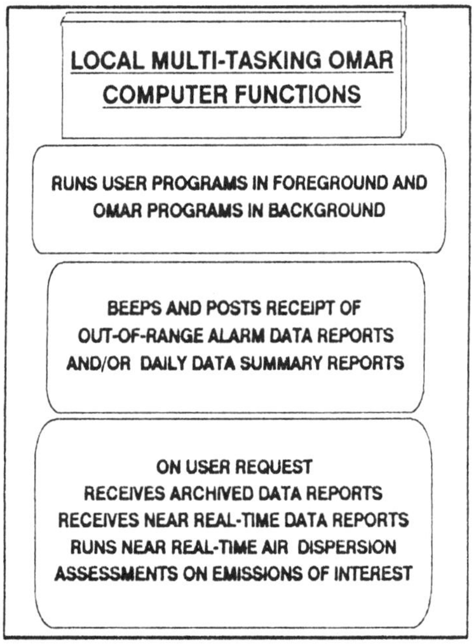

FIGURE 4. OMAR LOCAL COMPUTER FUNCTIONS

computers which operate in multi-tasking mode for the local and the remote computers. The remote computer is installed near enough to the CSI aerometric monitoring sites that efficient and cost effective frequent communication can be obtained. The remote computer is equipped with a CSI clock-SIO and power-up board which turns the computer off and on (cold boots) in cases where the computer hangs-up or in cases where communication links time-out.

The remote OMAR computer, Figure 3, serves as a remote node with communication links to each CSI data logger monitoring site and with communication links to the local OMAR computer users. The remote functions include receiving out-of-range alarm reports from the CSI monitoring sites, compiling out-of-range alarm reports, archiving short term (nominally 1 hour) near real-time data reports and archiving monitoring data in report form. When data is found to be out-of-range an alarm report is immediately sent to each local OMAR computer which beeps and posts upon receipt.

The local OMAR multi-tasking computers, Figure 4, can be used to run general purpose programs as well as running OMAR programs in the background. Programs automatically answer incoming calls from the remote OMAR computer and receive data reports, alarm reports (which beep on receipt) and long term archived data reports. On command, programs display numerical and graphical data reports and perform air dispersion assessments. Each OMAR installation has been designed to monitor conditions at and near receptors of concern and at sites of meteorological interest. The system design provides the necessary real-time data needed to run air dispersion assessments for emission sources of concern (Goddard, 1988). Ridge top and/or mountain top stations are used to determine the state of atmospheric stability, the presence of marine air intrusion and/or subsidence capping temperature inversions.

When a user requests an air dispersion assessment from the local OMAR computer, the first operation that is performed automatically is the retrieval of the latest real-time data report from the remote OMAR computer. The data is then processed to determine the present meteorological conditions and whether good or poor air dispersion conditions exist. The user enters the location of the necessary venting operations in UTM coordinates, the elevation and the type of emission source. The OMAR air dispersion program then estimates the incremental impact of the venting above ambient concentrations of H2S and the cumulative impact base on the monitored current ambient H2S levels.

The user can run several venting scenarios including decreases in existing venting from bleeding or testing wells, or other emission sources. While users must obtain local air district permission to surpass venting emission limits, they can use OMAR as a management tool to demonstrate that venting will not result in substantial increased H2S levels. OMAR will assist in avoiding poor air dispersion impacts when emergency venting break-downs occur since users can immediately determine what the ambient levels of H2S are and reduce venting of manageable emissions so that cumulative impacts are maintained well below AAQS.

REFERENCE

Goddard, W.B. and C.B. Goddard, Oct 1987, Micrometeorological Air Dispersion Assessment Methodology (MADAM), A Geothermal Air Quality Impact Assessment Toolbox Available As Shareware, GRC, TRANSACTIONS, Vol. 11, Davis, CA

Products mentioned include Desqview from Quarter Deck Office Systems, Santa Monica, CA; Procomm from Datastorm Technologies, Columbia, MO; Campbell Scientific, Logan UT; and Quatro and Pascal,Borland, Scotts Valley, CA

A DECISION SEQUENCE FOR SELECTING PUBLIC DOMAIN MODELS FOR ACCIDENTAL RELEASES

J. Ivor John

Santa Barbara County Air Pollution Control District
5540 Ekwill St, Suite B, Santa Barbara, Ca 93110

1. INTRODUCTION

Recent legislation in the United States has created a need for comprehensive assessment of the off-site consequences that may occur from potential accidental releases of toxic vapours. This paper provides a brief summary of public domain models which are available for this purpose. Models selected by the Santa Barbara County Air Pollution Control District (SBCAPCD) for use in the regulatory process are identified.

2. SELECTION OF PUBLIC DOMAIN MODELS

A conceptual modelling sequence for accidental releases is outlined in Figure 1. The sequence includes modelling of the release rates, the source term calculations, atmospheric dispersion and assessment of exposure.

A wide range of publicly available models have been reviewed by the SBCAPCD to identify the most appropriate models to use for risk management prevention planning. The selected models are categorized by type in Table 1.

Table 1. Public Domain Models Selected by SBCAPCD for Modelling Accidental Releases.

Model Type	First Choice	Alternative
Release Rates	OME [1]	ARCHIE [2]
Passive Releases	AFTOX [3]	INPUFF [4]
Dense Gases	SLAB [5]	DEGADIS [6]

References for the selected models are summarized in Section 3.

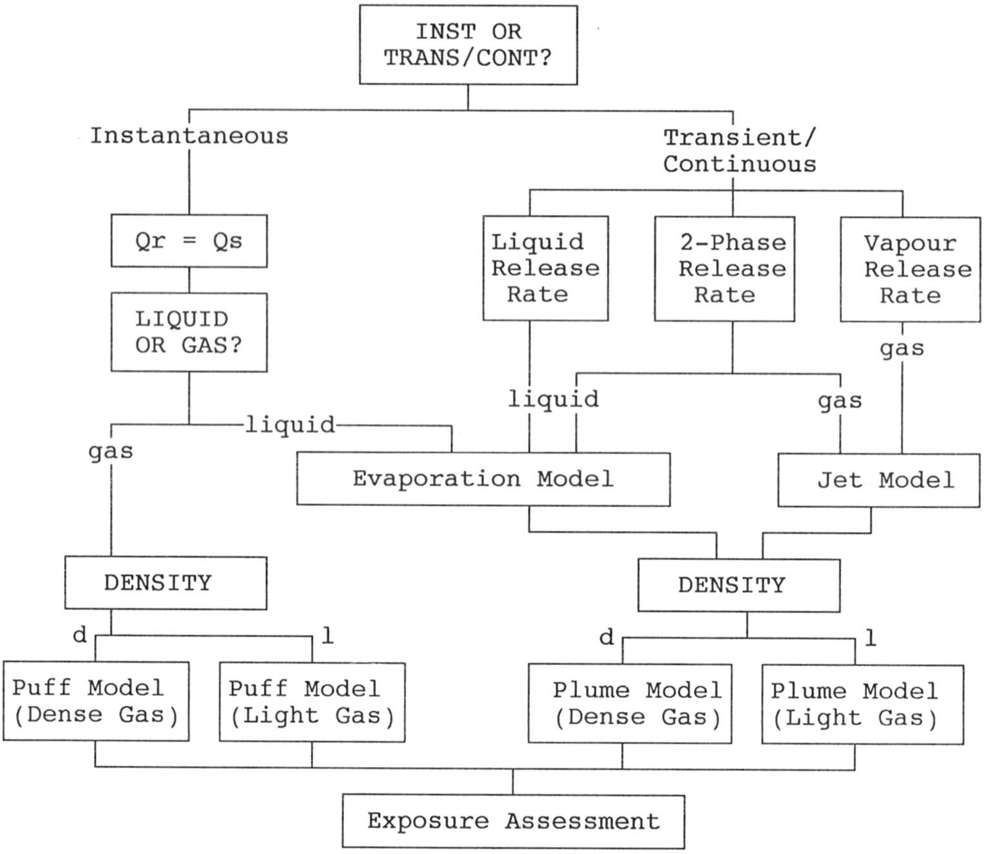

Figure 1. Modelling Sequence for Accidental Releases

3. REFERENCES

1. A Portable Computing System for Use in Toxic Gas Emergencies, Ontario Ministry of the Environment (OME) Model, March 1988.
2. Handbook of Chemical Hazard Analysis Procedures, (ARCHIE) FEMA/DOT/EPA, 1989.
3. Air Force Toxic Chemical Dispersion Model (AFTOX), U.S. Air Force Geophysics Laboratory, Hanscom AFB, MA 01731, AFGL-TR-88-0009, 1988.
4. INPUFF 2.2 - A Multiple Source Gaussian Puff Dispersion Algorithm, by William Petersen and Leonidas Lavdas, Meteorology and Assessment Division, EPA, Research Triangle Park.
5. SLAB: An Atmospheric Dispersion Model for Denser-Than-Air-Releases, Atmospheric and Geophysical Sciences Division, Lawrence Livermore National Laboratory, 1989.
6. DEGADIS: A Dispersion Model for Elevated Dense Gas Jet Chemical Releases, EPA-450/4-88-006a, 1988.

OBJECTIVE METEOROLOGICAL ANALYSIS FOR USE ON PERSONAL COMPUTERS TO PROVIDE AIR QUALITY MODEL INPUTS

F. L. Ludwig

SRI International
Menlo Park, CA 94025

Peter Lester

Department of Meteorology
San Jose State University
San Jose, CA 95192

INTRODUCTION

Typically, meteorological data are available from only a few sites in an area where air quality model applications are to be made. The most sophisticated air quality models require three-dimensional, gridded values of winds and turbulence. Even simple Gaussian plume models will benefit from the use of local conditions, rather than inputs from some remote observation site. The computer analysis code described here can provide gridded wind, temperature and turbulence values to satisfy the following air quality modeling needs:

1. For on-line, real time applications of transport and diffusion models using current observations as inputs

2. For predictive applications, using outputs of large scale weather service models as inputs

3. To provide more detailed input for analytical applications, by using the outputs of prognostic boundary layer models as inputs.

The spatial scale of this model's gridded outputs can be from about one-half kilometer to several kilometers. The code has two main sections which are described below. Those sections are a wind generation module and a turbulence parameter module.

MAJOR COMPONENTS

Wind Module

The wind module uses two physical constraints to achieve a realistic interpolation of observed winds. The vertical motions are constrained by application of the critical dividing streamline concept to define surfaces within which the flow should be confined. The second constraint is the frequently used one of mass conservation. In using the critical streamline concept to define flow surfaces, those surfaces are assumed to be lowest over the lowest terrain. The highest level to which they can rise is estimated from one or more temperature profiles and the initial estimates of wind speeds over the low terrain (e. g. Hunt and Snyder, 1980). Further constraints are applied to account for the effects caused by elevated stable layers and other special cases. A maximum rise is imposed for neutral and unstable conditions.

The flow surfaces are allowed to intersect the underlying terrain, so that when the second physical constraint, mass conservation, is imposed, the flow will be forced around obstacles as appropriate. This reduces the original three-dimensional problem to several two-dimensional problems that are easily solved with Endlich's (1967) fast iterative adjustment procedure.

Turbulence Module

A survey of turbulence parameterization schemes (Ludwig et al, 1990) showed that they all rely on some combination of the following basic information:

- Lapse rate
- Surface heating
- Surface roughness
- Wind shear.

Except for the surface characteristics, these factors are derived from the inputs and outputs of the wind module. Surface characteristics, cloud cover and solar angle are used to define surface roughness and heating. Solar angle, corrected for local terrain slope, is estimated from local time, latitude and longitude. The above information is used with the various parameterizations that have been published (e.g. Holtslag and van Ulden, 1983; Sorbjan, 1986; Stull, 1988; Sutherland et al, 1986) to generate both two- and three-dimensional turbulence indices.

STATUS OF THE COMPUTER CODE

A version of the computer code has been written (Ludwig et al,1990) to provide guidance to Army pilots operating at low altitudes over complex terrain. It is currently being tested against observations (Hansen, 1990). The turbulence indices in that version may not be the most suitable for air quality model applications, but the code is modular and all the basic parameters have been determined, so changes to provide more suitable measures should be easy. The model is written in FORTRAN77 and has been run MacIntosh II and IBM-AT personal computers. Calculations for a 20X20X5 grid required about 2 and 3 minutes, respectively, so the code is quite suitable for all the earlier cited applications.

ACKNOWLEDGEMENTS

We gratefully acknowledge the support of the U. S. Army (Contract DAAL03-86-001/068) and helpful suggestions of Roy Endlich, and Sue and Frank Hansen.

REFERENCES

Endlich, R.M., 1967: An iterative method for altering the kinematic properties of wind fields, J. Appl. Meteorol., 6:837-844.

Hansen, S.,1990: personal communication.

Holtslag,A.A.M., and van Ulden,1983: A simple scheme for daytime estimates of surfaces fluxes from routine weather data, J. Clim. and Appl. Meteorol.,22:517-529.

Hunt,J.C.R. and Snyder, W.H., 1980: Experiments on stably stratified flow over a three-dimensional hill, J. Fluid Mech.,96:671-704.

Ludwig, F.L., Endlich, R.M. and Lester,P., 1990: A computer code for objective analysis of turbulence and winds in the lower atmosphere, final report U.S. Army contract DAAL03-86-0001/0680, SRI Internat., Menlo Pk. CA, 151 pp.

Sorbjan, Z., 1986: On similarity in the atmospheric boundary layer, Bound. Lay. Meteorol.,34:377-397.

Stull, R.B., 1988: "An Introduction to Boundary Layer Meteorology," Kluwer Academic Publishers,Dordrecht, 666 pp.

Sutherland, R.A., Hansen, F.V. and Bach, W.D., 1986: A quantitative method for estimating Pasquill stability class from windspeed and sensible heat flux density, Bound. Lay. Meteorol.,37:357-369.

van Ulden, A.P. and Holtslag,A.A.M., 1985: Estimation of atmospheric boundary layer parameters for diffusion applications, J. Clim. and Appl. Meteorol., 24:1196-1207.

AN INVESTIGATION INTO TWO-PHASE JETS

J.S. Kukkonen[1] and D.M. Webber[2]

1. Finnish Meteorological Institute, Air Quality Dept. Sahaajankatu 22 E, SF-00810 Helsinki, Finland
2. UKAEA, Safety and Reliability Directorate Wigshaw Lane, Culcheth, Warrington, WA3 4NE, UK

INTRODUCTION

Hazard analysis of storage or transport of pressurised chemicals requires mathematical models of the behaviour of two-phase jets. We have reviewed the most important factors in modelling two-phase jets, illustrating the problems with a simple analytical model and with results from the computer code TRAUMA (Webber and Kukkonen, 1988 and 1990).

THE MODELS

We have examined a set of jet equations for species, mass, and momentum conservation. It has been shown that these equations can be solved entirely analytically for the both Ricou-Spalding and the Morton-Taylor-Turner entrainment models. The analytic model gives also explicit estimates of length scales over which gravity and wind effects will be significant.

The computer code TRAUMA was developed by Wheatley (1987) to model the behaviour of ammonia jets. The thermodynamic behaviour of two-phase ammonia is fully represented, as is its interaction with atmospheric water vapour. More recently, this model has been generalised to apply to other substances. We have made sensitivity analyses of the model, to estimate the relative significance of different factors on jet evolution.

THE RESULTS AND DISCUSSION

The analytic results give useful insight on the effect of entrainment models, gravity, ambient air flow and liquid deposition. These are supported in the presence of more complex thermodynamic effects by the results from the computer code TRAUMA. Among the most important factors to be modelled are gravitational spreading, deposition of liquid, ambient air flow and 'transition' to dense gas dispersion.

The sensitivity analyses have shown that the effect of different entrainment models and the effects of water vapour do

Air Pollution Modeling and Its Application VIII, Edited by H. van Dop and D. G. Steyn, Plenum Press, New York, 1991

not have larger direct effect than about 10 % each in the concentration values. Liquid deposition may cause a direct decrease in concentration by a factor of 2 to 4 in typical conditions. It has also been shown that the choice of transition criterion to a heavy gas dispersion model may have a substantial effect on predictions.

Analytic and numerical calculations have shown the existence of a region where gravity effects have become significant, but where jet momentum effects have not decayed. This has important consequences for the way in which one models the transition from jet to cloud. Gravitational effects may be substantial even at small distances from the source.

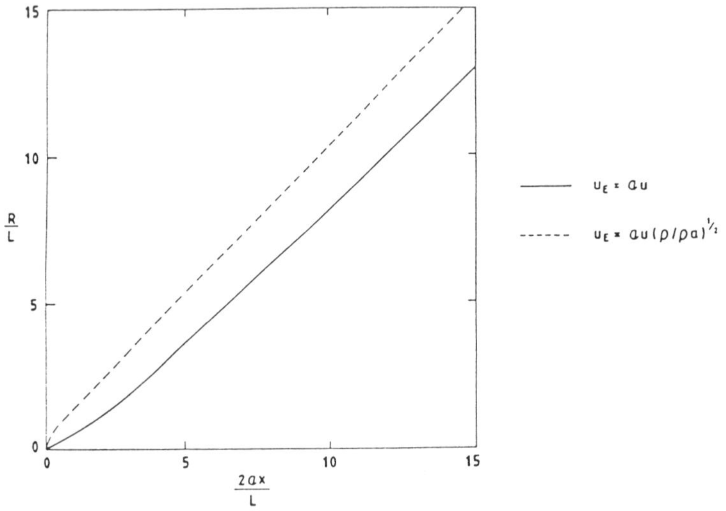

Fig. 1. Results of the analytic model for dimensionless jet radius against dimensionless downstream distance, for two entrainment submodels (Webber and Kukkonen, 1988). Notation: R = jet radius, L = jet length scale, α = entrainment coefficient, x = downstream distance, u_E = entrainment velocity, ρ = jet density, ρ_a = density of the ambient air.

REFERENCES

Webber, D.M. and Kukkonen, J.S., 1988, Modelling two-phase jets. In: 1988 European seminar on the pressurised storage of flammable liquids. 24th-26th October 1988, London. IBC Technical Services Ltd and the Health and Safety Executive

Webber, D.M. and Kukkonen, J.S., 1990. Modelling two-phase jets for hazard analysis. J. Hazard. Materials (accepted).

Wheatley, C.J., 1987. Discharge of liquid ammonia to moist atmospheres - survey of experimental data and model for estimating initial conditions for dispersion calculations. UKAEA, Safety and Reliability Directorate, SRD R 410.

DISPERSION MODELING WITHIN RISK ASSESSMENT STUDIES

Michael Schatzmann and Klaus Marotzke

Meteorological Institute
University of Hamburg
Hamburg, W. Germany

INTRODUCTION

In recent years, several mathematical models for the prediction of heavy gas dispersion have been developed. Although the models have been reasonably successful in some cases, they are still limited in scope. Especially in cases where obstacle effects dominate the dispersion of the cloud, physical models turn out to be a more powerful tool than mathematical models for, e.g., the determination of lower flammability distances.

The paper briefly describes a method for estimating the dispersion of gases heavier than air released under unfavorable atmospheric conditions. The method is based on the results of a wind tunnel study investigating the dispersion of instantaneous as well as continuous releases into a boundary layer shear flow disturbed and undisturbed by surface obstacles. The method has recently been adopted by the German Engineering Society (VDI) and forms the basis for risk assessment studies for flammable and toxic heavy gases in the Federal Republic of Germany (VDI 3783, part 2).

MAIN FEATURES OF THE METHOD

Range of applicability

A gas in the sense of the guidline is considered as heavy if the initial density excess ratio $(\rho_o - \rho_a)/\rho_a > 0,16$ and if, for instantaneous release, the source volume $V_o > 0,1m^3$ or, for continuous release, the source volume flux $\dot{V}_o > 10^{-3}m^3/s$. Furthermore, it must be reasonable to assume the gas disperses near to the ground. If one of these criteria is not fulfilled the corresponding guideline for neutrally-dense or buoyant gas releases (VDI 3783, part 1) has to be applied.[1]

Determination of initial density

In the determination of the intitial density ρ_o, the guideline distinguishes between four cases. In case of gas/air mixtures, only the densitiy of the heavy gas fraction is considered.

[1]The application of the guideline is facilitated by a menu-controlled program which runs on personal computers with operating system MS-DOS. When using this program, all criteria mentioned are automatically asked for and the user is guided to the correct final result. The program is available on a diskette from the VDI-Kommission 'Reinhaltung der Luft', P.O. Box 11 39, D-4000 Düsseldorf.

Case	initial density ρ_o
a) gases, general, as long as not to be classified under b), c), or d).	ρ_o is the density of the gas at process temperature and ambient pressure
b) gases liquified without pressure	ρ_o is the density of the gas at process temperature (max. boiling point temp.) and ambient pressure
c) gases liquified by pressure (see also d)	ρ_o is the density of the gas at boiling point temperature and ambient pressure
d) NH_3 liquified by pressure, and other gases behaving similar to NH_3	ρ_o is assumed to be 1.4

Determination of initial volume

The source volume for an instantaneous release V_o resp. the source volume flow rate for a continuous release $\dot{V}_o$ are calculated from the released mass M_o resp. the mass flow rate $\dot{M}_o$ of the substance in its gaseous phase after the release, including the aerosol suspended in the cloud. Ambient air which is entrained before and during the release is not considered. The volume increase due to temperature increase during dispersion is anticipated and leads to an increase of V_o or $\dot{V}_o$, respectively.

Determination of release type

For very short release times ($t \to 0$) a spill is considered as instantaneous, for very long release times ($t \to \infty$) as continuous, respectively. Spills with finite release times or discontinuous flow rates must be assigned to one of these two types of release. In case of doubt which release type is more appropriate, the computer program [1)] should be applied which tests both alternatives and chooses the best approximation.

Determination of characteristic quantities

From the initial conditions of the cloud characteristic quantities can be deduced which are subsequently used to scale up the results from the small scale experiments to real world proportions. For details of the theory behind see König-Langlo and Schatzmann (1990).

	instantaneous release	continuous release
length scale	$L_{ci} = V_o^{1/3}$	$L_{cc} = (\dot{V}_o^2/g')^{1/5}$
time scale	$T_{ci} = (L_{ci}/g')^{1/2}$	$T_{cc} = (\dot{V}_o/g'^3)^{1/5}$
velocity scale	$U_{ci} = (L_{ci} \cdot g')^{1/2}$	$U_{cc} = (\dot{V}_o \cdot g'^2)^{1/5}$

with effective gravity acceleration $g' = g(\rho_o - \rho_a)/\rho_a$.

Choice of ambient wind velocity

According to German law, an unfavorable dispersion situation must be considered within a risk assessment study. Largest lower flammability distances were found for an ambient wind velocity (measured at the height of 1 characteristic length scale) of about $u_a(L_{ci}) = U_{ci}$ and $u_a(L_{cc}) = U_{cc}$ for instantaneous and continuous spills, respectively.

Quantities of evaluation

For flammable heavy gases, the mean lower flammability distance, LFD_{mean}, and the worst lower flammability distance, $LFD_{maximum}$, are required. LFD_{mean} represents the maximum distance from the source within which the mean cloud concentration exceeds the lower flammability concentration LFC. $LFD_{maximum}$ is accordingly defined but based on maximum concentration values (defined as mean $+2RMS$ in order to ensure reproducibility).

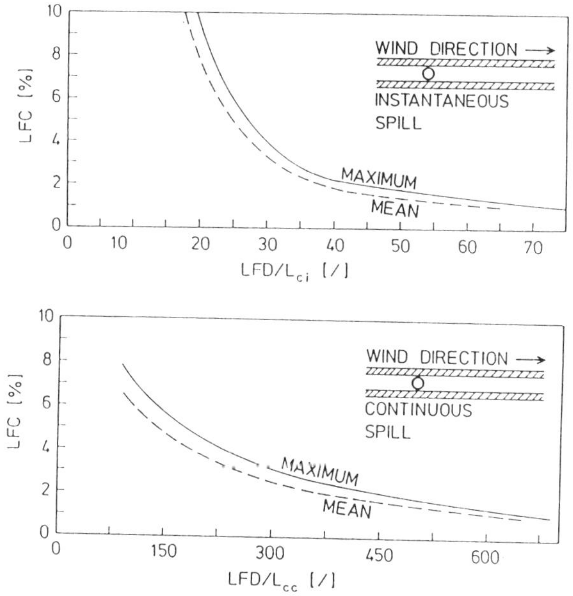

Fig. 1. Lower flammability distances LFD/L_c for instantaneously (top) and continuously (bottom) released heavy gases as function of lower flammability concentration LFC.

In the case of toxic heavy gases, the quantities of evaluation are concentration versus time traces as well as dose values at fixed receptor points downstream from the source.

Determination of quantities of evaluation

To meet the demand of authorities and industry for an easily applicable method for the determination of the quantities of evaluation defined above, systematic wind tunnel experiments with a variety of obstacle configurations have been carried out. These obstacle configurations include street canyons with different orientations towards the wind, street crossings, down slopes, dikes, ditches and mitigating walls. Some of these obstacles lead to a significant increase, others to a significant reduction of the extent of the area within which hazardous gas concentrations can occur. The obstacle configurations have been chosen such that both extremes were covered in the experiments, in order to obtain quantitative information to which degree obstacle effects can modify the dispersion of a heavy gas cloud.

Subsequently, as an example for an unfavourable obstacle configuration, the results from a street canyon oriented parallel to the wind direction will be presented. The canyon was infinitely long, its width was $2L_{ci}(14L_{cc})$ and the height of the bordering buildings was $1L_{ci}(7L_{cc})$. Flow visualization indicated that the height of the buildings was already sufficient to trap the heavy gas cloud completely. Instantaneous as well as continuous releases were realized. The wind velocity $\overline{u}_{aR}$ measured at height L_c was chosen to be the characteristic wind speed U_c. Ground level concentration versus time traces have been recorded at 8 locations downstreams from the source. Mean and maximum (mean $+2RMS$) values of the local concentrations were determineded.

The results obtained have been transformed into graphs as presented in Fig. 1. They allow the direct determination of lower flammability distances as a function of the lower flammability concentration for all major flammable gases.

For toxic heavy gases, the area of interest significantly exceeds the lower flammability distances. For such gases, a procedure in two steps has been elaborated. Diagrams as shown in Fig. 1 (or alternatively the computer program [1)]) are used to cover the gravity spreading phase of the cloud which is assumed to extend from the source down to 1 % (by volume) of the initial concentration. At the end of the gravity spreading phase, initial conditions for a numerical model utilized to calculate concentrations and doses at far-field receptor points are determined. The model chosen is described in guideline VDI 3783, part 1. In the Federal Republic, this model is routinely used for dispersion calculations within risk assessment studies for accidentally released neutrally-dense and buoyant gases.

Acknowledgements

We are grateful to the German Ministry of Research and Technology (BMFT) and to the German Environmental Protection Agency (UBA) for financial support.

References

König-Langlo, G., and Schatzmann, M. (1990): Wind tunnel modeling of heavy gas dispersion. Atmospheric Environment (in print).

VDI 3783, Part 1 (1987): Dispersion of emissions by accidental releases - safety study. Beuth Verlag, Berlin

VDI 3783, Part 2 (1990): Dispersion of heavy gas emissions by accidental releases - safety study. Beuth Verlag, Berlin

THE SEVEX PROJECT, AN INTEGRATED DISPERSION AND CONTAMINATION CALCULATION FOR COMPLEX TERRAIN AND ACCIDENTAL RELEASES

G. Schayes+, J. Smitz*, P. Thunis+, and L. Lesage*

+ Institut d'Astronomie et de Géophysique (Université de Louvain), Belgium
* Centre Environnement (Université de Liège), Belgium

1. Introduction

Following the EEC Seveso directive of 1982, industries using dangerous materials in large quantities and Public authorities are obliged to establish a distance of possible contamination occuring during hypothetical major accident. Three Belgian universities were charged of the practical study.

The SEVEX project (from SEVeso EXpert system) general layout is depicted on figure 1. First, an analysis of possible scenarii of major accidents is done by the Mons University group according to the data provided by the factory. This determines the source intensity and characteristics. In case of burning or explosion, simple conventional ways of calculating the effects are used. Explicit dispersion calculation as shown below is done only for toxic gas clouds released in the atmosphere.

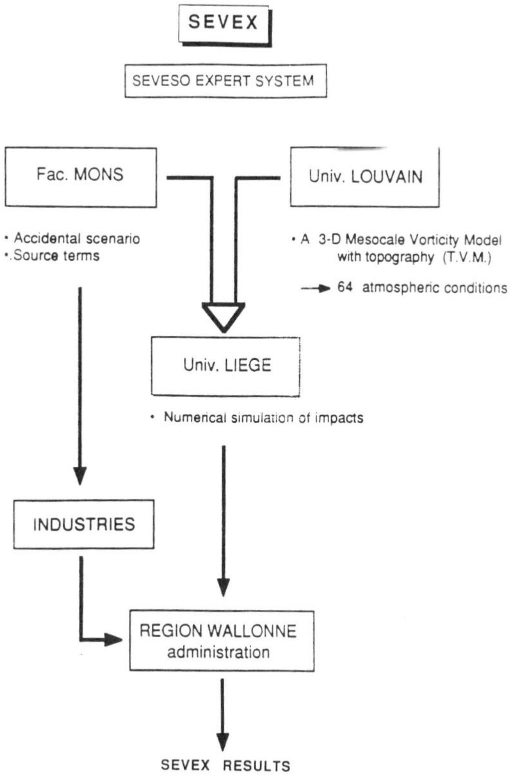

Fig. 1

In this particular case, the dispersion, concentration and deposition of the pollutant to the ground is obtained by using a three-dimensional mesoscale numerical model coupled with a Monte-Carlo particle dispersion scheme. The model is run several times with a series of prescribed typical meteorological conditions (3 different wind speeds, 8 wind directions and 2 stability classes) for each industrial site. For each run, an injury gravity function is deduced from the time integrated concentrations.

Finally, these results will be put in an expert system to be used by the administration whom goal is to provide efficient tools to public safety authorities.

First the mesoscale model will be described and compared with experimental data. Then the dispersion and concentration calculation scheme will be presented to finally show how all these steps are applied to a specific case.

2. Numerical mesoscale modelling

The three-dimensional numerical model of the planetary boundary layer TVM in use at the University of Louvain originated from the URBMET code developed at the San Jose State University (USA) and described in Bornstein (1987). The main characteristic of this model is that dynamic equations are written in vorticity (TVM for three-dimensional vorticity model) so that density and pressure are dropping off the equations. This model is hydrostatic and Boussinesq. The atmospheric part of the model domain is divided in two layers: a surface layer with constant fluxes where the variables are one dimensional and are supposed to obey the well known Monin-Obukhov similarity theory, and a transition layer where the hydrodynamic and thermodynamic equations are solved numerically with finite differences. The turbulence closure is the so-called one-and-a-half order closure based on the turbulent kinetic energy for which a prognostic equation is introduced in the model. The calculation of the diffusion coefficients is based on the mixing length formulation of Terry and Lacarrere (1983).

The temperature and humidity conditions at the ground level are computed by a soil submodel from the energy balance at the surface using the force-restore method of Deardorff (1978) for bare soil. The presence of the vegetation is accounted for indirectly by a Pennmann-Monteith formulation using resistances to compute the latent heat flux. The basic forcing comes from the solar radiation computed by the model itself and a correction is introduced when solar radiation is incident on an inclined surface.

A z-coordinate transformation is applied to take into account the topography but the hydrostatic assumption does not allow slopes steeper than 5 degrees. A filter is used in the uppermost 5 layers of the domain to attenuate gradually perturbations in each horizontal plane as they move up.

Other environmental factors (like the rugosity and the ground cover type) surrounding the factory are used as input in this model which generates the wind field, turbulence intensity and associated variables in the PBL.

Mesoscale model validation

In order to validate the model, we used the data collected at the sixth European Remote-Sensing campaign on Air Pollution that took place at Fos (near Marseille, France) from june 6 to 16, 1983. During this experiment, the meteorological conditions were dominated by a regular and strong sea-breeze phenomenon (due to the Mediterranean sea) interacting with the local topography (for details, see Rancillac and Despres, 1986). Continuous measurements of the sensible heat and momentum fluxes by means of a sonic anemometer have been made at a height of 22 m (see Cieslik & Hasenjager, 1985). These data are used to check the model.

The chosen day of june 9 was caracterised by relatively slow synoptic wind (3 m/s from WNW) allowing a full development of the sea-breeze and a clear sky (in order to avoid any radiative error due to clouds). The result of the time evolution of the N-S wind component at the observation point and computed by TVM for a 24 h cycle are presented in fig 2, where circles represent observations and continuous lines the simulation results. Similar good agreement was achieved for the temperature and the friction velocity.

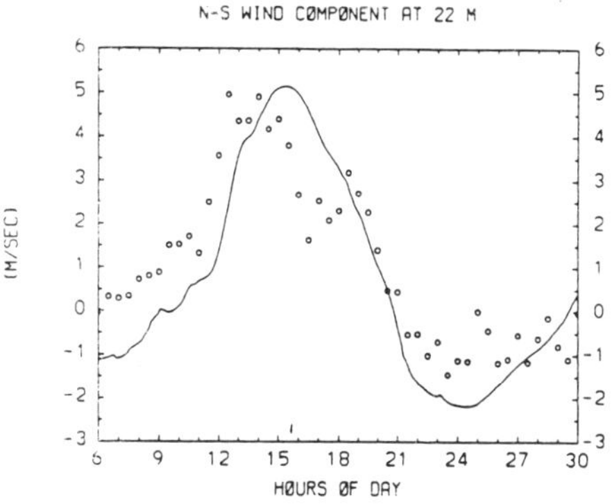

Fig. 2 Calculated (-) versus observed (°) wind velocity

A more complete description of the model and its validation can be found in Schayes and Thunis (1990).

3. Lagrangian particle diffusion and pollutant concentration

As a second step, we use a lagrangian particle diffusion to compute the concentration of the effluent in the air. Following the scheme of Yamada and Bunker (1988), the position x_i of the particles are computed for each time step δt from

$$x_i(t+\delta t) = x_i(t) + U_{pi}\ \delta t,$$

where

$$U_{pi} = U_i + u_i,$$

$$u_i(t + \delta t) = a\ u_i(t) + b\ \sigma_{ui}\Phi + \delta_{i3}\ (1-a)\ t_{Lxi} \frac{\partial}{\partial X_i} (\sigma_{ui})^2, \quad (1)$$

$$a = \exp(-\ \delta t/t_{Lxi}) \quad \text{and} \quad b = (1 - a^2)^{0.5}.$$

In the above expressions, U_{pi} is the particle velocity in the x_i direction; U_i the mean velocity; u_i the turbulence velocity; Φ a random number from a Gaussian distribution with a zero mean and unit variance; t_{Lxi} the Lagrangian integral time for velocity u_i and σ_{ui} the standard deviation of velocity fluc-tuation. The last term on the right-hand side in eq.(1) was introduced by Legg and Raupach (1982) in order to correct accumulation of particles into lower energy regions. The mean velocity U_i and standard deviation δ_{ui} of

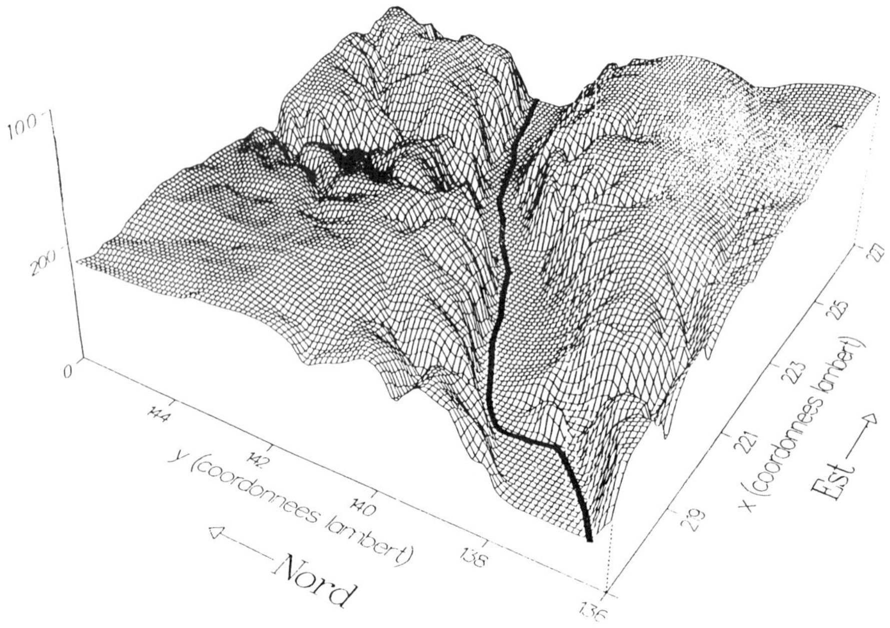

Fig. 3. Example case topography (Vertical scale is exagerated).

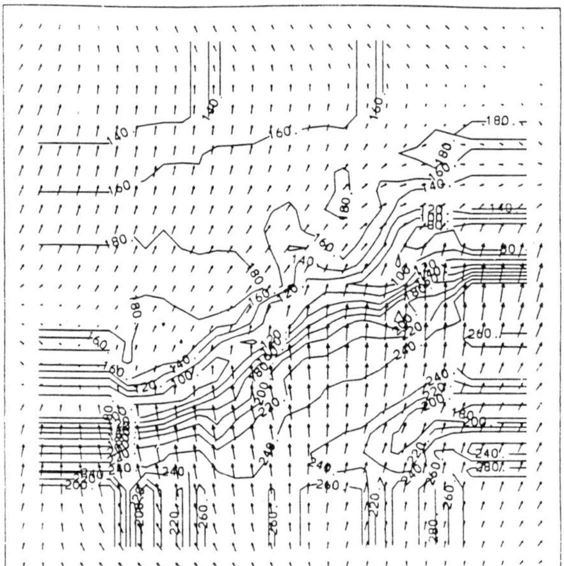

Fig. 4. Wind field close to the ground for a 2m/s south-west geostrophic wind applied to the example case of fig. 3.

the wind are obtained from the mesoscale model discussed before. The Lagrangian time scales have been fixed to t_{Lz} = 20 s, $t_{Lx} = t_{Ly} = 10^4$ s.

Outdoor concentrations

Following Yamada and Bunker scheme (1988), each particle represents a center of a puff. The concentration distribution in this puff is expressed by a Gaussian where variances are determined as the time integration of the velocity variances encountered along the history of the puff. The total outdoor concentration level at a given space and time is then the sum of the contributions of all particles.

4. Injury gravity and example case

Probable indoor concentrations are then calculated with different ventilation coefficients for buildings. Finally, contamination (injury) data are estimated from the outdoor and indoor concentrations by using the Probit functions calculated as the time integration of the concentrations at a given point.

The final result of the simulation gives a series of maps showing the zones affected by the accident to various degrees of contamination under outdoor and indoor exposure.
All the simulated data are to be included in an intelligent data base (so called expert system) in order to be used by the public authorities for (i) planning emergency responses and (ii) quick decision making in case of accident.

Injury level code:

black >99% death
hatched > 1% death
cross > 1% lesion

The thin lines represent the topography (m). The thick line is the river.

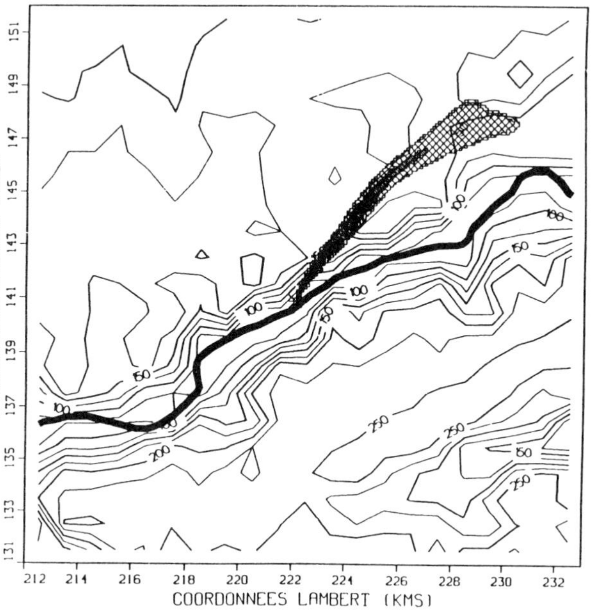

Fig.5 Example map of injury level result.

As a typical example, the figure 3 shows the location of an industry in a valley; the mesoscale numerical modelling provides the wind field in the case of a 2 m/s geostrophic wind from the South-West (fig. 4) and finally fig. 5 shows the probable outdoor injury zone.

Acknowledgements

The SEVEX project has been financed by the "Ministère de l'Environnement" of the Walloon Region of Belgium and has received a support from IBM. Part of this work has been done under Collaborative Grant nr. 86/0430 of NATO Scientific Affairs Division. One of us (G. Schayes) has been visiting the European Community Joint Research Centre of Ispra where many developments of the code have been made. Finally, the authors thank R. Bornstein for providing the original URBMET code and have appreciated the helpful discussion with Dr. D. Steyn.

References

Bornstein R. et al. (1987). Modeling the Polluted Coastal Urban Environment. Vol 1, EPRI Report EA-5091.

Cieslik S. and Hasenjager H. (1985). Sensible and latent heat fluxes and temperature gradients deduced from sonic anemometer and radiometric data obtained during the VIth ECC Campaign at Fos. EUR Report 9829 EN1985.

Deardorff J.,(1978). Efficient prediction of ground surface temperature and moisture, with inclusion of a layer of vegetation. J. Geoph. Res. 83, 1198-1903.

Hasenjager H. (1985). Micrometeorological observations and comparison of Pasquill to Smith classes using data obtained with an ultrasonic anemometer during the VIth ECC campaign at FOS, 1983. EUR Report 10127 EN-1985.

Legg R. and Raupach M. (1982). Markov-chain simulation of particle dispersion in inhomogeneous flows: The mean drift velocity induced by a gradient in Eulerian velocity variance. Bound. Layer Met. 24, 3-13.

Rancillac F. and Despres A.(1986). Résultats de la VIeme Campagne Européenne de mesure de la pollution atmosphérique à Fos (juin 1983). Rapport final. Commissariat à l'Energie Atomique (France).

Schayes G. and Thunis P. (1990). A three-dimensional mesoscale model in vorticity mode. Contribution nb. 60, Institut d'Astronomie et de Géophysique, UCL, Louvain-la-Neuve, 42 pp.

Terry G. and Lacarrère P. (1983). Improving the Eddy Kinetic Energy Model for Planetary Boundary Layer Description. Bound. Layer Met. 25, 63-88.

Yamada T. and Bunker (1988). Development of a nested grid second moment turbulence closure model and application to the 1982 Ascot Brush Creek data simulation. J. Atm. Sci. 27, 562-578.

AN OPERATIONAL EULERIAN DISPERSION MODEL

APPLIED TO DIFFERENT SCALES

C. Persson and L. Robertson

Swedish Meteorological and
Hydrological Institute (SMHI)
601 76 Norrköping, Sweden

A three-dimensional Eulerian dispersion model has been adapted to different air pollution problems on meso and regional scale. With some modifications the model can be applied to continuous as well as accidental releases. All meteorological information needed is based on routine weather observations. The purpose is to have three different options of the dispersion model for operational applications on
- meso-γ scale (different subregions within Sweden)
- meso-β scale (Sweden)
- European scale (Europe).

DISPERSION MODEL

The Eulerian model has on
- meso-γ scale about 1000 gridsquares in the horisontal, each with a size of 5 x 5 km^2
- meso-β scale about 2000 gridsquares of 20 x 20 km^2
- European scale about 4000 gridsquares of 55 x 55 km^2.

The vertical extension of the model, which basic concept is similar to van Egmond and Kesseboom (1983), is separated into the following 3 different levels: A surface layer which deph is kept constant in time, the rest of the mixed boundary layer, and above that a reservoar layer. The thicknesses of layers 2 and 3 and the vertical fluxes within the model are continuously variable in time and space depending on the state of the boundary layer. The variation in space of the boundary layer height causes some numerical disturbances, which however can be dealt with, and they are of very little importance. The advantage with this approach is that it makes it possible to obtain rather detailed information about the physical dispersion processes within the frame of a limited computing volume.

An operator split time integration scheme is used. The horizontal flux is calculated by means of a finite element method (the Weighted Residual, Bubnow-Galerkin, Method). The vertical flux is obtained from time and space variations of the boundary layer height, the vertical wind and the turbulent vertical flux (K-theory). Wet deposition is calculated with regard to the present precipitation intensity in each grid square and dry deposition is obtained with regard to physiography, topography and vertical turbulent flux within each grid square. In order to improve the description of the initial dispersion, a Gaussian puff model is used, for the major sources.

Air Pollution Modeling and Its Application VIII, Edited by H. van Dop and D. G. Steyn, Plenum Press, New York, 1991

METEOROLOGICAL INFORMATION

The meteorological information needed is provided by the so called objective analysis technique. The basic concept of this technique is to find the most likely state of the atmosphere by a statistical approach in terms of the weather observations available. For each of the three vertical levels information such as horisontal and vertical winds, air temperature, vertical stability, surface heat flux and precipitation are provided.

Meso-γ scale: Analyses based on observations from 30-40 ground based routine weather stations and 1-2 radio sonde stations.

Meso-β scale: Analyses based on observations from about 200 ground based routine weather stations and about 10 radio sonde stations.

European scale: Analyses based on meteorological fields will be provided by a high resolution, state-of-the-art numerical weather prediction system developed within the HIRLAM project, which is a joint Nordic, Duth and Irish research project aiming at a complete system for high resolution objective analysis and short range forcasting. The HIRLAM system will at SMHI be based on an operational access to ECMWF analysed and forecasted meteorological fields.

RESULT

Meso-γ studies for SO_2 over Skåne, the southernmost part of Sweden with an area of 15000 km^2, are given as an example. All types of sources within the Skåne area are included. For the time being only a simple parametrisation of the chemistry is used. By combining daily measurements from a background air pollution station with model calculations, good agreement for total sulphur compounds are obtained over the studied area, see Figure 1. Within the cities 40-50% of all SO_2 in the air come from sources outside Skåne.

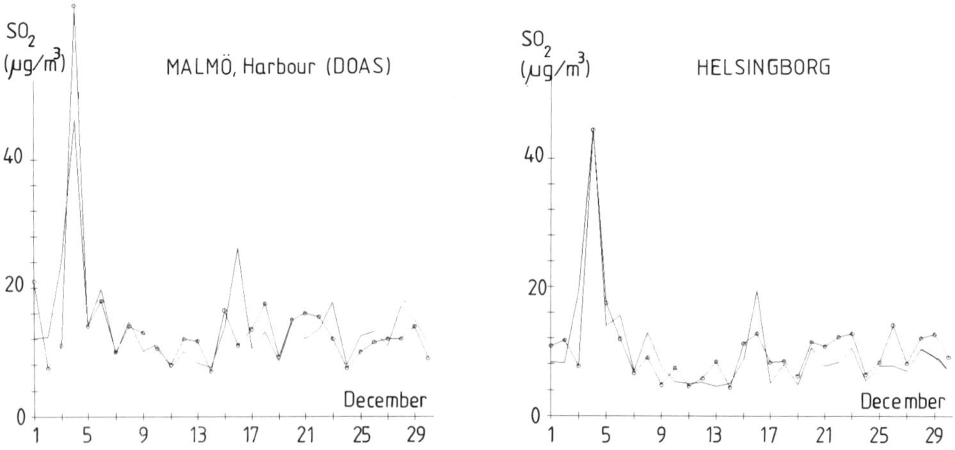

Figure 1. Observed (o—o) and total calculated (——) total 24-h mean values of SO_2 ($\mu g\ SO_2/m^3$) during December 1988 for some cities in southern Sweden.

REFERENCE

Van Egmond N.D., and Kesseboom H., 1983, Mesoscale air pollution dispersion models-I. Eulerian grid model. Atm.Env., Vol 17 No 2.

A LAGRANGIAN PARTICLE DISPERSION MODEL FOR LONG-RANGE SIMULATION OF ACCIDENTAL RELEASES OF RADIOACTIVITY AT THE GERMAN WEATHER SERVICE

Barbara Fay, Hubert Glaab, Ingo Jacobsen

Deutscher Wetterdienst
Frankfurter Straße 135
6050 Offenbach, Federal Republic of Germany

Following the Chernobyl reactor accident, the German Weather Service (DWD) is bound by national legislation to perform air pollution forecasts in case of accidental releases of radioactivity involving long-range transport of contaminants. As part of the national radioactive monitoring and alert system (IMIS), a Lagrangian particle dispersion model (LPDM) was chosen due to physical and numerical advantages.

The LPDM calculates the trajectories of a multitude of contaminated particles from a point source. The model is fully threedimensional and covers the whole of the troposphere. It uses model output of the mesoscale "Europa-Modell" (EM) (horizontal grid length 50 km) at the DWD (Müller et al., 1986). Transport and dispersion depend on EM grid-scale winds ($\overline{v}$) and on turbulent velocities (v'). The latter consist of a Markov chain-formulation consisting of a time-correlated part (A) and a stochastic part (B, Monte Carlo model), and a drift velocity (C) (a. o. Legg and Raupach, 1982; de Baas, 1988). They are parameterized by EM turbulent kinetic energy derived diagnostically from the equations of second turbulent moments (Müller, 1981; Mellor and Yamada, 1974).

$$v_i(t) = \overline{v_i}(t) + \underbrace{R_i(\Delta t) \cdot v_i'(t-\Delta t)}_{A} + \underbrace{(1-R_i^2)^{1/2} \cdot \sigma_i \cdot \lambda_i}_{B} + \underbrace{(1-R_i(\Delta t)) \cdot T_i \cdot \partial\sigma_i^2/\partial x_i \cdot \delta_{i3}}_{C} \qquad i = 1, 2, 3$$

T: Lagrangian time scale
R: Lagrangian autocorrelation coefficient
σ: standard deviation of turbulent velocity
λ: random number.

Standard microscale turbulence parameterization is not appropriate to model long-range horizontal transport. Therefore, different diffusion parameterizations have been tested, also taking into account horizontal deformation and shear of grid scale winds and allowing for longer time steps (Gifford, 1988; McNider, 1989).

Dry and wet deposition are crucial for cleansing the atmosphere and for contaminating soil, vegetation, animals, and humans. Dry deposition comprises the removal of gases and particles from the atmosphere by various processes such as sedimentation, and turbulent transfer and uptake at the surface. The latter are parameterized using a deposition velocity determined by resistance analogy (Chamberlain and Chadwick, 1966). It is based upon EM atmospheric and surface resistances including spatial, diurnal and seasonal variations.

Wet deposition results from the interception of the plume by precipitation. It is a very complex process depending on the cloud macro- and micro-physics as well as the physical and chemical properties of the aerosols involved. The EM uses an elaborate micro-physical cloud parameterization scheme for precipitation that allows a detailed computation of rain-out and wash-out processes (Müller, 1986). The preliminary version of the LPDM calculates wet deposition employing a scavenging coefficient depending on precipitation intensity (ApSimon, 1987).

The model will be optimized and made operational on the CRAY Y-MP/432, using the DWD's future forecast suite of EM and global model.

REFERENCES

ApSimon, H. M., and Wilson, J. J. N., 1987, Modeling atmospheric dispersal of the Chernobyl release across Europe, Bound.-Lay. Met., 41:123-133

Chamberlain, A. C., and Chadwick, R. C., 1966, Transport of iodine from atmosphere to ground, Tellus XVIII, 2:226-237.

de Baas, A. F., 1988, Some properties of the Langevin model for dispersion, Report No. Risø-M-2627, January, Risø Nat. Lab., Roskilde.

Gifford, F. A., Barr, S., Malone, R. C., and Mroz, E. J., 1988, Tropospheric relative diffusion to hemisphere scales, Atm. Environm. Vol. 22, 9:1871-1879.

Glaab, H., 1986, Lagrangesche Simulation der Ausbreitung passiver Luftbeimengungen in inhomogener atmosphärischer Turbulenz, PhD thesis, Technische Universität, Darmstadt.

Legg, B. J., and Raupach, M., 1982, Markov-chain simulation of particle dispersion in inhomogeneous flows: the mean dirft velocity induced by a gradient in Eulerian velocity variance, Bound.-Lay. Met., 24:3-13.

McNider, R. T., 1989, Mesoscale and long range pollutant transport and diffusion, IAMAP 1989, abstracts Vol. 1, Reading.

Mellor, G. L., and Yamada, T., 1974, A hierarchy of turbulence closure models for planetary boundary layers, J. Atmos. Sci., 31:1791-1806.

Müller, E., 1981, Turbulent flux parameterization in a regional-scale model, ECMWF Workshop on planetary boundary layer parameterization, 193-220.

Müller, E., Frühwald, D., Jacobsen, I., Link, A., Majewski, D., Schwirner, J.-U., and Wacker, U., 1986, Results and prospects of mesoscale modeling at the Deutscher Wetterdienst, Collection of papers presented at the WMO/IUGG NWP Symposium, Tokyo.

Smith, F. B., and Clark, M. J., 1989, The transport and deposition of airborne debris from the Chernobyl nuclear power plant accident with special emphasis on the consequences to the United Kingdom, U. K. Met. Office Scientific paper No. 42.

METEOROLOGICAL MONITORING NEAR THE PICKERING NUCLEAR GENERATING STATION

Peter A. Taylor[1], James R. Salmon[2] and Paul G. Stalker[2]

[1]Department of Earth and Atmospheric Science
York University
North York, Ontario, Canada, M3J 1P3.

[2]1435, Colonsay Drive
Burlington, Ontario, Canada, L7P 2T8

INTRODUCTION

A demonstration network of surface meteorological stations was operated for two years (Jan 1988 - Dec 1989) in the vicinity of the Pickering NGS. The study was supported by the Atmospheric Environment Service, the Atomic Energy Control Board, the Ontario Ministries of the Solicitor General and of the Environment and by Ontario Hydro, with most of the work being undertaken by the authors and by AES personnel. The main objective of the exercise was to demonstrate the feasibility of deploying permanent, low cost but reliable, mesoscale meteorological networks (Mesonets) around Nuclear Generating Stations and other sites with the potential for the accidental release of toxic material into the atmosphere.

THE MONITORING NETWORK

The Pickering Mesonet consisted of a network of 8 (9 originally) instrumented 10m posts (PAQMOS-10m's) and one instrumented 26m tower (PAQMOS-26m). A full description of the network, which covered an area of about 50 km x 20 km, and of the standard PAQMOS-10m towers and their instrumentation are contained in Salmon and Taylor (1989). Each PAQMOS-10m post supported an anemometer (at 10m) and a T/H sensor installed on a mounting arm at the 1.5m level within a vaned, self-ventilating radiation shield. A barometer was attached to the post at the 1.5m level and a datalogger (within its environmental enclosure) was mounted directly on the post as were a solar panel and an auxiliary battery.

The Pickering NGS is located on the shoreline of Lake Ontario about 30km to the east-north-east of the central core of the city of Toronto. Oshawa lies 20km further east-north-east and the rapidly expanding town of Pickering is about 5km north of the NGS. We should stress that the network was installed as a demonstration and was not intended as a permanent network. We would also emphasize the difficulties encountered in finding reasonably secure and representative sites in this heavily populated area and the restricted number of stations that were available.

AN ILLUSTRATIVE CASE - 12 MAY 1988

The early morning hours were characterised by light winds and strong radiational cooling. Light winds then developed, associated in part with urban heat island and lake breeze effects. There were distinct differences in wind direction with the shoreline stations reporting easterly winds and others being almost southerly. Predictions of the trajectory of a plume based on data from any single station would be highly suspect and even with data from all of the stations this would be a difficult case to assess. Figure 1 shows the vector winds for a typical 10 min period. Wind symbols have one full tick for every 10 knots (approx 5 ms^{-1}). The shoreline of Lake Ontario is also shown and coordinates are relative to the King City Radar. The winds show substantial direction differences between stations. From this information it might be reasonable to predict that a plume released from Pickering at surface level would follow the subjectively drawn trajectory indicated, somewhat tentatively, on the plot. The wind directions recorded at Toronto International and Toronto Island Airports at this time were both south-southwest.

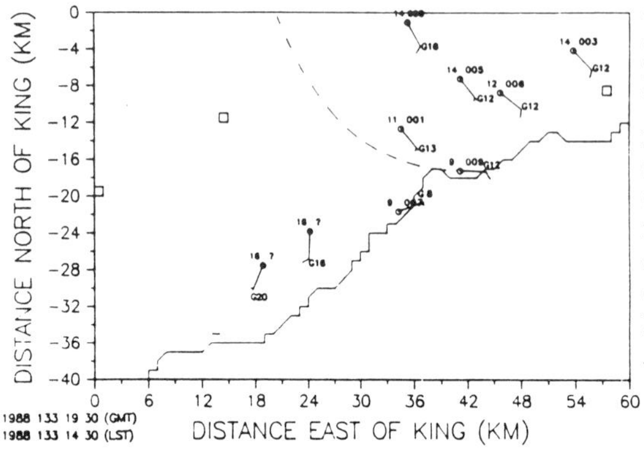

Figure 1 Sample data from Mesonet

STATISTICAL DATA

Basic yearly and monthly statistics have been compiled for the data collected during 1988 and 1989 by the PAQMOS-10m network and the PAQMOS-26m. These are presented in Salmon and Taylor (1989) and in the 1989 data report with a limited amount of discussion. As one of the reasons for installing the Pickering Mesonet was to gather data for a local climatology of the region, we calculated joint-frequency tables to relate wind speed, wind direction and atmospheric stability class. The atmospheric stability is represented by Pasquill class (A to F) inferred from measured temperature and velocity profile data from the 26m tower.

CONCLUSIONS AND RECOMMENDATIONS

We should stress that the network deployed in this demonstration project is a limited one and that a more extensive network of on-line stations would clearly do a better job in defining the mesoscale surface wind field. As a long term goal we would seriously recommend that all nuclear generating stations near urban areas should be provided with permanent surface networks of this type, perhaps comprising between 10 and 20 stations. In the vicinity of a major centre such as Toronto this could be part of a more broadly based network covering the urban and surrounding areas, with several users of the data. Multiple use, for both routine and emergency situations would spread the costs and make the deployment quite feasible.

REFERENCE

Salmon, J.R. and Taylor, P.A.,1989, The Pickering Mesonet, 1988 Data Report, to be published as an Atomic Energy Control Board Report.

AN IMPROVED GAUSSIAN MODEL FOR REAL TIME EVALUATION OF HEAVY GAS DISPERSION

C. Borrego, P. Leitão, J. M. Martins and M. Conceição

Department of Environment, University of Aveiro
3800 Aveiro Portugal

THE MODEL

This model was developed to be used in the real time prediction of a toxic or flammable cloud position, in case of industrial accidents. It considers two main features: a **slumping phase** with a top-hat model for concentration, and a **gaussian phase**. The main difficulty in the first phase is the correct assessment of the gas escape characteristics. To avoid this initial conditions problem a database with most probable and worst case accidents was built. At least a first estimate can be done using database accident cases. A pre-processor calculates the evolution of the emission rate and simulates the flow variation by a series of short time, constant flow rate, point sources with the corresponding meteorological data. In case of prediction of cloud path, most recent meteorological data are used. The accident pre-processor is used in both continuous and instantaneous releases: a liquid pool of evaporating gas is considered as a continuous source, with emission rate depending on a complex way from parameters such as surrounding temperature and ground type. To reduce to a minimum the number of operator inputs, the meteorological data are acquired automatically from a meteorological mast. This data include 10, 30 and 100 m measurements of temperature and wind speed and direction. A pre-processment of this data allows the calculation of the shear velocity, the Monin-Obukov length and the atmospheric stability, necessary for the determination of the σ parameters of the gaussian equation.

During the **slumping phase**, after a brief initial mixture with the surrounding air, the heavy gas cloud falls down due to gravity and spreads around following an hydrostatic equation (Kaiser et al. 1978; Chlorine Institut Pamphlet 1982; Zerefos et al. 1988). During this gravitic spread air entrainment occurs due to turbulent mixing, increasing the cloud volume, although the height remains more or less constant (Picknett, 1981). The point of transition to the second phase is calculated comparing cloud spreading speed with the atmospheric shear velocity, a parameter of atmospheric turbulence, calculated by the pre-processor refered above from the meteorological mast data (Chlorine Institut Pamphlet 1982; Zerefos et al. 1988).

The **second phase** follows a gaussian model, using a virtual source calculated after Turner model for area sources as a function of the dispersion during the initial stage (Turner 1988). Continuous decaying releases, typical of the evaporation of liquified gases, are divided into

Air Pollution Modeling and Its Application VIII, Edited by H. van Dop and D. G. Steyn, Plenum Press, New York, 1991

several short step emissions as refered before and the Palazzi correction is applied to the gaussian model (Pallazzi et al. 1982). For each step emission, updated meteorological data can be used and the cloud trajectory is corrected. Roughness effects will also be included (Borrego et al. 1989). The concentration at each point of the calculation grid is obtained by the summation of the contribution of each step source, active or extinguished. A final routine calculates the isoconcentration lines, which define the cloud, normally corresponding to values as the flammable limits or the detection, irritation and lethal concentrations. These lines are plotted on the screen over a map of the region.

CONCLUSION

This model is suitable to be used in case of an accidental release of a heavy gas to give a real-time evaluation of the endangered area. For this purpose the model must have access to on-time meteorological data, as well as chemical products, geographic and demographic data bases. Its results are presented on the computer screen as isoconcentration contours over the geographic map of the region.

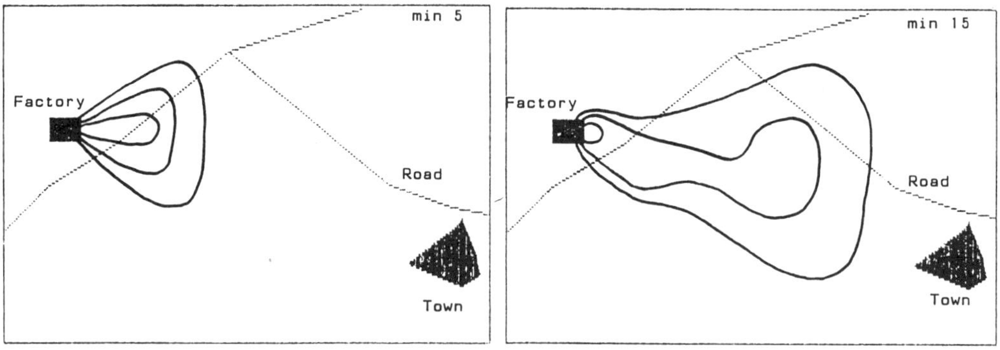

Fig.1 Cloud path: at minute 10 wind changed from 270° to 290°.

REFERENCES

Borrego,C., Coutinho,M. and Costa,M.J., 1989, Terrain roughness effects on a Gaussian dispersion, to be published in The Science of the Total Envir.

Chlorine Institute Pamphlet, 1982, Calculating the Area Affected by Chlorine Releases, edition 1 revision 1, the Chlorine Institute, inc.

Kaiser,G.D. and Walker,B.C., 1978, Releases of Anhydrous Ammonia from Pressurized Containers - the Importance of Denser-then-air Mixtures, Atm. Envir., 12:2289.

Palazzi,E., De Faveri,M., Fumarola,G. and Ferraiolo,G., 1982, Diffusion from a Steady Source of Short Duration, Atm. Envir., 16:2785.

Picknett,R.G., 1981, Dispersion of Dense Gas Puffs Released in the Atmosphere at Ground Level, Atm. Envir., 15:509.

Turner, D.B., 1970, "Workbook of atmospheric dispersion estimates", AP-26,US, Environmental Protection Agency, Research Triangle Park, NC.

Zerefos,C.S., Ziomas,I.C. and Bais,A.F., 1988, Heavy Gas Dispersion With Demonstration of Computerized Predictive Tools, ISPRA Course "Major Hazard Analysis", Thessaloniki, 88.10.10-14.

MODEL-PREDICTED OZONE TRANSPORT PATTERNS IN THE NORTHEAST UNITED STATES

Norman C. Possiel* and Thomas E. Pierce*

U.S. Environmental Protection Agency
Research Triangle Park, North Carolina 27711

INTRODUCTION

Photochemical smog often enshrouds metropolitan areas throughout the Northeastern United States during the summer. Ozone levels exceeding the level of the National Ambient Air Quality Standard (NAAQS), 120 ppb, occur during such smog episodes. Ambient monitoring data indicate that high ozone levels persist despite efforts to reduce ozone precursor emissions from automobiles and industrial sources. Also, field studies have shown that ozone pollution often spreads over several States to impact areas across hundreds of miles. The Environmental Protection Agency has responded to the need to understand this complex problem of long-range ozone transport in part by sponsoring the development of the Regional Oxidant Model (ROM). The purpose of this video is to display ozone concentrations predicted by ROM in a manner which graphically depicts the formation and transport of ozone within the Northeast. As shown in the video, patterns of interurban and interstate transport of moderate to high ozone levels are clearly evident across this region.

TECHNICAL APPROACH

The region modeled extends from Ohio to the East Coast and from Virginia to Maine and includes the major metropolitan areas of Washington, DC, Baltimore, Philadelphia, New York, and Boston (i.e., the Northeast Corridor). This Corridor along with other major metropolitan areas such as Pittsburgh, Cleveland, and Detroit emit moderate to high levels of ozone precursors. In addition to man-made emissions of nitrogen oxides, hydrocarbons, and carbon monoxide, hydrocarbon emissions from vegetation are included.

Description of Transport Patterns

This video was created from hourly ozone concentrations calculated by ROM for a 15-day ozone episode, July 12 - 27, 1980. The portion of the video which displays the movement over time of predicted ozone concentra-

---

* On assignment from the National Oceanic and Atmospheric Administration, United States Department of Commerce

Air Pollution Modeling and Its Application VIII, Edited by H. van Dop and D. G. Steyn, Plenum Press, New York, 1991

tions for the July episode is divided into five segments, each comprised of 2 to 3 days. The remainder of this paper highlights the transport patterns for two segments with the highest predicted ozone concentrations.

July 12 - 16, 1980. On July 12, a cold front moved off the East Coast and a high pressure system moved into the Northeast with northwesterly wind flow across the region. By July 14, this high was centered over the Northeast Corridor and moved offshore on the following 2 days. Wind flow was generally from the west to southwest and afternoon temperatures were above 90°F. On July 14, ROM predictions show the build-up in ozone during midday with concentrations exceeding the NAAQS near the major Corridor cities and also near Pittsburgh, Cleveland, and Detroit. Overnight, plumes from these cities move to the east and northeast. By sunrise on July 15, the remains of the Baltimore-Washington plume from the previous day is predicted to be over southern New Jersey. Similarly, the New York City plume is over eastern Connecticut and Rhode Island. Thus, at the start of the day, ozone is already greater than the NAAQS in some areas due to transport from the day before. By midday, one continuous plume of high ozone extends from near Philadelphia to Boston with values greater than 160 ppb covering most of Connecticut. On July 16, ozone levels again form along the Corridor with values greater than 160 ppb over eastern Long Island by late afternoon.

July 20 - 22, 1980. During this segment, a large Bermuda high pressure system was located off the East Coast. Skies were generally fair and afternoon temperatures above 90°F prevailed. Wind flow was from the west to southwest. By late afternoon on July 20, plumes of high ozone greater than the NAAQS are located downwind of Pittsburgh as well as in the Northeast Corridor. Overnight, moderately high ozone (~80 ppb) is transported eastward across Pennsylvania. By 0900 EST on July 21, this area of ozone has reached eastern Pennsylvania and concentrations increase rapidly after sunrise as the photochemical process begins; ozone and precursors are fumigated from aloft; and, fresh emissions are injected into the plume. This same area merges with the Corridor plumes during the late afternoon and ozone increases rapidly to above 160 ppb in the Philadelphia plume located just south of New York City and also across Long Island. Ozone levels diminish overnight, but the remains of these urban plumes are located over Rhode Island and Cape Cod by 0900 EST on the following day, July 22. Ozone in this aged air mass increases to 160 ppb over Cape Cod by 1300 EST. By mid-afternoon, the New York City plume is evident as a large area of high ozone covering Connecticut. Although ozone levels diminish during the evening, high concentrations persist over portions of New England during the night.

SUMMARY AND CONCLUSIONS

The ozone predictions of ROM for a 15 day episode in July of 1980 reveal patterns of high ozone formation and transport over several days across hundreds of miles in the Northeast. On most days during this episode, interstate transport is evident along the Northeast Corridor as well as from urban areas in the western part of the region toward the East Coast. These results suggest that regional strategies which reduce transported ozone concentrations will be necessary to attain the NAAQS along the Northeast Corridor.

V: NEW DEVELOPMENTS IN DISPERSION MODELLING AND THEORY

chairman: T.R. Oke

rapporteur: B. Artíñano

COMPREHENSIVE MODELS FOR ENVIRONMENTAL SYSTEMS

Akula Venkatram

ENSR Consulting and Engineering
1220 Avenida Acaso
Camarillo, CA 93012

INTRODUCTION

The fate of pollutants released into the environment is governed by complex systems. Acid deposition is governed by such a system. It consists of a large number of interacting processes, which include gas-phase and aqueous-phase chemistry of the relevant pollutants, meteorology, cloud physics, and the dry and wet scavenging of pollutants. Each of these processes is itself very complex. For example, the gas-phase chemistry that leads to the formation of sulfuric and nitric acids from their precursors, SO_2 and NO_x, involves at least 100 reactions among 50 species. Ideally, a model of the acid deposition system would incorporate all our understanding of the component processes. In practice, this is not possible for a number of reasons, the most important of which are the large manpower and the computational resources required to develop such a model.

There are two possible approaches to constructing a practical model for an environmental system. The first approach, followed in most existing models, assumes that only the net effect of individual processes needs to be considered in modeling the behavior of the system. Then, a process such as dry deposition can be represented by a simple equation or a parameter. The model of the system consists of equations that describe the interaction between these simplified processes. The mathematical formulation of the processes relies on measurements of their effects on the system being modeled. Because this approach to modeling relies directly on observations, the resulting model is often referred to as semi-empirical.

In the second approach, referred to as comprehensive modeling, every attempt is made to include all the processes in as much detail as practically possible. The underlying philosophy is that there is no justification for neglecting any of our understanding of the processes. This paper discusses the development and application of comprehensive models. To be specific in our discussion, we will focus our attention on the system that governs acid deposition.

The majority of currently available acid deposition models are semi-empirical. In spite of their inadequacies, we have continued to use them because the alternative, comprehensive models, represents such a major investment in both labor and computational resources. However, recent advances in computing hardware have led to preliminary steps towards using comprehensive models as more than research tools. This paper discusses these steps. In addition, it provides a preliminary blueprint for the development of a new generation of comprehensive models. For the sake of completeness, we will briefly discuss some of the ideas that underlie semi-empirical models, which often complement comprehensive models in understanding environmental systems.

Air Pollution Modeling and Its Application VIII, Edited by H. van Dop and D. G. Steyn, Plenum Press, New York, 1991

SEMI-EMPIRICAL MODELS

Semi-empirical models rely on observations to complete their formulation of the processes; the values of some of the model parameters are obtained by fitting model predictions to corresponding observations. For example, the value of the parameter used to represent the conversion of SO_2 to sulfate in the gas-phase could be derived this way. The values of the derived parameters are considered valid if they are consistent with our prior understanding of the process in question. However, in certain cases, this probing of observations might actually provide insight that might suggest re-examination of this prior understanding. For example, the role of aqueous-phase oxidants in enhancing the wet removal of SO_2 was first suggested by this fitting procedure. Venkatram and Pleim (1985) and Venkatram et al. (1990) provide examples of this application.

The validity of the semi-empirical approach, to a large extent, depends upon the model's ability to describe the behavior of the environmental system as a whole rather than its components. In such a model, the range of observations used to develop the model constrains the range of predicted responses of the model that can be considered valid. However, our knowledge of the processes, whose effects are parameterized in the semi-empirical models, can help in determining the range of applicability of the model. Thus, it is possible to justify the application of semi-empirical models to situations that are different from those used to derive model parameters.

COMPREHENSIVE MODELS

A comprehensive model is designed to be a numerical surrogate for the system being studied. After such a model has been evaluated against observations, it can be used to conduct experiments that would be impossible in the real world. For example, an acid deposition model can be used to examine the effect of SO_2 emission reductions in the midwestern United States on acid deposition at sensitive receptor sites in eastern North America.

In building a comprehensive model, one attempts to describe the governing processes in as much detail as possible. This description is in the form of mathematical equations, which then have to be translated into code that directs a computer to perform the associated computations. Because the model is a representation of a complex environmental system, its mathematical formulation and the associated code are necessarily complex. Thus, developing a comprehensive model is very much like designing and manufacturing a complex piece of machinery. The science in the comprehensive model is only one of the factors that determine the adequacy of the model.

The first comprehensive models for air pollution were developed in the late 1960s to examine the oxidant problem in Los Angeles. Tesche (1989) provides a review of these models. These urban scale models tend to emphasize the gas-phase chemistry that controls ozone formation. Because ozone is produced only during the daytime, when the boundary layer is well-mixed, the treatment of meteorology in these models has been kept simple by assuming a vertically homogeneous boundary layer.

Over the past ten years, concerns about the long-range transport of pollutants have given rise to the need for comprehensive models that incorporate other processes besides gas-phase chemistry in detail. A model for acid deposition has to incorporate processes that affect the transport and deposition of acidifying pollutants and their precursors. These processes include transport by three-dimensional flows over a domain with horizontal scales of a few thousand kilometers and a vertical scale of ten kilometers, transport and scavenging by cloud processes, transformation by gas-phase and aqueous-phase chemistry, and interaction of pollutants with the ground. The development of a model that includes these processes is best illustrated by considering a specific model (Venkatram et al., 1988), the Acid Deposition and Oxidant Model (ADOM), which is being sponsored by a group consisting of the Ontario Ministry of the Environment (OME), the Canadian Atmospheric Environment Service (AES), the Umweltbundesamt of the Federal Republic of Germany, and the Electric Power Research Institute (EPRI). A similar model (Chang et al., 1987), the Regional Acid Deposition Model (RADM), is being developed under the auspices of the National Acid Precipitation Assessment Program (NAPAP).

THE DEVELOPMENT OF ADOM

The eight-year development of ADOM, which was initiated in 1982, was divided into three phases. Phase I involved the design of the model. In this phase, the acid deposition system was divided into a number of components, which are illustrated in Figure 1. Then, these components were described with formulations that reflected our current understanding of the corresponding processes. In selecting the appropriate formulations, it was necessary to make sure that they were consistent with the available computational resources. Thus, it was often necessary to choose a formulation that a specialist in the relevant field would not support wholeheartedly. However, we made sure that the design of ADOM was overseen by a peer review committee consisting of experts in fields relevant to acid deposition. This review ensured that we were aware of the uncertainties in the parameterizations incorporated in ADOM. Furthermore, the committee made sure that the best science was brought to bear on the model. It is important to point out that even in a comprehensive model, limitations in computational resources impose simplifications of current understanding of the science in developing modules for the processes.

Phase I included the development of a logical framework and model architecture to incorporate submodels. We also delineated the data requirements for operating and validating the long-range transport model. Available data bases were then examined to determine their suitability for this task.

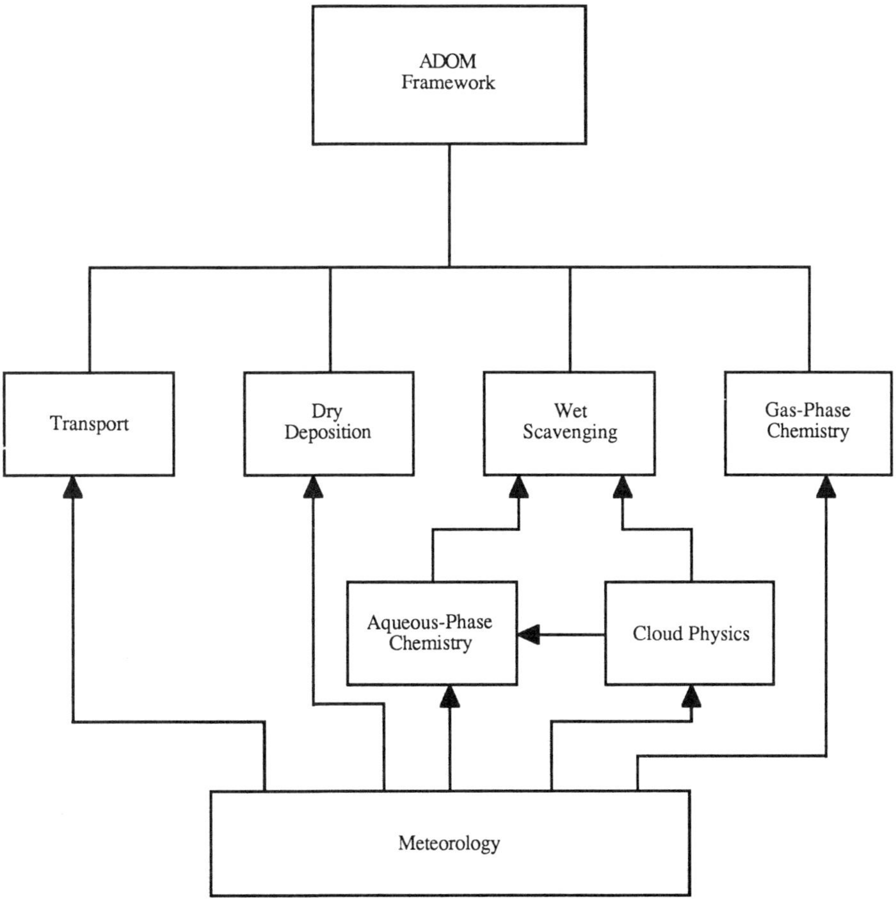

Figure 1. Structure of the Acid Deposition and Oxidant Model (ADOM)

Phase II of the project involved the implementation and refinement of the Phase I design. Submodels of physical processes were constructed for transport, dispersion, dry deposition, cloud processes, gas-phase chemistry, aqueous-phase chemistry and wet deposition. Sensitivity studies were performed to determine the relative importance of controlling variables in each model. These studies helped us to debug the codes and understand the processes involved in each module. During this process, we also made sure that the codes for the individual submodels and the overall model were vectorized to take advantage of the capabilities of supercomputers such as the Cray 1S and the Cray X-MP.

The assembly of the modules to construct ADOM involved an evolutionary process, which we believe is critical to the successful development of any comprehensive model. We essentially constructed a hierarchy of models, in which each hierarchy represented an increase in the complexity of the modules. The first stage of the hierarchy was constructed by combining modules that essentially represented parameterizations derived from semi-empirical models. Specifically, the wind field was based on surface geostrophic winds, the transformation rates of SO_2 to sulfate and NO_x to nitrate were described by constant rates, the wet scavenging was described by a scavenging coefficient, and the dry deposition velocities were specified. The model was debugged by conducting several sensitivity tests, which also provided insight into the responses of the complete model.

This relatively simple model was tested against observations of sulfur in rain collected by a network of stations operated by the Multistate Atmospheric Power Production Pollution Study (MAP3S) during the period July 1-14, 1980 in the eastern United States. The emission inventory used in the study was compiled for the Sulfate Regional Experiment (SURE), which was conducted during the period 1977-1980 (Mueller and Hidy, 1983). Because the model was based on empirically derived formulations for the processes, its predictions were in reasonable agreement with observations of sulfur in rain averaged over the two-week period of the study.

In the subsequent evolution of the model, each module of one generation of the model was replaced by a more complete version. These modules were replaced one at a time, and the results of the two versions of ADOM compared. Because this process of replacement of modules was designed to be evolutionary, substantial differences between the results of successive generations of ADOM indicated one of several possible problems: the new module had a bug, it was incorrect in its formulation, or it was implemented incorrectly in ADOM. We proceeded to the next version of ADOM only after we had identified and corrected such problems. In most cases, the incorporation of a more complete module resulted in better model performance. For example, we found that replacing the empirical wet scavenging coefficients by a cloud physics module that included aqueous-phase chemistry resulted in better agreement of model estimates of sulfur in rain with observations from the MAP3S network (ERT, 1985). This version of ADOM also included a detailed gas-phase chemistry mechanism (Lurmann et al., 1986).

The assembly of the modules using these evolutionary procedures was necessary for several reasons. First, it facilitated the assembly of the extremely complex model; the codes were debugged, and the interfaces between the modules were implemented in the most efficient manner. Second, examination of the results obtained from several generations of ADOM provided valuable understanding of the rich repertoire of responses that a complex model like ADOM can exhibit. This understanding is vital to the evaluation of the model against observations, as well as in understanding the results of simulations conducted with the model. We believe that the proper application of a comprehensive model requires the type of understanding of the model's responses that can only be obtained by participating in the evolution of the model. In a sense, the modeler is part of the comprehensive model.

The version of ADOM (ADOM I) that incorporated all the designed modules became available two years after the project was initiated. In Phase III of the project, ADOM I was tested with observations of sulfur and nitrogen in rain collected during the Oxidation and Scavenging Characteristics of April Rains (OSCAR) field study in 1981 (this study, designed to understand the wet removal of pollutants by cyclonic storms, operated a precipitation chemistry network of 37 stations spaced at 100-200 km intervals across eastern North

America). This testing indicated the need for improvements in several of the modules. For example, the cloud physics module was replaced by a more complete formulation to improve the performance of the model. The dry deposition module was improved when ADOM estimates of ozone concentrations were tested against observations made during an oxidant episode in Ontario, Canada, that occurred during June 9 to June 17, 1983 (Karamchandani et al., 1989).

We also improved modules when we could incorporate developments in the understanding of the processes without incurring large penalties in computational resources. We point out that the testing of model predictions against observations invariably results in improvements in the model. These improvements are possible because, in most cases, our understanding of the processes far exceeds what is incorporated in the model at any stage of its development.

EVALUATION OF ADOM AND RADM

The data used in Phase III were not complete enough to test ADOM or RADM adequately. For example, they did not include ambient air quality measurements coincident with the precipitation chemistry data. In response to the need for a high quality data set to evaluate RADM and ADOM, the U.S. Environmental Protection Agency (EPA), Environment Ontario, Environment Canada, and the Electric Power Research Institute embarked upon a two-year monitoring program, which included two intensives, to collect the necessary data. The program, which started in mid-1988, is collecting both ambient and precipitation chemistry data at about 70 stations distributed over the U.S. and Canada, with the greatest density of stations in the eastern U.S. The data set includes ambient concentrations of SO_2, sulfate, NO, NO_2, HNO_3, total nitrate, and ozone; it also includes measurements of sulfur, nitrogen, ammonium, and pH in rain. A small number of sites measured hydrogen peroxide and organics. The intensives, the first of which was conducted during the period from August through September 1988, had aircraft measurements of several species of relevance to acid rain.

Both ADOM and RADM are currently being evaluated with data from this measurement program, which is referred to as the Eulerian Model Evaluation Field Study (EMEFS). Here, we present some preliminary results from the evaluation of ADOM with data collected during the period from August through September 1988. Our intention here is to show how this evaluation can provide the type of understanding of the acid deposition system that would be impossible to obtain from a semi-empirical model.

Figure 2 shows a comparison of the model estimates of sulfur species with corresponding observations averaged over the 35 days of the experiment. We see that the model overpredicts ambient SO_2 concentrations, and underpredicts sulfate concentrations; there is little bias in the model estimates of total sulfur concentrations in air. This behavior of the model predictions indicates that the model is not producing enough oxidants, particularly the hydroxyl radical, OH, to convert SO_2 to sulfate. Note that, in spite of this obvious problem, the model estimates of sulfur in rain show little bias when compared with observations.

Figure 3, which compares model estimates of nitrogen species with observations, shows that the model overestimates all three species — NO_2 and nitrate in air, and nitrate in rain. The model also overpredicts total nitrogen (NO_2 + nitrate) concentrations in air. This strongly suggests the possibility that the NO emissions used by the model are too high. This overestimation of NO could also suppress oxidant levels in high emission density regions, which in turn would explain the low conversion of SO_2 to sulfate. The underestimation of oxidants is confirmed by examining the levels of ozone, a major precursor of OH. ADOM underestimates ozone in urban regions, and overestimates ozone in rural areas (not shown here). This indicates that the excess NO is scavenging ozone in urban areas, and leading to excess production of NO_2 in rural areas.

This diagnosis of the problem in ADOM led to an examination of the NO emissions inventory. It turned out that there was an error in the inventory that was equivalent to a 30% overestimate. ADOM has since been rerun with the corrected inventory. Preliminary results indicate that the model estimates of the nitrogen species now show little bias when compared

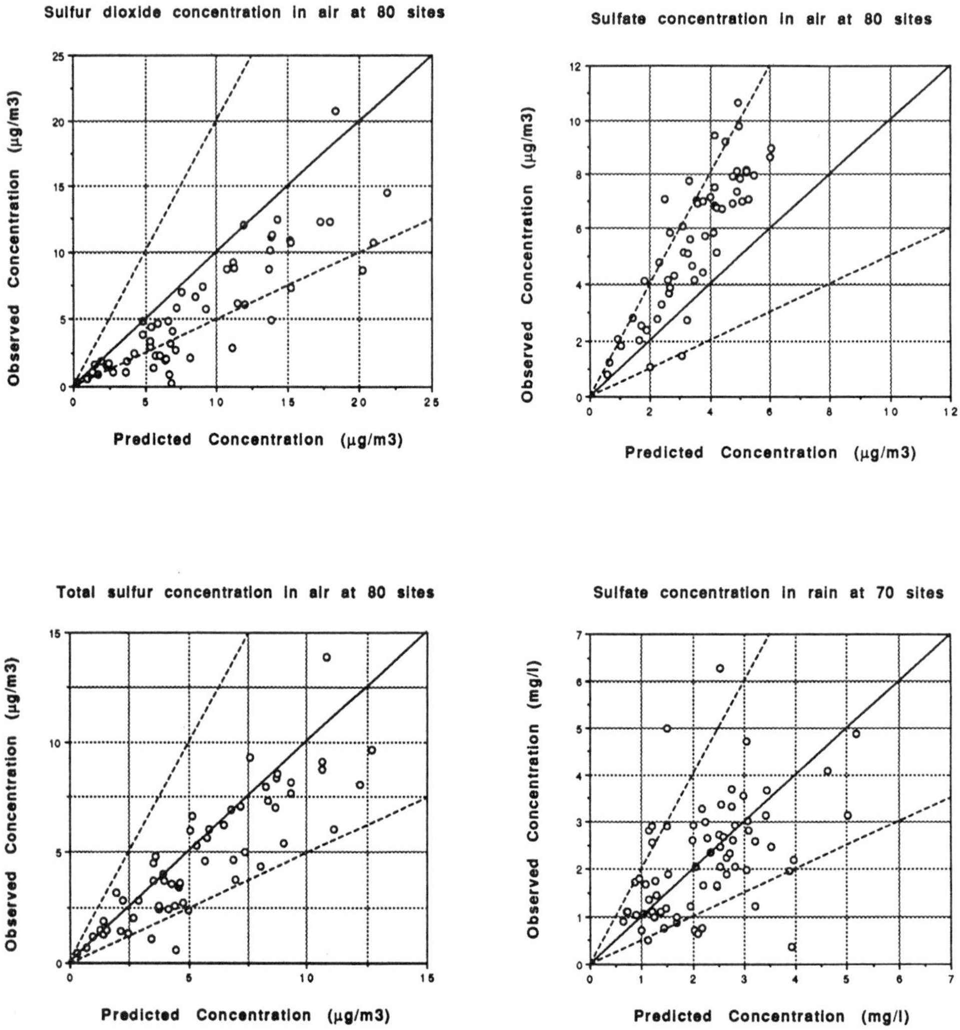

Figure 2. Comparison of ADOM predictions of concentrations of sulfur species in air and rain with observations made in eastern North America. Concentrations are averaged over the period August 25, 1988 to September 28, 1988. The solid line indicates perfect agreement between observations and predictions and the dashed lines show agreement within a factor of two.

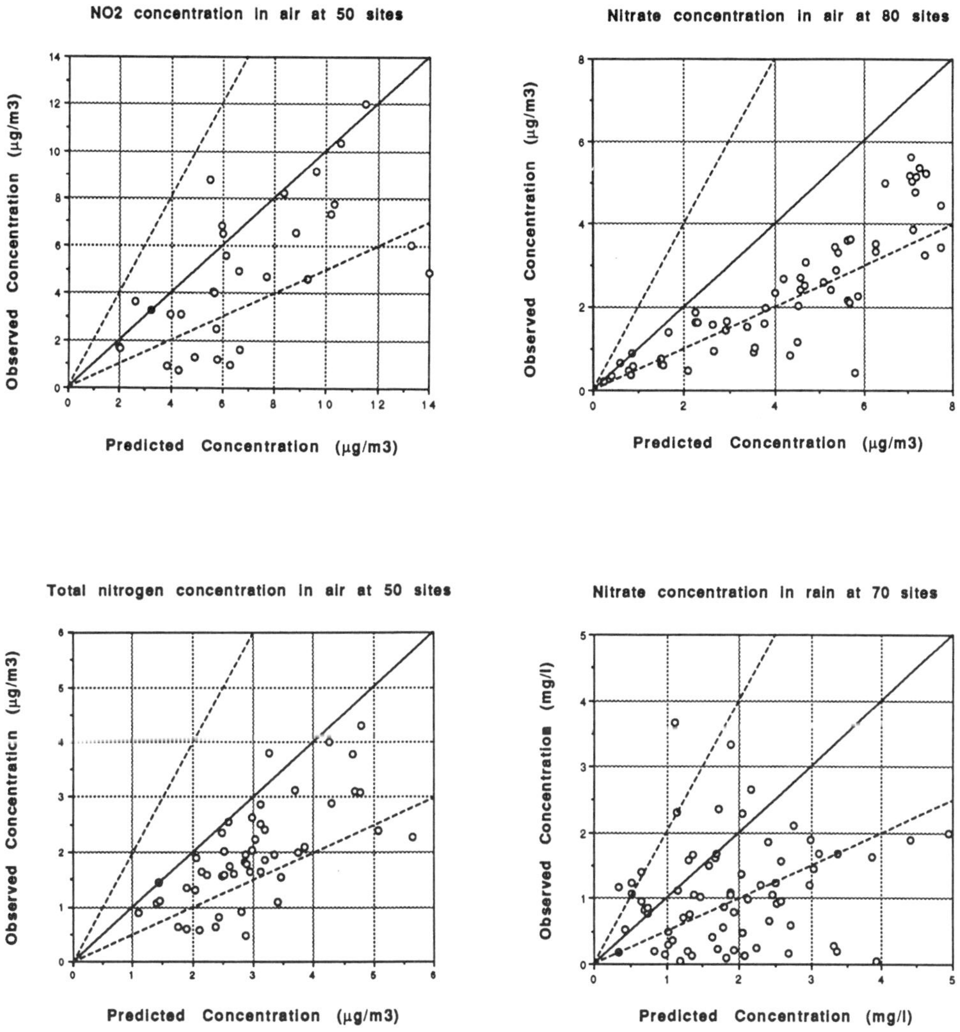

Figure 3. Comparison of ADOM predictions of concentrations of nitrogen species in air and rain with observations made in eastern North America. Concentrations are averaged over the period August 25, 1988 to September 28, 1988. The solid line indicates perfect agreement between observations and predictions and the dashed lines show agreement within a factor of two.

with observations. However, the sulfate and ozone are still being underpredicted in high emission density areas. The ozone is overpredicted in rural areas. This suggests that the reactive hydrocarbon emissions in the model might be too low; hydrocarbons enhance the production of ozone in urban areas, and scavenge ozone and NO_2 (the precursor of ozone) in rural areas. Sensitivity studies with higher emissions of hydrocarbons are currently being conducted to examine this theory. Another possible reason for the underprediction of sulfate could be related to the treatment of non-precipitating clouds, which can produce sulfate (in the aqueous-phase) that can make substantial contributions to ambient concentrations.

This preliminary evaluation teaches us several lessons. The acid deposition system consists of complex interacting components; for example, nitrogen and organic chemistry have important bearing on the sulfur chemistry of acid rain. Estimates of the effect of SO_2 emission reductions on sulfur deposition have to account for any concurrent changes in the NO_x and hydrocarbon emissions; semi-empirical models cannot account for this interaction. The evaluation also shows that the comprehensive model can throw important light on current estimates of emissions of the precursors of acid rain. Without accurate estimates of these emissions, strategies to reduce acid deposition are likely to be less than effective.

THE FUTURE OF COMPREHENSIVE MODELING

The experience of the past ten years indicates that a comprehensive model represents the only viable tool for understanding the complex processes that govern the fate of pollutants released into the ecosystem. This theme is underlined in a recent paper by Hansen (1989). With the expected advances in the science as well as the computational hardware, we are poised to develop comprehensive models that can address a wide range of environmental problems. In order to achieve this objective, we have to modify the approach we have taken thus far in developing comprehensive models.

We have to treat the comprehensive model as a complex system, rather than as an assembly of algorithms that can be designed independently of the host computer. The structure and the code of the model have to be given as much attention as the science in the model. Quality control of the computer code is likely to become more important with the expected increase in the complexity of new models. We have to design and manufacture the model as carefully as we would a complex piece of machinery, such as an automobile. All this suggests that building a comprehensive model will require an organization that will support a team of scientists committed full-time to the task of developing, improving and maintaining comprehensive models for environmental systems. The multi-disciplinary team will have to include systems analysts and computer scientists in addition to experts in the relevant physical sciences.

Computer science is likely to become increasingly important in model development because the computational demands of the new generation of comprehensive models will only be met by computer systems designed by experts. Even now, the main deterrent to the routine use of currently available comprehensive models is the computing time required by them. ADOM, which is typical of other comprehensive models, runs at about 3.5% of real time on a Cray X-MP, which has a peak performance of about a gigaflop (a billion floating point operations per second, abbreviated as gflop). This means that a simulation of a 30-day episode requires about a day of computing, and the simulation of a year, required for calculations of long-term effects, will run for 13 days. Our experience has shown that we cannot evaluate or use the model effectively at these simulation time to real time ratios. At a typical cost of 1,000 dollars per hour for CPU time, running an annual simulation can be prohibitively expensive; one such simulation would cost close to half a million dollars.

An order of magnitude increase in computing speed over current levels, which is possible with a Cray-3, will be matched by the likely increases in the computational demands of new models. To illustrate this, consider ADOM, which has over 10,000 grid points with a horizontal resolution of 100 km; it handles 29 species. Future models will have to use much finer spatial resolution in the horizontal to avoid sub-grid scale effects. Using even a 50 km spacing will result in a factor of four increase in computing time. Increasing the modeling domain or the number of species will easily use up more than the factor of ten increase in performance afforded by the new generation of supercomputers such as the Cray-3 (16 gflops) or the NEC SX-X (22 gflops). (It is useful to put the computing speeds we are talking about in perspective. If we sat down to make a billion calculations, taking one second per calculation, we would only complete the task in about 32 years!).

The most commonly used supercomputers, such as the Cray series, obtain their high performance using two methods. One is to use several high speed processors to perform pipelined vector operations concurrently. The other is to decrease memory access time by reducing the physical distance between the memory and the processing units of the computer, which is equivalent to reducing the size of the computer. There is a price to pay for increasing performance by packing circuit elements in a small space — high heat dissipation per unit volume of the computer. Removing this heat is a major problem in present day supercomputers. In the Cray-1, Freon is circulated through channels in each board carrying the chips; to carry heat away in the more densely packed Cray-2, the entire system is immersed in a bath of liquid fluorocarbon.

Computers, such as the Cray and the Alliant, that link relatively few high performance computational elements are referred to as "coarse grained" machines. The main advantage of this architecture is that it is not necessary to design the code to take advantage of the multiple processors. The performance increases with the degree of vectorization of the code. Compilers for such machines can optimize codes to take advantage of concurrent vector operations. As mentioned earlier, the main disadvantage of this approach is that we are rapidly approaching the limits of performance.

There appears to be consensus in the computer science community (Denning, 1989) that large increases in performance — approaching teraflop speeds — can only be obtained through the massive parallelism of "fine grained" computers, which link a great many processors. Because of several constraints, these processors have to be relatively weak. An example of such a computer is the Connection Machine 2 (Hillis, 1987), which can deliver a peak performance of 30 gigaflops. It obtains such speeds by linking 65,536 processors, each of which has its own memory of 4,096 bits. Because these processors are relatively slow, the performance of the machine is highly dependent on the degree of concurrent operations demanded by the code. Even small fractions of serial code can have a disproportionately large adverse impact on machine performance (See discussion of Amdahl's law in Worlton, 1981). This means that the problem being solved has to allow translation into code that consists of predominantly concurrent operations.

"Fine grained" machines appear to be particularly suited to the analysis of environmental systems because their architecture is consistent with the topology of the system. Each of the processors can be assigned to different parts (perhaps, at every grid point) of the modeling domain (domain decomposition, as described in Fox and Messina, 1987) of the system; they would perform simultaneous identical calculations, such as chemistry and scavenging, during a calculation cycle and store the results. Processes such as advection and dispersion would be represented by transfer of information between neighboring processors followed by simple calculations. It is easy to see that this type of computational architecture has the potential of increasing computational speeds over that of serial machines by factors approaching the number of grid points used to represent the domain of the environmental system.

The use of such a machine would also alleviate another problem associated with comprehensive models, which is the handling and analysis of large data sets. Here, visual display of data, made possible by computer graphics, becomes essential to the process of data management. Hillis (1987) points out that computer graphics is a field in which massively parallel machines are likely to be most useful. It is easy to appreciate this when we notice that it is possible to assign a processor to every pixel.

Fine grained machines are particularly suited for the analysis of large volumes of data because the processors can "see" the entire domain during a calculation cycle; the processors can be instructed to perform operations that would help in discovering spatial patterns in the data. One such operation is convolution, which blurs an image by averaging each point with its nearest neighbors. This blurring, which is also believed to be carried out in the human visual system, is useful for removing fluctuations in the data to bring out significant patterns. This analysis is most efficiently carried out during the calculations that evolve the system in time, because the required input is available to the processors during this time. In principle, the analyzed patterns can be displayed on a graphics terminal at regular intervals during the computation. This type of visual display is invaluable in understanding the behavior of complex environmental systems.

The use of explicitly parallel computers, such as the Connection Machine, will require codes that are compatible with the architecture of these computers (Gelernter, 1987). It appears unlikely that the currently available codes can be readily modified for the new architectures. Thus, for the time being, we might have to rely on implicitly parallel machines such as the Cray and the Alliant. However, it is clear that any new initiatives in comprehensive modeling will have to take a very careful look at massively parallel machines.

Dramatic increases in hardware performance are not likely over the next few years. However, over the same period, the costs of supercomputing at gigaflop speeds are expected to go down subtantially. For example, while the Cray goes for a few million dollars, we can now obtain machines with slightly lower performance, such as the Alliant, for around hundred thousand dollars. This development suggests a new strategy in the development of comprehensive models. We should perhaps view the computer as an integral part of the model itself, because the cost of the computer is now only a small fraction of the total model development costs. By doing so, we are in a position to decrease simulation times by paying careful attention to all the components of the model — the "science", the software and the hardware. In the past, we have perhaps concentrated on the science at the expense of the software and the hardware.

Software results from processes at three levels. The first level is the translation of partial differential equations into algebraic equations using "numerical" methods. At the second level these algebraic equations are translated into algorithms for efficient computation. At the third level, these algorithms are converted into computer code. We can improve numerical methods, particularly those used to represent the complex chemistry of environmental systems. We can also afford to pay greater attention to algorithms. McRae (1989) points out that most existing codes have not taken advantage of advances in the algorithms for elementary operations such as matrix multiplication. The computer code can be improved in several ways. It can be made readable through structured programming; this will facilitate quality control, which is a major problem in large computer codes. Worlton (1981) suggests that some of the subroutines should be converted into generic modules that can be certified for use in a wide variety of applications. The code should also be designed to take advantage of the special features of the host computer.

Realistic analysis of environmental problems will depend on comprehensive models becoming more than research tools. The experience gained in developing models such as ADOM and RADM will be valuable in developing a new generation of models that can be used routinely. However, further progress will require a new paradigm, which is that the computational framework is an integral part of the model; it cannot be separated from the science.

ACKNOWLEDGEMENTS

The Electric Power Research Institute, Environment Ontario, and Environment Canada supported the research described in this paper. I would like to thank P. Karamchandani for valuable comments on this paper. I am also grateful to G. Kuntasal for reviewing this paper, and to K. Kashanian for his assistance in preparing the figures.

REFERENCES

Chang, J. S., Brost, R. A., Isaksen, I. S. A., Madronich, S., Middleton, P., Stockwell, W. R. and Walcek, C. J., 1987, A three-dimensional Eulerian acid deposition model: physical concepts and formulation, J. Geophys. Res., 92:14681-14700.

Denning, P. J., 1989, Massive parallelism in the future of science, American Scientist, 77:16-18.
ERT, 1985, Summary of recent sensitivity analyses with the ADOM/TADAP prototype model, available from A. Venkatram, ENSR Corp., 1220 Avenida Acaso, Camarillo, CA 93012, USA.
Fox, G. C. and Messina, P. C., 1987, Advanced computer architectures, Scientific American, October, 1987, 67-74.
Gelernter, D., 1987, Programming for advanced computing, Scientific American, October, 1987, 91-98.
Hansen, D. A. 1989, A concept paper on tropospheric model development and evaluation, available from the author at Electric Power Research Institute, 3412 Hillview Avenue, Palo Alto, CA 94303, USA.
Hillis, W. D., 1987, The Connection Machine, Scientific American, June, 1987, 108-115.
Jenkins, R. A., 1986, "Supercomputers of Today and Tomorrow; The Parallel Processing Revolution", Tab Books Inc., Blue Ridge Summit, PA.
Karamchandani, P. K., Kuntasal, G. and Venkatram, A., 1989, Results from a comprehensive acid deposition model: predictions of regional ozone concentrations, in: "Air Pollution Modeling and Its Application VII", Han van Dop, ed., Plenum Press, New York.
Lurmann, F. W., Lloyd, A. C., and Atkinson, R., 1986, A chemical mechanism for use in long-range transport/acid deposition computer modeling, J. Geophys. Res., 91:10905-10936.
Mueller, P. K., and Hidy, G. M., 1983, The sulfate regional experiment: report of findings, Volume 1, report EA-1901, available from Electric Power Research Institute, 3412 Hillview Avenue, Palo Alto, CA 94304, USA.
McRae, G. J. 1989, Chemical process modeling and simulation using advanced computational architectures, paper prepared for Foundations of Computer Aided Process Design Meeting, Snowmass, Colorado, 9-14 July, 1989, sponsored by AIChE and the National Science Foundation.
Tesche, T. W., 1989, Ozone air quality models: how effective are they?, Seminar on the Contemporary Oxidant Problem, 27-28 February, Vancouver, BC, Canada.
Venkatram, A., and Pleim, J., 1985, Analysis of observations relevant to long-range transport and deposition of pollutants, Atmos. Environ., 19:659-667.
Venkatram, A., Karamchandani, P. K. and Misra, P. K., 1988, Testing a comprehensive acid deposition model, Atmos. Environ., 22:737-747.
Venkatram, A., Kashanian, K., Karamchandani, P. K., and Avitzur, T., 1990, Probing the acid deposition system with a semi-empirical model: the role of oxidant limitation, Atmos. Environ., 24A:125-131.
Worlton, J., 1981, A philosophy of supercomputing, presented at Conference on High-Speed Computing, Glenenden Beach, Oregon, March 30, 1981.

DISCUSSION

F.L. LUDWIG

The system that you describe sounds much like the 200 inch telescope (or a cyclotron), where researchers sign up for time to do their experiments. Is this what you envision?

A. VENKATRAM

Yes, to some extent. This will also allow continuous improvement of the model.

H.J.A. VAN JAARSVELD

Do you agree that for the development of a comprehensive model, a comprehensive measurement programme is also required?

A. VENKATRAM

Yes. Nearly 20 million dollars were spent on a field study to collect the data for evaluating comprehensive models such as RADM and ADOM.

C.J. NAPPO

Please give me some examples of the use of comprehensive models.

Comprehensive models can be used to conduct experiments that would be impossible in the real world. Specifically, one could use the comprehensive model to examine the effect of SO_2 emission controls in Ohio on acidic deposition in the Adirondacks.

PLUME MODELS WHICH INCORPORATE UNCERTAINTY

W. S. Lewellen, R.I. Sykes
S. F. Parker, and D. S. Henn

ARAP Div., California Research and Technology, Inc.
P.O. Box 2229, Princeton, N.J., 08543

INTRODUCTION

Atmospheric turbulence forces random fluctuations in plume dispersion which generally imply that any observable concentration value will have significant uncertainty. This makes the rigorous comparison between observed samples and any plume predictions at the same point in time and space difficult. The Electric Power Research Institute's extensive project on plume model evaluation found that the existing models could predict no more than 15 % of the one hour surface samples within a factor of 2 at the plains site (Bowne et al., 1983). Subsequently, ARAP was asked by EPRI to develop and test a new family of dynamic plume models incorporating uncertainty.

In this paper we wish to review the resulting models (Lewellen et al., 1988) which span the range from a large-eddy-simulation of the detailed dispersion to a steady state, Gaussian plume model. At all levels the models provide the capability to estimate the probability distribution of expected concentration samples. The high resolution LES results provide details on the expected concentration fluctuations generally unavailable from field data. The lowest resolution model, SCIMP (Second-order Closure Integrated Model Plume) provides an engineering tool which has been evaluated by comparison with extensive field data taken by EPRI downwind of power plants, demonstrating the capability of providing valid frequency distributions for both five-minute and one-hour, time-averaged concentration samples at a network of locations. The predicted probability distribution may be used to specify probabilities of interest to the user, such as the probability of exceeding a threshold value for a particular time period at a given location.

STOCHASTIC FRAMEWORK FOR DISPERSION MODELS

In addressing any concept of a statistical distribution of possible concentration observations, it is necessary to define what particular ensemble of possible states is being considered. We believe the most useful ensemble to consider is that composed of all possible concentration fields which could be produced by the observed wind and emissions data, as suggested by Venkatram (1982). The ensemble of concentration values thus depends directly on the observations used by the model. A high resolution meteorological data set combined with precise emissions might produce a narrowly defined ensemble of possible concentration fields with little inherent uncertainty. However, the detailed data required becomes prohibitively large as the desired sampling time of the concentration field is reduced.

For any practical model, it seems highly desirable to consider the concentration as a random variable and seek to predict probabilistic quantities, such as ensemble mean values, variances and probability distributions. Lewellen and Sykes (1983) proposed a method by which a prediction of the mean and variance of the concentration fluctuations could be used as a means of representing the full probability distribution. The variance model is based on second-order closure (Sykes et al., 1984, 1986). The probability distribution function (pdf) is then estimated from a two-parameter family of truncated Gaussian functions, which replace any unrealizable negative values with a delta function at zero. The resulting clipped normal pdf was found to provide a very reasonable description of the fluctuations in lidar plume observations (Lewellen and Sykes, 1986).

The prediction of the ensemble mean and variance of the concentration fluctuations can be attempted at any resolution, but the most useful appears to be at a resolution which is compatible with the resolution of the ensemble defining meteorological and emissions data. For this paper, we will consider both the high resolution and low resolution members of the hierarchy of plume models generated by ARAP for EPRI. At the upper end, we use a large-eddy simulation (LES) of the boundary-layer-scale velocity fluctuations to provide a high resolution meteorological data set, and at the low end, a Gaussian plume model. These two models are briefly described in the following sections. An intermediate resolution puff model SCIPUFF (Sykes et al., 1989a) which is most appropriate when unsteady effects are important will not be discussed here.

A major difficulty with evaluating any statistical model's prediction, is the dearth of reliable experimental data on the concentration field downwind of a tracer source. Atmospheric experiments are obviously difficult since the defining conditions cannot be sufficiently fixed to obtain reliable ensemble average statistics. It is possible to compare models with some good laboratory experimental data, but this does not satisfy the requirements for a good field test. Even though the statistical variance in concentration fluctuation has rarely been measured, it is possible to determine whether a set of observed samples are a consistent realization from the predicted ensemble. When the model has predicted a pdf as a function of time and space, repeated sampling can determine possible realizations of any sampler data.

LARGE EDDY SIMULATION OF DISPERSION

The LES model has the capability to provide relatively complete information about the flow field, but the price is high. Not only are the computational requirements great, but we must also restrict the ensemble conditions more than is usually possible in physical experiments. Within these restrictions Sykes and Henn (1990) have shown that it is possible to numerically resolve most of the variance in a fluctuating concentration field downwind of a localized source of passive tracer under the laboratory boundary-layer conditions used by Fackrell and Robins (1982). Their results for the ratio of the maximum r.m.s. fluctuation to the maximum mean concentration are shown in fig. 1. Sykes and Henn (1990) discuss reasons for believing that the lower data point at $x/H = 5$ is anomalous.

Attempts have been made to simulate the conditions which occurred at certain times during the EPRI PMV&D project at the Kincaid power plant. Geostrophic wind conditions, mixed layer heights, and surface roughness and heat flux are set, and the turbulent boundary layer eddies are computed for a domain of approximately 10 km on a side. The periodic boundary conditions imposed do not allow any lower frequency variations to enter the problem. Figure 2 shows a comparison between the vertical distributions of turbulence as simulated and the observations at 100 meters. The simulation is believed to capture all of the vertical velocity variance but miss some of the low frequency horizontal variance.

Qualitatively, the resulting plume has a very realistic look, particularly when viewed in animation. The buoyant plume rises into the stable inversion above the boundary layer and bounces back to become mixed down to the surface by the large boundary layer eddies. Figure 3 shows how the observed cumulative distribution function (CDF) of SF_6 surface samples compares with 6 different simulation realizations. There is significant scatter between the realizations, demonstrating that the boundary-layer turbulence captured by the LES simulation does contribute to the uncertainty in hourly samples. The observed CDF has somewhat smaller maximum values, and fewer zeros. This is consistent with the LES missing some of the low frequency horizontal turbulence as noted in fig. 2.

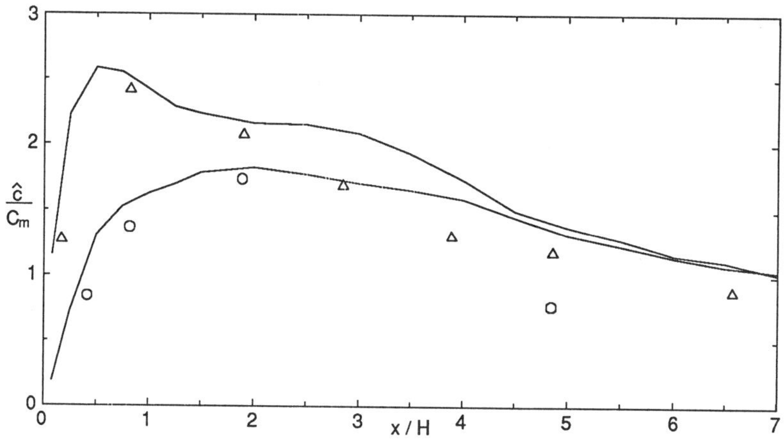

Fig. 1. Centerline relative concentration fluctuations downstream of a point release for two source sizes. Symbols show the observations of Fackrell and Robins (1982); Model results are from LES results of Sykes and Henn (1990).

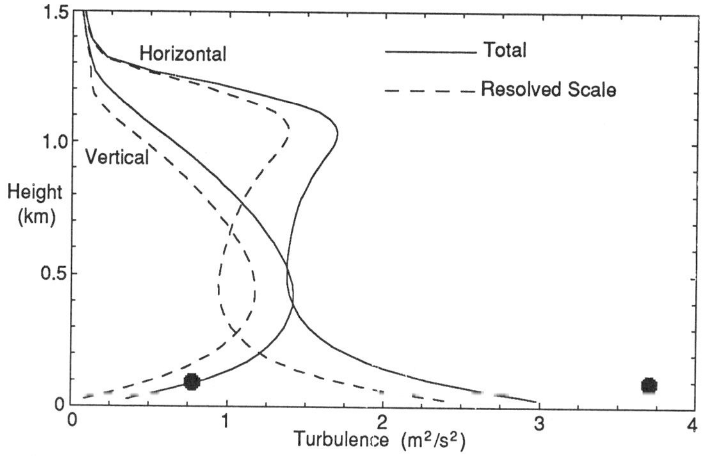

Fig. 2. Vertical distributions of turbulence used to simulate 1200-1300 on 5/28/81. Symbols represent observations from the 100 meter tower.

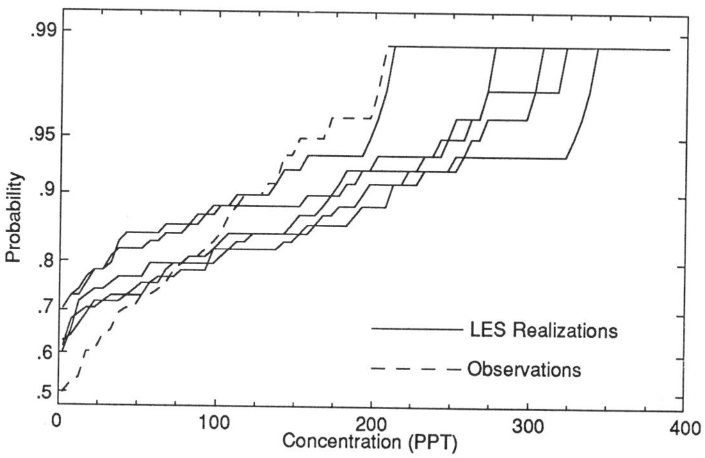

Fig. 3. Observed CDF of SF6 samples compared with results of 6 LES simulations.

The greatest utility of an LES dispersion model is to provide fundamental data on concentration statistics. Such results have already expanded the present experimental data base and more results are rapidly becoming available.

A GAUSSIAN PLUME DISPERSION PREDICTABILITY MODEL

Our lowest resolution model SCIMP (Lewellen and Sykes, 1989; Sykes et al., 1989a) provides a simulation of the probability distribution of concentration fluctuations at the level of a steady state Gaussian plume model. It is derived by assuming a steady plume and integrating the turbulent transport equations over the plane transverse to the plume to obtain a set of ordinary differential equations describing the downstream evolution of the plume. In addition to predicting the mean concentration, the model predicts the variance of the concentration fluctuations. As noted in the earlier section, knowledge of the mean and the variance of an ensemble of concentration fluctuations can be used to provide a reasonable description of the full pdf as a clipped normal distribution.

The meteorological input data required for SCIMP are an estimate of the vertical distribution of the wind, and the turbulent correlations of the wind and temperature fluctuations. The model requires both the average of the ensemble of possible conditions represented by the available meteorological data, and an estimate of the variance in all three components of the wind. Our approach has been to break the wind variation into two components: low frequency fluctuations, and high frequency turbulence. The break between the high and low is set at the lowest frequency of the boundary layer eddies. This division permits us to use different approaches to calculating the concentration variance that arises from these two components of the velocity fluctuations. The contributions from the high frequency are computed from an evolution equation for the mean square scalar, which provides for a balance between advection and dissipation (Sykes et al., 1986). The contributions from the low frequency fluctuations are computed from repeated runs chosen randomly from the predetermined ensemble of possible meteorological conditions.

Figure 4 shows the SCIMP predictions for the same hour as the LES results of fig. 3. These SCIMP predictions not only include a parameterization of the boundary layer scale fluctuations computed by the LES model of fig. 3, but also include an estimate of the contributions from the energy left out of the LES calculation and the low frequency meteorological uncertainty. The domain of the SCIMP calculation is somewhat larger than that for the LES in order to include all of the sampler data for that hour. Thus the observed distributions of fig. 2 and fig. 3 are slightly different.

SCIMP was developed using half of the EPRI Plume Model Validation and Development data set, and then evaluated against the other half of the data set. Figure 5 shows the comparison for the evaluation data set. Two versions of the observed

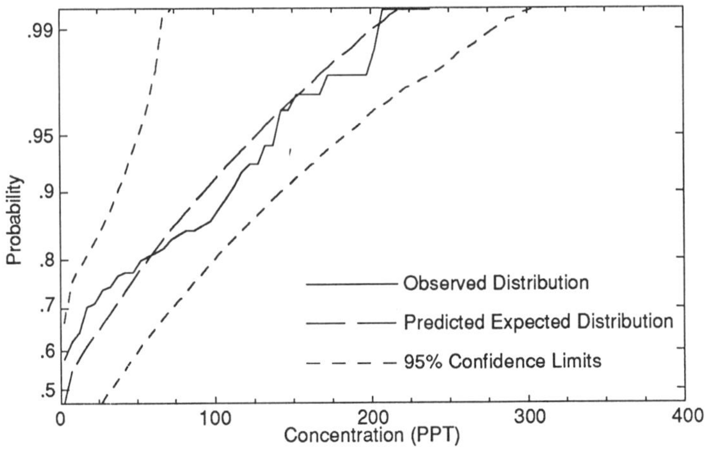

Fig. 4. SCIMP predictions of CDF for the same hour as the simulation of Fig. 3.

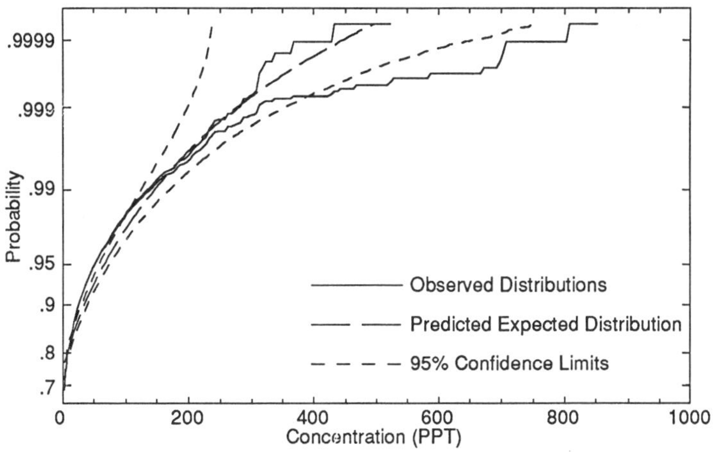

Fig. 5. Statistical comparisons of the SCIMP simulations with the 20,000 hourly samples of the Kincaid Evaluation Data Set.

distributions are shown because of some questions about a few of the highest reported samples. The actual distribution of the 20,000 sample points is not as close to the model predicted expected-value distribution as it was for the development data set (Sykes et al., 1989a), but generally falls within the 95% confidence bands predicted by the model.

The LES results provide a hint at what may be part of the cause of the less than perfect agreement in fig. 5. Sykes and Henn (1990) show that the pdfs obtained from the LES results show some consistent departures from the SCIMP assumed clipped normal distribution. Although, the clipped normal distribution provides the best overall fit, there appears to be a consistent under prediction of large values along the centerline relatively far downstream. When the plume has travelled several boundary layer thicknesses downstream from its source, the pdf along the plume centerline starts to look more like a log normal than the assumed clipped normal.

The format of figs. 4 and 5 is aimed at providing a direct measure of validity. Once the validity of the model is established, other, more useful, presentations of the statistical output of the model can be envisioned. For emergency response type applications it may be desired to provide a map of the probabilities of exceeding a given critical value for a given sampling time, as illustrated in fig. 6. Alternatively, it may be useful to fix a minimum level of probability which is of interest, and exhibit isopleths of concentration values which would be expected to occur at this minimal probability level.

SAMPLING TIME INFLUENCE

The role of uncertainty is increased as the sampling time is reduced (Sykes et al., 1989b). Time averaging provides the opportunity to filter out the fluctuation variances induced by velocity fluctuations at frequencies much higher than the sampling frequency. The reduction factor can be computed directly by the LES results. Sykes and Henn (1990) show that the shape of the temporal autocorrelation is almost constant with its scale increasing downstream in proportion to the plume width over the wind speed. This implies that longer time averages are necessary to reduce the variance with increasing downstream distance.

The influence of the sampling time on the cumulative distribution functions as predicted by SCIMP and as observed is given in Fig. 7 for the 250,000 five minute SO_2 samples included in the EPRI PMV&D data set. Both 5 minute (Fig.7a), and one hour (Fig. 7b) sampling times are shown. Although there is a consistent tendency for the model predicted expected curve to be somewhat larger than the observations, the influence of sampling time is well represented by the SCIMP results . The modest decrease in the number

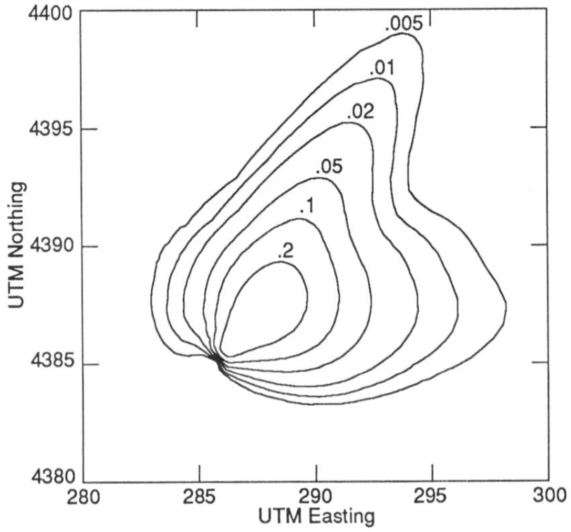

Fig. 6. Examples of isopleths of the probability that the concentration of SF6 exceeds 100 ppt for a single hour on 5/28/81.

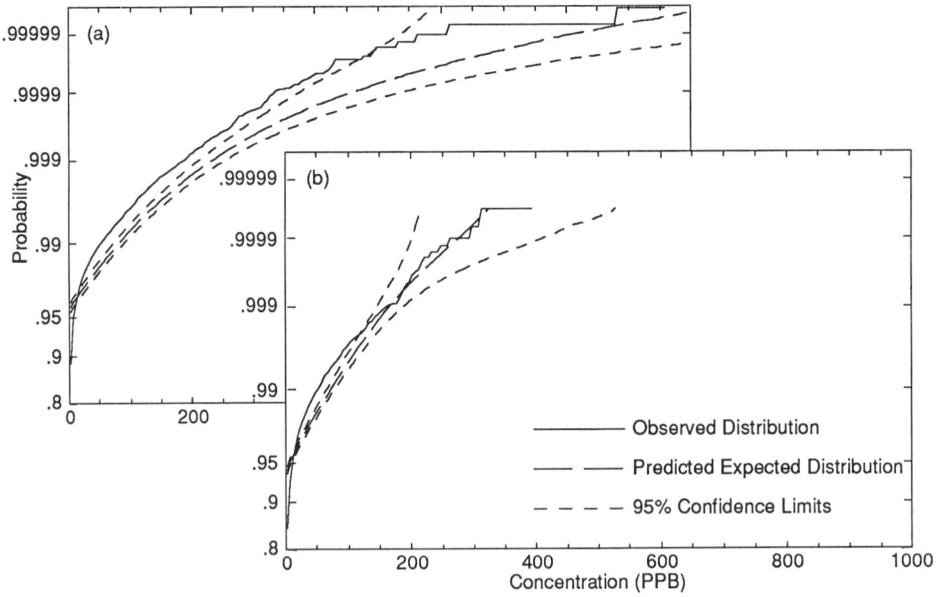

Fig. 7. Comparison of predicted CDF for the Kincaid (a) 5-minute SO_2 data, and (b) one-hour averaged data.

of zeros and large reduction in extreme values caused by increasing the sampling time from 5 minutes to 1 hour are both very well represented by the model.

CONCLUDING REMARKS

We have shown that uncertainty which is an inherent part of plume dispersion is amenable to simulation at different levels of resolution. High resolution LES simulations can provide fundamental data on the level of uncertainty under different conditions, on the spatial and temporal correlations, and on the detailed probability distribution functions which should be expected.

At the engineering level, a Gaussian plume model (SCIMP) has been shown to do a credible job of statistically representing the uncertainty expected in power plant plumes under a number of different conditions. We believe that utilization of this model could provide a major step forward in assessing the risk of exceeding threshold values over different time periods for general plume emissions.

ACKNOWLEDGEMENT

This work was supported by EPRI under Project No. RP1616-28.

REFERENCES

Bowne, N.E., R.J. Londergan, D.R. Murray and H.S. Borenstein, (1983) "Overview, results, and conclusion for the EPRI PMV&D Project: Plains site." EPRI Report EA 3074, RP 1616-1.

Fackrell, J.E. and A.G. Robins, (1982) "Concentration fluctuations and fluxes in plumes from point sources in a turbulent boundary layer." J. Fluid Mech. 117, 1-26.

Lewellen, W.S. and R.I. Sykes, (1983) "On the use of concentration variance predictions as a measure of natural uncertainty in observed concentration samples. "Proc. Sixth Sym. on Turb. and Diff., Boston, MA, Amer. Meteor. Soc.

Lewellen, W.S. and R.I. Sykes, (1986) "Analysis of concentration fluctuations from lidar observations of atmospheric plumes." J. Clim. & Appl. Met. 25, 1145-1154.

Lewellen, W.S. and R.I. Sykes, (1989) "Meteorological data needs for modeling air quality uncertainties." J. Atmos. & Oceanic Tech. 6, 759-768.

Lewellen, W.S., R.I. Sykes, S.F. Parker, D.S. Henn, N.L. Seaman, D.R. Stauffer, T.T. Warner and H.R. Chang, (1988) "A hierarchy of dynamic plume models incorporating uncertainty, Vol. 1, EPRI Report EA-6095, RP 1616-28.

Sykes, R.I. and D.S. Henn, (1990) "Large-Eddy Simulation of Concentration Fluctuations in a Dispersing Plume." Submitted to J. Fluid Mech.

Sykes, R.I., W.S. Lewellen and S.F. Parker, (1984) "A turbulent-transport model for concentration fluctuations and fluxes." J. Fluid Mech. 139, 193-218.

Sykes, R.I., W.S. Lewellen and S.F. Parker, (1986) "A Gaussian plume model of atmospheric dispersion based on second-order closure." J. Clim. & Appl. Met. 25, 322-331.

Sykes, R.I., W.S. Lewellen, S.F. Parker and D.S. Henn (1989a) "A hierarchy of dynamic plume models incorporating uncertainty, Vol. 3 and 4, EPRI Report EA-6095, RP 1616-28.

Sykes, R.I., W.S. Lewellen and S.F. Parker, (1989b) "Prediction of the effect of sampling time on concentration probabilities." Air & Waste Man. Assoc. 82nd Mtg., Paper No. 89-114.1.

Venkatram, A., (1982) "A framework for evaluating air quality models," Bound.-Layer Meteor., 24, 371-385.

DISCUSSION

G.A. BRIGGS

We found large variations in vertical velocity (w) distributions during very convective conditions of NOAA's Boulder Atmospheric Observatory 300-m meteorological tower, during the CONDORS diffusion experiment. During one 60-min period, out of 7 hours collected, $\overline{w}$ at the upper tower levels was -0.4 w^* = -0.8 m/s! Such large deviations from the usual modeling assumption, $\overline{w}$ = 0, should have a large effect on extreme value distributions of surface concentration. Do your models for uncertainty consider variations in $\overline{w}$?

W.S. LEWELLEN

The tuned SCIMP model used the developmental half of the EPRI data to determine reasonable magnitudes for modeled uncertainties in wind speed, mixing depth, etc. By definition, our ensemble average, $\overline{w}$ = 0, but this does not mean that we expect hour-averaged values of w well above the surface to be zero. Quite the contrary, we expect hour-averaged w's to differ significantly from $\overline{w}$, just as hour-averaged concentrations differ significantly from $\bar{c}$. Your example of a large deviation from zero in an hour-averaged w at 300 m is quite consistent with our high-resolution LES results. Under free convection conditions in a 1 km mixed layer with $w^* = 1$ m/sec, hour-averaged w's were found to vary from -.6 m/sec to +1.2 m/sec. Although, we expect these to be reduced by the addition of a mean wind, the reduction may be modest under conditions which lead to cloud streets.

ENTRAINMENT AND PLUME RISE IN TURBULENT WINDS

Dennett D. J. Netterville

Research Department, Syncrude Canada Ltd.
P.O. Box 5790, Station L
Edmonton, Alberta, Canada T6C 4G3

INTRODUCTION

Most modern closed-form plume models treat several important physical phenomena as separate and distinct sequential processes, even though the phenomena have a smooth and continuous interdependence. Some examples of this are the artificial decoupling of dispersion from plume rise by use of the 'effective source height' point-source dispersion model, the 'semi-impermeable boundary' used to simulate the effect of plume penetration into and dispersion below elevated 'capping' inversions, and the abrupt termination of plume rise produced by semi-empirical theories used to simulate the effect of atmospheric turbulence on plumes.

Model simplifications such as these have been justified in the past because data to drive the models was normally limited to surface observations. The recent introduction of remote sensing techniques for measuring wind speed, turbulence and temperature aloft means that such limitations can finally be removed. This paper briefly describes a new analytical plume model developed by Netterville (1990), which combines within one theoretical framework several aspects of the phenomena of plume rise, dispersion, thermal stratification and atmospheric turbulence.

REVIEW OF THEORY

'Mainstream' Models

The popular '$x^{2/3}$ law' provides a useful and widely-used closed-form description of plume rise in near-neutral conditions. The model assumptions were chosen to embody the most important physical processes affecting plume rise. Of necessity, some relevant physics had to be excluded, either to simplify the mathematics or because the processes themselves were not adequately understood. Thus the $x^{2/3}$ law is strictly valid in the mathematical sense only for a neutrally stable laminar atmosphere, and, not surprisingly, gives predictions which sometimes do not agree with measurements. For example, real plumes in adiabatic turbulent conditions eventually level off (Hanna et al. (1982)), but $x^{2/3}$ plumes keep rising steadily. It is generally accepted that this discrepancy between theory and observation is due mainly to the model's neglect of atmospheric turbulence effects (Briggs (1975)).

Semi-empirical adjustments to this 'laminar' plume model have been the method of choice for evaluating rise in turbulent winds. These adjustments usually involve an abrupt transition from a buoyancy-dominated $x^{2/3}$ phase of plume rise to an atmosphere-dominated phase in which further rise is either very slow or is entirely absent (Briggs (1969, 1975), Slawson and Csanady (1967, 1971)). With this approach, the two (or sometimes three) plume phases differ only in the entrainment assumption used to derive the plume mass continuity equation. In most models of this type the plume buoyancy and momentum equations are unaffected by atmospheric turbulence.

In reality, the buoyancy-dominated and atmosphere-dominated phases of plume rise exist only as extremes of behavior close to the source and far downwind, respectively. Within the intermediate region, the plume is subject to both influences at the same time, and transition between extremes is a gradual process. The phases are kept separate in most models because mathematical difficulties prevent finding a closed form solution when several entrainment mechanisms act simultaneously. A discontinuity in the physics of entrainment is of course artificial, and, as pointed out by Briggs (1975), the methods recommended to determine the downwind location of such an artifact are therefore somewhat arbitrary.

<u>Other Models</u>

One of the earliest models incorporating atmospheric turbulence is due to Priestley (1953, 1955) and Priestley and Ball (1954), who introduced the idea that buoyant parcels lose momentum and buoyancy to their surroundings at a rate governed partly by parcel size and partly by the level of ambient turbulence. Priestley divided the parcel trajectory into a traditional buoyancy-dominated growth phase followed by an atmosphere-dominated 'steady loss' phase during which the parcels gradually leveled off.

The first single formulation was presented by Turner (1963) in his study of rising spherical thermal updrafts. He proposed that ambient turbulence causes the thermal to lose mass by inducing a constant fluid removal velocity, proportional to the ambient rms turbulence velocity, which acts over the thermal's systematically growing surface area. The thermal's buoyancy and momentum are slowly carried off in the lost mass, weakening the thermal core so that eventually it stops rising.

Turner did not describe the mechanism by which turbulence causes mass to be removed from the plume. A physical mechanism for turbulence induced mass loss was suggested by Benjamin (1975). He assumed that, in convective turbulence, chimney plumes in the second, atmosphere-dominated phase of plume rise lose mass and buoyancy (apparently overlooking momentum) by entrainment into thermal updrafts which penetrate and pass through the plume.

Djurfors (1977) recognized that perhaps the mathematically simplest form of leveling-off behavior was one in which the vertical distance between the plume centerline and its final height would decrease exponentially with time. Thus he investigated the mathematical circumstances that could cause such asymptotic behavior. By working backwards from the assumed trajectory, Djurfors showed that plume buoyancy and momentum decay terms of the form $-aF$ and $-bM$ (F and M being local buoyancy and momentum, respectively, with a and b as constants) would be required in the plume conservation equations.

Development of 'turbulent' plume models currently rests here, with the question of which approach is most realistic still undecided. Each model contains some good ideas, and these taken together provide the conceptual roots from which the present model has grown. Beyond this, however, we must look outside the familiar plume rise literature to obtain further guidance.

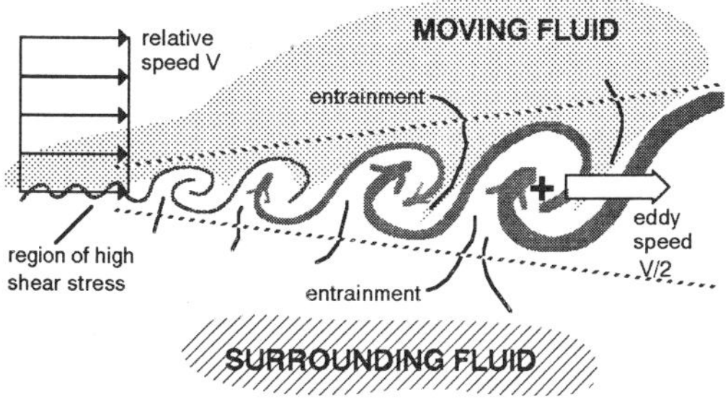

Fig. 1. Shear-induced vortex roll-up [adapted from Hernan & Jimenez (1982)]

THE PHYSICS OF ENTRAINMENT

The main difference between the Priestley and Turner types of model is in the way the plume mass conservation equation is perceived to respond to atmospheric turbulence. The plume mass equation models the mass transfer process between plume and atmosphere, and the mechanism of entrainment is central to the transfer of mass between fluids. Thus the two approaches may be reconciled if we look closely at what is known about entrainment.

Recent studies of fluid mixing within free shear layers are relevant to our problem of understanding plume/air entrainment when we make the key assumption that the interface zone between any moving fluid parcel and its surroundings acts as a free shear layer. This working hypothesis seems reasonable, so long as the interface zone between moving fluid parcels is thin compared with the distance to the nearest solid flow boundary.

Entrainment Events

The deterministic eddy structures discovered by Brown and Roshko (1974) within the apparently random turbulence of a free shear layer provide our starting point. Based on their observations and the work of others, it appears that shear forces arising from the relative motion between adjacent fluid regions cause small vortices to form systematically and grow at the interface between the fluids. This process is sketched in Figure 1.

High-speed movies of shear-layer mixing taken by Hernan and Jimenez (1982) show that entrainment is not a smoothly continuous process. Instead, shear-layer mixing occurs in short, quick gulps, through the quasi-regular formation of small vortices. The close interleaving of fluids within each vortex is essentially irreversible. This means entrainment is virtually complete as soon as the eddy has formed (a suggestion made by Hernan and Jimenez), even though the mixture still has an organized structure. This observation supports our first assumption that an entrainment 'event' can be defined as the formation of an eddy at the shear layer interface between two mixing fluids. Thus the amount of mass m_e entrained per unit eddy length during each event is approximately proportional to the eddy cross-sectional area,

$$m_e \sim r_e^2 \times \rho_e , \qquad (1)$$

where r_e is the eddy radius and ρ_e is the density of the entrained fluid. Here we take the usual convention that moving fluid parcels entrain material from their larger surroundings.

Hernan and Jimenez (1982) found that, for shear layers with a specific origin, eddy size r_e grows linearly with distance d from the origin. They also observed that eddy life span (the length of time t_e over which an eddy maintains its physical integrity) varies inversely with eddy speed v_e, and grows linearly with the downstream position d at which the eddy was born.

From their observations we draw the simple conclusion that eddy life span grows linearly with the eddy's diameter at birth.

For a free shear layer in which turbulence is stationary and homogeneous, the eddy population must be in equilibrium, i.e. the eddy birth and death rates are in balance at every eddy scale. This means the eddy birthrate (frequency f_e) has to be inversely proportional to the eddy life span t_e. Then the frequency of entrainment events in the shear layer is

$$f_e \sim 1/t_e = \text{constant} \times v_e/r_e \; . \tag{2}$$

<u>Rate of Entrainment</u>

A fluid parcel of radius R moving at speed v_r through a relatively stationary fluid medium generates eddies of radius r_e and speed v_e in the shear layer at the fluid interface. It is reasonable to suppose that the number of eddies N_e that form on the parcel surface (each eddy formation being an entrainment 'event') is proportional to the ratio of the parcel circumference to eddy diameter,

$$N_e = \text{constant} \times R/r_e \; . \tag{3}$$

Since the parcel radius R and speed v_r are the only length and velocity scales in the flow, the eddy radii and speeds must scale proportionally. Using this simple model, the rate (per unit eddy length) $\dot{m}_{dm\to p}$ at which mass is directly entrained into a fluid parcel of radius R moving with relative speed v_r through the surrounding medium of density ρ_m is equal to the shear eddy mass multiplied by the shear eddy birthrate, multiplied by the number of shear eddies on the parcel surface,

Direct Entrainment:

$$\dot{m}_{dm\to p} = m_e \times f_e \times N_e = \text{constant} \times R\, v_r \times \rho_m \; , \tag{4}$$

where the subscript $_{dm\to p}$ indicates mass <u>d</u>irectly entrained from the <u>m</u>edium into the <u>p</u>arcel. The name 'direct' entrainment distinguishes this type of entrainment from the indirect entrainment discussed in the next section.

Eq. (4) is identical to the well-known Taylor entrainment assumption proposed for plumes on empirical and dimensional grounds by Taylor (1945) and Morton et. al. (1956). What is special about Eq. (4), however, is that here it has been derived for the first time from basic principles of fluid mixing within free shear layers. The derivation shows that Eq. (4) should be valid for all fluid parcels moving in uniformly dense free shear layers.

TURBULENT MASS TRANSFER

In a turbulent atmosphere, plumes are exposed to moving fluid parcels in the air around them. These moving fluid parcels are retarded by slower fluid around them, and shear force instabilities cause the birth of small shear layer eddies on their surface. Hence atmospheric turbulence eddies must entrain the material through which they move, at a rate prescribed by Eq. (4), even when they are inside a plume. Figure 2 shows a simple schematic diagram of this process.

Each atmospheric and plume turbulence eddy, and the plume itself, is surrounded by its own family of small shear-generated eddies. This elementary picture of parcel interpenetration is consistent with the data of Fackrell and Robins (1982) that show the presence of clean air eddies on the plume centerline. Thus the air eddies also become 'spongy' due to penetration by the plume's internal eddies, and these invasive plume eddies themselves become spongy when exposed to penetration by small scale air eddies within the larger air eddy. This 'endless chain' of interpenetration

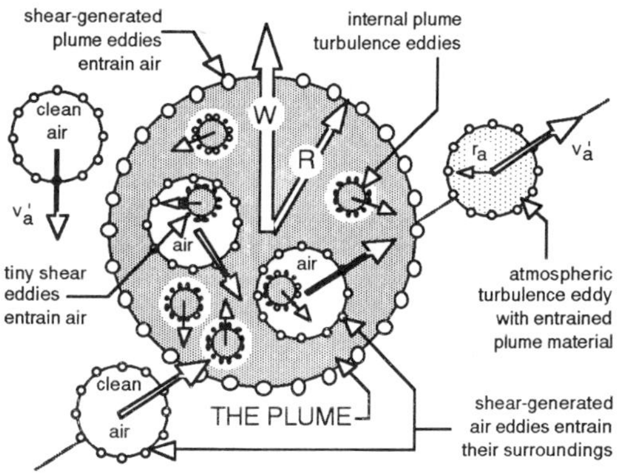

Fig. 2. Cross-sectional schematic of plume/atmosphere interaction.

extends down through ever-smaller scales until eddies approach the size of the Kolmogorov microscale, where viscosity smooths out and eliminates the steep velocity gradients that produce shear layer eddies.

Rate of Extrainment

The amount of plume material cumulatively entrained by all atmospheric eddies in the plume, including the smallest, is simply the sum of the separate amounts entrained by each specific size of air eddy. The entrainment rate of a single air eddy of scale s is found from Eq. (4) to be

$$\dot{m}_{dp\to a_s} = c_{\dot{m}} \times r_{a_s} v'_{a_s} \times \rho_p \, , \tag{5}$$

where the subscript $_{dp\to a_s}$ indicates direct entrainment of plume mass into air eddies, $c_{\dot{m}}$ is a proportionality constant and v'_{a_s} is the rms velocity difference between an air eddy of scale s and the plume. The total plume mass entrained per unit eddy length by all air eddies of scale s is simply the product of the number of penetrating air eddies multiplied by the entrainment rate of a single air eddy. The number $N_{a\to p}$ of air eddies per unit length inside any larger plume eddy is assumed proportional to the ratio of plume eddy volume to air eddy volume,

$$N_{a\to p} = c_N \times r_p^2 / r_a^2 \, , \tag{6}$$

where c_N is also a constant. The lowercase subscript $_{a\to p}$ indicates an air eddy that has penetrated a plume eddy. It follows that the total rate per unit eddy length at which plume mass is directly entrained and carried off by atmospheric eddies of arbitrary scale s is

$$\dot{m}_{dP\to a_s} = \dot{m}_{dp\to a_s} \times N_{a_s\to p} = (c_N \, c_{\dot{m}})_s \times r_p^2 \, v'_{a_s} / r_{a_s} \times \rho_p \, , \tag{7}$$

where the capital $_P$ in the subscript $_{P\to a_s}$ indicates the total plume mass entrained by all air eddies of scale s.

Because the rms turbulence velocity measured in the atmosphere is basically the same in either a fixed (Eulerian) or moving (Lagrangian) coordinate system, the insensitivity of measured rms turbulence levels to coordinate system translation means the radius r_{a_s} and rms speed v'_{a_s} (relative to the moving plume) of all air eddies larger than the Kolmogorov microscale will scale with the atmosphere's Lagrangian integral length scale Λ_L and rms turbulence velocity u'_L respectively.

Summing the plume mass entrained per unit eddy length by all dispersive eddy scales yields the expression

$$\dot{m}_{dP\to A} = \sum_{\text{microscale}}^{\text{integral scale}} (\dot{m}_{dp\to a_s} \times N_{a_s\to p}) = c \times R^2\, u'_L / \Lambda_L \times \rho_p \,, \tag{8}$$

where the uppercase subscript $_{P\to A}$ indicates total plume mass extrained to the atmosphere by all air eddies.

The Boussinesq Approximation for Extrained Mass. G. Briggs (1975) originally coined the word 'extrainment' to describe negative entrainment (mass loss) to the atmosphere. We could calculate plume extrainment with Eq. (8), wherein the extrained mass has density ρ_p. However, the development of closed-form solutions for the plume motion equations is greatly simplified by use of the Boussinesq approximation, in which density differences are neglected everywhere except in the buoyancy terms. When the density ratio ρ_p/ρ_a is set approximately equal to unity, the plume density ρ_p in Eq. (8) is replaced with the ambient density ρ_a, and the extrainment is approximated by

Extrainment:

$$\dot{m}_{dP\to A} \cong c \times R^2\, u'_L / \Lambda_L \times \rho_a \,. \tag{9}$$

Eq. (9) is important in the analysis which follows, as it describes the overall rate at which plume mass is entrained and carried off by atmospheric eddies. It will play a key role in the derivation of the plume conservation equations.

Indirect Entrainment

The extrainment mechanism that causes the loss of plume mass given by Eq. (7) is really just the familiar entrainment of Eq. (4) in another guise. Eq. (4), and hence Eq. (7), have a general validity that applies to all free shear layer eddies. It is reasonable to expect air turbulence eddies to intermix with their surroundings in just the same way as do plume turbulence eddies. Hence an atmospheric eddy moving through the turbulent plume must lose some of its mass to the plume just as the plume loses some of its mass by extrainment into the surrounding turbulent atmosphere.

By taking into consideration the time required for an air eddy to pass through the plume, Netterville (1990) has shown that the total amount of air entrained into a unit length of plume by this secondary mass transfer route (called indirect entrainment) is given by

Indirect Entrainment:

$$\dot{m}_{iA\to P} = \dot{m}_{da\to p} \times N_{p\to a} \times N_{A\to P} = \text{constant} \times R^2\, v_r/r_a \times \rho_a \,. \tag{10}$$

Like the direct entrainment Eq. (4) used in its derivation, indirect entrainment depends on the velocity and length scale of the entraining eddy. However, indirect entrainment also depends explicitly on the velocity and length scale of turbulent motion in the fluid that is being entrained.

The results derived in this section are valid not only for mass transfer between an air and a plume eddy, but they also apply to mass transfer between two plume eddies or two air eddies. The general form of Eq. (10) is, for a "Fluid A" eddy that has been penetrated by eddies from "Fluid B",

Generalized Entrainment:

$$\dot{m}_{dB\to A} = \dot{m}_{db\to a} \times N_{a\to b} \times N_{B\to A} = \text{constant} \times r_A^2\, v_{BrelA}/r_B \times \rho_B. \tag{11}$$

where $\dot{m}_{dB\to A}$ is the total rate of direct entrainment of fluid B into eddy A, $\dot{m}_{db\to a}$ is the rate of direct entrainment of fluid B into small internal eddies of fluid A, $N_{a\to b}$ is the number of small internal fluid A eddies that pass through each penetrating fluid B eddy, and $N_{B\to A}$ is the number of fluid B eddies that penetrate the main fluid A eddy.

It can be shown that the classic Taylor entrainment assumption derived as Eq. (4) is a special case of this general expression. We see that direct and indirect entrainment must depend on the same basic entrainment constant, since they are really just two different manifestations of the same underlying generalized entrainment relation given by Eq. (11).

The Turbulence Buffet Frequency

The main difference between plume rise in laminar and turbulent conditions is that in the latter case the plume is exposed to a constant barrage of turbulent gusts that erode plume mass. Thus it is useful to define the atmospheric turbulence 'buffet' frequency f as

$$f \equiv c_{nx}(u'_L/\Lambda_L) = \text{constant}/\tau_L \,, \tag{12}$$

where τ_L is the Lagrangian integral time scale of atmospheric turbulence, and f has the units of frequency, sec^{-1}. Following common practice, we assume that the ratio of Lagrangian to Eulerian time scales is given by $\tau_L/\tau_E = \text{constant}/i_E$, where $i_E = u'_E/U$ is the intensity of Eulerian turbulence. Then Eq. (12) can be written as

$$f = \text{constant} \times i_E/\tau_E = \eta\, i_E U/\Lambda_E \,, \tag{13}$$

where the turbulent entrainment coefficient η is an unknown constant. Netterville (1990) has shown that the classic entrainment constant β and the newly defined turbulent entrainment constant η are related by

$$\eta = 2\beta \,. \tag{14}$$

Then the turbulence buffet frequency f becomes

$$f = \text{constant}/\tau_L = \eta\, i_E U/\Lambda_E = 2\beta\, i_E U/\Lambda_E \,. \tag{15}$$

THE PLUME RISE EQUATIONS

The complete set of governing equations for plume motion in a turbulent wind is derived by balancing fluxes of mass, momentum and buoyancy through a two-dimensional plume slice of radius R. Consider the truncated conical plume control volume sketched in Fig. 3. Mass from the surrounding atmosphere is directly entrained into the control volume at a rate $\dot{m}_{dA\to P}\,\Delta s$, where $\dot{m}_{dA\to P}$ is given by Eq. (4). Mass is indirectly entrained from penetrating air eddies at a rate $\dot{m}_{iA\to P}\,\Delta s$, with $\dot{m}_{iA\to P}$ calculated either from Eq. (10) or the more general Eq. (11). The entrained mass comes from the surrounding atmosphere, and by definition carries a momentum, temperature and species excess of zero. The only quantity added by entrainment into the plume is mass. Mass is removed from the plume control volume by atmospheric turbulence at a rate $\dot{m}_{dP\to A}\Delta s$, where $\dot{m}_{dP\to A}$ is given by Eq. (9). The extrained mass carries with it the values of momentum, temperature and species concentration that pertain to the plume at the point where the mass is removed.

Time-Dependent Plume Equations

Netterville (1990) has derived equations for the balance of mass, momentum, heat and species in the plume control volume sketched in Fig. 3. With the well-known Brunt-Väisälä frequency N characterizing the air temperature gradient, the conservation equations for a "bent-over" plume moving downwind at speed U are:

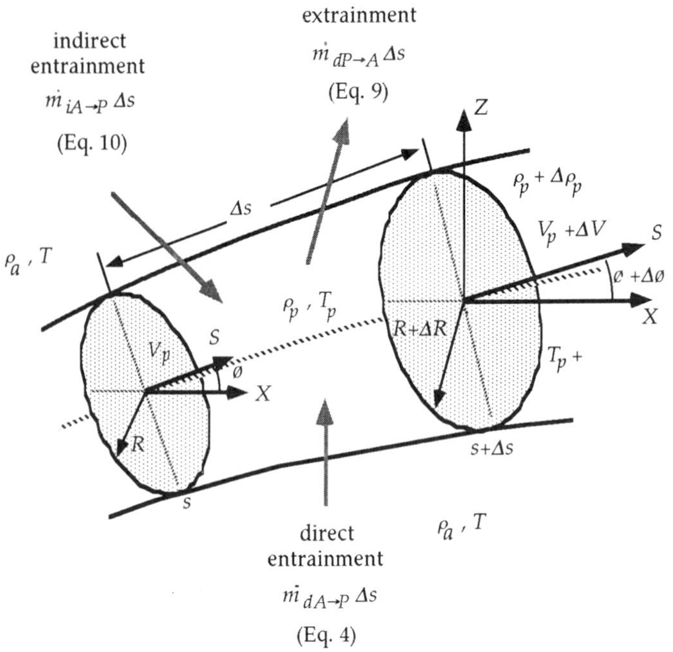

Fig. 3. The plume control volume and coordinate system.

Mass Equation:

$$\frac{d}{dt}(UR^2) = \eta URW + f\,UR^2 - f\,UR^2 = 2\beta URW, \qquad (16)$$

rate of mass accumulation — inflow by direct entrainment — inflow by indirect entrainment — outflow by extrainment

Vertical Momentum Equation:

$$\frac{d}{dt}(UR^2W) = UR^2 g\left(\frac{T_p - T_a}{T_a}\right) - f\,UR^2W, \qquad (17)$$

rate of momentum accumulation — momentum created by buoyancy force — momentum carried off in extrained mass

Heat Equation:

$$\frac{d}{dt}\left[UR^2 g\left(\frac{T_p - T_a}{T_a}\right)\right] = -UR^2WN^2 - f\,UR^2 g\left(\frac{T_p - T_a}{T_a}\right), \qquad (18)$$

rate of heat loss from plume — effective heat loss due to ascent through thermal stratification — heat carried off in extrained mass

Species Equation:

$$\frac{d}{dt}(UR^2\chi) = -f\,UR^2\chi, \qquad (19)$$

rate of pollutant loss from plume — pollutant carried off in extrained mass

Equations with the form of (17) and (18) were first obtained by Djurfors (1977) by working backwards from his assumption that plumes in neutral conditions approach final rise at an exponentially decreasing rate. Here we have shown that Djurfors' decay terms for momentum and buoyancy are part of a family of new removal terms that originate in the two-way mass exchange between a turbulent plume and a turbulent atmosphere.

Solution of the Plume Equations. For constant wind speed U, the solution of the mass equation (16) has the familiar form

$$R = R_o + \beta z \, , \qquad (20)$$

where R_o is the plume radius at the end of the "bending-over" phase.

Following Csanady (1973) and Djurfors (1977), it is convenient to introduce the flux of vertical momentum, $M \equiv U R^2 W$, and the flux of buoyancy, $F \equiv U R^2 g (T_p - T_a)/T_a$. Then, following Netterville (1990), the solution of Eqs. (16) - (18) for plume height $z(\tau)$ is found to depend on the degree of ambient thermal stratification as indicated by the squared Brunt-Väisälä frequency N^2 being greater than, less than or equal to zero:

Stable Atmosphere ($N^2 > 0$):

$$z(\tau) = \left\{ \frac{3}{\beta^2 U \left(f^2 + N^2\right)} \left[F_o + f M_o + \left\{ N M_o \left[\sin\left(\frac{N}{f}\tau\right) - \frac{f}{N}\cos\left(\frac{N}{f}\tau\right) \right] - F_o \left[\cos\left(\frac{N}{f}\tau\right) + \frac{f}{N}\sin\left(\frac{N}{f}\tau\right) \right] \right\} e^{-\tau} \right] + \left(\frac{R_o}{\beta}\right)^3 \right\}^{1/3} - \frac{R_o}{\beta} \qquad (21)$$

Neutral Atmosphere ($N^2 = 0$):

$$z(\tau) = \left\{ \frac{3}{\beta^2 f^2 U} \left[F_o + f M_o - \left\{ f M_o + F_o(\tau + 1) \right\} e^{-\tau} \right] + \left(\frac{R_o}{\beta}\right)^3 \right\}^{1/3} - \frac{R_o}{\beta} \qquad (22)$$

Unstable Atmosphere ($N^2 < 0$):

$$z(\tau) = \left\{ \frac{3}{\beta^2 U \left(f^2 + N^2\right)} \left[F_o + f M_o - \left\{ N M_o \left[\sinh\left(\frac{N}{f}\tau\right) + \frac{f}{N}\cosh\left(\frac{N}{f}\tau\right) \right] + F_o \left[\cosh\left(\frac{N}{f}\tau\right) + \frac{f}{N}\sinh\left(\frac{N}{f}\tau\right) \right] \right\} e^{-\tau} \right] + \left(\frac{R_o}{\beta}\right)^3 \right\}^{1/3} - \frac{R_o}{\beta} \qquad (23)$$

where $N \equiv + \sqrt{|N^2|}$. The quantities M_o, F_o and R_o are initial values of plume momentum, buoyancy and radius defined by Djurfors (1979, 1983) for the plume after it has bent over.

As $t \to \infty$, all three solutions asymptotically approach the same functional form for final rise,

$$z_{final} = \left[\left(\frac{3}{\beta^2 U}\right) \frac{F_o + f M_o}{\left(f^2 + N^2\right)} + \left(\frac{R_o}{\beta}\right)^3 \right]^{1/3} - \frac{R_o}{\beta} \qquad (24)$$

For stable conditions ($N^2 > 0$) the final rise is always finite. For neutral conditions ($N^2 = 0$) the final rise is finite only if the atmosphere is turbulent ($f > 0$). For unstable conditions ($N^2 < 0$) the final rise is finite or infinite depending on the sign of ($f^2 + N^2$), i.e. on whether thermal instability or atmospheric turbulence dominates plume motion.

COMPARISON WITH EXPERIMENT

The predictions of the neutral atmosphere model, Eq. (22), have been tested with 192 LIDAR measurements of plume height obtained by Edwards and Misra (1984) at Ontario Hydro's Nanticoke power station. Fig. 4 shows a plot of Eq. (22) against the Nanticoke data, for which

$$U = 8\,m/s,\ T_a = 10\,°C,\ i_E = 0.25,\ \Lambda_E = 250\,m,\ R_s = 2.5\,m,\ W_s = 30\,m/s,\ T_s = 127\,°C.$$

Marquardt Estimation was used to obtain a value for the entrainment constant that provided the best fit of the 'turbulent' neutral trajectory to the Nanticoke data. The 'best-fit' entrainment constant was $\beta = 0.65$.

In laminar winds ($f = 0$), and for negligible initial plume radius and momentum, Eq. (22) reduces to the classic '2/3' power-law trajectory,

$$Z_{lam}(t) = 1.14\ (F_o/U_s\ \beta^2)^{1/3}\ t^{2/3}. \qquad (25)$$

This curve is plotted in Fig. 4, using Briggs' (1975) recommended value of $\beta = 0.6$. Briggs assumed that plume rise terminates when the effective rate of dissipation of in-plume turbulence equals the ambient turbulence dissipation rate. With this assumption he obtains for final rise,

$$\Delta h = 1.3\ (F/U\,u*^2)\ (1 + h_s/\Delta h)^{2/3}\ . \qquad (26)$$

Care was taken to eliminate any chance that the trajectory from Eqs. (25) and (26) could be at a disadvantage because of a possibly incorrect $U/u*$ ratio and non-optimized value for β. Since Eq. (25) is really the laminar special case of Eq. (22), the entrainment constant in Eq. (25) was adjusted to $\beta = 0.65$ to agree with the optimized value used in Eq. (22), and $U/u*$ was arbitrarily set to $U/u* = 8.23$ so the final rise from Eq. (26) would agree with the final height predicted by Eq. (22).

The two-tailed paired T-Test was used to test the two models. The analysis showed that, at the 95% confidence level, there was no significant difference between the Nanticoke data and the trajectory predicted by Eq. (22) ($P = 0.088$), but a significant difference did exist between the observations and the trajectory predicted by Eqs. (25) and (26) ($P \le 0.0001$).

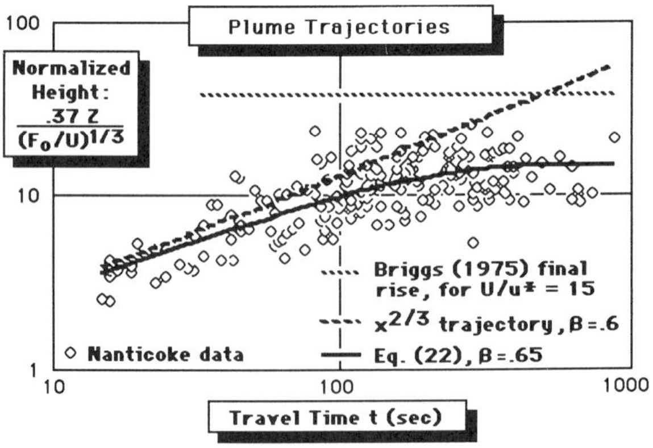

Fig. 4. Normalized Plume Height as a Function of Travel Time

DISPERSION

The species Eq. (19) describes the rate at which a pollutant is removed from the plume by atmospheric turbulence. It has the simple solution

$$X(\tau) = X_o(1 - e^{-\tau}), \qquad (27)$$

where $X = UR^2\chi$ is the species mass flux at any point along the plume centerline, χ is the species concentration governed by Eq. (19), and X_o is the species mass flux emitted from the source. The plume acts along its length as a distributed line source of pollutant, which diffuses passively from whatever (x,z) coordinate it has when it leaves the plume. Dispersion is explicitly coupled to plume rise because the local source term $X(\tau)$ gets smaller as the plume rises with increasing travel time. In the traditional 'effective source height' dispersion model, the plume must wait until it reaches final rise before it can disperse. In the model developed here, the least buoyant parts of the plume begin diffusing immediately, as soon as the plume leaves the stack. Gradually other portions of the plume are entrained into and carried off by atmospheric eddies, until eventually all the plume has been caught up and dispersed by the turbulent wind.

CONCLUSIONS

A new model for entrainment has been developed, based on an understanding of several aspects of turbulent mixing within free shear layers. The entrainment model predicts that a turbulent atmosphere entrains plume material just as the plume entrains the atmosphere. The model also shows that Taylor's entrainment assumption corresponds to entrainment by shear-generated eddies on the plume's surface, and is only the first in a series of entrainment terms acting on the plume at ever-decreasing scales.

The two-way entrainment model has led to development of a new model for plume rise in turbulent winds, which combines Turner-type plume mass loss with solutions consistent with the 'decay constant' approach of Priestley, Csanady and Djurfors. The plume rise model was tested with almost 200 LIDAR measurements of plume height obtained at Ontario Hydro's Nanticoke power station. The new turbulent model significantly outperformed the laminar plume rise model at distances beyond 1 km from the source.

The new plume model also has implications for dispersion, since it shows that the plume acts along its length as a distributed line source of pollutant, rather than as a single point source as is commonly assumed in current dispersion models.

ACKNOWLEDGMENTS

The author is indebted to G. Briggs, J. Deardorff and D. J. Wilson for their helpful comments, and to Grant Edwards of Ontario Hydro for providing the plume rise data. The study of plume rise in turbulent winds was suggested to the author by S. G. Djurfors, who has a long-standing interest in the subject. His constant encouragement and helpful discussions were vital to the completion of this work, and are gratefully acknowledged.

REFERENCES

Benjamin, S. F., 1975, The behavior of hot plumes from tall chimneys under conditions of strong thermal convection, in: "Proc. APCA/PNWIS Annual Meeting," Vancouver, B.C., 1 - 41.

Briggs, G. A., 1969, "Plume Rise," USAEC Report, 25 - 37.

Briggs, G. A., 1975, Plume Rise Predictions, in: "Lectures on Air Pollution and Environmental Impact Analyses, AMS, 59 - 111.

Brown, G. L. and Roshko, A., 1974, On density effects and large structure in turbulent mixing layers, J. Fluid Mech. 64, 755 - 816.

Csanady, G. T., 1964, Turbulent diffusion in a stratified fluid, J. Atmos. Sci. 21, 439 - 447.

Djurfors, S. G., 1977, "On the rise of buoyant plumes in turbulent environments," Syncrude Canada Ltd. Professional Paper 1977-4.

Djurfors, S. G., 1983, Discussion of effective source flux parameters for use in analytical plume rise models. Atmospheric Environment 17, 197 - 198.

Edwards, G. C. & Misra, P. K., 1984, The effects of mesoscale turbulence on plume rise in a land/lake environment, Boundary-Layer Meteorol. 29, 155 - 165.

Fackrell, J. E. & Robins, A. G., 1982, Concentration fluctuations and fluxes in plumes from point sources in a turbulent boundary layer, J. Fluid Mech. 117, 1 - 26.

Hanna, S. R., Briggs, G. A. and Hosker, R. P. Jr., 1982, "Handbook on Atmospheric Diffusion," US DOE/TIC-11223.

Hernan, M. A. & Jimenez, J., 1982, Computer analysis of a high-speed film of the plane turbulent mixing layer, J. Fluid Mech. 119, 323 - 345.

Morton, B. R., Taylor, G. I. & Turner, J. S., 1956, Turbulent gravitational convection from maintained and instantaneous sources, Proc. Roy. Soc. (London) A, 234, 489 - 512.

Netterville, D. D. J., 1990, Plume rise, entrainment and dispersion in turbulent winds, Atmospheric Environment, in press.

Priestley, C. H. B., 1953, Buoyant motion in a turbulent environment, Australian J. Physics 6, 279 - 290.

Priestley, C. H. B., 1955, A working theory of the bent-over plume of hot gas, Quart. J. Roy. Met. Soc. 82, 165 - 176.

Priestley, C. H. B. and Ball, F. K., 1954, Continuous convection from an isolated source of heat, Quart. J. Roy. Met. Soc. 81, 144 - 157.

Slawson, P. R. and Csanady, G. T., 1967, On the mean path of buoyant, bent-over chimney plumes, J. Fluid Mech. 28, 311 - 322.

Slawson, P. R. and Csanady, G. T., 1971, The effect of atmospheric conditions on plume rise, J. Fluid Mech. 47, 33 - 49.

Taylor, G. I., 1945, "Dynamics of a Mass of Hot Gas Rising in the Air," USAEC Report MDDC-919 (LADC-176), Los Alamos Scientific Laboratory.

Turner, J. S., 1963, The motion of buoyant elements in turbulent surroundings, J. Fluid Mech. 16, 1 - 16.

DISCUSSION

G.A. BRIGGS

The most unique aspect of this plume rise model is the introduction of a time scale f^{-1} to characterize ambient turbulence. When conditions are near neutral and $f^2 >> N^2 = (g/\theta)\partial\theta/\partial z$, when f is assumed constant, the model gives a final rise equation identical in form to the usual stable rise equation, except that f^2 replaces N^2. Thus, in the application to the Nanticoke data shown in Fig. 4, it is assumed that f = const. and N = 0, but the same good fit could be obtained using f = 0 and N > 0 with $\partial\theta/\partial z \approx 0.3°/100$ m - only a slightly stable stratification (close to the "Standard Atmosphere"). Are you certain that $\partial\theta/\partial z$ was closer to zero than the above value?

D.D.J. NETTERVILLE

I was unable to obtain $\partial\theta/\partial z$ for these data. According to Edwards and Misra (personal communication) who first reported the data, 60% of the cases are neutral to unstable, and 40% are slightly stable to stable. Because the individual data points could not be grouped by stability, I compared all the observations with model predictions for the neutral case. With almost two-thirds of the data being taken in non-stable conditions, it is more likely that the data would be better represented by an average $\partial\theta/\partial z$ of zero than with a stable $\partial\theta/\partial z$ of 0,3°C/100 m. In any case, I strongly support the idea of testing the model with data for which plume level turbulence intensity and scale, and $\partial\theta/\partial z$, have all been measured.

G.A. BRIGGS

It is more realistic to assume that f varies with height, although it could be evaluated at a specific height for a specific case. For instance, for the neutral surface layer it is commonly agreed that turbulence intensity $\propto u^*/U$ and the Eulerian scale of turublence $\propto z$. This gives $f \propto u^*/z$. Applying this to the present model, with $z = h_s + \Delta h$, we get a final rise equation very similar to Eq. 26, quoted from Briggs (1975), except for the exponent in the $(1 + h_s/\Delta h)$ term.

D.D.J. NETTERVILLE

I agree it is most realistic to use a turbulence buffet frequency f that varies with height. However, for many years we have used "laminar wind" plume models that in effect assume f to be zero. My plume model splits the difference, so to speak, and assumes that f is constant but non-zero through the plume layer. Even so, as pointed out in my paper in Atmospheric Environment (1990), this restriction can be worked around by dividing the atmosphere into several layers, each one having its own value for f.

Your raise the point that substituting a variable $f \alpha u^*/z$ into my final rise Eq. (24) gives a formula having the same functional dependence on F_o/Uu^{*2} as in the final rise Eq. (80) of Briggs (1975). My theory was developed for the case of a <u>constant</u> f, so this functional similarity may be a coincidence that should not be pushed too far. For example, using a variable f in my Eq. (24) leads to a final rise dependence on $(1+h_s/\Delta h)^2$ rather than on $(1+h_s/\Delta h)^{2/3}$ as in Briggs' Eq. (80), and this difference is very important for tall stacks with $h_s/\Delta h \sim O[1]$. It would be interesting to rederive the model for a variable f and see how that changes the final rise equation.

<u>Reference</u>

Netterville, D.D.J.; Plume rise, entrainment and dispersion in turbulent winds, *Atmospheric Environment* Vol. 24A, No. 5, pp. 1061-1081, 1990.

A COMPARISON OF THE MONTE CARLO AND THE FLUX GRADIENT METHOD FOR ATMOSPHERIC DIFFUSION*

Rolf Lange

Atmospheric and Geophysical Sciences Division
Lawrence Livermore National Laboratory
Livermore, California 94550

INTRODUCTION

In order to model the dispersal of atmospheric pollutants in the planetary boundary layer, various methods of parameterizing turbulent diffusion have been employed. These approaches differ greatly in sophistication and complexity (Monin and Yaglom, 1971). Historically, the Gaussian plume models were the first dispersion models. The Gaussian formula is based on statistical theory and empirical observation of the horizontal and vertical standard deviation of the wind speed σ_y and σ_z. In K-theory, the Eulerian diffusion-advection equation closure problem is circumvented by assuming a gradient transport paramaterization and the postulation of the turbulent diffusivity parameter K, which must be provided empirically. The stochastic Markov chain (Monte Carlo) method employs generally the Langevin equation to model dispersion with the use of very many particles. The approach needs the empirical prescription of the wind velocity variances σ_u and the Lagrangian integral time scales T_L. All three methods have two things in common: they rely on some mathematical scheme and they need empirically derived diffusion parameters. (Higher order closure models are not part of this discussion.)

The purpose of this paper is to use a three-dimensional particle-in-cell transport and diffusion model to compare the Markov chain (Monte Carlo) method of statistical particle diffusion with the deterministic flux gradient (K-theory) method. The two methods are heavily used in the study of atmospheric diffusion under complex conditions, with the Monte Carlo method gaining in popularity partly because of its more direct application of turbulence parameters. The basis of comparison is a data set from night-time drainage flow tracer experiments performed by the U.S. Department of Energy Atmospheric Studies in Complex Terrain (ASCOT) program at the Geysers geothermal region in northern California. The domain of this complex terrain study is about 7×10 kilometers. The Atmospheric Diffusion Particle-in-Cell (ADPIC) model used is the main model in the Lawrence Livermore National Laboratory emergency response program: Atmospheric Release Advisory Capability (ARAC). As a particle model, it can simulate diffusion in both the flux gradient and Monte Carlo modes.

* Work performed under the auspices of the U.S. Department of Energy by the Lawrence Livermore National Laboratory under contract No. W-7405-Eng-48.

MODEL DESCRIPTION

ADPIC (Lange, 1978 and 1989) is a three-dimensional numerical particle-in-cell diffusion and transport model capable of simulating the time and space varying dispersal of atmospheric pollutants under complex conditions. It is a particle-in-cell model in which Lagrangian "mass" particles are transported inside a fixed Eulerian grid. The model solves the three-dimensional flux conservation equation,

$$\frac{\partial \chi}{\partial t} + \nabla \cdot (\chi \vec{U}_p) = 0 \quad , \tag{1}$$

χ is the pollutant concentration obtained by summing the particles in a given cell. $\vec{U}_p$ is a pseudo transport velocity defined as the sum of the mean advection velocity $\vec{U}_a$ and a diffusive velocity $\vec{U}_d$, all interpolated to individual particle positions.

$$\vec{U}_p = \vec{U}_a + \vec{U}_d \tag{2}$$

$\vec{U}_a$ is provided by a gridded mean wind field model MATHEW (Sherman 1978) which uses interpolated meteorological data and terrain information to produce a mass conservative wind field. The diffusion velocity $\vec{U}_d$ is computed in ADPIC depending on the type of diffusion theory chosen: either Monte Carlo or K-theory.

For K-theory $\vec{U}_d$, the diffusion velocity, is obtained by the mathematical model of gradient transport,

$$\vec{U}_d = -K \cdot \nabla \chi / \chi \tag{3}$$

where K is the diffusion coefficient, empirically derived for the vertical as

$$K_z = \frac{k\ u_* z}{\phi(z/L)}\ \epsilon^{-\left(\frac{V_g z}{U_* h}\right)} \quad . \tag{4}$$

where k is the Von Karman constant, u_* is the friction velocity, z is the height, above terrain $\phi(z/L)$ is an atmospheric stability function based on z and the Monin-Obukhov scale length L, V_g is the geostrophic wind and h the height of the mixing layer (Businger and Arya 1974).

The horizontal diffusivity, K_h, is based on the semi-empirical expression (Draxler 1976)

$$\sigma_y = \sigma_\theta U t (1 + \frac{t}{\tau})^{-1/2} \quad , \tag{5}$$

combined with the analytical relationship

$$K_h = \sigma_y d\sigma_y / dt \quad , \tag{6}$$

where σ_y is the horizontal standard deviation of the plume, σ_θ the standard deviation of the fluctuation of the wind direction, U the local mean wind speed, t is time, and τ is an empirical time constant. Because of the term $\nabla \chi / \chi$ in Eq. 3, the gradient theory approach to diffusion needs a grid-cell mesh.

For Monte Carlo statistical diffusion, the diffusion velocity $\vec{U}_d$ is computed using the Langevin equation (Legg and Raupach 1982). The vertical component equation being,

$$w(t + \Delta t) = a w(t) + b\ \sigma_w \zeta + (1 - a)\ T_L\ \partial(\sigma_w^2)/\partial z, \tag{7}$$

where $w(t)$ is the vertical velocity at time t, ζ a random number from a Gaussian distribution with zero mean and unit variance, T_L the Lagrangian integral time scale, $a = exp(\Delta t/T_L)$, and $b = (1 - a^2)^{1/2}$. The last term in Eq. 7 allows for inhomogeneous turbulence in which σ_w and T_L can vary with height, z. σ_w and T_L are empirical parameters which are computed from the following equations (Hanna 1981).

$$\sigma_w = 1.3\ u_*(1 - z/h);\ T_L = 0.08\ h/u_* \frac{(z/h)^{0.8}}{1 - z/h} \quad . \tag{8}$$

The two horizontal components of $\vec{U}_d$ are computed using equations of the same form as Eq. 7 without the last term which deals with the vertical inhomogeneity of turbulence. Similarly, the prescriptions for the horizontial wind variances σ_u and σ_v and time scales T_L follow the same functional forms as shown in Eq. 8. With the Monte Carlo method, the particles are totally independent from one another and no grid is needed to calculate the diffusion velocities.

DESCRIPTION OF THE EXPERIMENT

In 1978 the Department of Energy developed a program aimed at atmospheric studies in complex terrain (ASCOT). The ASCOT program was designed to develop the technology needed to assess atmospheric properties and the impact of energy sources on air quality in areas of complex terrain. The initial study site was the Geysers geothermal resource area located approxiamtely 120 km north of San Francisco, California, and the purpose was to study nocturnal air drainage flow.

The data set used here is one of five tracer release experiments in the Anderson Springs Valley in the Geysers in September 1980 (Gudiksen, 1983). The experiments consisted of obtaining meteorological and tracer release data over the area of the Anderson Springs Valley during nighttime drainage flow. The experiments began at 2300 PST and ended at approximately 0500 PST. Anderson Springs Valley is surrounded by Boggs Mountain to the northeast, by Cobb Mountain to the north and by Pine Mountain to the southwest, and forms a bowl shaped drainage regime for the Putah, Gunning, Anderson and Bear Creeks, opening towards the east. It has rugged topography and a ground cover varying from bare soil to forest canopy. Figure 1 shows the geography of the Anderson Springs Valley, the meteorological measurement, tracer release, and sampler locations.

The experiment chosen was on the night of September 19/20, during which 3 one-hour simultaneous and distinguishable surface tracer releases were conducted (2 perfluorocarbons, 1 sulfur hexafluoride) from the Anderson, Gunning and Putah creeks. Average concentrations were sampled at some 60 sampler locations and two vertical profile stations throughout the area. Averaging times varied from 10 minutes to 2 hours. In this study, the tracer data from only the Anderson Creek (PMCH) was used providing some 170 samples.

The extensive meteorological data were collected by surface anemometers, tower anemometers, tethersondes accoustic sounders, and tetroons, providing M/A with 47 surface stations and 8 vertical profiles of wind speed and direction as input (Fig. 1), plus the necessary information on stability and mixing height. All six hours during this night were stable with good air drainage flow developing.

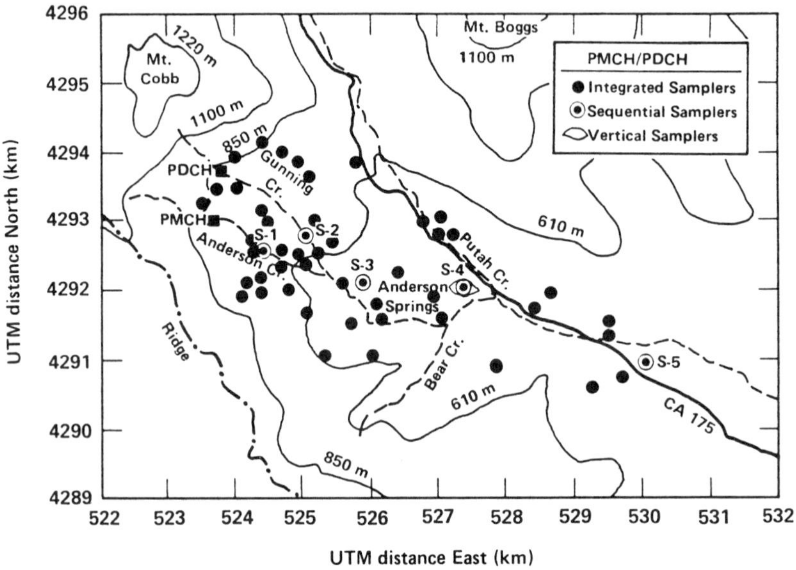

Fig. 1. Sampling and traces release locations for the 1980 Geysers Anderson Creek Valley experiments. The solid squares designate tracer release points.

RESULTS AND DISCUSSION

Figure 2 shows a typical advection wind field at 40 m above terrain for the Anderson Valley at 0100 PST. The convergence of the drainage winds towards the center of the bowl-shaped valley and the outflow to the southeast are quite evident. The heavy dashed lines are the four main drainage creeks and the dot-dashed line indicates the mountain ridge line. The wind vectors are drawn at every other grid point. Figure 3 shows an example of the ADPIC particle plume representing the tracer distribution at 0100 PST, two hours after the start of the one-hour release. The distribution shows little similarity with a well-behaved plume. The pattern is short and disorganized with a shape intermediate between that of a plume and a puff. This heterogeneity in the plume structure is well-supported by the wide fluctuations shown by the measured tracer concentrations, which at times varied by a factor of 5 over a distance of 50 m.

In the evaluation of the quantitative differences between K-theory and the Monte Carlo method of diffusion in comparison to measured tracer concentrations, Figs. 4 and 5 show the sequential surface air tracer concentrations for 20-minute averages at the three locations S-1, S-4, and S-5 as identified in Fig. 1. Figure 4 depicts the comparison between the plume modeled by the K-theory method and measurements, and Fig. 5 the same for the Monte Carlo method. In each case, the arrival time of the modeled plume is late by 10 to 40 minutes as a result of lower wind speed coming from the advection wind field model. Also, in both cases, the modeled plume is "shorter," an effect that can be traced to the lack of ability of the model to take into account the drag effects of terrain and vegetation (trees) on the real tracer plume. Overall the differences between the two diffusion methods as shown in Figs. 4 and 5 are small.

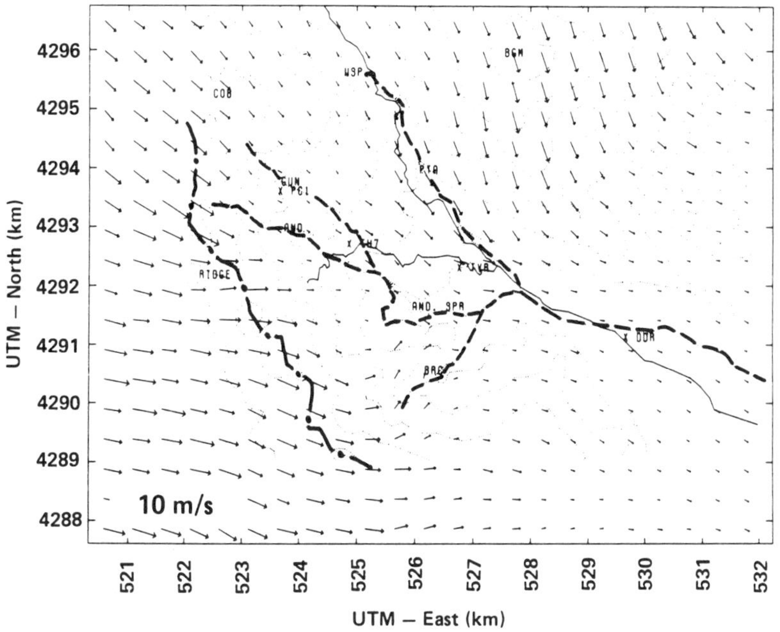

Fig. 2. MATHEW advection wind field 40 m above terrain for the Anderson Valley at 0100 PST.

It is difficult to devise a statistical method that in all cases adequately describes a model's performance in complex terrain. The standard correlation coefficient is used sometimes, although a few points can overly influence the entire results. Such points are typically located close to the tracer release point, where high concentrations are present, and the plumes are narrow so that spatial resolution is critical. Especially in complex terrain, a difference of 5° in the angle of the computed plume trajectory can produce outlying points that completely overshadow the statistics and the performance of the model in the rest of the domain under study. Thus, another method, which is less biased by a few high concentrations, was selected for comparing the relative accuracy of the K-theory and Monte Carlo diffusion schemes.

This method involves a band analysis which equally weights the model performance over the entire spatial domain of interest by using the ratio of measured to computed tracer concentrations within the entire domain. A factor R is computed for each pair of measurements (C_m) and model calculations (C_c) which represents the ratio between the two. The percentage of sample comparisons within a factor R are plotted as a function of R. The definition of R is

$$R = C_m/(C_c + B), \quad \text{except when R} < 1, \text{ then R } \equiv 1/\text{R} \ . \qquad (9)$$

Here B is background which is added to C_c. If both C_m and C_c are zero, the sample is disregarded. The average tracer background concentrations ranged from 0.01 to 0.03 ppt, which are small in comparison with most measurements. Background is included in the computed concentrations because the model does not provide background, and to avoid division by zero in the calculation of R when C_c is zero at some sampler location.

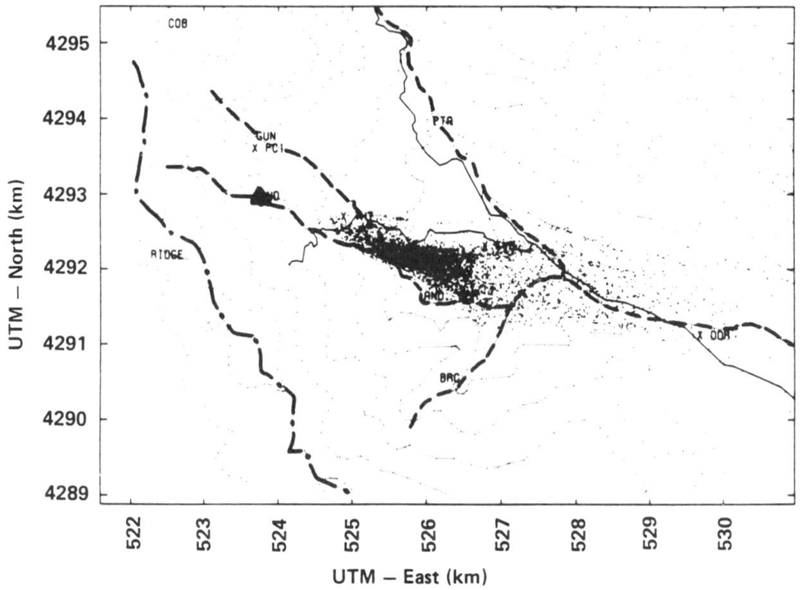

Fig. 3. ADPIC plume simulation example representing the tracer release in the Anderson Valley at 0100 PST one hour after the end of release. Solid triangle designates location of release.

Figure 6 shows the percentage of measured samples agreeing with those computed to within a factor R as defined above. The curves show that the Monte Carlo method gives consistently slightly better results. It does considerably better when looking at the residual defined as the difference between measured and computed concentrations shown in the figure. This is due to the Monte Carlo method providing a better simulation of the high measured concentrations near the source.

CONCLUSIONS AND COMMENTS

In comparing the simulations derived from the three-dimensional particle-in-cell diffusion model ADPIC, in both its gradient diffusion (K-theory) and Monte Carlo modes, against the results from a tracer release conducted in complex terrain during night time stable flow, the Monte Carlo method of diffusion gave slightly, but consistently better results when comparing measured and computed concentrations. This is encouraging because the Monte Carlo method of statistical diffusion provides advantages over the K-theory method. It represents both mathematically and physically a more fundamental approach to diffusion because (1) it is stochastic and, hence, does not depend on the concentration distribution, (2) it does not require a grid mesh to calculate gradients since the stochastic particles are independent of each other, and (3) it relies on diffusion parameters like σ_w and T_L which are more fundamentally linked to the turbulent diffusion regimes in the atmosphere than K profiles.

The following comments made for the case of vertical diffusion are equally valid for its horizontal components. Both the Monte Carlo and the K-theory methods

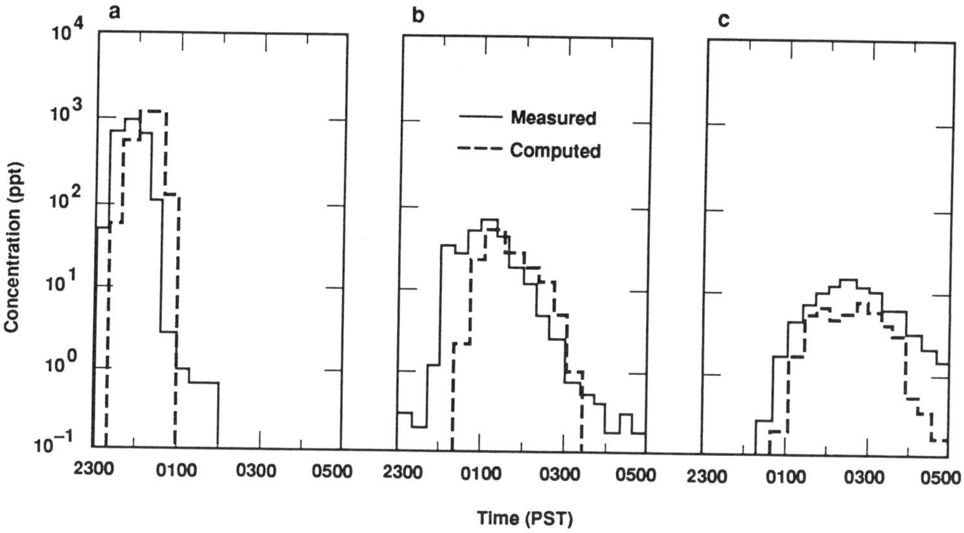

Fig. 4. K-theory computed vs. measured concentrations of the Anderson Creek tracer release in parts-per-trillion (ppt) at the sequential sampler locations: (a) S-1, (b) S-4, (c) S-5, plotted against time (see Fig. 1 for locations).

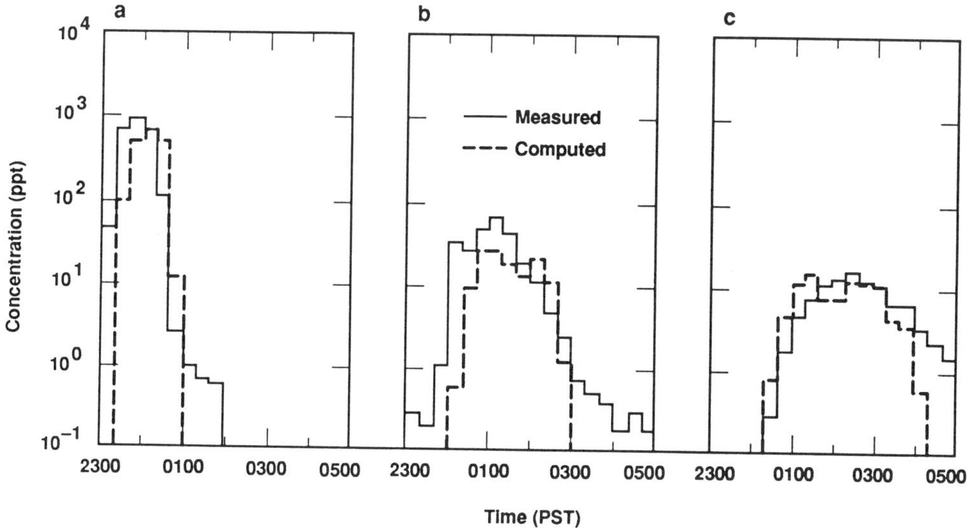

Fig. 5. Same as Fig. 4 except for Monte Carlo method.

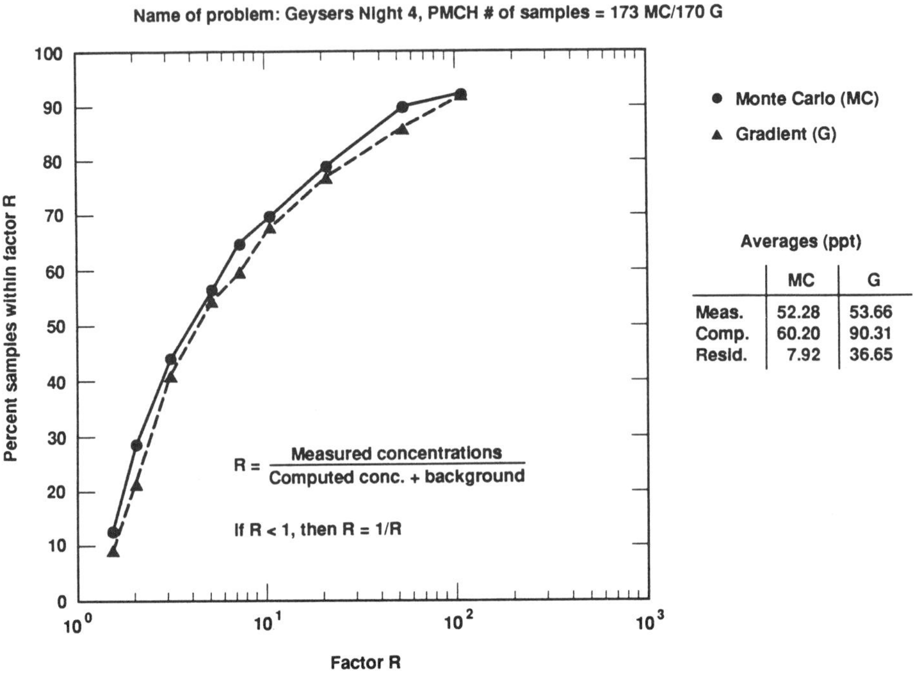

Fig. 6. Percentage of measured samples agreeing with those computed to within a factor R (as defined in the figure) for the Anderson Creek tracer plume comparison between K-theory and Monte Carlo diffusion. Also shown are the average measured and computed concentrations on the residuals in (ppt).

(as is also true for the Gaussian method mentioned in the introduction), require diffusion coefficient profiles of one kind of another, i.e., K or σ_w and T_L. The profiles used in this study are shown in Eqs. 4 and 8. The strong, proportional dependence of the diffusion velocity $\vec{U}_d$ on K is evident from Eq. 3. The sensitivity of $\vec{U}_d$ to σ_w and T_L is not so immediately clear (Eq. 7). Initial numerical experiments with ADPIC, which are outside the scope of this paper, indicate a strong sensitivity of $\vec{U}_d$ to σ_w and a rather weaker one to T_L. Presently, there is very little information on σ_w and T_L profiles available. Data on σ_w are becoming more frequent through field experiments, but T_L remains a rather elusive quantity. More research is needed to provide these profiles because numerical experiments indicate that the differences in K, σ_w and T_L parameterizations may well outweigh the difference between the K-theory and Monte Carlo diffusion methods.

With regard to the relative computer time requirements, an ADPIC calculation with the Monte Carlo method showed a 10-percent savings over the K-theory gradient method in this limited study. While the Monte Carlo method saves the need to calculate concentration gradients in a three-dimensional grid, it makes frequent calls to Gaussian distributed random number generators which is time consuming.

ACKNOWLEDGEMENTS

I would like to thank Leonard A. Lawson for programming the Monte Carlo diffusion into ADPIC, which was previously exclusively a K-theory code, and for doing most of the computations.

REFERENCES

Businger, J.A. and S.P.S. Arya, 1974, "Height of the Mixed Layer in the Stably Stratified Planetary Boundry Layer," *Advances in Geophysics*, Academic Press, **18A,** 73–92.

Draxler, R.R., 1976, "Determination of Atmospheric Diffusion Parameters," *Atmos. Environ., 10,* 99–105.

Gudiksen, P.H., 1983, *ASCOT Data from the 1980 Field Measurement Program in the Anderson Creek Valley, California,* Lawrence Livermore National Laboratory Report #UCID-18874-80.

Hanna, S.R., 1981, "Applications in Air Pollution Modeling," *Atmospheric Turbulence and Air Pollution Modelling,* F.T.M. Nieuwstadt and H. Van Dop, Editors, D. Reidel Publishing Co., Dortrecht/Boston/Lancaster, 1981.

Lange, R., 1978, "ADPIC—A Three-Dimensional Particle-in-Cell Model for the Dispersal of Atmospheric Pollutants and Its Comparison to Regional Tracer Studies, *J. Appl. Meteoro.* **17,** 320–329, 1978.

Lange, R., 1989, "Transferability of a Three-Dimensional Air Quality Model Between Two Different Sites in Complex Terrain," *J. of Appl. Meteoro.*, **Vol. 28, No. 7,** July 1989.

Legg, B.J., and M.R. Raupach (1982). "Markov Chain Simulation of Particle Dispersion in Inhomogeneous Flows: The Mean Drift Velocity Induced by a Gradient in Eulerian Velocity Variance," *Boundary-Layer Met.* **24**, 3–13.

Monin, A.S., and A.M. Yaglom, 1971, *Statistical Fluid Mechanics: Mechanics of Turbulence,* **1**, MIT Press.

Sherman, C.A., 1978, "A Mass-Consistent Model for Wind Fields Over Complex Terrain," *J. Appl. Meteor.* **17,** 312–219.

DISCUSSION

H. VAN DOP

1. In the Lagrangian model which you applied, you used a Lagrangian timescale which increased with height to a constant value. Would it not be more appropriate to let the timescale decrease near the top of the atmospheric boundary layer?

2. Though the Lagrangian concept of diffusion certainly has distinct advantages over K-diffusion theory, it should be noted that the Legg-Raupach model you applied has the undesired property that it "demixes" a uniform concentration distribution.

R. LANGE

1. Probably. The parameterisation for the timescale was taken from a compilation by Steve Hanna of timescales for the unstable, neutral and stable PBL. ("Atmospheric Turbulence and Air Pollution Modelling," F.T.M. Nieuwstadt and H. van Dop, D. Reidel Publishing Co., Dordrecht, Holland, 1984.) For the unstable case, the timescale actually decreases near the top of the PBL, but not so for the stable case. There is definitely a need to develop better parameterisations for these scales.

2. We have made many calculations for simple two-dimensional cases applying the Legg and Raupach drift term to correct for inhomogeneous turbulence in the vertical and found that for cases of moderate inhomogeneity the drift term preserved uniform mixing in the vertical as well as other drift correction terms found in the literature.

SPORADIC TURBULENCE AND DISPERSION IN THE STABLE BOUNDARY LAYER

Carmen J. Nappo

Atmospheric Turbulence and Diffusion Division
Air Resources Laboratory
National Oceanic and Atmospheric Administration
P.O. Box 2456, Oak Ridge, TN 37831

INTRODUCTION

It is commonly assumed that the levels of turbulence and dispersion in the stable planetary boundary layer (PBL) are low. This assumption strongly affects the modeling of air quality and plume dispersion in the nocturnal atmospheric surface layer. However, beginning with the early observations by Durst (1933), Gifford (1952), and Lyons et al. (1964) and latter by, for example, Bean et al. (1973), Kondo et al. (1978), Mahrt (1985), and Gossard et al. (1985) it is now clear that sporadic outbreaks of turbulence and enhanced vertical diffusion are common in the stable PBL. These effects have not been accounted for in conventional PBL plume diffusion theory and, because of this, model predictions of surface deposition may be underestimated. For long-term air quality considerations, the assumption that vertical transfer is small in the stable PBL is generally acceptable; however, in treating the problem of short-term exposure to hazardous materials consideration of the non-stationary aspects of the PBL is crucial. For example, during a breakdown event pollutants released in the upper levels of the PBL can be brought downward to the surface layer with relatively little dilution. Harrison et al. (1978) and Winkler (1980) show that sudden increases in ozone concentration at the ground surface are correlated with disturbances in the stable PBL as indicated by sodar traces. They relate these concentration increases to turbulent transport from higher levels where ozone concentrations are higher than at the ground surface.

In this paper we present the results of a statistical analysis of PBL breakdowns over simple and complex terrains. We shall show that sporadic turbulent breakdowns are a consistent feature of the nighttime boundary layer, and that a significant portion of the nocturnal exchanges of heat, momentum, and atmospheric pollutants between the atmosphere and the ground surface occurs during these breakdown episodes. Some of the results presented in this paper as well as an extended discussion of the possible causes of breakdowns appear in Nappo (1990).

THE DATA

Data which we consider representative of simple terrain environments were taken from the EPA Regional Air Monitoring System (RAMS) which is described in detail by Schiermeier (1978). These data were collected at 25 stations within and around St. Louis, MO from 1975 through 1977. One-minute averages of meteorological variables were formed at each station from half-second samples. In this study we use horizontal wind speed, temperature, and ozone mixing ratio measured at 30 m above the ground surface at stations 1 (urban) and 23 (rural). Station 1 was located north of the city center in a area of one- and two-story buildings; station 23 was located east of the city in open country and farm land.

Data over complex terrain were taken at the Walker Branch Watershed field station near Oak Ridge, TN from February 1987 to June 1988. Figure 1 illustrates the terrain contours in the Oak Ridge region. In this part of the eastern Tennessee River Valley, the terrain consists mostly of parallel ridges aligned along the valley axis. The ridges are typically 50 m high, 1 km wide, and 1 to 2 km apart. The region is mostly forested except for towns and small farms in the valleys. The monitoring system at the Walker Branch site consisted of a three-component Gill anemometer,* an aspirated bead thermistor, and a Meloy continuous-measuring ozone monitor all mounted at the 40 m level of the Walker Branch tower. This level is some 17 m above the forest canopy. The tower is on the crest of a ridge which has an average elevation of about 60 m above the floor of the Tennessee River Valley. Observations were taken every half-second, and from these, one-minute averages of horizontal wind speed, temperature, and ozone mixing ratio were formed. In addition to these measurements, a monostatic sodar was in operation in a clearing located about 400 m away from the tower. The number of nights of useful data are 165 for the Walker Branch, 103 for RAMS station 1, and 76 for RAMS station 23.

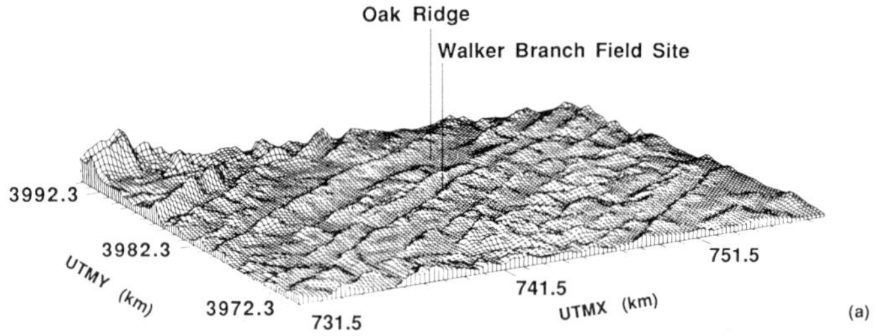

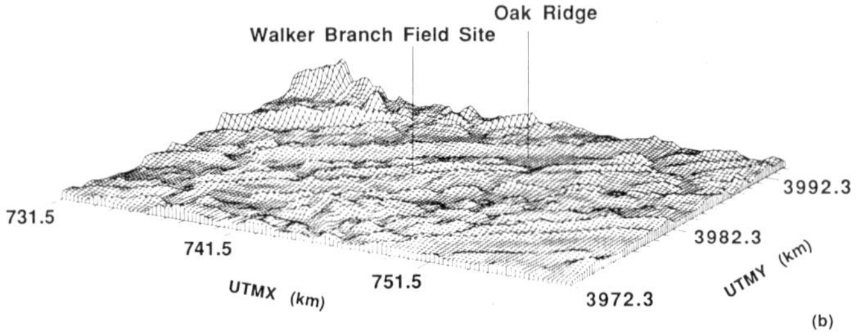

Fig. 1. Topography of the eastern Tennessee River Valley in the Oak Ridge area viewed from the southwest (a) and the southeast (b).

METHOD OF ANALYSIS

The breakdown event is pictured as an overturning of the PBL air. In the presence of vertical wind shear and a stable thermal stratification, such an overturning transfers downward high-speed warm air toward the ground surface, and lifts low-speed cool air upward away from the ground surface. These actions will result in a positive value of the covariance of wind speed and temperature, C_{UT}. We calculate the covariance time series using digital filtering techniques. The details of these calculations are

* Mention of a commercial company or product does not constitute an endorsement by NOAA. Information from this publication concerning proprietary products or the test of such products for publicity or advertising purposes is not authorized.

given in Nappo (1990). Basically, the raw wind speed and temperature time series are first smoothed using a Gaussian-weighted, 5-minute running mean. This reduces the amplitudes of the minute-to-minute fluctuations which we consider too short-lived to be breakdown events. Next, we smooth this time series using a Gaussian-weighted, 30-minute running mean which removes long-term trends. A bandpass filtering is accomplished by subtracting to two smoothed time series. The covariance is obtained by forming the products of bandpassed wind speed and temperature time series. An example of this covariance is shown in Figure 2. We see that C_{UT} is mostly positive throughout the night, but periods of negative covariance also exist. We assume a breakdown begins when C_{UT} exceeds 10 percent of the average nighttime covariance, and ends when C_{UT} drops below this value. We take the nighttime period to run from 2000 hours to 0600 hours LST; however, the final results do not appear to be overly sensitive to these times. The statistics of interest are the number of breakdowns per night, the durations of the breakdowns, and the length of time between breakdowns. This latter quantity is significant because it relates to the relaxation time of the PBL.

During stable conditions, the ozone mixing ratio generally increases with height above the ground surface (Harrison et al., 1978). Overturning of the PBL air will bring high-speed, ozone-rich air down toward the ground surface, and lift low-speed, ozone-depleted air away from the ground surface. The result is a positive covariance of wind speed and ozone, C_{UO}. However, Nappo et al. (1988) show that the nighttime vertical distribution of ozone-mixing ratio can have zones of significant ozone depletion due to scavenging by nitrogen, sulfur, or hydrocarbon compounds. If air from these zones is brought downward to the ground surface, then a negative covariance of wind speed and ozone will result.

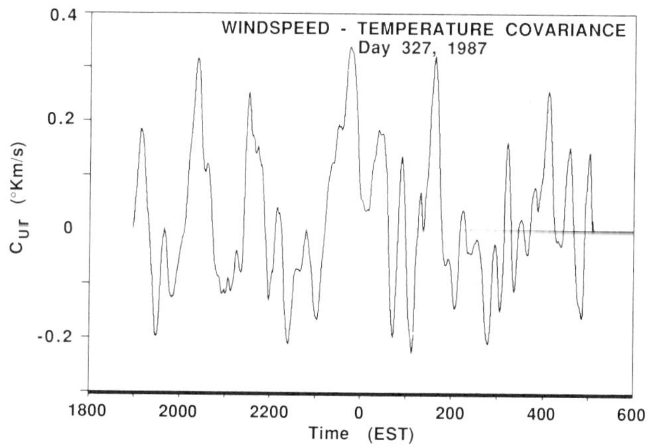

Fig. 2. Wind speed-temperature covariance at the Walker Branch site.

RESULTS

The average values of the breakdown characteristics based on the analysis of the wind speed-temperature covariance are listed in Table 1. We have performed similar calculations using the wind speed ozone covariance and obtained essentially the same results. Figure 3 shows the relative frequencies of breakdowns per night measured over contrasting environments, i.e., urban, simple terrain (RAMS station 1), and rural, complex terrain (Walker Branch). The distributions appear somewhat Gaussian but slightly skewed toward lower values. A greater range of breakdown frequency is seen over the complex terrain site than over the simple terrain site. The relative frequencies of the durations of breakdown events observed at these two stations are shown in Figure 4. The distributions appear somewhat lognormal. All breakdowns longer than 60 minutes are grouped into the 60-minute cell. Note that because the resolution of the covariance time series is one minute, we cannot detect breakdowns which last less than two minutes. Figure 5 shows the relative frequencies of the intervals between

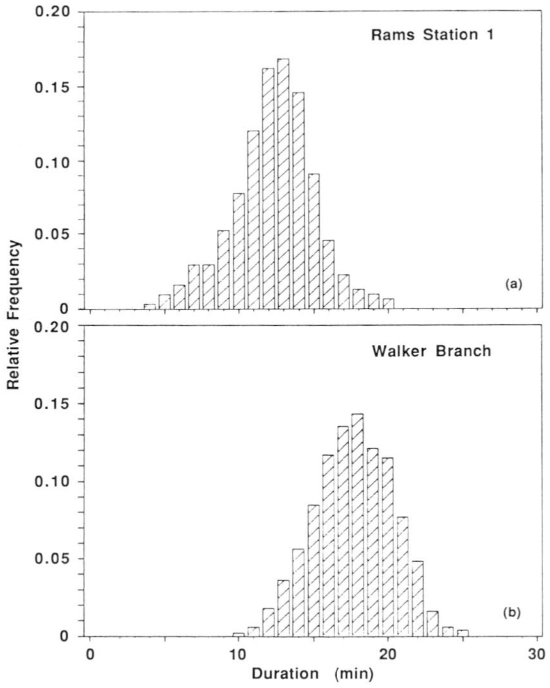

Fig. 3. Relative frequencies of breakdowns per night.

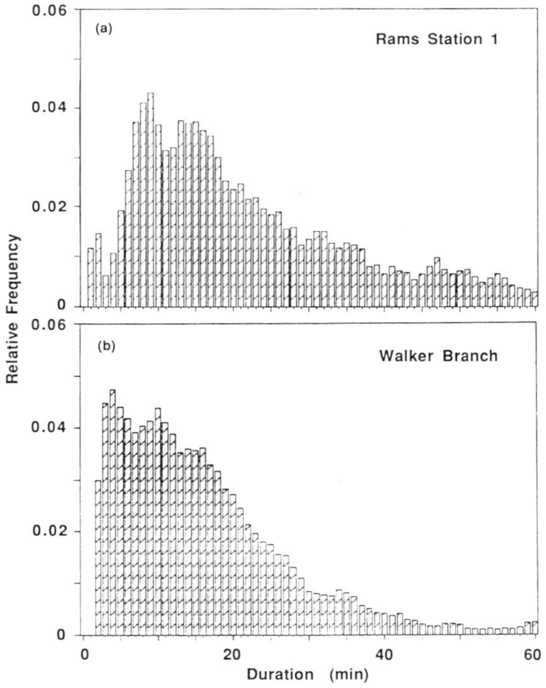

Fig. 4. Relative frequencies of the durations of breakdowns.

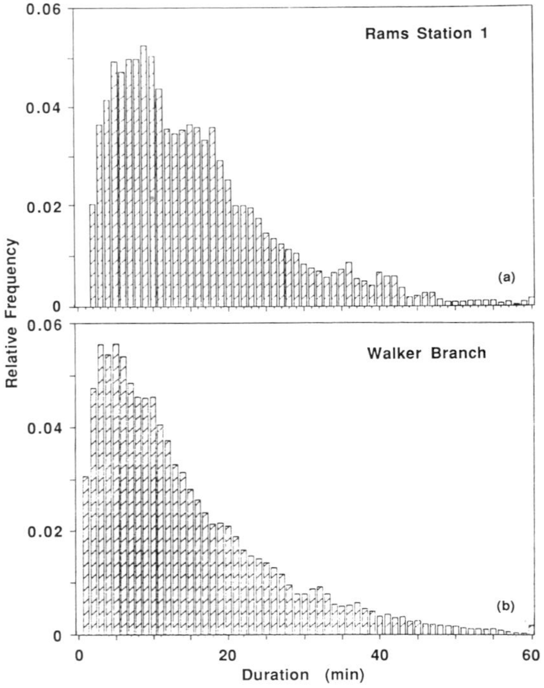

Fig. 5. Relative frequencies of the intervals between breakdowns.

breakdown events for the two contrasting environments. These quiescent periods are skewed more toward higher values for the urban station than for the rural station. The mode of a probability distribution marks the most likely value of the variant, and for this reason is a more meaningful statistic than the average. The modal values listed in Table 1 were estimated by inspection of the various histograms.

Table 1. Average Values of Breakdown Characteristics

	RAMS #1 (urban)	RAMS #23 (rural)	Walker Branch (complex terrain)
Breakdowns per night	13	15	18
Duration of breakdowns (min)	15(8)[a]	19(10)	16(6)
Interval between breakdowns (min)	21(12)	17(6)	14(4)
Covariance per breakdown (°K-ms^{-1})	3.2×10^{-2}	5.5×10^{-2}	4.8×10^{-2}
Nighttime average covariance (°K-ms^{-1})	0.6×10^{-3}	$12. \times 10^{-3}$	-7.8×10^{-5}

[a]Numbers in parentheses are the modal values of the distributions.

DISCUSSION

The values of the individual breakdown characteristics listed in Table 1 are seen to be not very different over the three sites, yet these sites are located in quite contrasting environments. This is significant because it suggests that breakdowns are not exceptional events limited to a certain terrain type but rather are a regular feature of the stable PBL flow. However, we see that the nighttime average covariances differ considerably between the sites. Recall that the Walker Branch site is located atop a 60-m ridge. Air flow directed over the ridge will tend to speed up as the flow streamlines converge above the ridge, and this will create a positive wind speed perturbation. However, because this flow originates at lower elevations, it will be colder than the ambient air at the instrument height some 100 m above the valley floor, and this will create a negative temperature perturbation. The result is a negative covariance. These types of events occur throughout the night and lead to the low values of nighttime average covariance. At the RAMS station 1 the urban heat-island effect and the large surface roughness often result in near-neutral or even unstable temperature gradients at night. We believe these conditions result in many prolonged periods of negative covariance, and these in turn lead to the small value of the nighttime average covariance shown in Table 1.

The relationship between breakdown events and turbulence is illustrated in Figure 6, where the monostatic sodar trace and C_{UT} are plotted for one night at the Walker Branch site. The elevation of the tower instruments is about 40 m above the sodar. The covariance is essentially zero until just before 2200 hrs when a sharp rise occurs. This jump in C_{UT} corresponds with the sudden appearance of turbulence in the PBL as indicated by the sodar trace at the 40 m height. After this initial period, the covariance and the turbulence are seen to vary with time; however a correlation exists between these quantities if examined on the 40 m level of the sodar trace. At about 0400 hrs, the turbulence and C_{UT} both fall to zero and remains there. Large negative values of the covariance such as those seen at 0215 and 0330, we attribute to the effects of flow over the ridge as described above.

This analysis of breakdown events is based on fluctuations of one-minute averages of horizontal wind speeds and temperatures, which we assume to arise from overturning of the PBL air. However, we have seen that horizontal advection can also lead to fluctuations, but this effect may be more pronounced over complex terrain. If we take the covariance as defined here to represent, in some sense, a vertical flux, then a heat flux can be defined as $-Dc_pC_{UT}$, where D is the air density and c_p is the specific heat of air at constant pressure. Using the values from Table 1, and the standard density for air, the average heat fluxes per breakdown are about -40 Wm^{-2} for the urban station, -68 Wm^{-2} for the rural station, and -59 Wm^{-2} for the complex terrain station. These values are larger than the heat flux values normally associated with the stable PBL; however these are intermittent values. For example, if at the rural station 15 breakdowns occur during a night, and if each breakdown is 10 minutes long, then the heat flux averaged over on 8-hour night is about -21 Wm^{-2}, which is a reasonable value.

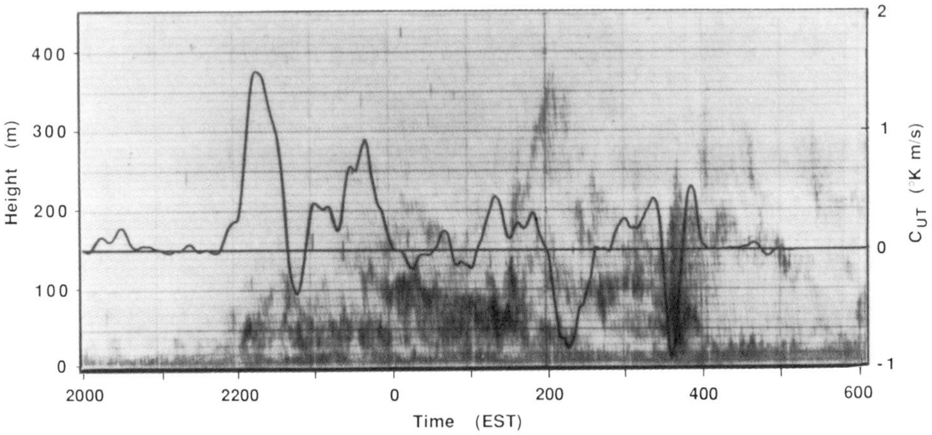

Fig. 6. Sodar trace and wind speed-temperature covariance for a typical night at the Walker Branch site. The covariance is measured at about the 40 m height of the sodar record.

We have examined the wind speed-ozone covariances; however these results are not easily interpreted. For example, if we associate the covariance with a vertical flux of ozone, then we can define a deposition velocity as the ratio of the covariance to the concentration; however we find that the average deposition velocities during breakdown events at the RAMS stations are about two orders of magnitude greater than what we would expect from eddy correlation measurements. These results suggest that we cannot simply relate the covariance of horizontal wind speed and ozone to the vertical flux of ozone. Rather, we would expect the vertical flux to be some small fraction of the horizontal flux.

CONCLUSION

The results of this study demonstrate that sporadic turbulent breakdowns are a consistent feature of the stable PBL over all types of terrain. We speculate that a significant portion of the nocturnal exchanges of heat, momentum, and atmospheric pollutants between the atmosphere and the ground surface occur during these breakdown episodes. Conventional PBL diffusion theory does not account for these episodes, and because of this, model predictions of nighttime vertical fluxes may be inaccurate. This may not be a serious problem for long-term air quality considerations; however, for short-term predictions of ground surface concentrations of hazardous materials, failure to account for PBL breakdowns is potentially dangerous.

ACKNOWLEDGEMENTS

This research was performed under an agreement with the U.S. Army Research Office, the U.S Department of Energy, and the National Oceanic and Atmospheric Administration under DOE interagency agreement no. 40-833-84.

References

Bean, B. R., A. S. Frisch, L. G. McAllister, and J. R. Pollard, 1973: Planetary boundary-layer turbulence studies from acoustic echo sounder and in situ measurements. *Boundary-Layer Meteorol.* **4**, 449-474.

Durst, C. S., 1933: The breakdown of steep wind gradients in inversions. *Quart. J. Roy. Meteorol. Soc.* **54**, 131-136.

Gifford, F. A., 1952: The breakdown of a low-level inversion studied by means of detailed soundings with a modified radiosonde. *Bull. AMS.* **33**, 373-379.

Gossard, E. E., J. E. Gayner, R. J. Zamora, and W. D. Neff, 1985: Fine structure of elevated stable layers observed by sounder and in situ tower sensors. *J. Atmos. Sci.* **42**, 2156-2169.

Harrison, R. M., C. D. Holman, H. A. McCortney, and J. F. R. Mclvenn, 1978: Nocturnal depletion of photochemical ozone at a rural site. *Atmos. Environ.* **12**, 2012-2026.

Kondo, J., O. Kanechika, and N. Yosuda, 1978: Heat and momentum transfers under strong stability in the atmospheric surface layer. *J. Atmos. Sci.* **35**, 1012-1021

Lyons, R., H. A. Panofsky, and S. Wallaston, 1964: The critical Richardson number and its implications for forecast problems. *J. Appl. Meteorol.* **3**, 136-142.

Mahrt, L., 1985: Vertical structure and turbulence in the very stable boundary layer. *J. Atmos. Sci.* **42**, 2333-2349.

Nappo, C. J., J. A. Herewehe, and A. M. Thompson, 1988: Observations of ozone profiles in a developing boundary layer. Proceedings of the Quadrennial Ozone Symposium, 1988, Gottengen, FRG, 4-6 August.

Nappo, C. J., 1990: Sporadic breakdown of the stable planetary boundary layer over simple and complex terrain. Submitted to *Boundary-Layer Meteor.*

Schiermeier, F. A., 1978: RAPS field measurements are in. *Environ. Sci. and Tech.* **12**, 644-651.

Winkler, K., 1980: Storungen der nachtlichen Grenzschicht. *Meteorol. Rdsch.* **33**, 90-94.

DISCUSSION

H. VAN DOP: Did you try to correlate ozone and temperature data? This may give some insight in the apparent discrepancy in deriving vertical fluxes of heat and ozone.

C.J. NAPPO: We have not looked at temperature-ozone co-variances. It will be interesting to do so especially if it will clear up the large discrepancy between the estimated fluxes of ozone and temperature.

G.A. DALU: Did you try to approach the problem with wave and wave-stability theory?

C.J. NAPPO: It is felt that the PBL breakdown process is intimately related to gravity waves and wave instabilities in the boundary layer. In this study, we have concentrated on evaluating breakdown characteristic. We are now satiesfied that breakdowns are a regular feature of the stable PBL. In future work we shall try to identify the causes of these events. At that time we shall use linear wave theory to examine PBL stability.

J.I. JOHN

The proposed model assumes a nocturnal surface inversion. If the atmosphere is not stable how would this affect the covariance parameter as an indicator of vertical flux?

C.J. NAPPO

For neutral and unstable thermal stratifications, the covariance of wind speed and temperature will be negative indicating a heat flux away from the surface. However, in such conditions the question of PBL breakdown is moot. We have examined covariances, as calculated in this paper, during convective conditions. During these conditions the signatures of discrete convective events are clear, and we assume these to be due to convective eddies.

DIFFUSION OF BUOYANT PLUMES FROM TALL STACKS IN THE CONVECTIVE LAYER

L. E. Venegas and N. A. Mazzeo

Dept. of Meteorology, University of Buenos Aires
Ciudad Universitaria-Pab II
1428- Buenos Aires, Argentina

INTRODUCTION

Since 1970, there has been significant progress in our knowledge of Planetary Boundary Layer structure and diffusion, specially for the daytime convective boundary layer (CBL) (see i.e. Briggs, 1985; Weil, 1988). The basis of most this progress is convective scaling which provides a means for describing the more important features of boundary layer turbulence and diffusion in convective conditions (Deardorff, 1970). Applications of convective scaling to diffusion in the CBL have also proliferated, beginning with Deardorff and Willis (1974) laboratory studies. The improved knowledge of CBL structure has motivated the development of dispersion models. Dispersion models in the CBL can be divided into: a) for passive releases (see i.e. Briggs, 1985; Weil, 1988) and b) for buoyant plumes (see i.e. Weil, 1988; Venegas and Mazzeo, 1989).

We derive alternative formulas for the lateral dispersion coefficient (σ_Y), the dimensionless cross-wind integrated concentration (Cy*) and the centreline ground-level concentration (Cc) following the empirical method applied by, for example, Briggs (1985), Weil et al. (1986) and Hanna and Paine (1987). These dimensionless scaling relations are developed using data from Kincaid and Bull Run power plants (Hanna and Paine, 1987). We obtain different expressions for different ranges of the dimensionless buoyancy flux parameter (F*). The obtained formulas are tested with data from Dickerson (Venkatram, 1980), Morgantown (Venkatram, 1980) and INCO (Venkatram and Vet, 1981).

FIELD DATA

Hanna and Paine (1987) present meteorological and power plants data corresponding to 187m stack at Kincaid and 244m stack at Bull Run. The available data are: the source emission rate (Q), the mean wind speed (u), the convective velocity scale (w*), the CBL depth (Zi), the plume buoyancy flux (F), the maximun concentration (Cc) and the cross-wind integrated concentration (Cy) on arcs at downwind distance (X) ranging from 1.0 to 50.0 km. Venkatram (1980) presents measurements corresponding to two 122m stacks, 60m apart at Dickerson and two 213m stacks, 76m apart at Morgantown. The available data are: Q, u, w*, Zi, F, Cc and X. Venkatram and Vet (1981)

Air Pollution Modeling and Its Application VIII, Edited by H. van Dop and D. G. Steyn, Plenum Press, New York, 1991

present the following data of 381m stack of International Nickel Company (INCO) at Sudbury: Q, u, w*, Zi, F, Cc and X.

From these available data we have calculated the following dimensionless parameters: the plume buoyancy $F^*=F/(uw^{*2}\ Zi)$ (Briggs, 1985); the downwind distance $t^*=xw^*/(uZi)$ and the ground-level crosswind integrated concentration $Cy^*=Cy.u.Zi/Q$. Since the crosswind integrated concentration at Morgantown, Dickerson and INCO is not reported, we have estimated it assuming a Gaussian lateral distribution and following Briggs (1985).

ANALYSIS

The first step has been to rank the Cy* data using an objective ranking scheme in which the buoyant effect on ground-level concentration distribution is considered. The Cy* data is classified according to F*. We have separated the cases that verify both, $F^* \leqslant 0.02$ and $Zs \leqslant 0.1Zi$ (18 cases) (Zs is the emission height). We have considered them as "not bouyant" because the emission takes place below the mixed layer and F* is too small.

From the remainder data, we have first analyzed the following groups: A: $0.02 < F^* \leqslant 0.06$ (110 cases); B: $0.06 < F^* \leqslant 0.1$ (71 cases)) and C: $0.1 < F^* \leqslant 0.2$ (72 cases). We have applied the discriminatory analysis between them and we have considered the Fisher Test of significance. The $\hat{F}$-value computed between A and B is 3.51 with 2 and 178 degrees of freedom (see Table 1). Since the $\hat{F}$-value is 4.61 for 1% significance level and 3.0 for 5% significance level, there is no clear difference between groups A and B. The analysis of groups B and C has given an $\hat{F}$-value of 12.53 with 2 and 144 degrees of freedom. It shows that these classes can be considered statistically different. In this way, we have obtained a limit of class at $F^*=0.1$.

The data with $F^*>0.2$ has been grouped in the following way: group D: $0.2 < F^* \leqslant 0.3$ (51 cases); E: $0.3 < F^* \leqslant 0.5$ (38 cases); F: $0.5 < F^* \leqslant 1.0$ (31 cases) and G: $1.0 < F^*$ (37 cases). The $\hat{F}$-values and the degrees of freedom obtained from the analysis of different combinations of the classes defined above are also shown in Table 1. The critical $\hat{F}$-values are the same as considered before. The $\hat{F}$-value obtained from the analysis of groups C and D is not enough to affirm that data of these classes belongs to different regimes. The $\hat{F}$-value between the groups D and E is smaller than both considered as limits. The difference between groups E and F is not clear, because the $\hat{F}$-value is between the two limits of significance. As group D has almost twice the number of cases of groups E and F, we have joint these groups into one named (E+F). Then we have applied the discriminatory analysis between D and (E+F) finding there is no significant difference between them (see Table 1). Finally we have analyzed groups F and G. It can be seen in Table 1 that they can be considered statistically different. This comparative analysis of the results given in Table 1 shows that groups C, D, E, and F can be joint together.

In this way we have found three well differentiated groups: Group I: $0.02 < F^* \leqslant 0.1$; Group II: $0.1 < F^* \leqslant 1.0$ and Group III: $1.0 < F^*$. These limits are similar to those mentioned by other authors (see i.e. Briggs, 1985).

Table 1. Values of $\hat{F}$-Fisher test of significance

Groups considered	A-B	B-C	C-D	D-E	E-F	D-(E+F)	F-G
$\hat{F}$-value	3.51	12.53	4.38	1.99	3.53	0.96	20.49
degrees of freedom	2;178	2;144	2;124	2;86	2;66	2;117	2;65

Ground-level Cy* parameterizations

The second step has been to derive alternative formulas for predicting dimensionless cross-wind integrated ground-level concentration (Cy*) for each Group (I, II and III). We have used only the data from Kincaid and Bull Run for the model development. The data from Morgantown, Dickerson and INCO have been left untouched until final model evaluation has been performed.

The general form of the empirical expression proposed for Cy*, include not only the main variables (F* and t*) in buoyant convective scaling but also the source height (Zs). It is given by:

$$Cy^* = \exp(- A\ (t^{*a}\ F^{*b}\ (Zs/Zi)^c\ )^B\ ) \qquad (1)$$

where a, b, c, A and B are parameters to be determined.

We give different values to a, b, and c and use the regression analysis to find A and B. A considerable number of different approaches has been comparatively evaluated in order to find the best fit. We have calculated the following statistical quantitative indices (Willmott, 1982): the mean bias error (MBE); the variance of the distribution of differences (S_d^2); the mean square error (MSE); the mean absolute error (MAE); the index of agreement (d); the mean fractional error (MFE); the dimensionless mean square error (DMSE) and the determination coefficient (r^2). We also report the magnitude and degrees of freedom associated with the $\hat{F}$-Fisher. The values of the parameters (a, b, c, A and B) included in (1) that belong to a selection of the most interesting expressions that better fit data of groups I, II and III are presented in Tables 2-4. The statistics associated to those fits and the number of cases are also summarized in Tables 2-4.

We should note that we have found $Cy^* > 1$ among the observed data of Group I ($0.02 < F^* \leqslant 0.1$) only. As Briggs (1985) has pointed out, because the smallest F* cases would be associated with smallest t*, it is not surprising that Cy* can be higher than 1. The analysis described here is based only on Cy* data less than unity.

Table 2. Parameters and quantitative measures of Cy* parameterization performance. Group I ($0.02<F^*\leqslant0.1$) N=98 cases (degrees of freedom: 1; 96).

a	b	c	A	B	MBE	S_d^2	MSE	MAE	d	MFE	DMSE	$\hat{F}$	r^2
1	0	0	0.765	-0.820	0.05	0.05	0.05	0.18	0.75	0.19	0.14	44.2	0.32
1	-1	0	40.755	-1.257	-0.01	0.04	0.04	0.16	0.85	-0.03	0.13	65.6	0.44
1	1	0	0.150	-0.492	0.06	0.06	0.07	0.21	0.58	0.23	0.18	18.2	0.16
1	-1	-1	40.004	-0.844	0.04	0.04	0.05	0.17	0.79	0.17	0.13	54.5	0.36

Table 3. Parameters and quantitative measures of Cy* parameterization performance. Group II ($0.1<F^*\leqslant1.0$) N=181 cases (degrees of freedom: 1; 179).

a	b	c	A	B	MBE	S_d^2	MSE	MAE	d	MFE	DMSE	$\hat{F}$	r^2
-1	1	0	2.360	0.370	-0.03	0.14	0.14	0.28	0.43	0.22	0.60	19.8	0.10
1	-2	0	2.410	-0.264	-0.03	0.14	0.14	0.29	0.42	0.22	0.60	18.4	0.09
1	-1/6	-2	15.915	-0.709	-0.04	0.10	0.11	0.23	0.70	0.11	0.45	115.2	0.38
1	1	0	0.778	-0.108	-0.03	0.17	0.17	0.31	0.14	0.24	0.70	1.9	0.01

Table 4. Parameters and quantitative measures of Cy* parameterization performance. Group III (1.0<F*) N=37 cases (degrees of freedom: 1; 35).

a	b	c	A	B	MBE	S_d^2	MSE	MAE	d	MFE	DMSE	$\hat{F}$	r^2
-1	1	0	3.397	0.244	-0.02	0.02	0.02	0.09	0.52	0.17	0.83	9.2	0.21
1	0	0	4.064	-0.282	-0.02	0.02	0.02	0.09	0.52	0.17	0.81	10.2	0.23
1	-1	-6	6.152	-0.253	-0.01	0.01	0.01	0.08	0.82	0.08	0.42	37.4	0.52
1	0	-1	5.537	-0.352	-0.01	0.01	0.01	0.08	0.66	0.15	0.67	19.0	0.35

We have also included in the analysis the form: $Cy^*=\exp[-A(F^*/t^*)^{B}]$ (a=-1; b=1; c=0) which has been considered by other authors (see i.e. Weil, 1988)(Tables 3 and 4).

The following expressions show the best fit to the observed data:

Group I: $Cy^*= \exp(-40.76\ (t^*/F^*)^{-1.257})$

Group II: $Cy^*= \exp(-15.915\ (t^*/F^{*1/6}(Zs/Zi)^2\ )^{-0.709})$ (2)

Group III: $Cy^*= \exp(-\ 6.152\ (t^*/F^*\ (Zs/Zi)^6\ )^{-0.253})$

This set of formulas verifies the boundary conditions for Cy*. The values of Cy* fall to zero at the point source (t*=0). Far from the source (large t*), Cy* approaches its 'well-mixed' limit of unity.

Lateral dispersion coefficient parameterizations

Since the observed lateral distributions of average plume concentration are roughly Gaussian, we can estimate the lateral dispersion coefficient (σ_y) from:

$$\sigma_y = Cy\ (2\pi)^{-1/2}\ Cc^{-1}$$

We have considered the observed data from Kincaid and Bull Run to estimate σ_y/Zi. Empirical σ_y/Zi parameterizations have been developed using the same procedure as that included in previous analysis of Cy* data. However, the general form proposed for σ_y/Zi is given by:

$$\sigma_y/Zi = D\ (t^{*f}\ F^{*g}\ (Zs/Zi)^h\ )^E \qquad (3)$$

where f, g, h, D and E are parameters to be determined. The expression (3) is similar to the forms of σ_y/Zi that have been found at present (see Briggs, 1985; Weil, 1988). We have given different values to f, g and h , and we have determined D and E by the regression analysis. According to previous results (see Weil, 1988), we have also tried f=2/3, g=1/3 and E=1. We have done a considerable number of combinations. Tables 5-7 present the values of the parameters (f, g, h, D and E) included in (3) of a selection of the forms that fit the observed data. The statistics associated to each fit and the number of cases considered are also included in Tables 5-7. It should be noted that the whole Group I has been considered in the analysis. It means that we have included every σ_y/Zi data from Bull Run and Kincaid that verifies $0.02 < F^* \leq 0.1$. This is the reason of having 144 cases in Group I instead of the 98 cases that have been considered before.

Table 5. Parameters and quantitative measures of σ_y/Zi parameterization performance. Group I ($0.02<F^*\leqslant 0.1$) N=144 cases (degrees of freedom: 1; 142).

f	g	h	D	E	MBE	S_d^2	MSE	MAE	d	MFE	DMSE	$\hat{F}$	r^2
1	0	0	0.332	0.740	-0.07	0.18	0.18	0.24	0.74	-10^{-3}	0.24	404.6	0.74
2/3	1/3	0	1.069	1.000	0.04	0.19	0.19	0.27	0.89	0.13	0.22	309.0	0.69
1	-1	0	0.055	0.650	-0.09	0.29	0.30	0.33	0.78	-10^{-3}	0.42	224.0	0.61
1	-1	-1	0.035	0.533	-0.13	0.41	0.41	0.38	0.65	10^{-3}	0.59	120.5	0.46
1	0	-1	0.107	0.656	-0.10	0.30	0.31	0.32	0.77	-10^{-3}	0.43	215.7	0.60
2/3	1/3	1	5.867	1.000	-0.03	0.19	0.19	0.29	0.90	-0.01	0.24	307.0	0.69

Table 6. Parameters and quantitative measures of σ_y/Zi parameterization performance. Group II ($0.1<F^*\leqslant 1.0$) N=181 cases (degrees of freedom: 1; 179).

f	g	h	D	E	MBE	S_d^2	MSE	MAE	d	MFE	DMSE	$\hat{F}$	r^2
1	0	0	0.331	0.874	-0.19	1.55	1.57	0.78	0.85	10^{-3}	0.36	475.8	0.73
2/3	1/3	0	0.918	1.000	0.08	1.52	1.52	0.82	0.84	0.20	0.31	246.2	0.48
1	0	1	1.058	0.640	-0.21	1.53	1.56	0.78	0.84	-10^{-3}	0.36	399.0	0.69
8	2	2	0.664	0.100	-0.05	1.43	1.42	0.75	0.87	0.06	0.31	275.0	0.61

Table 7. Parameters and quantitative measures of σ_y/Zi parameterization performance. Group III ($1.0<F^*$) N=37 cases (degrees of freedom: 1; 35).

f	g	h	D	E	MBE	S_d^2	MSE	MAE	d	MFE	DMSE	$\hat{F}$	r^2
1	0	0	0.292	1.000	-0.13	14.7	16.1	2.58	0.81	-0.27	0.48	15.0	0.98
2/3	1/3	0	0.926	1.000	-0.07	11.0	10.7	2.38	0.83	0.14	0.26	40.0	0.58
0.8	0.2	1	0.843	1.000	-0.56	11.3	11.3	2.34	0.85	-0.03	0.29	37.2	0.81
2/3	1/3	2	1.698	1.000	-0.41	11.4	11.3	2.52	0.85	-10^{-4}	0.28	37.1	0.54
2/3	1/3	1	1.305	1.000	-0.11	10.1	9.8	2.41	0.85	0.11	0.23	48.0	0.64
1	0	1	0.929	0.769	-0.43	11.5	11.4	2.33	0.85	10^{-3}	0.29	88.3	0.72

Analysis of the statistics associated with each attempt reveals that the following expressions may be considered as the best fits obtained:

Group I: $\sigma_y/Zi = 1.069\ (F^{*1/3} t^{*2/3})$ and $\sigma_y/Zi = 0.332\ t^{*0.74}$

Group II: $\sigma_y/Zi = 0.664\ F^{*.2}\ t^{*.8}\ (Zs/Zi)^{.2}$ (4)

Group III: $\sigma_y/Zi = 1.698\ F^{*1/3}\ t^{*2/3}\ (Zs/Zi)^{2}$

We have found little difference between the performance of both σ_y/Zi expressions included in (4), for Group I (see Table 5). Both expressions will be included in the model evaluation analysis for low buoyancy fluxes ($0.02 < F^* \leqslant 0.1$) in order to find if F^* may be removed. Another interesting result obtained from Table 5 is that including the source height into the expressions mentioned above does not clearly improve their statistic analysis.

We have derived the following expressions for the dimensionless centreline ground-level concentration (Cc*) assuming a Gaussian lateral distribution and combining expressions (2) and (4):

Group I:

$$Cc^{*}=((2\pi)^{1/2}\, 0.332\ t^{*0.74})^{-1} \exp(-40.76(t^{*}/F^{*})^{-1.257}) \qquad (5a)$$

$$Cc^{*}=((2\pi)^{1/2}\, 1.069\ F^{*1/3}\ t^{*2/3})^{-1} \exp(-40.76(t^{*}/F^{*})^{-1.257}) \qquad (5b)$$

Group II:

$$Cc^{*}=((2\pi)^{1/2}\, 0.664(F^{*}\ t^{*4}\ Zs/Zi)^{.2})^{-1} \exp(-15.915(t^{*}/F^{*1/6}(Zs/Zi)^{2})^{-0.709}) \qquad (5c)$$

Group III:

$$Cc^{*}=((2\pi)^{1/2}\, 1.698\ F^{*1/3}\ t^{*2/3}(Zs/Zi)^{2})^{-1}\exp(-6.152(t^{*}/F^{*}(Zs/Zi)^{6})^{-0.253}) \qquad (5d)$$

where Cc*= Cc.u.Zi /Q.

EVALUATION

Ground-level centreline concentrations (Cc) estimated using (5) have been evaluated using the reserved data from Morgantown, Dickerson and INCO, which have been untouched up to this point. For low buoyancy fluxes (Group I) we have also evaluated a Cc-expression which does not include F*. The expression for Cc* has been found from the Cy* (Table 2) and σ_y/Zi (Table 5) better fits from which F* has been removed. The third Cc* obtained for Group I is given by:

$$Cc^{*} = ((2\pi)^{1/2} 0.332\ t^{*0.74})^{-1} \exp(-0.765\ t^{*-0.82}) \qquad (6)$$

Figures 1a-1c contain the values of Cc(pred) predicted by expressions (5a), (5b) and (6) for Group I plotted as a function of Cc(obs) observed, respectively. Solid line represents one-to-one correlation and dashed lines represent a factor of 2 agreement with observations. In the three cases the 68% of the model predictions are within a factor of 2 of the observed concentration. The variance (r^2) explained by the model is 0.49 for expression (5a); 0.37 for (5b) and 0.35 for the predictions which do not include the buoyancy flux. We find no clear difference between the performance of the last two. The expression (5a) seems to give slightly better predictions.

Evaluation of Cc-expression for Group II (Figure 1d) shows that 73% of the predictions are within a factor of 2 of the observations. The limited amount of data and the large scatter have resulted in a relatively low variance ($r^2 \sim 0$). We have found no value of F*>1.0 within the available data of Morgantown, Dickerson and INCO.

Although we have found an acceptable magnitude of the "percent within a factor of 2" statistic in the model evaluation for low and moderate buoyancy fluxes, further work is necessary to explain the low variances and the large scatter. In this regard, emphasis must be placed on concentration uncertainties due to stochastic variability, input data uncertainties, potential model errors and the limited amount of data. We also think that the proposed expressions require further testing before making final recommendations.

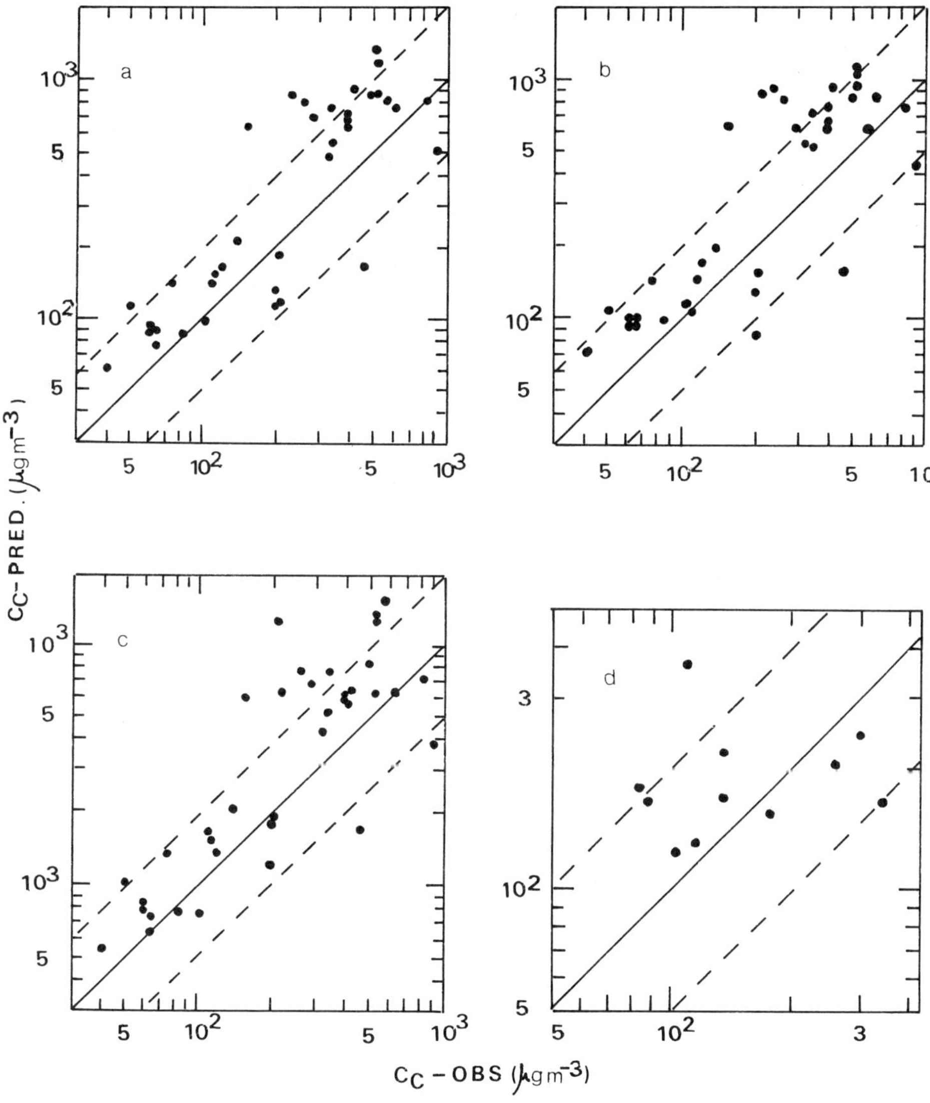

Fig. 1. Observed and predicted ground-level centreline concentrations ($\mu g/m^{-3}$)

a) Group I - expression (5a)
b) Group I - expression (5b)
c) Group I - expression (6)
d) Group II - expression (5c)

CONCLUSIONS

The model evaluation described in the last section suggest that for low buoyancy fluxes the main scaling parameters in the parameterization of ground-level cantreline concentration are t* and F*. The values estimated using the expression (5a) show a fair to good agreement with the observed data. However, at moderate ($0.1 < F^* \leqslant 1.0$) and high ($F^* > 1.0$) bouyancy fluxes, statistic analysis of the observed data reveals that including the source height in the parameterization of ground-level concentration may improve the predictions. For $0.1 < F^* \leqslant 1.0$ we have obtained an expression for the ground-level centreline concentration as a function of t*, F* and Zs/Zi (equation (5c)). The limited amount of data available for the evaluation does not allow us to make further recommendations. As mentioned before, for $F^* > 1.0$, ground-level centreline concentration also depends on t*, F* and Zs/Zi, and is given by equation (5d). However, the test of this expression is not included in the evaluation analysis because we have not found any cases with $F^* > 1.0$ within the power-plant data left for the evaluation. A better test of the proposed expressions is necessary.

REFERENCES

Briggs, G. A., 1985, Analytical Parameterizations of Diffusion: The Convective Boundary Layer, J. Climate Appl. Met, 24, 1167-1186.

Deardorff, J. W., 1970, Convective velocity and temperature scales for the unstable boundary layer for Rayleigh convection, J. Atmos. Sci., 27,1211-1213.

Deardorff, J. W. and Willis, G. E., 1974, Computer and laboratory modeling of the vertical diffusion of nonbuoyant particles in the mixed layer, Adv. in Geophysics,Vol. 18B, H. E. Landsberg and J. van Mieghem, eds., Acad. Press, 187-200.

Hanna, S. R. and Paine, R. J., 1987, Convective scaling applied to diffusion of buoyant plumes from tall stacks, Atmos. Environ., 21,2153-2162.

Venegas, L. E. and Mazzeo, N. A., 1989, Dispersion of buoyant particles within the convective boundary layer, in: Poceedings of the 8th World Clean Air Congress, The Hague, 11-15 September 1989, Elsevier Sci. Pub., Amsterdam.

Venkatram, A., 1980, Dispersion from an elevated source in a convective boundary layer, Atmos. Environ.,14,1-10.

Venkatram, A. and Vet, R., 1981, Modeling of dispersion from tall stacks, Atmos. Environ.,15,1531-1538.

Weil, J. C., 1988, Dispersion in Convective Boundary Layer, in: Lectures on Air Pollution Modeling. Venkatram, A. and Wyngaard, J. C.,Eds., Amer. Meteor. Soc., Boston., 167-227.

Weil, J. C., Corio, L. A. and Bower, R. P., 1986, Dispersion of buoyant plumes in the convective boundary layer, in: Fifth Joint Conf. Application of Air Pollution Meteorol. Amer. Meteor. Soc, 335-338.

Willmott, C. J., 1982, Some Comments on the Evaluation of Model Performance, Bull. Amer. Meteor. Soc., 63, 1309-1313.

LAGRANGIAN PARAMETERS FROM DIFFUSION EXPERIMENTS

Osvaldo L. L. Moraes and Gervasio A. Degrazia

Departamento de Fisica
Universidade Federal de Santa Maria
97119 Santa Maria, RS, Brasil

ABSTRACT

In this work we present an analysis of the measurements done in the Hanford-67 series. From Statistical Theory of Taylor to large and short diffusion times we obtain the Lagrangian parameters which are essential in diffusion studies. These parameters allow us to derive an expression for the Lagrangian autocorrelation function.

INTRODUCTION

Turbulent diffusion of a tracer which is emitted continously from a fixed point source is properly described by Taylor's diffusion equation

$$\overline{Y^2}(T) = 2\,\sigma_{vL}^2 \int_0^T\!\!\int_0^t R_{vL}(\xi)\, d\xi\, dt \ , \qquad (1)$$

where $\overline{Y^2}(T)$ is the mean square lateral displacement of a large number of particles from the mean particle trajectory, σ_{vL} is the Lagrangian variance of the lateral velocity components, R_{vL} is the Lagrangian autocorrelation coefficient and T is the time-diffusion. In order to derive Taylor's equation one assumes that the turbulence is homogeneous and stationary. A similar expression for the mean square vertical displacement $(\overline{Z^2})$ can be write too.

This equation can be used in two complementary ways to study of atmospheric diffusion. The first one has been to assume reasonable expressions for σ_{vL} (σ_{wL}) and R_L in order to predict $\overline{Y^2}$ $(\overline{Z^2})$; the other has been to deduce the Lagrangian expressions from measurements of $\overline{y^2}$ $(\overline{Z^2})$.

In this work we use the asymptotic behavior of the Taylor's model (to large and short diffusion times) to obtain, from several experiments data, collected from stable to near neutral conditions, σ_{vL} and t_L, where

$$t_L = \int_0^\infty R_{vL}(\xi)\, d\xi \ , \qquad (2)$$

is the Lagrangian integral time scale.

Using these parameters we derive an empirical expression for $\overline{Y^2}$ in terms of T/t_L. In consequence we obtain $R_L(\xi)$.

Air Pollution Modeling and Its Application VIII, Edited by H. van Dop and D. G. Steyn, Plenum Press, New York, 1991

THE PRACTICAL METHODS

For practical use to be made of diffusion formulas numerical values for the diffusion coefficients Y^2 and Z^2 must be determined. On the basis of available data, including Prairie Grass experiments, and guided by theoretical expectations, Pasquill sugested in an unpublished note in 1958 a practical scheme for the estimation of difussion which is particularly suitable for practical applications.

Pasquill's method of estimating diffusion can be sumarized in a families of curves to $\sigma_y^2(=\overline{Y^2})$ and $\sigma_z^2(=\overline{Z^2})$ for various stability categories, based on the smoke-plume elevation, H, and an angular spread, Θ. Furthemore the experimental data collected since the Prairie Grass experiments demonstrate that the standard deviation of the horizontal wind direction, σ_θ, for a short averaging time, can be related empirically to the measured values of plume width or to normalized average concentrations. On the basis of these data, Pasquill's stability categories can be related approximately in terms of measured values of σ_θ.

In the earlys seventy Pasquill (1971) sugested expressions for the diffusion parameters from Taylor's equations which can be write as

$$\sigma_y / \sigma_\theta U = f(T/t_L) , \tag{3}$$

where U is the mean wind speed and f is an universal function, submited as the same conditions of eq.(1), and with values sugested by Pasquill. In eq.(3) we assume that $\sigma_v T \simeq \sigma_\theta X$, where σ_v is the standard deviation of the turbulent wind speed in the transversal direction and X=UT. A similar expression for σ_z can be obtained. This equation is well suited to field use because a simple recording wind vane and anemometer erected at a proposed site can furnish climatological useful estimates of σ_θ and U rapidly.

PREDICTIONS OF σ_θ DONE WITH LAGRANGIANS PARAMETERS

Kampe de Feriet (1939) rewrite the Taylor's equation in the form

$$\overline{Y^2} = 2\,\sigma_{vL}^2 \int_0^T (T-\xi)\, R_{vL}(\xi)\, d\xi , \tag{4}$$

which with further simple transformation may be write as

$$\sigma_y^2 = 2\,(\sigma_{vL}^2/U^2)\, U^2 t_L^2 \int_0^{X/Ut_L} \{(x/Ut_L)-(\xi/t_L)\}\, R_L(\xi/t_L)\, d(\xi/t_L) , \tag{5}$$

and adimensionalized as

$$\frac{\sigma_y^2}{2\,\sigma_{vL}^2\, t_L^2} = \int_0^{T/t_L} \{(T/t_L) - (\xi/t_L)\}\, R_L(\xi/t_L)\, d(\xi/t_L) , \tag{6}$$

or

$$\sigma_y^2 / (2\sigma_{vL}^2 t_L^2) = f(T/t_L) , \tag{7}$$

which can be compared with eq.(3).

In eq.(7) the left-hand side is normalized by the intensity of turbulence and its scales. The right-hand side has f which is a function of T/t_L. In this form the Taylor's model suggest that the parameters to be used in the determination of σ_y are the Lagrangean quantities σ_{vL} and t_L.

The limits of eq.(7) for large and short diffusion times are

$$\sigma_y^2 = \sigma_{vL}^2 \; T^2 \; , \qquad T << t_L \; , \tag{8}$$

and

$$\sigma_y^2 = 2\;\sigma_{vL}^2 \; t_L \; T \; , \quad T >> t_L \; . \tag{9}$$

Eq.(8) corresponding to the limiting case of Taylor's formula for small diffusion times has been used by Frenkiel (1953). It undoubtedly gives realiable predictions for diffusion times up to at least a few minutes (More et al., 1985). At the opposite limit, eq.(9), corresponding to the case of large diffusion times, has been used to solve the problem of diffusion on scale ranging from continental to global (Smith 1983, Degrazia et al., 1988, Degrazia and Moraes, 1990).

The Hanford-67 series gives the values of the lateral dimension of the plume in several distances from the release point (σ_y vs X). The transformation of σ_y vs X for σ_y vs T is obtained by the relation X=UT. To estimate σ_{vL} and t_L we used the asymptotics relations (8) and (9) and the σ_y vs T curves. The importance of (8) in the diffusion atmospheric process was showed by Hay and Pasquill (1959) and corresponds to the linear region in the curve of σ_y vs T. The use of (9) was done in the runs which the parabolic behavior was observed in the graphics. With (8), (9) and the transformation X=UT we obtain

$$\sigma_{vL}^2 = \sigma_i \; U^2 \; / \; X_i \; . \tag{10}$$

and

$$t_L = \sigma_f^2 \; U \; / (2 \; \sigma_{vL}^2 \; X_f) \; , \tag{11}$$

where (X_i,σ_i) and (X_f,σ_f) are the initial and final points in each experiment where (8) and (9) were valid. With these equations it is possible to estimate σ_{vL} and t_L. These values are presented in table 1.

The comparision between the values of σ_y predicts by $\sigma_\theta U$ (Eulerian) and by σ_{vL} (Lagrangean) is observed in figures 1 and 2. Although there is considerable scatter in the data we can see that the predictions done by σ_{vL} (Fig.2) are better than the prediction done by $\sigma_\theta U$ (Fig.1).

The importance of the time scale t_L in the determination of σ_y is showed in Figure 3. When we ploted $\sigma_y/\sqrt{2}\sigma_{vL}t_L$ vs T/t_L the scattering in the foregoing figures disappears.

Table 1. Values of σ_{vL} and t_L as obtained from eq.(10) and (11). his the height of release (m), σ_{vL} is given in m/s and t_L in s.

Run	h	σ_{vL}	t_L	Run	h	σ_{vL}	t_L	Run	h	σ_{vL}	t_L
D2	2	0.288	251	U61	26	0.407	332	U75	26	0.261	223
U56	2	0.525	335	U62	26	0.645	70	U76	26	0.744	81
U59	2	0.281	670	U64	26	0.326	341	U78	26	0.476	87
U61	2	0.227	695	U68	26	0.684	295	U82	26	0.403	258
U62	2	0.332	200	U69	26	0.635	135	V5	26	0.503	167
U70	2	0.694	256	U70	26	1.189	156	V6	26	0.480	98
U58	26	0.564	168	U71	26	0.759	181	V7	26	1.155	91
U59	26	0.509	478	U72	26	0.691	249	V8	26	0.208	299

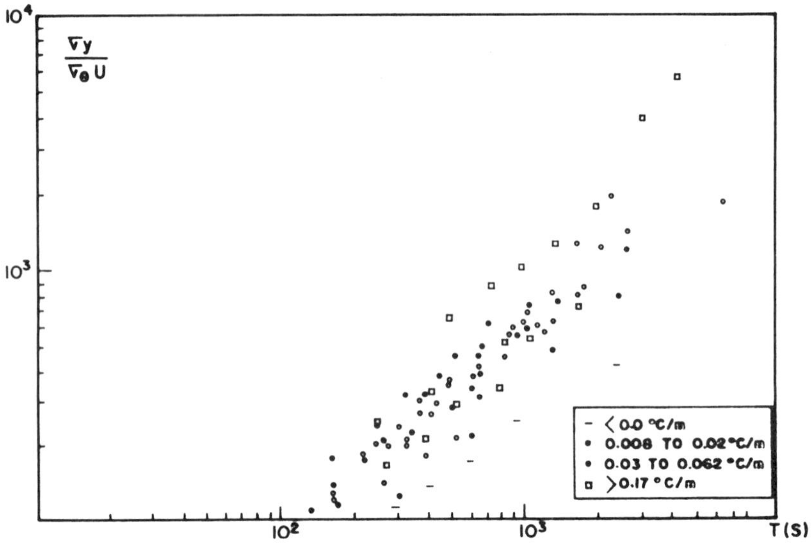

Fig.1 - Values of σ_y, normalized by $\sigma_\theta U$, as a function of the travel time.

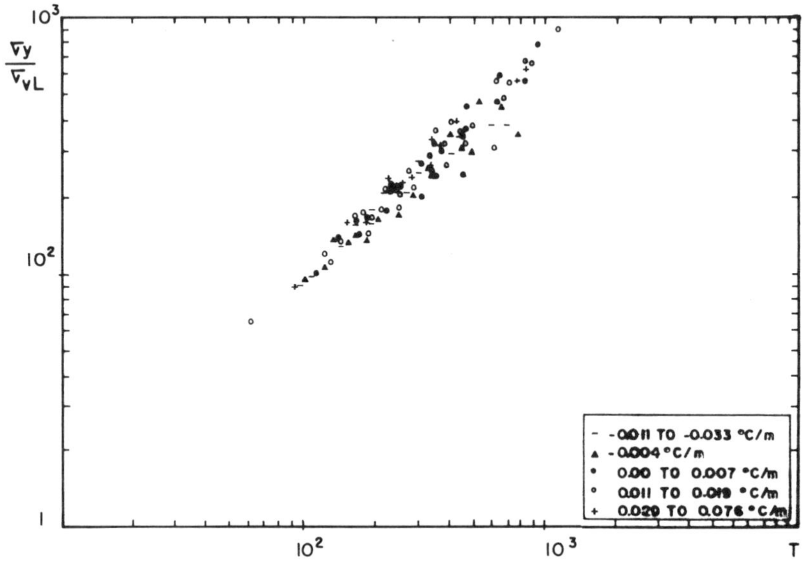

Fig.2 - Values of σ_y, normalized by σ_{vL}, as a function of the travel time.

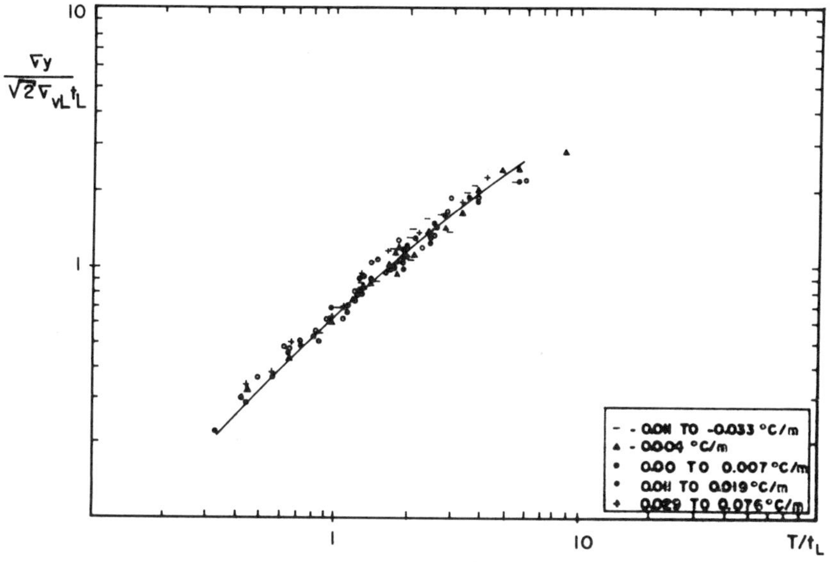

Fig.3 - Values of σ_y, normalized by $\sqrt{2}\sigma_{vL}t_L$, as a function of the travel time.

The observed behavior on the experimental points allow us to draw a fitting curve, represented in the Figure 3 by the full line, which is write as

$$f(T/t_L) = 1.6 \{ (T/t_L) + [\exp -1.03(T/t_L)] - 1\} \tag{12}$$

We are now able to derive the Lagrangian autocorrelation function. Using

$$(2\sigma_{vL}^2 t_L^2)^{-1} \frac{d^2\sigma_y^2}{d\xi^2} = R_L(\xi) , \tag{13}$$

and the eq.(12) we obtain

$$R_L(T/t_L) \simeq 1.7 \exp -1.03(T/t_L) , \tag{14}$$

an expression suitable different from that usually adoppt in the literatura.

CONCLUSIONS

Using the limits of Taylor's diffusion theorem for small and large times we are able to derive an empirical formula for the lateral diffusion parameter, adimensionalized by the Lagrangian velocity and the Lagrangian integral time scale. We can observe the important roles represnted by these quantities to specifies the diffusion parameters by analysis of atmospheric turbulence measurements. We can also verify that the practical methods based on Eulerian parameters are from doubts applicability to predict the diffusion parameters, at least to experiments analysed in this work.

REFERENCES

Degrazia, G.A. and Hugelman, C.P., 1988, Studio della dispersione di rilasci in quota nel PBL instable, Bollet. Geof., Marzo:98.

Degrazia, G.A. and Moraes, O.L.L., 1990, Numerical simulation of dispersion of pollutants in planetary boundary layers, Rev. Bras. Meteor., in press.

Frenkiel, F.N., 1953, Application of the statistical theory of turbulent diffusion to micrometeorology, J. Met., 9:52.

Hay, J.S. and Pasquill, F., 1949, Diffusion from a continuous source in relation to the spectrum and scales of turbulence, in: Adv. in Geoph., F.N. Frenkiel and P.A. Sheppard, ed., Academic Press.

Kampet de Feriet, M.J., 1939, Les fonctions aléatoires stationnaires et la théorie statisque de la turbulence homogène, Ann. Soc. Sci. Brux., 59:145

Moore, G.E. and Liu, M.K., 1985, Estimates of integral time scales from a 100m meteorological tower at a plain site, Bound.-layer Meteorol., 31:349.

Pasquill, F., 1971, Atmospheric dispersion of pollution, Quart. J. Roy. Met. Soc., 97:369.

Smith, F.B., 1983, Discussion on 'Horizontal diffusion on the atmosphere: A Lagrangean-dynamical theory' by F.A. Gifford, Atmosp. Environm., 17:194.

EDDY DIFFUSIVITY IN THE STABLE BOUNDARY LAYER

Gervásio A. Degrazia and Osvaldo L. L. Moraes

Departamento de Fisica
Universidade Federal de Santa Maria
97119 Santa Maria, RS, Brasil

ABSTRACT

A model for the vertical eddy diffusivity K_{zz} in the stable boundary layer is developed using the statistical diffusion theory and the concept of local scaling. The functional form of K_{zz} for large travel time is in agreement with the relations for K_M and K_H given by Mizuno et al.,(1983) and Sorbjan (1985).

INTRODUCTION

Turbulence and dispersion in stable nocturnal boundary layer are well less understood than in day-time convective boundary layer (CBL). The main reasons are the low intensity of turbulence fluctuations and the presence of gravity waves which make it difficult to interpret the data.

Despite this situation, data analysed by Sorbjan (1985) from the Minnesota experiment (Caughey et al., 1979) as well data presented by Nieuwstadt (1984) from Cabauw showed similarity structure of the stable boundary layer (SBL).

In the presence of moderate to strong geostrophic winds, when turbulence is continuous in the entire boundary layer, the use of the Local Similarity Theory (LST) developed by Nieuwstadt and Sorbjan allows us to describe various important turbulent parameters in the SBL.

This study assumes that Statistical Theory of Taylor describe vertical diffusion in the SBL and uses the LST to obtain the eddy diffusivities. The results here derived can be applied to the dispersion of pollutants as well as other species as for example humidity.

VERTICAL VELOCITY SPECTRUM

According to Olesen et al. (1984) the w-spectrum under stable conditions can be approximated by a smooth curve with one maximum only (when plotted as log nS(n) vs log n), where n is the frequency and S(n) is the spectral density. A suitable relationship for such spectra is

$$\frac{nS(n)}{u_*^2} = \frac{A\, f}{1 + Bf^{5/3}} \tag{1}$$

where $f(nz/U)$ is the reduced frequency, z is the height above the ground, U is the mean speed of the wind and u_* is the friction velocity. For the particular case of the Minnesota experiment, wher this w-spectrum was measured, Sorbjan (1986) obtained, for the reduced frequency at the peak $(f_m)_w$, the relation

$$(f_m)_w = 0.33(1+ 3.7z/L) \ , \qquad (2)$$

where L is the Monin-Obukhov parameter. Using the criteria $B/A=2.5\phi_\varepsilon^{2\,3}$ (Olesen et al., 1984), where ϕ_ε is the dimensionless dissipation rate, and $(f_m)_w=(1.5/B)^{3\,5}$ the constants A and B can be determined. With this procedure eq.(1) is rewriten as

$$\frac{nS_w(n)}{u_*^2} = \frac{3.807\ (nz/qU)}{1 + 9.518\ (nz/qU)^{5/3}}\left(\frac{\phi_\varepsilon}{q}\right)^{2/3} \ , \qquad (3)$$

where q is defined as $1+3.7(z/L)$.

The dimensionless vertical velocity variance can be calculated from the integral of $S_w(n)$ over all frequencies, resulting in

$$\sigma_w^2 = 2.3\ (1-z/h)^2\ u_*^2 \ , \qquad (4)$$

In the above expression, we use for the dimensionless dissipation rate the relation given by Sorbjan (1986) for the Minnesota data, i.e.

$$\phi_\varepsilon = 1.26\ (1+3.7z/L)(1-z/h)^3 \ . \qquad (5)$$

The eq.(4) is in good agreement with those given by Sawford (1985).

It is desirable for our future purposes to obtain the vertical velocity spectrum normalized by σ_w^2. From relations (3), (4) and (5) we find

$$\frac{nS_w(n)}{\sigma_w^2} = \frac{0.644\ f/(f_m)_w}{1 + 1.5\left[f/(f_m)_w\right]^{5/3}} \qquad (6)$$

which is the form of the spectral densities given by Sorbjan (1986). Fig. 1 is a plot of eq.(3) for six different heights and for the stability parameter h/L=2. The curves show that the function (3) shifts towards lower frequencies

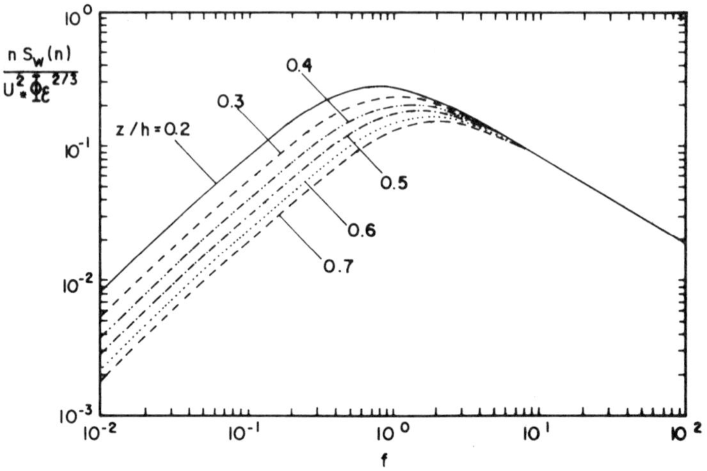

Fig.1 - Vertical velocity spectra for six differents z/h at a fixed h/L (=2) as obtained from eq.(3).

and as z/h decreases the area under the curves increases. This is in agreement with the fact that the vertical turbulent energy should decrease from the ground to the top of the SBL.

The parameter h/L is an indicator of the level of turbulence in the SBL. For a given Monin-Obukhov lenght $L > 0$ with h/L small ($h/L < 1$), the stability is close to neutral, while for large values of h/L ($h/L > 6$) the fully turbulent state of the atmosphere tends to dissapear with the the increasing of the intermitency phenomena as poited out by Holtslag and Nieuwstadt (1986).

TIME-DEPENDENT VERTICAL DIFFUSIVITY

Taylor (1921) poited out that turbulent diffusion differs in the near and far regions of an elevated source. In the proximity of the source the particles of pollutants retain memory of their initial turbulent environment. For far regions this memory effect is lost with the particles following only the local properties of the turbulence.

The gradiente-transfer theory (K-theory) which is used in the diffusion equation for the description of mean concentration field of contaminants fails close to point sources and is valid only for large travel time. The introduction of a diffusivity as a function of time travel permits to describe diffusion in the near and far fields of an elevated source. Runca (1982) has already suggested a diffusivity coeficient K_{zz} as a function of distance of travel and of the vertical coordinate.

A K_{zz} time-dependent can be obtained from the equation for the standard deviation σ_z given by Pasquill and Smith (1983)

$$\sigma_z^2 = \sigma_w^2 t^2 \int_0^\infty S_w^N(n) \left[\frac{\sin(n\pi t/\beta_w)}{(n\pi t/\beta_w)}\right]^2 dn \ . \tag{7}$$

Here $S_w^N(n)$ denote the normalized vertical energy spectra for σ_w^2 and β_w is defined as the ratio of the Lagrangian to the Eulerian time scale. The quantity β_w increases withe the increasin of the stability (Moore et al., 1985). Wantel and Kofoed-Hansen (1962) have show that $\beta_w = U\sqrt{\pi}/4\sigma_w$. The statistical diffusion theory shows that an expression for the time-dependent vertical diffusivity coefficient K_{zz} is equal to $(1/2)(d\sigma_z^2/dt)$, so that from eq.(7) we obtain

$$K_{zz} = \frac{\sigma_w^2 \beta_w}{2\pi} \int_0^\infty S_w^N(n) \frac{\sin(2\pi nt/\beta_w)}{n} dn \ . \tag{8}$$

Substituing eq.(6) in (8) with σ_w from (4), eq.(8) becomes

$$\frac{K_{zz}}{u_* h} = \frac{0.207(1-z/h)(z/h)}{1 + 3.7(z/h)(h/L)} \int_0^\infty \frac{\sin\left\{5.564\frac{(1-z/h)}{(z/h)}\left[1+3.7(z/h)(h/L)\right]X'n'\right\}}{\left[1 + n'^{5/3}\right]n'} dn' \tag{9}$$

where $X' = xu_*/hU$ represents a non-dimensional travel distance and h/u_* is a time scale of turbulence. We notice that the adimensional vertical diffusivity becomes a function of the quantities X', z/h and h/L.

For large travel times, when the concentration distribution is just a function of turbulence, the filter function in eq.(8) is very narrow and selects $S_w^N(0)$ so that for $t \to \infty$ eq.(9) can be approximated as

$$K_{zz}/u_*h = \frac{0.33(z/h)(1-z/h)}{1+3.7(z/h)(h/L)} \; . \tag{10}$$

Eq.(10) can be compared with the relation for momentum and heat eddy viscosities given by Sorbjan (1985), i.e.

$$K_M/u_*h = \frac{0.35(z/h)(1-z/h)}{1 + 4.7(h/L)(z/h)} \; , \tag{11}$$

$$K_H/u_*h = \frac{0.35\ (z/h)(1-z/h)}{0.74+4.7(h/L)(z/h)} \; . \tag{12}$$

Eq.(10) has the same functional form as eqs.(11) and (12), which was used by Mizuno et al., (1983) to calculate the concentration of contaminants in the SBL.

A SIMPLER EQUATION FOR K_{zz}

Venkatram et al., (1984) adopt the following model for the vertical dispersion of elevated releases in the SBL

$$\sigma_z = \frac{\sigma_w t}{\left[1 + t/(2t_{Lw})\right]^{1/2}} \; , \tag{13}$$

where t_{Lw} is the vertical Lagrangian time scale.

Hanna (1981) showed that the Eulerian time scale is related to λ_m through the relation $t_E = 0.166\ \lambda_m/U$. For the Minnesota case we can get an expression for t_{Ew}. Using the eq.(4) a formula for t_{Lw} in the SBL is

$$t_{Lw} = \frac{0.145(z/h)}{\left[(1-z/h)\right]\left[1+3.7(h/L)(z/h)\right]u_*} \; . \tag{14}$$

If the functional form (13) is derived with respect to the time, and the eq.(14) is used, a relation for K_{zz} simpler than eq.(9) can now be obtained

$$\frac{K_{zz}}{u_*h} = \frac{0.667\left[(1-z/h)\right]^2 X'(z/h)\left\{0.29(z/h)+0.5X'\left[1+3.7(h/L)(z/h)\right](1-z/h)\right\}}{\left\{0.29\ (z/h) + X'\ \left[1+3.7(z/h)(h/L)\right](1-z/h)\right\}^2} \; . \tag{15}$$

For large travel time, eq.(15) gives

$$K_{zz}/u_*h = \frac{0.33(1-z/h)(z/h)}{1+3.7(z/h)(h/L)} \; , \tag{16}$$

in excellent concordance with eq.(10).

DISCUSSION

The behavior of the time and space dependent vertical diffusivity calculated from eq.(9) is exhibited in Fig. 2 for the particular heights z/h=0.2;0.5;0.8. The decrease of K_{zz} with height should be interpreted in the light of the gradual decrease of turbulence towards zero near the top of the SBL as demonstraded by σ_w-profile.

As can be seen from Fig.3, the curves of K_{zz} given by eq.(9) compare well with the curves given by eq.(15) for all range of X'. The comparision of profile (10) with (11) and (12) as illustrated in Fig. 4 indicates that in the SBL a good correlation exist between turbulent fluctuation of concentration and those of wind speed and temperature.

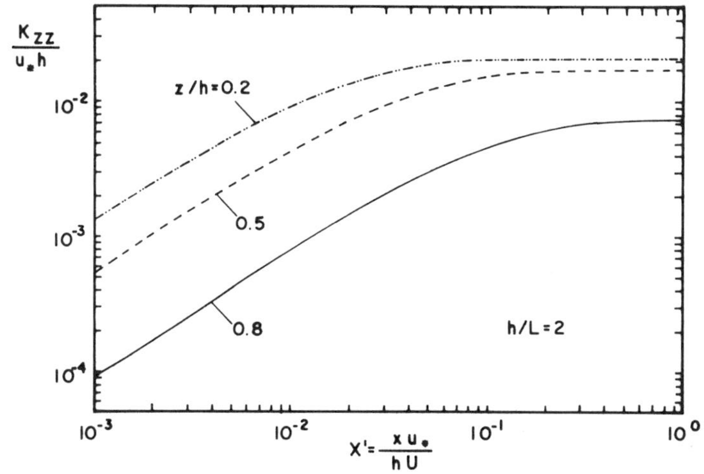

Fig.2 - Computed diffusion coefficient as given by eq.(9)

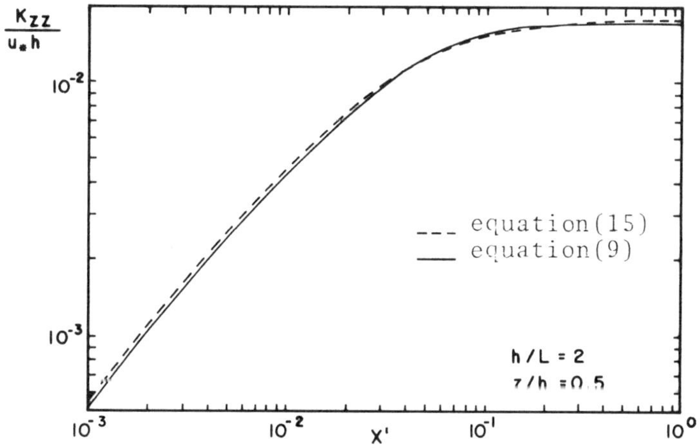

Fig.3 - Comparision between diffusion coefficient as given by eqs.(9) and (15)

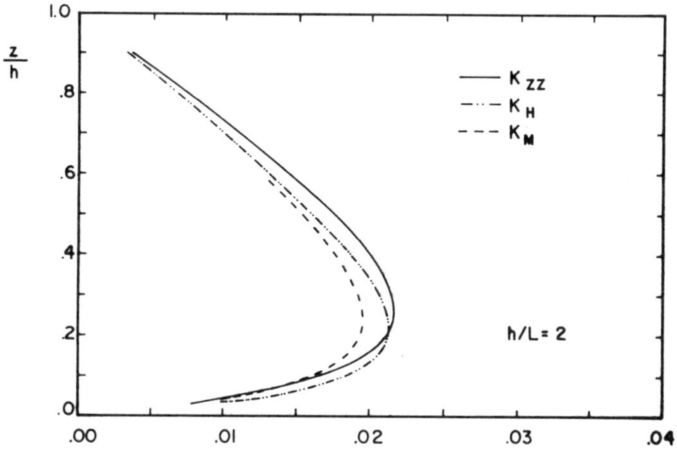

Fig.4 - Comparision among the profiles predicted by (10) and those given by Sorbjan for K_M and K_H.

CONCLUSIONS

We have presented a model for K_{zz} in the SBL valid in the near and far fields of an elevated source using the statistical theory and local similarity theory. The expression for K_{zz} besides being dependent on z/h and xu_*/Uh, is also dependent on z/L.

One partial test for this model is the comparision of the values of K_{zz} for large X' with the values of K_M and K_H as given by Sorbjan and Mizuno et al. This comparision showed good agreement between the model here presented and those proposed from experimental results.

One important problem in the first-order closure problem is finding a rational parametrization for K_{zz}, for example as a function of height, stability and others parameters. The K_{zz}-profile shape (9) is a proof that statistical theory can yields this parametrization when turbulence is continuous in the SBL and its structure is described by similarity coordinates as h/L, z/h and h/u_*.

REFERENCES

Caughey, S.J., Wyngaard, J.C. and Kaimal, J.C., 1979, Turbulence in the evolving stable boundary-layer, J. Atmos. Sci., 36:1041.

Hanna, S.R., 1981, Lagrangian and Eulerian time-scale relations in the day-time boundary layer, J. Appl. Meteor., 20:242.

Holtslag, A.A.M. and Nieuwstadt, F.T.M., 1986, Scaling the atmospheric boundary layer, Bound.-Layer Meteorol., 36:201.

Mizuno, T., Yokoyama, O. and Yasuraoka, A., 1983, On the vertical diffusion from elevated sources in the atmospheric boundary layer, J. Meteor. Soc. Japan, 61:414.

Moore, G.E. and Liu, M.K. 1985, Estimates of integral time scales from a 100m meteorological tower at a plain site, Boun.-Layer Meteorol., 31:349.

Nieuwstadt, F.T.M., 1984, The turbulent stucture of the stable nocturnal boundary layer, J. Atmos. Sci., 41:2202.

Olesen H.R., Larsen, S.E. and Hojstrup, J., 1984, Modelling velocity spectra in the lower part of the planetary boundary layer, Bound.-Layer Meteorol., 29:285.

Pasquill, F. and Smith, F.B., 1983, Atmospheric Diffusion, Ellis Horwood Limited, Chichester.

Runca, E., 1982, A practical numerical algorithm to compute steady-state level concentration by a K-model, Atmos. Environm., 16:75.

Sawford, B.L., 1985, Lagrangian statistical simulation of concentration mean and fluctuations fields, J. Clim. Appl. Meteor., 24:1152.

Sorbjan, Z., 1985, On similarity in the atmospheric boundary layer, Bound.-Layer Meteorol., 34:377.

Sorbjan, Z., 1986, Local similarity of spectral and cospectral characteristic in the stable-continuous boundary layer, Bound.-Layer Meteorol., 35:257.

Taylor, G.I., 1921, Diffusion by continuous movements, Proc. London Math. Soc., 20:196.

Venkatram, A., Strimatis, D. and Dicristofaro, D., 1984, A semiempirical model to estimate vertical dispersion of elevated releases in the stable boundary layer, Atmos. Environm., 18:923.

Wandel, C.F. and Kofoed-Hansen, O., 1962, On the Eulerian-Lagrangean transform in the statistical theory of turbulence, J. Geophys. Res., 67:3089.

DEVELOPMENT OF THE ADAPTIVE VOLUME PUFF MODEL

R. Bornstein and
Z. Boybeyi

Department of Meteorology
San Jose State University
San Jose, CA, U.S.A.

F. Ludwig

SRI International
Menlo Park, CA
U.S.A.

INTRODUCTION

The Lagrangian Adaptive Volume Plume Model (AVP) of Ludwig et al (1989) approximates a continuous plume by a series of discrete volumes that change shape in response to vertical wind shear and changes of atmospheric stability with height (Fig. 1). Each volume is described by five Lagrangian points at different altitudes. Horizontal and temporal changes in meteorological parameters are accounted for as the five points move and diffuse independently according to local meteorological conditions along their trajectories; thus each adaptive volume element consists of four segmented parts.

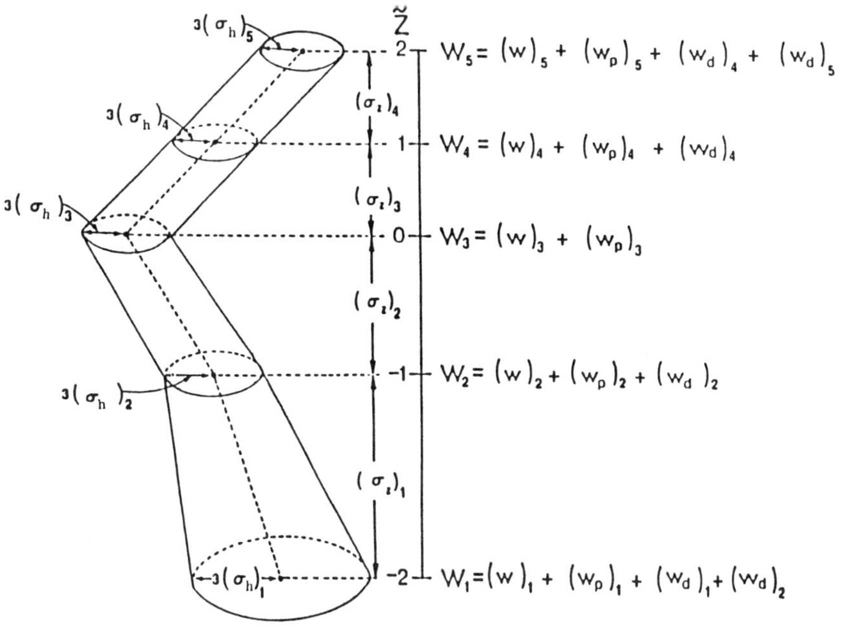

Fig. 1. Schematic representation of one adaptive volume; see text for explanation of symbols.

Air Pollution Modeling and Its Application VIII, Edited by H. van Dop and D. G. Steyn, Plenum Press, New York, 1991

Each volume has the following distribution of concentration:

$$C = \frac{Q}{(2\pi)^{3/2}\,\sigma_h^2\sigma_z} \exp\left[-\frac{1}{2}\left(\frac{s}{\sigma_h}\right)^2\right] \exp\left(-\frac{\tilde{z}^2}{2}\right), \tag{1}$$

where σ_y and σ_x have been assumed equal to an horizontally isotropic value of σ_h, and thus where horizontal concentration is a function of horizontal distances from the puff center. The dimensionless height $\tilde{z}$ is the number of "equivalent" standard deviations between the center of the adaptive volume and height z as follows:

$$\tilde{z}(z) = \int_{z_0}^{z} \frac{d\eta}{\sigma_z(\eta)}. \tag{2}$$

METHOD

Original Model

The horizontal motion of each of the five points in one adaptive volume is calculated using spatially interpolated meteorological conditions at two-minute time-steps. Vertical advection (Fig. 1) arises from the sum of the specified vertical air motion w, calculated rate of plume rise w_p (as discussed in Ludwig et al, 1989), and vertical diffusion velocity w_d. To calculate w_d, the model incorporates the parameterization of Ludwig et al (1977) for σ_z in terms of both the horizontal travel distance d of each of the five points defining an adaptive volume and Pasquill-Gifford stability class j as follows:

$$\sigma_z(d,j) = \alpha_j\, d^{\gamma_j}. \tag{3}$$

The derivative of (10) with respect to d is used to obtain w_d as follows:

$$(w_d)_i = -V_h\left(\frac{\partial\sigma_z}{\partial d}\right)_i, \quad i \neq 3.$$

The horizontal diffusion coefficients σ_h at the five points are also functions (via Ludwig et al, 1977) of the total horizontal travel distance of its corresponding point, i.e.,

$$\sigma_h(d,j) = \lambda_j\, d^{0.9}. \tag{4}$$

Values obtained from (3) and (4) approximate the Turner (1970) diffusion curves.

Modified Model

The main modifications made to the AVP model involve techniques developed to locate systematically (at each time step) a previously emitted plume or puff. These locations are used to establish a fine mesh concentration computational grid. Definition of the grid requires several steps, including determination of an estimate of the "absolute" concentration field based on a "relative" concentration field.

An approximate concentration is calculated for each cell by taking a weighted sum of all points within that cell. The weight F assigned each point is approximately proportional to the mass associated with it and is defined by

$$F = \frac{f(\sigma_z)}{\sigma_h^2},$$

where σ values are those associated with the point and f is the value of the Gaussian normal function at 0, 1, or 2 σ from the mean, i.e., 1.0, 0.37, and 0.13, respectively. The relative concentration for a grid cell is then the sum of the weights associated with each point within it.

RESULTS

The current study only considers a specified constant prescribed for effective stack height (for ease of comparison with analytical Gaussian solutions), so some source parameters need not be specified. The following constant source information was specified for all simulations presented here: x_o = 100m, y_o = 2.5 km, Q_e = 4.8 kg/s , and H = 250m.

Case A. Steady-State Uniform Meteorology

An arbitrary specified uniform steady-state wind with a westerly flow of 0.4 ms was specified for a "short term" (10 minutes) release of a single adaptive volume into a moderately unstable (Class B) temperature profile (γ = 1.8 C/100m); results were compared with analytical Gaussian solutions give by using σ_z and σ_h values approximated by (3) and (4), respectively. A low wind speed was chosen, as such conditions are difficult for finite-difference Eulerian models to simulate, and the AVP Model will ultimately be embedded in such a model.

Comparison of Lagrangian AVP surface concentration values with those from the analytical solution given shows exact agreement beyond the 1 km distance. The lack of agreement near the stack arises because AVP puffs only extend $\pm 2\sigma_z$ vertically about its center point. A puff thus does not contribute to surface concentrations until σ_z exceeds half the centerpoint height.

Case B. Speed Shear

To study the effect of a speed shear with height, the u component of an Ekman profile was specified.

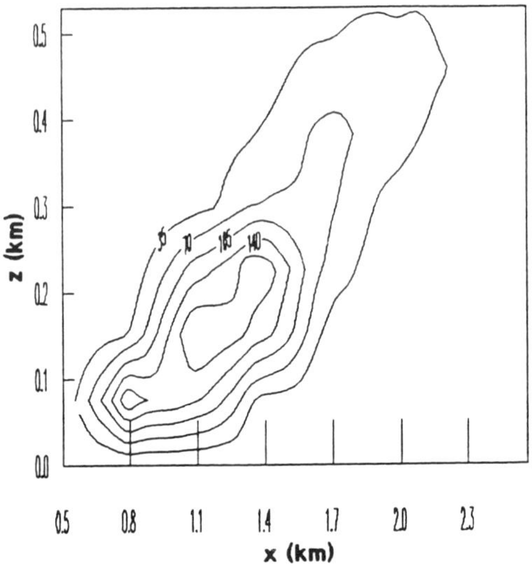

Fig. 2. Longitudinal axial concentration (10^{-4}g/m^3) cross section after 60 minutes travel with Case B positive wind speed shear.

The advection and diffusive growth of each of the four segments of the adaptive volume after 40 minutes travel reflects the effects of the specified speed shear. The puff axis is no longer vertical, but curved and tilted downstream with height. While the middle point of the puff advects at the same rate as in Case A, the shear now causes the upper segments to move faster, and as σ is dependent on d to diffuse more rapidly than the lower two segments. If a more realistic stability stratification was used, so that turbulent diffusion proceeded more slowly aloft, away from the heated surface, this effect would not be so pronounced. The upwind displacement of the upper two center points relative to the tilted dashed line that represents constant shear results from the less than linear nature of the shear produced by an Ekman spiral.

While wind shear produces a downstream (longitudinally) tilted puff with a peak concentration at H, a boundary problem ultimately creates an erroneous peak concentration (before surface reflection) below H (Fig. 2). The vertical gradient from this erroneous maximum concentration is larger downward than upward, as concentration normally increases towards H. A time-dependent σ formulation would correct this artifact.

Case C. Directional Shear

To produce a directional shear with height, the v component of the Ekman spiral of Case B was evaluated. All other input parameters are again identical to those of Case A, including a uniform u-component.

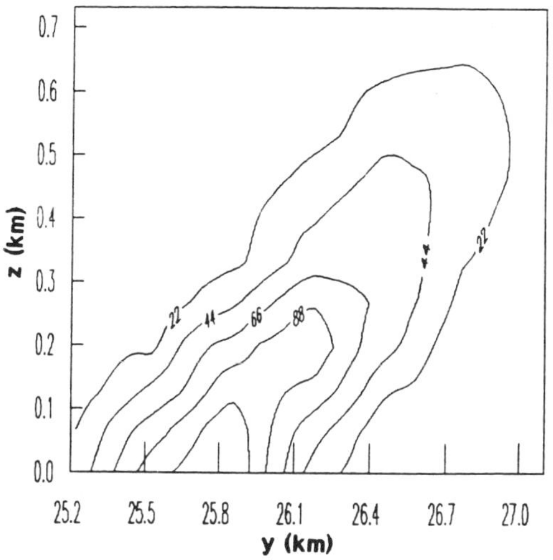

Fig. 3. Lateral concentration (10^{-4} g/m^3) cross section after 60 minutes travel (1.6 km east of source) with Case C positive wind directional shear.

All concentration values for Case C (before effects of surface reflection) at any given travel time are less than those of Case A, because the total horizontal travel distance of the puff center is increased by the lateral displacement. Increased travel distance increases both σ_h and σ_z, thus producing the lower concentrations and producing an earlier surface impaction (almost 40 versus almost 50 min).

While concentration patterns (not shown) are still symmetric in the horizontal (as in Cases A and B) and longitudinal planes (as in Case A), the directional shear produces a tilted (in the direction of the v-component of the flow) vertical puff axis. After 40 minutes, the peak concentration is at H, when the puff first impacts the surface. After 60 minutes, the peak concentration has moved to the surface (Fig. 3), due to initiation of the well mixed layer (as in Case A). In this case, however, the layer forms earlier due to the increased travel distance.

Case D. Subsidence Velocity

The effects of subsidence on a dispersing puff has been studied by specifying vertical wind components that vary linearly with height. All other variables are again identical to those of Case A.

The nonlinear subsidence of the puff center point is a combination of the constant u and the linearly decreasing w(z), as the point approaches the surface. The puff, of course, hits the ground earlier than in Case A (within 40 rather than 50 minutes). The concentration distribution (not shown) in both the horizontal and vertical planes through the puff center is symmetric before surface reflection, as in Case A, but concentrations are greater. The decreasing subsidence with height compresses the volumes of the four puff segments unequally as they sink to the surface. Subsidence produces a well mixed pollutant layer before it appears in Case A, because surface reflection is initiated earlier. Surface concentrations, and those at the level of the maximum, are both higher after 60 minutes in Case D than in Case A for the reasons cited above.

Case E. Fumigation

Fumigation conditions were investigated by creating a growing layer of depth $h_c(t)$ in which the Pasquill-Gifford stability class B conditions of the previous cases were replaced by class A conditions. The change was initiated after 20 minutes. The depth of the extremely unstable layer increased uniformly 50m every 10 minutes until reaching 250m after 70 minutes. All other variables are identical to those of Case A.

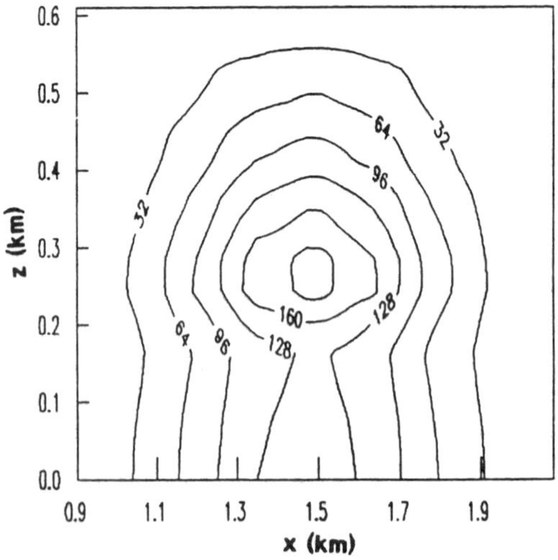

Fig. 4. Longitudinal axial concentration (10^{-4}g/m^3) cross section after 60 minutes travel with Case E fumigation conditions.

The Longitudinal axial concentration cross section at 60 minutes (Fig. 4) also shows that Case E concentrations are greater than those of Case A. Figure 4 also shows the effects of fumigation on puff shape through its lowest 200m. The cross section after 110 minutes likewise shows Case E concentrations to be less at H relative to those shown for Case A. Fumigation also dispersed the peak concentrations that had been found near the surface for Case A, producing peak concentrations at H for Case E.

REFERENCES

Ludwig, F., Salvador, R., and Bornstein, R., 1989: An adaptive volume plume model. Atmos. Environ., 23:127.

Ludwig, F., Gasiorek, L., and Ruff, R., 1977: Simplification of a Gaussian puff model for real-time mini-computer use. Atmos. Environ., 11:431.

Turner, D.B., 1970: Workbook of Atmospheric Dispersion Estimates. Environmental Protection Agency, Pub. No. AP-26.

THE OPERATIONAL STREET POLLUTION MODEL (OSPM)

Ole Hertel and Ruwim Berkowicz

Environmental Research Institute, Frederiksborgvej 399
4000 Roskilde, Denmark

Steinar Larssen

Norwegian Institute for Air Research, Postbox 64
N-2001 Lillestrøm, Norway

1. INTRODUCTION

A new model for estimating dispersion of air pollution from traffic in street canyons has recently been developed (Hertel and Berkowicz, 1989a). This model is named the Operational Street Pollution Model (OSPM). OSPM contains also a chemical submodel describing formation af NO_2 in street canyons (Hertel and Berkowicz, 1989b).

The dispersion model was tested on data from Denmark, Norway and the Netherlands. The results showed good correlation with the observed concentrations.

The main improvements in prediction of dispersion in street canyons have been achieved by the introduction of a better description of wind direction and wind speed dependence, compared with e.g. the widely used STREET-model (Johnson et al., 1973).

In this paper, only a short description of the model is presented. For details, the readers are referred to Hertel and Berkowicz (1989a,b,c). The performance of OSPM, considering both dispersion and chemistry, is illustrated in the present report by several examples. Model implications are also discussed.

2. DESCRIPTION OF THE MODEL

The concentration of a pollutant in a street canyon consist of two main contributions: the street contribution, which is due to the emission from traffic in the street; and the background contribution, which is the contribution from all other sources in the area, such as other streets, industry and power plants etc. The street contribution is in OSPM described by two subcontributions: the direct contribution from the traffic and the contri-

Air Pollution Modeling and Its Application VIII, Edited by H. van Dop and D. G. Steyn, Plenum Press, New York, 1991

bution from recirculation in the street canyon. The total concentration is thus

$$C_{tot} = C_d + C_r + C_{bk} \quad (1)$$

where C_d is the direct contribution, C_r is the contribution from recirculation in the street canyon and C_{bk} is the background contribution from other sources in the area.

We model the dispersion of traffic pollution by assuming a straight street with buildings on both sides. We define line sources such that they are perpendicular to the wind direction in the street. The basic equation for computing the direct contribution is the Gaussian formula for ground level concentrations from an infinite line source:

$$\delta C_d = 2/\pi \frac{\delta Q}{u_b \cdot \sigma_z(x)} \quad (2)$$

where u_b is the wind speed at street level, $\sigma_z(x)$ is the vertical dispersion parameter at a downwind distance x from the line source. δQ and δC_d ifinitisimal emissions and concentrations from a number of line sources in the street. Equation (2) is integrated along the wind path. When the distance to the receptor is larger than the distance at which σ_z exceeds the height of the buildings in the street, an extra exponential decay term is added in order to describe the removal of pollutants from the street canyon.

The vertical dispersion parameter is given by

$$\sigma_z(x) = \sigma_w \frac{x}{u_b} + h_o \quad (3)$$

where σ_w is the standard deviation of the turbulent velocity, and h_o is the initial dispersion. We use h_o = 2 m.

The turbulence in the street consists of two contributions: the turbulence generated by wind and the turbulence generated by traffic in the street. The ambient turbulence, i.e. that generated by wind, is in OSPM modelled as a certain fraction of wind speed at street level. We use 10%.

$$\sigma_w = \left[(\alpha \cdot u_b)^2 + \sigma_{wo}^2 \right]^{1/2} \quad (4)$$

where α = 0.1 and σ_{wo} is the traffic generated turbulence.

The traffic generated turbulence is an important parameter. At low wind speeds (below ca. 2 m/s) it makes the main contribution to the turbulence in the street and thus determines the concentration of pollutants. A simple model for σ_{wo} is presented in Hertel and Berkowicz (1989c).

The contribution from recirculation is computed using a simple box model. The width of the recirculation zone is set equal to the least of the street width and twice the height of the characteristic buildings in the street. Ventilation of the recirculation zone is modelled by turbulent exchange at the top of the recirculation zone and an advective exchange at the sides of the zone. The latter is only active when the width of the recirculation zone is smaller than the width of the street canyon (Hertel and Berkowicz, 1989a).

There is no contribution from recirculation for parallel wind directions. Recirculation disappears also at low wind speeds. This is modelled by a linear decrease of the width of the recirculation zone for wind speeds below 2 m/s.

In order to simulate the effect of fluctuations in wind direction, an integration of the resulting concentrations with respect to the average hourly wind direction fluctuation around the mean value is introduced (Hertel and Berkowicz, 1989c). The integration over wind direction has a pronounced effect on the calculated concentrations for low wind speeds and wind directions close to parallel.

In Fig. 1, model results for different wind speeds are shown as functions of the wind direction. Calculations are performed with parameters for Vesterbrogade in Copenhagen. It is seen from the figure that for low wind speeds (< 2 m/s) the dependence on the wind direction is small. For higher wind speeds the wind direction is very important, and the highest concentrations are found for parallel wind flow (70 and 250 degrees). High concentrations are found also in the leeward side of the street. Lowest concentrations are found in the windward side. The difference between concentrations on the leeward side and the windward side predic-

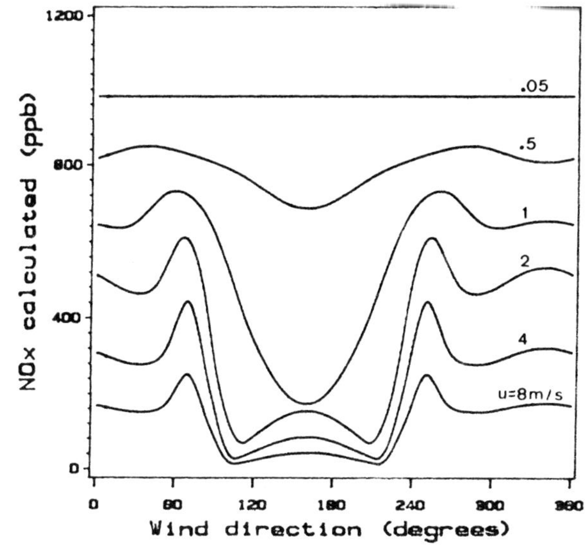

Fig. 1. NO_x concentrations as a function of wind direction for different wind speeds (0.05, 0.5, 1.0, 2.0, 4.0 and 8.0 m/s). Calculations with OSPM for Vesterbrogade in Copenhagen. Receptorpoint is in windward side for wind directions from south.

ted by OSPM is much larger than e.g. that calculated by the STREET-model. On the other hand, OSPM predicts that this difference decreases with decreasing wind speed. In the STREET-model the ratio between the leeward and windward concentrations is independent of wind speed (Johnson et al., 1973).

2.1 Chemical submodel for NO_2

The primary assumption in the chemical submodel used in OSPM for describing formation of NO_2 in a street canyon is that only the reaction between NO and O_3 is important. Also other reactions can lead to formation of NO_2, e. g. reactions involving hydrocarbons, but these reactions are too slow to significantly influence the formation of NO_2 inside street canyons. Such reactions may have influence on background concentrations of NO_2. The NO_2 concentration in street canyons is calculated using a set of equations describing both the chemical reactions and exchange with the background air. Steady-state conditions are assumed (Hertel and Berkowicz, 1989b):

$$\frac{d[NO]}{dt} = -k[NO][O_3] + J[NO_2] + \frac{[NO]_v}{\tau} + \frac{[NO]_b - [NO]}{\tau} = 0$$

$$\frac{d[NO_2]}{dt} = k[NO][O_3] - J[NO_2] + \frac{[NO_2]_v}{\tau} + \frac{[NO_2]_b - [NO_2]}{\tau} = 0 \quad (4)$$

$$\frac{d[O_3]}{dt} = - k[NO][O_3] + J[NO_2] + \frac{[O_3]_b - [O_3]}{\tau} = 0$$

where J and k are chemical reaction rates.

The first two terms in (4) account for chemical reactions. Thereafter follow the contribution from the direct emission in the street, and the exchange rate between the street canyon and the background air. There is no emission term for ozone.

The exchange rate is governed by the time constant τ. $[NO]_v$ and $[NO_2]_v$ are concentrations of NO and NO_2 due to the direct emission from the vehicles in the street. The terms with index b are the background concentrations of the respective components.

The time constant τ for exchange with background air is calculated as the ratio between the depth of the street canyon and the turbulent exchange velocity, σ_w. From (4) it follows that

$$[NO_2] = 0.5\left[b - (b^2 - 4([NO_x][NO_2]_a + [NO_2]_n D))^{1/2}\right] \quad (5)$$

where

$b = [NO_x] + R + [NO_2]_a + D$
$D = 1/(\tau \cdot k)$ is the exchange coefficient
$R = J/k$ is the photochemical equilibrium constant
$[NO_2]_n = [NO_2]_b + [NO_2]_v$
$[NO_2]_a = [NO_2]_n + [O_3]_b$

It is shown in Hertel and Berkowicz (1989b) that the background concentrations of ozone ($[O_3]_b$) play a very significant role in formation of NO_2 in streets. The contribution from the NO_2 directly emitted in the street ($[NO_2]_v$) is usually much less significant. The rate of exchange with background air (τ or the exchange coefficient D) is also important, at least at low to modest concentrations of NO_x.

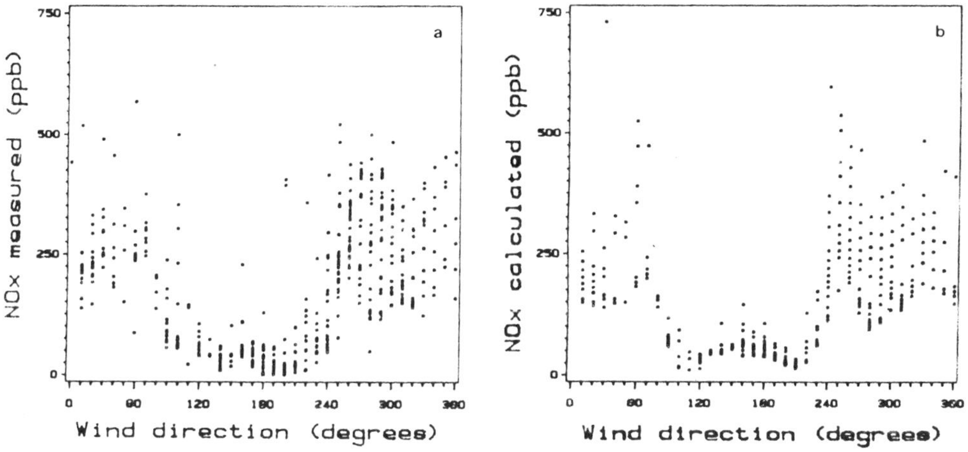

Fig. 2. Dependence of the NO_x street contribution, normalised with respect to emission, on the wind direction: a) measurements; b) computed by OSPM.

3. COMPARISON WITH MEASUREMENTS

Model calculations have been performed for a number of streets. Here we will present results from Vesterbrogade in Copenhagen. About four month of measurements during 1985 and 1986 are used (Høg and Sørensen, 1986). In Fig. 2 modelled and measured concentrations of NO_x are shown as functions of the wind direction. It is seen that OSPM reproduces the observed dependence on wind directions very well.

Comparison between the modelled and the observed concentrations is shown in Fig. 3. The correlation is good. Much of the scatter can be explained by uncertainty in the used wind directions. It should be noted here that the meteorological data are from the Kastrup Airport, 20 km southeast from Vesterbrogade.

Emissions of NO_x are estimated using the method reported in Larssen (1984). Due to a very limited amount of traffic data, the estimated emissions are very uncertain. The measurements shown in Figs. 2 and 3 are for the hours 8 - 16. Within this time period the emission from traffic is roughly constant.

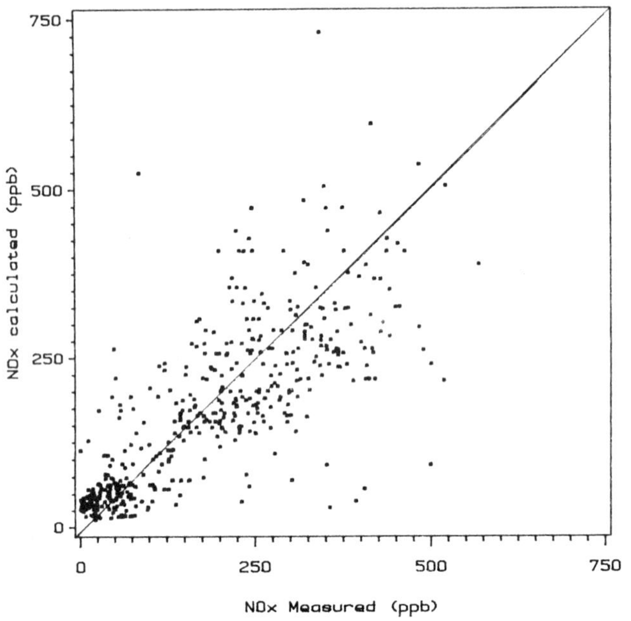

Fig. 3. Comparison of measured and computed NO_x street contributions. Calculations with OSPM for Vesterbrogade in Copenhagen. Data are for the hours 8 - 16.

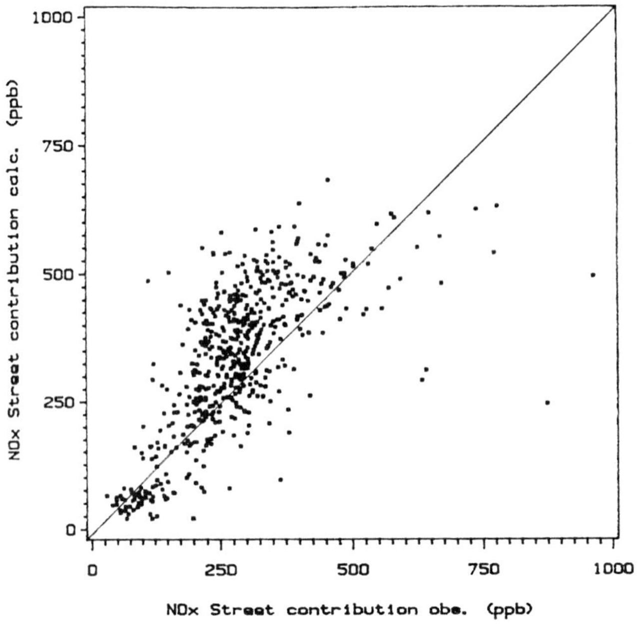

Fig. 4. Comparison of measured and computed NO_x street contributions. Calculation with OSPM for St. Olavs Street in Oslo. Data are for the hours 8 - 16.

Another data set used for evaluation of OSPM consists of measurements of NO_x in St. Olavs Street, Oslo, Norway (Larssen, 1988). These measurements are more detailed than those from Vesterbrogade. Hourly traffic counts were carried out simultaneously with the air pollution measurements. Also on-site meteorological measurements were available. Frequent occurrence of conditions with low wind speeds provided the opportunity to test performance of OSPM for these situations. Low wind speeds (below 2 m/s) were practically absent during the measurements in Vesterbrogade.

A comparison of the measured and modelled concentrations of NO_x for St. Olavs Street in Oslo is shown in Fig. 4. A detailed discussion of the results is presented in Hertel and Berkowicz (1989c).

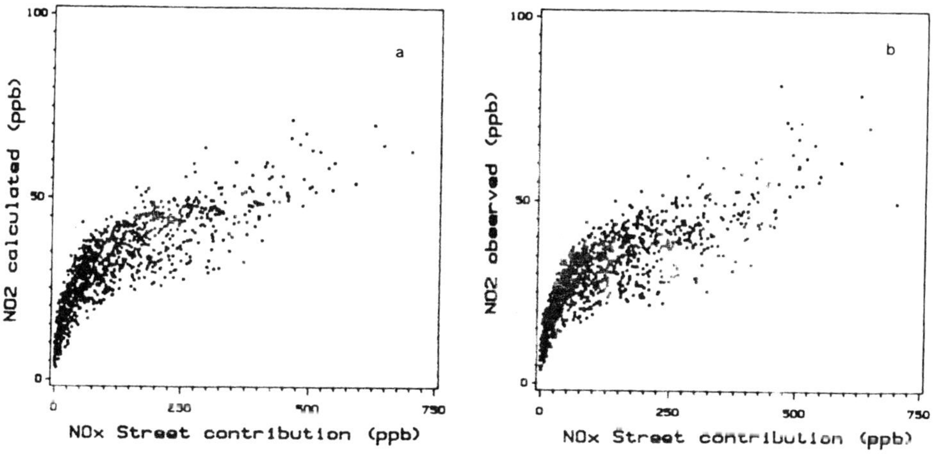

Fig. 5. Dependence of NO_2 concentrations on NO_x street contributions for St. Olavs Street in Oslo. a) computed NO_2; b) measured NO_2.

3.1 Test of the chemical submodel

The chemical submodel in OSPM was previously tested on data from Wittevrouwenstraat in Utrecht, the Netherlands (Hertel and Berkowicz, 1989b). It was shown that the model performs well for prediction of ozone concentrations in the street. Comparison with the available NO_2 data was, however, hampered by significant uncertainties in measurements of NO_2.

Here we present a new test of the model on data collected in January and February 1989 in St. Olavs Street in Oslo. The data are believed to be of very high quality, so the uncertainties in measurements of NO_2 are significantly reduced. During the measuring campaign both the street and background concentrations of NO/NO_2 were monitored. Also background measurements of ozone were available.

In Fig. 5 the NO_2 concentrations are shown as a function of the NO_x street contribution. Fig. 5a shows model results and 5b the observed concentrations. Because we are only interested here in the performance of the chemical submodel, not modelled but measured NO_x concentrations are used here. It is worth noticing how well the model reproduces the observed nonlinearity in the NO_2-NO_x relationship.

A comparison of the measured and calculated NO_2 concentrations is shown in Fig. 6a. It is seen that the agreement is very good. For comparison we show in Fig. 6b also results of calculations, but without taking the chemical reactions into account. In this case a significant underestimation is evident. This example illustrates the importance of chemical production for concentrations of NO_2 in street canyons. It should be noted that all the measurements presented here are performed in winter when concentrations of ozone are usually small (less than 20 ppb). It is expected that the importance of chemical production of NO_2 will be much greater in a summer period, because concentrations of ozone then are higher.

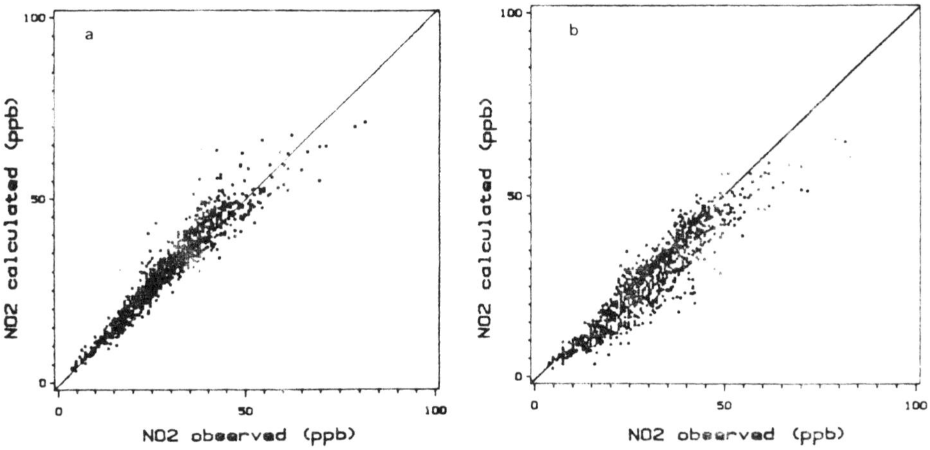

Fig. 6. Comparison between computed and observed NO_2 concentrations in St. Olavs Street in Oslo. a) with chemistry; b) without chemistry.

4. CONCLUSION

A new model for estimating dispersion of air pollution from traffic in street canyons - the Operational Street Pollution Model - is presented here. The dispersion part consists of a combined Gaussian formula for diffusion from an infinite line source and a box model describing recirculation in the street canyon. OSPM also contains a submodel for the chemical reactions leading to formation of NO_2 in the street.

Dispersion in a street canyon is governed by turbulence created by both wind and by the traffic in the street. The traffic created turbulence makes a significant contribution to dispersion when wind speed is low.

OSPM reproduces well the observed dependence of pollution concentrations on wind direction. In the case of Vesterbrogade in Copenhagen, both the measurements and model results show that the highest concentrations occur for parallel wind directions. The dependence on wind direction decreases with decreasing wind speed.

The tests of the chemical submodel on data from St. Olavs Street in Oslo show that the model is able to reproduce well the observed relationship between NO_2 and NO_x. This comparison was possible due to availability of high quality NO_2 measurements. The chemical formation of NO_2 in a street makes a significant contribution to the observed concentrations.

ACKNOWLEDGEMENTS

This work has been supported by the Danish National Agency of Environmental Protection and the Nordic Council of Ministers. Part of the work was conducted within the framework of the internordic project Nordic Computational Method for Air Pollution from Car Exhaust (NBB). We appreciate the cooperation with Sten Laurin from the Swedish Meteorological and Hydrological Institute and other members of the Nordic group. The Dutch National Institute of Public Health and Environmental Protection (RIVM) and especially Hans Eerens is acknowledged for the permission to use their data in the project. Nicky Brown is acknowledged for his help with preparation of this paper.

REFERENCES

Hertel, O. and Berkowicz, R., 1989a. Modelling pollution from traffic in a street canyon. Evaluation of data and model development. DMU Luft A-129.

Hertel, O. and Berkowicz, R., 1989b. Modelling NO_2 concentrations in a street canyon. DMU Luft A-131.

Hertel, O. and Berkowicz, R., 1989c. Operational Street Pollution Model (OSPM). Evaulation of the model on data from St. Olavs Street in Oslo. DMU Luft A-135.

Høg, I. and Sørensen, I. S., 1986. Gadeluftforurening i Københavns kommune. Målinger af gadeluftforurening i København 1985. Miljøkontrollen, Miljøkontrollens skriftserie nr. 2/1986.

Johnson, W. B., Ludwig, F. L., Dabbert, W. F. and Allen, R. J., 1973. An urban diffusion simulation model for carbon monoxide. **JAPCA, 23,** 490-498.

Larssen, S., 1984. Nordisk beregningsmetode for bilavgasser. NILU Lillestrøm (project NILU 0-8216).

Larssen, S., 1988. Overvågning av luftforurensning fra biltrafikk. Målinger i Oslo 1980-1986. NILU Lillestrøm (projekt NILU 0-8413).

DISCUSSION

M.E. BERLYAND

Do you take into account the background ozone concentration?

O. HERTEL

We do take into account the background ozone concentration in our model for NO_2 concentrations in street canyons. Due to the low NO_2 concentrations in car exhausts, the formation of NO_2 from reactions between NO and O_3 is one of the main contributions to the NO_2 concentrations in a street canyon. Thus, for highly trafficked streets, the background concentration of O_3 is the limiting factor for the NO_2 concentrations.

For developing the chemical submodel in OSPM, we have used background measurements of NO_2 and O_3 and street measurements of NO, NO_2 and O_3 from Utrecht in the Netherlands, Oslo in Norway and Copenhagen in Denmark.

In a recent work (in Danish) we have developed a model for urban concentrations of NO_2 and O_3. As input we used urban emission factors for NO_x and measured regional O_3 concentrations. In the future we plan to use a long-range model to calculate regional O_3 concentrations. Our urban model uses the same simple chemical scheme as the street canyon model does.

REVISION OF THE HYBRID PLUME DISPERSION MODEL (HPDM) FOR APPLICATION TO URBAN AREAS

Steven R. Hanna and Joseph C. Chang

Sigma Research Corporation
234 Littleton Road, Suite 2E
Westford, MA 01886 USA

ABSTRACT

The technical components of the urban Hybrid Plume Dispersion Model (HPDM) are summarized and the results of its evaluation with the Indianapolis field data are presented. The model contains state-of-the-art formulations for the atmospheric boundary layer and for dispersion in convective, neutral, and stable conditions. It is shown to be slightly conservative, by about 10-15% at both the middle and high range of the distribution, and yields a correlation coefficient of about 0.8. It is seen to be a significant improvement (at the 95% confidence level) over two other models (the EPA's RAM model and the rural HPDM).

OBJECTIVES AND BACKGROUND

The impetus for this study was the uncertainty that exists in available procedures to model the dispersion from tall stacks in urban areas. Until the Indianapolis tracer study was carried out in 1985 there had been no comprehensive field study of dispersion from tall stacks located within the urban boundary layer (TRC, 1986). The U.S. Environmental Protection Agency (EPA) arbitrarily set the requirement that urban tall stacks be modeled using the McElroy-Pooler (1968) dispersion coefficients, which are derived from near-surface tracer data. These coefficients imply relatively large vertical dispersion, and hence, greater concentrations close to the stack than would be produced by the coefficients used in rural areas. There was a concern whether the McElroy-Pooler dispersion coefficients would apply to plumes from tall stacks in urban areas, since the excess turbulence observed over urban surfaces is known to decay to lower values characteristic of the rural surface at elevations of a few hundred meters. The McElroy-Pooler dispersion coefficients have been included by the EPA in their RAM model (Catalano et al., 1987), which is the standard model recommended for application to urban areas. The RAM model predictions are known to produce poor correlations and large biases when compared with observations taken during field studies in St. Louis (Wackter and Minott, 1984) and Indianapolis (TRC, 1986).

The authors previously developed the "rural" Hybrid Plume Dispersion Model (HPDM), for application to tall stack plumes dispersing over nearly flat terrain in rural areas (Hanna and Paine, 1989). Emphasis was on convective conditions, because those conditions were observed to cause the highest

Air Pollution Modeling and Its Application VIII, Edited by H. van Dop and D. G. Steyn, Plenum Press, New York, 1991

concentrations at flat terrain sites with stack elevations of about 200 m. The algorithms in the model that deal with stable or neutral conditions were not adequately tested, since observed ground-level concentrations were quite low during those stability conditions at the two sites used for model evaluation. Furthermore, the validity of the model at locations with large, complex roughness elements (urban areas or complex terrain areas) was never established.

When the field data from the urban power plant stack in Indianapolis became available (TRC, 1986) it was of interest to test the existing models (RAM and rural HPDM) with the new urban database. Based on the results of that model evaluation exercise, a plan for the development of an urban version of HPDM was drawn up. The urban HPDM includes the following components: 1) specification of an urban surface energy balance and boundary layer parameterization scheme, 2) derivation of urban dispersion coefficients, and 3) inclusion of a new "lofting" model for buoyant plumes that rise up to the mixing lid. The remainder of this paper describes the development of the urban HPDM model and its evaluation using the Indianapolis database.

DESCRIPTION OF INDIANAPOLIS EXPERIMENT

During the Indianapolis field experiment, tracer gas was released from the 83 m stack of a power plant located near the downtown area of Indianapolis, Indiana (TRC, 1986). 19 experiments with duration 8 to 9 hours were carried out, covering a wide range of stabilities. Ground level concentrations were observed at about 160 monitoring sites located at downwind distances ranging from 0.25 to 12.0 km, and several types of meteorological data were obtained concurrently. Meteorological data were taken at five sites, including four short towers and one tower on a tall downtown building. Slow rise radiosondes provided profiles of wind, temperature, and relative humidity from an urban and rural site at intervals of about one hour during each experiment. Routine National Weather Service data were available from the Indianapolis airport.

Measurements of plume rise and plume concentration distribution aloft were made by two types of lidar instruments. SRI International's Differential Absorption Lidar (DIAL), which measures SF_6 in the plume, monitored the plume's cross-section and location at approximately 0.5 to 1.0 km downwind of the stack. The SRI Airborne Lidar for Plume and Haze Analysis (ALPHA-1) measured plume location and cross-sectional distribution at distances out to approximately 5 km from the stack.

All Indianapolis data were checked for errors and were compiled into a comprehensive database available on magnetic tape from the experiment's sponsor, the Electric Power Research Institute. This database also includes derived parameters such as the mixing depth, h, and the friction velocity, u_*.

RESULTS OF INDIANAPOLIS DATA ANALYSIS

The intent of this project was to use half of the Indianapolis data (the "developmental" database) to test and develop new theories, and to reserve half of the data (the "evaluation" database) for independent final evaluation of the new model. The data were randomly divided, by day, into these two halves. It was found that the observed maximum concentrations at Indianapolis were nearly equal during three types of meteorological conditions: 1) light-wind convective conditions, 2) high-wind neutral conditions, and 3) nighttime conditions when a nocturnal jet is formed.

General boundary layer theories were shown to be valid in Indianapolis as long as the surface boundary conditions (i.e., surface roughness, albedo, moisture availability) were properly described. The observed components of the urban surface energy balance could be simulated within ± 10 to 20% by theories during mid-day periods. However, during the night and near sunrise and sunset, the errors in the predictions are larger than the components themselves. The urban sensible heat flux, H, and the urban friction velocity, u_* could be just as accurately estimated using simple semi-empirical formulas as they could be simulated using more complex formulas. The typical root mean square error (rmse) in H was 50 w/m^2 and the rmse in u_* was 0.1 m/s (i.e., about 20% of their mid-day values).

During clear nights, a low-level jet was observed at elevations of 100-300 m in the wind profile. This jet consistently developed at about 11 p.m., and had an important effect on vertical dispersion, causing a downward mixing of the plume to the ground.

TECHNICAL IMPROVEMENTS TO HPDM

Based on the analysis of the developmental database from Indianapolis, several changes were made to the rural version of HPDM in order to improve its performance in urban areas. Details on all of these improvements are given in the project report by Hanna and Chang (1989). A few of the more significant changes are briefly reviewed in this section.

Minimum Monin-Obukhov Length, L

Most urban dispersion models employ a modified stability classification scheme to account for the increased surface roughness and the heat input in urban areas. The net result of these modifications is increased vertical mixing over urban areas. In contrast to most models, HPDM employs the Monin-Obukhov length, L, as an indicator of stability. The Monin-Obukhov length is thought to represent the depth of the boundary layer that is dominated by mechanical mixing processes. As pointed out by Uno et al. (1988), surface obstacles such as buildings generate their own mechanical turbulence up to heights of two to three times the obstacle height. For obstacles such as grass blades or gravel, this effect is unimportant, since the calculated $|L|$ seldom falls below 2 to 5 m. However, for urban areas where the obstacle height is over 10 m, it is necessary to correct for the effects of mechanical turbulence.

Intuitively, it is reasonable to assume that the minimum L is proportional to the building height during stable conditions. It is suggested that the EPA/Auer (1978) land usage classification methods be applied to the area within 3 km of the site of interest. Once the dominant land usage is determined, the urban HPDM assumes that the following minimum L values are assigned:

	Minimum L
Large City Downtown Area	150 m
Urban/Commercial	50 m
Dense Residential	50 m
Rural	2 m

When these minimum L values are applied in urban areas, very stable conditions will never occur, and conditions will always be either nearly-neutral or unstable. These values are clearly somewhat arbitrary and should be further refined in future studies.

Vertical Turbulence Intensity in Nocturnal Jet

Relatively high ground-level concentrations are observed in Indianapolis during clear nights after the nocturnal jet forms. The boundary-layer parameterizations in rural HPDM, as in most other models, do not account for this increased vertical mixing, since they assume that a stable layer with little vertical turbulence at elevations of 100-200 m, would persist all night. Consequently, urban HPDM contains empirical parameterizations of this increased turbulence. The vertical turbulence, σ_w, is assumed by urban HPDM to show different variation with height, depending on time of night and cloudiness. Analysis of nocturnal jet development at several sites shows that the jet develops at a height of 100 to 300 m on clear nights after the hour of 10 or 11 p.m. The nocturnal jet consistently develops after the hour of 10 p.m. at Indianapolis. Tjemkes and Duynkerke (1989) report its formation about 10 p.m. in the Netherlands, and Gaffen and Bornstein (1988) come to the same conclusion concerning New York City. Consequently σ_w is assumed to be larger after the jet forms than before. The following parameterizations are used in urban HPDM:

If cloudiness is less than 0.8 and the hour is 10 p.m. or earlier,

$$\sigma_w = 0.15\sigma_{wo} \qquad \text{for } z \geq 100 \text{ m} \tag{1}$$

and

$$\sigma_w = \sigma_{wo}(1 - 0.0085z) \qquad \text{for } z < 100 \text{ m} \tag{2}$$

These formulas are used in rural HPDM (Hanna and Paine, 1989).

If cloudiness is less than 0.8 and the hour is 11 p.m. or after, or if cloudiness is greater than 0.8,

$$\sigma_w = 0.70\sigma_{wo} \qquad \text{for } z \geq 100 \text{ m} \tag{3}$$

and

$$\sigma_w = \sigma_{wo}(1 - 0.0030z) \qquad \text{for } z < 100 \text{ m} \tag{4}$$

Unfortunately, there are no theoretical boundary layer models and very few sets of field data that deal with turbulence profiles above the inversion capping the surface mixed layer (Ching et al., 1983). It is expected that the parameterizations can be further refined as more field data become available.

Lofting Model for Buoyant Plumes

The new lofting model is based on a theory by Weil and Corio (1988) for the capped convective boundary layer, and has been extended in this project to the nearly-neutral boundary layer. The lofting model assumes that when a buoyant plume rises or "lofts" against the top of the mixed layer, it is mixed down to the surface only by eddies whose downward vertical speeds exceed the upward buoyant force due to the plume's excess temperature. They define a dimensionless parameter, ϕ, that expresses this force ratio:

$$\phi = 0.8(F/x)^{1/3}\sigma_w^{-1} \tag{5}$$

where F is the plumes initial buoyancy flux and x is downwind distance. The constant, 0.8, in Equation (5) provides a good fit to field data from the Indianapolis developmental data base. The ground-level plume centerline concentration is given by the formula:

$$C/Q = (1-\mathrm{erf}(\phi))/(\sqrt{2\pi}h\sigma_y u) \quad (6)$$

where h is the mixing depth, u is the wind speed, erf is the error function, and

$$\sigma_y = 1.6\ F^{1/3}x^{2/3}u^{-1} \quad \text{if } L > 0 \text{ or } (L < 0 \text{ and } u/w_* \geq 2) \quad (7)$$

$$\sigma_y = 0.8\ F^{1/3}x^{2/3}u^{-1} \quad \text{if } L < 0 \text{ and } u/w_* < 2 \quad (8)$$

The formulas in Equations (7) and (8) are based on the assumption that x is small enough that σ_y is completely determined by the component due to plume buoyancy.

The lofting model should be used only if the plume rises up near the mixing lid. This condition is met if the dimensionless buoyancy flux, F_*, exceeds the following limits, determined from the plume rise formulas for "break-up" conditions.

$$\text{Convective} \quad F_* = F/uw_*^2h > 0.1 \quad (9)$$

$$\text{Neutral} \quad F_{*n} = F/uu_*^2h > 0.5 \quad (10)$$

This model has more scientific credibility at high F_* than other available models, but has received only limited testing. It is now included in urban HPDM as an option, but should ultimately become the preferred model as evaluation proceeds and as the constant in Equation (5) becomes better known.

EVALUATION OF HPDM URBAN AND RURAL VERSIONS AND RAM

Methods

In this preliminary exercise, an evaluation of the HPDM urban and rural models and the RAM model has been carried out using the 79 hours of the "developmental" database from Indianapolis. In the future, the models will be evaluated using the "evaluation" database and the results will be reported later. Statistical tests are applied to the sets of observations and model predictions in order to determine whether the differences are significant between the models and the observations and between the various model options. In addition, the individual model errors are plotted versus meteorological parameters such as wind speed in order to determine whether the errors are independent of these variables.

This model evaluation exercise employs a statistical computer code currently being applied in several other model evaluation exercises (Hanna, 1989).The code uses a blocked bootstrap or jackknife resampling method to estimate whether there are significant differences in fractional bias (FB) normalized mean square error (NMSE), correlation (R), and fraction within a factor of two (FAC2), defined by:

$$FB = (\bar{C}_o - \bar{C}_p)/(0.5(\bar{C}_o + \bar{C}_p)) \quad (11)$$

$$NMSE = \overline{(C_o - C_p)^2}/\bar{C}_o\bar{C}_p \quad (12)$$

$$R = \overline{(C_o - \bar{C}_o)(C_p - \bar{C}_p)}/\sigma_{C_o}\sigma_{C_p} \quad (13)$$

$$FAC2 = \text{Fraction of data for which } 0.5 \le C_p/C_o \le 2.0 \quad (14)$$

where C_o and C_p are the observed and predicted maximum concentrations for that hour anywhere on the monitoring network.

In this evaluation exercise, the RAM model is evaluated using wind speeds observed above a downtown building at 94 m elevation. The HPDM models are given two wind speeds, one from the 10 m tower at the urban site, and the other from the 94 m bank tower. The 10 m wind data are used by the HPDM meteorological preprocessor to calculate surface fluxes of sensible heat and momentum. The 94 m wind data are used in the dispersion code to calculate plume rise and dilution. The same file of observed mixing depths is provided as input to all models. The lateral turbulence observations, σ_θ, from the 10 m tower at the urban site, are used directly in the urban and rural versions of HPDM. The roughness length, z_o, is assumed to be 1.0 m at the location of the 10 m urban tower. The urban albedo is assumed to be 0.18. The moisture availability parameter, α, is assumed to be 0.5 in the urban HPDM, and the Bowen Ratio is assumed to be 2.0 in the rural HPDM.

Results

The highest concentrations are of great importance for regulatory purposes, even though they do not have as much weight in the overall statistical tests. The maximum observed normalized concentration, C_o/Q, in the units 10^{-9} s/m^3, is 1149. The maximum predicted normalized concentrations, C_p/Q, are 1276 for HPDM-urban, 878 for HPDM-rural, and 1100 for RAM. The predicted maxima by the HPDM versions occur under convective conditions, in agreement with the observation. However, the predicted maximum by RAM occurs under stable conditions, in disagreement with the observation.

The results of the model evaluation using all 79 hours of data and the performance measures in equations (11)-(14) are given in Table 1. The HPDM-urban version is seen to show the best performance for nearly all performance measures, with a fractional (relative) bias, FB, of -0.08, a normalized mean square error, NMSE, of 0.12, a correlation of 0.82, and the fraction within a factor of two (FAC2) of 0.91. These performance measures are exceptionally good for any air quality model (Hanna, 1989) and are definite improvements over most of those for the HPDM-rural and RAM models. Note that the fractional bias (-0.08) for HPDM-urban is not significantly different from zero and that the correlation (0.82) is significantly different from zero. In fact, over 60% of the variance in C_o/Q is explained by HPDM-urban.

The performance measures in Table 1 for RAM are generally poor, and in all cases, there are significant differences (at 95% confidence) between the performance measures for RAM and for HPDM-urban. The relatively poor performance for RAM is chiefly due to its large overpredictions during stable conditions.

The performance of the older HPDM-rural model tends to be in between that for HPDM-urban and RAM. However, all performance measures for HPDM-urban are significantly better than those for HPDM-rural. HPDM-rural tends to predict maximum concentrations that are about 30% lower than those predicted by HPDM-urban during convective conditions, and underpredicts by an even larger amount during neutral and stable conditions.

Table 1. Performance Measures (Defined in Eqs. (11)-(14) for HPDM-urban, HPDM-rural and RAM for the Developmental Data Set at Indianapolis. Normalized Concentrations, C/Q, in Units of 10^{-9} s/m^3 are Used.

Model	$\overline{C/Q}$	$\sigma_{C/Q}$	$\overline{C_o/Q}-\overline{C_p/Q}$	NMSE	R	FAC2	FB
Observed	426.58	235.39					
HPDM-urban	460.20	261.70	-33.62	0.12	0.820	0.911	-0.076
HPDM-rural	313.17	199.66	113.41	0.39	0.594	0.570	0.307
RAM	580.37	270.14	-153.79	0.58	0.065	0.544	-0.305

Analysis of Variation of Residuals

The previous discussion provided information on the overall performance of the models, when applied to the 79 hours in the Indianapolis developmental data set. It is also important to know whether the performance of each model varies with input parameters. Knowing this relative performance, one can better assess the problems with a model and can decide which technical algorithms are not performing well. Figure 1 presents the ratio C_p/C_o, for the HPDM-urban model plotted on a logarithmic scale as a function of 10 m urban wind speed, stability, on-site urban mixing depth, and hour of day. The figure shows that the residuals C_p/C_o are not strong functions of any of the four parameters that are studied, and therefore that the HPDM-urban model does not have any major problems with its physical formulations.

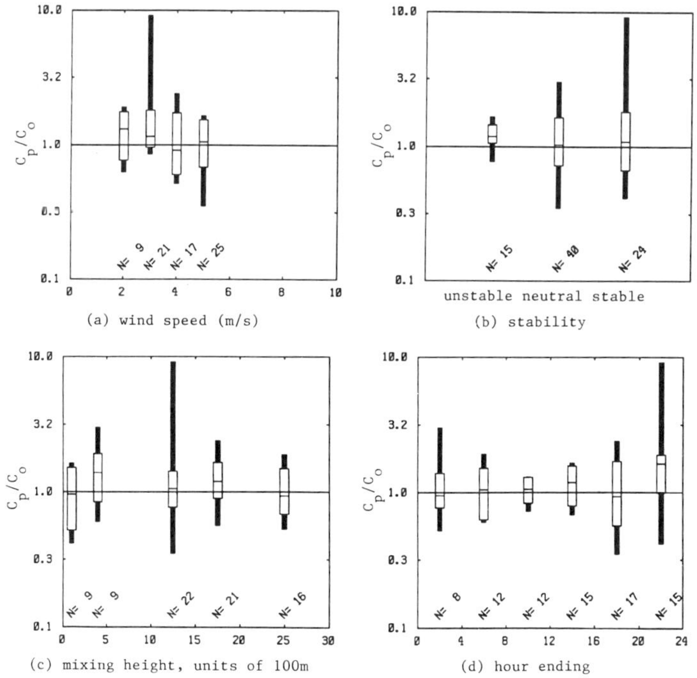

Figure 1. Residuals of concentration predictions (C_p/C_o) plotted as a function of wind speed, stability index, mixing height, and hour of day, for HPDM-urban. Significant points on each box plot represent the average, the average ± one standard deviation, and the average ± two standard deviations. N is the number of points in each box.

RECOMMENDATIONS FOR FUTURE MODEL DEVELOPMENT AND TESTING

As a result of the analyses of model performance at the Indianapolis site, we have gained considerable insight into the problems associated with the HPDM models. In the future, the HPDM-urban model should be tested at a number of sites using at least one year of hourly data, since its current evaluation experience is limited to 79 hours at Indianapolis. In a few cases this testing should include evaluation of individual model algorithms, such as the boundary layer formulations, the nocturnal jet parameterizations, the plume lofting algorithm, and the plume rise and dispersion coefficient formulas.

Acknowledgements: This research was sponsored by the Electric Power Research Institute, with Dr. Gordon Beals and Dr. Charles Hakkarenin as technical project advisers. The lofting model was formulated by Dr. Jeffrey Weil.

References

Auer, A.H., 1978, Correlation of land use and cover with meteorological anomalies. *J. Appl. Meteor.*, **17**, 636-643.

Catalano, J.A., D.B. Turner and J.H. Novak, 1987, User's guide for RAM--Second edition. EPA/600/8-87/046, NTIS Access. No. PB 88-113 261/AS. US EPA, Research Triangle Park, NC.

Ching, J.K.S., J.F. Clarke and J.M. Godowitch, 1983, Temporal behavior of turbulence parameters relevant to elevated point source dispersion. Proceedings of Sixth Symp. on Turbulence and Diffusion. American Meteorological Society, Boston, 309-312.

Hanna, S.R., 1989, Confidence limits for air quality models, as estimated by bootstrap and jackknife resampling methods. *Atmos. Envir.*, **23**, 1385-1398.

Hanna, S.R. and J.C. Chang, 1989, Modification of HPDM for Urban Conditions and Its Evaluation using the Indianapolis dataset. Report A089 for EPRI, 3412 Hillview Avenue, Palo Alto, CA, Three Volumes.

Hanna, S.R. and R.J. Paine, 1989, Hybrid Plume Dispersion Model (HPDM) development and evaluation. *J. Appl. Meteorol.*, **28**, 206-224.

McElroy, J.L. and F. Pooler, Jr., 1968, *St. Louis Dispersion Study, Vol. II-Analysis.* U.S. Dept. of Health, Education and Welfare, NAPCA Pub. AP-53.

Tjemkes, S.A. and P.G. Duynkerke, 1989, The nocturnal boundary layer: model calculations compared with observations. *J. Appl. Meteorol.*, **28**, 161-175.

TRC Environ. Cons. Inc., 1986, Urban power plant plume studies. EPRI Report EA-5468, Research Project 2736-1, prepared by TRC, 800 Conn. Blvd., East Hartford, CT for EPRI, 3412 Hillview Ave., Palo Alto, CA.

Uno, I., S. Wakamatsu, H. Ueda and A. Nakamura, 1988, An observational study of the structure of the nocturnal urban boundary layer. *Boundary Layer Meteorol.*, **45**, 59-82.

Wackter, D.J. and D.H. Minott, 1984, Evaluation of six urban air quality simulation models. Conference Volume: Fourth Joint Con. on Applic. of Air Poll. Meteorol., AMS, Boston, MA, 83-86.

Weil, J.C. and L.A. Corio, 1988, A modification of the PPSP dispersion model for highly buoyant plumes. Maryland Power Plant Research Program, Maryland Dept. of Natural Resources, Annapolis, Report No. PPRP-MP-60.

THE OFFSHORE AND COASTAL DISPERSION (OCD) MODEL:

REVISIONS AND EVALUATIONS

Donald C. DiCristofaro and Steven R. Hanna

Sigma Research Corporation
234 Littleton Road, Suite 2E
Westford, MA 01886

INTRODUCTION

The Offshore and Coastal Dispersion (OCD) Model simulates plume dispersion and transport from offshore point, area, or line sources to receptors on land or water. The OCD model is an hour-by-hour steady state Gaussian model with enhancements that consider the differences between overwater and overland dispersion characteristics, the sea-land interface, and platform aerodynamic effects. A revised OCD model (Version 4) (or OCD/4) has been developed that includes modifications to boundary layer parameterizations and dispersion model assumptions. An overview of the physical and theoretical aspects of the model is given, along with the results of a statistical model evaluation exercise that compares the revised model (OCD/4) with the earlier OCD model (Version 3) (or OCD/3). Data from four field experiments (Ventura, Pismo Beach, and Carpinteria, California; Cameron, Louisiana) are used to evaluate the model performance. The reader is referred to the report by DiCristofaro and Hanna (1989) for detailed discussions of the points that are only briefly reviewed in this paper.

MODEL OVERVIEW

A technical discussion of the original OCD/3 model is given by Hanna et al. (1985). The modifications in the OCD/4 model include: the addition of area and line source emissions, a revised complex terrain algorithm, the assessment of annual impacts from non-permanent activities, a revised shoreline fumigation algorithm, a revised TIBL (Thermal Internal Boundary Layer) growth formula for all stabilities and over complex terrain, a revised platform downwash algorithm, a revised mixing depth routine, changes to the default meteorological parameterization, a revised calculation of predicted lateral turbulence intensity, and a revised lateral dispersion algorithm. Additional modifications included streamlining the code, restructuring the algorithms to more realistically represent the impact of the plume on shoreline terrain, and a standardization of the grid-cell sizes used in the shoreline mapping routine.

Gaussian Equation

The OCD concentration equation is based on the Gaussian diffusion equation (Gifford, 1968) and accounts for the multiple eddy reflections from both the ground and the top of the mixed layer (Turner, 1969):

$$C = \frac{Q}{2\pi u \sigma_{\theta_c} x \sigma_z} \exp\left(-\frac{1}{2} \left(\frac{\theta' - \theta}{\sigma_{\theta_c}} \right)^2 \right) \left[\exp\left(-\frac{1}{2} \left(\frac{z - h_e}{\sigma_z} \right)^2 \right) + \exp\left(-\frac{1}{2} \left(\frac{z + h_e}{\sigma_z} \right)^2 \right) \right.$$

$$+ \sum_{N=1}^{J} \left(\exp\left(-\frac{1}{2} \left(\frac{z - h_e - 2Nz_i}{\sigma_z} \right)^2 \right) + \exp\left(-\frac{1}{2} \left(\frac{z + h_e - 2Nz_i}{\sigma_z} \right)^2 \right) \right.$$

$$\left. \left. + \exp\left(-\frac{1}{2} \left(\frac{z - h_e + 2Nz_i}{\sigma_z} \right)^2 \right) + \exp\left(-\frac{1}{2} \left(\frac{z + h_e + 2Nz_i}{\sigma_z} \right)^2 \right) \right) \right] \qquad (1)$$

where Q is the emission rate, u is the wind speed, σ_z represents the lateral dispersion parameter, z is the receptor height above ground level, h_e is the effective plume height, and z_i is the mixing height. The lateral terms make use of a polar coordinate system where θ' is the direction from the source to receptor, θ is the wind direction, and $\sigma_{\theta c}$, the standard deviation (in radians) of the lateral angular distribution of plume material at the downwind distance, x, is given by

$$\sigma_{\theta c} = \frac{\sigma_\theta}{1 + 0.9(T/1000)^{1/2}} , \qquad (2)$$

where σ_θ is the standard deviation (in radians) of the wind direction fluctuations measured at the source and T is the travel time, which is equal to x/u. Equation (2) represents the Draxler (1976) lateral dispersion equation as formulated for a polar system, where it is assumed that the lateral dispersion coefficient, σ_y, is equal to $\sigma_{\theta c} x$. Irwin (1983) of the U.S. EPA recommends the Draxler (1976) formulation of σ_y, although the original OCD/3 model (Hanna et al., 1985) used the Briggs (1973) formulation. Recent studies have shown that Equation (2) provides better agreement with over-water data (Hanna and DiCristofaro, 1988). If observed values of the horizontal wind fluctuations, σ_θ, or turbulence intensity (i_y or σ_v/u) are not available, the OCD/4 model defaults to $\sigma_\theta = i_y = 0.37/u$ for all stability classes for plumes above and below the mixing depth. In the earlier OCD/3 model, $\sigma_\theta = i_y = 0.5/u$ was used only for stable conditions and for plumes below the mixing depth.

Vertical Dispersion

The vertical dispersion parameter (σ_z) is approximated by $\sigma_z = i_z \times f_z(x)$, where the vertical turbulence intensity, i_z equals σ_w/u, and for overland sources, the Briggs (1973) f_z formulation as a function of overland stability is used. For overwater sources, the f_z formulation is corrected such that the Class D formula is used for overwater classes A, B,

C, and D. This assumption reflects the fact that vertical dispersion is less intense over water than land, and can be approximated by the vertical shape factor appropriate for neutral conditions.

The use of observed vertical turbulence intensity ($i_z=\sigma_w/u$) is not recommended, since experience by a wide variety of users has shown that these data are highly uncertain, especially for stable conditions. The default value of overwater i_z for neutral and unstable conditions is assumed to be $i_z=0.2/u$ where u is in m/s, and 0.2 m/s is the median value of σ_w observed at the Carpinteria experiments for these stabilities.

The OCD/4 model allows for very stable conditions (Stability Class G) which are triggered when the observed dθ/dz in the lowest 100 m of the marine boundary layer is greater than or equal to 0.04°C/m. The earlier version of the model used a trigger of 0.05°C/m. This change provides better agreement with data from very stable conditions at the experiment sites analyzed by Hanna et al. (1985).

Thermal Internal Boundary Layer (TIBL)

The thermal internal boundary layer (TIBL) is formed when stable air flows from over the water to over a heated land surface (Lyons, 1975). The TIBL develops near coastlines because of the roughness and temperature difference between land and water. The underlying physical principle is that the heat added to the boundary layer as it flows over the land is used to warm the air and form an adiabatic layer at the base of the initial stable overwater temperature profile. When a plume travelling in a stable layer intersects the TIBL, fumigation occurs that could lead to potentially high ground-level concentrations. The OCD/4 model uses the following empirical formulas for the TIBL height:

$$h_T = 0.1\ x \qquad (x \le 2000\ \text{m}) \qquad (3)$$

$$h_T = 200 + 0.03\ (x-2000) \qquad (x > 2000\ \text{m}) \qquad (4)$$

where x is the distance inland from the shoreline. Hanna (1987) used observations of the height of the daytime coastal internal boundary layer at several "flat" coastal sites to justify this empirical formula. There are no observations of TIBL height over coastal areas with complex terrain. Consequently, the simplest assumption is made that the TIBL is terrain following, i.e., the TIBL height above the local terrain at a given distance inland equals that over flat terrain at the same distance inland.

It is also possible that marine air with a neutral or unstable stratification can flow onto the land on a clear night, resulting in a stable layer that develops at the surface. At the base of this stable layer is a mixed layer that begins at the shoreline and deepens with increasing distance inland. The depth h of this mixed layer will approach a constant value derived by Zilitinkevich (1975), $h=0.4\ (u_*L/f)^{1/2}$ where f is the Coriolis parameter, and the parameters u_* and L refer to the friction velocity and Monin-Obukhov length characteristic of the overland boundary layer. Therefore, for stable conditions over land, the OCD/4 model assumes that the stable internal boundary layer (SIBL) is capped at the height defined by Zilitinkevich, such that $h=\min(h_T, h)$.

Fumigation

In the OCD/4 model, fumigation, or the vertical dispersion of the plume to the ground after it passes through the TIBL, is calculated as the maximum of concentrations predicted by (1) the Turner (1969) complete vertical mixing assumption and (2) the OCD virtual source assumption. The Turner formulation assumes that after the plume centerline intersects the TIBL it is uniformly mixed vertically between the surface and the mixing depth h_T. The virtual source method assumes that the vertical dispersion over land proceeds as if the atmosphere is unstable. However, a virtual source distance is calculated as the distance upwind from the point of TIBL intersection where the source would be if an unstable dispersion rate were present in the overwater atmosphere, and σ_z had its given value at the point of TIBL intersection.

The OCD/3 model did not use the Turner (1969) complete vertical mixing assumption. This version of the model used the Deardorff and Willis (1982) fumigation algorithm, which is a convective-scaling formulation based on observations of mixed layer growth in a laboratory convective tank. After testing this empirical algorithm in a wide range of real-world scenarios, it was discovered that the TIBL slopes for which the model was derived were usually less than those that would occur in most coastal areas. Consequently, in the revised version of the OCD model this option is replaced with the Turner assumption.

Plume Behavior in Complex Terrain

The revised version of the OCD model uses recent theory and observations of transport over simple hill shapes to calculate deflection around or over terrain. Much theoretical, laboratory, and field research over the past few years supports the use of the concept of the critical dividing streamline height (Snyder and Hunt, 1984; DiCristofaro et al., 1984). Given the hill height, H, the wind speed, u, and the temperature gradient at plume height, $d\theta/dz$, a critical streamline height, H_c, is calculated:

$$H_c = H - u/((g/T)d\theta/dz)^{1/2} \quad . \tag{5}$$

If the initial plume elevation, H_e, is below H_c, then no lifting of the plume is assumed to occur and the plume path coefficient (PPC) equals zero. The PPC, which varies from 0.0 to 1.0, describes the relative amount of vertical deflection of the plume centerline by the terrain. Consequently, the plume experiences no vertical deflection if PPC=0.0 and is "terrain following" if PPC=1.0. If H_e is above H_c, then PPC equals 0.5 and partial lifting of the plume occurs. This lifting is calculated with respect to H_c rather than with respect to the ground surface under H_c.

If H_{eo} is the effective plume plume height over the water, then the adjusted plume height above local terrain (referred to msl), H_a, is given by:

$$H_a = PPC \cdot (H_{eo} - H_c) \qquad \text{for } H(x) > H_{eo} \tag{6}$$

$$H_a = H_{eo} - (1 - PPC)\cdot(H(x) + H_c) \qquad \text{for } H(x) \le H_{eo} \tag{7}$$

where H(x) is the local terrain height.

Area/Line Source Formulation

The OCD/4 model Gaussian plume equation given by Equation (1) is used for point, area, and line source emissions. Adjustments are made to the lateral dispersion parameter, σ_y, for area and line sources.

For an area source, the source region is approximated by the user by a series of just touching circular areas, such that the emission rate in each area is Q_i. Each area source, as approximated by an effective circle with radius r_i, is modeled with Equation (1) using a virtual source assumption with $\sigma_y = r_i/\sqrt{2}$ to calculate the virtual source distance for the appropriate overwater stability class.

The emissions from a line source or moving ship are represented as a series of point sources for N segments along the path of the ship. A virtual source procedure with initial σ_{yi} defined by the equation,

$$\sigma_{yi} = \frac{x' - x''}{\sqrt{2}} \quad , \tag{8}$$

is used to calculate the virtual source distance. The parameters, x′ and x′′, are the successive midpoint distances along the path of the ship. The concentration at each receptor is represented by the sum of each concentration using Equation (1) calculated for the midpoint of each of the N line segments. The emission rate used in Equation (1) is adjusted to account for the time averaging.

MODEL EVALUATION PROCEDURES AND RESULTS

Observations from four separate offshore and coastal diffusion experiments are available for evaluating both versions of the OCD model. There were three California experiments in Ventura, Pismo Beach, and Carpinteria and one Gulf of Mexico experiment in Cameron, Louisiana. Each experiment is further divided into subsets with similar source conditions or meteorological conditions. The Ventura, Pismo Beach, and Cameron studies were similar in that the coastal terrain was not steep and the tracer source was positioned at about 5 to 8 km offshore. At Carpinteria, the coastal terrain was moderately steep and the source was a distance of about 0.5 to 1.5 km offshore. Two sets of OCD model runs are compared: 1) using observed lateral turbulence intensity i_y and 2) using predicted lateral turbulence intensity i_y. Statistical tests (Hanna, 1989) are applied to the sets of observations and model predictions in order to determine whether the differences are significant between the models and the observations and between the various model options.

The observed and predicted overall maximum normalized concentrations (C/Q) for each experiment are summarized in Table 1. The ratio of predicted to observed (C_p/C_o) maximum concentrations are also presented in parenthesis following the C_p/Q entry in the table. The model performance varies from experiment to experiment, with underpredictions by a factor of two at one experiment and overpredictions of a factor two at another. For the OCD/3 model, the maximum concentrations are overpredicted at four of the nine experiments using i_y (Observed data) and the maximum concentrations are underpredicted at three of the nine experiments using i_y (Observed data). Using i_y (Predicted), the OCD/4 model overpredicts and underpredicts an equal number of experiments and one experiment has a maximum predicted concentration within one percent of the maximum observed concentration. Moreover, the average (C_o Max)/(C_p Max) for the nine experiments is 0.87 for the OCD/3 model and 1.04 for the OCD/4 model using i_y (Predicted). Based on this simple statistical analysis, the OCD/4 model is better than the OCD/3 model using the i_y (Predicted) option, which is the recommended method for running the model. When using i_y (Observed data), the OCD/4 model overpredicts while the OCD/3 model underpredicts with an average (C_o Max)/(C_p Max) of 0.94 for the OCD/3 model and 1.20 for the OCD/4 model.

Table 2. Evaluation of Fractional Bias ($FB = (\bar{C}_o/Q - \bar{C}_p/Q)/(0.5(\bar{C}_o/Q + \bar{C}_p/Q))$) For Maximum C/Q During Each Experiment

Model Number:		1	2	3	4	Models with FB Not Sig.	Model Pairs with ΔFB Not Sig.
		OCD/3		OCD/4			
Data Set	N	i_y (Obs)	i_y (Pred)	i_y (Obs)	i_y (Pred)	Dif. from 0	Dif. from 0
Cameron-Winter	17	0.159	0.809	-0.208	0.325	1, 3, 4	1-4
Cameron-Summer	9	0.900	0.878	-0.456	-0.544	3	1-2, 3-4
Carpinteria - SF_6	18	0.569	0.391	-0.217	-0.413	3, 4	1-2, 3-4
Carpinteria - Fumigation	9	0.949	-0.235	0.161	-0.327	2, 3, 4	2-3, 2-4
Carpinteria - CF_3Br	9	0.075	-0.265	-0.320	-0.577	1, 2, 3	1-2, 2-3, 3-4
Pismo - Winter	15	-0.198	0.245	0.023	0.025	1, 2, 3, 4	1-3, 1-4, 2-3, 2-4, 3-4
Pismo - Summer	16	-0.108	0.480	-0.324	-0.027	1, 3, 4	1-4, 3-4
Ventura - Fall	9	-0.211	-0.396	-0.351	-0.491	1, 2, 3, 4	1-2, 1-3, 1-4, 2-3, 2-4, 3-4
Ventura - Winter	8	0.360	0.081	0.216	-0.119	1, 2, 3, 4	1-3, 1-4, 2-3, 2-4, 3-4

REFERENCES

Briggs, G.A., Diffusion Estimates for Small Emissions, ATDL Cont. 79, ATDL, Oak Ridge, TN (1973).

Deardorff, J.W. and G.E. Willis, Ground Level Concentrations Due to Fumigation Into an Entraining Mixing Layer. *Atmos. Environ.*, **16**, 1159-1170 (1982).

DiCristofaro, D.C., D. Strimaitis, B. Greene, R. Yamartino, D. Godden, T. Lavery, and B. Egan, EPA Complex Terrain Model Development Program: Fifth Milestone Report-1984. EPA600/3-84-110, U.S.EPA, RTP, NC (1986).

DiCristofaro, D.C. and S.R. Hanna, OCD: The Offshore and Coastal Dispersion Model Volume I: User's Guide. U.S. MMS, Herndon, VA (1989).

Gifford, F.A., An Outline of Theories of Diffusion in the Lower Layers of the Atmosphere in: "Meteorology and Atomic Energy-1968," D.H. Slade ed. USAEC Report TID-24190 (1968).

Hanna, S.R., An Empirical Formula for the Height of the Coastal Internal Boundary Layer. *Bound. Lay. Met.*, **40**, 205-207 (1987).

Hanna, S.R., Confidence Limits for Air Quality Models as Estimated by Bootstrap and Jacknife Resampling Methods. *Atmos. Environ.*, **23**, 1385-1395 (1989).

Hanna, S.R. and D.C. DiCristofaro, Development and Evaluation of the OCD/API Model. Final Report, API Pub. 4461, API, Washington, D.C. (1988).

Hanna, S.R., L.L. Schulman, R.J. Paine, J.E. Pleim, and M. Baer, Development and Evaluation of the Offshore and Coastal Dispersion Model. *JAPCA*, **35**, 1039-1047 (1985).

Irwin, J.S., Estimating Plume Dispersion--A Comparison of Several Sigma Schemes. *J. Clim. and Appl. Meteorol.*, **22**, 92-114 (1983).

Lyons, W.A., *Lectures on Air Pollution and Environmental Impact Analysis.* (D. Haugen ed). AMS, Boston, 136-208 (1975).

Snyder, W.H. and J.C.R. Hunt, Turbulent Diffusion from a Point Source in Stratified and Neutral Flows Around a Three-Dimensional Hill; Part II - Laboratory Measurement of Surface Concentrations. *Atmos. Envir.*, **18**, 1969-2002 (1984).

Turner, D., *Workbook of Atmospheric Dispersion Estimates.* Office of Air Programs Publ. No. AP-26. U.S. EPA, Research Triangle Park, NC (1969).

Zilitinkevich, S.S., Resistance Laws and Prediction Equations for the Depth of the Planetary Boundary Layer. *J. Atmos. Sci.*, **32**, 741-752 (1975).

DISCUSSION

J.L. WALMSLEY

You indicated that the TIBL height should depend on the surface roughness and land-sea temperature difference. Yet, your empirical formulae are functions of distance from the coastline alone. Can you explain?

D.C. DICRISTOFARO

In the original development of the OCD/3 model in 1983, a search was conducted for an equation for the TIBL height (h_T) that would be theoretically correct, would agree with available data, and would be capable of producing reasonable results for all hours of the year. Like the EPA regulatory models, the OCD model was intended to be applied to a year or more of hourly data. Several of the formulae tested tended to produce reasonable results only for a limited range of conditions. These formulae produced very small or very large values of h_T during other hours. For example, if the overwater potential temperature gradient approaches zero, the TIBL height prediction becomes very large. Similarly, if the overland sensible heat flux is small, the TIBL height prediction is very small. It was concluded that these models were not robust, since they permitted large variations in the value of h_T and were quite sensitive to uncertainties in input parameters such as the overwater potential temperature gradient. It was discovered during the course of the OCD investigation that many of the available h_T data could be fit by a model that permits no variation in h_T with meteorological conditions at a given downwind distance. Most of the data tested are within a factor of two of the prediction.

PARTICIPANTS

The 18th NATO/CCMS International Technical Meeting on Air Pollution Modeling and its Application
Vancouver, Canada, 13-17 May, 1990

AUSTRIA

Alcamo J.	International Institute for Applied Systems Analysis Schlossplatz 1 A - 2361 Laxenburg
Bartnicki J.	International Institute for Applied Systems Analysis Schlossplatz 1 A - 2361 Laxenburg
Züger J.	Österreichisches Forschungszentrum Seibersdorf GmbH A-2444 Seibersdorf

BELGIUM

Rojas C.M.	Universitaire Instelling Antwerpen Universiteitsplein 1 B-2610 Antwerpen-Wilrijk
Schayes G.	UCL Institut d'Astronomie et de Géophysique Chemin du Cyclotron, 2 B-1348 Louvain-la-Neuve
Stief-Tauch P.H.	Commission of the European Communities - SEPC Rue de la Loi, 200 B-1049 Brussel
Thunis P.	UCL Institut d'Astronomie et de Géophysique Chemin du Cyclotron, 2 B-1348 Louvain-la-Neuve

CANADA

Ayotte K.	166 Indian Grove Rd. Toronto, Ontario M6P 2H2
Bennett R.	BC Ministry of Environment 810 Blanshard Street Victoria, B.C. V8V 1X5

Cheng L.	Alberta Research Council 250 Karl Clark Road Edmonton, Alberta T6H 5X2
Coligado M.	BC Ministry of Environment 810 Blanshard Street Victoria, B.C. V8V 1X5
D'Amours R.	Canadian Meteorological Centre 2121 Trans Canada Highway 404 North Service Road Dorval, Quebec H9P 1J3
Fung C.	Ontario Ministry of the Environment Air Resources Branch P.O. Box 213 125 Resources Road, East Wing Rexdale, Ontario M9W 5L1
Hrebnyk B.	Senes Consultants #506 - 21 Water Street Vancouver, B.C. V6B 1A1
Kotturi M.S.	Pacific Air Consultants 2440 Alpine Crescent Victoria, B.C. V8N 4B6
Matthias C.S.	Atmospheric Environment Service 4905 Dufferin Street Downsview, Ontario M3H 5T4
McBean G.	The University of British Columbia Department of Geography 217-1984 West Mall Vancouver, B.C. V6T 1W5
McCormick W.	Plaza 400 1011 Fourth Avenue Prince George, B.C. V2L 3H9
Netterville D.D.J.	Syncrude Canada Ltd. Research Department P.O. Box 5790 Edmonton, Alberta T6C 4G3
Oke T.R.	The University of British Columbia Department of Geography 217-1984 West Mall Vancouver, B.C. V6T 1W5
Pudykiewicz J.	Atmospheric Environment Service 2121 Trans Canada Highway Suite 534 Dorval, Quebec H9P 1J3
Ramsay S.R.	The University of Western Ontario Faculty of Engineering Science London, Ontario N6A 5B9

Rawlings M. | Senes Consultants
#506 - 21 Water Street
Vancouver, B.C. V6B 1A1

Roy S. | Environment Canada
100 Alexis-Nihon, Room 300
St. Laurent, Quebec H4M 2N8

Sakiyama S.K. | Alberta Environment
Air Quality Branch
9820 - 106 St
Edmonton, Alberta T5K ZJ6

Stalker P. | York University
Box 170
Ballinifad, Ontario N0B 1H0

Steyn D.G. | The University of British Columbia
Department of Geography
217-1984 West Mall
Vancouver, B.C. V6T 1W5

Thomson B. | Atmospheric Environment Service
1200 West 73rd Avenue, Suite 700
Vancouver, B.C. V6P 6H9

Voigt C. | P.O. Box 58281, STN L.
Vancouver, B.C. V6P 6E3

Wallace D.J. | Containment Analysis Branch
White Shell Nuclear Research Establishment
Pinawa, Manitoba R0E 1L0

Walmsley J.L. | Atmospheric Environment Service
Boundary Layer Research Division
4905 Dufferin Street
Downsview, Ontario M3H 5T4

Wilson D. | University of Alberta
Department of Mechanical Engineering 4-9 Bldg.
Edmonton, Alberta T6G 2G8

Zelt B. | University of Alberta
Department of Mechanical Engineering 4-9 Bldg.
Edmonton, Alberta T6G 2G8

DENMARK

Christensen J.H. | National Environmental Research Institute
Frederiksborgvej 399
DK-4000 Roskilde

Hertel O. | National Environmental Research Institute
Frederiksborgvej 399
DK-4000 Roskilde

FEDERAL REPUBLIC OF GERMANY

Brost R.A.	Max-Planck-Institut für Chemie Abt. Chemie der Atmosphäre Postfach 3060 D-6500 Mainz
Dittmann E.	Deutscher Wetterdienst Frankfurter Strasse 135 D-6050 Offenbach
Fay B.	Deutscher Wetterdienst Kaiserleistrasse 42 Postfach 100465 D-6050 Offenbach
Feichter J.	Meteorologisches Institut Universität Hamburg Bundesstrasse 55 D-2000 Hamburg 13
Geiss H.	Institut für die Chemie der belasteten Atmosphäre Forschungszentrum Jülich GmbH Postfach 1913 D-5170 Jülich
Hass H.	Institut für Geophysik und Meteorologie der Universität zu Köln Projekt EURAD Salierring 48 D-5000 Köln 1
Jakobs H.J.	Institut für Geophysik und Meteorologie der Universität zu Köln Projekt EURAD Salierring 48 D-5000 Köln 1
Klug W.	Institut für Meteorologie Technische Hochschule Darmstadt Hochschulstrasse 1 D-6100 Darmstadt
Laube M.	Institut für Geophysik und Meteorologie der Universität zu Köln Albertus-Magnus-Platz D-5000 Köln 41
Löbel J.	VDI - Commission on Air Pollution Prevention P.O. Box M39 Düsseldorf
Massmeyer K.	Gesellschaft für Reaktorsicherheit mbH Schwertnergasse 1 Postfach 101650 D-5000 Köln 1
Petersen G.	GKSS Forschungszentrum Geesthacht GmbH Max-Planck-Strasse 1 Postfach 1160 D-2054 Geesthacht

Schaller E.	Fraunhofer-Institut für Atmosphärische Umweltforschung - IFU Kreuzeckbahnstrasse 19 D-8100 Garmisch-Partenkirchen
Schatzmann M.	Meteorologisches Institut Universität Hamburg Bundesstrasse 55 D-2000 Hamburg 13
Scheich G.	Landesanstalt für Immissionsschutz NRW Wallneyer Strasse 6 D-4300 Essen 1
Walter H.	Bundesgesundheitsamt Institut für Strahlenhygiene Ingoldstädter Landstrasse 1 D-8042 Neuherberg
Weber H.	German Military Geophysical Office Mont Royal 5580 Traben-Trarbach

FINLAND

Kukkonen J.S.	Finnish Meteorological Institute Air Quality Department Sahaajankatu 22E SF-00810 Helsinki 81
Nickmo J.	Finnish Meteorological Institute Air Quality Department Sahaajankatu 22E SF-00810 Helsinki 81
Savunen T.	Finnish Meteorological Institute Air Quality Department Sahaajankatu 22E SF-00810 Helsinki 81

FRANCE

Albergel A.	Electricité de France DER, Départment EAA 6, quai Watier BP 49 78401 Chatou CEDEX
Chaumerliac N.	LAMP/OPGC 12, avenue des Landais 63000 Clermont-Ferrand
Musson-Genon L.	Direction de la Météorologie Nationale SCEM/D/ES 2, avenue Rapp 75340 Paris CEDEX 07

GREECE

Kallos G. — Laboratory of Meteorology
University of Athens
Department of Applied Physics
33, Ippocratous Street
10680 Athens

ITALY

Baldi M. — Area Ricerca Frascati - CNR
Via Tiburtina 770
00159 Roma

Brusasca G. — ENEL/DSR/CRTN
Via Rubattino 54
20134 Milano

Riva G.M. — CISE
Via Reggio Emilia 39
20090 Segrate (Milano)

Salerno R. — ARS S.p.A. (ENI group)
Via Medici Del Vascello 26
20138 Milano

JAPAN

Kitada T. — Toyohashi University of Technology
Department of Regional Planning
Tempaku-Cho
Toyohashi 440

Wakamatsu S. — National Institute for Environmental Studies
Tsukuba
Ibaraki 305

MEXICO

Ruiz M. — Instituto Mexicano del Petroleo
Apdo Postal 14-805
Mexico 07730 DM

NORWAY

Mylona S. — Norwegian Meteorological Institute
P.O. Box 43
Blindern
N-0313 Oslo 3

Saltbones J. — Norwegian Meteorological Institute
P.O. Box 320
Blindern
N-0314 Oslo 3

PORTUGAL

Borrego C. — Departamento de Ambiente e Ordenamento
Universidade de Aveiro
3800 Aveiro

Coutinho M. — Departamento de Ambiente e Ordenamento
Universidade de Aveiro
3800 Aveiro

SPAIN

Artíñano B. — Ministerio de Industria y Energia
CIEMAT-ipryma
Avenida Complutense 22
E-28040 Madrid

Millán M.M. — Ministerio de Industria y Energia
CIEMAT-ipryma
Avenida Complutense 22
E-28040 Madrid

Serrano Garcia F. — Instituto Nacional de Meteorologia
Sección de Contaminación Ambiental
Apartado de Correos No. 285
E-28040 Madrid

SWEDEN

Andrén A. — Department of Meteorology
Uppsala University
Box 516
S-75120 Uppsala

Enger L. — Department of Meteorology
Uppsala University
Box 516
S-75120 Uppsala

Persson C. — Swedish Meteorological and Hydrological Institute (SMHI)
S-60176 Norrköping

Robertson L. — Swedish Meteorological and Hydrological Institute (SMHI)
S-60176 Norrköping

SWITZERLAND

Schneiter D. — Swiss Meteorological Institute
Les Invuardes
CH-1530 Payerne

Van Dop H. — World Meteorological Organization
Case Postale No. 2300
CH-1211 Genève

THE NETHERLANDS

De Valk J.P.J.M.M. — Institute of Meteorology and Oceanography
University of Utrecht
Princetonplein 5
3584 CC Utrecht

Evers C.W.A. — Head of the Emission Inventory Department
Dr. Van der Stamstraat 2
2260 MB Leidschendam

Holtslag A.A.M. — Royal Netherlands Meteorological Institute
P.O. Box 201
3730 AE De Bilt

Roelofs G.J.H. — Institute of Meteorology and Oceanography
University of Utrecht
Princetonplein 5
3584 CC Utrecht

Van Jaarsveld H.J.A. — National Institute of Public Health and Environmental Protection
P.O. Box 1
3720 BA Bilthoven

Van Mameren A.C. — Organization for Applied Scientific Research TNO
Ruskesslatweg 4
7157 BK Rekken

Van Rheineck Leyssius H.J. — National Institute of Public Health and Environmental Protection
P.O. Box 1
3720 BA Bilthoven

UNITED KINGDOM

Dickson D.A. — HM Industrial Pollution
Inspectorate for Scotland
27, Perth Street
Edinburgh EH3 5RB

Hunt J.C.R. — University of Cambridge
Department of Applied Mathematics and Theoretical Physics
Silver Street
Cambridge CB3 9EW

U.S.A.

Barchet W.R. — Batelle Pacific Northwest Laboratories
P.O. Box 999
Richland, WA 99352

Barry R.J. — State University of New York at Albany
Atmospheric Sciences Research Center
100 Fuller Road
Albany, NY 12205

Bornstein R.D. — San Jose State University
Department of Meteorology
One Washington Square
San Jose, CA 95192-0104

Bowen B. — Los Alamos National Laboratory
Environmental Protection (HSE) Group
MS K490
P.O. Box 1663
Los Alamos, NM 87545

Briggs G.A. — U.S. Environmental Protection Agency
Mail Drop 80
Research Triangle Park, NC 27711

Bullock, Jr. O.R.	U.S. Environmental Protection Agency Atmospheric Research and Exposure Assessment Laboratory Mail Drop 80 Research Triangle Park, NC 27711
Carmichael G.R.	University of Iowa Department of Chemical & Biochemical Engineering 125A CB Iowa City, IA 52242
Cederwall R.T.	University of California Lawrence Livermore National Laboratory P.O. Box 808 Livermore, CA 94550
Chang Joseph C.	Sigma Research Corporation 234 Littleton Road, Suite 2E Westford, MA 01886
Chang Julius S.	State University of New York at Albany Atmospheric Sciences Research Center 100 Fuller Road Albany, NY 12205
Dalu G.A.	CIRA-CSU Foothill Campus Fort Collins, CO 80523
Daye R.L.	US EPA VIII, ARTX 726 Minnesota Avenue Kansas City, KA 66101
Dennis R.L.	U.S. Environmental Protection Agency Atmospheric Sciences Modeling Division Mail Drop 80 Research Triangle Park, NC 27711
DiCristofaro D.C.	Sigma Research Corporation 234 Littleton Road, Suite 2E Westford, MA 01886
Gillani N.V.	Washington University Department of Mechanical Engineering Campus Box 1185 One Brookings Drive St. Louis, MO 63130
Goddard W.B.	Goddard & Goddard Engineering Environmental Studies 6870 Frontage Road Lucerne, CA 95458
John J.I.	Santa Barbara County Air Pollution Control District 5540 Ekwill Street Suite B Santa Barbara, CA 93111
Karamchandani P.K.	ENSR Consulting and Engineering 1220 Avenida Acaso Camarillo, CA 93010

Lange R.	University of California Lawrence Livermore National Laboratory P.O. Box 808 Livermore, CA 94550
Leone J.M.	University of California Lawrence Livermore National Laboratory P.O. Box 808 Livermore, CA 94550
Lewellen W.S.	ARAP Division of California Research & Technology Inc. 50 Washington Road P.O. Box 2229 Princeton, NJ 08543
Ludwig F.L.	SRI International 333 Ravenswood Avenue Menlo Park, CA 94025
Mathur R.	University of Kentucky Department of Chemical Engineering 160 Anderson Hall Lexington, KY 40506-00461
Meroney R.N.	Colorado State University Engineering Research Center B-217 Fort Collins, CO 80523
Moran M.D.	Colorado State University Department of Atmospheric Science Fort Collins, CO 80523
Nappo C.J.	Atmospheric Turbulence and Diffusion Division ARL/NOAA 456 South Illinois Avenue Oak Ridge, TN 37830
Nasstrom J.S.	EG&G Energy Measurements P.O. Box 9051 Pleasanton, CA 94566
Nickerson E.C.	NOAA Forecast Systems Laboratory 325 Broadway Boulder, CO 80303
Overstreet R.	NOAA/HAZMAT 7600 Sand Point Way NE Seattle, WA 98115
Pielke R.A.	Colorado State University Department of Atmospheric Science Fort Collins, CO 80523
Pierce T.E.	U.S. Environmental Protection Agency Mail Drop 14 Research Triangle Park, NC 27711

Rao S.T.

New York State Department of Environmental Conservation
Division of Air Resources
50 Wolf Road
Albany, NY 12233-3259

Renné D.S.

Council of Environmental Quality
Executive Office of the President
722 Jackson Place, NW
Washington, DC 20503

Rodriguez D.J.

University of California
Lawrence Livermore National Laboratory
P.O. Box 808
Livermore, CA 94550

Saylor R.D.

University of Kentucky
Center for Applied Energy Research
3572 Iron Works Pike
Lexington, KY 40511-8433

Scheffe R.D.

U.S. Environmental Protection Agency
Mail Drop 80
Research Triangle Park, NC 27711

Schiermeier F.A.

U.S. Environmental Protection Agency
Meteorology Division
Mail Drop 80
Research Triangle Park, NC 27711

Shannon J.D.

Argonne National Laboratory
Environmental Research Division
9700 South Cass Avenue
Argonne, IL 60439

Uliacz M.

Colorado State University
Department of Atmospheric Science
Fort Collins, CO 80523

Venkatram A.

ENSR Consulting and Engineering
1220 Avenida Acaso
Camarillo, CA 93010

Walcek C.J.

State University of New York at Albany
Atmospheric Sciences Research Center
100 Fuller Road
Albany, NY 12205

U.S.S.R.

Berlyand M.E.

Main Geophysical Observatory
7, Karbyshev Street
Leningrad 194018

AUTHOR INDEX

SUBJECT INDEX